st Quiz - 1-6 chapters
└ general course information
└ study guide

Need Skills book

Pg 9-11
Jobsite
Tank storage
Confined space safety rules
PPE - 18rr

Welding Skills

Fourth Edition

AMERICAN TECHNICAL PUBLISHERS
ORLAND PARK, ILLINOIS 60467-5756

B. J. Moniz

R. T. Miller

Welding Skills contains procedures commonly practiced in industry and the trade. Specific procedures vary with each task and must be performed by a qualified person. For maximum safety, always refer to specific manufacturer recommendations, insurance regulations, specific job-site and plant procedures, applicable federal, state, and local regulations, and any authority having jurisdiction. The material contained is intended to be an educational resource for the user. American Technical Publishers, Inc. assumes no responsibility or liability in connection with this material or its use by any individual or organization.

American Technical Publishers, Inc., Editorial Staff

Editor in Chief:
    Jonathan F. Gosse
Vice President—Production:
    Peter A. Zurlis
Art Manager:
    James M. Clarke
Technical Content Development and Review:
    Stephen V. Houston
Technical Editor:
    Scott C. Bloom
Copy Editor:
    Catherine A. Mini
Cover Design:
    James M. Clarke
Cover Photo:
    Christopher T. Proctor
Illustration/Layout:
    Nicole D. Bigos
    William J. Sinclair
Multimedia Coordinator:
    Carl R. Hansen
CD-ROM Development:
    Gretje Dahl
    Daniel Kundrat
    Nicole S. Polak

4 5 6 7 8 9 – 10 – 9 8 7 6 5 4 3 2

Printed in the United States of America

ISBN 978-0-8269-2992-1

 This book is printed on recycled paper.

Acknowledgments

The author and publisher are grateful to the following companies, organizations, and individuals for providing information, photographs, and technical assistance.

Airco
American Welding Society
ASI Robicon
Bacharach, Inc.
Baker Testing
Bernard Welding Equipment Company
Bobcat Company, a Unit of Ingersoll-Rand
Boeing Commercial Airplane Group
Buehler Ltd.
CK Worldwide, Inc.
Chrysler Corporation
Cleaver-Brooks
Columbus McKinnon Corporation, Industrial
    Products Division
The Duriron Co, Inc.
E.I. du Pont de Nemours and Company
ESAB Welding and Cutting Products
Exxon Company
Fanuc Robotics North America
Faxitron X-Ray Corporation
Fel-Pro Chemical Products
G.A.L. Gage Company
Harrington Hoists, Inc.
Haynes International, Inc.
Hobart Welders
Hobart Welding Products

Ironworkers, Local Union 378
Jackson Safety, Inc.
Kamweld Technologies, Inc.
LECO Corporation
The Lincoln Electric Company
LOCK-N-STITCH, Inc.
LPS Laboratories, Inc.
Mathey Dearman
Miller Electric Manufacturing Company
Motoman, Inc.
Nederman, Inc.
Osborn International
Pandjiris, Inc.
Rath Gibson, Inc.
Sciaky, Inc.
Sellstrom Manufacturing Co.
SIFCO Selective Plating, Cleveland, OH
Smith Equipment
SPM Instrument, Inc.
Stork Technimet, Inc.
Thermadyne Industries, Inc.
Thermo GasTech
Tinius Olsen Testing Machine Co., Inc.
Victor, a Division of Thermadyne Industries, Inc.
Wall Colmonoy Corporation
Weld Tooling Corp.

Dennis Klingman, AWS – CWI/CWE,
formerly with the Lincoln Electric Company

Thomas J. Clark
    Ironworkers, Local Union 378
Charlie R. Cramlet
    E. I. du Pont de Nemours and Company
Dave Doner
    Prairie State College
Dave Heidemann
    Miller Electric Manufacturing Company

Thomas P. Heraly
    Milwaukee Technical College
Pipe Fitters Training Center, Local Union 597
Gary Reed
    SIFCO Selective Plating
Glen Schulte
    Joliet Junior College
Mark Schumann
    Miller Electric Manufacturing Company

Contents

section . one
Introduction to Welding

section . two
Oxyacetylene Welding (OAW)

Contents

section . **four**
Gas Tungsten Arc Welding (GTAW)

section . **five**
Gas Metal Arc Welding (GMAW)

section . **six**
Other Welding and Joining Processes

section . **seven**
Weld Evaluation and Testing

section . **eight**
Welding Technology

Interactive CD-ROM Contents

- **Using This Interactive CD-ROM**
- **Quick Quizzes®**
- **Illustrated Glossary**
- **Flash Cards**
- **Welding Resources**
- **Media Clips**
- **ATPeResources.com**

Introduction

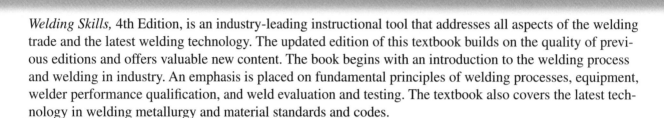

Welding Skills, 4th Edition, is an industry-leading instructional tool that addresses all aspects of the welding trade and the latest welding technology. The updated edition of this textbook builds on the quality of previous editions and offers valuable new content. The book begins with an introduction to the welding process and welding in industry. An emphasis is placed on fundamental principles of welding processes, equipment, welder performance qualification, and weld evaluation and testing. The textbook also covers the latest technology in welding metallurgy and material standards and codes.

The chapters have been organized into eight sections to progressively enhance knowledge and skills. Safety procedures are covered in context with appropriate cautions and warnings. Competencies specified in the American Welding Society (AWS) Schools Excelling through National Skills Standards Education (SENSE) program are included throughout the book. This comprehensive textbook has been updated and expanded to include the following:

- Welding safety and personal protective equipment (PPE)
- Welding equipment
- Surfacing and hardfacing
- Dissimilar metal welding
- Distortion control
- Welding symbols and printreading

The author, Bert J. Moniz, has over 43 years of experience in metallurgy and in many facets of welding, Mr. Moniz currently serves as Materials Engineering Consultant with the DuPont Company. In his current position, he is involved with selecting materials for construction, fabrication, and failure analysis worldwide. He has taught related courses, authored books, and written and presented several papers. His hands-on knowledge and expertise are reflected throughout the book.

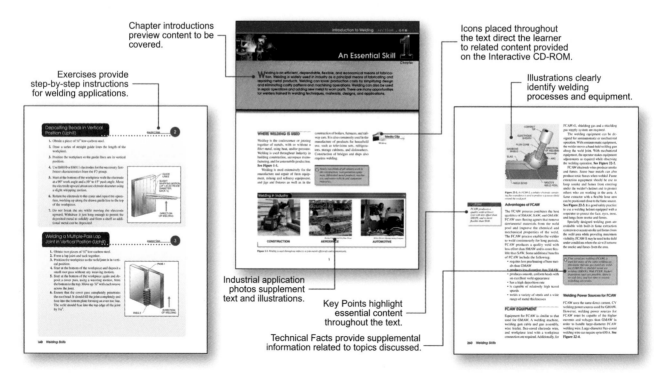

Chapter introductions preview content to be covered.

Icons placed throughout the text direct the learner to related content provided on the Interactive CD-ROM.

Exercises provide step-by-step instructions for welding applications.

Illustrations clearly identify welding processes and equipment.

Industrial application photos supplement text and illustrations.

Key Points highlight essential content throughout the text.

Technical Facts provide supplemental information related to topics discussed.

The Interactive CD-ROM included with this textbook is a study aid with the following features:

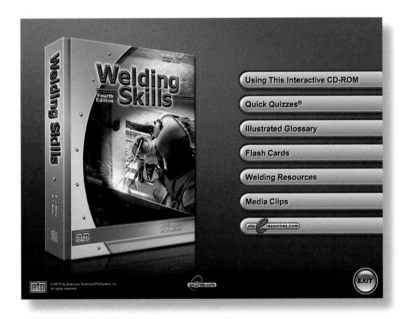

- Quick Quizzes® that reinforce fundamental concepts, with 10 questions per quiz
- An Illustrated Glossary of industry terms, with links to illustrations, video clips, and animated graphics
- Flash Cards that enable a review of common welding terms, definitions, and symbols
- Welding Resources that consist of reference tables and illustrations of welding concepts presented in the text
- Media Clips that depict welding processes through video clips and animated graphics
- Access to ATPeResources.com, which provides a comprehensive array of instructional resources

To obtain information on related training products, visit the American Tech web site at www.go2atp.com

The Publisher

An Essential Skill

Welding is an efficient, dependable, flexible, and economical means of fabrication. Welding is widely used in industry as a principal means of fabricating and repairing metal products. Welding can lower production costs by simplifying design and eliminating costly patterns and machining operations. Welding can also be used in repair operations and adding new metal to worn parts. There are many opportunities for welders trained in welding techniques, materials, designs, and applications.

WHERE WELDING IS USED

Welding is the coalescence or joining together of metals, with or without a filler metal, using heat, and/or pressure. Welding is used throughout industry in building construction, aerospace manufacturing, and for automobile production. **See Figure 1-1.**

Welding is used extensively for the manufacture and repair of farm equipment, mining and refinery equipment, and jigs and fixtures as well as in the construction of boilers, furnaces, and railway cars. It is also commonly used in the manufacture of products for household use, such as television sets, refrigerators, storage cabinets, and dishwashers. Construction of bridges and ships also requires welding.

Media Clip

Welding

 Nearly two-thirds of all welders work in the construction, transportation equipment, fabricated metal products, machinery, and motor vehicle and equipment industries.

Welding in Industry

CONSTRUCTION

Boeing Commercial Airplane Group

AEROSPACE

Miller Electric Manufacturing Company

AUTOMOTIVE

Figure 1-1. *Welding is used throughout industry to join metals efficiently and economically.*

DEVELOPMENT OF WELDING PROCESSES

Modern welding processes evolved from discoveries and inventions dating back to the year 2000 BC when forge welding was first used as a means of joining two pieces of metal. Forge welding was a crude process of joining metal by heating and hammering until the objects were fused together. Today, forge welding is used only in limited applications.

Acetylene gas was discovered in 1836 by Edmund Davy. When combined with oxygen, acetylene produced a flame suitable for welding and cutting, and oxyacetylene welding (OAW) was developed.

The application of heat generated from an electric arc between carbon electrodes was the basis for the arc welding process. Resistance welding was developed in the late 1800s and first used in the early 1900s.

One of the most significant developments at that time was the invention of shielded metal arc welding (SMAW). In SMAW, an electrode could be consumed into the weld while providing heat from an electric arc. The ability to modify the electrode coating to suit different applications expanded the use of arc welding in industry.

Another addition to the family of arc welding processes occurred when it was discovered that a nonconsumable tungsten electrode could be used to generate a welding arc. The process was originally called heliarc welding, because helium was used as an inert shielding gas to protect the electrode and weld area from atmospheric contamination. The process proved to be especially important for welding magnesium and aluminum on World War II fighter planes. The process was renamed tungsten inert gas (TIG) welding because it was discovered that argon could also be used as an inert shielding gas. Today the process is referred to as gas tungsten arc welding (GTAW).

Gas metal arc welding (GMAW) was introduced in the late 1940s. GMAW soon became the process of choice for high-production welding because it used continuous solid-wire electrodes.

Flux-cored arc welding (FCAW) was introduced in the 1950s and combined the versatility of SMAW with the high production rates of GMAW.

Developments in the field continue to influence welding requirements and applications in industry. Current welding processes are the product of continued refinements and variations of the welding processes discovered in the 1800s.

WELDING PROCESSES

The demands of a growing industrial economy during the 1800s spurred the development of modern welding processes. The welding process to be used for a particular job is determined by the following:
- type of metals to be joined
- costs involved
- nature of products to be fabricated
- production techniques used
- job location
- material appearance
- equipment availability
- welder experience

Welding processes used today are commonly classified as oxyfuel welding, arc welding, and resistance welding. **See Figure 1-2.**

In 1810, Sir Humphrey Davy discovered that an electric arc could be maintained at will by bringing two terminals of high voltage electricity near each other. The length and intensity could be varied by adjusting the voltage of the circuit.

OXYFUEL WELDING

Oxyfuel welding (OFW) is a type of welding that uses heat from the combustion of a mixture of oxygen and a fuel gas such as acetylene, methylacetylene-propadiene (MAPP), propane, natural gas, hydrogen, or propylene. The heat is obtained from the burning of a combustible gas in the presence of oxygen. OFW welding processes are used with or without filler metal. If filler metal is not used in the joint, the weld is autogenous. An *autogenous weld* is a fusion weld made without filler metal.

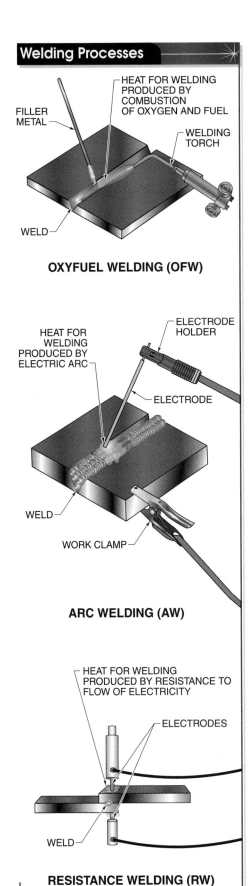

Welding Processes

HEAT FOR WELDING PRODUCED BY COMBUSTION OF OXYGEN AND FUEL

FILLER METAL

WELDING TORCH

WELD

OXYFUEL WELDING (OFW)

HEAT FOR WELDING PRODUCED BY ELECTRIC ARC

ELECTRODE HOLDER

ELECTRODE

WELD

WORK CLAMP

ARC WELDING (AW)

HEAT FOR WELDING PRODUCED BY RESISTANCE TO FLOW OF ELECTRICITY

ELECTRODES

WELD

RESISTANCE WELDING (RW)

Figure 1-2. *Welding processes are commonly classified as oxyfuel welding, arc welding, and resistance welding.*

Oxyacetylene welding is the most commonly used oxyfuel process. *Oxyacetylene welding (OAW)* is an oxyfuel welding process that uses acetylene as the fuel gas. Because of its flexibility and mobility, oxyacetylene welding is used in all metalworking industries, but is most commonly used for maintenance and repair work.

ARC WELDING

Arc welding (AW) is a type of welding that produces coalescence in metals by heating them with an electric arc. The arc is struck between a welding electrode and the base metal. The welding electrode is a component of the welding circuit that terminates at the arc. The joint area is shielded from the atmosphere until it is cool enough to prevent the absorption of harmful impurities from the atmosphere.

AW is the most common method of welding metals. AW processes include shielded metal arc welding (SMAW), gas tungsten arc welding (GTAW), gas metal arc welding (GMAW), flux-cored arc welding (FCAW), submerged arc welding (SAW), and plasma arc welding (PAW).

Shielded Metal Arc Welding. Shielded metal arc welding (SMAW) is an arc welding process that produces an arc between a consumable, coated electrode and the workpiece, creating a weld pool. The arc and molten weld pool are protected by shielding gas produced by the electrode coating. The electrode coating also contains flux that forms a protective slag over the weld. Alloying elements can also be added to the electrode coating to produce a weld bead with different chemical and mechanical properties to suit a variety of applications, making SMAW a versatile process. **See Figure 1-3.**

Common applications of SMAW include fabrication, construction, repair welding, and surfacing for abrasion resistance (hardfacing). SMAW does not require external shielding gas, which makes SMAW more portable than other welding processes.

✓ **Point**

SMAW electrodes can be modified to allow for wider application of SMAW processes.

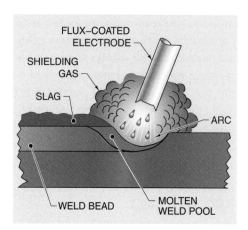

Figure 1-3. *SMAW is a versatile process that does not require external shielding gas.*

Gas Tungsten Arc Welding. *Gas tungsten arc welding (GTAW)* is an arc welding process that produces an arc between a nonconsumable tungsten electrode and the workpiece, creating a weld pool. The electrode, the arc, and the weld pool are protected from the atmosphere by a shielding gas. **See Figure 1-4.** Argon and helium can be used for shielding. Argon is most commonly used because it costs less than helium.

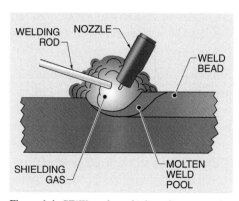

Figure 1-4. *GTAW produces high quality, spatter-free welds on a variety of thin-gauge ferrous and nonferrous metals.*

GTAW can be used with or without filler metal. It produces high quality, spatter-free welds on a variety of metals. It is used in the aerospace, medical equipment, food equipment, and racing industries. GTAW is widely used for joining thin-wall tubing and depositing the root pass in pipe joints.

Gas Metal Arc Welding. *Gas metal arc welding (GMAW)* is an arc welding process that produces an arc between a continuous wire electrode and the workpiece, creating a weld pool in the presence of a shielding gas. GMAW can be used to weld a variety of ferrous and nonferrous metals. **See Figure 1-5.**

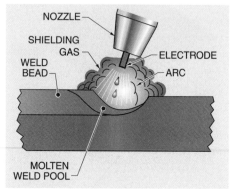

Figure 1-5. *GMAW is a continuous wire process that can be used semiautomatically or automatically. It is well suited for high-production welding.*

The type of shielding gas used depends on the type of metal to be welded and the type of metal transfer desired. Metal transfer refers to the way the molten filler metal crosses the arc to become part of the weld. The three types of metal transfer are short circuiting transfer, globular transfer, and spray transfer.

GMAW can be applied semiautomatically. This means the welder sets the parameters and manipulates the welding gun while a wire feeder delivers filler metal to the weld. GMAW can also be fully automated as part of a robotic cell. GMAW is better for high-production welding than SMAW or GTAW.

Flux-Cored Arc Welding. *Flux-cored arc welding* (FCAW) is an arc welding process that produces an arc between a continuous tubular electrode and the workpiece, creating a weld pool. The core of the electrode is similar in function to the outside coating on an SMAW electrode. It contains powdered alloying elements, shielding gas, and flux. It also contains deoxidizers and scavengers that remove oxygen and impurities from the molten weld pool. **See Figure 1-6.**

Electrodes for FCAW can be self-shielded or gas shielded. Self-shielded electrodes produce sufficient shielding gas to protect the weld. Gas shielded electrodes require auxiliary shielding gas to produce a sound weld.

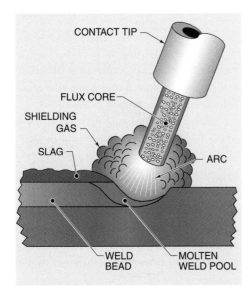

Figure 1-6. *FCAW is a continuous tubular wire process that can be used semiautomatically or automatically. It produces high deposition rates and is typically used on thicker materials.*

FCAW produces an easy-to-remove slag coating to protect the weld from the atmosphere. It is a fast process with high deposition rates, which lowers production costs. FCAW produces welds with excellent appearance. It is commonly used to weld carbon, low-alloy and stainless steels, and cast iron. Typical applications include field and shop fabrications. Like GMAW, FCAW can be applied semiautomatically or automatically.

Submerged Arc Welding. *Submerged arc welding (SAW)* is an arc welding process that uses an arc between a bare metal electrode and the workpiece, creating a weld pool. The electrode, arc, and weld pool are submerged in a granular flux poured on the base metal. SAW is limited to flat or low-curvature base metals. SAW produces high-quality weld metal with fast deposition rates. The weld surface is smooth with no spatter. SAW is automated and most often used to join thick metals requiring deep penetration, such as in heavy steel plate fabrication.

Plasma Arc Welding. Plasma arc welding (PAW) is an arc welding process that uses a constricted arc between a non-consumable tungsten electrode and the weld pool (transferred arc) or between the electrode and constricting nozzle (nontransferred arc). Transferred arc PAW produces a deep, narrow, uniform weld zone and is suitable for almost any metal. Transferred arc PAW is used for welding high-strength, thin metal. Nontransferred arc PAW is typically used for thermal spraying.

Resistance Welding

Resistance welding (RW) is a type of welding in which welding occurs from the heat obtained by resistance to the flow of current through the joined metals. A resistance welding machine fuses metals together by heat and pressure. RW is used to make localized (spot) or continuous (seam) joints. An advantage of resistance welding is its adaptability to rapid fusion of seams.

RW uses special fixtures and automatic handling equipment for the mass production of automobile bodies, electrical equipment, hardware, or other domestic goods. RW can be used for joining almost all steels, stainless steels, aluminum alloys, and some dissimilar metals.

OCCUPATIONAL OPPORTUNITIES IN WELDING

The widespread use of welding in American industry provides a constant source of employment for welders. According to the U.S. Department of Labor, there are more than 500,000 persons employed as welders. Over half of these work in industries that manufacture durable goods such as transportation equipment, machinery, and household products. Many others work for construction firms and repair shops. A growing number of welders are required to operate automated and robotic welding machines.

✓ **Point**

The need for certified welders continues to grow in the welding industry.

Employment Outlook

Opportunities for those who desire to become welders differ by occupational specialty. The need to replace skilled welders who are retiring, combined with the need for additional skilled trades-workers to repair and replace failing infrastructure, creates a demand for welders. Certified welders, especially those certified in more than one process, have better employment opportunities than noncertified welders.

Although many companies have automated some tasks traditionally performed manually, qualified welders are still required. Many automated welding machines and robots require a single operator overseeing multiple operations. However, fabrication and repair applications are still common in the welding industry. **See Figure 1-7.**

The Lincoln Electric Company

Figure 1-7. *Robotic welding machines are programmed to perform repetitive welds on mass-produced products and require supervision by a skilled welding machine operator.*

Training

Welder training is available from different sources. Many schools offer comprehensive welder training programs. Company training programs can vary from a few months of on-the-job training to several years of formal training. Apprenticeship programs that include welder training are also available through unions such as the United Association of Journeyman and Apprentices of the Plumbing and Pipe Fitting Industry of the United States and Canada (UA), the International Association of Bridge, Structural, Ornamental, and Reinforcing Iron Workers and the International Union of Operating Engineers (IUOE). Most employers prefer applicants who have some welding experience or who have completed welder training in one or more welding processes, along with courses in trade-related math and printreading.

Welders must have good manual dexterity, eyesight, and hand-eye coordination. They should be able to concentrate on detailed work for long periods and must be free of physical disabilities that would prevent them from bending, stooping, or working in awkward positions. Welders must also be able to lift 50 lb regularly and 100 lb occasionally.

Before being assigned to work where the quality and strength of the weld are critical, a welder generally has to pass a certification test given by an employer, government agency, or inspection authority. **See Figure 1-8.** Typically, welders are certified by an employer to perform specific welds.

Figure 1-8. *Welders are certified by an employer, government agency, or inspection authority to perform specific welds.*

The American Welding Society (AWS) was commissioned by the federal government to develop a program called *Schools*

Excelling through National Skill Standards Education (SENSE). The objective of SENSE is to provide minimum standards with corresponding guidelines for welding education. Individuals receiving welding training at public or private training organizations participating in SENSE can receive AWS SENSE certification in one or more welding processes. SENSE certification indicates to employers that individuals have attained a minimum level of knowledge and skill to weld limited-thickness plate (entry welder) or plate and pipe (advanced welder).

Job Classifications

Welding jobs differ in the degree of skill required. Welding machine operators can learn the required procedures in several hours, while welders may need years of on-the-job training to master their craft. A beginning welder usually starts on simple production jobs and gradually works up to higher levels of skill with experience.

Welders must have a working knowledge of metal properties and effects of heat on welded structures. They must also have an understanding of how materials are fabricated. Welders must be able to read detailed drawings and interpret welding symbols, prepare the work area, control distortion, recognize weld defects, and perform all tasks required to finish the welding job.

Skilled welders have a variety of career options open to them. If they enjoy the challenges associated with welding, they can enjoy a rewarding career under the hood in manufacturing, construction, aerospace, racing, or any other satisfying jobs that involve welding. If they enjoy the sciences, welding is an excellent training ground for a career in metallurgy or welding engineering. If they find reward in teaching others, there is always a need for committed welding educators. Welding can also lead to a career in man-agement, quality control, or welding inspection. The opportunities for careers involving welding are many. Some of the principal job titles for welders include the following:

- Welder helper—entry-level welder, cleans slag for welder, positions workpieces, helps move materials
- Welder—person who performs welding using the required process
- Welder operator—welder who operates automatic welding equipment, such as that found on automobile assembly lines
- Pipe welder—welder with additional training and certification in welding pipe
- Welding layout and set-up person—welder with print reading experience; must prepare workpieces for welding

Some welding personnel are required to oversee welder certification, instruction, and quality control. The following supervisory positions require additional training:

- Welding inspector—person who has undergone training and certification to work as an inspector
- Welding supervisor—person with good management skills who can effectively run a weld shop and maintain the required welding schedule and quality of workmanship; welding supervisors must be knowledgeable about company standards and procedures
- Welding instructor—person employed by a high school, community college, vocational program, or apprenticeship program; instructors must be certified to meet AWS standards
- Welding engineer—person with a college degree and professional certification qualified to specify necessary weld requirements

 Refer to Quick Quiz® on CD-ROM

Points to Remember

- The combustion of a mixture of acetylene and oxygen produces a flame that is suitable for welding and cutting.
- Oxyfuel welding is a type of welding that uses heat from the combustion of a mixture of oxygen and a fuel gas.
- SMAW electrodes can be modified to allow for wider application of SMAW processes.
- FCAW uses a tubular electrode with flux in its core.
- The need for certified welders continues to grow in the welding industry.
- Certified welders that are certified in more than one process have better employment opportunities than noncertified welders.

Questions for Study and Discussion

1. What manufacturing applications is welding commonly used for?
2. What is the basis of the arc welding process?
3. What are the three common classifications of the welding processes used today?
4. What is an autogenous weld?
5. What is the most common method used for welding metals?
6. What is one difference between FCAW and GMAW?

Refer to Chapter 1 in the *Welding Skills Workbook* for additional exercises.

Welding Safety

2

Chapter

Appropriate safety precautions are effective in reducing the occurence of accidents on the job site. Every year, thousands of welders are injured due to carelessness, lack of knowledge, or indifference to safety regulations. Welders need to understand the hazards associated with welding, take appropriate safety precautions against them, and use common sense to protect themselves and others from injury and long-term health problems.

JOB-SITE SAFETY

Welding involves intense heat and can be very dangerous. Looking at the welding arc is like looking into the sun. Sparks, bits of slag, and dross (molten metal from cutting) can cause eye injury and serious burns. The arc produces ultraviolet and infrared radiation that can burn exposed skin on the face, neck, arms, and hands. Fumes from base materials, contaminants, surface coatings, filler metals, and shielding gases can be hazardous. Exposure to noise in the work area can damage hearing, and working in confined spaces and in hot work zones add another layer of potential hazard.

Industry places a strong emphasis on safety in the workplace. A tremendous amount of time and effort is spent on safety training and awareness. *The Occupational Safety and Health Administration (OSHA)* in the United States and the *Canadian Centre for Occupational Health and Safety (CCOHS)* in Canada are federal agencies that standardize safety practices for most types of work environments. OSHA and the CCOHS require all employers to provide a safe and sanitary environment for their employees. **See Figure 2-1.**

Employers are responsible for safety training at the job site and for ensuring that their employees are familiar with, and follow, OSHA or CCOHS regulations. Most companies have a comprehensive new-hire training program that covers the overall safety requirements of the company. Many companies require employees to attend weekly safety meetings or "toolbox talks." These meetings are designed to reinforce company safety regulations, to discuss safety concerns, and to make employees and supervisors aware of potential hazards.

✓ **Point**

Weekly safety meetings are used by employers to address job site safety issues and concerns.

Figure 2-1. *The Occupational Safety and Health Administration (OSHA) and the Canadian Centre for Occupational Health and Safety (CCOHS) require employers to provide a safe and sanitary work environment for employees.*

Reporting Accidents

Welders are exposed to potential injuries and health risks every day. The ultraviolet rays of the welding arc can burn the eyes and skin. Fumes produced by welding may be toxic and can cause health problems if inhaled. Welding or cutting near flammable materials or welding on containers that have held combustible materials poses the risk of fire or explosion.

While precautions must be taken to prevent injuries, accidents do happen.

All accidents should be reported and documented, regardless of how minor they may be. A small scratch might lead to a serious infection, or a tiny particle lodged in the eye could result in a serious eye injury. Prompt attention can help to minimize the seriousness of the injury.

Most companies have an established accident reporting procedure. It is in the best interest of employees to become familiar with the reporting procedure and to report accidents as soon as they occur. **See Figure 2-2.**

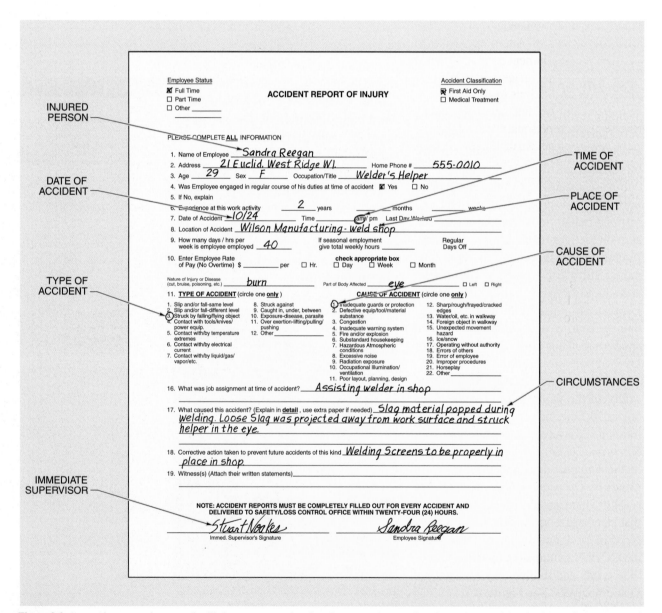

Figure 2-2. *An accident report form must be filled out to accurately reflect the events of an accident, list injuries, and detail job hazards that may need attention.*

Work Behavior

Welders must act responsibly in the workplace. Welding is inherently dangerous and welders must proceed with caution in everything they do. Any focus away from the job may result in an accident, causing injury to the welder and/or damage to property and equipment. Welders should not play games, use cell phones, or engage in any activity that draws their attention away from their work.

SAFE EQUIPMENT OPERATION

All welding equipment must be maintained and used in the proper manner. Manufacturer recommendations should be followed at all times. Attempting to operate a piece of equipment without instruction may not only damage the equipment, it could result in serious injury.

Welders and welding operators should wear appropriate personal protective equipment, properly maintain the equipment they operate, and should never override safety features on the equipment. Malfunctioning welding equipment should be repaired by a trained service technician.

Gas Cylinder Safety

Gas cylinders contain gas under high pressure and can be dangerous if not handled and stored properly. For example, a fully charged oxygen cylinder has a pressure of about 2200 psi at 70°F. The following are some essential safety procedures for working with gas cylinders:

- Always use a cylinder cart to transport gas cylinders. Cylinders valve caps should be in place, and cylinders should be secured by a chain or by other means while in transit.
- Cylinder valves should be fully closed when not in use, and valve caps should be in place when cylinders are not connected.
- Cylinders should be separated from combustible materials by at least 20″, and oxygen cylinders should be separated from fuel gas cylinders by at least 20″ unless there is a firewall between them. The firewall must be a minimum of 5″ tall with a half-hour burn rate.
- Cylinder valves on shielding gas and oxygen cylinders should be fully open when in use to prevent gas from leaking around the valve stem.
- Cylinder valves on acetylene cylinders should be opened a maximum of one (1) turn, and the valve wrench should be left on the cylinder valve so it can be closed quickly in case of emergency.
- The maximum safe operating pressure for acetylene is 15 psi.
- Oil should never be used on gas cylinders, regulators, connections, or hoses. Oil can deteriorate seals, gaskets, and rubber hoses.
- Cylinders should be secured in the upright position at all times, whether in use or in storage. This is especially true of acetylene cylinders because acetone can leak into valves, regulators, and hoses.
- Cylinder valves should be closed and lines should be bled before leaving for the day or at the end of a job.

Confined Spaces

A confined space permit is necessary when welding is carried out in a work space that has any of the following features:

1. It is large enough and so configured that a person can bodily enter it and perform assigned work.
2. It has limited or restricted means for entry or exit.
3. It is not designed for continuous occupancy.

Examples of confined spaces include tanks, silos, storage bins, hoppers, vaults, pits, and trenches. Specific safety precautions required when working in a confined space include having a standby person available, guarding openings, using adequate ventilation, and performing oxygen content checks.

> Confined space permits are issued for a specific period of time. Work must be completed in the allotted time or a new permit must be obtained.

⚠ WARNING
Any welding equipment malfunctions shall be reported to the supervisor.

✓ Point
A fully charged oxygen cylinder has a pressure of about 2200 psi at 70°F. A fully charged acetylene cylinder has a pressure of about 250 psi.

✓ Point
Gas cylinders should be secured in the upright position at all times.

⚠ WARNING
The maximum safe operating pressure for acetylene is 15 psi. Above 15 psi, acetylene becomes unstable and may cause a fire or explosion.

Gas Cylinder Safety

Welding and cutting operations performed in confined spaces create specific safety hazards. For instance, shielding gas or other gas leaks can displace life-supporting oxygen. **See Appendix.** Some gases, such as argon, cannot be detected by smell and can build to toxic levels in confined or low-ventilated areas. Safety precautions should be taken before welding or cutting in a confined space. The welder must be satisfied that the confined space entry procedure and paperwork are satisfactory. If not, the welder has the right to refuse to perform the work until remedial actions are taken.

A permit is required when a confined space contains atmospheric hazards that have the potential to cause serious physical harm to a welder. A *permit-required confined space* is a confined space with one or more of the following characteristics:

- It contains or has the potential to contain a hazardous atmosphere.
- It contains a material that has the potential to engulf the entrant.
- It has an internal configuration such that the entrant could be trapped or asphyxiated by inwardly converging walls or by a floor that slopes downward and tapers to a smaller cross section.
- The confined space contains any other recognized serious safety or health hazard.

Under any of these conditions, a permit system is required in which worker entry into the confined space is regulated. The employer must develop procedures for preparing and issuing permits to enter, work inside, and return the confined space to service at the end of the job. Entry procedures for permit-required confined spaces must be assessed to ensure that they are in compliance with OSHA standards.

A *non-permit confined space* is a confined space that does not contain, or have the potential to contain, any hazards capable of causing death or serious physical harm. Conditions can change as tasks such as welding occur.

Hot Work Safety

Hot work is any operation that involves open flames, heat, and or/sparks and has the potential to cause a fire or explosion. Thousands of hot work fires occur every year causing millions of dollars of damage and even death. One way to control the number and seriousness of hot work accidents is to require hot work permits for any welding or cutting operations that pose potential fire hazards. A hot work permit must be posted at the job site. It specifies the name of the company responsible for the hot work, the job location, the welding or cutting processes used, and the dates that the work is to be performed.

Typically, at least one person trained in the use of fire extinguishers is assigned as a "fire watch" and has the authority to stop work if problems arise. The fire watch continues to monitor the work area for at least one-half hour after the work is completed. The National Fire Protection Association (NFPA) has a standard for covering fire prevention: NFPA 51B is the *Standard for Fire Prevention During Welding, Cutting and Other Hot Work*.

Ventilation

Welding should only be performed in well-ventilated areas. There must be sufficient movement of air to prevent an accumulation of toxic fumes or, possibly, a deficiency of oxygen. All wind or air movement (ventilation) should be across the body, not from in front or from behind. Front- and rear-directed air movement causes wind tunnels (rolling) in front of the body and into the respiratory tract.

Adequate ventilation becomes extremely critical in confined spaces where dangerous fumes, smoke, and dust are likely to collect. When working in a shop, the installed ventilation system may not be adequate to vent the toxic fumes generated by welding. Additional ventilation is required through the use of a respirator, fans, or an exhaust system. Ventilation should be placed as close to the fume source as practical. **See Figure 2-3.**

Nederman, Inc.

Figure 2-3. *A ventilation system is required to remove toxic fumes, smoke, and dust caused by welding.*

An exhaust system is necessary to keep toxic gases below their prescribed limits in areas where welding is performed. Areas under 10,000 cu ft per welder, or that have ceilings lower than 16′, require forced ventilation. An adequate exhaust system is especially necessary when welding or cutting brass and bronze or metal containing zinc, lead, cadmium, or beryllium. This includes galvanized steel and metals painted with lead-based paint. These materials are toxic.

Stainless steels containing chromium as well as paints, primers, and other surface coatings that contain chromates or chromic acids also require special attention to ventilation. Chromium, chromates, and chromic acids break down in the heat of the arc to form hexavalent chromium. Long-term exposure to hexavalent chromium can damage the nose, throat, lungs, eyes, and skin.

Even when ventilation is provided, a welding helmet equipped with a respirator should be worn when welding metals that produce toxic fumes. This protects the welder from inhaling fumes before they can be extracted by the ventilation system. **See Figure 2-4.**

PERSONAL PROTECTIVE EQUIPMENT (PPE)

Personal protective equipment (PPE) is any device worn by welders to prevent burns and protect against hazardous radiation, fumes, and gases, as well as eye injuries and hearing loss. All personal protective equipment must meet requirements specified in OSHA 29 CFR and other applicable safety standards. A welder must be aware of the dangers associated with welding and cutting operations and follow safe operating practices at all times. This includes wearing suitable personal protective equipment. **See Figure 2-5.**

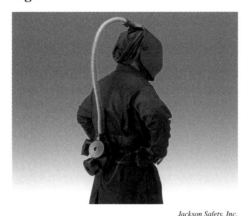

Jackson Safety, Inc.

Figure 2-4. *A respirator should be worn when welding metals that produce toxic fumes.*

Eye and Face Protection

Eye and face protection are essential for welders. Ultraviolet and infrared radiation produced by arc welding and cutting may cause flash burn, which is a burn to the eyes. Arc radiation can also cause burns to the face. The effects depend on the radiant energy wavelengths, intensity, and amount of exposure.

In addition, sparks, slag, and dross can be projected from the work toward the welder. Therefore, welders must wear a suitable helmet with the proper filter plate shade number and safety glasses when arc welding or cutting. They must also take appropriate safety precautions when oxyfuel welding and cutting by wearing goggles or a face shield with the appropriate shade number.

Welders should always be alert for welding arcs that can pose hazards to others near the welding operation. Passersby can be protected by welding screens, curtains, or by remaining an adequate distance from the job. Anyone observing the welding operation should wear a welding helmet or use a hand-held face shield with a filter plate that has the appropriate shade number.

⚠ WARNING

Ultraviolet and infrared radiation produced by arc welding and cutting may cause a burn to the eyes called "flash burn."

✓ Point

Safety glasses should always be worn under the helmet when welding.

Media Clip

Personal Protective Equipment

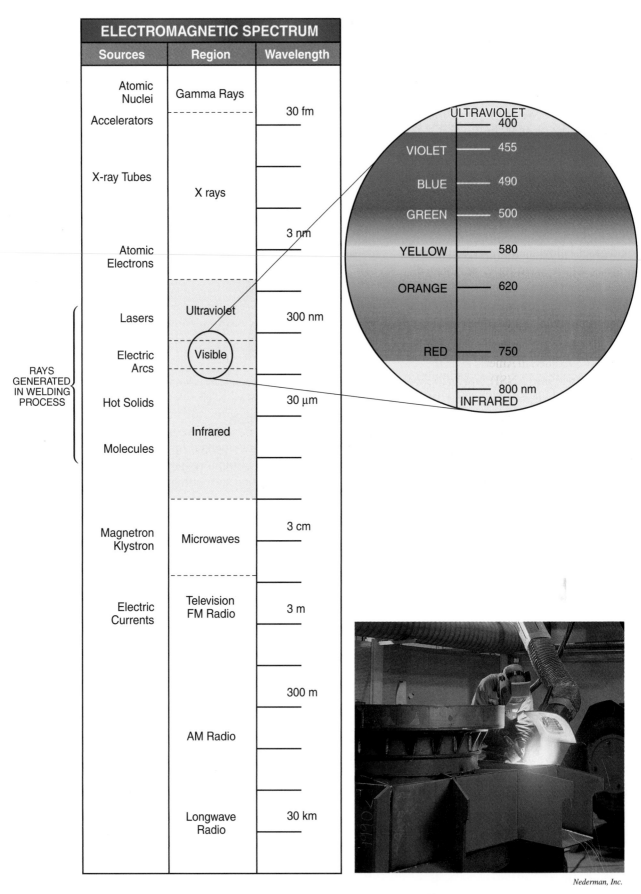

ELECTROMAGNETIC SPECTRUM

Sources	Region	Wavelength
Atomic Nuclei	Gamma Rays	
Accelerators		30 fm
X-ray Tubes	X rays	
Atomic Electrons		3 nm
Lasers	Ultraviolet	300 nm
Electric Arcs	Visible	
Hot Solids	Infrared	30 µm
Molecules		
Magnetron Klystron	Microwaves	3 cm
Electric Currents	Television FM Radio	3 m
	AM Radio	300 m
	Longwave Radio	30 km

RAYS GENERATED IN WELDING PROCESS

ULTRAVIOLET
400
VIOLET — 455
BLUE — 490
GREEN — 500
YELLOW — 580
ORANGE — 620
RED — 750
800 nm
INFRARED

Nederman, Inc.

Figure 2-5. *The rays generated by welding are harmful to workers. A welder should always wear suitable personal protective equipment to protect against the ultraviolet and infrared rays generated during welding.*

Helmets and Filter Plates. Welding helmets should be in good condition. Openings or cracks in a helmet can expose the welder to harmful light from the arc. Helmets may have fixed filter plates, or they may be equipped with auto-darkening technology. Fixed filter plates come in different shades for different degrees of eye protection. Auto-darkening helmets darken within a ten-thousandth of a second when arc light strikes the filter. **See Figure 2-6.**

In arc welding and cutting, the filter-plate shade number should be matched to the type of process and the welding current. In oxyfuel welding and cutting, the filter plate should be matched to the thickness of the material. Filter plate shade numbers for arc and oxyfuel processes are specified in American National Standards Institute (ANSI), Z49.1, *Safety in Welding, Cutting and Allied Processes.* **See Figure 2-7.** Filter plates are marked showing the manufacturer, shade number, and the letter H, which indicates they have been treated for impact resistance and will not shatter.

A filter plate is protected from spatter by a clear glass or plastic safety plate. The safety plate should be replaced before it becomes coated with enough spatter to obstruct the welder's view. Safety plates are inexpensive and can be purchased from any welding supply dealer. A magnifying plate can be inserted behind the filter plate for detailed work or to aid with vision. Also, the filter plate should be checked regularly and replaced if cracked. The welder should follow manufacturer recommendations and use appropriate filter plates and safety plates.

Safety Glasses, Goggles, and Face Shields. Safety glasses must be worn in the work area at all times and must be worn when welding to protect the eyes from slag and other debris that might be deflected under the helmet. Eye protection is also available with prescription lenses for welders who normally wear glasses, and magnifying plates can be installed behind the filter plate to aid with vision.

✓ **Point**

For most arc welding and arc cutting operations, a minimum filter-plate shade number of 10 is suitable. For most oxyfuel welding and cutting operations, a minimum shade number of 5 can be used.

Welding Helmets

Hobart Welders

AUTO DARKENING

Sellstrom Manufacturing Co.

FLIP FRONT

Hobart Welders

FIXED SHADE

Sellstrom Manufacturing Co.

COMPACT

Figure 2-6. *A welding helmet protects the welder from infrared rays, ultraviolet rays, and hot sparks.*

GUIDE FOR FILTER-PLATE SHADE NUMBERS

Operation	Electrode Size 1/32 in. (mm)	Arc Current (A)	Minimum Protective Shade	Suggested Shade No.
Shielded metal arc welding (SMAW)	Less than 3 (2.5)	Less than 60	7	—
	3–5 (2.5–4)	60–160	8	10
	5–8 (4–6.4)	160–250	10	12
	More than 8 (6.4)	250–550	11	14
Gas metal arc welding (GMAW) and flux cored arc welding (FCAW)		Less than 60	7	—
		60–160	10	11
		160–250	10	12
		250–500	10	14
Gas tungsten arc welding (GTAW)		Less than 50	8	—
		50–150	8	11
				12
		150–500	10	14
Arc carbon	Light	Less than 500	10	
Arc cutting	Heavy	500–1000	11	
Plasma arc welding (PAW)		Less than 20	6	6 to 8
		20–100	8	10
		100–400	10	12
		400–800	11	14
Plasma arc cutting (PAC)	Light	Less than 300	8	9
	Medium	300–400	9	12
	Heavy	400–800	10	14
Torch blazing		—	—	3 or 4
Torch soldering		—	—	2
Carbon arc welding		—	—	14

Operation	Plate thickness		Minimum Protective Shade	Suggested Shade No.
	in.	mm		
Gas welding	Under 1/8	Under 3.2		4 to 5
Light	1/8 to 1/2	3.2 to 12.7		5 to 6
Medium	Over 1/2	Over 12.7		6 to 8
Heavy				
Oxygen cutting	Under 1	Under 25		3 to 4
Light	1 to 6	25 to 150		4 to 5
Medium	Over 6	Over 150		5 to 6
Heavy				

Figure 2-7. *Filter plates come in different shades for different degrees of protection. For arc welding and cutting, filter plates are selected based on amperage. For oxyfuel cutting and welding, filter plates are selected based on material thickness.*

Goggles with the proper filter-plate shade number and face shields should be used for oxyfuel welding and cutting operations. Face shields should be worn to protect the face from metal particles when grinding. **See Figure 2-8.** All eye and face protection should meet the requirements of ANSI Z87.1, *Occupational and Educational Eye and Face Protection Devices.*

Protective Clothing

Welders and workers in the work area are required to wear protective clothing made of fire-resistant materials to prevent burns from sparks, spatter, and dross. The clothing should be heavy enough to protect against the intense heat of the arc, and it should prevent infrared and ultraviolet arc light from reaching the skin.

Welding jackets and shirts should not have open pockets, and pants should not have cuffs that could trap slag or molten metal. A welding cap with a brim should be worn under the helmet to protect the head and the back of the neck. The helmet protects the face and neck and should have the proper filter plate shade to protect the eyes. Leather gloves should be worn to protect the hands. Pant legs should be worn over the outside of boots. Work boots should be steel-toed to protect the feet from impact.

Protective clothes for welders should be made of natural materials such as leather, wool, cotton, or denim. These materials have a higher resistance to burning than synthetic materials. Synthetic materials such as polyester should never be worn while welding because they melt and burn easily and can cause severe injury. Loose fitting clothes pose a hazard because they can become caught in rotating equipment. Likewise, long hair should be kept under a welding cap or under a welding jacket.

The best way to decide what kind of protective gear to wear is to determine what kind of welding the job involves. Oxyfuel welding and cutting, GTAW, and GMAW short circuiting transfer on thin gauge material can be considered light-duty work. Protective clothing should be selected accordingly.

SMAW, FCAW, and GMAW spray transfer are examples of heavy-duty welding. Heavy-duty welding requires heavier protective clothing such as a leather welding jacket, a leather apron, leather sleeves, leather leggings, gauntlet-type welding gloves, and an insulated hand shield to protect the back of the hand from intense heat. **See Figure 2-9.**

Gloves. Gloves should be worn at all times to prevent the hands from being burned by hot metal, weld spatter, and arc radiation. Several types of gloves are available for welding, ranging from light-duty leather gloves for light-duty welding, to heavy-duty gauntlet-type gloves. Gloves should be flexible enough to let the hands move freely while providing adequate protection against heat and radiation. **See Figure 2-10.**

The chromium in stainless steels requires special attention to ventilation because it forms hexavalent chromium, which can damage the nose, throat, lungs, eyes, and skin.

Safety Glasses, Goggles, and Face Shields

SAFETY GLASSES
Jackson Safety, Inc.

GOGGLES
Hobart Welders

FACE SHIELD
Jackson Safety, Inc.

Figure 2-8. *Safety glasses should always be worn under the helmet during arc welding and cutting operations to prevent damage to the eyes resulting from deflected slag or other debris. Safety goggles or a face shield should be worn during oxyfuel welding or cutting.*

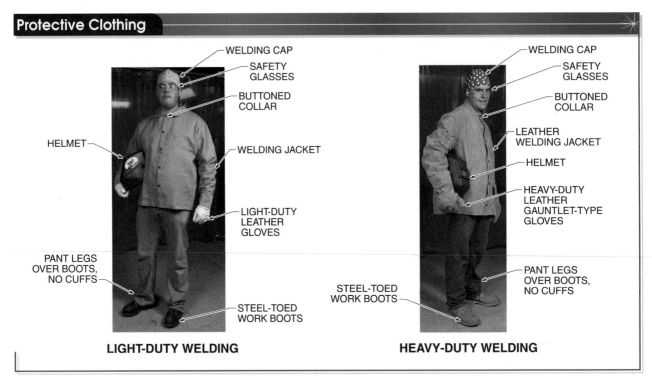

WELDING CAP
SAFETY GLASSES
BUTTONED COLLAR
HELMET
WELDING JACKET
LIGHT-DUTY LEATHER GLOVES
PANT LEGS OVER BOOTS, NO CUFFS
STEEL-TOED WORK BOOTS

LIGHT-DUTY WELDING

WELDING CAP
SAFETY GLASSES
BUTTONED COLLAR
LEATHER WELDING JACKET
HELMET
HEAVY-DUTY LEATHER GAUNTLET-TYPE GLOVES
STEEL-TOED WORK BOOTS
PANT LEGS OVER BOOTS, NO CUFFS

HEAVY-DUTY WELDING

Figure 2-9. *The proper protective clothing is required to prevent burns and injuries during welding.*

Smith Equipment

WORK

GAUNTLET-TYPE

Figure 2-10. *Gloves must be worn when welding and cutting to protect the hands from arc radiation, intense heat, and spatter.*

✓ Point

Never place jackets or shirts over oxygen or gas cylinders. Gas can leak under clothing and start a fire.

Leather Jackets and Aprons. A leather jacket or apron is recommended when welding as spatter can cause injury. A leather apron offers the best protection from hot spatter. In situations where there may not be an excessive amount of metal spatter, suitable fire-retardant coveralls may be worn to protect the clothing.

Work Boots. Work boots must be approved safety shoes or boots made of leather or other approved material with steel toes to prevent impact injuries. Metatarsal (instep) protection should also be worn to prevent slag, sparks, and dross from dropping into the laces. Street shoes must never been worn, regardless of the material they are made from.

Ear Protection

Some welding operations, such as chipping, peening, air carbon arc gouging, and plasma arc cutting, produce high levels of noise. Engine-driven generators can also be noisy. Exposure to excessive noise on a regular basis can result in permanent hearing loss. Hearing protection such as earplugs or earmuffs must be worn to prevent hearing impairment.

Earplugs and ear muffs are supplied by the employer in situations where workers are exposed to excessive noise. *Earplugs* are a device inserted into the ear canal to reduce the level of noise reaching the eardrum. Earplugs are made of moldable rubber, foam, or plastic. *Earmuffs* are a device worn over the ears to reduce the level of noise reaching the eardrum. **See Figure 2-11.**

Safe noise levels and levels at which hearing protection is required are covered in regulations developed by the Environmental Protection Agency (EPA) and OSHA. Ear protection devices are rated for noise reduction to maintain permissible noise levels. A *noise reduction rating (NRR) number* is a number that indicates the noise level reduction in decibels (dB). **See Figure 2-12.**

For example, an NRR of 27 means that the noise level is reduced by 27 dB when tested under factory conditions. If a factory has a noise level of 95 dB, the exposure limit without ear protection is 4 hr. For workers exposed to those noise levels for an 8 hr shift, ear protection is required. Earplugs commonly have an NRR of about 27, which would reduce the noise level from 95 dB to 68 dB. Sixty-eight decibels is a moderate intensity and well within the permissible exposure limit for an 8 hr shift, thus reducing the danger of hearing impairment.

HAZARDOUS SUBSTANCES

Fumes and solid particles originate from the welding process. Gases are generated during welding or are produced by the effects of welding radiation on the surrounding environment. Many of the fumes and gases produced by welding and cutting are hazardous to a welder's health. Hazardous substances include substances that are combustible, toxic, or corrosive.

Adequate ventilation must be available to remove fumes and gases from the work area. Where ventilation is inadequate, air sampling should be used to determine where corrective measures are to be applied.

Ear Protection

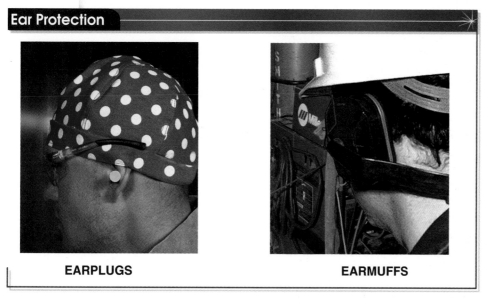

EARPLUGS   EARMUFFS

Figure 2-11. *Ear protection is required for tasks that expose workers to high noise levels.*

SOUND LEVELS		
Intensity of Noise	**Decibels (dB)**	**Example**
Deafening	120 110	thunder, artillery, nearby riveter
	100	
Very Loud	90	loud street noise, noisy factory, unmuffled truck
	80	
Loud	70	noisy office, average street noise, average factory
	60	
Moderate	50	noisy home, average conversation, quiet radio
	40	
Faint	30	quiet home, private office, quiet conversation
	20	
Very Faint	10 0	whisper, soundproof room, threshold of audibility

NOISE LEVEL INTENSITY	
Duration per Day*	**Noise Level†**
8	90
6	92
4	95
3	97
2	100
1½	102
1	105
¾	107
½	110
¼ or less	115

* in hrs
† in decibels (dB)

PERMISSIBLE EXPOSURE TIMES

NOISE REDUCTION RATING: ‡
Earplugs = 27 dB
Earmuffs = 32 dB
‡ typical, varies by manufacturer

Example: A noisy factory has a decibel level of 95 dB that with earplugs can be lowered to 68 dB, which is of moderate intensity but well within permissible exposure times.

Figure 2-12. *Ear protection lowers the decibel level to which the eardrums are exposed, reducing the risk of hearing loss.*

Hazardous Substance Containers

Problems may be compounded by welding or cutting on surfaces contaminated by chemicals or corrosive products. Hazardous substances may be present in a container having previously held any of the following:

- a volatile liquid that releases potentially hazardous, flammable, and/or toxic vapors at atmospheric conditions
- an acid or alkaline material that reacts with metals to produce hydrogen
- a nonvolatile liquid or solid that at ordinary temperatures does not

release potentially hazardous vapors but does so if the container is heated

- a dust cloud of finely divided airborne particles that may still be present in an explosive concentration
- a flammable or toxic gas

Cleaning Hazardous Substance Containers

For maximum safety, only qualified personnel shall designate the container cleaning method. The cleaning method used depends upon the substance previously held in the container. The water method of cleaning is used when the substance is known to be readily soluble in water. The residue can be removed by completely filling the container with water and draining it several times.

When the substance originally held in the container is not readily soluble in water, additional methods of cleaning the container are available, including the hot chemical solution, steam, mechanical cleaning, or chemical cleaning methods. Occasionally, combinations of all methods of cleaning must be used prior to welding or cutting. Care must be taken to protect personnel and to prevent hazardous reactions when combining cleaning methods.

Hot Chemical Solution Method. The hot chemical solution method uses trisodium phosphate (a strong washing powder) or a commercial caustic cleaning compound dissolved in hot water. The cleaning agents are mixed with hot water and added to the container to be cleaned. The container is then filled with water and stirred until the chemicals have been cleaned from the container.

Steam Method. The steam method for cleaning containers uses low-pressure steam and a hot soda or soda ash to remove substances. The cleaning agents are added to the container and the container is filled with live steam and stirred until the chemicals have been removed from the container.

Mechanical Cleaning Method. The mechanical cleaning method is generally used when scaly, dry, or insoluble residues have been left on the surface of the container. Mechanical cleaning may be performed by scraping, sand or grit blasting, high-pressure water washing, brushing, filling the container one-quarter full of clean dry sand and rolling it on the floor, or any method in which the contaminant can safely be dislodged. During mechanical cleaning, the container should be grounded to minimize the possibility of static charge buildup and spark charges.

Chemical Cleaning Method. The chemical cleaning method is generally used when the container has insoluble deposits or when it cannot be mechanically cleaned. Care must be used in selecting a chemical solvent because some solvents may be as hazardous as the deposits they are intended to remove. When selecting chemical solvents, consult the manufacturer of the material to be removed.

Containers should be checked carefully after any cleaning method to ensure that all chemicals have been thoroughly removed. As a final precaution after cleaning, a container should be vented and filled with water before welding or cutting. The container should be arranged so that it can be kept filled to within a few inches of the point where the welding or cutting is to take place, but not interfere with welding. **See Figure 2-13.** When welding or cutting on containers, the following safety precautions should be observed:

- Vent the container to allow for the release of air pressure or steam during welding.
- Use a spark-resistive tool to remove heavy sludge or scale when scraping or hammering.
- Never use oxygen to vent a container as it may start a fire or cause an explosion.
- Never rely on sight or smell to determine the safety of welding or cutting a closed container. A small amount

of residual flammable liquid or gas may not be detectable, but it could cause an explosion.

- Never weld or cut drums, barrels, or tanks until the danger of fire or explosion has been eliminated.

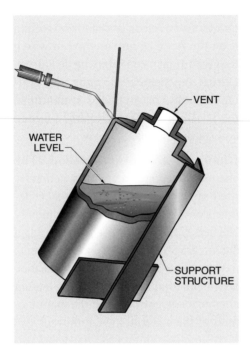

Figure 2-13. *Containers should be partially filled with water before cutting or welding.*

Material Safety Data Sheets

Before any container is cleaned, the hazardous characteristics of the substance previously held by the container must be determined. Information about the substance and safety precautions to follow when working with the substance are contained in a material safety data sheet. A *material safety data sheet (MSDS)* is a document that includes data about every hazardous component comprising 1% or more of a material's content and is used by a manufacturer, importer, or distributor to relay chemical hazard information to the employee.

MSDSs are obtained from the suppliers of welding filler metals, fluxes, and gases. They should be kept on file at a designated location in the workplace. The information is used to inform and train employees on the safe use of hazardous materials. **See Figure 2-14.** If an MSDS is not provided, the employer must contact the manufacturer, distributor, or importer to obtain the missing MSDS.

An MSDS has no prescribed format but must contain certain information about the material including identification; material, physical and chemical characteristics; fire hazard, reactivity, and health hazard data; handling; and control measures. MSDS files must be kept up-to-date and well organized to allow quick access to information in an emergency situation. Employees should become familiar with the MSDS for chemicals commonly encountered on the job.

OXYFUEL AND ARC CUTTING SAFETY

Fires associated with cutting operations generally occur because the proper safety precautions were not taken. Sparks and dross can travel up to 30′ or more and can pass through cracks in walls and floors out of sight. All of the safety precautions associated with gas cylinders, hot work, ventilation, and PPE apply to oxyfuel and arc cutting safety. The following is a quick checklist that welders and supervisors should follow for any cutting operation:

- Never use a cutting torch where sparks or flames pose a fire hazard without first obtaining a hot work permit.
- When cutting in an area that poses a fire hazard, have the appropriate fire extinguisher on hand and make sure that at least one person trained in the use of fire extinguishers serves as a fire watch during the cutting operation. The fire watch should monitor the area for at least 30 min after the work is complete.
- Sweep shop floors clean and use a partially filled bucket of water or sand to catch dross from the cutting operation.

- Clean grease and dirt away from the work area as much as possible and make sure that any flammable gases are purged from the area to prevent fires and explosions caused by sparks or flames.
- Keep combustible materials at least 35′ away from flame or arc cutting operations. If flammable materials cannot be moved, cover them with fire-resistant blankets or use fire-proof partitions or screens to protect them from sparks and dross.
- Ensure that oxygen cylinders, gas cylinders, and gas hoses are protected from sparks, dross, and flames.

- Never exceed 15 psi operating pressure for acetylene.
- Identify the material to be cut, make sure that ventilation is adequate to remove toxic fumes, and use a respirator if necessary.
- Never use oxygen as a substitute for compressed air (shop air).

> Welders are frequently exposed to hazardous situations. PPE such as gloves, safety shoes, safety glasses, helmets with protective filter plates, welding jackets, and other safety equipment should always be used to prevent injury.

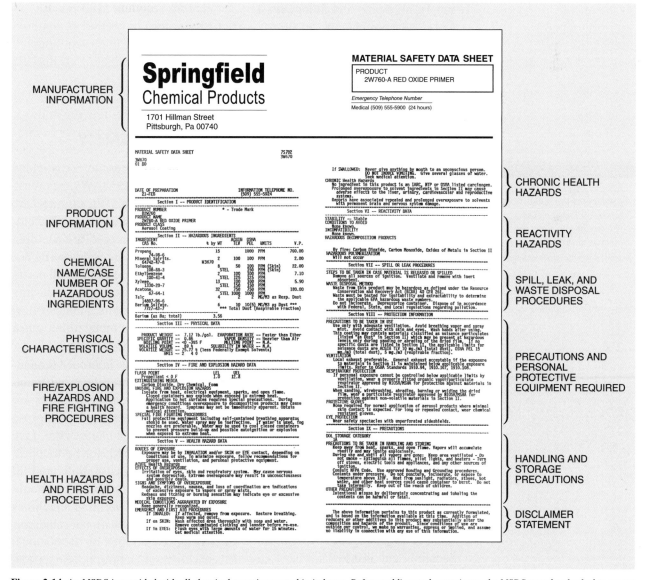

Figure 2-14. *An MSDS is provided with all chemical containers used in industry. Before welding such containers, the MSDS must be checked to ensure that chemicals have been properly removed from the container.*

OXYFUEL WELDING SAFETY

Most of the safety precautions that apply to cutting operations also apply to oxyfuel welding, including gas cylinder safety, hot work, ventilation, and PPE. The two primary fuel gases used for welding are acetylene and methylacetylene-propadiene (MAPP gas). Acetylene produces flame temperatures in the range of 5800°F to 6300°F when mixed with oxygen, while MAPP gas produces a flame temperature in the 5300°F range. The following are important safety considerations for oxyfuel welding:

- When gases are supplied through a manifold system from a central supply, all piping and fittings must withstand a pressure of 150 psi.
- Oxygen piping may be carbon steel, brass, or copper, and only oil-free compounds should be used on threaded connections.
- Fuel gas piping must be carbon steel and must be purged with shop air or nitrogen before first use to remove any foreign material.
- When using cylinders, crack the valves before attaching regulators. *Cracking* is the opening and closing of cylinder valves quickly to clear any debris.
- As an added safety precaution, oxyfuel hoses are color coded. Fuel gas hoses are red and have left-hand threads, while oxygen hoses are green with right-hand threads to prevent them from being connected to the wrong supply or to the wrong side of the torch.
- Keep oxyfuel equipment clean, free of oil, and in good operating condition.
- Never handle cylinders with oily gloves and never use oil on regulators, connections, or hoses.
- Install flashback arrestors at the torch or at the regulator to prevent backfires and flashbacks.
- Test all regulator and hose connections with soapy water to ensure that there are no leaks before lighting the torch.
- Purge oxygen and fuel gas hoses to clean out any dirt before lighting the torch.
- If flammable materials are in the welding area and cannot be moved to a safe distance, cover them with fire-resistant blankets or use fireproof partitions or screens to protect them, keep a fire extinguisher at hand, and follow the safety procedures for hot work.
- Always use a striker (sparklighter) to light a torch. Never use matches or a lighter.
- The maximum safe working pressure for acetylene is 15 psi.
- Proper eye protection is required for oxyfuel welding even though it produces lower light levels than arc welding.

ARC WELDING SAFETY

Arc welding processes include shielded metal arc welding (SMAW), gas tungsten arc welding (GTAW), gas metal arc welding (GMAW), flux cored arc welding (FCAW), submerged arc welding (SAW), and plasma arc welding (PAW). Equipment may range from small portable SMAW machines to industrial-duty power sources. Welders should follow manufacturer recommendations when using equipment. **See Figure 2-15.**

Faulty insulation, improper grounding, and incorrect operation and maintenance of electrical equipment are typical sources of danger from electric shock. Electric shock can be fatal. Live electrical parts should not be touched, and manufacturer instructions and all recommended safety practices should be followed. Only welding machines that meet recognized national standards, such as those identified by the National Electrical Manufacturers Association, NEMA EW-1, *Electric Arc Welding Power Sources,* should be used.

LOCATE POWER DISCONNECT NEAR WELDING MACHINE

WEAR GLOVES AND USE PLIERS TO HANDLE HOT METAL

ATTACH WORK LEAD SECURELY

The Lincoln Electric Company

USE OPERATING CURRENT LEVELS WITHIN RATED CAPACITIES

Figure 2-15. *Following common safety precautions reduces the chances of an accident occurring during welding.*

The following are important safety considerations when working with arc welding equipment:

- Properly ground all electrical equipment and connect the work lead to the workpiece, or as close as possible to the workpiece, to complete the welding circuit. Never attach the work lead to piping carrying gases or flammable liquids.
- Ensure that the electrical leads are correctly sized for the welding application. Sustained overloading causes failure of the welding leads and may result in electric shock or fire.
- Check all electrical connections to make sure they are tight, clean, dry, and in good condition. Poor connections can overheat and melt, or produce dangerous arcs and sparks.

Thermadyne Industries, Inc.

The work lead should be attached securely to the workpiece, or as close as possible to the workpiece, to complete the welding circuit and to prevent personal injury due to electric shock.

- Inspect welding cables for cuts, nicks, or abrasions in the insulation and make appropriate repairs if necessary. Make sure there are no cable splices within 10' of the work.
- Ensure that all plugs, sockets, and electrical units are dry and clean.
- Use insulated electrode holders (stingers) for SMAW.
- Be aware that open circuit voltage can be amplified by the number of welders working on the same weldment, which increases the severity of the electrical shock hazard.
- Never coil electrode leads, gun cables, or work leads around welding machines or allow them to clutter the work area where they can create a tripping hazard for the welder.
- Place electrodes leads and gun cables on the proper hangers when not in use so they do not accidentally short out against metal surfaces.
- Remove SMAW electrodes from electrode holders when not in use.
- Keep welding leads away from the power cable supplying power the welding machine.

- Keep the welding power source dry. If moisture penetrates into the electrical components, have it dried properly by electrical maintenance personnel.
- Never operate arc welding equipment or power tools while standing in water or in damp areas. Water conducts electricity and increases the potential for electrical shock. If the area cannot be dried, stand on boards or on an insulated platform.

The following are additional safety practices common to most arc welding operations:

- Install welding equipment according to the appropriate electrical code.
- Ensure that a power disconnect switch is located at or near the welding machine so that the machine can be shut off quickly in case of emergency.
- Never override safety features on welding equipment.
- Wear appropriate welding gloves and use pliers to handle hot metal.
- Turn power OFF before making repairs to welding equipment. High open circuit voltage can cause serious injury or death.
- Turn power OFF before changing polarity. Otherwise, an arc could damage the switch and burn the welder.
- Do not weld on hollow (cored) castings unless they have been properly vented to prevent explosions.
- Never weld against concrete. Welding against concrete can weaken the concrete due to overheating and cause trapped gas and moisture to expand, resulting in explosion.
- When welding is completed for the day, close the valve on the shielding gas cylinder and bleed the gas line (or remove the SMAW electrode from the electrode holder), turn the welding machine OFF, and hang the electrode holder, welding gun, or torch cable in the designated place.

Welders who follow all safety requirements ensure a safer work environment for themselves and others in the work area.

PREVENTING FIRES

Welding and cutting operations expose welders to intense heat, sparks, and flame. Precautions should be taken to ensure that the job site is safe and that adequate fire prevention strategies are in place. A fire extinguisher should be kept near cutting and welding operations at all times. A hot work permit should be obtained for jobs that pose a high risk of fire, and hot work procedures should be followed carefully, including the use of fire watchers trained in the operation of fire extinguishers. If possible, the work area should be enclosed with portable, fire-resistant screens. Welding or cutting should not be done where dangerously reactive or flammable gases are present.

The NFPA classifies fires into five types: A, B, C, D, and K. The classifications are based on the combustible material and the type of extinguisher required to put out the fire. Extinguisher classifications can also be identified by color and shape. **See Figure 2-16.**

Common dry chemical extinguishers should be available in case sparks from welding set other materials on fire. The two basic types of dry chemical extinguishers are the stored-pressure and the cartridge operated. A fire extinguisher labeled either A, B, or C can only extinguish the fire for which it is labeled. A fire extinguisher labeled ABC is composed of dry chemicals and is capable of extinguishing class A, B, and C fires. Using the wrong fire extinguisher can make the fire worse rather than extinguishing it.

Welders must be particularly aware of the fire hazards associated with the metals they are welding as well as the environment in which they are working and ensure that the proper type or types of extinguishers are available. If a fire erupts, the nozzle of the fire extinguisher should be aimed at the base of the fire.

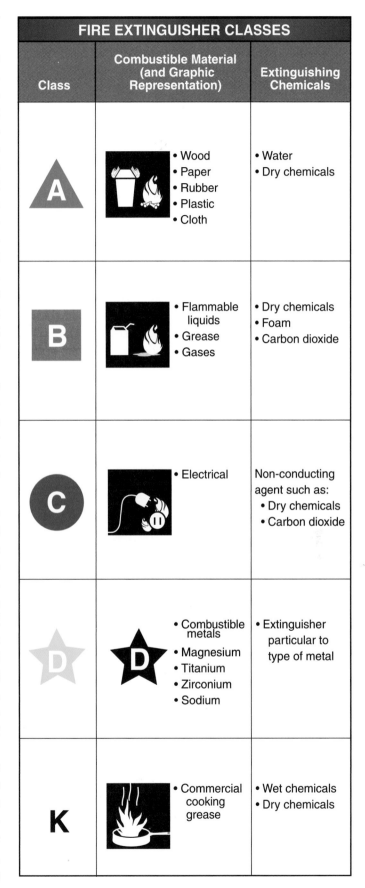

Class	Combustible Material (and Graphic Representation)	Extinguishing Chemicals
A	• Wood • Paper • Rubber • Plastic • Cloth	• Water • Dry chemicals
B	• Flammable liquids • Grease • Gases	• Dry chemicals • Foam • Carbon dioxide
C	• Electrical	Non-conducting agent such as: • Dry chemicals • Carbon dioxide
D	• Combustible metals • Magnesium • Titanium • Zirconium • Sodium	• Extinguisher particular to type of metal
K	• Commercial cooking grease	• Wet chemicals • Dry chemicals

Figure 2-16. *Fire extinguishers are classified as A, B, C, D, and K. When using a fire extinguisher, it should be aimed at the base of the fire.*

Class A

A Class A fire may be caused by most combustible materials, such as wood, paper, rubber, plastic, and cloth. Class A fires are the most common type of fire. A Class A fire extinguisher is identified by the color green inside a triangle shape. Class A fires can be extinguished with water or dry chemicals. Carbon dioxide, sodium bicarbonate, and potassium bicarbonate should not be used on a Class A fire.

Class B

A Class B fire is caused by flammable liquids, gases, or grease. A Class B fire extinguisher is identified by the color red inside a square. Class B fires can be extinguished with dry chemicals. Foam and carbon dioxide extinguishers may also be used.

Class C

A Class C fire is an electrical fire. A Class C fire extinguisher is identified by the color blue inside a circle. Electrical fires require a non-conducting agent, such as carbon dioxide or dry chemicals, to extinguish them. Foam extinguishers or water should never be used on an electrical fire.

Class D

A Class D fire is caused by combustible metals, such as magnesium, titanium, or sodium. A Class D fire extinguisher is identified by the color yellow inside a star. Class D fires cannot be extinguished with common A, B, or C extinguishers. The chemicals in common extinguishers can intensify the fire, rather than put it out. Dry powder extinguishers are available that are made specifically for metal hazards.

Class K

A class K fire is caused by grease in commercial cooking equipment. Class K fire extinguishers coat the fire with wet or dry chemicals.

Miller Electric Manufacturing Company

Welding shops should be equipped with fire extinguishers located near the work area for easy access in case of fire.

> *Explosion, fire, or other health hazards may result if welding or cutting is performed on containers that are not free of hazardous substances. No container should be presumed to be clean or safe. Containers can be made safe for welding and cutting provided the necessary steps and safety precautions are followed.*

 Refer to Quick Quiz® on CD-ROM

- Weekly safety meetings are used by employers to address job-site safety issues and concerns.
- A fully charged oxygen cylinder has a pressure of about 2200 psi at 70°F (21°C). A fully charged acetylene cylinder has a pressure of about 250 psi.
- Gas cylinders should be secured in the upright position at all times.
- When working in a confined space, have a standby person available to ensure a safe environment.
- Ventilation should be placed as close to the fume source as practical.
- Safety glasses should always be worn under the helmet when welding.
- For most arc welding and arc cutting operations, a minimum filter-plate shade number of 10 is suitable. For most oxyfuel welding and cutting operations, a minimum shade number of 5 can be used.
- In addition to approved work clothes, heavy-duty welding requires a leather jacket or apron and leather gauntlet-type gloves.
- Never place jackets or shirts over oxygen or gas cylinders. Gas can leak under clothing and start a fire.
- Ear protection should be worn to protect against noise from welding, cutting, grinding, and other activities.
- A material safety data sheet (MSDS) includes data about every hazardous component comprising 1% or more of a material's content.
- Always use a striker to light a welding or cutting torch.
- Always use insulated electrode holders when welding with SMAW to prevent electric shock and injury.
- Never weld against concrete. Welding against concrete can cause trapped gas and moisture to expand, resulting in explosion.

1. What are some of the main causes of accidents?
2. Why should all accidents be reported immediately?
3. Why should welders always stay focused while working in the shop?
4. What may happen if welding equipment is used without proper instruction?
5. What should be done if a malfunction occurs in any welding equipment?
6. What general practice should be followed regarding ventilation when performing welding?
7. Why should used containers be thoroughly cleaned and safety-processed before any welding or cutting is done?
8. Why do fires occur during a cutting operation?
9. What are some precautions that should be taken when using a cutting torch?
10. Why must welders wear proper PPE when welding?
11. A dry chemical fire extinguisher can be used to extinguish which class or classes of fire?
12. What class of fire extinguisher should be used for a fire involving burning metal?
13. What is the purpose of an MSDS?
14. Why should a welder never look at an electric arc without eye protection?
15. What determines the correct shade of lens for use during welding?
16. Why should shaded lenses be covered with clear plastic lenses?
17. Why are safety glasses required when welding?
18. Why should leather gloves be worn when welding?

Refer to Chapter 2 in the *Welding Skills Workbook* for additional exercises.

Joint Design and Welding Terms

Chapter

Engineers and designers consider all factors in the design of a weld joint to ensure safety and efficiency. These factors include load requirements of the weld, adaptability of the joint for the product being designed or welded, accessibility of the weld, intended function of the structure, governing codes and specifications, and economic considerations such as the cost of preparing the joint.

Welded joints are used in virtually every industry. In the building industry, welds are used to join structural elements such as columns, trusses, girders, and other structural components. A complete knowledge of welding requires an understanding of the terms associated with welding.

WELDING TERMS

Before proceeding with any welding operation, welders must understand common welding terms. The *base metal* is the metal or alloy that is to be welded. An *electrode* is a component of the welding circuit that conducts electrical current to the weld area. Electrodes may be consumable or nonconsumable, depending on the welding process. Consumable electrodes are depleted during welding, non-consumable electrodes are not. Some electrodes, such as those used in shielded metal arc welding, are covered with a flux coating.

A *weld bead* is a weld that results from a weld pass. A *weld pass* is a single progression of welding along a weld joint. **See Figure 3-1.** A single-pass weld requires only one weld pass. When laying a bead in a multiple-pass weld, each weld pass builds on the previous pass. The movement of the heat source creates ripples as the weld bead is deposited. A *ripple* is the shape within the deposited bead caused by the movement of the welding heat source.

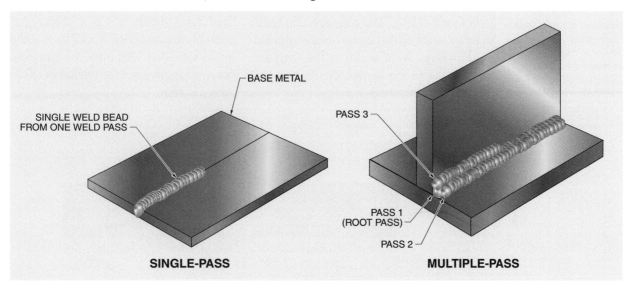

Figure 3-1. *When depositing a weld, each pass builds on the previous pass and should overlap it by one-third to one-half to ensure complete fusion. The movement of the electrode by the welder creates ripples as the bead is deposited.*

A *crater* is a depression at the termination of the weld bead. *Depth of fusion* is the distance that fusion extends into the base metal from the surface melted during welding. **See Figure 3-2.** *Joint penetration* is the distance the weld metal extends from the weld face into the joint. Joint penetration does not include the face or root reinforcement.

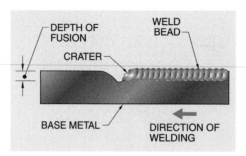

Figure 3-2. *A crater is a depression at the termination of a weld bead. Depth of fusion is the distance that fusion extends into the base metal from the surface melted during welding.*

Weld reinforcement is the amount of weld metal in excess of that required to fill the joint. *Root reinforcement* is reinforcement on the side opposite the one on which welding took place. *Face reinforcement* is reinforcement on the same side as the welding.

The *root face* is the portion of the groove face within the joint root. The *root opening* is the distance between joint members at the root of the weld before welding. The root opening must be accurate so that excess welding is not necessary. *Weld width* is the distance from toe to toe across the face of the weld. **See Figure 3-3.**

The *weld toe* is the junction of the base metal and the weld face. The *weld face* is the exposed surface of the weld, bounded by the weld toes on the side on which welding was done. The face may be either concave or convex. The *weld root* is the area where filler metal intersects the base metal and extends the furthest into the weld joint.

The *actual throat* is the shortest distance from the face of a fillet weld to the weld root after welding. **See Figure 3-4.** The *effective throat* is the minimum distance, minus convexity, between the weld face and the weld root. The *theoretical throat* is the distance from the joint root, perpendicular to the hypotenuse of the largest right-angle triangle that can be inscribed within the cross section of a fillet weld. A *weld leg* is the distance from the joint root to the toe of a fillet weld.

Filler metal is metal deposited in a welded, brazed, or soldered joint during the welding process. *Fusion welding* is the melting of metal and filler metal, or filler metal only, to make a weld. Fusion welding is the most common method of joining metals.

Welding progression refers to the addition of filler metal in a weld joint root and beyond. A *joint root* is the portion of a weld joint where joint members are the closest to each other. In cross section, the joint root may be either a point, a line, or an area. A *root bead* is a weld bead that extends into or includes part or all of the joint root. A *root pass* is the initial weld pass that provides complete penetration through the thickness of the joint member.

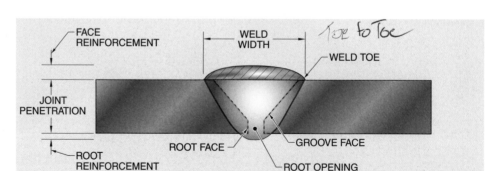

Figure 3-3. *A root opening must be prepared properly to prevent excess welding. Weld reinforcement is the convex surface of the weld. The joint may also require a root face to prevent excess penetration.*

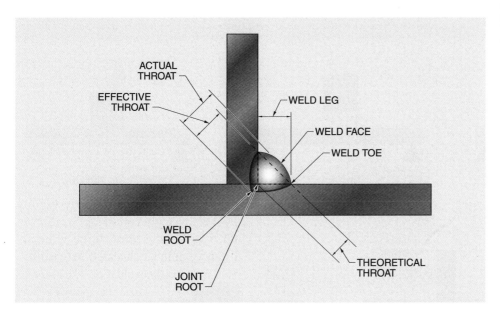

Figure 3-4. *Parts of a weld are identified for specifying weld characteristics.*

Several weld beads (multiple-pass weld) may be required to complete a weld. A multiple-pass weld contains two or more weld beads.

WELD JOINTS

A *weld joint* is the physical configuration at the juncture of the members to be welded. Weld joints must be correctly designed and have adequate root openings to support the loads transferred from one member to another through the welds. **See Figure 3-5.** The following are some basic considerations in the selection of any weld joint:

- the type and thickness of material to be welded
- whether the load will encounter tension, compression, bending, fatigue, or impact stresses
- how the load is to be applied to the joint, i.e., whether the load is a static, impact, cyclic, or variable
- the displacement of the load in relation to the joint
- the direction from which the load is to be applied to the joint
- whether the joint can be welded from both sides or only from one side
- the cost of preparing the joint

Weld joint design is based on the strength of the joint, safety requirements, and the service conditions under which the joint must perform. Joints are also designed for economy or accessibility during construction and inspection. The five basic weld joints designs are the butt, T, lap, corner, and edge joints. **See Figure 3-6.**

> When designing weld joints for buildings, consideration must be given to the effects of transverse shrinkage, which occurs in support columns as a building is constructed. Shrinkage can accumulate if unaccounted for in the weld designs.

The Lincoln Electric Company

Welders must understand welding terminology, the performance characteristics of different welding processes, and how to read welding procedure specifications and prints.

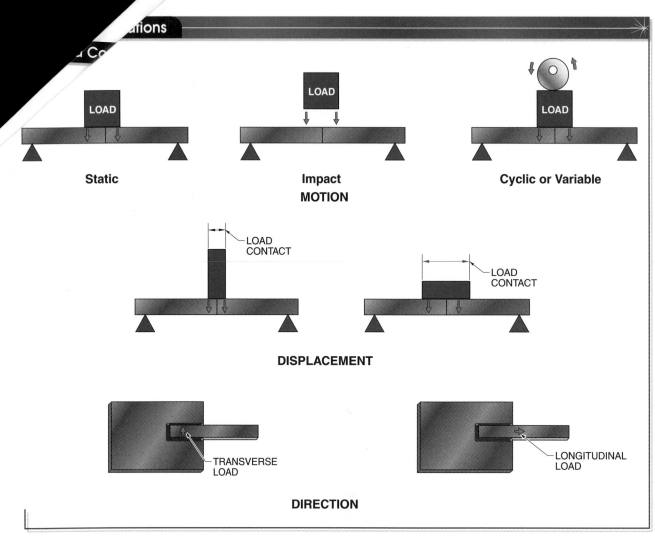

Static | Impact | Cyclic or Variable

MOTION

LOAD CONTACT

LOAD CONTACT

DISPLACEMENT

TRANSVERSE LOAD

LONGITUDINAL LOAD

DIRECTION

Figure 3-5. *Welders should be familiar with how loads impact welded joints and with the criteria for selecting the proper joint configuration.*

Basic Weld Joints

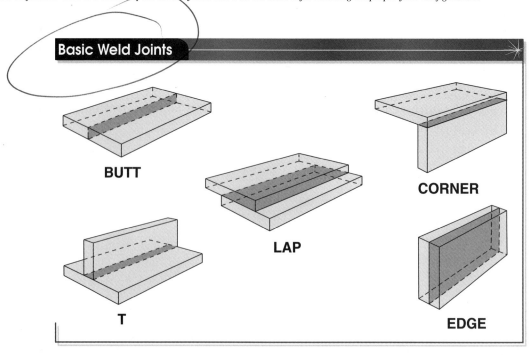

BUTT

LAP

CORNER

T

EDGE

Figure 3-6. *The five basic joint configurations used in welding are the butt, T, lap, corner, and edge.*

(handwritten at top) Fillet weld — Each leg = lengths. Equal Legs

Butt Joints

A *butt joint* is a weld joint in which two members are set approximately level to each other and are positioned edge-to-edge. **See Figure 3-7.** The weld is made between the edge surfaces of the two sections. Butt joints include square-groove, single-bevel-groove, double-bevel-groove, single-J-groove, double-J-groove, single-V-groove, double-V-groove, single-U-groove, and double-U-groove. Butt joints are commonly used in fabricating vessels and subassemblies and for repair operations.

Square-Groove Butt Joints. The square-groove butt joint is intended primarily for materials that are ³⁄₁₆″ thick or less. Square-groove joints require complete fusion for optimum strength. Square-groove joints in materials up to ³⁄₈″ thick with a minimum root opening of ⅛″ can be welded with submerged arc welding, The square-groove joint is reasonably strong in static tension but is not recommended when the joint will be subjected to fatigue or impact loads, especially at low temperatures. A square-groove joint

is the least expensive weld joint option because it requires little or no preparation other than aligning the joint edges.

Single-Bevel-Groove Butt Joints. A single-bevel-groove butt joint is welded from one side and is generally used on metals no more than ½″ thick. Single-bevel-groove joints are similar to single-V-groove joints in terms of properties and applications, but they are more difficult to weld than V-groove and U-groove joints because one edge of the joint is vertical. However, a single-bevel-groove joint requires less filler metal than most joint configurations, except the single-J-groove. Also, a single-bevel-groove joint requires less preparation than the single-J-groove, the single-V-groove, or the single-U-groove.

Double-Bevel-Groove Butt Joints. Double-bevel-groove joints have the same characteristics as double-V-groove joints, but they take less preparation, and they require less filler metal. Like single-bevel-groove joints, however, it is more difficult to achieve complete fusion because one side of the joint has a perpendicular face.

> ✓ **Point**
> *Square-groove butt joints are typically used with materials ³⁄₁₆″ thick or less.*

> ✓ **Point**
> *A double-V-groove butt joint is suitable for all load conditions.*

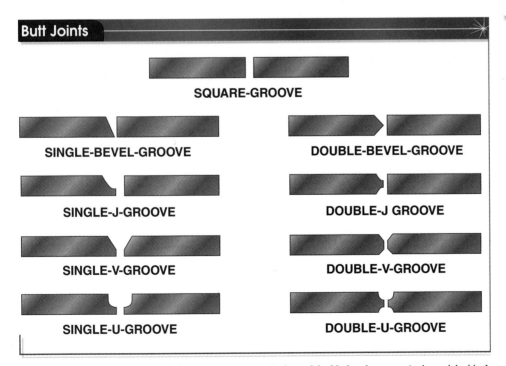

Butt Joints

SQUARE-GROOVE

SINGLE-BEVEL-GROOVE

DOUBLE-BEVEL-GROOVE

SINGLE-J-GROOVE

DOUBLE-J GROOVE

SINGLE-V-GROOVE

DOUBLE-V-GROOVE

SINGLE-U-GROOVE

DOUBLE-U-GROOVE

Figure 3-7. *Common butt joints include the square-groove, single- and double-bevel-groove, single- and double-J-groove, single- and double-V-groove, and single- and double-U-groove.*

Single- and Double-J-Groove Butt Joints. Single- and double-J-groove joints require less filler metal than single- and double-bevel-groove joints because the groove faces are steeper. But filler metal savings have to be compared to the cost of joint preparation. Like single- and double-bevel-groove joints, it is more difficult to achieve complete fusion into the perpendicular face of the joint.

Single-V-Groove Butt Joints. A single-V-groove butt joint is used on metal from ⅜″ to ¾″ thick. Preparation for a single-V-groove joint is more costly than a single-bevel-groove joint because both members to be joined require beveling, and more filler metal is required to fill the joint. A single-V-groove joint is strong in static loading but, like square-groove and single-bevel-groove joints, it is not particularly suitable when subjected to fatigue or impact loads at the weld root.

Double-V-Groove Butt Joints. A double-V-groove butt joint is suitable for all load conditions. The double-V-groove is typically specified for metals that are ¾″ thick or greater and where both sides of the joint can be welded.

It costs more to prepare a double-V-groove joint compared to a single-V-groove. But the groove angles are smaller, which means less filler metal is required to fill the joint. To keep distortion to a minimum, welding should be done first on one side and then the other, with the welder alternating sides until the joint is filled.

Single-U-Groove Butt Joints. A single-U-groove butt joint meets all ordinary load conditions and is used for work requiring high-quality welds. The single-U-groove works well on metals ½″ to ¾″ thick. The single-U-groove joint needs less filler metal than the single-V-groove joint, and generally, less distortion occurs.

Double-U-Groove Butt Joints. A double-U-groove butt joint is intended for metals ¾″ thick or greater that can be welded from both sides. The double-U-groove joint can meet all regular load conditions. However, preparation costs are higher than for the single-U-groove joint.

T-Joints

A *T-joint* is a weld joint formed when two members are positioned at approximately 90° to one another in the form of a T. **See Figure 3-8.** Depending on metal thickness and service conditions, the vertical member can be machined to produce different T-joint configurations. Basic T-joint designs are square, single-bevel, double-bevel, single-J, and double-J.

Square T-Joints. A square T-joint can be welded on one or both sides with fillet welds. Square T-joints can be used for thin or reasonably thick materials where applied loads subject the weld to longitudinal shear. Since the stress distribution of the joint may not be uniform, this factor should be considered where severe impact or heavy transverse loads are encountered. For maximum strength, considerable weld metal is required. This requirement can be met by depositing multiple fillet welds on both sides of the joint.

Single-Bevel T-Joints. A single-bevel T-joint can withstand a more severe load than the square T-joint because it allows for better distribution of stresses. It is generally confined to plates ½″ thick or less where welding can be performed from one side only.

Double-Bevel T-Joints. A double-bevel T-joint is intended for use where heavy loads are applied in both longitudinal and transverse directions. A double-bevel T-joint is welded from both sides.

Single-J T-Joints. A single-J T-joint is used on plates 1″ thick or more where welding is limited to one side of the joint. It is especially suitable for weldments that are exposed to severe loads.

Double-J T-Joints. A double-J T-joint is particularly suitable for heavy plates 1½″ thick or more where unusually severe loads are encountered. Joint location should permit welding on both sides.

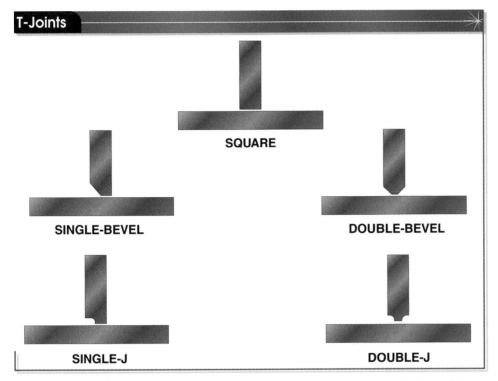

SQUARE

SINGLE-BEVEL

DOUBLE-BEVEL

SINGLE-J

DOUBLE-J

Figure 3-8. *T-joints are used on all standard metal thicknesses and include square, single-bevel, double-bevel, single-J, and double-J.*

Lap Joints

A *lap joint* is a weld joint between two overlapping members in parallel planes. A lap joint is one of the strongest joints available despite the lower unit strength of the filler metal. Lap joints are typically welded on both sides. An overlap greater than three times the thickness of the thinnest member is recommended. Depending on the service conditions, fillet welds may be deposited on one side of a lap joint or on both sides. **See Figure 3-9.**

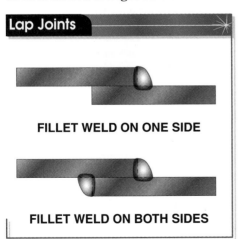

FILLET WELD ON ONE SIDE

FILLET WELD ON BOTH SIDES

Figure 3-9. *Fillet welds, whether on one side or on both sides of a lap joint, are the strongest welds.*

Fillet Weld on One Side of a Lap Joint. Lap joints on metal up to ½″ thick can be welded with a fillet weld on one side of the joint if the loading is not too severe. The strength of the fillet weld depends on the size of the weld.

Fillet Welds on Both Sides of a Lap Joint. A lap joint with fillet welds on both sides is one of the more widely used joints in welding. The joint can withstand greater loads than a lap joint with a fillet weld on one side only.

Corner Joints

A *corner joint* is a joint formed when two members are positioned at an approximate right angle in an L shape. Corner joints are used in many applications where weldments are exposed to general service loads. Common corner joints are flush, half-open, and full-open. **See Figure 3-10.**

Flush Corner Joints. A flush corner joint is designed primarily for welding sheet metal 12 gauge or thinner. It is restricted to thin materials because complete penetration is sometimes difficult to achieve, and the joint is able to support only moderate loads.

> ✓ Point
>
> *A lap joint is usually welded on both sides of the joint.*

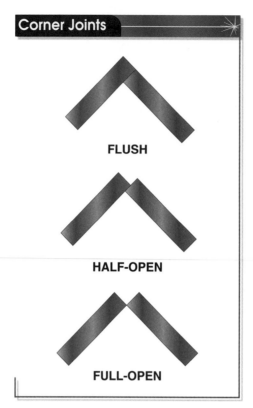

Corner Joints

FLUSH

HALF-OPEN

FULL-OPEN

Figure 3-10. *Corner joints are generally used for light to moderate loads.*

Half-Open Corner Joints. A half-open corner joint is usually more adaptable for materials thicker than 12 gauge. It is suitable for loads where fatigue or impact are not too severe, and where the joint can be welded from one side only. The two edges of the joint are shouldered together so there is less tendency to burn through the plates.

Full-Open Corner Joints. A full-open corner joint permits welding on both sides, so it produces a strong joint capable of carrying heavy loads. All metal thicknesses can be welded with full-open corner joints. A full-open corner joint provides good stress distribution.

Edge Joints

An *edge joint* is a weld joint formed when the edges of two or more parallel, or nearly parallel, members are joined. The edge joint is suitable for plates ¼″ thick or less and can sustain only light loads. Edge joints can be combined with butt joints or corner joints, and the edges can be squared or beveled. **See Figure 3-11.**

ESAB Welding and Cutting Products

The weld joint must be designed to withstand the service conditions and the load requirements of the weldment.

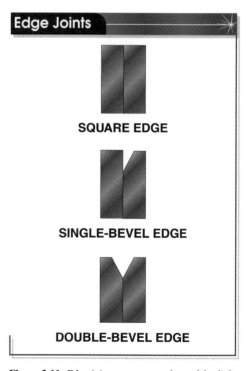

Edge Joints

SQUARE EDGE

SINGLE-BEVEL EDGE

DOUBLE-BEVEL EDGE

Figure 3-11. *Edge joints are commonly used for light load applications. Many combinations of joint edges are possible since edge joints can be square or beveled.*

An edge joint is commonly used to join support structures and short lengths of structural steel. A *flanged joint* is an edge joint in which one of the joint members has a flanged edge at the weld joint.

WELD TYPES

A *weld type* is the cross-sectional shape of the weld. The weld type is determined by the weld joint design and depends on the load requirements of the weld. To maximize weld strength and economy, the following basic rules are observed:

- Minimize edge preparation—Minimizing edge preparation reduces cutting and machining costs.
- Provide access to the joint for welding—Taking the welding equipment needed for the job into consideration and ensuring the welder has enough room to deposit the required welds is essential.
- Minimize filler metal—Minimizing filler metal reduces costs.
- Reduce excess heat—Reducing the amount of excess heat applied to the weld area during welding minimizes metallurgical changes in the weld and in the heat affected zone (HAZ).
- Minimize the number of welds—Minimizing the number of welds reduces the amount filler metal required. It also helps to minimize distortion by controlling heat input.
- Size the weld for the thinnest joint member—The size of the weld should not exceed the strength of the thinnest joint member.

Good joint design uses root openings and groove angles that require the least amount of weld metal, yet still provide accessibility to the joint. Joint design is also influenced by the type of metal to be welded, the location of the joint in the weldment, and the required performance of the weld. Weld types are selected for specific applications and include fillet welds, groove welds, plug and slot welds, surfacing welds, stud welds, spot and seam welds, projection welds, and back welds. **See Figure 3-12.**

Fillet Welds

A *fillet weld* is a weld of approximately triangular cross section that joins two surfaces at approximately right angles. Fillet welds may be used in lap, T, or corner joints. Fillet welds are the most commonly used weld and are preferred over groove welds because they are easier to prepare and are less expensive to complete. Fillet welds may be made from one side or both sides of the joint. Fillet weld size is the length of the largest side of a right-angle triangle that can be inscribed within the fillet weld cross section.

Fillet welds are commonly used when load stresses are low and the required effective throat is less than $\frac{5}{8}''$. The strength of the fillet weld is based on the effective throat of the weld. If the load requires an effective throat of $\frac{5}{8}''$ or larger, a groove weld should be used, possibly in combination with a fillet weld to provide the required size.

Groove Welds

A *groove weld* is a weld made in the groove between the two members to be joined. A groove weld may be square groove, single-groove, or double-groove. A square-groove weld is the most economical to produce because it requires no edge preparation, but its use is limited by the thickness of the joint and the service load. A groove weld is adaptable for a variety of joints, most commonly the butt joint. A groove weld should use the smallest root opening and groove angle possible for the job to produce a sound weld with the least amount of filler metal.

With a suitable opening and backing strip, square-groove weld joints up to $\frac{1}{4}''$ thick can be made with SMAW. Square-groove weld joints up to $\frac{3}{8}''$ thick can be made with GMAW spray transfer, FCAW, or SAW. The root of a square-groove weld should not be under tension when the weld is bent under load.

> **✓ Point**
>
> *A groove weld should use the smallest root opening and groove angle possible for the job to produce a sound weld with the least amount of filler metal.*

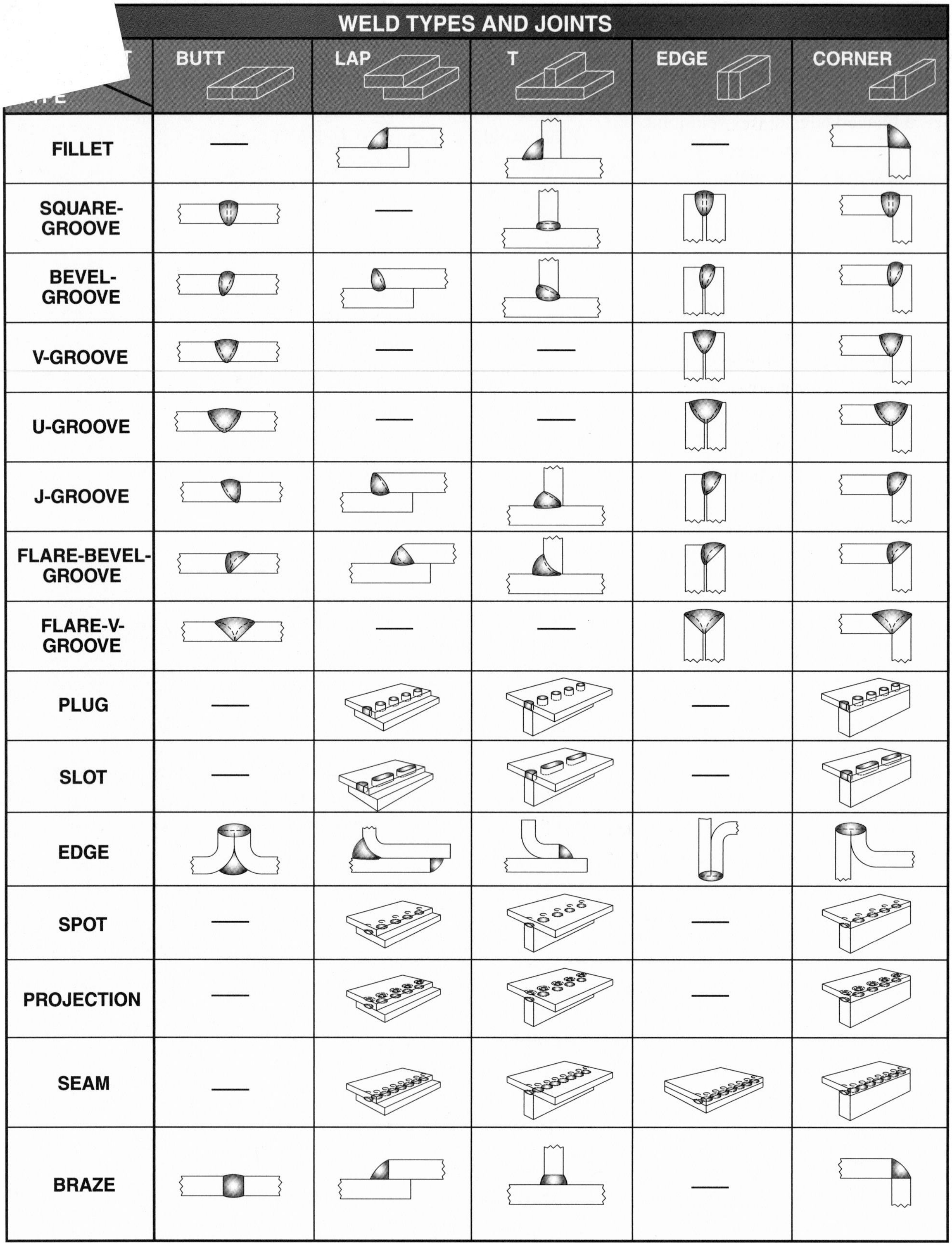

	BUTT	LAP	T	EDGE	CORNER
FILLET	—			—	
SQUARE-GROOVE		—			
BEVEL-GROOVE					
V-GROOVE		—	—		
U-GROOVE		—	—		
J-GROOVE					
FLARE-BEVEL-GROOVE					
FLARE-V-GROOVE		—	—		
PLUG	—			—	
SLOT	—			—	
EDGE					
SPOT	—			—	
PROJECTION	—			—	
SEAM	—				
BRAZE				—	

Figure 3-12. *The basic weld joints are used with applicable weld types to meet load requirements.*

Single-groove and double-groove welds are normally used for thick joints. A *single-groove weld* is a groove weld made from one side of the joint only. However, in some cases, a back weld may be applied on the root side of the joint. Single-groove welds include the following welds:

- square-groove
- single-bevel-groove
- single-V-groove
- single-U-groove
- single-J-groove
- single-flare-bevel-groove
- single-flare-V-groove

A *double-groove weld* is a groove weld made from both sides of the joint. Double-groove welds include the following welds:

- double-square-groove
- double-bevel-groove
- double-V-groove
- double-U-groove
- double-J-groove
- double-flare-bevel-groove
- double-flare-V-groove

The selection of a single-groove weld over a double-groove weld is dictated by material type and thickness, service conditions, whether the joint is accessible for welding from one side or both sides, filler metal cost, time required to deposit the weld, and joint preparation cost.

Plug and Slot Welds

Plug welds and slot welds may be used to join two overlapping members by welding through circular holes or slots. A *plug weld* is a weld made in a circular hole in one member to fuse it to the other member. A *slot weld* is a weld made in an elongated hole in one member to fuse it to the other member. Welding is done by completely filling the circular hole or slot to join the two members.

Plug welds were originally used during design transitions from riveted to welded structures. They are often used for joining sheet metal to a substrate to provide protection such as corrosion resistance.

Surfacing Welds

A *surfacing weld* is a weld applied to a surface, as opposed to making a joint, to obtain the desired properties or dimensions. Surfacing welds are commonly used to modify the properties of selected surfaces of a single component, such as an extruder. When a surfacing weld is used to increase wear resistance, this is known as hardfacing or fusion hardfacing.

A surfacing weld is different from thermal spraying. *Thermal spraying (THSP)* is a term that refers to a group of processes in which finely divided metallic or nonmetallic materials are deposited in a molten or semi-molten condition to form a coating. The surfacing material may be in the form of a powder, rod, cord, or wire. THSP includes arc spraying, flame spraying, and plasma spraying. Thermal spray hardfacing (non-fusion hardfacing) is the application of a thin layer of materials to the surface such that local melting does not occur.

Stud Welds

A *stud weld* is a weld produced by joining a metal stud or similar part to a member. During the welding process, part of the stud is melted, providing weld reinforcement at the base of the stud. Welding may be done with heat and pressure.

Spot and Seam Welds

A *spot weld* is a weld with an approximately circular cross section made between or upon overlapping members. Spot welds are specified in pounds per weld. A *seam weld* is a continuous weld made between or upon overlapping members that produces a continuous seam or series of overlapping spot welds. Seam welds are specified in pounds per inch of joint strength. In both spot and seam welds, coalescence (joining) may start and occur on the faying surfaces or may proceed from the outer surface of one member. The *faying surface* is the surface of a joint member that is in contact with, or in close proximity to, the member to which it will be joined.

> **✓ Point**
>
> *A plug weld or slot weld is used to join overlapping members through circular holes or slots made in one member.*

Projection Welds

A *projection weld* is a resistance weld produced by the heat obtained from the resistance to the flow of welding current. The resulting welds are localized at predetermined points by projections, embossments, or intersections. Projection welds are specified in pounds per weld. A weld strength greater than the strength of the minimum nugget size should be specified in the design. A *nugget* is the weld metal that joins the members in spot, seam, and projection welds.

Back Welds

A *back weld* is a weld made in the weld root opposite the face of the weld. A back weld is deposited after welding on the face side of the joint is completed. Back welds are usually made to improve the quality of the root pass. This is achieved by gouging or grinding out imperfections in the root pass, followed by depositing the back weld.

WELD DESIGNS

Weld joint designs are governed by American Welding Society (AWS) codes and other appropriate codes. For example, in building construction, AWS codes govern structural and welding materials, weld details, processes and techniques, weld quality, and inspection. The design of the structural elements is governed by American Institute of Steel Construction (AISC) specifications. The weld joint design selected must factor in wind forces, loads, seismic conditions, and other conditions that can cause fatigue. Additional codes such as the Uniform Building Code and other appropriate state and local codes may also apply.

The designer or engineer is responsible for determining the proper weld design to use. However, a welder should be aware of joint design requirements in order to produce a weld that better meets the established specifications for the job.

Weld Joint Selection

Welded joints provide strength and efficiency, and welding is quicker than other joining methods. Welded joints have replaced many parts and structures that previously used fasteners or the casting process. Most welded joints are subjected to loads that require strength and rigidity to prevent failure. Loads in a structure are transferred from member to member through the welds. Welds in joints subjected to minimum loads are considered "no-load" welds. For example, access covers, panels, and safety guards require "no-load" welds.

AWS Welding Positions. Weld joint selection is also affected by welding position. The four basic welding positions are flat, horizontal, vertical, and overhead. Each position has a number designation: flat = 1, horizontal = 2, vertical = 3, and overhead = 4. Each type of weld has a letter designation: groove weld = G and fillet weld = F. **See Figure 3-13.** Therefore, a 1G weld is a groove weld in the flat position, and a 2F weld is a fillet weld in the horizontal position.

Flat position is the most widely used welding position because welding can be done quickly and easily. Also, flat position welding allows for the greatest control of the welding process.

In the horizontal, vertical, and overhead positions, gravity affects the molten weld pool, making it difficult to control. This may produce weld discontinuities or defects like undercut, overlap (cold lap), excessive reinforcement, and incomplete or excessive penetration, resulting in weak or unacceptable welds.

Horizontal welding is difficult because the molten pool has a tendency to sag. Vertical welding is done in a vertical line from the bottom to the top (uphill) or from the top to bottom (downhill) of the joint. On thin material, a downhill welding technique helps to reduce heat input. Overhead welding is difficult because the molten metal sags due to heat build-up. A good welder can read the weld pool and make appropriate adjustments to the welding technique and to the power source to produce a uniform bead with the proper penetration.

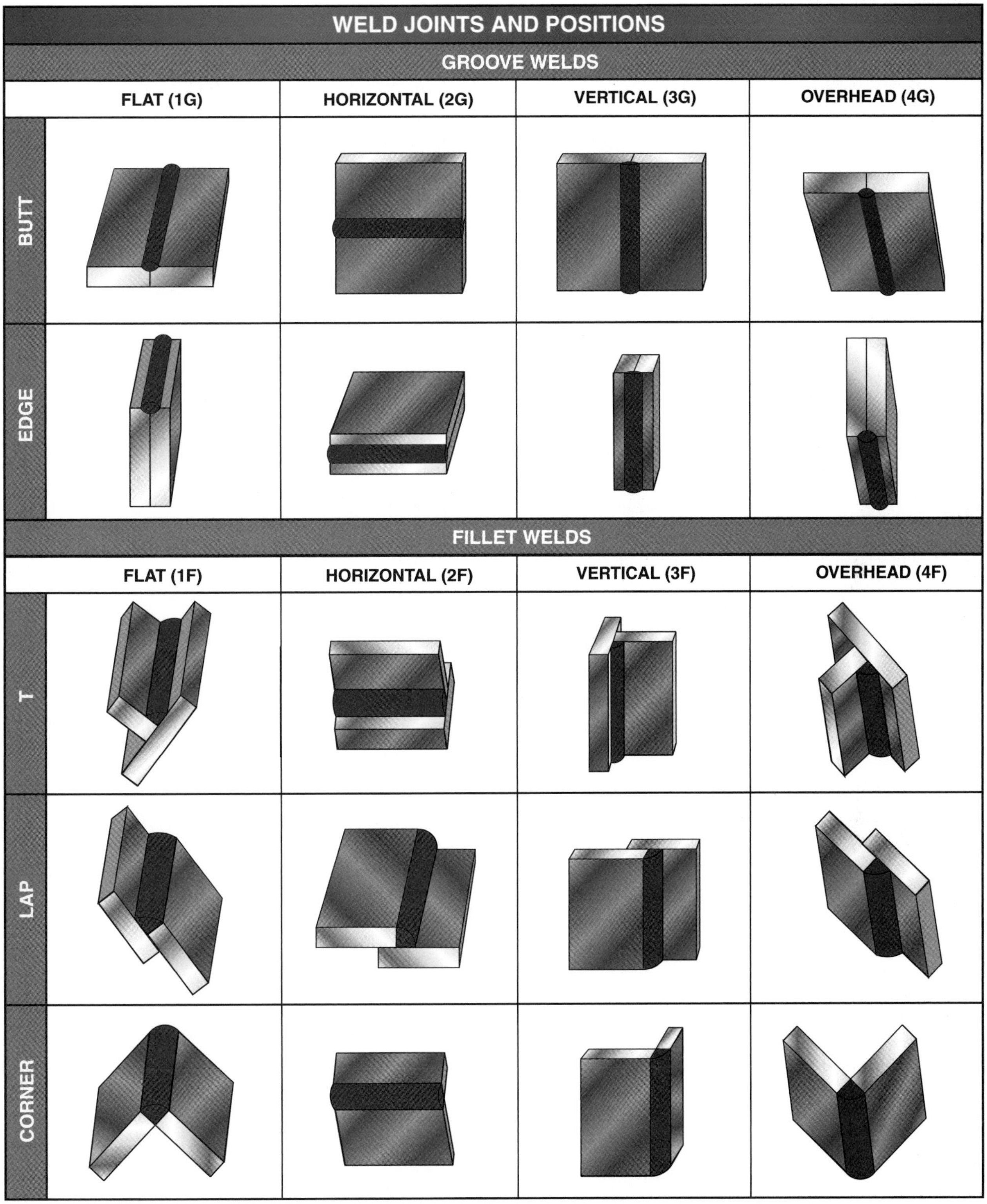

WELD JOINTS AND POSITIONS			
GROOVE WELDS			
FLAT (1G)	HORIZONTAL (2G)	VERTICAL (3G)	OVERHEAD (4G)

Figure 3-13. *The four common welding positions are flat, horizontal, vertical, and overhead. Each position has a number designation, with groove and fillet welds designated by "G" and "F" respectively.*

Small parts are commonly welded in the flat position for efficiency. Large parts that cannot be fixtured for flat position welding can be controlled using jigs, tack welds, spacers, or consumable inserts.

A *tack weld* is a weld used to hold members in proper alignment until the welds are made. Spacers provide a consistent root opening while the members are tack-welded in place. After tack welding, the spacers are removed and welding is performed. In some joints, consumable inserts are used as spacers. Consumable inserts are melted during the welding process and become part of the filler metal added to the weld joint.

Joint Preparation. For a quality weld, proper joint preparation is essential. Edges are commonly cut, sawed, or machined to provide good fit-up. Fit-up must be consistent along the entire joint. Proper joint preparation requires the following:

- Lap joints on sheet metal must be clamped tightly along the entire length of the joint to be welded. Root openings and bevels must be consistent along the entire length joint. The welder must compensate for any variations in joint preparation or fit-up by adjusting travel speed, arc length, or electrode manipulation to produce an acceptable weld.
- A correct groove angle is required for good bead shape and penetration. **See Figure 3-14.**

Groove Angle

60° — ELECTRODE

FULL PENETRATION

CORRECT

45°

INCOMPLETE PENETRATION

INCORRECT

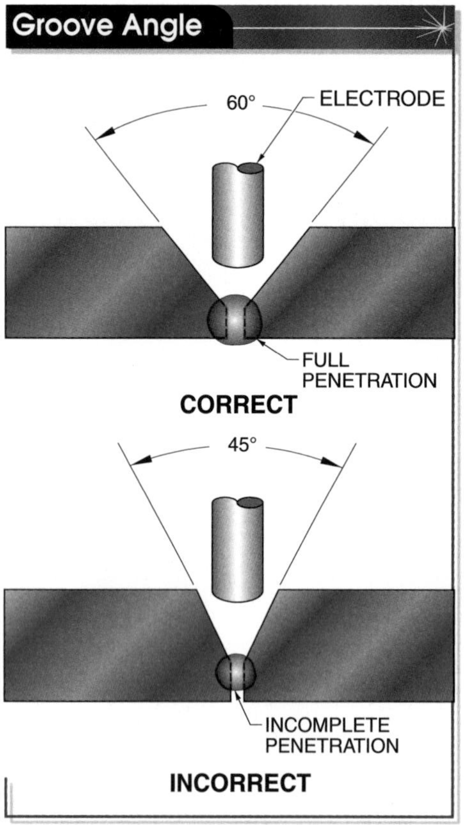

Figure 3-14. *The correct groove angle is essential for a good weld as it allows the electrode to access the root of the joint, which ensures complete penetration.*

- An insufficient bevel prevents the electrode from penetrating into the joint. Deep, narrow beads may lack penetration, and they have a tendency to crack. A large groove angle should be used in pipe welding to ensure complete penetration. **See Figure 3-15.**

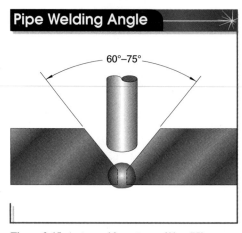

Pipe Welding Angle

60°–75°

Figure 3-15. *A pipe weld requires a 60° to 75° groove.*

- A proper groove angle of 60° to 75° should be maintained. A sufficient bevel is necessary for a quality bead; however, any excess bevel creates additional work for the welder and wastes filler metal. Filler metal is expensive, and any variation from the recommended groove angle contributes to excess cost, in both material and time. **See Figure 3-16.**

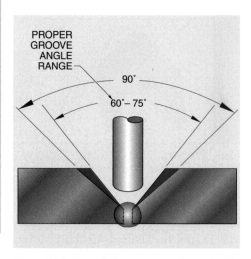

PROPER GROOVE ANGLE RANGE

90°

60°–75°

Figure 3-16. *An overly large groove angle wastes filler metal and time, resulting in greater welding costs.*

- Sufficient root opening is needed for full penetration. Without adequate penetration, a welded joint cannot withstand the loads imposed on it. Although penetration is affected by electrode manipulation, it is essential that the root opening is adequate to achieve full penetration. **See Figure 3-17.**

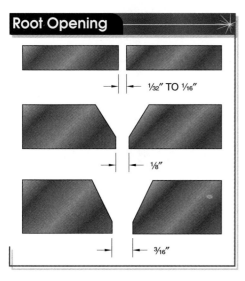

Figure 3-17. *Proper root opening size is required in order to make a sound weld. The root opening size is determined by the thickness of the metal.*

- Either a ⅛″ root face or a backing strip is required for a good quality weld on thicker materials. Feather-edge preparations require a slow, costly root bead. However, double-V-groove butt joints without a root face are practical when the root bead is offset by easier edge preparations and when the root opening is limited to about ³⁄₃₂″. **See Figure 3-18.**

Proper fit-up ensures that joint members are in correct alignment, have the correct edge preparation, and have the required root opening for proper penetration and sufficient weld reinforcement. **See Figure 3-19.** Members should be aligned edge-to-edge and end-to-end, and should lie in the same plane.

Joint Access. Sufficient joint access is required for the welding equipment and for the deposition of filler metal. For example, joints located in tight areas provide limited access to the large welding gun required for flux-cored arc welding. Fabricating a weldment in subassemblies can eliminate some access problems.

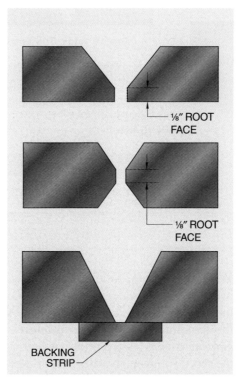

Figure 3-18. *The proper root face must be provided for a quality weld unless a backing strip is used.*

Joint design must take into consideration access to the joint by the welder.

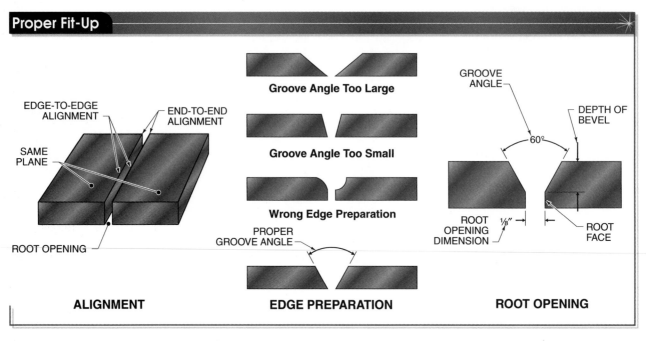

Figure 3-19. *Correct alignment, edge preparation, and root opening are necessary for proper fit-up.*

Welding Location. Welding is performed in the shop or in the field, depending upon the size and fabrication requirements of the structure. Small parts, structures, and subassemblies are often welded in the shop. The shop provides a controlled environment in which welding variables can be closely controlled. Additionally, fixtures and positioners can be used to move a part or hold a part in position for improved welding productivity.

A *fixture* is a device used to maintain the correct positional relationship between weldment components as required by print specifications. A *positioner* is a mechanical device that supports and moves weldments for maximum loading, welding, and unloading efficiency.

Positioners can be used with hand- and machine-controlled welding machinery. In production settings, positioners and welding equipment are used together for maximum welding efficiency.

Welding is performed in the field when equipment breaks down and it is impractical or impossible to transport it, or when the size of a fabrication prohibits assembly in the shop, such as a pipeline. Welding in the field often results in a decrease in welding productivity because additional variables are introduced that can influence the finished weld. Ambient temperature, weather, welding conditions, and welder efficiency in the field all affect welding productivity.

 Refer to Quick Quiz® on CD-ROM

- The root pass is the initial weld pass that produces the root bead, with complete or partial penetration into the root of the joint.
- Square-groove butt joints are typically used with materials ³⁄₁₆″ thick or less.
- A double-V-groove butt joint is suitable for all load conditions.
- A T-joint is formed when two members are positioned approximately 90° to one another.
- A lap joint is usually welded on both sides of the joint.
- A groove weld should use the smallest root opening and groove angle possible for the job to produce a sound weld with the least amount of filler metal.
- A plug weld or a slot weld is used to join overlapping pieces of metal through circular holes or slots made in one member.
- AWS codes and other applicable codes (API 1104 for pipe and ASME section IX for pressure vessels) provide guidelines and specifications for designing quality weld joints.

Questions for Study and Discussion

1. What factors must be considered when determining the type of joint to use in welding any structural unit?
2. What is a fillet weld?
3. In what type of joints are groove welds made?
4. What is a plug weld?
5. When is a surfacing weld used?
6. Why are grooved butt joints better for welding thick plates than square butt joints?
7. What are the basic types of T-joints?
8. Describe a double-fillet lap joint.
9. Which type of corner joint is the strongest?
10. What is the toe of a weld?
11. What is the root of a weld?
12. What are some of the basic principles that contribute to good joint-geometry?
13. When are double-bevel T-joints normally used?
14. Which butt joint requires the least amount of preparation before welding?
15. What is reinforcement of the weld?
16. How is the root opening size determined?
17. Why is a proper groove angle required?
18. How is the size of a weld leg determined?

Refer to Chapter 3 in the *Welding Skills Workbook* for additional exercises.

OAW–Equipment

Oxyacetylene welding does not require electricity and is typically used for maintenance, in body shops, and in the repair of small parts where other welding processes are too expensive.

Oxyacetylene welding can be used to join iron, steel, cast iron, copper, brass, aluminum, bronze, and other metals. Often, dissimilar metals such as steel and cast iron, brass and steel, copper and iron, and brass and cast iron can be joined with oxyacetylene welding. Oxyacetylene welding equipment can also be used for preheating, cutting metal, case hardening, annealing, soldering, and brazing.

OXYACETYLENE WELDING

Oxyacetylene welding (OAW) is one of the oldest welding processes. Its use has declined somewhat in popularity in recent years, but it is still widely used for welding pipes and tubes, as well as for repair work.

Oxyacetylene welding uses the heat from the combustion of a mixture of oxygen and acetylene (the fuel gas) to fuse base metals and filler metal together. This mixture produces flame temperatures in the range of 5800°F to 6300°F, which can be used to weld a variety of metals. Other fuel gases can be used, including methylacetylene-propadiene (MAPP), propane, natural gas, hydrogen, or propylene, but acetylene is the most common of the oxyfuel welding gases. Much of the same equipment and many of the same techniques and precautions apply when welding with these other mixtures.

An oxyacetylene torch is the primary welding component of the system. The oxygen and acetylene are connected to separate valves on the torch, which can be adjusted to vary the mixture

for the optimal heat production. The flammable gas mixture is lit with a sparklighter and the flame is directed at the weld joint with one hand. Typically, filler metal in the form of a thin rod is directed into the weld pool with the other hand, requiring the welder to coordinate the welding process with both hands. For this reason, the skills required for gas tungsten arc welding (GTAW) are often compared with those developed by practicing oxyacetylene welding.

Thermadyne Industries, Inc.

A mixture of oxygen and acetylene is used for most welding and cutting operations.

Proper eye protection is required for oxyacetylene welding, even though it produces lower light levels than arc welding. Unlike arc welding, the oxyacetylene welding process requires no electricity, which makes it well suited for welding in remote locations, such as farms or construction sites. However, it does require a set of cylinders to hold the compressed gases. The hoses that connect the valves to the torch, and often the cylinders themselves, are color-coded for safety. Acetylene hoses are red, and have left-hand threads, while oxygen hoses are green with right-hand threads to prevent them from being connected to the wrong supply or to the wrong side of the torch.

Oxyacetylene welding equipment is relatively portable, which is one of its advantages, though the gas cylinders must be transported securely in special carts. There are special hazards involved with acetylene gas. When installed in a facility, the gas cylinders must be stored in separate designated areas. For high-volume welding, multiple cylinders of each gas can be connected together with manifold systems that supply the welding gases to other areas of the facility.

Oxyacetylene welding gases and equipment can also be used for cutting, with a different type of torch. Most of the safety precautions that apply to oxyacetylene welding also apply to cutting operations, including gas cylinder safety, hot work safety, ventilation, and personal protective equipment.

OXYGEN FOR WELDING

The atmosphere (air) is comprised of approximately 20% oxygen. The majority of the atmosphere is made up of nitrogen with a percentage of rare gases such as helium, neon, and argon. For oxygen to be usable for welding, it must be separated from the other gases. The two methods that can be used to isolate oxygen are the liquid-air and the electrolytic methods.

The liquid-air method of producing oxygen draws air from the atmosphere into huge containers called washing towers. In the washing towers, the air is washed and purified of carbon dioxide. A solution of caustic soda is circulated through the towers by means of centrifugal pumps to wash the air.

As the air moves out of the washing towers, it is compressed and passed through oil-purging cylinders. In the oil-purging cylinders, oil particles and water vapor are removed. From the oil-purging cylinders, the air moves into drying cylinders. The drying cylinders contain dry, caustic potash that dries the air and removes any remaining carbon dioxide and water vapor. At the top of each drying cylinder are special cotton filters to prevent particles of foreign matter from being carried into the high-pressure lines.

The dry, clean, compressed air then goes into rectifying or liquefaction columns where the air is cooled and expanded to approximately atmospheric pressure. As the pressure is lowered, the extremely high-pressure, cold air cools and liquefies.

The separation of the nitrogen from the oxygen is possible once the air has liquefied because nitrogen and oxygen have different boiling points. Nitrogen boils at $-320°F$ and oxygen at $-296°F$. The nitrogen, having a lower boiling point, evaporates first, leaving the liquid oxygen at the bottom of the condenser. The isolated liquid oxygen passes through a heated coil, which changes the liquid oxygen into a gaseous form. After the gas moves through the heated coil, it is stored in a storage tank. A gas meter mounted between the heating coil and the storage tank registers the amount of gas entering the storage tank. The stored oxygen gas can then be drawn from the storage tank and compressed into receiving cylinders.

The electrolytic method is a process that uses water and electricity to isolate oxygen. Water is a chemical compound consisting of oxygen and hydrogen. By sending an electrical current through a solution of water containing caustic soda, oxygen is given off at one terminal plate, and hydrogen at the other. The oxygen, having been separated from the hydrogen, is suitable for welding. The electrolytic method is a very expensive method of producing oxygen; for this reason the liquid-air method is more commonly used to produce commercial oxygen.

Oxygen Cylinders

Oxygen cylinders are made from seamless drawn steel and tested with a water (hydrostatic) pressure of 3360 psi. The cylinders are equipped with a high-pressure valve that can be opened by turning the handwheel on top of the cylinder. The valve handwheel should always be opened by hand and not with a wrench. The handwheel must be turned slowly to permit a gradual pressure load on the regulator. The valve handwheel is turned to full open position. This provides a seal to reduce leakage from the valve. A protector cap screws onto the neck ring of the cylinder to protect the valve from damage. The protector cap must always be in place when the cylinder is not in use. **See Figure 4-1.**

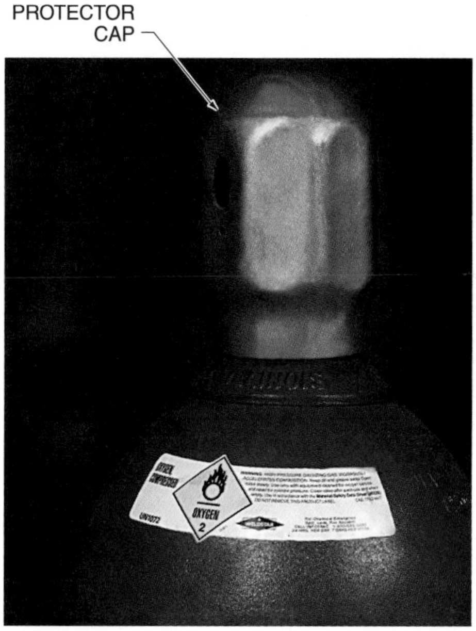

There are three common sizes of oxygen cylinders. The large cylinder holds 244 cubic feet (cu ft) of oxygen. The large size is commonly used in industrial plants and shops that require large quantities of gas. A medium-size cylinder can contain 122 cu ft of oxygen and a small cylinder can hold 80 cu ft.

Cylinders are charged with oxygen at a pressure of 2200 psi at a temperature of 70°F. Gases expand when heated and contract when cooled, so the oxygen pressure will increase or decrease as the temperature changes. For example, if a full cylinder of oxygen is allowed to stand outdoors in near-freezing temperatures, the pressure of the oxygen will register less than 2200 psi. However, none of the oxygen has been lost; cooling has only reduced the pressure of the oxygen.

Since the pressure of gas varies with the surrounding temperature, all oxygen cylinders are equipped with a safety nut that permits the oxygen to drain slowly if the temperature increases the cylinder pressure beyond its rated safety load. If a cylinder were exposed to a hot flame, the safety nut would relieve the pressure before the cylinder reached its exploding point.

250 lbs Acytline

ACETYLENE FOR WELDING

Acetylene is a colorless gas with a very distinctive odor. Acetylene is highly combustible when mixed with oxygen. A fully charged acetylene cylinder has a pressure of about 250 psi. However the operating pressure for acetylene should never exceed 15 psi. Above 15 psi, acetylene becomes unstable.

Acetylene gas is formed by the mixture of calcium carbide and water. The commercial generator in which the gas is produced consists of a large tank containing water. A specified quantity of carbide is put into a hopper and raised to the top of the generator. The carbide is then allowed to fall into the water. As the carbide meets the water, bubbles of gas are given off. The gas is collected, purified, cooled, and slowly compressed into cylinders.

PROTECTOR CAP

Figure 4-1. *A protector cap screws onto the neck ring of the cylinder to protect the valve from damage when not in use.*

Acetylene Cylinders

To ensure the safe storage of acetylene, the cylinder is packed with a porous material that is saturated with acetone. Acetone is a liquid chemical that stabilizes acetylene and allows the cylinder to be pressurized to 250 psi. The acetylene cylinder is equipped with a fusible plug that melts, relieving excess pressure, if the cylinder is subjected to any mechanical pressure or undue heat, such as from a fire. **See Figure 4-2.** Acetylene cylinders should never be laid down as the corrosive nature of the acetone can erode the seals in the tanks.

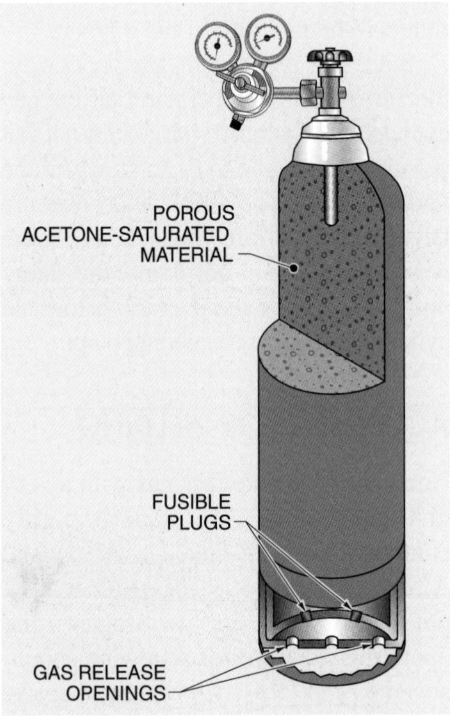

Figure 4-3. *The acetylene cylinder valve should be opened with the valve handle (if installed) or with a proper valve wrench.*

When a considerable amount of welding is to be performed in an area, as in industry or in a school welding shop, acetylene cylinders are frequently connected to a manifold system with pipelines carrying the gas to the welding stations. **See Figure 4-4.** The demand for acetylene is usually higher than can be supplied by a single cylinder, so a manifold system is commonly needed. A multiple cylinder manifold system allows the necessary volume of acetylene to be supplied to the work area. Acetylene can be drawn off no faster than one-seventh the total volume of the cylinder per hour, which is the quickest the acetylene can be released from the acetone lining in the cylinder. A flash arrestor is also used in the manifold system to prevent a flashback from reaching the stored cylinders.

POROUS
ACETONE-SATURATED
MATERIAL

FUSIBLE
PLUGS

GAS RELEASE
OPENINGS

Figure 4-2. *An acetylene cylinder is packed with a porous material that is saturated with acetone to allow the safe storage of acetylene.*

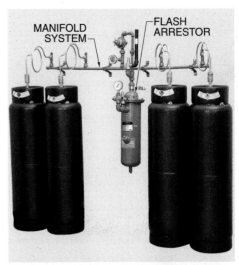

MANIFOLD
SYSTEM

FLASH
ARRESTOR

ESAB Welding and Cutting Products

Figure 4-4. *Acetylene cylinders are connected to a manifold system in areas where a high volume of welding is to be performed.*

The cylinder valve is opened with an installed valve handle. **See Figure 4-3.** The cylinder valve should never be opened more than one complete turn. It is advisable to open the cylinder only slightly so the valve can be closed quickly in case of an emergency.

Flash Arrestors. A *flash arrestor* is a safety device that prevents an explosion or a backfire in the torch or torch head from reaching the regulator and the acetylene cylinder. Two types of flash arrestors are the torch-mounted and the regulator-mounted.

The torch-mounted flash arrestor is a check valve that prevents a reverse gas flow from reaching the cylinder. The regulator-mounted flash arrestor is a combination check valve and flame barrier. The barrier metal is a porous flame-retardant material that allows gas to flow through, but blocks out a flame. Torch-mounted and regulator-mounted flash arrestors should always be used on fuel hoses and oxygen hoses. Regulator-mounted flash arrestors prevent backfires and flashbacks from entering the hoses, and possibly the cylinders.

A backfire is caused by the flame going out suddenly on the torch. A backfire may occur when the tip is touched against the workpiece; if the flame settings are too low; if the tip is dirty, damaged, or loose; or if the tip is overheated.

When a torch backfires, it could cause a flashback. A flashback is a condition in which the flame burns inside the tip, the torch, or the hose. In case of a flashback, the oxygen and fuel valves must be immediately closed to prevent possible explosion of the cylinders. Flashbacks are typically caused by malfunctioning equipment. If a flashback occurs, the equipment should be removed from service and a service technician called to correct the problem or replace the equipment. Hoses should be discarded after a flashback. The torch tip is reusable, but it should be removed from the torch and thoroughly blown out with air to remove any soot or residue.

SAFE HANDLING OF CYLINDERS

Cylinder carts should be used to move cylinders. Cylinders should be chained in place, and the protector caps should be in place over the valves before the cylinders are moved. **See Figure 4-5.** Follow these safety precautions when handling oxygen and acetylene cylinders:

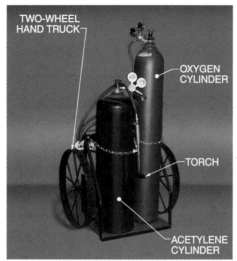

Figure 4-5. *To safely move a cylinder, always use a cylinder cart with the cylinder chained in place, and make sure the protector cap is installed.*

- Never lift a cylinder by the protector cap.
- Always keep cylinders in a vertical position.
- Do not allow grease or oil to come in contact with cylinder valves. Although oxygen is in itself nonflammable, it quickly aids combustion if exposed to flammable materials.
- Avoid exposing cylinders to furnace heat, radiators, open fire, or sparks from a torch.
- Never transport a cylinder by dragging, sliding, or rolling it on its side. Avoid striking it against any object that might create a spark, as there may be just enough gas escaping from the cylinder to cause an explosion.
- Close cylinder valves completely before moving cylinders.
- Do not tamper with or attempt to repair cylinder valves. If valves leak or do not function properly, notify the supplier immediately.
- Keep valves closed on empty cylinders.
- Do not use a hammer or wrench to open cylinder valves. If they cannot be opened by hand or with a T-wrench, notify the supplier.
- Keep cylinders covered with valve protector caps when not in use.
- Cylinders should be chained in position at all times during use and when stored. Cylinders in use should be securely attached to a hand cart, or chained near the work station. **See Figure 4-6.**

ACETYLENE
CYLINDERS

Figure 4-6. *Cylinders should be chained at all times during use and when stored.*

WELDING EQUIPMENT

Welding equipment consists of a torch with an assortment of different-sized tips; two lengths of hose, one red for acetylene and the other green for oxygen; two pressure regulators; two cylinders, one containing acetylene and the other oxygen; a welding sparklighter; and a pair of goggles. **See Figure 4-7.**

Cylinders are typically chained to a two-wheel hand truck to permit moving the equipment to a desired location. If the cylinders are positioned near the workbench, they should be chained to a fixed object.

Welding Torches

The welding torch, or blowpipe, is a tool that mixes acetylene and oxygen in the correct proportions and permits the mixture to flow to a tip, where it is burned. Although torches vary to some extent in design, they are made to provide complete control of the flame during the welding operation. **See Figure 4-8.**

The two primary types of torches are the medium-pressure and the injector. The medium-pressure torch requires acetylene pressures of 1 psi to 10 psi. The injector torch is designed to use acetylene at very low pressures (0 up to 1 psi). Both types of torches operate when acetylene is supplied from cylinders or medium-pressure generators.

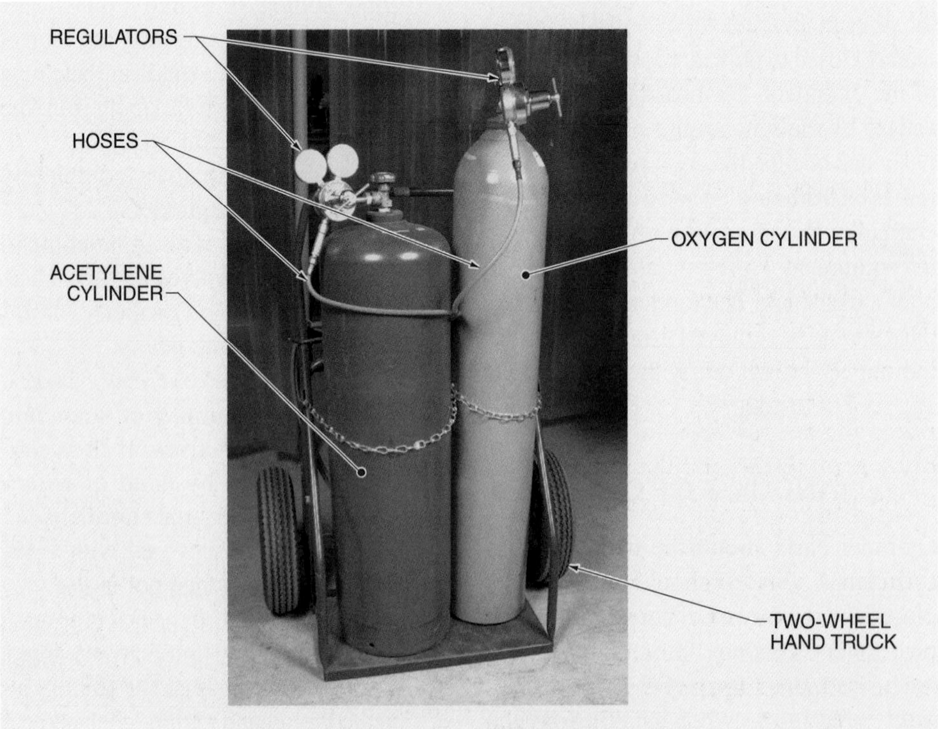

REGULATORS

HOSES

ACETYLENE
CYLINDER

OXYGEN CYLINDER

TWO-WHEEL
HAND TRUCK

Figure 4-7. *Cylinders can be chained to a two-wheel hand truck for easy transportation.*

Figure 4-8. *An oxyacetylene welding torch directs the flame during welding.*

In a medium-pressure torch, the oxygen and acetylene are fed independently to a mixing chamber, after which they flow out through the tip. In an injector torch, the oxygen, as it passes through a small opening in the injector nozzle, draws acetylene into the oxygen stream. When small fluctuations in the oxygen supply occur, a corresponding change occurs in the amount of acetylene drawn, maintaining consistent proportions of the two gases while the torch is in operation. The medium-pressure torch is the most commonly used torch.

Both types of torches are equipped with two needle valves; one regulates the flow of oxygen at the torch and the other regulates the flow of acetylene at the torch. At the base of the torch are two fittings for connecting each hose. To eliminate any chance of interchanging the hoses, the oxygen fitting is made with a right- hand thread and the acetylene fitting is made with a left-hand thread.

Care of Torches. When welding is completed, the torch should be properly secured to prevent it from falling and becoming damaged. Needle valves are especially delicate, and if the torch drops and strikes a hard object, the needle valves can break easily. Needle valves may loosen and turn too freely, making it difficult to keep the proper adjustment for the required mixture. When the needle valves loosen, the packing nuts on the stem of the needle valves should be tightened with a slight turn of a wrench. **See Figure 4-9.**

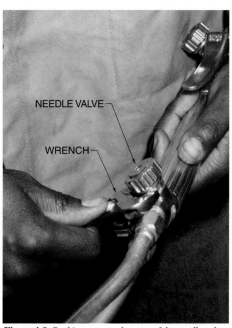

NEEDLE VALVE

WRENCH

Figure 4-9. *Packing nuts on the stem of the needle valves should be tightened with the correct wrench.*

Welding Tips

Welding on different thicknesses of metal is possible because torches are equipped with an assortment of different size heads, or tips. The size of the tip is governed by the diameter of its opening, which is marked on the tip.

Care of Welding Tips. A welding tip is designed to be installed and removed by hand. Frequent torch use causes carbon to form in the passage of the tip. Carbon must be removed from the tip regularly to ensure the free flow of gas. **See Figure 4-10.** To clean a torch tip, follow the procedure:

1. File the end of the tip flat with a metal file.
2. Insert a properly-sized tip cleaner into the tip and pull it straight out. Repeat until the tip is clean.

Cleaning a Torch Tip

1 FILE TIP FLAT

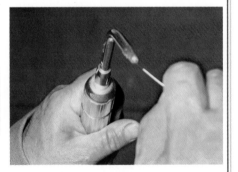

2 INSERT TIP CLEANER

Figure 4-10. *A tip cleaner is used to clean a torch tip.*

Regulators

Oxygen and acetylene pressure regulators perform two functions. They control the flow of gas from the cylinder to maintain the required working pressure, and they produce a steady flow of gas under varying cylinder pressures.

Regulators are equipped with two gauges—a cylinder pressure gauge, which indicates the actual pressure in the cylinder, and a working-pressure gauge, which shows the working, or line, pressure used at the torch.

The oxygen cylinder pressure can be as high as 2200 psi. The required working pressure for oxygen is from 1 psi to 25 psi. The acetylene cylinder pressure can be as high as 250 psi. The working pressure for acetylene must be between 1 psi and 12 psi. It should never exceed 15 psi. The regulator must maintain the proper working pressure, even as the cylinder pressure changes. If the oxygen in the cylinder is below 1800 psi and a pressure of 6 psi is needed at the torch, the regulator must maintain a constant pressure of 6 psi even if the cylinder pressure drops to 500 psi.

The oxygen cylinder pressure gauge is a graduated scale up to 4000 psi. A second scale on the gauge is calibrated to register the contents of the cylinder in cubic feet. The oxygen working-pressure gauge is graduated in divisions from 0 psi to 60 psi and the acetylene working-pressure gauge is graduated in divisions from 0 psi to 30 psi. The acetylene working-pressure gauge is usually marked with a warning color above 15 psi. The acetylene cylinder pressure gauge is graduated up to 350 psi or 400 psi. **See Figure 4-11.**

The two types of regulators are the single-stage and the two-stage. The single-stage regulator is typically less expensive than the two-stage type. With the single-stage regulator, there is no intermediate chamber through which gas passes before it enters the low-pressure chamber. The gas from the cylinder flows into the regulator and is controlled entirely by the adjusting screw.

Regulators

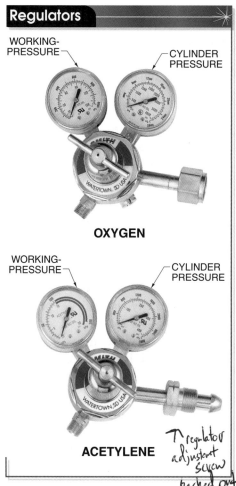

T regulator adjustment screw backed out

Figure 4-11. *Oxygen and acetylene regulators control the flow of gas for welding. An acetylene regulator has a red line above 15 psi, indicating a dangerous operating range.*

A single-stage regulator must be continually adjusted to maintain correct working pressure. The adjusting screw on a regulator must be released (turned out) before the cylinder valve is opened. If the adjusting screw is not released and the cylinder valve is opened, the tremendous pressure of the gas in the cylinder, forced onto the working-pressure gauge, may blow out the screw and damage the regulator.

The adjusting screw is turned to increase or decrease the gas pressure from the torch to the regulator by controlling the force of a spring on the flexible diaphragm. The diaphragm moves a valve, allowing gas to flow into the regulator. As the gas pressure in the regulator increases, it bends the diaphragm back, closing the valve. During welding, the

regulator reduces the gas pressure behind the diaphragm and the spring opens the valve, allowing gas to flow. The change in internal pressure is registered on the working-pressure gauge.

With the two-stage regulator, the reduction of the cylinder pressure to that required at the torch is accomplished in two stages. In the first stage, the gas flows from the cylinder into a high-pressure chamber. A spring and diaphragm keep a predetermined gas pressure in the chamber. For oxygen, the pressure is usually 200 psi, and for acetylene, 50 psi. From the high-pressure chamber, the gas passes into a reducing chamber. Control of the pressure in the reducing chamber is governed by an adjusting screw.

When acetylene and oxygen are mixed correctly and ignited, the flame can reach temperatures of 5700°F to 6300°F, which melts commercial metals so completely that they flow together to form a complete bond without the application of any mechanical pressure or hammering. Filler metal is usually added to the molten metal to build up the joint for greater strength. On very thin metals, the edges are generally flanged and melted together. In either case, if the weld is performed correctly, the section where the bond is made is as strong as the base metal.

Care of Regulators. Regulators are sensitive instruments and must be treated as such. A slight jolt can render a regulator useless. Regulators should be handled extremely carefully when being removed from the cylinder. Never leave a regulator on a bench top or floor for any length of time as it could be moved and damaged. General guidelines for the care of regulators include the following:

- Check the adjusting screw before the cylinder valve is turned ON and release it when welding has been completed.
- Never use oil on a regulator. Use only soap or glycerin to lubricate the adjusting screw.

✓ **Point**

Be sure the adjusting screw on the regulator is fully released before opening the cylinder valve.

1st d last step "Back out regulator"

- Do not attempt to interchange the oxygen and acetylene regulators.
- If a regulator does not function properly, shut OFF the gas supply and have a qualified service technician check the regulator.
- Check the regulator regularly for creeping. If the regulator creeps (does not remain at set pressure), have it repaired immediately. Creeping can be seen on the working-pressure gauge after the needle valves on the torch are closed. A creeping regulator usually requires that the valve seat or stem be changed.
- Check the mechanisms regularly. If the gauge pointer fails to go back to the pin when the pressure is released, the mechanism is likely sprung, caused by pressure entering the gauge suddenly. This condition should be repaired.
- Always keep a tight connection between the regulator and the cylinder. If the connection leaks after tightening,

close the cylinder valve and remove the regulator. Clean both the inside of the cylinder valve seat and the regulator inlet-nipple seat. If the leak persists, the seat and threads are probably marred, and the regulator must be returned to the manufacturer for repair.

Check Valves

A *check valve* is a valve that allows the flow of liquid or gas in one direction only. **See Figure 4-12.** In welding equipment, the pressure in the supply hose is higher than the pressure in the torch, allowing a valve disk in the check valve to open and release the gas into the torch. If the pressure in the torch becomes higher than that in the supply hose, such as when a flashback occurs, the valve disk closes, shutting OFF the supply of gas to the torch. A check valve must be positioned at the torch inlet, and can also be placed at the regulator outlet. The check valve must be replaced if a flashback occurs.

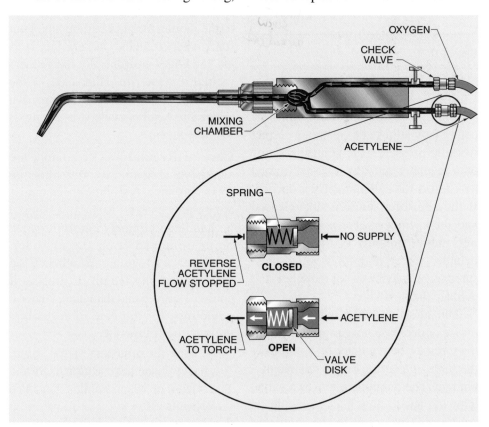

Figure 4-12. *A check valve is connected to the torch and to each hose to prevent acetylene or oxygen from traveling back through the hoses in the event of a flashback.*

Oxygen and Acetylene Hoses

A special nonpermeable hose is used for welding. To prevent the hoses from being misconnected, the oxygen hose is always green in color and the acetylene hose is red. Hoses must be properly marked because if oxygen were to pass through a hose that had previously contained acetylene, a dangerous combustible mixture might result.

A standard connection is used to attach the hose to the regulator and torch. The connection consists of a nipple that is forced into the hose and a nut that connects the nipple to the regulator and the torch. The acetylene nut can be distinguished from the oxygen nut by the notch that runs around the center, indicating a left-hand thread. **See Figure 4-13.** A clamp is used to squeeze the hose around the nipple to prevent it from working loose.

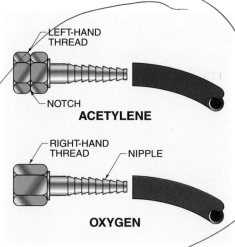

Figure 4-13. *The nut on the acetylene connection has a notch that runs around the center, distinguishing it from the nut on the oxygen connection.*

In North America, green (oxygen) and red (acetylene) are the standard colors used for hoses. In Europe, blue is used for oxygen and orange for acetylene. Some parts of the world use black for oxygen hoses.

Care of Welding Hoses. All hose connections must be tight. The connections should be tightened with a close-fitting wrench to prevent damage to the nuts. Do not drag the hose across a greasy floor, as grease or oil can eventually soak into, and erode, the hose. The hose should not be pulled around sharp objects or across

hot metal, and should be positioned so that it cannot be stepped on or damaged. When welding has been completed, the hose should be rolled up and suspended so that it will not drop to the floor. Also note these additional precautions:

- All new hose is dusted with talcum powder inside. The powder should be blown out with dry, grease-free shop air before first use.
- Long lengths of hose tend to kink. Use the shortest length of hose to properly service the shop.
- Do not try to repair a leaking hose, replace it with a new hose assembly.

Sparklighters

A sparklighter, or striker, is a tool used for igniting the torch. **See Figure 4-14.** A sparklighter should always be used to light a torch. Never use matches or lighters to light a torch because the flame produced by acetylene can cause serious burns.

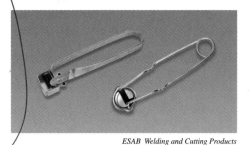

ESAB Welding and Cutting Products

Figure 4-14. *A sparklighter (striker) is used for lighting a torch.*

Goggles and Face Shields

Oxyacetylene welding produces sparks, spatter, and intense light that may injure the eyes if they are not properly shielded. Goggles or face shields with the proper filter-plate shade number should always be worn when gas welding or cutting to provide adequate eye protection. For most oxyfuel welding operations, filter-plate shade numbers of 4, 5, and 6 are recommended. **See Figure 4-15.** The American Welding Society (AWS) produces standards for eye protection that have additional information on the correct shielding for each welding operation.

Point

Never interchange oxygen and acetylene hoses. Avoid dragging them over greasy floors.

Point

Always use a striker (sparklighter) to light a torch. Never use a match or a lighter.

Point

Wear goggles or a face shield with the proper filter plate along with other PPE appropriate for the job.

TYPE OF GAS WELDING		PLATE THICKNESS	SHADE NUMBER
Gas Welding			
	Light	less than 1/8″	4 or 5
	Medium	1/8″ to 1/2″	5 or 6
	Heavy	over 1/2″	6 or 8
Oxygen Cutting			
	Light	less than 1″	3 or 4
	Medium	1″ to 6″	4 or 5
	Heavy	over 6″	5 or 6

Figure 4-15. *Goggles or a face shield with the recommended filter-plate shade number should always be worn during welding.*

Protective Clothing

An apron, shop coat, or coveralls should always be worn when welding with oxyacetylene equipment. Sparks commonly shoot away from the molten metal and, unless suitable covering is worn, will burn holes in clothes. Sparks that burn through clothes may also burn the skin. Under no circumstances should flammable garments be worn when welding. A small spark that falls on flammable garments may burst into a rapidly spreading flame. A welding cap should also be worn to prevent hot metal particles from falling on the hair.

A pair of lightweight gloves should be worn to prevent burns. Occasionally the hot end of filler metal or a piece of metal that has been set down to cool is picked up by mistake, and without gloves, serious burns may result.

OTHER WELDING GASES

Although acetylene is commonly used for certain types of welding, other gases may be used. The most common of these are methylacetylene-propadiene stabilized, more commonly known as MAPP gas, and hydrogen. The principal difference between these gases and acetylene is in the properties of the gas used in the burning mixture; the welding technique is the same.

MAPP Gas

Acetylene produces a very high flame temperature but is very unstable. MAPP gas has many of the physical properties of acetylene, but lacks the shock sensitivity of acetylene. MAPP gas is the result of a rearrangement of the molecular structures of acetylene and propane. When the two gases are combined, their molecular structure is changed and a very stable fuel results, with a flame temperature nearly comparable to acetylene.

Although propane itself is very stable, its low flame temperature limits its capabilities for welding. MAPP gases can be used for welding if the fuel-to-oxygen ratio is increased to raise the temperature of the flame. Deoxidized filler metal must also be used to ensure a sound weld when using MAPP gas for welding.

Generally, a slightly larger welding tip is required with MAPP gas because of its greater gas density and slower flame propagation rate. The only significant difference is in the flame appearance. A neutral flame for welding will have a longer inner cone than with oxyacetylene gas.

Since MAPP gas is not sensitive to shock, it can be stored and shipped in lighter cylinders. Because acetylene must be stored in cylinders filled with a porous filler material saturated with acetone, empty acetylene cylinders weigh about 220 lb. Empty MAPP cylinders weigh only 50 lb. Normally, a filled cylinder of acetylene weighs 240 lb while a filled cylinder of MAPP gas weighs 120 lb.

Hydrogen

The combination of oxygen and hydrogen generates a low-temperature flame used primarily for welding thin sections of metal, usually aluminum, on which low temperatures are required. One of the unusual characteristics of an oxyhydrogen flame is that the flame is difficult to see. Consequently, it is often difficult to adjust for a neutral flame. To avoid welding with an oxidizing flame, the regulator should be adjusted for an accurate hydrogen flow before adjusting the oxygen. Oxyhydrogen welding is commonly used for underwater welding as it can be used at higher pressures than acetylene.

  Refer to Quick Quiz® on CD-ROM

⚠ Points to Remember

- Handle oxygen and acetylene cylinders with care. Never expose them to excessive heat and prevent contact with oil and grease.
- Always hang up a torch when not in use to prevent it from dropping to the floor and being bent or damaged.
- Be sure the adjusting screw on a regulator is fully released before opening a cylinder valve.
- Do not lubricate the adjusting screw on a regulator with oil. Use soap or glycerin.
- Never interchange oxygen and acetylene hoses. Avoid dragging them over greasy floors.
- Wear goggles or face shield with the proper filter plate along with other PPE appropriate for the job.
- Always use a striker (sparklighter) to light a torch. Never use a match or a lighter.
- Always use shop air to remove dirt and dust from clothing or to clean the work area. Never use air or oxygen blown through the torch.

1. What safety devices are used to prevent cylinders from exploding when subjected to intense pressure?
2. What is the purpose of the protector cap on a cylinder?
3. How much should the cylinder valve be opened on an acetylene cylinder? On the oxygen cylinder?
4. Why is it dangerous to allow grease or oil to come in contact with the oxygen cylinder valve?
5. What is the function of the needle valves on a welding torch?
6. Why are the oxygen and acetylene hose fittings made with different screw threads?
7. How is the size of a welding tip indicated?
8. What could happen if pliers are used when removing welding tips?
9. What is a tip cleaner? When and why should it be used?
10. What is a two-stage pressure regulator?
11. What precautions should be observed in handling a pressure regulator?
12. Why is it dangerous to light a torch with a match or a lighter?
13. What are the advantages and disadvantages of using MAPP gas?
14. Hydrogen is often used instead of acetylene for what operation?
15. What welding goggle shade numbers are commonly used for most oxyacetylene welding?
16. What type of protective clothing is commonly worn when oxyacetylene welding?
17. Name three ways of distinguishing oxygen hoses from acetylene hoses. Name three ways of distinguishing oxygen fittings from acetylene fittings.
18. Who is responsible for repairing a damaged regulator?
19. How are oxygen and acetylene cylinders moved safely?

Refer to Chapter 4 in the *Welding Skills Workbook* for additional exercises.

OAW–Setup and Operation

Chapter

Welding equipment must be assembled in the correct sequence to ensure safe operation. Once the equipment is assembled, the torch can be lit and adjusted.

Oxygen and acetylene cylinders must be safely stored when not in use. When stored, cylinders must be chained in an upright position, with the oxygen cylinders separated from the acetylene cylinders. When in use, cylinders can be secured on a hand truck, chained to a secure object such as a bench in the shop, or secured in position adjacent to a manifold system.

WELDING EQUIPMENT ASSEMBLY

Before assembling welding equipment, cylinders must be securely fastened to a hand truck or some fixed object. Remove the protector cap from each cylinder and examine the outlet nozzles closely. Make sure the connection seat and screw threads are not damaged. A damaged screw thread may ruin the regulator nut, while a poor connection seat causes the gas to leak. **See Figure 5-1.** To assemble welding equipment follow the procedure:

1. Crack the cylinder valves by opening and closing them quickly to remove foreign matter. Particles of dirt can collect in the outlet nozzle of the cylinder valve. Wipe out the connection seat with a clean cloth. If not cleaned out, the dirt can work into the regulator when the pressure is turned ON.

2. Connect oxygen regulator and hose. Connect the oxygen regulator to the oxygen cylinder and the oxygen hose to the oxygen regulator. Use a cylinder wrench to tighten the nuts and avoid stripping the threads.

Always use the proper size wrench to tighten the nuts. A loose-fitting wrench will eventually wear the corners of the regulator nuts.

3. Connect the acetylene regulator and hose. Connect the acetylene regulator to the acetylene cylinder and the acetylene hose to the acetylene regulator.

4. Purge hoses. Check the adjusting screw on each regulator to ensure that it is released, then open the cylinder valves. Blow out any dirt that may be lodged in the hoses by opening the regulator adjusting screws. Opening the adjusting screws slightly will also purge the hoses of any residual gases. Promptly close the regulator adjusting screws.

5. Connect check valves and hoses to torch. To prevent the reverse flow of gases that would result in a combustible mixture in the welding hose, check valves are mounted to the welding torch. Under normal conditions, gases flow toward the welding torch.

✓ **Point**

Cylinders must be properly secured to prevent damage and possible injury.

Welding Equipment Assembly

1 CRACK THE CYLINDER VALVE OPEN AND CLOSED TO CLEAR OUT DEBRIS

OXYGEN REGULATOR

2 ATTACH OXYGEN REGULATOR AND HOSE

ACETYLENE REGULATOR

3 ATTACH ACETYLENE REGULATOR AND HOSE

4 PURGE HOSES

[handwritten: 1½ wrench]

[handwritten: Get 1½ wrench]

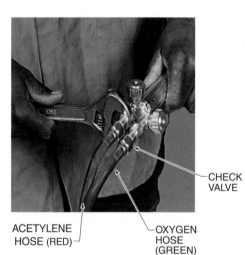

CHECK VALVE

ACETYLENE HOSE (RED)

OXYGEN HOSE (GREEN)

5 CONNECT CHECK VALVES AND HOSES TO TORCH

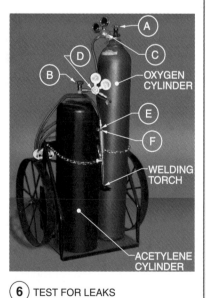

A
C
D
B

OXYGEN CYLINDER
E
F
WELDING TORCH
ACETYLENE CYLINDER

6 TEST FOR LEAKS

Figure 5-1. *Welding equipment must be properly assembled to ensure proper and safe operation during welding.*

Any condition that might cause a reverse flow of gas will close the valve. Check valves should be left in place on the torch when the hose is detached.

Connect the hoses to the check valves mounted on the torch. The red hose is connected to the acetylene check valve mounted on the needle valve fitting marked AC. The green hose is connected to the check valve mounted on the needle valve fitting marked OX. Acetylene hose connections always have left-hand threads as indicated by the notched nut, and oxygen hose connections have right-hand threads.

6. Test for leaks. All new welding equipment must be tested for leaks before being operated. It is advisable to periodically test equipment in service to ensure that no leaks have developed. Leaking gas may be exposed to a spark and develop into a fire. Additionally, leaks mean that gas is wasted.

To test for leaks, open the oxygen and acetylene cylinder valves and, with the needle valves on the torch closed, adjust the regulators to approximately normal working pressure. Apply soapy water with a brush on the following points:

A—Oxygen cylinder valve

B—Acetylene cylinder valve

C—Oxygen regulator inlet connection

D—Acetylene regulator inlet connection

E—Hose connections at the regulators and torch

F—Oxygen and acetylene needle valves

Inspect each point carefully. Any noises, such as a hissing sound or bubbles, are an indication of leakage. If a leak is detected at a connection, use a wrench to properly tighten the fitting. If tightening does not remedy the leak, shut the gas pressure OFF, open the connections, and examine the screw threads.

To check for leakage in the welding hose, adjust the regulators to working pressure. Submerge the hose in clean, clear water. Check for any bubbles indicating a leak. On sections of welding hose that cannot be submerged, brush on soapy water and check for bubbles. Welding hoses should be routinely inspected for cuts and worn areas that could eventually leak.

Using the correct size welding tip provides sufficient heat to melt the base metal for the required welding process.

Selecting Welding Tips

The size of the welding tip used depends on the thickness of metal to be welded. If very light sheet metal is to be welded, a tip with a small opening is used, while a large-sized tip is needed for thick metal.

A numbering system is used to identify tip sizes. The number system ranges from 000 to 15, with the most common tip sizes between 000 and 10. **See Figure 5-2.** With this system, the higher the number, the larger the tip diameter.

The correct welding tip must be used with the proper working pressure. If too small a tip is used, the heat will not be sufficient to fuse the metal to the proper depth. When the welding tip is too large, the heat is too great and burns holes in the metal.

Smaller # = smaller hole

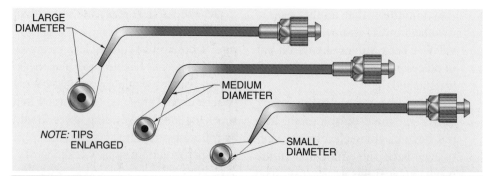

LARGE DIAMETER

MEDIUM DIAMETER

NOTE: TIPS ENLARGED

SMALL DIAMETER

COMMON WELDING TIP SIZES													
Tip Number	000	00	0	1	2	3	4	5	6	7	8	9	10
Thickness of Metal*	up to ¹⁄₆₄	¹⁄₆₄	¹⁄₃₂	¹⁄₁₆	³⁄₃₂	⅛	³⁄₁₆	¼	⁵⁄₁₆	⅜	½	⅝	¾ and up
Oxygen Pressure†	1	1	1	1	2	3	4	5	6	7	7	7½	9
Acetylene Pressure†	1	1	1	1	2	3	4	5	6	7	7	7½	9

* in in.
† in psi

Figure 5-2. *The size of the welding tip is determined by the thickness of metal welded. The proper tip size and working pressure must be selected to provide a quality weld.*

A satisfactory weld must have the right amount of penetration and smooth, even, overlapping ripples. Unless conditions are optimized, it is impossible for the torch to function the way it should, and a poor weld will result. Ensure that the equipment, including the hoses, regulators, check valves, torch, and welding tip are properly connected before lighting the torch.

Lighting Torches

1. Select the correct welding tip size for the metal to be welded and connect it to the torch.
2. Stand to one side and open the oxygen and acetylene cylinder valves slowly. **See Figure 5-3.** Open the acetylene cylinder valve approximately one complete turn and open the oxygen valve all the way. Do not face the regulator when opening the cylinder valve. Oxygen and acetylene are stored under high pressure. If the gas is permitted to come against the regulator suddenly, it may cause damage to the equipment. In addition, a defect in the regulator may cause the gas to blow through, shattering the glass and causing injury to the welder.

✓ **Point**

Stand to one side before opening a cylinder valve of the outlet nozzle and be sure the regulator adjusting screw is fully released.

Figure 5-3. *Stand to one side of the regulator when opening a cylinder valve.*

3. Set the working pressure of the oxygen and acetylene regulator adjusting valves to correspond to the required working pressure of the welding tip being used.
4. Turn the acetylene needle valve on the torch approximately one-half turn.
5. With the sparklighter (striker) held about 1″ away from the end of the welding tip, ignite the acetylene as it leaves the tip. Adjust the acetylene until the smoke disappears. **See Figure 5-4.**

When igniting a torch, keep the tip of the torch facing downward. Lighting the torch while it is facing outward or upward could cause injury to workers nearby.

Figure 5-4. *Hold the sparklighter (striker) approximately 1″ from the tip when lighting the torch.*

Adjusting the Welding Flame

With the acetylene ignited, gradually open the oxygen needle valve until a well-defined white cone appears near the tip surrounded by a second, bluish cone that is faintly luminous. This is known as a neutral flame because there is an approximate one-to-one mixture of acetylene and oxygen, which results in a flame that is chemically neutral. A *neutral flame* is a flame that has neither oxidizing nor carburizing characteristics. The brilliant white cone should be approximately 1⁄16″ to 3⁄4″ long, depending on the welding tip size. **See Figure 5-5.** A neutral flame is used for most welding operations.

Any variation from the one-to-one oxygen-acetylene mixture will alter the flame characteristics. When excess oxygen is forced into the oxyacetylene mixture, the resulting flame is said to be oxidizing. An *oxidizing flame* is a flame in which there is an excess of oxygen. The oxygen-rich zone extends around and beyond the cone. An oxidizing flame resembles the neutral flame slightly, but has an inner cone that is shorter and more pointed with an almost purple color rather than brilliant white. It is sometimes used for brazing.

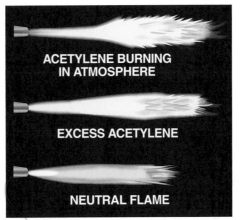

Figure 5-5. *With the acetylene burning, gradually open the oxygen needle valve to obtain a neutral flame.*

If the oxyacetylene mixture consists of a slight excess of acetylene, the flame is carburizing, or reducing. A *carburizing flame* is a reducing flame in which there is an excess of fuel gas. The carbon-rich zone extends around and beyond the cone. This flame can be easily identified by the existence of three flame zones instead of the usual two found in the neutral flame. The end of the brilliant white cone is no longer as well defined, and it is surrounded by an intermediate white cone, which has a feathery edge in addition to the usual bluish outer envelope. **See Figure 5-6.**

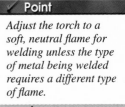

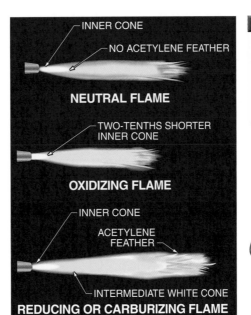

Figure 5-6. *An oxidizing flame is the result of an excess of oxygen in the mixture. A slight excess of acetylene produces a carburizing flame.*

✓ Point

Adjust the torch to a soft, neutral flame for welding unless the type of metal being welded requires a different type of flame.

[handwritten notes: Weld w/ neutral flame ☆ B = part htd + O₂ (Quiz)]

Flame Characteristics. A flame may be harsh or quiet. A harsh flame is produced by too much pressure of both gases to the welding tip. A harsh flame is undesirable, since it has a tendency to depress the molten surface and cause the metal to spatter around the edges of the weld pool. A harsh flame is noisy and makes it extremely difficult to achieve complete fusion with smooth, uniform ripples.

A quiet flame is just the opposite of a harsh flame and is achieved by the correct pressure of gases flowing to the tip. The flame is not a harsh, noisy flame but one that permits a continuous flow of the weld pool without any undue spatter.

To ensure a soft, quiet, neutral flame, the welding tip must be clean and the correct oxyacetylene mixture used. Even with the proper proportion of acetylene and oxygen, a good weld is difficult to achieve unless the opening in the tip allows a free flow of gases. Any foreign matter in the welding tip restricts the heat necessary to melt the metal.

Flame Control. As welding progresses, the flame cone should be observed to ensure that the mixture remains consistent. Changes in the flame occur as a result of slight fluctuations in the flow of the gases from the regulators. A slight adjustment to either the oxygen or the acetylene will readjust the flame.

During welding, the torch may occasionally "pop." Popping is an indication that there is an insufficient amount of gases flowing to the welding tip. Popping can be stopped by further opening both the oxygen and acetylene needle valves on the torch. Another cause of popping is overheating of the weld pool by lingering, or keeping the flame too long in one position and not melting enough filler metal into the weld pool.

Backfire and Flashback. When the flame goes out with a loud pop, it is called a backfire. A *backfire* is a quick recession of the flame into the welding tip, typically followed by extinction of the flame.

A backfire may be caused by operating the torch at lower pressures than required for the welding tip used; touching the welding tip against the work; overheating the welding tip; or by an obstruction in the welding tip. If a backfire occurs, shut the needle valves and, after remedying the cause, relight the torch.

A *flashback* is a recession of the flame into or back of the mixing chamber in a flame torch or flame spray torch. A flashback flashes quickly into the torch and burns inside with a shrill hissing or squealing noise. If a flashback occurs, close the needle valves immediately. A flashback generally is an indication that something is wrong. A welding tip may be clogged, the needle valves may be functioning improperly, or the acetylene or oxygen pressure may be incorrect. The malfunction must be corrected and damaged equipment replaced before relighting the torch.

Shutting Off Torches

When welding is completed, the torch must be properly shut off. After the torch is shut off, it must be stored properly. The hoses must be removed from the cylinders and hung out of the way. Protector caps must be screwed onto the cylinders to protect the handwheels and valves. Cylinders must be chained and stored safely. Following is the correct sequence of steps for shutting off a torch:

1. Close the oxygen needle valve.
2. Close the acetylene needle valve.
3. If the entire welding unit is to be shut down, shut off both the acetylene and the oxygen cylinder valves.
4. Open the needle valves until the lines are drained to remove pressure from the working pressure gauges. Then promptly close the needle valves.
5. Release the adjusting screws on the pressure regulators.

- Cylinders must be properly secured to prevent damage and possible injury.
- Point the valve outlet nozzle away before cracking the cylinder.
- Periodically test connections and hoses for leaks. Use soapy water only.
- The welding tip size is determined by the thickness of metal welded.
- Stand to one side before opening a cylinder valve of the outlet nozzle and be sure the regulator adjusting screw is fully released.
- Adjust the torch to a soft, neutral flame for welding unless the type of metal being welded requires a different type of flame.
- Prevent conditions that may cause a backfire or flashback.
- Keep the passage in the welding tip clean and flowing freely.

 Refer to Quick Quiz® on CD-ROM

 EXERCISES

Testing the Flames

exercise **1**

The characteristics of the carburizing and oxidizing flames must be understood for correct adjustment of the neutral flame. To become familiar with the effects of the various flames, complete the following exercise:

1. Obtain a piece of scrap metal. Light the acetylene and turn on the oxygen until a white cone appears on the end of the welding tip enveloped by another fan-shaped cone that has a feathered edge.

2. While wearing goggles, apply the carburizing flame to the metal, holding the point of the white cone close to the metal. Notice that as the metal melts, it has a tendency to boil. This is an indication that carbon is entering the molten metal. After the metal has cooled, the surface will be pitted and very brittle.

3. Open the oxygen needle valve completely. The white cone becomes short and the color changes to a purplish hue. The flame burns with a roar.

4. Apply the oxidizing flame to the piece of metal, allowing the cone to come in contact with the surface. As the metal melts, numerous sparks are given off and a white foam forms on the surface. After the piece cools, the metal will be shiny.

5. Adjust the needle valve until the flame is balanced. Apply the neutral flame to the piece of metal. The molten metal flows smoothly, with very few sparks.

1. Why should cylinders be securely fastened before being used?
2. Why should the cylinder outlet nozzles be examined closely?
3. What is the proper order for setting up the welding equipment?
4. Why are check valves used?
5. What is the proper method of testing for gas leaks?
6. What governs the size of the welding tip that should be used?
7. Describe the process for lighting and adjusting the flame for a cutting torch.
8. How far should the acetylene needle valve be opened when lighting the torch?
9. What is an oxidizing flame?
10. What is a carburizing flame?
11. What is the difference between a neutral flame and a carburizing flame?
12. What are the characteristics of a neutral flame?
13. What is the difference between a harsh flame and a quiet flame?
14. What are some of the conditions that may cause a backfire?
15. What is meant by a flashback when one is using an oxyacetylene torch?
16. Why are hoses purged after being connected to the regulators?
17. Why should the welder stand to one side when opening cylinder valves?
18. What is the last step done to the regulator when shutting off the torch?
19. What kind of mixture of oxygen and acetylene is required to achieve a neutral flame?
20. What happens when an oxidizing flame is used to melt the metal?

Refer to Chapter 5 in the *Welding Skills Workbook* for additional exercises.

OAW–Flat Position

Chapter

Welding with an oxyacetylene torch requires practicing a series of operations in a prescribed order. These operations involve carrying a weld pool, depositing a weld bead with filler metal, and welding various types of joints. In flat position welding, the torch and filler metal are held with the weld joint in the flat position.

CARRYING A WELD POOL

Before performing welding operations, beginning welders should learn the proper technique for forming and maintaining a uniform weld bead. A consistent weld bead can be formed and maintained using an oxyacetylene torch to create and carry a weld pool. The weld pool must be carried along the joint at a consistent width and depth. How the torch is held, the torch position in relation to the joint, and the motion used to carry the weld pool have a direct effect on the quality of the weld bead.

Holding the Torch

A torch should be held like a hammer, with the fingers lightly curled underneath the torch. **See Figure 6-1.** To prevent fatigue, the torch should balance easily in the hand.

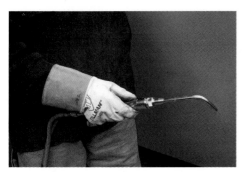

Figure 6-1. *When welding light-gauge metal in flat position, grasp the torch like a hammer.*

Positioning and Moving the Torch

The torch should be held so that the flame points in the direction of welding and at an angle of about 45° to the weld joint. If right-handed, start the weld at the right edge of the metal. The left-handed welder should start welding at the left edge of the metal, working in the reverse direction. **See Figure 6-2.**

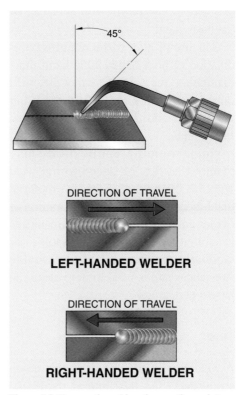

Figure 6-2. *To move the weld pool across the workpiece, hold the torch at a 45° angle and manipulate it in a circular motion.*

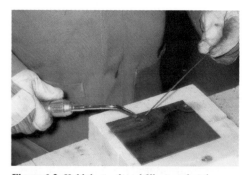
Bring the inner cone of the neutral flame to within ⅛″ of the surface of the workpiece. Hold the torch still until a molten weld pool forms, then move the weld pool across the work-piece. As the weld pool travels forward, rotate the torch in a circular pattern to form a series of overlapping ovals.

Do not move the torch ahead of the weld pool, but slowly work forward, giving the heat a chance to melt the metal. If the flame is moved forward too rapidly, the heat fails to penetrate far enough into the metal and the metal does not melt sufficiently. If the torch is kept in one position too long, the flame will burn a hole through the metal.

ADDING FILLER METAL

On some joints, it is possible to weld two workpieces without adding filler metal. For most welding jobs, however, filler metal is advisable because it builds up the weld, adding strength to the joint. The strength of a weld depends largely on the skill with which the filler metal is blended, or interfused, with the edges of the base metal.

The use of filler metal requires coordination of both hands. One hand must manipulate the torch to carry the weld pool across the plate, while the other hand must add the correct amount of filler metal.

Some welding applications may require that flux be added to the weld with the filler metal. *Flux* is a material that hinders or prevents the formation of oxides and other undesirable substances in molten metal. Flux also dissolves or facilitates the removal of undesirable substances and is used to help clean the base metal.

Selecting Filler Metal

A welded joint should possess as much strength as the base metal itself. To achieve the required strength, it is necessary to use filler metal that has the same properties as the base metal. Inferior filler metals may contain impurities that make them difficult to use and that create a weak or brittle weld. A good filler metal flows smoothly and readily unites with the base metal without excessive sparking.

A poor quality filler metal sparks profusely, flows irregularly, and leaves a rough surface filled with punctures, like pinholes.

Filler metals come in a variety of sizes ranging from ⅟₁₆″ to ⅜″ in diameter. The size filler metal to use depends largely on the thickness of the base metal. The general rule is to use filler metal with a diameter equal to the thickness of the base metal. For example, if a ⅟₁₆″ thick metal is to be welded, a ⅟₁₆″ diameter filler metal should be used.

Many types of filler metal are available for welding a variety of metals. For example, cast iron filler metal is used to weld cast iron, and high strength carbon steel is used as filler metal for welding carbon steel. Copper or copper-silicon filler metal are used for copper, and various aluminum filler metals are used for aluminum. Filler metals must be selected for compatibility with the welding process and the metal being joined.

Manipulating Filler Metal

Hold the filler metal at approximately the same angle as the torch but slanted away from the torch. The filler metal should be moved at a consistent rate and speed as it is fed into the weld pool. **See Figure 6-3.**

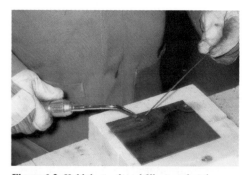

Figure 6-3. *Hold the torch and filler metal at the same angle and maintain a consistent travel angle and feed speed when adding filler metal to the weld.*

Melt a small pool of the base metal and then insert the tip of the filler metal into the weld pool. To ensure proper fusion, the correct diameter filler metal must be used.

If the filler metal is too large, the heat of the weld pool will be insufficient to melt it. If the filler metal is too small, the heat of the weld pool cannot be absorbed by the filler metal, and a hole will be burned in the workpiece.

As the filler metal melts in the weld pool, advance the torch forward. Concentrate the flame on the base metal and not on the filler metal. Do not hold the filler metal above the weld pool, as the molten metal will have to drip down to the weld pool. When molten metal falls, it combines with the oxygen of the air and part of it burns up, causing a weak, porous weld. Always dip the filler metal in the center of the weld pool.

A beginning welder may have trouble holding the filler metal steady, which can cause the filler metal to stick to the base metal. Instead of inserting the filler metal in the middle of the weld pool where the heat is sufficient to melt it readily, the beginning welder may insert it near the edge of the weld pool where the temperature is lower. However, the heat at the edge may not be hot enough to melt the filler metal. If the filler metal is not melted sufficiently it may stick to the weld. Do not try to jerk filler metal loose, since such an action will simply interrupt the welding. Instead, to loosen the filler metal, play the flame directly on the tip and the filler metal will be loosened. While the filler metal is being freed, the weld pool will likely solidify. Therefore, the weld pool must be re-formed before moving forward.

Depositing Weld Beads

Rotate the torch to form overlapping ovals, and keep raising and lowering the filler metal as the weld pool is moved forward. Advance the weld pool about 1/16″ with each complete motion of the torch. An alternate torch movement is a semicircular motion. **See Figure 6-4.** When the filler metal is not in the weld pool, keep the tip just inside the outer envelope of the flame.

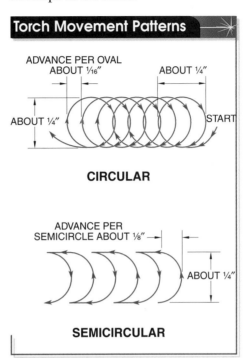

Figure 6-4. *The torch can be moved in a circular or semicircular motion when depositing beads in flat position.*

Maintaining Travel Speed. To secure weld beads of uniform width and height, keep the forward movement of the torch consistent. If the travel speed is too slow, the weld pool is carried forward too slowly, it becomes too large, and may burn through the metal. If the travel speed is too rapid, the filler metal does not fuse thoroughly with the base metal but merely sticks on the surface. It will also be impossible to form even ripples.

When the weld pool appears to be getting too large, withdraw the flame slightly so that only the outer envelope of the flame is touching the weld pool. Do not move the flame to one side, since such a movement allows air to strike the hot metal, oxidizing the metal.

WELDING BUTT JOINTS

Once the task of carrying a weld pool across the surface of a workpiece while

✔ **Point**

Do not hold the filler metal so high above the weld pool that the molten metal drips onto the weld pool.

✔ **Point**

When welding with filler metal, move the torch in a semicircular or circular motion.

adding filler metal is mastered, the next task is to fuse two workpieces together using a butt joint.

Tack Welds

Workpieces must be tacked at regular intervals before welding to maintain the root opening. **See Figure 6-5.** To make a tack weld, apply the flame to the workpiece until it melts and then add filler metal.

Progressive spacing may be used to allow for closing of the root opening. Progressive spacing between the edges of a seam is not commonly used, but if it is specified, allow a gap of about $\frac{1}{16}''$ at the starting end of the joint and approximately $\frac{1}{8}''$ at the other end. The space permits the flame to melt the edges all the way through to the bottom of the workpieces, allowing for complete fusion.

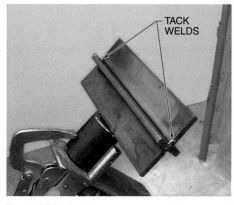

Figure 6-5. *Tack welds restrict expansion forces in metal that is to be welded.*

Butt Joint Defects

The first few welds a beginning welder makes may easily break. A beginning welder should practice until a straight, smooth weld that does not open when bent can be made. Some common defects that may occur when first learning to weld are:

- holes in the joint, caused by holding the flame too long in one spot
- a brittle weld, resulting from improper flame adjustment during welding or dripping filler metal
- excessive metal hanging underneath the weld, as the result of too much

penetration caused by moving the torch forward too slowly
- insufficient penetration, caused by moving the torch forward too rapidly. When penetration is correct, the underside of the seam should show that fusion has taken place completely through the joint. **See Figure 6-6.**

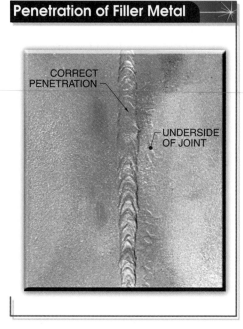

Penetration of Filler Metal

CORRECT PENETRATION

UNDERSIDE OF JOINT

Figure 6-6. *When penetration is correct, complete fusion is evident on the underside of the joint.*

- hole in the end of the joint, caused by not lifting the torch when the end of the weld has been reached
- uneven weld bead, caused by moving the torch too slowly or too rapidly

Often, a joint appears to have the correct penetration but still cracks open when tested. Cracks may be caused by one of several problems, such as:

- improper space allowances between the edges of the workpieces
- filling the space between the workpieces with molten filler metal without sufficiently melting the edges of the workpieces, which results in a poor bond between the base metal and the filler metal
- holding the torch too flat, causing the weld pool to lap over an area that has not been properly melted

WELDING OTHER JOINTS

When the ability to weld a correct butt joint is mastered, other joints may be welded using techniques similar to those used on the butt joint. A flange joint is used a great deal in sheet metal work, particularly on material that is 20 gauge or less. The flange portion should extend above the surface of the sheet a distance roughly equal to the thickness of the sheet.

A corner joint is used extensively in fabricating products such as tanks and vessels, as well as in repair work. The edges are fused without filler metal, as in welding a flange joint.

A lap joint is formed when one piece of metal is laid on top of another. Careful control of the direction of heat is needed for a lap joint weld.

A T-joint is made by laying one workpiece flat and standing the other workpiece on top to form a T. The T-joint requires a greater amount of filler metal than other joints. Correct filler metal usage is critical.

OAW – CAST IRON

Gray cast iron may be welded; however, extreme caution must be taken to offset expansion and contraction forces. Since gray cast iron is brittle, it is susceptible to rapid temperature changes (thermal shock), making preheating and postheating necessary when welding cast iron.

To maintain the gray iron structure throughout the weld area, the weld must be made with the correct filler metal. All weld parts must be cooled slowly. If the casting is cooled too rapidly, the weld area is likely to turn into white cast iron, making the weld section extremely brittle and so hard that machining may be impossible.

Preparing Edges

The edges of the casting should be beveled at 45° to form a 90° groove angle. The V should extend only to ⅛″ from the bottom of the break. Beveling makes it easier to build up a sound weld near the bottom and

lessens the likelihood of melt-through. Placing carbon backing bars underneath the joint also helps to prevent the molten cast iron from running out the seam.

Precautions must be taken to clean the surfaces of the joint before welding. The weld area should be cleaned on both sides of the joint. Improperly cleaned surfaces result in porosity in the weld, even if sufficient flux is used.

Preheating and Postheating

When welding cast iron, the entire casting must be preheated to a dull red. Uniform preheating equalizes expansion and contraction forces and minimizes the possibility of cracks. On a small section of cast iron, preheating can be carried out by playing the flame over the casting. A large casting may have to be placed in a preheating furnace. The temperature must be monitored carefully on a heavy casting, especially if it has thin members, to prevent overheating.

After welding is completed, postheat the cast iron by bringing the entire casting up to a uniform temperature. Use the same techniques as for preheating the casting.

Filler Metal and Flux

A cast iron filler metal that has the same composition as the base metal is used to weld cast iron. The cast iron filler metal contains silicon to ensure flowability. Correct preheating and postheating allow for machinability.

Using flux is also essential when welding cast iron to keep the weld pool fluid. Otherwise, infusible slag mixes with the iron oxide that forms on the weld pool. If infusible slag mixes with iron oxide, the weld will contain inclusions and porosity.

OAW – ALUMINUM

Although the gas shielded arc welding processes (GTAW and GMAW) are the most practical for welding commercially pure aluminum, oxyacetylene welding is occasionally used. If oxyacetylene

✓ Point
If possible, use carbon backing bars when welding cast iron.

✓ Point
Clean all welding surfaces around the joint to be welded.

✓ Point
Preheat cast iron to a dull red before welding.

✓ Point
Postheat cast iron after the weld is completed and then allow it to cool slowly.

welding must be used on aluminum, care must be taken not to overheat the aluminum, weakening the metal.

The following considerations must be kept in mind when welding aluminum with an oxyfuel process:

- Aluminum has a relatively low melting point compared to other metals. Pure aluminum melts at 1220°F.
- The thermal conductivity of aluminum is high—almost four times that of steel.
- Aluminum collapses suddenly into liquid when heated. Since it is light in color, there is practically no indication when the melting point is reached.
- Molten aluminum oxidizes very rapidly. A heavy coating forms on the surface of the seam, which necessitates the use of a good flux.
- Aluminum is very flimsy and weak when hot. Care must be taken to support it adequately during welding.
- Aluminum welds should be made in a single pass if possible.

Joint Designs

In general, the same principles of joint design for welding steel apply to aluminum. Aluminum from ¹⁄₁₆″ to ³⁄₁₆″ thick can be welded using a butt joint, provided the edges are notched with a saw or chisel. Notching minimizes the possibility of burning holes through the joint, permits full penetration, and prevents local distortion. Permanent backings and fillet welded lap joints should not be used when welding aluminum as they may cause the flux to become entrapped in the weld, which leads to a greater likelihood of corrosion.

When welding heavy aluminum plate ³⁄₁₆″ to ³⁄₈″ thick, the edges should be beveled to form a 90° to 100° V. Allow a ¹⁄₁₆″ to ¹⁄₈″ notched root face. Aluminum that is greater than ³⁄₈″ thick should be prepared as a double-V butt joint with a notched root face. The edges should be beveled to form a 100° to 120° V. **See Figure 6-7.**

As a rule, the lap joint is not recommended for aluminum welding because flux and oxide may become trapped between the surfaces of the joint, causing the aluminum to corrode.

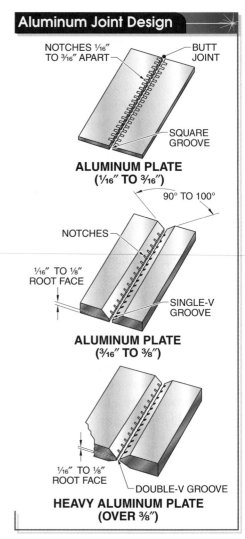

Aluminum Joint Design

Figure 6-7. *Aluminum joint design is similar to that for most other metals. The angles, notches, and flanges are dependent on the thickness of the aluminum.*

Using Flux

The edges of aluminum to be welded must be thoroughly clean. All grease, oil, and dirt must be removed with an appropriate solvent or by rubbing the surface with steel wool or a wire brush.

Since aluminum oxidizes rapidly, a layer of flux must be used to ensure a sound weld. Flux is sold as a powder, which can be mixed with water to the consistency of a thin paste (approximately two parts flux to one part water). If filler metal is not required, the flux is applied to the joint by means of a brush.

When filler metal is used, it is coated with flux by first heating the filler metal and then dipping it into the flux.

On thick sections of metal, it is advisable to coat the base metal as well as the filler metal to ensure complete fusion. When welding is complete, all traces of flux must be washed away. Flux that remains on the weld can cause corrosion. Flux is removed by washing the workpiece in hot water or by immersing in a 10% cold solution of sulfuric acid, followed by rinsing in hot or cold water.

Selecting Filler Metal

The proper filler metal must be used when welding aluminum. The filler metal composition should be comparable to that of the aluminum to be welded. The three most common filler metals for welding nonheat-treatable aluminum are 1100, 4043, and 5356. The 4043 and the 5356 filler metals are recommended when greater strength is required.

Filler metals are available in $\frac{1}{16}''$, $\frac{1}{8}''$, $\frac{3}{16}''$, and $\frac{1}{4}''$ diameters. Generally, a filler metal whose diameter equals the thickness of the aluminum to be welded should be used.

Preheating. All aluminum to be welded, including thin sheet, should be preheated to minimize the effects of expansion and cracking. Aluminum $\frac{1}{4}''$ thick or more should be preheated to a temperature of 300°F to 500°F. Preheating to these temperatures can usually be done by playing the flame of the oxyacetylene torch over the work. For large or complicated parts, preheating is done in a furnace.

The preheating temperature must not exceed 500°F. If the temperature rises above 500°F, the alloy may be weakened or the aluminum may collapse under its own weight. The correct preheating temperature may be determined with a temperature-indicating crayon or by one of the following methods:
- A mark made on the metal with a carpenter's blue chalk will turn white.
- A pine stick rubbed on the metal will leave a char mark.
- No metallic ringing sound is heard if the metal is struck with a hammer.

Selecting Torches

Since aluminum has high thermal conductivity, a welding tip slightly larger than one used for steel of the same thickness should be used. **See Figure 6-8.**

Many welders use hydrogen instead of acetylene when welding aluminum, and in many cases this is preferable, especially for welding light-gauge material. In either case, the torch should be adjusted to a neutral flame. Some authorities recommend a slightly reducing flame, but usually a neutral flame is satisfactory for producing a clean, sound weld. Whether using acetylene or hydrogen, the flame should be adjusted to a low gas velocity to permit a soft flame.

The torch angle has much to do with welding speed. Instead of lifting the flame from time to time to avoid melting holes in the metal, the welding torch should be held at a flatter angle to increase the welding speed. The welding speed should also be increased as the edge of the metal is approached.

OAW – STEEL

Heavy steel is rarely welded with oxyacetylene unless other types of welding equipment are not available. Welding heavy steel with oxyacetylene is much slower and less cost-efficient than other methods. Occasionally, it may be necessary to use oxyacetylene welding to weld or repair a structure. When welding steel using OAW, maintain the proper oxygen and acetylene pressures. **See Figure 6-9.**

Single-V Groove Butt Joints

Complete penetration of the weld is necessary to achieve maximum weld strength. On steel $\frac{1}{8}''$ thick or less, complete penetration is reasonably easy to achieve. On thicknesses over $\frac{1}{8}''$, penetration is not possible unless the edges are beveled. Edges can be beveled using a torch, a beveling machine, or a grinder.

✓ Point
When welding aluminum, use a slightly larger welding tip than is used for steel.

✓ Point
Use an 1100, 4043, or 5356 filler metal for welding aluminum.

✓ Point
Use a neutral or slightly reducing flame for all aluminum welding.

✓ Point
When welding aluminum, keep the preheat temperature below 500°F.

✓ Point
Steel thicker than $\frac{1}{8}''$ should be beveled before welding.

Points to Remember

- Move the torch just fast enough to keep the weld pool active and flowing forward.
- Use filler metal with a diameter equal to the thickness of the base metal.
- Do not hold the filler metal so high above the weld pool that the molten metal drips onto the weld pool.
- When welding with filler metal, move the torch in a semicircular or circular motion.
- When using progressive spacing, allow a space between work-pieces to compensate for expansion forces.
- If possible, use carbon backing bars when welding cast iron.
- Clean all welding surfaces around the joint to be welded.
- Preheat cast iron to a dull red before welding.
- Postheat cast iron after the weld is completed and then allow it to cool slowly.
- Always use the recommended flux and filler metal when welding.
- When welding aluminum, use a slightly larger welding tip than is used for steel.
- Use an 1100, 4043, or 5356 filler metal for welding aluminum.
- Use a neutral or slightly reducing flame for all aluminum welding.
- When welding aluminum, keep the preheat temperature below 500°F.
- Steel thicker than ⅛″ should be beveled before welding.
- When using a single-V-groove joint on steel, the groove angle should be 60°.
- A double-V joint must be used with steel ½″ thick or more.
- On steel ½″ thick or more, do not try to fill the joint in a single pass. Deposit the weld in layers with two or more welds per layer.
- When backhand welding, do not swing the torch. Instead, move the filler metal.

 EXERCISES

Carrying a Weld Pool without Filler Metal

exercise **1**

1. Obtain a piece of low-carbon steel ¹⁄₁₆″ to ⅛″ thick, approximately 3″ wide, and 5″ long.

2. Be sure the surface is free of oil, dirt, and scale.

3. Light the torch and adjust it for a neutral flame.

4. Hold the inner cone of the flame approximately ⅛″ from the work and position the torch at a 45° angle to the workpiece. Move the torch from the right side of the workpiece to the left side, using a circular manipulation. Left-handed welders should reverse the direction of travel.

5. Maintain a consistent travel speed to prevent melt-through in the workpiece.

6. Practice depositing beads without filler until properly formed beads are consistently produced.

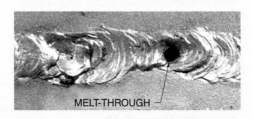
MELT-THROUGH

Depositing Beads with Filler Metal

1. Obtain a piece of low-carbon steel 1/16″ to 1/8″ thick, approximately 3″ wide, and 5″ long.

2. Be sure the surface is free of oil, dirt, and scale.

3. Light the torch and adjust it for a neutral flame.

4. Practice running consistent straight beads while manipulating the torch and the filler metal at the correct angles.

5. As the torch is withdrawn at the end of the pass, fill the crater by adding filler metal.

Welding a Butt Joint in Flat Position

1. Obtain two pieces of metal 1/16″ to 1/8″ thick, approximately 1½″ wide, and 5″ long.

2. Place the workpieces on two firebricks. Space for progressive spacing or tack weld the workpieces together.

3. Begin welding at the right end (or the left end if left-handed), using the same torch and filler motion as when depositing beads with filler.

4. Work the torch slowly to give the heat a chance to penetrate the joint. Add sufficient filler metal to build up the weld about 1/16″ above the surface. Be sure the weld pool is large enough and the metal is flowing freely before dipping the filler metal.

5. Maintain a molten weld pool approximately 1/4″ to 3/8″ wide.

6. Advance the weld pool about 1/16″ with each complete motion of the torch while maintaining a uniform bead width.

7. Uniform torch motion will produce smooth, even ripples.

Welding a Flange Butt Joint in Flat Position

exercise 4

1. Obtain two pieces of metal with flanged edges.

2. Place the pieces so the flanged edges are touching. Tack weld the edges.

3. Hold the torch on one end until a weld pool is formed.

4. Carefully manipulate the torch to maintain the pool as the pool is carried along the entire joint.

5. Withdraw the torch at the end of the joint to prevent burning a hole in the joint.

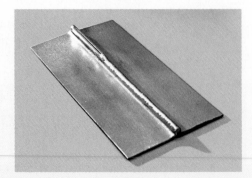

Welding a Corner Joint in Flat Position

exercise 5

1. Obtain two pieces of metal and tack weld to form a corner joint.

2. Hold the torch on the end of the joint until a weld pool is formed.

3. Manipulate the torch to maintain the weld pool along the entire joint using a technique similar to that used on the flange joint.

4. Withdraw the torch at the end of the joint to prevent burning a hole in the joint.

5. If additional buildup is required, filler metal may be added as the weld pool is carried along the joint.

Welding a Lap Joint in Flat Position

1. Obtain two pieces of metal 1/16″ to 1/8″ thick, approximately 1½″ wide, and 4″ to 5″ long.

2. Lay one workpiece on top of the other, slightly offset, and tack in place to form a lap joint.

3. Tilt the tacked workpiece 45° to the work surface and place a firebrick under one side for support.

4. Weld the workpieces using a semicircular motion of the torch.

5. While manipulating the torch and filler metal, direct more of the heat to the bottom plate. This may be accomplished by increasing the duration of the torch motion on the bottom plate. The top plate requires less heat and may overheat if too much heat is applied.

6. Weld one side of the workpiece and then practice on the reverse side.

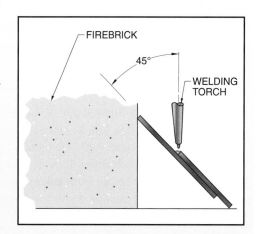

Welding a T-Joint in Flat Position

1. Obtain two pieces of metal approximately 1/16″ to 1/8″ thick, 1½″ wide, and 4″ to 5″ long.

2. Lay one plate flat and stand the other on top to form a T-joint. Tack weld the plates.

3. Tilt the tacked workpiece 45° to the work surface and place a firebrick under one side for support.

4. Hold the torch so the welding tip forms an approximately 45° angle to the bottom workpiece.

5. Using the same technique used when welding a butt joint, keep the inner cone of the flame about 1/8″ away from the deepest part of the weld.

6. Manipulate the torch constantly while adding filler metal to produce a consistent weld free from undercuts.

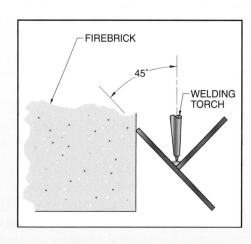

1. Obtain two pieces of cast iron and prepare the edges to be welded. Bevel the joint if necessary and remove all foreign matter from the surface.

2. Slowly heat the entire workpiece to a dull red.

3. Concentrate the flame near the starting point of the weld until the metal begins to melt. Keep the torch in the same position as in welding mild steel, with the inner cone of the flame about ⅛″ to ¼″ from the root of the joint.

4. When the bottom of the V is thoroughly fused, move the flame from side to side, melting down the sides so the molten metal runs down and combines with the fluid metal in the bottom of the V. Rotate the torch in a circular motion to keep the sides and bottom of the V in a molten condition. If the metal gets too hot and tends to run, raise the torch slightly.

5. Once the weld pool is molten, bring the filler metal into the outer envelope of the flame and keep it there until it is fairly hot. When the filler metal is hot, dip it into the flux. Insert the fluxed end of the filler metal into the molten pool. The heat of the weld pool will melt the filler. The filler metal should remain in the weld pool. Do not dip it into and out of the pool. As the filler metal melts, the molten metal will rise in the groove. When the metal has been built up slightly above the top surface of the workpiece, move the weld pool forward about 1″ and repeat the operation. Be sure not to move the weld pool before the sides of the V have been broken down, as this will force the molten weld pool ahead onto the cold metal.

6. When gas bubbles or white spots appear in the weld pool or at the edges of the seam, add more flux and play the flame around the specks until the impurities float to the top. Skim these impurities off the weld pool with the filler metal. Tapping the filler metal against the bench will remove impurities.

7. After the weld is completed, postheat the entire workpiece to a dull red. Allow the casting to cool slowly by covering with a blanket.

8. To test the weld sample, place it in a vise. The weld should be flush with the top of the jaws. Wearing proper eye protection, strike the upper end of the workpiece with a heavy hammer until the workpiece breaks. If the metal has been welded properly, the break should occur in the base metal, not along the welded line.

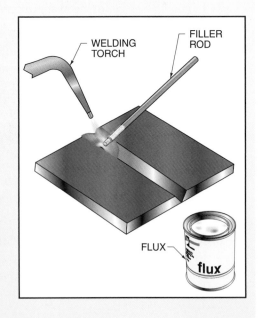

Welding Aluminum

1. Form a square groove joint from two pieces of 0.125″ aluminum plate with a ⅛″ root opening.

2. Preheat the workpiece to the proper temperature.

3. Flux the workpieces using the recommended flux.

4. Pass the flame over the starting point until the flux melts.

5. Scrape the surface with the filler metal at about 3- or 4-second intervals, permitting the filler metal to come clear of the flame each time. Otherwise, it will melt before the base metal. The scraping action indicates when welding should begin without overheating the aluminum.

6. Using the forehand welding technique, angle the torch at a low angle (less than 30° above horizontal when welding thin material). The torch should be moved forward without any side-to-side motion.

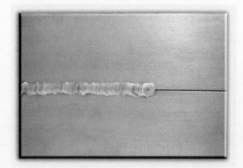

7. While moving the torch forward, dip the filler rod into the weld pool and withdraw it. The dip technique closes the weld pool, prevents porosity, and assists the flux in removing the oxide film.

8. Maintain the same procedure throughout welding.

9. A correct oxyacetylene weld on aluminum will have the necessary penetration with the correct bead ripple and contour.

Welding a Backhand Weld

1. Form a single-V groove joint with a 60° groove angle, and a ¹⁄₁₆″ root opening. Tack weld the plates at each end.

2. Position the workpiece so the weld joint is in horizontal position.

3. Start the weld at the left edge of the workpiece if right-handed (the right edge if left-handed) and form a weld pool by melting the edges at the root of the joint. Hold the end of the filler metal in the outer envelope of the flame so it melts as soon as the weld pool forms.

4. At the start, concentrate the flame slightly more on the bottom of the joint. Once the weld pool is fluid, dip the filler metal into it. As the weld pool moves, direct the flame more on the filler metal and build up the weld pool to the top of the V. As the molten metal fills up the V, move the filler metal slightly from side to side to ensure that the weld metal fuses evenly with the edges of the base metal.

5. To test the weld, cut two 1½″ specimens from the center of the workpiece. Grind off the face reinforcement so that the top of the weld (face) is flush with the top of the base metal. The grind marks should run lengthwise on the specimen to prevent premature failure during testing. Bend each specimen in the guided bend tester. Make one face bend and one root bend. If the weld is satisfactory, there should be no indications of cracking or fracturing. Use proper PPE when testing each specimen.

Welding a Single-V Butt Joint in Flat Position

exercise **11**

1. Form a single V-groove joint from two pieces of ¼″ low-carbon steel with a ¹⁄₁₆″ root opening.

2. Tack weld the plates at each end.

3. Position the workpiece so the weld joint is in flat position.

4. Use ³⁄₁₆″ filler metal. Use the correct size tip for the weld.

5. Hold the torch at an angle 60° from the vertical, rather than the 45° angle used for other steels.

6. Direct the flame onto the V and, as the edges begin to melt, dip the tip of the filler metal into the weld pool. Before adding filler metal, ensure that the sides of the V are thoroughly molten to the bottom of the V. Fill in the bottom of the V about ½″, with the weld pool extending upward to one-half the depth of the V. While the weld pool is still molten, move the torch in a semicircular motion and fill the V. The completed bead should be between ³⁄₈″ and ½″ wide with a slight face reinforcement. Return the flame to the bottom of the V, advance another ½″, and again raise the bead section to the top of the V. Continue until the weld is finished.

7. To test the weld, cut two 1½″ specimens from the center of the workpiece. Grind off the face reinforcement so that the top of the weld (face) is flush with the top of the base metal. The grind marks should run lengthwise on the specimen to prevent premature failure during testing. Bend each specimen in the guided bend tester. Make one face bend and one root bend. If the weld is satisfactory, there should be no indications of cracking or fracturing. Use proper PPE when testing each specimen.

❓ Questions for Study and Discussion

1. Why is a filler metal used in welding?
2. What determines the size of the filler metal that should be used?
3. Where is the filler metal inserted when depositing beads with filler metal?
4. What happens if the filler metal is too large for the base metal that is being welded? If it is too small?
5. How should the torch be manipulated when using filler metal on a butt weld?
6. If the metal does not melt readily, what is the probable cause?
7. What happens if the torch is moved forward too slowly?
8. Why should cast iron pieces be preheated before welding?
9. Why is flux necessary when welding cast iron?
10. What is the melting point of aluminum?
11. What type of filler metal is recommended for welding aluminum?
12. How can it be determined when aluminum has reached its preheating temperature?
13. How are smooth, even ripples formed in the weld bead?
14. How is the flux manipulated in order to deposit it in the weld?
15. If cast iron has been properly welded, where should the break occur when the completed weld is tested?
16. Why should the edges be beveled when heavy steel is ⅛″ thick or more?
17. At what angle should the torch be held when welding heavy steel?
18. How should the torch and filler metal be handled in backhand welding? At what angle?
19. When welding heavy steel over ½″ thick, why use more than one pass?

Refer to Chapter 6 in the *Welding Skills Workbook* for additional exercises.

OAW–Other Positions

Welding in the flat position is easier and somewhat faster than other positions, but welding cannot always be done in the flat position. Occasionally, welding must be performed in the horizontal, vertical, or overhead positions. The main challenge associated with producing a sound weld in the horizontal, vertical, or overhead positions is managing the gravitational pull on the molten weld pool.

HORIZONTAL AND VERTICAL WELDING

When welding in horizontal or vertical position, a jig or positioner may be used to hold the workpieces in position. A semicircular torch movement should be used for horizontal and vertical welding. Maintaining a consistent size weld pool helps control the weld and prevent sagging.

For horizontal position groove welds, the workpiece should be fixed in the vertical position with the joint in the horizontal position. As the weld progresses, heat builds up in the plate, and the weld metal tends to build up on the lower edge of the joint. To overcome this tendency, hold the torch so the tip forms an angle of 45° above the workpiece and in line with the joint. Angle the filler rod about 30° above the joint with no side angle. Direct the flame over both sides of the joint, but hold the flame longer on the bottom edge of the joint without allowing the weld pool to sag. To prevent undercutting, add filler near the top edge of the joint.

For horizontal position fillet welds in T-joints and lap joints, place the workpiece flat on the work surface so the joint is in the horizontal position. Angle the torch 45° from the bottom plate and 45° from the vertical member (T-joint) or the top plate (lap joint). Angle the filler rod

about 30° above the joint with a 15° to 20° side angle.

Vertical welding is performed uphill or downhill. *Uphill welding* is welding performed with an uphill progression. **See Figure 7-1.** *Downhill welding* is welding performed with a downhill progression. When vertical welding, do not allow the weld pool to become too large. If the weld pool gets too big or too fluid, it could get out of control and run down the face of the weld. If the weld gets too fluid, pull the flame away slightly so that it does not play directly on the weld pool.

> ✓ **Point**
>
> *Use a semicircular torch movement for horizontal, vertical, and overhead welding.*

> ✓ **Point**
>
> *In horizontal welding, direct the flame more on the lower edge of the joint.*

Figure 7-1. *Uphill welding is performed with an uphill progression. Do not allow the weld pool to get too large.*

OVERHEAD WELDING

Overhead welding is more difficult to perform than horizontal or vertical welding because of the unusual working position and the skill needed to keep the molten weld pool from sagging.

Overhead welding is possible because molten metal has cohesive (sticky) qualities, as long as the weld pool does not get too large.

The amount of heat directed on the joint must be carefully regulated, since excessive heat increases the fluidity of the weld pool and can cause it to fall out of the joint.

Use the same semicircular motion of the torch for overhead welding as for other welding positions. Move the filler metal slowly in a circular or side-to-side motion to help keep the weld pool shallow. The movement of the filler metal distributes the molten weld pool and prevents it from forming large drops and falling out of the joint. **See Figure 7-2.** If the weld pool becomes too fluid and starts to run, move the torch slightly away from the joint.

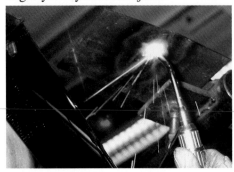

Figure 7-2. *The filler metal moves ahead of the torch and distributes the molten weld pool as it is moved.*

 Refer to Quick Quiz® on CD-ROM

Points to Remember

- Use a semicircular torch movement for horizontal, vertical, and overhead welding.
- In horizontal welding, direct the flame more on the edge of the lower joint.
- On overhead welds, move the filler metal slowly in a circular or side-to-side motion.

EXERCISES

Welding a Butt Joint in Horizontal Position

exercise **1**

1. Obtain two pieces of ¹⁄₁₆≤ or ⅛≤ low-carbon steel.

2. Form a square groove butt joint, with a ⅛≤ root opening to allow complete penetration, and tack together.

3. Position workpiece so the weld joint is in the horizontal position.

4. Start welding at the right edge if right-handed (or the left, if left-handed), using a semicircular torch motion. As welding progresses, gravity can cause metal to build up on the bottom workpiece. To overcome this tendency, direct the flame longer on the bottom edge of the joint and keep the tip of the filler metal closer to the top of the joint.

Welding a T-Joint in Horizontal Position

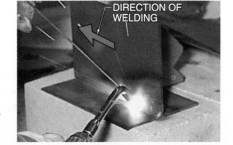

DIRECTION OF WELDING

1. Obtain two pieces of ¹⁄₁₆″ or ⅛″ low-carbon steel.

2. Form a T-joint with the pieces at a 90° angle to each other and tack together.

3. Position the workpiece so the weld joint is in horizontal position.

4. Start welding at the right edge if right-handed (or the left, if left-handed), using a semicircular torch movement.

5. Hold the torch so the tip forms a 45° angle to the bottom plate, and a 45° angle from the vertical member.

6. Point the filler metal toward the welding tip at an angle of approximately 30° above the joint with a 15° to 20° side angle from the vertical member.

7. Direct the flame evenly over the workpiece. To prevent undercutting, add filler metal closer to the top edge of the joint (along the bottom of the vertical member).

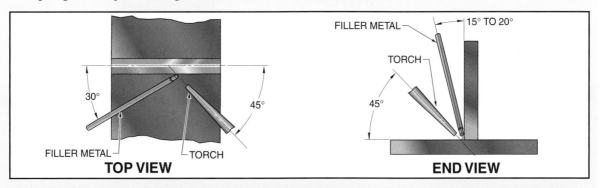

30°   45°   FILLER METAL   TORCH   **TOP VIEW**

FILLER METAL   15° TO 20°   TORCH   45°   **END VIEW**

Welding a Butt Joint in Vertical Position

1. Obtain two pieces of ¹⁄₁₆″ or ⅛″ low-carbon steel.

2. Form a square groove butt joint, with a ⅛″ root opening to allow complete penetration, and tack together.

3. Position the workpiece so the weld joint is in vertical position.

4. Hold the torch and filler rod at the same angle as in flat position. As welding progresses, vary the torch angle as necessary to control the weld pool.

5. Weld uphill. Start the weld at the bottom of the joint and work upward, using a semicircular torch motion. Do not allow the weld pool to become too large or it will run down the face of the weld.

6. To prevent the weld pool from becoming too fluid, direct more of the flame on the filler rod. If the weld pool becomes too fluid, pull the flame away slightly.

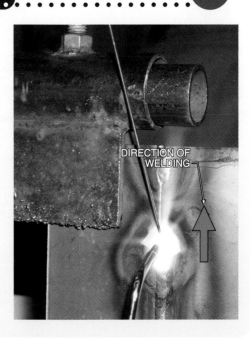

DIRECTION OF WELDING

Welding a Butt Joint in Overhead Position

1. Obtain two pieces of ¹⁄₁₆″ or ⅛″ low-carbon steel.

2. Form a square groove butt joint, with a ⅛″ root opening to allow complete penetration, and tack together.

3. Position the workpiece so the weld joint is in the overhead position. The weld joint should allow clearance for manipulating the torch.

4. Use the same semicircular motion of the torch as other welding positions. Move the filler metal slowly in a circular or side-to-side motion to help keep the weld pool shallow. The movement of the filler metal also distributes the molten weld pool to prevent it from falling out of the joint.

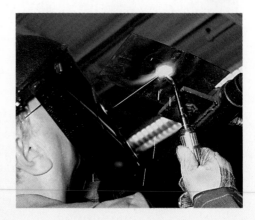

5. If the weld pool has a tendency to run, pull the torch slightly away from the surface.

? Questions for Study and Discussion

1. What can be done to prevent the weld pool from sagging when welding in vertical position?
2. At what angle should the torch be held for horizontal welding?
3. How should the torch be moved for vertical, horizontal, and overhead welding?
4. In horizontal welding of a butt joint, why should the flame be directed more on the edge of the lower workpiece?
5. What should be done when welding in vertical position to prevent the weld pool from becoming too fluid?
6. Why is overhead welding more difficult to perform than horizontal or vertical welding?
7. How can the weld pool be prevented from dropping off in overhead welding?
8. How should the filler metal be manipulated in overhead welding?
9. What can be done to prevent undercutting of the weld when welding a horizontal T-joint?
10. What can be done to maintain a shallow weld pool when welding in the overhead position?

Refer to Chapter 7 in the *Welding Skills Workbook* for additional exercises.

SMAW–Equipment

Shielded metal arc welding (SMAW), sometimes referred to as stick welding, is used in the fabrication of many products, including ships, pressure vessels, tanks, automobiles, and appliances. SMAW machines are used to weld light- and heavy-gauge metals of all kinds.

A constant-current welding machine is used for SMAW. Power to produce a welding arc can be static, such as is supplied by a transformer, transformer-rectifier, or inverter, or can be engine-driven. The power source design is selected based on the requirements of the welding task.

Proper personal protective equipment must be used during welding to protect the welder from injury and to prevent damage to the materials or structures being welded.

SHIELDED METAL ARC WELDING

Shielded metal arc welding (SMAW) is often the first welding process learned by new welders, with the possible exception of oxyacetylene welding (OAW). SMAW is extremely versatile and can be used for many types of applications. Though it requires an electrical power source, the equipment is fairly simple and inexpensive, making it well suited for welding practice, repair welding, and in-the-field welding.

SMAW is can weld carbon steel, low- and high-alloy steel, stainless steel, cast iron, and ductile iron. While it is less popular than other processes for welding nonferrous materials, it can be used on nickel, copper, and aluminum alloys.

SMAW uses special flux-coated electrodes to strike and maintain an arc with the workpiece. The electrode is held securely in a conductive clamp, which is supplied electrical current from the welding machine. The circuit is completed by a workpiece lead that is clamped to the workpiece itself or to a conductive welding table. The electrode is consumed in

the process, becoming shorter as the filler metal is added to the weld. Welders must practice the skills involved in maintaining a consistent weld while dealing with a constantly changing electrode length. When the electrode becomes too short to manipulate effectively, it is removed from the clamp and replaced.

Shielded Metal Arc Welding

Shielded metal arc welding can be used for many types of welding applications.

This process is referred to as "shielded" because as the electrode metal melts, the flux coating gives off vapors that protect the molten weld pool from oxygen and other atmospheric gases that can adversely affect the weld. These vapors remain in the weld area long enough to adequately shield the weld immediately behind the arc. Because it uses consumable flux for shielding, SMAW does not need separate shielding-gas supplies, which would add heavy cylinders and extra complexity to the equipment.

The remainder of the flux becomes slag in the weld metal, which floats to the top as it cools, forming a hard crust. This crust further protects the cooling metal and keeps it from cooling too quickly, which improves strength. When finished with a weld bead, the welder chips off the slag crust with a small hammer and brushes the weld clean. It is especially important to clean all slag from a weld before welding near the same bead so that bits of slag do not become embedded in the weld, which will weaken it.

Due to the frequent stops for electrode changes and slag removal, SMAW is one of the least efficient processes in terms of amount of time spent actually laying weld. SMAW has lost some popularity as newer welding processes permit higher productivity. However, because of its simplicity and versatility, SMAW remains common in many applications. It is particularly dominant in the maintenance and repair industry and is also heavily used in industrial fabrication and the construction of steel structures.

The electrodes are available in many filler-metal alloys and flux-coating combinations, which are designated with numerical codes. This variety is needed for welding different base metals in different positions and under different conditions. However, these variables affect the optimal welding machine settings and behavior of the arc. Therefore, welders must practice with each type of electrode likely to be used.

SMAW can be very dangerous because it involves intense heat, bright light, and radiation from the welding arc that can burn the eyes and skin. Sparks, bits of slag, and dross (molten metal from cutting) can cause eye injury and serious burns. SMAW power sources operate at high open-circuit voltages that can cause fatal electric shock. Exposure to noise from grinding, arc cutting, and other shop activities can cause permanent hearing loss. Therefore, safety should be a primary concern when working with SMAW or any welding or cutting process.

> *Shielded metal arc welding equipment or power tools should never be operated while standing in water or in damp areas. Water conducts electricity and increases the potential for electrical shock. If the area cannot be dried, stand on boards or an insulated platform.*

ELECTRICAL PRINCIPLES

When welding using SMAW, an electrical circuit is created. An *electrical circuit* is a path taken by electric current flowing from one terminal of the welding machine, through a conductor, and to the other terminal. *Current* is the amount of electron flow through an electrical circuit.

A *conductor* is any material through which electricity flows easily. Conductors can be found in the form of wire, cable, or busbars. A person can also act as a conductor of electricity when not following proper safe welding practices. When welding using the SMAW process, the welding leads serve as conductors in the circuit. *Resistance* is the opposition of the material in a conductor to the passage of electric current, causing the electrical energy to be transformed into heat. Resistance is measured in ohms. An *ohm* is the basic unit of measurement of resistance. One ohm is the result of 1 volt applied across a resistance that allows 1 ampere to flow through it.

Welding Current

When electrical current moves through a wire, heat is generated by the resistance of the wire to the flow of electricity. The greater the current flow, the greater the heat generated. The heat generated during the SMAW process comes from an arc that develops when electricity jumps across a gap between the end of an electrode and the base metal. The gap produces a high resistance to the flow of current. This resistance generates intense heat that can range from 6000°F to 10,000°F. Welding current can be direct current (DC) or alternating current (AC). Current has the most effect on the depth of penetration into the base metal.

Direct current (DC) is an electrical current that flows in one direction only. This means that there is no change in the direction of current flow. *Alternating current (AC)* is an electrical current that has alternating positive and negative values. In the first (positive) half-cycle, current flows in one direction. In the second (negative) half-cycle, current flows in the opposite direction. **See Figure 8-1.**

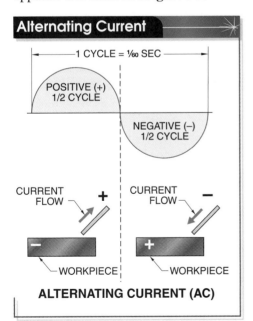

Figure 8-1. *Direct current flows in one direction only. Alternating current has positive values and negative values. Current flows in one direction for the positive half-cycle, and in the opposite direction for the negative half-cycle.*

Frequency is the number of cycles (AC sine waves) per second measured in hertz (Hz). In North America, AC operates at a frequency of 60 cycles per second (60 Hz), compared to 50 Hz in Europe.

Current is measured in amperes. An *ampere (A)* is the basic unit of measure for electrical current. One ampere is the amount of current that will flow through a resistance of 1 Ω in one second, when 1 V of electromotive force (pressure) is applied. An *ammeter* is an electrical instrument that measures amperage.

The primary voltage (input) to a welding machine may be 120 V, 230 V, 460 V, or 600 V. The welding machine frame (chassis) must be well grounded since primary voltages can be very dangerous.

Polarity. *Polarity* is the positive (+) or negative (–) state of an object. Polarity determines the direction of current flow in a DC circuit. Since current moves in one direction only in a DC circuit, polarity must be selected for some welding operations. DC current used for welding can be either direct current electrode negative (DCEN) or direct current electrode positive (DCEP). **See Figure 8-2.** The terminology DCEN and DCEP replaces the formerly used terms straight polarity (DCEN) and reverse polarity (DCEP).

Polarity is changed by connecting the electrode lead to either the positive or negative terminal. When the electrode lead is connected to the negative terminal of the welding machine and the workpiece lead is connected to the positive terminal, the polarity is DCEN. In DCEN welding, current flows from the electrode into the workpiece.

When the electrode lead is connected to the positive terminal of the welding machine and the workpiece lead is connected to the negative terminal, the circuit is DCEP. In DCEP welding, current flows from the workpiece to the electrode.

On some machines, polarity is changed by moving a switch or lever on the welding machine to DCEN (–) or DCEP (+). Polarity is of no consequence in AC welding machines because current is constantly changing direction.

Polarity determines whether current flows into the workpiece or toward the electrode. It also has an effect on penetration. In DCEN, current flows from the electrode into the workpiece. DCEN welding with SMAW produces shallow penetration compared to DCEP.

In DCEP, current flows from the workpiece into the electrode. DCEP welding with SMAW produces deeper penetration and a narrower weld bead profile than DCEN. Whether DCEN or DCEP is used for a particular application is determined by the welding to be performed, and the operating characteristics of the electrode.

Miller Electric Manfacturing Company

Figure 8-3. *Voltage and current values can be shown on a digital display on the front of a welding machine.*

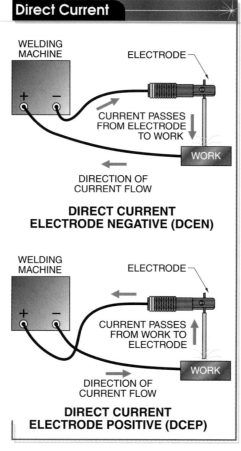

Figure 8-2. *When welding with DC current, polarity can be changed from DCEN to DCEP to control the amount of heat directed to the base metal.*

Voltage

The force (electromotive force, or emf) or pressure that causes current to flow in a circuit is called voltage. *Voltage* is the amount of electrical pressure in a circuit. Voltage does not flow, only current flows. Voltage is measured using a voltmeter. Voltage and current values are commonly shown with a digital display on the front of a welding machine. **See Figure 8-3.**

Voltage (force) is similar to the pressure used to make water flow in pipes. In a water system, a pump provides the pressure to make the water flow, whereas in an electrical circuit a power supply produces the force (voltage) that pushes the current through the wires. Voltage has the most effect on the height and width of the weld deposit.

Voltage drop is the voltage decrease across a component due to resistance to the flow of current. Just as the pressure in a water system drops as the distance from the water pump increases, so does voltage lessen as the distance from the generator increases. When there is too great a drop, the welding machine cannot supply enough current for welding. A voltage drop problem is usually associated with using welding cables that are too small, too long, or that have been damaged. Poor connection of the cables at the power source or work ends (electrode holder or workpiece clamp) can also contribute to voltage drop.

Open-circuit voltage is voltage produced when a welding machine is ON, but no welding is being done. Open-circuit voltage varies from 50 V to 100 V. *Arc voltage (working voltage)* is the voltage present after an arc is struck and maintained. Arc voltage is generally between 18 V and 36 V. **See Figure 8-4.** An adjustment is provided to vary the open-circuit voltage so that welding can be done in different positions. Arc voltage is measured as close to the welding arc as possible and to measure voltage loss in the circuit.

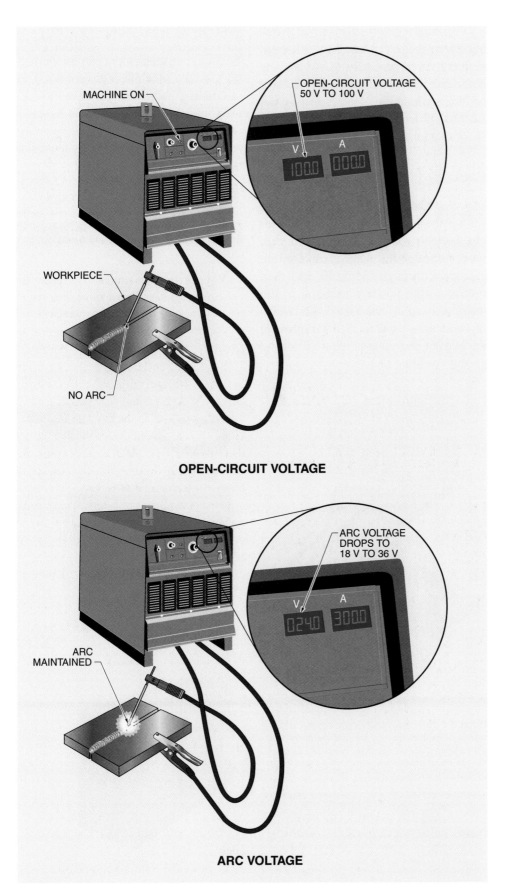

MACHINE ON

OPEN-CIRCUIT VOLTAGE
50 V TO 100 V

V   A
100.0 000.0

WORKPIECE

NO ARC

OPEN-CIRCUIT VOLTAGE

ARC VOLTAGE
DROPS TO
18 V TO 36 V

V   A
024.0 300.0

ARC
MAINTAINED

ARC VOLTAGE

Figure 8-4. *Open-circuit voltage (usually between 50 V and 100 V) is voltage produced when the welding machine is ON but no welding is being done. High-open circuit voltage aids in arc starting. Arc voltage is the working voltage (usually between 18 V and 36 V) after an arc is struck.*

The actual voltage used to provide welding current is low (18 V to 36 V), whereas high current is necessary to produce the heat required for welding. The low voltage and high current used for welding are not particularly dangerous if proper grounding and insulation are used.

Circuits

The electrical circuit used for welding starts at the negative terminal of the welding machine where current is produced, moves through the wire or cable to the electrode, through the work, and then returns to the positive terminal of the welding machine. **See Figure 8-5.** Welding machines used for SMAW provide the current and voltage required for the specific welding task.

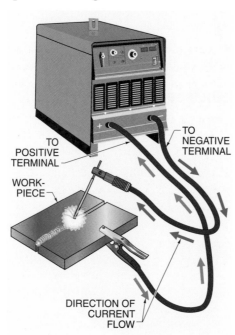

TO POSITIVE TERMINAL

TO NEGATIVE TERMINAL

WORK-PIECE

DIRECTION OF CURRENT FLOW

Figure 8-5. *Direct current electrode negative (DCEN) current flows from the negative terminal of the welding machine along the electrode lead to the electrode, through the work, and then returns to the positive terminal, and is commonly used for SMAW.*

> All welding equipment must be maintained and serviced. The welder is responsible for checking the fluid levels (water, oil, fuel) on all engine-driven machines. Electrode leads and holders should be checked regularly to ensure a tight connection and for proper grounding. Loose connections generate heat and burn leads and connections.

WELDING MACHINE OUTPUT

Welding machine output can be alternating current (AC), direct current (DC), or alternating current/direct current (AC/DC), depending on the welding task. **See Figure 8-6.** The electrode used must match the current produced by the welding machine.

Welding Machine Output

AC

DC

AC/DC

Figure 8-6. *Welding current is provided by AC, DC, or AC/DC output.*

Alternating Current (AC)

AC current output provides a constantly alternating current that can be used for SMAW welding. AC current allows a welder to easily maintain an arc during welding. Other features of AC current include low operating and maintenance costs, and high overall electrical efficiency. AC welding machines typically operate on single-phase (1ɸ) primary power. The electrode and work cables can be connected to either output terminal on AC-only welding machines.

Direct Current (DC)

DC current output for SMAW may use single-phase (1ɸ) or three-phase (3ɸ) primary electrical power. The most stable DC welding is provided by welding machines that provide 3ɸ transformers and full-wave rectifiers. DC output welding machines have output terminals identified as positive (+) and negative (–). If they have a polarity switch, the output terminals are identified as work and electrode.

Alternating Current/Direct Current (AC/DC)

AC/DC current output is available on constant-current welding machines that typically operate using a 1ɸ primary power source. The main difference between AC and AC/DC output for SMAW is that AC/DC contains a rectifier. Rectified 1ɸ welding power is not as stable as rectified 3ɸ DC welding power. AC/DC welding machines are commonly used for SMAW. **See Figure 8-7.** Most AC/DC welding machines have a polarity switch with the output terminals identified as work and electrode.

CONSTANT-CURRENT WELDING MACHINES

A *constant-current welding machine* is a welding machine that maintains a relatively constant current over a wide range of welding voltages caused by changes in arc length. Constant-current welding machines

are designed primarily for manual welding such as SMAW and gas tungsten arc welding (GTAW). All welding machines used for SMAW are constant current.

Miller Electric Manufacturing Company

Figure 8-7. *An AC/DC output welding machine is commonly used for SMAW.*

Constant-current welding machines have a steep volt-ampere (VA) curve. A *volt-ampere (VA) curve* is a curve showing the relationship between output voltage and output current. A steep VA curve is ideal for manual welding because it provides between 50 V and 100 V of open-circuit voltage to aid with arc starting. Open-circuit voltage is voltage produced when a welding machine is on, but no welding is being done. A steep slope also maintains relatively constant current over a range of arc lengths (voltages). **See Figure 8-8.**

When a welder strikes an arc, voltage steps down from open-circuit voltage to welding voltage (18 V to 36 V). If the welder lengthens the arc, voltage increases because it takes more voltage to push the arc across the larger gap, and the amperage decreases slightly, which produces a corresponding increase in heat. If the welder shortens the arc, less voltage is required to maintain the arc, and the amperage increases slightly, which produces a corresponding decrease in heat. Therefore, the steep slope of a constant current VA curve makes it possible for the welder to vary arc length to control the weld pool while maintaining relatively constant heat input and electrode burn-off rates.

Media Clip

Volt-Amp Curve

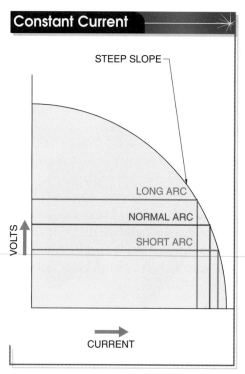

Constant Current

STEEP SLOPE

VOLTS

CURRENT

LONG ARC

NORMAL ARC

SHORT ARC

Figure 8-8. *A constant-current welding machine has a steep volt-amp curve that maintains relatively constant current over a range of arc lengths.*

STATIC POWER SOURCES

Static power sources used in a welding machine have no internal moving parts. They convert power from a utility line to the power needed for welding. Utility line power is typically supplied by a local utility company. Common static power sources include transformers, transformer-rectifiers, and inverters.

Based on the welding task, the welding leads are connected to the terminals on the front of the welding machine to supply the desired welding current. **See Figure 8-9.** Depending on how the leads are connected, current is supplied for electrode positive or electrode negative.

Transformers

Transformer welding machines are the most economical type of arc welding machine. When used with proper electrodes, they can produce quality welds for a large number of welding fabrications and repair applications. A *transformer* is an electrical device that changes AC voltage from one level to another. In welding machines, a transformer is used to transform a high-voltage and low-current input into a safe and useable low-voltage and high-current output. Typical values for a small transformer welding machine would be 230 V / 50 A AC input and 50 V to 75 V / 225 A AC output.

TERMINALS

Figure 8-9. *Based on the welding task, the welding leads are plugged into the terminals on the front of the welding machine to supply the desired welding current.*

In addition to a transformer, a reactor is needed to regulate output power. The two most common reactors for small AC welding machines are the tap and mechanical reactors. A tap reactor uses a click-type output current selector switch or a plug-in type output current selector. Each type is very simple, reliable, and maintenance free. The selectors provide only a specific output current at each setting or connection. There is no fine-tuning between output selections. A mechanical reactor uses a moveable iron or coil to regulate output current. A mechanical reactor can be fine-tuned to provide any desired output current within the capability of the machine. Since there are moving components, mechanical reactors must be properly maintained.

Transformer-Rectifiers

Transformer-rectifier welding machines are capable of producing constant DC. They contain a transformer, reactor, rectifier, and choke, or inductor. Single-phase input welding machines commonly produce AC/DC output. Three-phase welding machines are typically DC output. DC produces the highest quality SMAW welds. DC is also required for many electrodes, including most code-quality arc welding electrodes. **See Figure 8-10.**

A *rectifier* is an electrical device that changes AC into DC. Rectifiers are commonly used in a four-diode bridge called a full-wave bridge rectifier. Full-wave bridge rectifiers are typically used in 1φ transformer welding machines to produce DC output.

Three-phase transformer-rectifier welding machines typically use silicon-controlled rectifiers (SCRs). The 3φ welding machines use three transformers, one for each phase, and an SCR for each transformer. SCRs rectify and regulate the output of a welding machine, eliminating the need for a reactor to regulate the output.

A choke, or inductor, is an electrical device that smoothes out the rectified DC output. Chokes are used with full-wave bridge rectifiers of 1φ machines and with each SCR in 3φ machines.

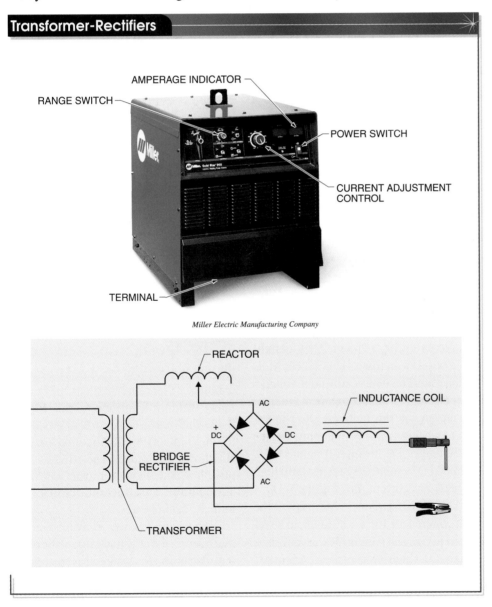

Transformer-Rectifiers

AMPERAGE INDICATOR

RANGE SWITCH

POWER SWITCH

CURRENT ADJUSTMENT CONTROL

TERMINAL

Miller Electric Manufacturing Company

REACTOR

AC

INDUCTANCE COIL

DC

DC

BRIDGE RECTIFIER

AC

TRANSFORMER

Figure 8-10. *A transformer-rectifier power source can be adapted for a variety of welding applications.*

Inverters

Inverter arc welding machines represent the latest technology in size, weight, efficiency, and control. An *inverter* is an electrical device that changes DC into AC. Inverter arc welding machines possess the following advantages:

- Size—Their small size saves space in schools, industry, and on construction sites.
- Weight—They are portable, which make them useful for maintenance welding, for the military, and on construction sites.
- Efficiency—They use less power, produce less heat, and produce less sound.
- Control—They have a very fast response to welding variables, are capable of multiple processes, contain memory settings, allow software updates, and some models are set up for wireless communication.

Basic inverter arc welding machines can be very simple to use, durable, and practical. High-end inverter arc welding machines have brought the capabilities of what has traditionally been associated with robotic, fully automatic welding centers and other high-end welding processes to arc welding.

Inverter components are very similar to transformer-rectifier components. However, they are in a different order, with a few devices added. The first component the incoming high-voltage, low-current AC meets is a rectifier. Here, 60 Hz AC is changed to DC, but is still high voltage, low current. The next component is a filter to smooth out the rectified DC ripples. Next is a high-speed switching device that changes DC to high-frequency AC (20,000 Hz or more). Again, it is still high voltage, low current. **See Figure 8-11.**

The high frequency AC allows the next component, the transformer, to be reduced in size and weight. The transformer changes the high-voltage, low-current AC to low-voltage, high-current AC. The next component is a silicon-controlled rectifier (SCR) that rectifies the AC to DC and regulates the low voltage and high current. The last component is a choke, or inductor, that smoothes out the DC. The result of these components is safe, high-quality DC output.

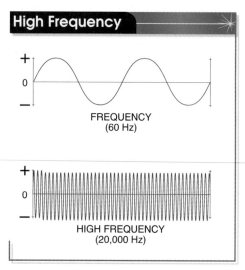

Figure 8-11. *Inverters can produce high frequency, so a smaller transformer can be used.*

Many basic inverters follow this process and are DC constant-current, output-only, arc welding machines, which are excellent for SMAW. Some welding machines add hot start and arc force controls to allow tailoring of the current specifically for SMAW.

More advanced welding machine models have these features for arc welding and are multi-process constant current and constant voltage.

ENGINE-DRIVEN POWER SOURCES

Engine-driven power sources provide welding current and auxiliary power for many welding applications. **See Figure 8-12.** They play a critical roll in producing the high-quality welds required on construction sites, bridges, and cross-country pipelines and for field repairs on heavy agricultural, construction, mining, and military equipment.

When utility power is available, it is much more cost effective to use static welding machines. An engine-driven welding machine's initial purchase, operating, and maintenance costs are higher than those of a static welding machine of the same current output per hour of operation. When selecting an engine-driven welding machine, the choice of engine, type of welding, and auxiliary power generation unit are important. The main advantages of each type of engine include the following:

- Gasoline air-cooled—light weight, less expensive
- Gasoline water-cooled—high current, long hours
- Propane air-cooled—in-plant welding, underground welding
- Diesel air and water cooled—high current, long hours, fuel efficient

In addition to the selection of engine type, the type of power generation equipment is taken into consideration.

The Lincoln Electric Company

Figure 8-12. *Engine-driven welding machines allow welding in areas where electricity is not available.*

An alternator produces AC. A generator produces DC. There are many advantages to engine-driven welding machines. **See Figure 8-13.**

Some of the largest welding machines are available with air compressors built in as compressed air is needed for weld chippers, arc gouging, and air-operated field tools. The fluid levels (oil, fuel, and coolant) of an engine-driven welding machine should always be checked before starting.

WELDING MACHINE RATINGS

Welding machines are rated (sized) according to their current at a voltage output at 60% duty cycle, such as 150 A, 200 A, 250 A, 300 A, 400 A, 500 A, or 600 A. The rating is the current output at the working terminal. Thus, a machine rated at 150 A can be adjusted to produce a range of power up to 150 A. The welding machine rating required is determined by the type of welding performed. A general guide to welding machine rating (size) and service is as follows:

- 150 A to 200 A. Light- to medium-duty welding. Excellent for all fabrication purposes, and rugged enough for continuous operation on light or medium production work.

ENGINE-DRIVEN WELDING MACHINES	
Type of Power Generation	**Welding and Auxiliary Current**
1φ alternator without rectifier	Lightweight; least expensive; produces AC only; high auxiliary output; high duty cycle
1φ alternator with rectifier	Lightweight; less expensive (except for units without rectifiers); produces AC and DC; high auxiliary output; some are CC/CV; high duty cycle
3φ alternator	Smoother DC output; many are CC/CV; higher welding amperage; 3φ auxiliary power; high duty cycle
Inverter	High-quality welding current; clean auxiliary power output; most are CC/CV with special features; high efficiency; high duty cycle
DC generator	Pipeline welding with CC; smooth DC/CC; dual control of welding current and voltage

Figure 8-13. *Engine-driven welding machines offer a number of choices to the welder.*

- 250 A to 300 A. Average welding requirements. Used in plants for production, maintenance, toolroom work, and general shop welding.
- 400 A to 600 A. Large-capacity, heavy-duty welding. Used extensively in heavy structural work, fabricating heavy machine parts, heavy pipe and tank welding, cutting scrap and cast iron, and for a wide range of welding applications.

Duty Cycle

Duty cycle is the percentage of time during a specified period that a welding machine can be operated at its rated load without exceeding the temperature limits of the insulation on the component parts. NEMA has set a standard based on a 10 min period. **See Figure 8-14.**

A welding machine rated at 300 A at 32 V, 60% duty cycle can put out the rated current at the rated voltage for 6 min out of every 10 min. The machine must idle and cool the other 4 min of every 10 min. Some welding machines used for automatic welding are rated at 100% duty cycle and can be run continuously without overheating.

WELDING EQUIPMENT

Welding equipment used for SMAW must be kept in good repair. Tools should be regularly inspected for signs of wear or damage. Required welding equipment includes welding leads, electrode holders, and workpiece leads. **See Figure 8-15.**

Welding Leads

Welding leads conduct current to and from the work. One lead runs from the welding machine to the electrode holder and the other is attached to the workpiece or the workbench. The lead connected to the electrode holder is called the electrode lead. The lead connected to the workpiece is called the workpiece lead.

When the welding machine is ON and the electrode in the electrode holder comes in contact with the workpiece, a circuit is formed, allowing current to flow.

The correct diameter welding lead for the length of cable specified for the welding machine output must be used. If welding leads are too small for the current, they overheat and power is lost. Larger leads are needed to carry the required current long distances from the welding machine; otherwise, there will be an excessive voltage drop. With smaller diameter welding leads, the recommended length must not be exceeded because voltage drop across the leads lowers the efficiency of the welding. Check with the welding machine manufacturer for the proper welding lead sizes, and for specific lengths and usage.

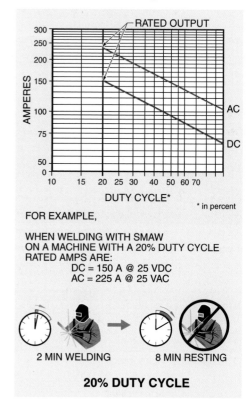

Figure 8-14. *Duty cycle is the amount of time that a welding machine can be operated at its rated load without overheating during a 10 min period. Duty cycle is expressed as a percentage.*

All welding lead connections should be tight because loose connections cause the lead to overheat. A loose connection may also produce arcing at the connection. Welding leads should be kept clean and should be handled so as to avoid damage to the insulation.

WELDING LEADS
The Lincoln Electric Company

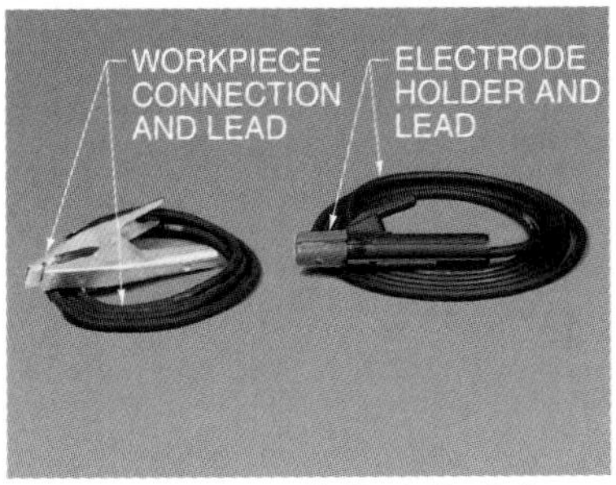

ELECTRODE HOLDER
ESAB Welding & Cutting Products

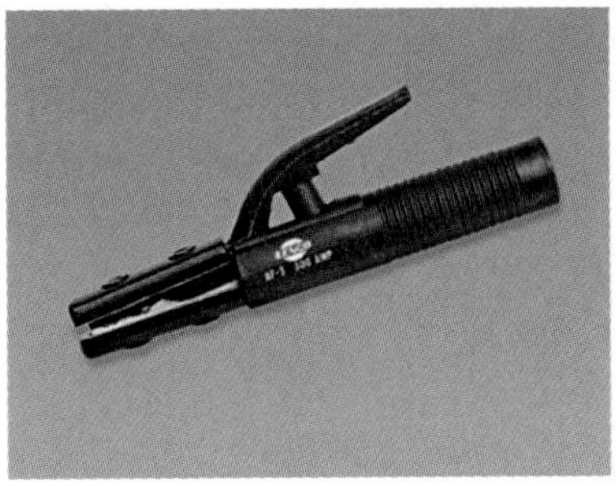

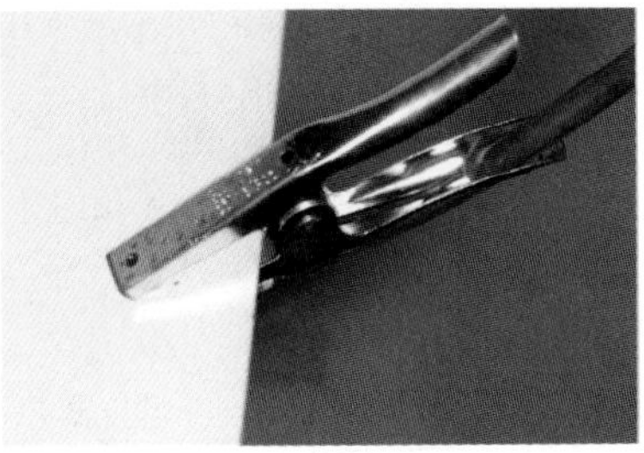

Removable

Clamped

WORKPIECE CONNECTIONS

Figure 8-15. *Proper welding requires welding equipment such as welding leads, electrode holders, and workpiece connections.*

Electrode Holders

An *electrode holder* (stinger) is a handle-like tool that holds the electrode during welding. The electrode holder is attached to the electrode lead during welding. The jaws of the electrode holder must be properly insulated. Laying an electrode holder with uninsulated jaws on the workbench while the machine is running may cause a flash. A well-designed electrode holder can be identified by the following features:

- It is reasonably light, to reduce excessive fatigue while welding.
- It does not heat too rapidly.
- It is well balanced.
- It secures and releases electrodes easily.
- It is properly insulated.

Work Leads

The work lead must be fastened to the workpiece or the workbench to provide a complete path for the electrical circuit. A workpiece connection is attached to the workpiece lead to complete the circuit. This type of workpiece connection is removable, making it easier for a welder to change locations. A workpiece lead can be attached or welded to the workbench using the lug on the end of the lead. Connections should be made as close to the welding location as possible.

SHOP TOOLS AND EQUIPMENT

Shop equipment, such as C-clamps, electrode ovens, tools, welding screens, and ventilation systems are required for a safe work area. **See Figure 8-16.** When welding workpieces that are too large to fit in a vise, C-clamps hold the workpieces in the proper position. Many electrodes must be stored at high temperatures to protect them from humidity. Electrode ovens maintain the required temperature and protect electrodes from damage. Floors should be constructed of fire-resistant materials and should be kept dry at all times to prevent possible shock.

✓ **Point**
If welding near other workers, set up welding screens so the arc does not harm workers nearby.

✓ **Point**
Use an electrode holder that is completely insulated.

✓ **Point**
Weld only in areas where there is adequate ventilation.

Hand Grinder

**Chipping Hammers/
Wire Brushes**

Ball Peen Hammer

TOOLS

Positioner

Welding Screen

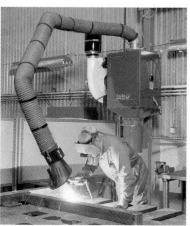

Nederman, Inc.

Ventilation System

C-Clamp

Electrode Oven

Figure 8-16. *Shop tools and equipment assist the welder in maintaining a safe work environment and producing quality welds.*

Tools

To produce a strong weld, the surface of the base metal must be free of foreign matter such as rust, oil, and paint. A wire brush (hand- or tool-powered) is used to clean metal surfaces.

After a bead is deposited, the slag that covers the weld is removed with a chipping hammer. The chipping operation is followed by additional wire brushing. Complete removal of slag is especially important when several passes must be made over a joint. If not properly removed, slag becomes trapped in the weld, forming inclusions that weaker the weld and may cause it to be rejected.

Welding Screen

Whenever welding is done in areas where other people may be working, the welding operation should be enclosed with screens so the ultraviolet rays cannot injure nearby workers. Welding screens can be easily constructed from fire-resistant canvas painted with black or gray ultraviolet-protective paint. When welding is done in a permanent location, a booth is desirable. A welding booth can be set up with a positioner, bench, ventilation, power source, and other equipment necessary for welding.

Ventilation System

Electrodes used for SMAW may emit a great deal of smoke and fumes, which should not be inhaled. Smoke and fumes should be properly vented. There should be a suction fan or other adequate fume extractor.

Permanent welding booths should be equipped with a sheet-metal hood with an extension arm mounted directly above the welding table and an exhaust system to draw out the smoke and fumes.

SMAW should not be performed without sufficient movement of air through the room. The general recommendation for adequate ventilation is a minimum of 2000 cu ft of air flow per minute per welding machine. If individual movable exhaust hoods can be placed near the work, the rate of air flow toward the hood should be approximately 100 linear feet per minute in the welding zone. The exhaust hood should never be placed in a manner that draws the gas and fumes across the face of the welder.

PERSONAL PROTECTIVE EQUIPMENT

An electric arc not only produces a brilliant light, but also gives off invisible ultraviolet and infrared rays that are extremely dangerous to the eyes and skin. Additionally, extreme heat is generated by welding, as well as slag and spatter, which may pop from the weld and strike a welder. Welders are required to wear personal protective equipment to prevent injury.

Approved work clothes, such as those made of leather, wool, or flame-resistant cotton; a headcap; safety glasses; approved work boots; and gloves are required for all welding and cutting operations. Light-duty welding requires leather gloves and a welding helmet with proper shading. Heavy-duty welding requires a leather jacket, leather gauntlet-type gloves, a leather apron, and a helmet with proper shading. **See Figure 8-17.**

The Lincoln Electric Company

Figure 8-17. *Proper protective clothing must be worn to protect the welder from ultraviolet and infrared rays, slag, and spatter produced during welding.*

Coveralls or work clothing should prevent exposure of the skin to infrared and ultraviolet rays. Synthetic materials such as polyester should never be worn. Pants should not have cuffs, and pockets should be covered to prevent molten metal from catching in the clothes. Sleeves and collars should be kept buttoned. Pant legs and shirt sleeves should be short enough that they do not bunch around the ankles or wrists.

The head and eyes must be protected from metal pieces or sparks that may be projected from a welding surface. Helmets with correct filter plate shades are required when performing any welding operation in order to prevent ultraviolet and infrared radiation from burning the eyes. Safety glasses should be worn under face shields, and helmets, and at all times when working in the shop. **See Figure 8-18.**

Eye Protection

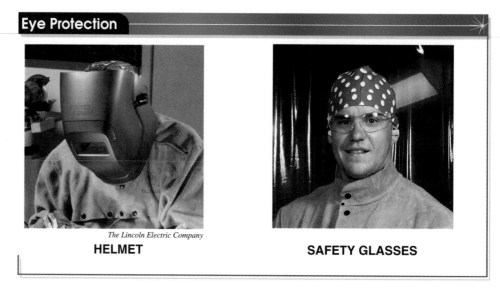

The Lincoln Electric Company
HELMET

SAFETY GLASSES

Figure 8-18. *Eye protection must be worn at all times when working in the shop and during welding.*

Refer to Quick Quiz® on CD-ROM

- The heat used for SMAW is generated from an arc that develops when electricity jumps across an air/gas gap between the end of an electrode and the base metal. The air/gas gap produces high resistance to the current flow, generating intense heat.
- The National Electrical Manufacturers Association (NEMA) has set a standard for duty cycle based on a 10 min period. The duty cycle standard expresses the actual operation time that a welding machine may be used at its rated load without exceeding the temperature limits of the insulation of the component parts.
- Keep welding leads orderly to prevent them from becoming a hazard. Fasten the welding leads overhead whenever possible. Never kink the welding leads.
- Use properly sized welding leads to prevent voltage drop.
- If welding near other workers, set up welding screens so the arc does not harm workers nearby.
- Use an electrode holder that is completely insulated.
- Weld only in areas where there is adequate ventilation.

Questions for Study and Discussion

1. What is an electrical circuit?
2. What is the difference between AC current and DC current?
3. What is polarity?
4. What determines whether the polarity of a welding machine is set for DCEN or DCEP?
5. What is voltage? What instrument is used to measure voltage?
6. What effect does welding polarity have on where heat is directed?
7. What is voltage drop? What effect does it have on welding current?
8. What is meant by open-circuit voltage and arc voltage?
9. What is meant by a constant-current welding machine?
10. What is a volt-amp curve?
11. Why is a transformer-rectifier often preferred for SMAW?
12. How are welding machines rated?
13. What is duty cycle when specifying welding machine ratings?
14. What are some of the requirements of an electrode holder?
15. Why is it important to weld only where there is adequate ventilation?

Refer to Chapter 8 in the *Welding Skills Workbook* for additional exercises.

SMAW–Selecting Electrodes

There are many different types and sizes of electrodes, and the correct one must be selected to ensure a quality weld. In general, electrodes are classified into five types: low-carbon steel, high-carbon steel, alloy steel, cast iron, and nonferrous. Electrodes are identified by American Welding Society (AWS) designations or manufacturer trade names where no corresponding AWS designation exists.

ELECTRODES

An *electrode* is a component of the welding circuit that conducts electrical current to the weld area. When current from a welding machine flows through the circuit to the electrode, an arc is formed between the end of the electrode and the work. The arc melts the base metal and filler metal to form the weld bead that joins the two pieces of metal. The arc also melts the electrode coating. **See Figure 9-1.** Depending on the composition of the coating, it breaks down to the following:

- a shielding gas that protects the molten weld pool from the atmosphere
- alloying elements that contribute to the chemical and mechanical properties of the weld
- a slag coating over the weld that protects the weld bead from the atmosphere as it cools, slows the cooling rate of the weld metal to help produce a more ductile bead, and may contain deoxidizers and scavengers that remove impurities from the weld

Electrodes are manufactured to weld different types and thicknesses of metal. Some electrodes are also designed to work with direct current (DC) only, some are designed to work best with alternating current (AC), and others work with DC and AC. Electrode selection also depends on welding position. Some electrodes are best suited for flat position and horizontal fillet welding due to the fluidity of the weld pool, while others may be used in all positions.

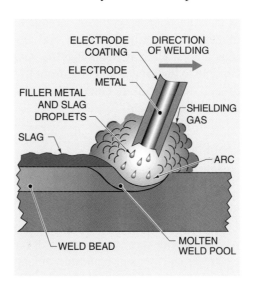

Figure 9-1. *Molten metal from the electrode combines with the base metal to form the weld bead. The arc melts the slag coating to form shielding gas, which protects the weld pool, and slag, which protects the weld bead as it cools.*

The coatings on SMAW electrodes contain various elements such as cellulose sodium, cellulose potassium, titania sodium, titania potassium, iron oxide, and iron powder, as well as several other ingredients. Each of the substances is intended to serve a particular function in the welding process, such as the following:

- act as a cleaning and deoxidizing agent to remove impurities and oxygen from the molten weld pool

✓ **Point**

Some electrodes are designed to work with direct current (DC) only, some are designed to work best with alternating current (AC), and others work with DC and AC.

✓ **Point**

Selecting the proper electrode for a job is a critical part of SMAW.

✓ **Point**

AWS A5.1/A5.1M, Specification for Carbon Steel Electrodes for SMAW, lists the requirements for carbon steel SMAW electrodes.

- produce shielding gas to prevent the molten weld pool from reacting with oxygen and nitrogen in the atmosphere that could cause porosity and other problems
- form a slag coating over the weld bead to protect it from oxygen and nitrogen in the atmosphere as it cools, as well as to produce a more ductile weld by slowing the cooling rate
- provide easier arc starting, stabilize the arc, and reduce spatter
- permit better penetration and help meet weld quality requirements

The coating on some electrodes contains iron powder, which becomes part of the weld deposit. It also helps increase travel speed and improve weld bead appearance.

Low-hydrogen electrodes have coatings high in limestone and other ingredients with low hydrogen content, such as calcium fluoride, calcium carbonate, magnesium-aluminum silicate, and ferrous alloys. Carbon steels may be susceptible to underbead cracking from hydrogen in the heat-affected zone (HAZ) adjacent to the weld.

Identifying Electrodes

Electrodes are referred to by manufacturer trade name and by American Welding Society (AWS) electrode classification. These classifications were established by AWS to provide formulation standards for electrodes and to ensure uniformity among manufacturers. AWS A5.1/A5.1M is the specification for carbon steel electrodes for SMAW. It establishes the requirements for the classification of electrodes, which includes mechanical properties of the weld metal, usability of the electrode, standard sizes, manufacturing, and packaging.

All electrodes in the same AWS classification should have similar welding characteristics regardless of manufacturer. Most electrodes are imprinted with an AWS classification number. This indicates that the manufacturer has qualified the electrode to the AWS specification for required mechanical and chemical properties. **See Figure 9-2.**

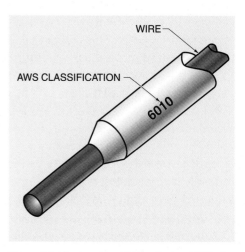

Figure 9-2. *The American Welding Society (AWS) electrode classification system identifies the tensile strength, welding position, type of coating, and operating characteristics of electrodes.*

Each type of electrode has a specific AWS classification number, such as E6010, E6012, or E7018. The prefix E stands for electrode. The first two digits of a four-digit number or the first three digits of a five-digit number specify the minimum tensile strength of the deposited weld metal in thousand pounds per square inch (ksi). For example, the number 60 indicates a minimum tensile (pull) strength of 60,000 psi (60 ksi). The number 70 indicates a minimum tensile strength of 70,000 psi (70 ksi). The number 110 indicates a minimum tensile strength of 110,000 psi (110 ksi).

The second to last digit in the classification number specifies welding position. The welding position is indicated by a 1, 2, or 4. The number 1 in the second to last position indicates that the electrode can be used in all positions. A number 2 indicates that the electrode is restricted to flat position groove (1G) and fillet welds (1F), and to horizontal fillet welds (2F) only, due to the fluidity of the weld pool. A number 4 indicates that the electrode can be used in the flat, horizontal, vertical down, and overhead positions.

The last digit in the classification (0 through 8) specifies the type of coating, welding current, and operating characteristics of the electrode. The last two digits are often viewed together, because the operating characteristics of the electrode determine the positions in which it can be used.

The E7018 low-hydrogen electrode may have a –1 suffix. This indicates that the weld deposit meets the requirements for improved impact strength of 20 ft-lb at –50°F, rather than 20 ft-lb at –20°F for a standard E7018. The suffix may also include an alphanumeric combination of H4, H8, or H16. An H4 indicates that the electrode meets the requirements of the diffusible hydrogen test with an average value not exceeding 4 mL of hydrogen per 100 grams of deposited metal. H8 and H16 indicate a hydrogen level of less than 8 mL and 16 mL/100 g of deposited metal respectively. The letter R may follow the hydrogen level if the electrode meets the requirements of the absorbed moisture test. **See Figure 9-3.**

An electrode for low-carbon steel with an AWS classification of E6010 produces a weld bead with 60,000 psi minimum tensile strength, can be used in all positions, can be used with DCEP only, and has a cellulose sodium coating. An E7024 electrode produces a weld bead with 70,000 psi minimum tensile strength, is restricted to the flat position and horizontal fillet welds only, can be used with AC, DCEP, or DCEN, and has an iron powder and titania coating. If an E7024 has a –1 suffix, it means that it has greater low temperature toughness than the standard E7024. **See Figure 9-4.**

Electrodes for Low-Alloy Steels

Electrodes for low-alloy steels may contain a variety of alloying elements that produce different chemical and mechanical properties in the weld bead. Other elements may be added to remove the oxygen and nitrogen, which can cause porosity, or to neutralize the sulfur and phosphorus, which can cause cracking. Phosphorus causes embrittlement and loss of toughness while sulfur causes embrittlement. Common elements added to electrodes include carbon, manganese, silicon, chromium, nickel, molybdenum, boron, copper, aluminum, titanium, columbium, and vanadium. **See Figure 9-5.** All carbon steels contain small amounts of phosphorus and sulfur as by-products of the steel-making process.

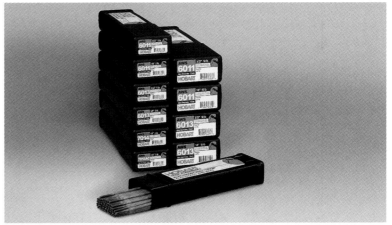

Hobart Welding Products

Electrodes contain different alloying elements for specific SMAW applications.

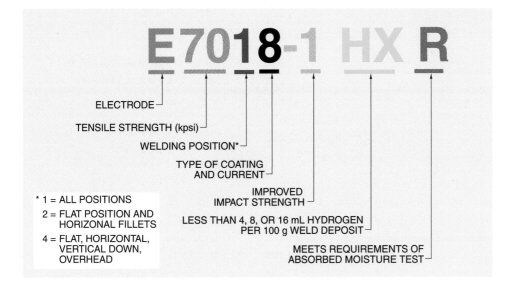

Figure 9-3. *The AWS electrode classification system uses a four- or five-digit number to identify electrodes. A suffix consisting of numbers and letters indicates additional operating characteristics.*

SMAW ELECTRODES FOR LOW-CARBON STEEL

AWS Class	Group	Coating	Polarity*	Characteristics	Position
EXX20	F1 (Fast Fill)	Iron oxide	DCEP	Shallow to medium penetration, high deposition smooth arc with low spatter, fluid weld pool, smooth bead with flat to slightly concave contour, heavy slag easy to remove Good for joining thick members	Flat and horizontal fillets only
EXX27		Iron powder, iron oxide	AC, DCEP, DCEN		
EXX24		Iron powder, titania	AC, DCEP, DCEN	Same as EXX20 and EXX27 except bead contour is flat to slightly convex	
EXX28		Iron powder, low-hydrogen potassium	AC, DCEN,	EXX28 produces weld deposits low in hydrogen for high-sulfur and high-carbon steels that tend to develop porosity and under-bead cracking	
EXX12	F2 (Fill Freeze)	Titania sodium	DCEN, AC	Shallow to medium penetration, medium deposition, evenly rippled bead with flat to slightly convex contour, low spatter, easy slag removal Good for sheet metal and poor fit-up	Flat, horizontal, vertical, overhead
EXX13		Titania potassium	AC, DCEP, DCEN		
EXX14		Iron powder, titania	AC, DCEP, DCEN		
EXX10	F3 (Fast Freeze)	Cellulose sodium	DCEP	Deep penetration, low deposition, crisp arc, bead solidifies rapidly with flat to slightly concave contour, coarse ripple pattern, thin slag	
EXX11		Cellulose potassium	AC, DCEP	Good for pipe welding, root passes, poor fit-up, general fabrication, and repair welding	
EXX16	F4 (Fill Freeze)	Low-hydrogen potassium	DCEP, AC	Medium penetration, medium to high deposition, soft arc, flat to slightly convex bead contour with smooth, even ripple pattern, easy slag removal, low hydrogen content in weld deposits, excellent ductility Good for high-sulfur and high-carbon steels that tend to develop porosity and under-bead cracking	
EXX18		Iron powder, low-hydrogen potassium	DCEP, AC		
EXX48		Iron powder, low-hydrogen potassium	DCEP, AC		Flat, horizontal, vertical down, overhead

* Preferred polarity for each electrode listed first

Figure 9-4. *The AWS has organized electrodes for low-carbon steel into F groups based on their performance characteristics.*

EFFECT OF ALLOYING ELEMENTS ON WELD DEPOSIT

Symbol	Element	Effect
C	Carbon	Increases hardenability and improves toughness
Mn	Manganese	Deoxidizes; increases hardness, strength and wear resistance; combines with sulfur and phosphorous to decrease embrittlement
Si	Silicon	Deoxidizes; improves oxidation resistance; increases hardenability, strengthens low-alloy steel
Cr	Chromium	Increases hardenability and strength; increases corrosion and oxidation resistance; improves abrasion resistance
Ni	Nickel	Adds ductility; increases strength and toughness at low temperatures
Mo	Molybdenum	Promotes uniform hardness and tensile strength; increases resistance to softening after tempering; increases strength and creep resistance at elevated temperatures
B	Boron	Increases hardenability in amounts from 0.0025% to 0.0030%; decreases ductility and causes cracking in amounts over 0.006%
Cu	Copper	Increases corrosion resistance, toughness, and ductility
Al	Aluminum	Deoxidizes; improves mechanical properties by reducing grain growth
Ti	Titanium	Deoxidizes; increases corrosion resistance; distributes sulfur in high-sulfur steel
Cb (Nb)	Columbium (Niobium)	Increases hardness; improves mechanical properties
V	Vanadium	Increases hardenability and corrosion resistance; controls grain growth; produces fine grain structure during heat treatment

Figure 9-5. *Electrodes for low-alloy steel contain a variety of alloying elements designed to produce different chemical and mechanical properties in the weld deposit.*

The AWS has developed a series of alphanumeric suffixes for electrodes (EXXXX-A1, EXXXX-B2, and EXXXX-C3, and so on). These suffixes indicate the alloying elements included in electrodes for low-alloy steels, stated as a percentage of the deposited weld. AWS A5.5 is the specification for low-alloy steel electrodes for SMAW. It lists the requirements for low-alloy steel SMAW electrodes for welding carbon and low-alloy steels. **See Figure 9-6.**

ELECTRODE CLASSIFICATION GROUPS

The AWS organizes electrodes for low-carbon and low-alloy steel into four F groups based on their performance characteristics. The F1 group (fast-fill electrodes) consists of high deposition electrodes such as the E7024. A *fast-fill (F1 group) electrode* is an electrode with a high iron powder coating that has a soft arc and high deposit rates. It produces a heavy slag that is easily removed and exceptionally smooth weld beads. It is generally used for production welding where all work can be performed in flat position.

The F2 group (fill-freeze electrodes) consists of shallow to medium penetrating electrodes such as the E6012. A *fill-freeze (F2 group) electrode* is an electrode that has a moderately forceful arc and a medium deposition rate. The arc and deposition rates are between those of the fast-fill and fast-freeze electrodes. Fill-freeze electrodes produce easy to remove slag and a weld bead with distinct, even ripples. They are general-purpose electrodes used in production shops and are particularly useful for sheet metal, repair work, and joints with poor fit-up. They can be used in all positions but are best suited for flat and horizontal welding.

The F3 group (fast-freeze electrodes) consists of deep penetrating electrodes like the E6010. A *fast-freeze (F3 group) electrode* is an electrode that produces a crisp, deep-penetrating arc and a fast-freezing weld bead. Fast-freeze electrodes produce a thin, tough slag and a flat to slightly concave weld bead. They are used for all-position welding for fabrication, repair work, and joints with poor fit-up. They are preferred for vertical and overhead positions.

Point

AWS A5.5, Specification for Low-Alloy Steel Electrodes for SMAW, lists the requirements for low-alloy steel SMAW electrodes for welding carbon and low-alloy steels.

Point

When welding with direct current, make sure the welding machine is set to the correct polarity for the electrode (DCEN or DCEP).

Point

The AWS organizes electrodes for low-carbon and low-alloy steel into four groups (F groups) based on their performance characteristics.

COMPOSITION OF UNDILUTED WELD METAL FOR LOW-ALLOY STEEL ELECTRODES

Suffix	C*	Mn*	Si*	Ni*	Cr*	Mo*	V*
A1	.12	.60–1.00†	.40–.80†	—	—	.40–.65	—
B1	.05-.12	.90	.60–.80†	—	.40–.65	.40–.65	—
B2L	.05	.90	.60–1.00†	—	1.00–1.50	.40–.65	—
B2	.05–.12	.90	.60–.80†	—	1.00–1.50	.40–.65	—
B3L	.05	.90	.80–1.00†	—	2.00–2.50	.90–1.20	—
B3	.05–.12	.90	.60–1.00†	—	2.00–2.50	.90–1.20	—
B4L	.05	.90	1.00	—	1.75–2.25	.40–.65	—
B5	.07–.15	.40–.70	.30–60	—	.40–.60	1.00–1.25	.05
B6	.05–.10	1.00	.90	.40	.40–.60	.45–.65	—
B8	.05–.10	1.00	.90	.40	8.0–10.5	.85–1.20	—
C1	.12	1.25	.60–.80†	2.00–2.75	—	—	—
C2	.12	1.25	.60–.80†	3.00–3.75	—	—	—
C3	.12	.40–1.25	.80	.80–1.10	.15	.35	—
D1	.12	1.00–1.75	.60–.80†	.90	—	.25–.45	—
D2	.15	1.65–2.00	.60–.80†	.90	—	.25–.45	—
D3	.12	1.00–1.80	.60–.80†	.90	—	.40–.65	—

* Percent by weight
† Actual percentage depends on electrode classification

Figure 9-6. *An alphanumeric suffix following an electrode classification, such as E7018-A1, indicates the alloy composition of the weld deposit.*

The F4 group (fill-freeze electrodes) consists of all position low-hydrogen electrodes like the E7018. The F4 group has operating characteristics similar to electrodes in the F2 group. A *fill-freeze (F4 group) electrode* is an electrode that produces sound welds with excellent notch toughness and high ductility. Low-hydrogen electrodes help prevent porosity because they produce welds with low hydrogen content. They also help to prevent underbead cracking on high-sulfur, high-carbon, and low-alloy steel. They have a soft arc and produce a slag coating that is easy to remove.

Selecting Electrodes

Seven factors should be considered when selecting an electrode—base metal properties, electrode diameter, joint design and fit-up, welding position, welding current and polarity, production efficiency, and service conditions.

Base Metal Properties. Electrodes are available for welding different classifications of metal. Some electrodes are designed to weld carbon steels, others are best suited for low-alloy steels, and some are intended for special purpose alloy steels such as chrome-moly. An electrode must produce a weld bead with the same or similar chemical and mechanical properties as the base metal. Therefore, the first step in selecting an electrode is to check the chemical analysis of the base metal and choose an electrode with the same or similar composition. Welding should never be performed on unidentified metal.

Electrode Diameter. Typically, the diameter of the electrode should not exceed the thickness of the base metal. Some welders prefer larger electrodes because they permit faster travel along the joint and thus speed up the welding operation, but this requires considerable skill.

For production efficiency, the largest diameter electrode practical for the job should be used. When making vertical or overhead welds, ³⁄₁₆″ is the largest diameter electrode that should be used

regardless of the base metal thickness. Larger electrodes make it too difficult to control the deposited metal. Typically, a fast-freeze (F3 group) electrode is best for vertical and overhead welding.

Joint Design and Fit-Up. The type and diameter of the electrode are also influenced by joint design. On thick metal with a narrow root opening, a small-diameter electrode is used to deposit the root bead. This ensures complete penetration into the root of the joint. Successive passes can be made with larger diameter electrodes from the appropriate electrode group.

Joints with insufficiently beveled edges require deep-penetrating, fast-freeze (F3 group) electrodes. These electrodes have a digging arc characteristic and may require more skillful electrode manipulation by the welder. Joints with wide root openings need a mild penetrating, fill-freeze (F2 group) electrode that rapidly bridges gaps and minimizes the possibility of burn-through.

Welding Position. The position of the weld joint must be considered when selecting an electrode. If the joint can be fixtured in the flat position, it might be possible to reduce welding time by using a high deposition fast-fill (F1 group) electrode. However, if the joint has to be welded out of position, an all-position fill-freeze (F2 group) or fast-freeze (F3 group) electrode would be the logical choice.

For vertical and overhead welding, all-position fast-freeze electrodes are typically limited to a ³⁄₁₆″ diameter, and fill-freeze electrodes are limited to a ⁵⁄₃₂″ diameter. These electrodes have lower deposition rates, which increase the time required to complete the job.

Welding Current and Polarity. Some SMAW electrodes are designed to operate with DCEP only, some with DCEP and AC, some with DCEN and AC, and some with AC, DCEP and DCEN. The type of welding current and polarity, along with the type of electrode coating determine the melting rate of the electrode. They also influence penetration. When selecting electrodes for SMAW, the following information is taken into consideration:

- Direct current (DC) produces a steadier arc and smoother metal transfer than AC because current travels in one direction only.
- DC is preferred for vertical and overhead welding because DC can sustain a shorter arc for better weld pool control.
- DCEP produces deeper penetration than DCEN or AC.
- DCEN produces higher electrode melting rates (deposition) than DCEP or AC.
- AC eliminates problems with arc blow on ferrous metals. This makes it possible to use larger diameter electrodes and operate at higher current levels, which increases deposition rates.

Production Efficiency. Deposition rate is important in production work. The faster a weld can be made, the lower the cost. Therefore, the largest diameter electrode practical for the job and from the appropriate electrode group should be used. The recommended amperage range for an electrode should never be exceeded in an attempt to increase deposition and travel speed. **See Figure 9-7.**

Figure 9-7. *The largest diameter electrode practical for the job from the appropriate electrode group should be used to maximize production efficiency.*

Service Conditions. The service conditions of the weldment may require high corrosion resistance, ductility, or high strength at extreme temperatures. The service conditions to which welds will be subjected must be taken into consideration when selecting electrodes.

Conserving and Storing Electrodes

Stub loss can add significantly to the cost of any job involving SMAW. Therefore, stub ends should not be discarded until they are down to the electrode classification number stamped on the coating, or about 1½″ to 2″ long. **See Figure 9-8.**

Figure 9-8. *Electrodes should be used until the stubs have been consumed down to the electrode classification number or until they are only 1½″ to 2″ long.*

Electrodes should be stored in a dry place, at normal room temperature, with a 50% maximum relative humidity. When exposed to moisture, electrode coatings disintegrate if they are not properly reconditioned. Low-hydrogen electrodes are especially vulnerable to moisture. Low-hydrogen electrodes, such as E7018, should be stored in drying ovens at 250°F to 300°F after the seal on their moisture-proof container is broken. Low-hydrogen electrodes should be stored separately from other types of electrodes. Stationary and portable drying ovens are designed for storing electrodes at specified holding temperatures. Care should be taken not to damage electrode coatings. Electrodes with chipped or damaged coatings are useless and must be discarded.

Low-Carbon Steel Electrodes

The most commonly used low-carbon steel electrode is E6010, because of its penetration and fast-freeze capabilities. The current settings for low-carbon steel electrodes are determined by the size of the electrode. Commonly used low-carbon steel electrodes include E6010, E6011, E6012, and E6013. **See Figure 9-9.**

LOW-CARBON STEEL ELECTRODE CURRENT SETTINGS		
Electrode	Diameter*	Amperes†
E6010	3/32	60 – 90
	1/8	80 – 120
	5/32	110 – 160
	3/16	150 – 200
	7/32	175 – 250
	1/4	225 – 300
	5/16	250 – 450
E6011	3/32	50 – 90
	1/8	80 – 130
	5/32	120 – 180
	3/16	140 – 220
	7/32	170 – 250
	1/4	225 – 325
E6012	3/32	40 – 90
	1/8	80 – 120
	5/32	120 – 190
	3/16	140 – 240
	7/32	180 – 315
	1/4	225 – 350
E6013	1/16	20 – 40
	5/64	25 – 50
	3/32	30 – 80
	1/8	80 – 120
	5/32	120 – 190
	3/16	140 – 240
	7/32	225 – 300
	1/4	50 – 350

* in in.
† ranges may vary depending on manufacturer

Figure 9-9. *Low-carbon steel electrodes are used for many general welding operations, and current settings vary depending on electrode diameter and type.*

E6010. The E6010 electrode is an all-position, fast-freeze (F3 group) electrode. It is suitable only with DCEP and is designed primarily for welding low-carbon and low-alloy steels. It is excellent for handling poor fit-up. The E6010 electrode has wide applications in shipbuilding, construction, bridges, tanks, and piping.

E6011. The E6011 is a fast-freeze (F3 group) electrode. It is similar in performance to the E6010 except that it can be used with DCEP and AC. Although the E6011 can be used with DCEP, it does not perform quite as well as the E6010. Current settings for the E6011 are slightly lower than for the E6010.

E6012. The E6012 electrode is a fill-freeze (F2 group) electrode. It is designed for use with DCEN or AC. An E6012 electrode provides medium penetration, a quiet arc, low spatter, and dense, easy-to-remove slag. Although it is considered an all-position electrode, it is more commonly used for flat and horizontal position welds. This electrode is especially useful for bridging wide root openings on work with poor fit-up. Higher currents can be used with the E6012 electrodes than with any other type of all-position electrode.

E6013. The E6013 electrode is a fill-freeze (F2 group) electrode. It is similar to the E6012, with a few exceptions. The E6013 can be used with AC, DCEP, or DCEN, but performs best with AC. The weld bead is noticeably flatter and smoother than the E6012 but produces shallower penetration. The E6013 has a stable arc, and slag is removed easily, especially with small-diameter electrodes, which permits better operation with lower open-circuit voltage. It performs well in all positions, and is a good choice for sheet metal and poor fit-up.

Iron Powder Electrodes

Iron powder electrodes have coatings that are high in iron powder. Iron powder electrodes are designed for welding low-carbon steel where high travel speeds and high deposition rates are required. AC is generally recommended for iron powder electrodes when used with higher currents to minimize arc blow. Iron powder electrodes produce shallow to medium penetration, low spatter, and a slag coating that is easy to remove. The three principal iron powder electrodes are the E6027, E7014, and E7024. Typical applications include railroad cars, earthmoving equipment, pressure vessels, piping, and ship building. The E7014 and E7024 are often used where higher strength joints are required. **See Figure 9-10.**

E6027. The E6027 is a fast-fill (F1 group) electrode. It produces high-quality welds with high travel speeds and high deposition rates. The E6027 is confined to the flat position and to horizontal fillet welds due to the fluidity of the weld pool. It can be used with AC, DCEN, or DCEP in the flat position.

IRON POWDER ELECTRODE CURRENT SETTINGS		
Electrode	Diameter*	Amperes†
E6027	3/16	225 – 300
	7/32	275 – 375
	1/4	350 – 450
E7014	3/32	80 – 110
	1/8	110 – 150
	5/32	140 – 190
	3/16	180 – 260
	7/32	250 – 325
	1/4	300 – 400
	5/16	400 – 500
E7024	3/32	90 – 120
	1/8	120 – 150
	5/32	180 – 230
	3/16	250 – 300
	7/32	300 – 350
	1/4	350 – 400
	5/16	400 – 500

* in in.
† ranges may vary depending on manufacturer

Figure 9-10. *Iron powder electrodes have various current settings depending on diameter and type and are commonly used for joints requiring high strength.*

In the horizontal position, it can be used with AC or DCEN to produce fillet welds with a flat to slightly concave contour. It is ideal for cover passes using DCEN or AC where a good bead appearance is required. A drag welding technique is recommended to prevent slag entrapment.

E7014. The E7014 electrode is an all-position, fill-freeze (F2 group) electrode. It is designed for use with AC, DCEP, or DCEN. The E7014 has higher deposition rates than an E6012 or E6013. It is particularly effective in downhill welding.

E7024. The E7024 electrode is a fast-fill (F1 group) electrode. It is designed for use with AC, DCEP, or DCEN. The E7024 is exceptionally economical for single- or multiple-pass welds as well as buildup applications because of its high deposition rate and easy slag removal. The E7024 is limited to the flat position and to horizontal fillet welds.

Low-Hydrogen Electrodes

Low-hydrogen electrodes are designed for welding high-sulfur and medium- or high-carbon steel, as well as thick sections. When such steels are welded, they tend to develop cracks under the weld bead because of hydrogen absorption from the arc atmosphere. Low-hydrogen electrodes were developed to prevent the introduction of hydrogen into the weld and heat affected zone (HAZ) adjacent to the weld. Low-hydrogen electrodes include the E7016, E7018, and E7028. **See Figure 9-11.**

LOW-HYDROGEN ELECTRODE CURRENT SETTINGS		
Electrode	Diameter*	Amperes†
E7016	3/32	75 – 105
	1/8	100 – 150
	5/32	140 – 190
	3/16	190 – 250
	7/32	250 – 300
	1/4	300 – 375
E7018	3/32	70 – 120
	1/8	100 – 150
	5/32	120 – 200
	3/16	200 – 275
	7/32	275 – 350
	1/4	300 – 400
E7028	5/32	175 – 250
	3/16	250 – 325
	7/32	300 – 400
	1/4	375 – 475

* in in.
† ranges may vary depending on manufacturer

Figure 9-11. *Low-hydrogen electrodes are recommended for steel with high-sulfur and high-carbon content. Current settings depend on electrode diameter and type.*

E7016. The E7016 electrode is an all-position, low-hydrogen, fill-freeze (F4 group) electrode. It is designed for use with AC and DCEP. The E7016 is recommended for welding hardenable steels where no preheat is used and where stress relieving normally would be required but cannot be performed.

E7018. The E7018 electrode is an all-position, low-hydrogen, fill-freeze (F4 group) electrode. The electrode coating on the E7018 contains iron powder, which produces higher deposition rates and faster travel speeds than the E7016. It is designed for use with AC and DCEP and produces high quality welds. The E7018 produces shallow to medium penetration with a smooth arc and low spatter. It is usually limited to small diameter electrodes for vertical and overhead welding.

E7028. The E7028 electrode is a low-hydrogen, fast-fill (F4 group) electrode. It produces a weld bead similar in physical properties and weld quality to the 7018, but it has unique characteristics that distinguish it from the E7018. It has a thicker coating because of its higher iron powder content. It produces higher deposition rates and allows faster travel speeds than the E7018, and it produces a spray transfer, not the globular transfer produced by the E7018.

ELECTRODE SELECTION VARIABLES

Although there are a variety of electrode classification charts that list the basic characteristics or differences in electrodes, many of the variables encountered in production often require testing to determine the suitability of an electrode for a specific application. Considerable time and effort can be saved by first analyzing the variables in terms of their importance to the quality of the finished weldment.

The suitability of an electrode for use with certain types of joints, such as groove butt welds and fillet welds, can be rated to help determine the proper electrode to use. Additionally, the type of metal to be welded, such as thin metal, heavy steel, or high-sulfur or off-analysis steel, is a determinant in the type of electrode to use. **See Figure 9-12.** Other variables associated with most types of welding include the following:

- deposition rate
- depth of penetration
- appearance and undercutting
- soundness
- ductility
- low-impact strength of the weld
- amount of spatter
- quality of fit-up required
- welder appeal
- ease of slag removal

The variables have a relative rating ranging from 1 to 10, with 10 as the highest value and 1 the lowest. These variables and their corresponding ratings are based on experience and are intended primarily as an aid in the electrode selection process. For example, if high-sulfur steel is to be welded, either E7016 or E7018 electrodes should be used. If poor fit-up is the problem, an E6012 is considered the best electrode. If the deposition rate is the primary factor, then the E6027 or the E7024 is the most suitable.

ELECTRODE SELECTION CHART*											
Variables	**Electrode Class†**										
	E6010	E6011	E6012	E6013	E6027	E7014	E7024	E7016	E7018	E7028	E6020
Groove butt welds, flat (< ¼″)	5	5	3	8	10	9	9	7	9	10	10
Groove butt welds, all positions (< ¼″)	10	9	5	8	(b)	6	(b)	7	6	(b)	(b)
Fillet welds, flat or horizontal	2	3	8	7	9	9	10	5	9	9	10
Fillet welds, all positions	10	9	6	7	(b)	7	(b)	8	6	(b)	(b)
Current (C)‡	DCEP	DCEP AC	DCEN AC	DC AC	DC AC	DC AC	DC AC	DCEP AC	DCEP AC	DCEP AC	DC AC
Thin material (¼″)	5	7	8	9	(b)	8	7	2	2	(b)	(b)
Heavy plate or highly restrained joint	8	8	8	8	8	8	7	10	9	9	8
High-sulfur or off-analysis steel	(b)	(b)	5	3	(b)	3	5	9	9	9	(b)
Deposition rate	4	4	5	5	10	6	10	4	6	8	6
Depth of penetration	10	9	6	5	8	6	4	7	7	7	8
Appearance, undercutting	6	6	8	9	10	9	10	7	10	10	9
Soundness	6	6	3	5	9	7	8	10	9	9	9
Ductility	6	7	4	5	10	6	5	10	10	10	10
Low-temperature impact strength	8	8	4	5	9	8	9	10	10	10	8
Low spatter loss	1	2	6	7	10	9	10	6	8	9	9
Poor fit-up	6	7	10	8	(b)	9	8	4	4	4	(b)
Welder appeal	7	6	8	9	10	10	10	6	8	9	9
Slag removal	9	8	6	8	9	8	9	4	7	8	9

* Rating is on a comparative basis of same-size electrodes with 10 as the highest value. Ratings may change with size
† AWS
‡ DCEP–direct current electrode positive; DCEN–direct current electrode negative; AC–alternating current; DC–direct current, either polarity
(b) Not recommended

Figure 9-12. *Electrodes are selected based on their suitability to different welding variables.*

 **Refer to Quick Quiz® on CD-ROM**

- Some electrodes are designed to work with direct current (DC) only, some are designed to work best with alternating current (AC), and others work with DC and AC.
- Selecting the proper electrode for the job is a critical part of SMAW.
- AWS A5.1/A5.1M, *Specification for Carbon Steel Electrodes for SMAW*, lists the requirements for carbon steel SMAW electrodes.
- AWS A5.5, *Specification for Low-Alloy Steel Electrodes for SMAW*, lists the requirements for low-alloy steel SMAW electrodes for welding carbon and low-alloy steels.
- When welding with direct current, make sure the welding machine is set to the correct polarity for the electrode (DCEN or DCEP).
- The AWS organizes electrodes for low-carbon and low-alloy steel into four groups (F groups) based on their performance characteristics.
- After a container of low-hydrogen electrodes has been opened, the electrodes must be stored in an electrode oven between 250°F and 300°F.
- Store electrodes in a dry place where the coating cannot be damaged by moisture.

Questions for Study and Discussion

1. What is the function of the coating on an SMAW electrode?
2. What has been done to ensure uniformity of electrode specifications?
3. What symbols have been adopted to identify different types of electrodes?
4. Explain the identifying symbols of the electrode classification E6010.
5. What is an all-position electrode?
6. How can the current and polarity an electrode is designed for be determined?
7. What factors should be taken into consideration when selecting an electrode for a job?
8. Why are smaller diameter electrodes used for overhead welding?
9. What precautions must be taken when storing electrodes?
10. What is the specific feature of electrodes with coatings containing powdered iron?
11. Why are low-hydrogen electrodes used?
12. What are some of the specific characteristics of electrodes designated as fast-freeze?
13. Some electrodes are classified as fill-freeze. What does this mean?
14. For what types of welding are fast-fill electrodes intended?
15. What is the function of slag in the welding process?
16. How does joint design affect the diameter of the electrode used?
17. What organization is responsible for establishing a standard numerical electrode classification?
18. Which F group includes deep penetrating electrodes?
19. How are electrodes marked for identification?
20. How should low hydrogen electrodes be stored once the seal on the moisture proof container is broken?

Refer to Chapter 9 in the *Welding Skills Workbook* for additional exercises.

SMAW–Striking an Arc

SMAW requires mastery of a specific series of operations through practice. Once these skills have been acquired, they can be applied on any welding job. The first basic SMAW operation is learning to strike an arc and deposit a straight (stringer) bead.

BASIC PRINCIPLES OF SUSTAINING A WELDING ARC

The success of any welding operation depends upon establishing and maintaining a stable arc. **See Figure 10-1.** To sustain a stable arc, four basic elements are necessary:

- Machine setting. The welding machine must be set to the correct type of current (DCEP, DCEN, or AC) and adjusted to the proper current setting. The type of current depends on the current requirements of the electrode as well as the type of base metal. The type of welding machine also influences the type of current. For example, some low-end machines produce AC only. Current, which is the flow of electricity through a circuit, is regulated by the current control on the welding machine. The current setting depends on electrode diameter and welding position.
- Electrode angles. *Electrode angles* are the angles at which the electrode is held in relation to the joint during welding. Using the correct electrode angles is critical to achieving complete penetration, proper bead formation, and preventing problems like undercut and overlap.
- Arc length. Arc length is the distance from the tip of the electrode core wire to the weld pool. Maintaining the proper arc length is essential to sustaining a strong, stable arc. In general, arc length should be equal to the diameter of the core wire of the electrode. An arc that is too long results in an unstable welding arc with excess spatter; produces flat, wide beads with reduced penetration; and prevents the gas shield from protecting the molten weld pool against atmospheric contamination, resulting in porosity. If too short an arc is used, weld beads will be uneven with irregular ripples. The arc may short out, causing the electrode to stick to the work.
- Travel speed. *Travel speed* is the rate at which the electrode moves along the weld joint. Factors such as size and type of electrode, current, welding position, and type of base metal affect travel speed.

> **✓ Point**
>
> *Set the welding machine to the recommended current (DCEP, DCEN or AC).*

Miller Electric Manufacturing Company
A stable arc is maintained during welding to ensure the required weld characteristics.

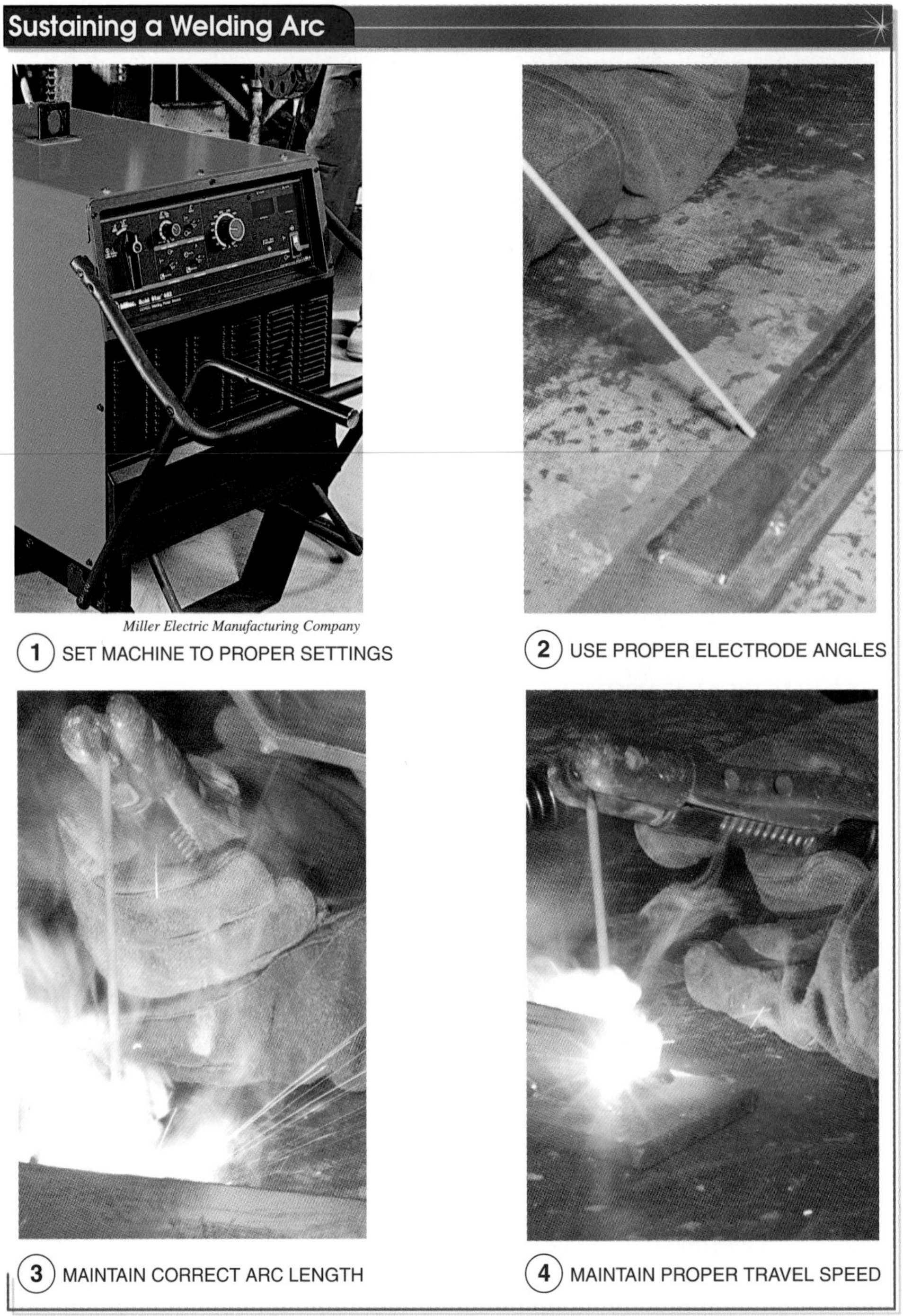

Miller Electric Manufacturing Company
(1) SET MACHINE TO PROPER SETTINGS

(2) USE PROPER ELECTRODE ANGLES

(3) MAINTAIN CORRECT ARC LENGTH

(4) MAINTAIN PROPER TRAVEL SPEED

Figure 10-1. *Proper machine settings, electrode angles, arc length, and travel speed are necessary in order to sustain an arc during welding.*

✓ **Point**

Inspect the equipment before starting to weld.

Checking and Adjusting Equipment

Equipment for SMAW includes a welding machine, electrode, electrode holder, electrode lead, workpiece lead, and workpiece connection. **See Figure 10-2.** Equipment must be checked regularly to ensure quality welding. Before starting the welding operation, apply the following procedure:

1. Check all electrical connections to make sure they are tight, clean, dry, and in good condition. Poor connections can overheat and melt, or produce dangerous arcs and sparks.
2. Inspect welding cables for cuts, nicks, or abrasions in the insulation, and make appropriate repairs if necessary. Make sure there are no cable splices within 10′ of the work.

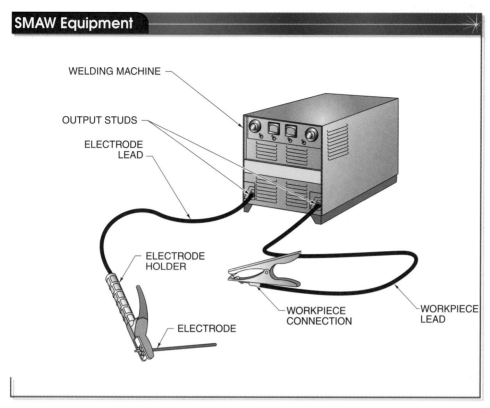

WELDING MACHINE

OUTPUT STUDS

ELECTRODE
LEAD

ELECTRODE
HOLDER

ELECTRODE

WORKPIECE
CONNECTION

WORKPIECE
LEAD

Figure 10-2. *Equipment for SMAW includes a welding machine, electrode holder, electrode lead, workpiece lead, and workpiece connection.*

3. Make sure the base metal is dry and free from dirt, rust, grease, and other contaminants.
4. Select the proper current (DCEP, DCEN, or AC).
5. Adjust the current to the recommended settings for the electrode diameter and welding position.
6. Make a few test welds on a piece of scrap metal and fine-tune the machine settings.

Gripping the Electrode

Clamp the bare end of the electrode in the jaws of electrode holder. **See Figure 10-3.** Keep the jaws of the electrode holder clean to ensure good electrical contact with the electrode.

Cradle the electrode holder in the hand. Do not grip it tightly. Gripping it too tightly will cause the hand and arm to tire quickly. It will also make it difficult to control the electrode. Do not touch metal surfaces with an un-insulated electrode holder while the power switch on the

welding machine is ON. This can cause an arc flash, damage the equipment, and may result in electric shock. When not in use, remove the electrode from the electrode holder. Coil the electrode and work leads loosely, and place on the hanger provided. Most welding machines have a bracket for storing cables. Do not leave electrode and work leads on the floor, and do not wrap them around the welding machine.

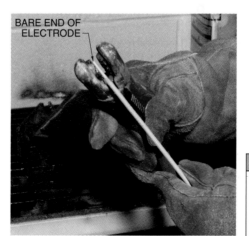

BARE END OF
ELECTRODE

Figure 10-3. *Place the bare end of the electrode in the electrode holder.*

✓ **Point**

Do not lay the electrode holder on a metal surface while the power to the welding machine is ON.

✓ **Point**

Always shut the welding machine OFF and remove the electrode from the electrode holder when leaving the welding area or when the job is completed.

⚠ **WARNING**

Keep combustible materials at least 35′ away from the welding area, cover them with nonflammable blankets, or protect them with screens to prevent flying sparks and spatter from igniting them.

Adjusting the Current

The recommended current settings for an electrode are only approximate. Final adjustment requires running test beads at different current setting within the range specified. The best way to do this is to use a piece of scrap metal. For example, if the current range for an electrode is 90 A to 100 A, set the current to 95 A. Run a few short welds and check the quality of the beads. Turn the current down to 90 A. Run a few more beads and compare them to first set of beads. Then turn the current down to 85 A and run another bead. Note that as the current is reduced, there is insufficient heat to melt the base metal. As the electrode burns off, it does not fuse with the base metal but piles up on the surface.

Reverse the process by gradually raising the current. Turn the machine up 5 A in several steps and run short test beads. As the current increases, the arc gets hotter, the electrode melts faster, and the weld bead exhibits better fusion into the base metal until the current exceeds the capabilities of the electrode.

Arriving at the proper current setting for a job is more art than science because many factors are involved, such as the skill of the welder, welding position, and type of base metal.

Arc welding equipment must be installed and grounded with the necessary disconnects, fuses, and incoming power lines in accordance with requirements of ANSI/NFPA 70, the National Electrical Code®, and relevant local codes.

 Refer to Quick Quiz® on CD-ROM

Points to Remember

- Inspect the equipment.
- Set the welding machine to the recommended current (DCEP, DCEN, or AC).
- Do not lay the electrode holder on a metal surface while the power to the welding machine is ON.
- Always shut the welding machine OFF and remove the electrode from the electrode holder when leaving the welding area or when the job is completed.
- Start within the recommended current range. Fine-tune the current by making small adjustments and depositing test welds on scrap metal to achieve the desired performance.

Striking the Arc

1. Obtain a piece of ¼″ low-carbon steel.

2. Position the workpiece in the flat position.

3. Use ⅛″ or 5⁄32″, E-6012, E-6013, or E-7024 electrode. Insert the electrode in the electrode holder and set the welding machine to the correct current.

4. Strike the arc. Either of two methods can be used to strike the arc—the tapping method or the scratching method. In the tapping method, the electrode is brought straight down to contact the workpiece and is withdrawn quickly. With the scratching method, the electrode is scratched on the base metal much like striking a match. Regardless of the motion used, upon contact with the workpiece, promptly raise the electrode a distance equal to the diameter of the electrode or the electrode will stick to the workpiece. Give it a quick twist to break it loose. If it does not break free, release it from the electrode holder.

5. Practice striking an arc until the operation can be performed quickly and easily without sticking the electrode.

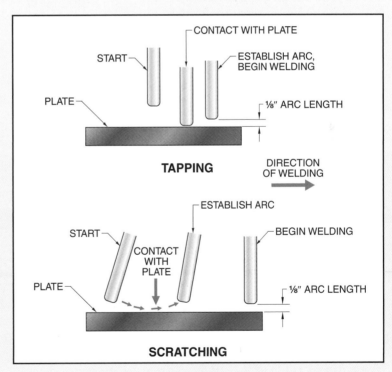

1. Obtain a piece of low-carbon steel.

2. With a soapstone, draw a series of lines on the workpiece, each line approximately 2″ in length and ⅜″ apart.

3. Position the workpiece so the lines are in the flat position.

4. Deposit a continuous bead over each line, moving the electrode from left to right. Hold the electrode in a vertical position and angle it slightly away from the direction of travel. This is called a drag travel angle.

5. Watch the weld pool and travel at a speed that produces a consistent weld bead and complete fusion into the workpiece. If the current is set properly and the arc is maintained at the correct length, there will be a continuous crackling or frying sound. An arc that is too long makes a humming sound. Too short of an arc makes a popping sound. Notice the action of the molten weld pool and how the trailing edge of the weld pool solidifies as the electrode moves along the workpiece.

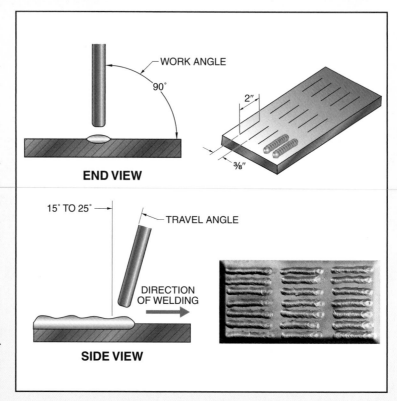

The molten weld pool is a liquid version of the finished weld. The weld pool should be clear and bright. If the edges of the weld bead have a dull, irregular appearance, it could mean that it contains slag inclusions.

❓ Questions for Study and Discussion

1. How does arc length affect a weld?
2. Why must the current be adjusted for a particular welding operation?
3. What equipment checks are made before proceeding to weld?
4. Why should the electrode holder never be placed on the workbench while the current is ON?
5. What two methods may be used to strike an arc?
6. When striking an arc, why should the electrode be withdrawn quickly?
7. What should be done if the electrode sticks to the plate?
8. The arc should be maintained at approximately what length?

Refer to Chapter 10 in the *Welding Skills Workbook* for additional exercises.

SMAW–Depositing a Continuous Bead

Chapter

To produce a quality weld, a welder must be able to manipulate the electrode and under-stand certain weld characteristics. A welder must be knowledgeable about factors that contribute to good quality and poor quality welds. Quality welds are produced by using the correct welding procedures, properly cleaning the weld, and preventing contamination of the weld.

ESSENTIALS OF ARC WELDING

The characteristics of the electrode must be known to ensure that the proper electrode is selected for each welding operation. The electrode must also be able to maintain its metal properties after deposition. Without the proper electrode, it is almost impossible to achieve the desired results, regardless of the welding technique used. As welding is performed, the heat from the arc melts the base metal, forming a crater into which the molten base metal and filler metal flow. As welding proceeds, the molten metal solidi-fies and a layer of slag forms on top of the weld. **See Figure 11-1.** The following fac-tors allow for a quality weld with the proper penetration: electrode selection, arc length, current, travel speed, and electrode angle.

Figure 11-1. *Heat from the arc melts the base metal, forming a crater into which molten base metal and filler metal can flow to create a quality bead.*

Electrode Selection

Electrode selection must take into account the position of the weld, the properties of the base metal, the type of joint, joint fit-up, and the type of welding current. Different types and diameters of electrodes are manufactured to meet various welding requirements.

> *The condition of electrode holders must be considered when setting arc welding variables. Electrode holders are exposed to extremely high heat on a regular basis, which causes them to deteriorate rapidly. Electrode holders should be cleaned regu-larly to ensure good electrical contact with the electrode and checked for wear.*

Arc Length

Arc length is the distance from the tip of the electrode to the weld pool. If the arc length is too long, the arc becomes unstable, and the electrode melts off in large globules that wobble from side to side. These large globules produce a wide, irregular bead with excessive spatter and incomplete fusion between the base metal and the deposited metal.

An arc length that is too short fails to generate enough heat to melt the base metal properly, producing high, uneven beads with irregular ripples. Depositing welds using too short an arc length also increases the possibility of the electrode sticking to the workpiece.

> ✓ **Point**
>
> *Use the proper elec-trode for each welding operation.*

The arc length required depends on the size of electrode used and the welding task. Small-diameter electrodes require a shorter arc length than large-diameter electrodes. For better control of the weld pool, the arc length should typically be approximately the diameter of the electrode core wire. For example, an electrode ⅛″ in diameter should have an arc length of about ⅛″. A shorter arc length is typically used for horizontal, vertical, and overhead welding because it gives better control of the weld pool.

The proper arc length also prevents impurities from entering a weld. A correct weld bead has the proper height and width and uniformly spaced ripples. A long arc length allows the atmosphere to flow into the weld area, permitting impurities of nitrides and oxides to form. Additionally, when the arc length is too long, heat from the arc stream is dissipated too rapidly, causing considerable metal spatter. **See Figure 11-2.** If the arc length is too short, the bead will have a narrow width and excessive height.

fast and the weld pool is large, irregular, and hard to control. Excessive spatter may also occur.

When the welding current is too low, there is not enough heat to melt the base metal and the weld pool will be too small. The result is poor fusion. Beads that pile up on the base metal and are irregular in shape can stick to the metal. The electrode can also stick to the metal. Too low a current setting also causes the arc to continually break.

Travel Speed

Travel speed is the rate at which an electrode moves along a joint. If the travel speed is too fast, the weld pool does not last long enough and impurities are locked in the weld. The resulting bead is narrow, with pointed ripples. If the travel speed is too slow, the metal piles up on the base metal. The bead is high and wide, with straight ripples, and excessive reinforcement. Traveling too slowly may also result in overlap. The correct travel speed produces a smooth weld bead with evenly spaced ripples. **See Figure 11-3.**

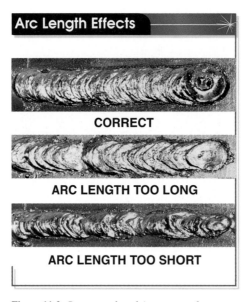

Arc Length Effects

CORRECT

ARC LENGTH TOO LONG

ARC LENGTH TOO SHORT

Figure 11-2. *Correct arc length is necessary for proper bead formation.*

Current Selection

For the desired weld characteristics, the correct current (AC, DCEP, DCEN) for a particular electrode must be used. If the current is too high, the electrode melts too

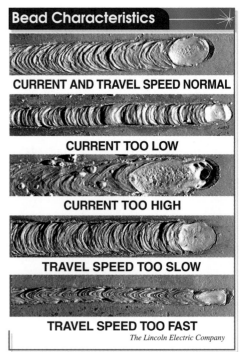

Bead Characteristics

CURRENT AND TRAVEL SPEED NORMAL

CURRENT TOO LOW

CURRENT TOO HIGH

TRAVEL SPEED TOO SLOW

TRAVEL SPEED TOO FAST

The Lincoln Electric Company

Figure 11-3. *Proper bead formation is dependent on many variables, which must be controlled to prevent a poor-quality bead.*

Electrode Angle

The electrode angle affects weld bead shape and penetration into the base metal. The electrode angle is determined by the travel angle and the work angle. **See Figure 11-4.**

The *travel angle* is an angle less than 90° between the electrode axis and a line perpendicular to the workpiece and in a plane determined by the electrode axis and the weld axis. The travel angle is along the weld axis and varies from 5° to 30° from vertical, depending on welder preference and conditions.

A *push travel angle* is a travel angle where the electrode points toward the direction of travel. It results in less heat input and shallow penetration, because the arc is directed away from the weld pool. A push travel angle is often used on thin material. A *drag travel angle* is a travel angle where the electrode points away from the direction of travel. It results in more heat input and deeper penetration because the arc is directed into the weld pool.

The *work angle* is an angle less than 90° in a line perpendicular to the workpiece and in a plane determined by the electrode axis and the weld axis. For example, the work angle is normally 90° when making a groove weld in flat position.

Ordinarily, a slight angle of the electrode in either direction from the work angle does not affect weld appearance or quality. However, when undercut occurs in the vertical plate of a fillet weld, the angle of the arc should be lowered and the arc directed more toward the vertical plate. Work angle is especially important in multiple-pass fillet welds.

CRATER FORMATION

As the arc comes in contact with the base metal, a crater is formed. A crater is a depression (weld pool or pocket) in the molten base metal made by the arc. The size and depth of a crater determine the amount of penetration. The depth of penetration should be one-third to one-half the total thickness of the bead, depending on the size of the electrode. **See Figure 11-5.**

Media Clip

Depth of Penetration

Electrode Angles

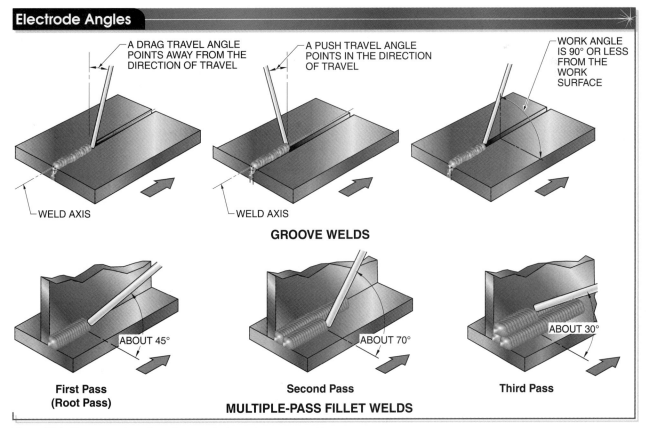

A DRAG TRAVEL ANGLE POINTS AWAY FROM THE DIRECTION OF TRAVEL

A PUSH TRAVEL ANGLE POINTS IN THE DIRECTION OF TRAVEL

WORK ANGLE IS 90° OR LESS FROM THE WORK SURFACE

WELD AXIS

WELD AXIS

GROOVE WELDS

ABOUT 45°

ABOUT 70°

ABOUT 30°

First Pass (Root Pass)

Second Pass

Third Pass

MULTIPLE-PASS FILLET WELDS

Figure 11-4. *The correct electrode angle is required to make a proper weld.*

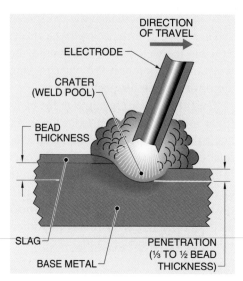

Figure 11-5. *The depth of the crater indicates the amount of penetration in the weld.*

✓ **Point**

Be sure the molten metal from the electrode fuses completely with the base metal.

In a whipping motion, the electrode is struck and held momentarily. It is then moved forward about ¼″ or ⅜″. Just as the weld pool begins to freeze, the electrode is moved back into the center of the weld pool and the sequence is repeated. The electrode is moved by pivoting the wrist and not moving the arm while making the pass.

Remelting Craters. When starting an electrode, there is a tendency for a large globule of metal to fall on the surface of the plate, resulting in little or no penetration. This is particularly a problem when restarting an electrode at the crater from a previously deposited weld. To remelt the existing crater and obtain proper fusion, strike the arc approximately ½″ in front of the crater. Maintain a slightly longer arc and move the electrode back through the crater, retracing it with the arc. The longer arc helps to reheat base metal and the crater. **See Figure 11-6.** As the crater is retraced, shorten the arc length to one electrode diameter and continue welding. Completely remove slag from the crater before restarting the arc to prevent problems with porosity and slag inclusions.

To obtain a sound weld, the metal deposited from the electrode must fuse completely with the base metal. Fusion results only when the base metal has been heated to a molten state and the molten metal from the electrode readily flows into it. If the arc length is too short, there is insufficient heat to form the correct size crater. When the arc length is too long, the heat is not centralized or intense enough to form the desired crater.

Controlling Craters

✓ **Point**

Restart the electrode ½″ from the front edge of the previously made crater, move the arc back through the crater to remelt the weld pool, and continue welding.

An improperly filled crater does not produce the required weld strength and may cause a weld to fail when a load is applied. Occasionally, the crater gets too hot and the molten metal has a tendency to run. When this happens, the electrode should be lifted slightly and quickly shifted to the side or ahead of the crater. Such a movement reduces the heat, allows the crater to solidify, and stops the deposit of metal from the electrode. The electrode is then quickly returned to the crater and the arc shortened.

Another method used by welders to control the temperature of the molten weld pool is a whipping motion of the electrode. The whipping technique is used with E6010 and E6011 electrodes and is especially helpful when welding pieces that have poor fit-up. It is also used in overhead and vertical welding to better control the weld pool.

Figure 11-6. *To continue a bead, strike the arc ½″ from the front edge of the previously deposited weld bead. Move the electrode back and retrace the crater to form a new weld pool, and continue welding.*

Crater Filling. An improperly filled crater at the end of a weld may result in crater cracking. To reduce the risk of crater cracks, back-step about ⅜″ to ¼″ to fill the crater before breaking the arc at the end of end of a weld.

Undercutting and Overlapping. *Undercutting* is creating a groove in the base metal that is not completely filled by weld metal during the welding process. Undercutting is the result of welding with excessive current, traveling too fast, or using an improper work angle. **See Figure 11-7.** Excessive current leaves a groove in the base metal along both sides of the bead, which greatly reduces the strength of a weld. Undercutting may also occur when there is insufficient deposition of metal on a vertical plate. Undercutting can be corrected by changing the electrode angle slightly.

Overlapping (cold lapping) is extending the weld metal beyond the weld toes. Overlapping occurs when the current is set too low, and the molten metal is deposited without actually fusing into the base metal, creating a poor quality weld. **See Figure 11-8.**

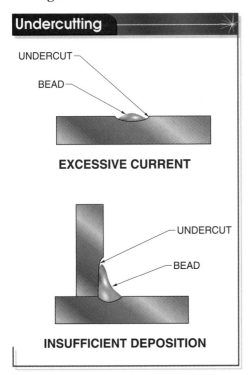

Figure 11-7. *Undercutting is caused by excessive current, insufficient metal deposition, traveling too fast, or using the improper work angle.*

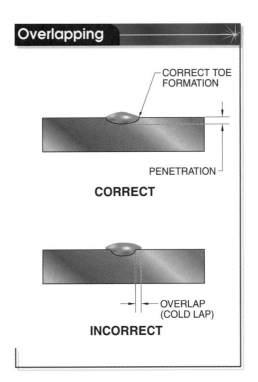

Figure 11-8. *Welding with too low current results in poor penetration and causes overlapping.*

Arc Blow

Arc blow is the deflection of the welding arc by magnetic forces that occur due to current flow. Arc blow is a common problem when welding with direct current. Current in a DC welding machine flows in one direction, which produces a strong magnetic force in the metal being welded. This magnetic force causes the arc to deflect from the weld area. Arc blow also breaks the continuity of the deposited metal, making it necessary for frequent restarts. This slows down the welding process and often leaves weak spots in the weld.

Arc blow typically occurs in carbon steels, but may be encountered in other metals as well. It is also more common in corners and near the ends of a joint when the workpiece connection is connected on only one side of the metal. Arc blow usually occurs forward or backward along the joint but may occasionally occur to the sides. **See Figure 11-9.**

Arc blow results in incomplete fusion and excessive spatter. If arc blow is severe enough, a satisfactory weld cannot be

Media Clip

Arc Blow

✓ Point

Avoid undercutting and overlapping of the weld joint by using the correct current and electrode angle.

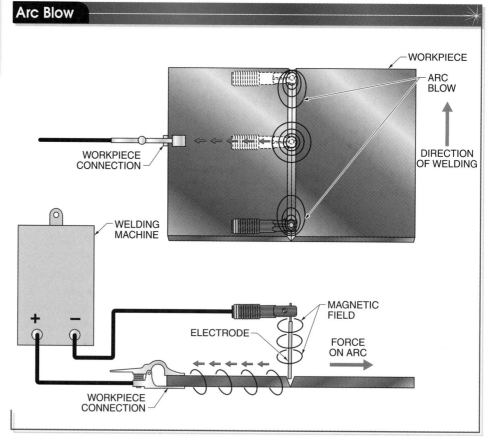
made. The best way to prevent arc blow is to use AC rather than DC on jobs where arc blow could be a problem. However, when AC is not an option, arc blow can be minimized by clamping the workpiece connection to the end of the workpiece, rather than on one side. Additional measures that may be taken to reduce arc blow include the following:

- welding away from the workpiece connection
- reducing the welding current
- using the back-step welding technique
- using the shortest possible arc to overcome the magnetic field
- using electrode angles to compensate for arc blow

CLEANING WELDS

The layer of slag that covers a deposited bead must be removed after welding. If a multiple-pass weld is required, the slag

must be removed between each pass. Slag allowed to enter the weld metal will weaken the weld. Additionally, finishing procedures, such as painting, should not be performed until all slag is removed. **See Figure 11-10.** To remove slag from a weld, use the following steps:

1. Strike the weld area with a chipping hammer. Hammer the bead so the chipping is directed away from the eyes, the face, and the body. Do not pound the bead too hard as the structure of the weld may be damaged. After the slag is loosened, drag the pointed end of the chipping hammer along the toes of the weld to loosen any remaining particles of slag. The flat end of the chipping hammer can also be used to remove spatter from the area around the weld.
2. Brush the weld with a stiff wire brush to remove residual slag particles after chipping.

Figure 11-9. *Arc blow is caused by the magnetism produced by a DC welding machine that causes the arc to deflect from the weld area.*

(1) STRIKE WELD AREA WITH CHIPPING HAMMER

(2) REMOVE REMAINING SLAG WITH WIRE BRUSH

Figure 11-10. *Strike the weld with a chipping hammer, and then rub with a wire brush to remove slag.*

Refer to Quick Quiz® on CD-ROM

Points to Remember

- Use the proper electrode for each welding operation.
- The arc length should be approximately the diameter of the electrode.
- Maintain a travel speed that is just fast enough to produce evenly spaced ripples.
- Use the correct current for a particular electrode.
- The depth of penetration should be one-third to one-half the total thickness of the weld bead.
- Be sure the molten metal from the electrode fuses completely with the base metal.
- Restart the electrode ½″ from the front edge of the previously made crater, move the arc back through the crater to form a new weld pool, and continue welding.
- Avoid undercutting and overlapping of the weld joint by using the correct current and electrode angle.
- Prevent arc blow during welding by using AC rather than DC on jobs where arc blow may be a problem.
- When cleaning slag from a weld, direct chipping away from the body, the eyes, and the face.

Depositing a Continuous Bead

1. Obtain a piece of ¼″ low-carbon steel, 4″ wide by 6″ long.

2. Position the workpiece in flat position.

3. With a soapstone, draw a series of lines approximately ¾″ apart and the length of the workpiece.

4. Use ⅛″ E6010, E6011, E6012, or E6013 electrode. Deposit continuous beads along the lines. Start from the left edge and work to the right.

5. After each line has been filled, remove the slag and examine the weld beads.

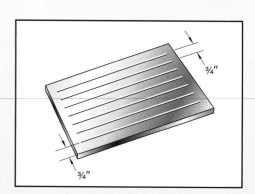

Moving the Electrode in Several Directions

1. Obtain a piece of ¼″ low-carbon steel.

2. Position the workpiece in flat position.

3. With a soapstone, draw a series of lines to form rectangles on the workpiece.

4. Deposit a continuous bead, moving the electrode from left to right, bottom to top, right to left, and top to bottom.

5. Maintain correct arc length, travel angle, and travel speed to control bead formation.

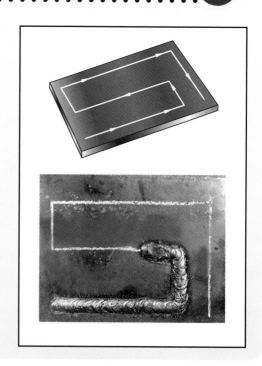

Restarting the Arc

1. Obtain a piece of ¼″ low-carbon steel.

2. Position the workpiece in flat position.

3. With a soapstone, draw a series of straight lines, divided into 2″ sections.

4. Deposit a bead over the first 2″ section, then break the arc.

5. Restart the arc and deposit a bead for another 2″ section.

6. Repeat the practice of breaking and restarting the arc by retracing the crater to form a new weld pool.

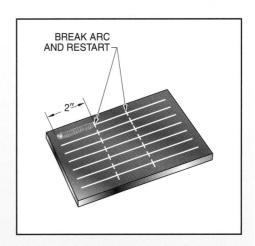

BREAK ARC AND RESTART

2″

Questions for Study and Discussion

1. What factors allow for a quality weld with the proper penetration?
2. What factors must be considered when selecting an electrode?
3. How is a crater affected when the arc length is too long? What happens when the arc length is too short?
4. When the arc length is too long, what happens to the metal as it melts from the electrode?
5. How is it possible to identify a weld that has been made with too long an arc length?
6. What is likely to happen to the electrode when the arc length is too short?
7. What are some characteristics of a weld made with too short an arc length?
8. What are some factors that must be considered when determining arc length?
9. In what way does the amount of current affect a weld?
10. What determines the travel speed at which an electrode should be moved?
11. What is a crater?
12. What should the depth of penetration be?
13. What should be done when the crater gets too hot and the metal has a tendency to run over the surface?
14. How should an electrode be restarted to fill a crater left from a previously deposited weld?
15. What causes undercutting? How can undercutting be prevented?
16. What are six ways to prevent arc blow?
17. How should slag from a weld be removed from a workpiece?

Refer to Chapter 11 in the *Welding Skills Workbook* for additional exercises.

SMAW–Flat Position

Chapter

The easiest position to weld in is flat position. When welding in flat position, the welding speed can be increased, molten metal has less tendency to run, better penetration of the base metal is possible, and the welding operation is less tiring for the welder. If possible, structures should be positioned so that they can be welded in the easier and more efficient flat position.

WELD PASSES

Some welds require more than one pass. In a multiple-pass weld, the first pass is the root pass. Additionally, an intermediate weld pass (or passes) and a cover pass are used for multiple-pass welding. Most welding operations require the workpieces to be tack welded. A tack weld is used to hold the workpieces in proper alignment until the final welds are made. Tack welds are spaced along the joint and must be consumed into the joint during welding. Once the joint is tacked, the necessary weld passes are made. **See Figure 12-1.**

Root Pass

The *root pass* is the initial weld pass that provides complete penetration through the thickness of the joint member. In groove welds, the root pass provides complete penetration through the thickness of the joint member. The root bead is made by moving a small-diameter electrode straight down into the groove without any weaving motion. The purpose of the root bead is to join the two members and fill the root opening.

The root bead serves as the base for subsequent passes, and it must produce complete penetration. Complete penetration is ensured if the root bead penetrates through to the back side of the joint and consumes all tack welds previously

made. Root reinforcement should not exceed fabrication code criteria or $\frac{1}{16}''$ beyond the bottom surface of the joint. Slag should be removed from the root bead before the next pass is deposited.

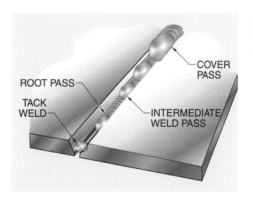

Figure 12-1. *Weld passes used for multiple-pass welding are the root pass, intermediate weld passes, and the cover pass.*

Intermediate Weld Passes

An *intermediate weld pass* is a single progression of welding subsequent to the root pass and before the cover pass. The term "intermediate weld pass" replaces terminology for both the hot pass and the filler pass. The first intermediate weld pass may be used to correct undercut or overlap. It uses a higher current setting to remove any remaining slag or inclusions and to create a quality weld surface for additional passes. It deposits a small amount of filler metal and, when

✓ **Point**

Tack welds are used to keep workpieces in position. They must be consumed into the joint during welding.

✓ **Point**

When depositing a root bead, advance a small-diameter electrode along the groove with no weaving motion.

completed, should form a concave bead. The first intermediate weld pass must be thoroughly cleaned before depositing additional weld passes.

Additional layers of intermediate weld passes may be needed to fill the groove depending on the thickness of the metal. When depositing intermediate weld passes, a slight weaving motion is generally used to ensure proper fusion with the previously deposited beads and the sides of the joint. When multiple passes are used, the beads should overlap by one-third to one-half to ensure a smooth surface. Each pass must be thoroughly cleaned of slag before additional passes are made. Intermediate weld passes must completely fuse into the previous passes, but should not penetrate too deeply as this can cause the previous passes to overheat and can weaken the weld. Time should be allowed for cooling between layers to prevent overheating.

Cover Pass

The *cover pass* is the final weld pass deposited. The cover pass provides additional reinforcement to a multiple-pass weld and provides a good appearance. The face reinforcement on the cover pass should not extend beyond the fabrication code criteria or more than 1/16″ above the base metal surface. A weaving motion can be used on the cover pass to obtain the necessary weld width when covering the intermediate passes. Restrictions to weave width are sometimes specified.

JOINTS WELDED IN FLAT POSITION

Flat position is the most efficient position for welding any joint. However, even in flat position, some joints require special techniques. For example, when a joint consists of two pieces of metal having different thicknesses, the work angle of the electrode should be adjusted so that the heat is concentrated on the thicker metal. **See Figure 12-2.** Joints commonly welded in flat position include lap, T, butt, and corner joints.

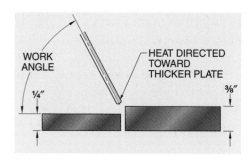

Figure 12-2. *When welding plates of different thicknesses, direct more heat to the thicker plate.*

Lap Joints

The lap joint is one of the most frequently used joints in flat position welding. It is a relatively simple joint, since no beveling or machining is necessary. Surfaces to be welded must be clean and evenly aligned. A lap joint is made by lapping one member over the other. **See Figure 12-3.** The amount the members overlap depends on the thickness of the metal and the strength required. Usually, the thicker the metal, the greater the amount of overlap needed. A fillet weld is used to join the two members. When the structure is subjected to heavy bending stresses, it is best to deposit welds on both sides of the joint.

A lap joint is adaptable for a variety of construction work as well as for many types of repairs. For example, a lap joint can be used when joining a series of metal plates together or when reinforcing another structural member. Since a lap joint stiffens the structure where the members are lapped, it is used a great deal in shipbuilding.

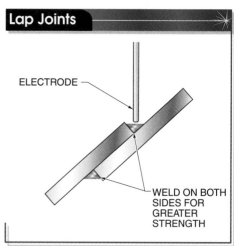

Figure 12-3. *A lap joint typically requires no edge preparation before welding. Welding both sides of the lap joint provides greater strength.*

When an exceptionally strong lap joint is required, especially on members ⅜″ thick or more, a multiple-pass fillet weld is recommended. This weld has two or more layers of beads, with each bead covering the adjacent bead by one-third to one-half.

T-Joints

A T-joint is frequently used in fabricating straight and rolled shapes. The strength of the joint depends on a close fit-up of the joint edges. A T-joint should not be used on structures subjected to heavy stresses from the opposite side of the welded joint. This weakness can be partially overcome by using a double fillet weld. **See Figure 12-4.** When welding thick metal, or when extra strength is required, a larger fillet is necessary. Fillet welds can be made larger by depositing several passes.

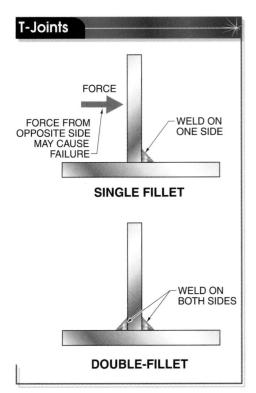

Figure 12-4. *The strength of a T-fillet depends on proper fit-up of the joint, the direction from which a force is applied, and whether a single-T or double-T fillet joint is used.*

Butt Joints

A butt joint may be closed, open, or prepared (such as beveled). **See Figure 12-5.** On a closed butt joint, the edges of the two workpieces are in direct contact with each other. A closed butt joint is suitable for welding steel that generally does not exceed ³⁄₁₆″ thick. Thicker metal can be welded, but only if the welding machine has sufficient current capacity and if larger diameter electrodes are used. On thicker metal, multiple passes are required because it is difficult to achieve enough penetration to produce a strong weld with one pass.

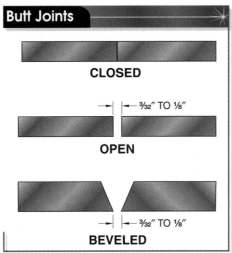

Figure 12-5. *A butt joint may be closed, open, or beveled.*

The edges of an open butt joint are spaced slightly apart, usually ³⁄₃₂″ to ⅛″, to allow for penetration of the filler metal and expansion of the base metal. Generally, a backing bar or block of scrap steel is placed under an open butt joint. **See Figure 12-6.** A backing bar prevents the bottom edges from burning through and retains filler metal in the joint.

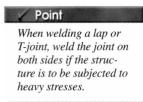

Point

When welding a lap or T-joint, weld the joint on both sides if the structure is to be subjected to heavy stresses.

Point

The edges of an open butt joint should be spaced about ³⁄₃₂″ to ⅛″ apart for penetration and joint expansion.

The five essentials of SMAW are selecting the correct electrode diameter for the job, matching welding current to the electrode type and diameter, using the proper arc length, adjusting travel speed to maintain a consistent weld bead, and maintaining the proper electrode angles.

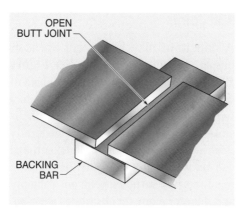

Figure 12-6. *A back-up strip prevents the bottom edges from burning through an open butt joint.*

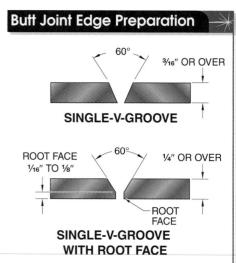

Figure 12-7. *The edges of a V butt joint are prepared in different ways, depending on the thickness of the metal.*

When the thickness of the metal exceeds ³⁄₁₆″, the edges of the butt joint should be beveled. Beveling the edges ensures better penetration and equalizes contraction forces. Edges can be beveled with a cutting torch or by grinding. The groove angle should not exceed 60° to limit the amount of contraction that usually results when the metal cools. The edges may be prepared in several ways, such as a single-V-groove, single-V-groove with root face, or double-V-groove with root face. **See Figure 12-7.** The thickness of the members determines the edge preparation required. For example, on metal ½″ thick or more, the edges should be beveled on both sides.

When welding thin stock with a single pass, as in a closed or open butt joint, move the electrode along the joint without any weaving motion. Move the electrode slowly enough to allow the arc sufficient time to melt the metal. Using too slow a travel speed can cause overheating and excessive penetration.

When a multiple-pass weld is to be made in a beveled joint, move the electrode down in the groove so that it almost touches both sides of the joint while depositing the root bead. Move the electrode fast enough to keep the slag flowing back on the finished weld. If the electrode is not moved rapidly enough, slag may become trapped in the weld causing inclusions and preventing complete fusion. After completing the root bead, proceed with the necessary intermediate weld passes. Complete the weld with a cover pass.

A butt joint is often used when joined structural pieces must have a flat surface, such as in tanks, boilers, and a variety of machine parts. **See Figure 12-8.**

Thermadyne Industries, Inc.

Construction equipment commonly requires surfacing welds with SMAW for building up worn parts and producing a hardened surface.

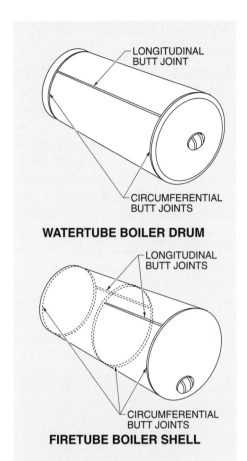

Figure 12-8. *Butt joints are often used for structural pieces that have flat surfaces, such as boilers.*

WATERTUBE BOILER DRUM

LONGITUDINAL BUTT JOINT

CIRCUMFERENTIAL BUTT JOINTS

FIRETUBE BOILER SHELL

LONGITUDINAL BUTT JOINTS

CIRCUMFERENTIAL BUTT JOINTS

dimensions. The operation consists of depositing several layers of welds, one on top of the other, to increase the dimensions of a part. **See Figure 12-9.**

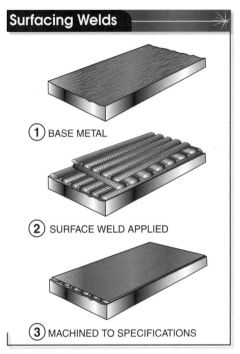

Surfacing Welds

(1) BASE METAL

(2) SURFACE WELD APPLIED

(3) MACHINED TO SPECIFICATIONS

Figure 12-9. *Surfacing welds are used to increase the dimensions of parts.*

Corner Joints

An outside corner joint may be used when constructing rectangular objects such as tanks, metal furniture, and other machine sections where the outside corner must have a smooth radius. A single pass is usually sufficient for welding corner joints.

A worn surface is ground down or machined to allow for two layers of surfacing to be deposited on the part. The first layer of deposited metal tends to become diluted and loses some of its alloying properties when mixed with the base metal. Additional layers of surfacing proved the required wear properties while maintaining the part thickness.

SURFACING

Surfacing is the application of a layer or layers of material to a surface to obtain desired dimensions or properties like abrasion resistance (hardfacing). Surfacing is commonly performed in flat position and is used to repair worn surfaces of shafts, wheels, and other machine parts. The worn base metal has a surfacing weld applied, which is then machined to specifications. A *surfacing weld* is a weld applied to a surface, as opposed to a joint, to obtain desired properties or

Surfacing is performed by depositing successive weld beads. Additional filler metal can be added by weaving. *Weaving* is a welding technique in which the electrode is moved transversely as it progresses along the surface. Weaving increases the bead width. Three common weave patterns are the crescent, figure eight, and rotary. The weave pattern used depends on the position and application. **See Figure 12-10.** Weaving is also used to provide a smooth weld finish on multiple-pass welds.

Weave Patterns

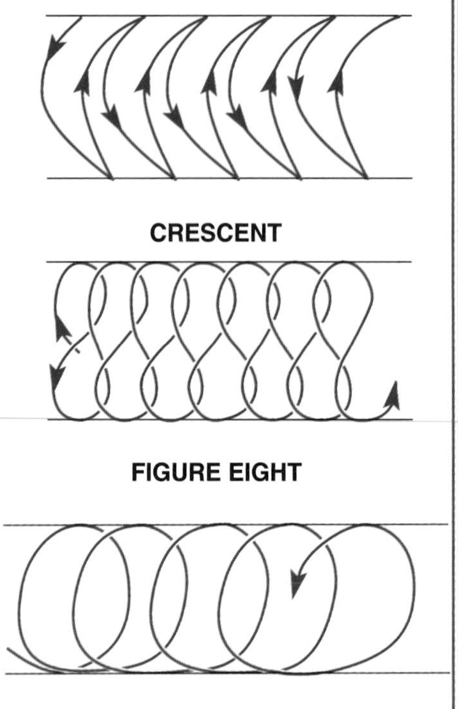

CRESCENT

FIGURE EIGHT

ROTARY

Figure 12-10. *Three common weaving patterns, crescent, figure eight, and rotary, are used to increase the width and volume of the bead.*

SMAW FLAT POSITION PROBLEMS

When welding in flat position, gravity helps direct the flow of molten weld metal. Welding is normally best performed in flat position, but problems that result in joint weakness and/or failure can still occur. Some of the problems that may be encountered when using SMAW in flat position include instability of the arc, poor penetration, a loud crackling noise when welding, difficulty striking an arc, weakness of the weld, and arcing at the workpiece connection. **See Figure 12-11.**

Problems that occur during SMAW are often the result of improper settings on the welding machine. Adjusting the current, changing polarity, or correcting a poor ground should solve the problem.

If there is no welding procedure specification (WPS) for a job, use the largest diameter electrode that the joint configuration will allow. Larger diameter electrodes produce higher deposition rates and faster travel speeds. This makes them more cost effective.

SMAW WELDING PROBLEMS		
Problem	**Cause**	**Remedy**
Unstable arc; arc goes out; excessive spatter	Arc too long	Shorten arc
Poor or no penetration; arc goes out often	Not enough current for size of electrode; wrong electrode	Increase current; use proper electrode
Loud crackling from arc; flux melts too rapidly; wide bead; spatter in large drops	Current too high for electrode; may be moisture in electrode coating or on base metal	Decrease current; change electrode; clean base metal
Difficulty in striking arc; poor penetration	Wrong polarity; current too low	Change polarity or increase current
Weak weld; arc hard to start; arc keeps breaking	Dirty work	Clean work; remove slag from previous weld
Arcing at workpeice connection	Poor ground	Properly attach workpiece clamp

Figure 12-11. *A welder should be alert to any signs of a problem during welding, such as instability of the arc or poor penetration, and remedy the situation quickly.*

Refer to Quick Quiz® on CD-ROM

- Tack welds are used to keep workpieces in position. They must be consumed into the joint during welding.
- When depositing a root bead, advance a small-diameter electrode straight along the groove with no weaving motion.
- The first intermediate weld pass may be used to remove slag inclusions or other defects from the root bead.
- Use a slight weaving motion when making intermediate weld passes.
- When welding a joint with base metals of different thicknesses, keep the heat concentrated on the thicker metal.
- When welding a lap or T-joint, weld the joint on both sides if the structure is to be subjected to heavy stresses.
- The edges of an open butt joint should be spaced about $3/32''$ to $1/8''$ apart for penetration and joint expansion.
- When welding a butt joint on metal more than $3/16''$ thick, bevel the edges to obtain the proper penetration.
- Always remove slag completely after each pass. Slag particles left on a bead can be incorporated into another weld as inclusions and can weaken it.
- Correct welding problems quickly when performing SMAW in flat position to prevent weakness or failure of the weld.

Welding a Single-Pass Lap Joint in Flat Position

exercise 1

1. Obtain two pieces of 3/16″ or 1/4″ low-carbon steel.

2. Form a lap joint and tack together.

3. Position the workpiece so the weld joint is in flat position.

4. Use a 1/8″ electrode and adjust the welding machine for the correct current.

5. Hold the electrode at a 45° angle with a 30° drag travel angle, and deposit a 1/4″ fillet weld along the joint.

6. Weave the electrode slightly, maintaining the arc for a slightly longer time on the bottom workpiece.

7. Make sure that fusion is complete at the joint root and prevent overlapping on the top workpiece. A weld made with a concave fillet is usually too weak because it lacks sufficient reinforcing metal. A weld with a convex bead has too much waste metal, which adds no strength to the weld.

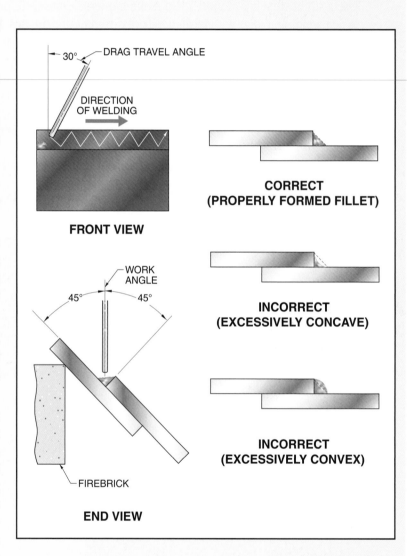

30° DRAG TRAVEL ANGLE

DIRECTION OF WELDING

FRONT VIEW

WORK ANGLE

45° 45°

FIREBRICK

END VIEW

CORRECT (PROPERLY FORMED FILLET)

INCORRECT (EXCESSIVELY CONCAVE)

INCORRECT (EXCESSIVELY CONVEX)

Welding a Multiple-Pass Fillet Lap Joint in Flat Position

exercise 2

1. Obtain two pieces of ¼" low-carbon steel.

2. Form a lap joint and tack together.

3. Position the workpiece so the weld joint is in flat position.

4. Deposit the root bead by moving the electrode straight down the joint without weaving.

5. Clean the weld carefully with a chip hammer and wire brush and deposit the second pass over the root bead.

6. While welding the second pass, weave the electrode, pausing for an instant at the top of the weave to deposit extra metal on the surface of the upper plate.

7. Maintain a consistent bead width along the joint.

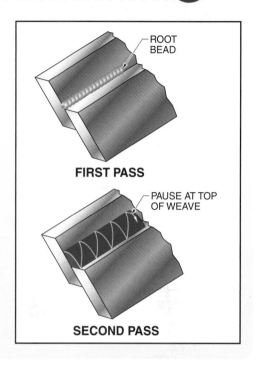

FIRST PASS

SECOND PASS

Welding a Single-Pass Fillet T-Joint in Flat Position

exercise 3

1. Obtain two pieces of ³⁄₁₆" or ¼" low-carbon steel.

2. Form a T-joint with the pieces at a 90° angle and tack together.

3. Position the workpiece so the weld joint is in flat position.

4. Hold the electrode at a work angle of 45° and a drag travel angle of 30° and advance it in a straight line without any weaving motion. Deposit a ¼" fillet weld.

5. Maintain travel speed to stay ahead of the weld pool. Concentrate the arc more on the bottom workpiece to prevent undercutting the top workpiece. Watch the crater closely to ensure that it forms a properly contoured bead.

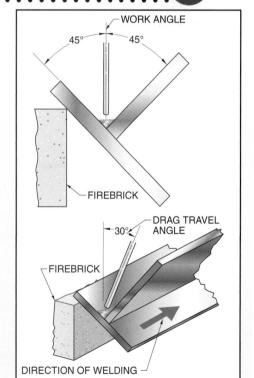

1. Obtain two pieces of ³⁄₁₆″ or ¼″ low-carbon steel.

2. Form a T-joint with the pieces at a 90° angle and tack together.

3. Position the workpiece so the weld joint is in flat position.

4. Deposit the root bead by moving the electrode straight down the joint without weaving. Remove slag completely.

5. Hold the electrode at a work angle of 70° and a drag travel angle of 30°. Deposit the first intermediate weld pass to partially cover the root bead. Remove slag completely.

6. Hold the electrode at a 30° work angle and a 30° drag travel angle. Deposit the second intermediate weld pass to cover the root bead and partially cover the second pass.

7. If more or less weld metal is required, make additional passes using different bead configurations.

8. Additional weld metal can be deposited on a multiple-pass fillet T-joint by weaving the electrode.

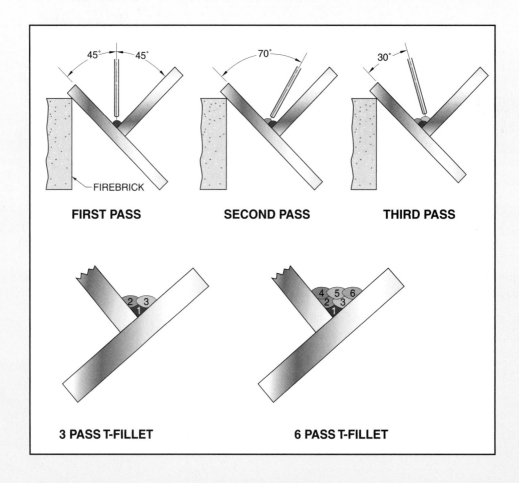

FIRST PASS   SECOND PASS   THIRD PASS

3 PASS T-FILLET   6 PASS T-FILLET

Welding a Butt Joint in Flat Position

1. Obtain two pieces of ³⁄₁₆″ or ¼″ low-carbon steel.

2. Form a butt joint with a root opening for expansion and tack together.

3. Position the workpiece so the weld joint is in flat position.

4. Hold the electrode at a work angle of 90° and a drag travel angle of 15° to 30°. Deposit a bead along the butt joint.

5. Let the workpiece cool and then repeat the procedure on the reverse side.

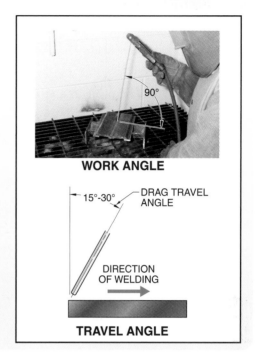

WORK ANGLE

TRAVEL ANGLE

Welding an Outside Corner Joint in Flat Position

1. Obtain two pieces of ³⁄₁₆″ or ¼″ low-carbon steel.

2. Form a corner joint with the pieces at a 90° angle and tack together.

3. Position the workpiece so the weld joint is in flat position.

4. Hold the electrode at a work angle of 45° and a drag travel angle of 30°. Deposit a bead along the outside of the joint.

5. For most corner joints, one bead is sufficient. Thick metals may require additional passes to fill the corner.

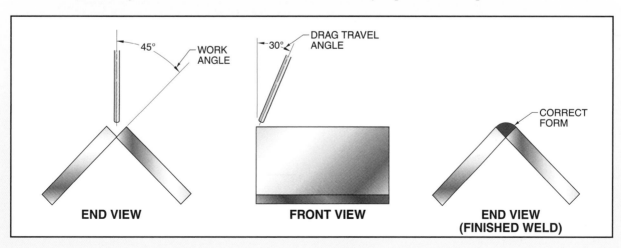

END VIEW   **FRONT VIEW**   **END VIEW (FINISHED WELD)**

Welding Round Stock in Flat Position

1. Obtain two pieces of round stock.

2. Form a butt joint with the beveled edges ground to the same groove angle.

3. Position the workpiece in a vise or section of angle iron so the weld joint is in flat position.

4. Deposit a small bead on one side. Then deposit a similar bead on the opposite side to prevent the shaft from warping.

5. Use a slight weaving motion on the last pass.

Weaving the Electrode

1. Obtain a piece of ¼″ low-carbon steel, 4″ wide and 6″ long.

2. Draw a series of straight lines on the plate.

3. Position the workpiece in flat position.

4. Deposit continuous beads along the guide lines. Remove slag completely.

5. Practice weaving by depositing a weld back and forth between the first pair of continuous beads.

6. Use a different weave motion to fill each section. Ensure that the short beads are fused into the long, straight beads.

7. Continue to practice weaving on several plates until a satisfactory plate is completed.

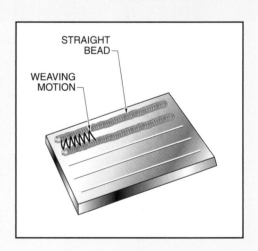

STRAIGHT BEAD

WEAVING MOTION

1. Obtain a piece of low-carbon steel, ¼″ thick or more, 3″ wide, and 5″ long.

2. Position the workpiece in flat position.

3. Deposit a layer of straight beads to completely cover the workpiece surface. Remove slag completely. A pass is often made around the edge of the plate before the layer of beads is deposited to maintain square edges.

4. Deposit a second layer of weaved beads about ½″ wide at right angles to the first layer. Remove slag completely.

5. Deposit a third layer of straight beads at right angles to the second layer. Remove slag completely.

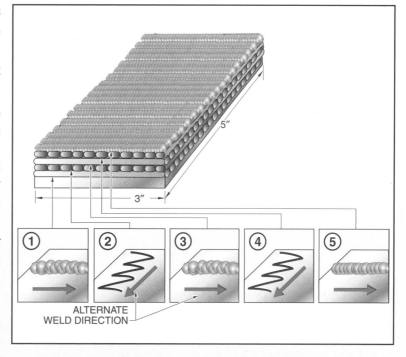

6. Deposit a fourth and fifth layer, in the same manner—each layer at right angles to the previous layer, with slag thoroughly removed before the subsequent layer is added.

1. What is an advantage of welding in flat position rather than in other positions?
2. How are tack welds used in welding?
3. What is the function of a root pass?
4. What is a cover pass and why is it used?
5. What procedures are followed when welding plates of different thicknesses?
6. When making a lap weld, what determines how much the workpieces should overlap?
7. How can undercutting be avoided when welding a lap joint?
8. Why should a double fillet be used on a lap joint?
9. When should multiple passes be used on a lap joint?
10. When welding a T-joint, why should the arc be directed more toward the bottom workpiece?
11. What is the difference between an open and closed butt joint?
12. When is a butt joint used in welding?
13. When should the edges of butt joints be beveled?
14. What determines the edge preparation required for welding a butt joint?
15. What are some common applications of outside corner welds?
16. How many passes should be made on an outside corner weld?
17. How should the edges of round stock be prepared for welding?
18. What work angle is used to weld a lap joint in flat position?
19. What work and travel angles are used when welding a butt joint in flat position?
20. What welding technique is recommended when making a root bead?
21. What work angles are used to weld a multiple-pass fillet T-joint weld?
22. What is the purpose of surfacing?
23. What is meant by weaving?
24. When is a weaving motion used?

Refer to Chapter 12 in the *Welding Skills Workbook* for additional exercises.

SMAW–Horizontal Position

On many jobs, welding cannot be performed in flat position. Occasionally, the welding operation must be done while the work is in horizontal position. Welds performed in horizontal position must have a uniform, consistent bead. A fill-freeze electrode from the F2 group or fast-freeze electrode from the F3 group should be used.

HORIZONTAL POSITION WELDING

A weld is in horizontal position when the workpiece is in a vertical position and the weld joint is approximately horizontal. **See Figure 13-1.** The weld bead must support the weld pool during welding to ensure sufficient buildup of the weld.

An overlap occurs when the weld pool runs down to the lower side of the bead and solidifies on the surface without actually penetrating the base metal. A sagging weld pool usually leaves an undercut on the top side of the weld seam and an improperly shaped bead. **See Figure 13-2.** Overlaps and undercuts can weaken a weld.

> ✓ **Point**
>
> *Do not allow the molten weld pool to sag and cause overlaps and undercuts.*

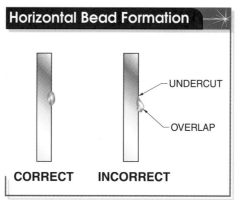

Horizontal Bead Formation

UNDERCUT

OVERLAP

CORRECT **INCORRECT**

Figure 13-2. *Using a short arc length minimizes the tendency of the weld pool to sag and cause overlapping. Sagging weld pools usually leave undercut on the top edge of the joint.*

Figure 13-1. *In horizontal welding, the weld joint is in horizontal position, and the workpiece is vertical.*

To weld in horizontal position, a short arc length should be used, with a slight reduction in current from that used for welding in flat position. The short arc length minimizes the tendency of the weld pool to sag and cause overlapping.

> ⊕ *Welding in horizontal position has a high failure rate on weld inspection. Many welders think horizontal position welding is easy and they don't pay close enough attention to the placement of the weld.*

> ✓ **Point**
>
> *When welding in horizontal position, use a lower welding current and shorter arc length than when welding in flat position.*

Point

*When welding in hori-
zontal position, hold
the electrode at a work
angle of 5° to 10° and a
travel angle of 20°.*

Welding Procedure

When welding a butt joint in horizontal position, hold the electrode at a work angle of 5° to 10° and a travel angle of 20°. **See Figure 13-3.**

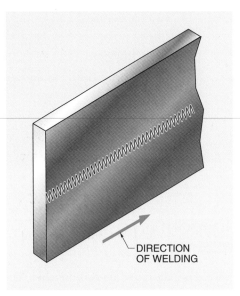

Electrode Position

WORK ANGLE

DIRECTION OF WELDING

5° TO 10°

20°

TRAVEL ANGLE

SIDE VIEW   **TOP VIEW**

Figure 13-3. *When welding in horizontal position, hold the electrode at a work angle of 5° to 10° and a travel angle of 20°.*

Point

*Use a slight weaving
motion when welding in
horizontal position.*

When depositing the bead, use a narrow weaving motion. Weaving the electrode distributes heat more evenly, further reducing any tendency for the weld pool to sag. **See Figure 13-4.** Keep the arc length as short as possible. If the force of the arc has a tendency to undercut the workpiece at the top of the bead, slightly tilt the electrode upward to increase the upward angle.

A fill-freeze electrode from the F2 group or fast-freeze electrode from the F3 group should be used for horizontal welding. As the electrode is moved in and out of the crater, pause slightly each time it is returned to the crater. This keeps the crater small and the bead is less likely to sag.

DIRECTION OF WELDING

Figure 13-4. *Use a slight weaving motion when welding in horizontal position to distribute heat more evenly.*

In horizontal position welding, the position of the electrode is changed for each pass. The number of passes required depends on the base metal, joint configuration, and thickness of the diameter of the electrode. Complete fusion into each adjacent pass and into the edges of the joint is necessary to produce a quality weld.

Refer to Quick Quiz® on CD-ROM

Points to Remember

- Do not allow the molten weld pool to sag, which can result in overlaps and undercuts.
- When welding in horizontal position, use a lower welding current and shorter arc length than when welding in flat position.
- When welding in horizontal position, hold the electrode at a work angle of 5° to 10° and a travel angle of 20°.
- Use a slight weaving motion when welding in horizontal position.

Depositing Straight Beads in Horizontal Position

1. Obtain a piece of ¼″ low-carbon steel.

2. Draw a series of guide lines ½″ apart and the length of the workpiece.

3. Position the workpiece so the guide lines are in horizontal position. The workpiece may be clamped onto a positioner, if available, or tack welded to another workpiece or the workbench.

4. Adjust the welding machine to the correct current and, with a slight weaving motion, deposit beads between the guide lines. Start at the left edge of the first guide line and deposit a bead, working to the right edge.

5. Move to the next guide line and reverse the direction of travel for the second bead.

6. Continue making beads in reverse directions until uniform beads can be made without overlapping and undercutting.

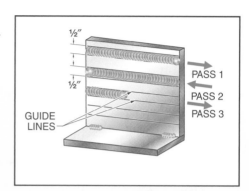

Welding a Single-Pass Lap Joint in Horizontal Position

1. Obtain two pieces of ¼″ low-carbon steel.

2. Form a lap joint and tack together.

3. Position the workpiece so the weld joint is in horizontal position.

4. Use a 45° work angle and deposit a single bead along the joint with a slight weaving motion. The weld should fuse into the top edge of the joint by ¹⁄₁₆″ and form an even top line along the bottom plate.

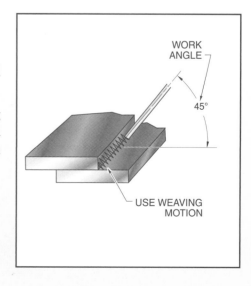

1. Obtain two pieces of ¼″ low-carbon steel.

2. Form a T-joint with the workpieces at a 90° angle and tack together.

3. Position the workpiece so the weld joint is in horizontal position.

4. Angle the electrode to 45° and deposit a root pass along the joint without any weaving motion. Remove slag completely to ensure proper penetration.

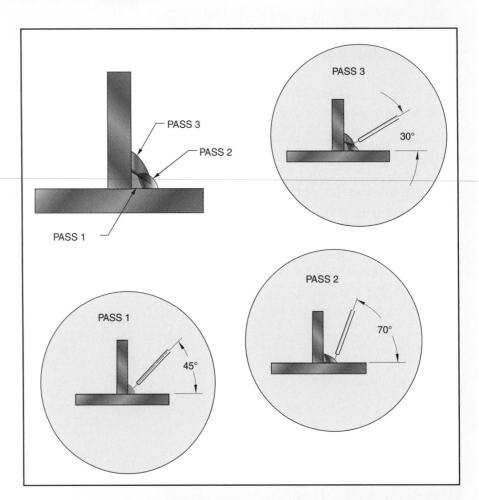

5. For the second pass, angle the electrode 70° from the bottom plate and deposit the weld along the bottom toe of the root pass. Use a slight weaving motion to control heat input, and make sure to penetrate the root bead and cover it by ⅓″ to ½″. The bead should fit into the base metal and form an even toe line. Remove slag completely.

6. For the third pass, angle the electrode 30° from the bottom plate and deposit the weld along the top toe of the root pass using a slight weaving motion. The third bead should penetrate into the first and second beads. It should cover the second pass by ⅓″ to ½″ and fuse into the vertical plate. The top toe line should be even with no undercut. Thoroughly clean the weld. Complete penetration of the weld passes must be obtained, otherwise a weak weld results and the layers may separate.

1. Obtain two pieces of ¼″ low-carbon steel and bevel the edge of one piece.

2. Form a butt joint, allowing a ¹⁄₁₆″ root opening, and tack together.

3. Position the workpiece so the weld joint is in horizontal position with the beveled piece on top and the piece that is not beveled on the bottom.

   The flat edge of the non-beveled workpiece serves as a shelf, helping to prevent the weld pool from running out of the joint.

4. Deposit the root pass deep into the joint. Remove slag completely.

5. Deposit two cover passes, clean the weld after each pass. Each bead should penetrate each previous pass, and fuse into the edges of the joint by ¹⁄₁₆″ to ⅛″.

On some welding jobs, both edges of the joint are beveled to form a 60° groove angle. This is a single-V-groove butt joint. Since a single-V-groove butt joint does not provide a retaining shelf for the bead, as does a single bevel butt joint, more skill is required to produce a satisfactory weld.

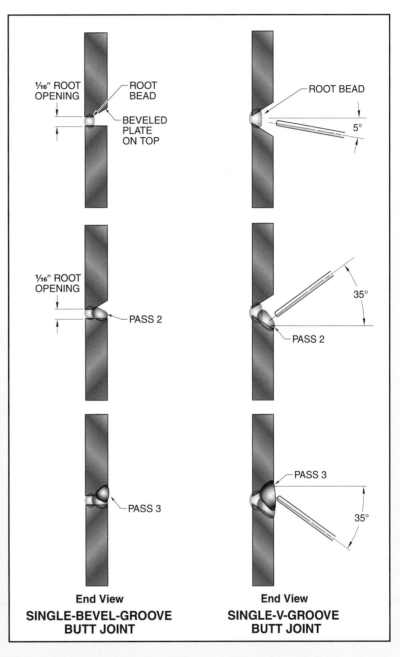

End View
SINGLE-BEVEL-GROOVE BUTT JOINT

End View
SINGLE-V-GROOVE BUTT JOINT

On a wide joint, the weld is sometimes finished with another pass to produce a smooth finish. A wide weaving motion that covers the entire area of the deposited beads is used to make the cover pass.

? Questions for Study and Discussion

1. Why must a low current and a short arc length be used when welding in horizontal position?
2. What can be done to prevent overlaps on horizontal welds?
3. In what position should the electrode be held for welding horizontal beads?
4. Why should a weaving motion be used when making horizontal welds?
5. What determines the number of passes that should be made on a weld?
6. What groove angle is used when beveling the edges for a butt joint?
7. When is a cover pass used?
8. What work angles are required for each pass of a multiple-pass fillet T-joint weld in horizontal position?
9. What must be done between passes of a multiple-pass fillet T-joint to ensure proper penetration?

Refer to Chapter 13 in the *Welding Skills Workbook* for additional exercises.

SMAW–Vertical Position

Chapter 14

Welding in vertical position is frequently used for the fabrication of structures such as steel buildings, bridges, tanks, pipelines, ships, and machinery.

When welding in vertical position, gravity tends to pull down the molten metal from the weld pool. To prevent this from happening, fast-freeze electrodes from the F3 group or fill-freeze electrodes from the F2 group should be used. Weld pool control can also be achieved with proper electrode manipulation. Vertical welding is done by depositing beads using one of two methods, downhill welding or uphill welding.

DOWNHILL WELDING

A *vertical weld* is a weld with the axis of the weld approximately vertical. *Downhill welding* is welding with a downward progression. Downhill welding is commonly used for welding light-gauge metal because penetration is shallow. Downhill welding can be performed rapidly, which is important in production work. Although generally recommended for welding light-gauge materials because it does not cause melt-through, downhill welding can also be used for other metal thicknesses.

In downhill welding, maintain a travel angle of 15° to 30°. **See Figure 14-1.** Start at the top of the joint and move downward with little or no weaving motion. If a slight weave is necessary, manipulate the electrode so the crescent of the weave is at the top. On metal ¼″ thick or more, uphill welding is more common.

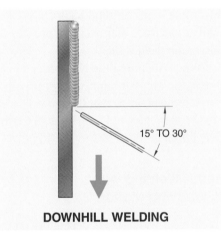

15° TO 30°

DOWNHILL WELDING

Figure 14-1. *Maintain a drag angle of 15° to 30° for downhill welding.*

UPHILL WELDING

Uphill welding is welding with an upward progression. Uphill welding is commonly used on metal more than ¼″ thick because deeper penetration can be obtained. Uphill welding also makes it possible to create a shelf for successive layers of beads.

For uphill welding, start with the electrode at a right angle to the workpiece. Position the electrode holder until the electrode forms a travel angle of 10° to 15°, pointing away from the direction of welding. **See Figure 14-2.**

> ✓ **Point**
>
> *When welding light-gauge metal in vertical position, downhill welding is used to control penetration and weld size to reduce distortion.*

> ✓ **Point**
>
> *On metal ¼″ thick or more, uphill welding is commonly used to obtain the required penetration.*

> *When welding in vertical position, molten weld metal has a tendency to run out of the weld pool. Using a small diameter electrode reduces this tendency by reducing the weld pool size.*

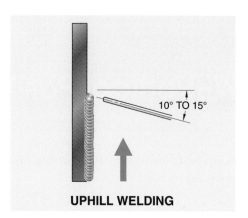

UPHILL WELDING

Figure 14-2. *Maintain a push angle of 10° to 15° for uphill welding.*

Uphill Welding with a Whipping Motion

Uphill welding commonly uses fast-freeze electrodes from the F3 group with a slight whipping motion. *Whipping* is a manual welding technique in which the arc is moved quickly forward about one electrode diameter, and back about one-half an electrode diameter as it progresses along the weld joint.

When whipping the electrode, do not break the arc, but simply pivot it with a wrist movement so that the arc is moved up ahead of the weld long enough for the weld pool in the crater to solidify.

Uphill Welding with a Weaving Motion

Weld joint width varies depending on the metal thickness and edge preparation. The weld bead width must be adjusted to completely fill the required joint width. The width of the weld bead can be controlled using a weaving motion, such as the figure eight, rotary motion, or crescent. Each weaving motion produces a bead approximately twice the diameter of the electrode.

The electrode is moved to allow penetration at the bottom of the stroke, and the upward motion momentarily removes the heat until the weld metal can solidify. **See Figure 14-3.**

The welder should pause at the toes of the weld. Pausing at the toes allows for complete fusion of the weld metal into

the joint. The electrode should be moved quickly across the center of the weld to prevent excessive heat buildup.

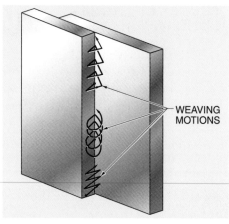

WEAVING MOTIONS

Figure 14-3. *A figure eight, rotary, or crescent weaving motion is used with uphill welding to control the width of the weld bead.*

LOW-HYDROGEN ELECTRODE WELDING TECHNIQUES

Although vertical welding techniques are generally applicable to all types of electrodes, a slight modification in procedure is advisable when using low-hydrogen electrodes.

The E7048 LS electrode is best suited for downhill welding. Drag the electrode lightly, using a short arc. Do not use a long arc since the weld depends on the molten slag for shielding. A single, narrow bead or small weave is preferred to wide weave passes. Use lower current when welding with DC as compared to AC. Point the electrode directly into the joint and tip it forward a few degrees in the direction of travel.

The E7018 electrode is best suited for uphill welding, a triangular weaving motion often produces better results. When using E7018 electrodes and a weaving motion, the width of the weave pattern should not exceed 2½ times the diameter of the electrode. Do not use a whipping motion or slag may become trapped in the weld, forming slag inclusions. Point the electrode directly into the joint and increase the travel angle slightly to permit the arc force to assist in controlling the weld pool. Current should be set toward the lower end of the recommended range.

Refer to Quick Quiz® on CD-ROM

 Points to Remember

- When welding light-gauge metal in vertical position, downhill welding is used to control penetration and weld size to reduce distortion.
- On metal ¼″ thick or more, uphill welding is commonly used to obtain the required penetration.
- On grooved joints, deposit the root pass deep into the root opening.

EXERCISES

Depositing Beads in Vertical Position (Downhill)

exercise **1**

1. Obtain a piece of ¼″ low-carbon steel.

2. Draw a series of straight guide lines the length of the workpiece.

3. Position the workpiece so the guide lines are in vertical position.

4. Use E6012 or E6013 electrodes.

5. Start at the top of the workpiece with the electrode pointed upward at about a 15° to 30° drag angle. Keep the arc short and move the electrode downward to form the bead.

6. Keep the arc on the leading edge of the weld pool and maintain a travel speed that is just fast enough to prevent the molten weld pool and slag from running ahead of the crater. Do not use any weaving motion at the start.

7. Once straight, single beads can be deposited, weave the electrode slightly, with the crest at the top of the crater.

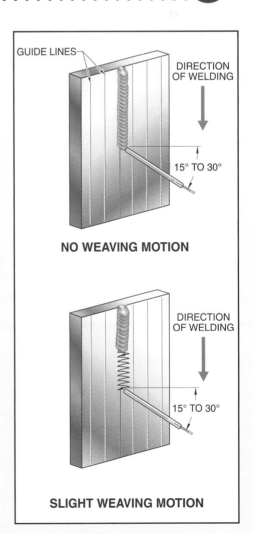

GUIDE LINES

DIRECTION OF WELDING

15° TO 30°

NO WEAVING MOTION

DIRECTION OF WELDING

15° TO 30°

SLIGHT WEAVING MOTION

Depositing Beads in Vertical Position (Uphill)

1. Obtain a piece of ¼″ low-carbon steel.

2. Draw a series of straight guide lines the length of the workpiece.

3. Position the workpiece so the guide lines are in vertical position.

4. Use E6010 or E6011 electrodes for the necessary fast-freeze characteristics from the F3 group.

5. Start at the bottom of the workpiece with the electrode at a 90° work angle and a 10° to 15° push angle. Move the electrode upward about one electrode diameter using a slight whipping motion.

6. Return the electrode to the crater and repeat the operation, working up along the drawn guide line to the top of the workpiece.

7. Do not break the arc while moving the electrode upward. Withdraw it just long enough to permit the deposited metal to solidify and form a shelf so additional metal can be deposited.

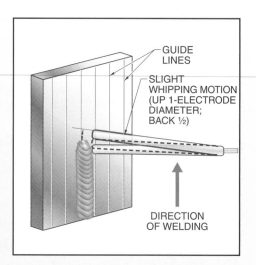

GUIDE LINES

SLIGHT WHIPPING MOTION (UP 1-ELECTRODE DIAMETER; BACK ½)

DIRECTION OF WELDING

Welding a Multiple-Pass Lap Joint in Vertical Position (Uphill)

1. Obtain two pieces of ¼″ low-carbon steel.
2. Form a lap joint and tack together.
3. Position the workpiece so the weld joint is in vertical position.
4. Start at the bottom of the workpiece and deposit a small root pass without any weaving motion.
5. Start at the bottom of the workpiece again and deposit a cover pass, using a weaving motion, from the bottom to the top. Move up ⅛″ with each weave across the joint.
6. Ensure that the cover pass completely penetrates the root bead. It should fill the joint completely and fuse into the bottom plate forming an even toe line. The weld should fuse into the top edge of the joint by ¹⁄₁₆″.

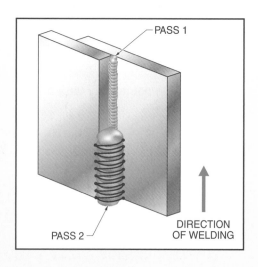

PASS 1

PASS 2

DIRECTION OF WELDING

Welding a Butt Joint in Vertical Position (Uphill)

1. Obtain two pieces of ¼″ low-carbon steel and bevel the edges to form a 60° groove angle.

2. Form a butt joint, with a ¹⁄₁₆″ root opening, and tack together.

3. Position the workpiece so the weld joint is in vertical position.

4. Start at the bottom of the workpiece and deposit a root pass. Remove slag completely.

5. Start at the bottom of the workpiece again and deposit an intermediate weld pass(es) as necessary to fill the joint about ¹⁄₁₆″ from the surface of the plates. Remove slag completely.

6. Finish the weld with a cover pass that fills the joint completely and fuses into the edge of the joint by ¹⁄₁₆″ to ⅛″. Remove slag completely. Face reinforcement should be ¹⁄₁₆″ maximum.

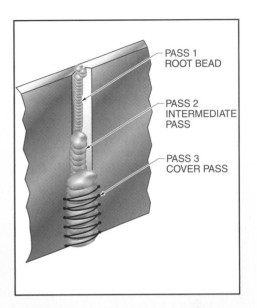

PASS 1
ROOT BEAD

PASS 2
INTERMEDIATE
PASS

PASS 3
COVER PASS

Welding a T-Joint in Vertical Position (Uphill)

1. Obtain two pieces of ¼″ low-carbon steel.

2. Form a T-joint with the pieces at a 90° angle and tack together.

3. Position the workpiece so the weld joint is in vertical position.

4. Start at the bottom of the workpiece and deposit a narrow root pass. Remove slag completely.

5. Start at the bottom of the workpiece again and, using a weaving motion, deposit an intermediate weld pass. Remove slag completely.

6. On the other side, start at the bottom of the workpiece and deposit a narrow root pass. Remove slag completely.

7. Deposit an intermediate weld pass on the second side. Remove slag completely.

8. Check for complete penetration of each pass. Deposit a cover pass on each side. Remove slag completely.

1. In vertical welding, what can be done to prevent the weld pool from sagging?
2. Why is downhill welding more applicable to light-gauge metal?
3. In what position should the electrode be held in downhill welding?
4. What motions should be used in downhill welding?
5. How should the electrode be held when making an uphill weld?
6. What is the advantage of using a whipping motion on a vertical weld?
7. How can the width of a bead be increased on an uphill weld?
8. What direction of travel provides the most penetration when welding in vertical position?
9. What types of electrodes are commonly used in vertical welding?
10. What kind of weaving motion is used when welding uphill using an E-7018 electrode?
11. Which is faster, uphill welding or downhill welding?
12. What determines if a weld is in vertical position?
13. What types of electrodes can be used with a whipping motion?
14. What is the advantage of using a weaving motion when welding in vertical position?

 Refer to Chapter 14 in the *Welding Skills Workbook* for additional exercises.

SMAW–Overhead Position

Welding in overhead position is one of the most difficult welding operations to master. Although overhead welding is similar to flat position welding in technique, overhead welding is done from an awkward position and is greatly affected by gravity. In overhead position the weld pool has a tendency to drop, making it harder to secure a uniform bead and correct penetration. With practice it is possible to secure welds with the same quality as those made in other positions

OVERHEAD WELDING

When overhead welding, the welder must be sure that the weld passes properly to fill the weld joint. Molten metal can easily drop from the weld pool, causing uneven, inconsistent weld beads and incomplete penetration. Keep the arc length as short as possible when welding in overhead position to prevent molten metal from falling out of the weld pool. Beginning welders should practice welding beads in overhead position until a consistent bead can be deposited repeatedly. When practicing welding in overhead position, a positioner is commonly used to secure workpieces. The positioner allows the welder to set the workpiece to any height or position. **See Figure 15-1.**

When overhead welding, personal protective clothing and equipment must be worn to protect against falling molten metal, slag, and sparks. A headcap, welding hood, and leather jacket or leather apron and leather sleeves should be worn to prevent slag and sparks from burning the skin. **See Figure 15-2.** Shirtsleeves should be rolled down and buttoned.

> ✓ **Point**
>
> *When welding in overhead position, keep the arc length as short as possible.*

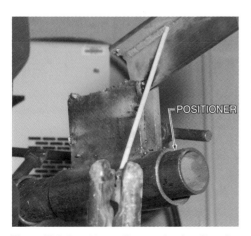

Figure 15-1. *A positioner allows work to be adjusted to any height or position.*

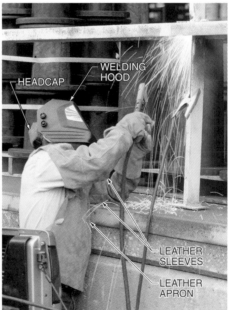

The Lincoln Electric Company

Figure 15-2. *Proper personal protective equipment must be worn when performing overhead welding to prevent injury.*

Overhead Welding Procedure

When welding in overhead position, use a fast-freeze electrode. To start welding, hold the electrode at a right angle to the joint. Hold the electrode at a work angle of 90° and a drag angle of 10° to 15°. **See Figure 15-3.**

Grip the electrode holder so the knuckles of the hand are up and the palm is down. This prevents particles of molten metal from being caught in the palm of the glove and allows spatter to roll off the glove. The electrode holder can be held in one hand; however, sometimes welding is easier if it is held with both hands. **See Figure 15-4.** To avoid hot metal spatter, stand to the side rather than directly underneath the arc. The weight of the electrode lead can be minimized by using a lightweight whip, or by supporting the cable with the other hand.

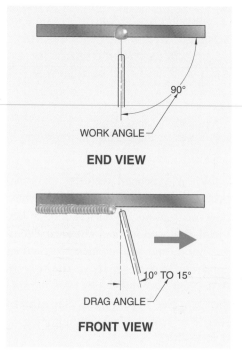

WORK ANGLE

END VIEW

DRAG ANGLE

FRONT VIEW

Figure 15-3. *For overhead welding, the electrode should be held at a work angle of 90° and a drag angle of 10° to 15°.*

⚠ **WARNING**

Molten metal can fall from the weld when welding in overhead position. Be sure sleeves are rolled down and a protective garment with a tight-fitting collar is zipped or buttoned up to the neck. Wear a headcap and heavy-duty shoes.

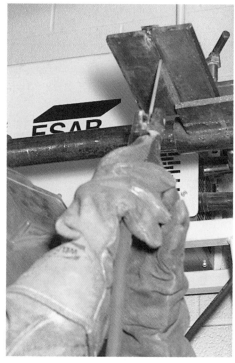

Figure 15-4. *A welder may use both hands to hold the electrode when welding in overhead position.*

 Refer to Quick Quiz® on CD-ROM

Points to Remember

- When welding in overhead position, keep the arc length as short as possible.
- A travel angle of 10° to 15° should be used for overhead welding.
- Grip the electrode holder so the knuckles of the hand are up and the palm is down.
- When welding in overhead position, stand to the side to avoid injury from hot metal spatter.
- To minimize the weight of the electrode lead, use a lightweight whip, or support the cable with the other hand.

Exercises

Depositing Beads in Overhead Position

exercise **1**

1. Obtain a piece of ¼″ low-carbon steel.

2. Draw a series of guide lines on the workpiece, each line approximately ½″ apart.

3. Position the workpiece so the guide lines are in overhead position.

4. Set current as recommended for overhead welding. Strike an arc and form a weld pool as in flat position welding. Move the electrode along the weld joint, keeping the arc as short as possible.

5. Deposit a series of straight beads with no weaving motion. If necessary to prevent the weld pool from dropping, reduce the current slightly.

6. Practice depositing beads in one direction, then reverse and practice in the opposite direction.

7. Deposit beads using a weaving motion to fill in the space between the beads.

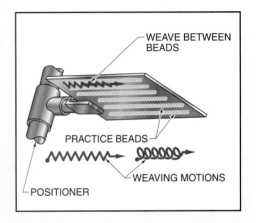

WEAVE BETWEEN BEADS

PRACTICE BEADS

WEAVING MOTIONS

POSITIONER

Welding a Multiple-Pass Single-V Butt Joint in Overhead Position

exercise **2**

1. Obtain two pieces of ¼″ low-carbon and bevel the edges to form a 60° groove angle.

2. Form a butt joint, with a ¹⁄₁₆″ root opening for expansion, and tack weld.

3. Position the workpiece so the weld joint is in overhead position.

4. Deposit a root pass in the root of the joint. Remove slag completely.

5. Deposit a cover layer consisting of two passes. The first pass should cover the root pass by ⅓ to ½. The second pass should overlap the first cover pass by ⅓ to ½. The cover layer should fill the joint completely and should fuse into the edges of the joint by ¹⁄₁₆″ to ⅛″. Remove slag between passes.

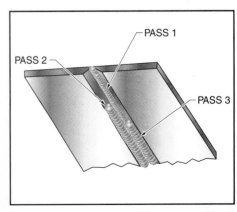

PASS 1

PASS 2

PASS 3

Welding a Multiple-Pass Lap Joint in Overhead Position

1. Obtain two pieces of ¼" low-carbon steel.

2. Form a lap joint and tack together.

3. Position the workpiece so the weld joint is in overhead position.

4. Hold the electrode with a 45° work angle and a 15° drag angle.

5. Deposit a root pass in the root of the joint. Remove slag completely.

6. Deposit a second weld on the bottom toe of the root pass, making sure the weld penetrates into the root bead and fuses into the bottom plate forming an even toe line. Remove slag completely.

7. Deposit a third weld on the top toe of the root pass, making sure the weld penetrates into the root bead and overlaps the first cover pass by ⅓ to ½. It should fuse into the top edge of the joint by ¹⁄₁₆".

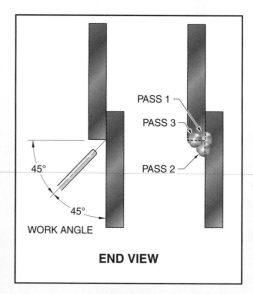

PASS 1
PASS 3
PASS 2
45°
45°
WORK ANGLE

END VIEW

Welding a Multiple-Pass T-Joint in Overhead Position

1. Obtain two pieces of ¼" low-carbon steel.

2. Form a T-joint with the pieces at a 90° angle and tack together.

3. Position the workpiece so the weld joint is in overhead position.

4. Deposit a root pass in the root of the joint. Remove slag completely.

5. Deposit a second pass on the bottom toe of the root pass, and a third pass on the top toe of the root pass. Remove slag completely between passes. Adjust the work angle of the electrode for each pass to ensure complete penetration.

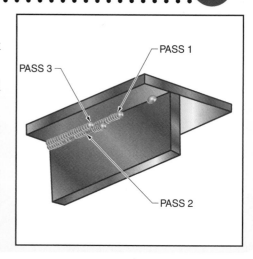

PASS 1
PASS 3
PASS 2

❓ Questions for Study and Discussion

1. Why is welding in overhead position more difficult than welding in other positions?
2. What is the recommended travel angle for overhead welding?
3. Why should the electrode holder be grasped so that the palm is facing down?
4. What should be done to prevent the weld pool from dropping?

Refer to Chapter 15 in the *Welding Skills Workbook* for additional exercises.

The gas tungsten arc welding (GTAW) process was developed in the late 1930s, primarily for welding aluminum and magnesium in the aircraft industry. A breakthrough in GTAW occurred during World War II, when alternating current (AC), with the addition of high frequency (HF) was found to produce high-quality welds on aluminum. Initially the process was referred to as heliarc welding because it used helium for shielding. It later became known as tungsten inert gas welding (TIG) because it was discovered that argon provided excellent shielding at much lower cost. The word "inert" was used in the name because helium and argon are inert gases. Today the process is known as gas tungsten arc welding (GTAW).

GAS TUNGSTEN ARC WELDING

Gas tungsten arc welding (GTAW) is an arc welding process in which shielding gas protects the arc between a nonconsumable tungsten electrode and the base metal. A *nonconsumable electrode* is an electrode that does not melt and become part of the weld. **See Figure 16-1.** GTAW is a versatile process that can be used to weld a wide variety of ferrous and nonferrous metals. It is commonly used to weld thin gauge materials and to deposit root passes on pipe. GTAW produces high-quality, spatter-free welds that require very little post-weld cleaning. However, GTAW has low deposition rates compared to other arc welding processes.

> Gas tungsten arc welding (GTAW) can weld more types of metal and metal alloys than any other arc welding process.

On certain types of joints, like lap joints and outside corner joints, GTAW can be performed without filler metal. On other types of joints such as T-joints and butt joints, filler metal is required. Filler metal is also required on thick metals to fill the joint. When welding with filler metal, the heated end of the filler metal should be kept in the flow of shielding gas to prevent oxidation.

Media Clip
Gas Tungsten Arc Welding

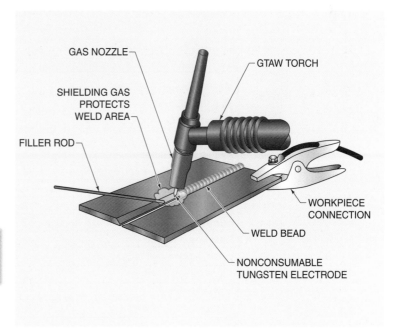

Figure 16-1. *Although gas tungsten arc welding (GTAW) can be performed without using filler metal, filler metal is required on thick metal and on butt and T-joints to provide reinforcement and prevent cracking.*

✓ Point

High frequency (HF) is used for arc starting with DC, and to start and maintain the arc with AC.

WELDING POWER SOURCES FOR GTAW

GTAW requires a constant-current (CC) welding power source capable of producing direct current (DC), alternating current (AC), and high frequency (HF). **See Figure 16-2.** The high open-circuit voltage associated with constant-current machines aids in arc starting. *High frequency (HF)* is high-voltage, low-current pulses over 16,000 cycles per second (16 kHz). HF is used to start an arc without touching an electrode to the work. When the welding machine is set to DCEN or DCEP, HF is used for arc starting only. For AC applications with certain types of welding power sources, high frequency is also used to maintain the arc. The choice of current depends on the type of metal to be welded. **See Figure 16-3.**

Direct Current

Direct current (DC) is an electrical current that flows in one direction only. DC must be electrode negative (DCEN) or electrode positive (DCEP). **See Figure 16-4.**

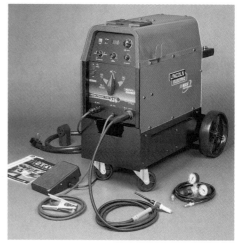

The Lincoln Electric Company

Figure 16-2. *A constant-current welding power source maintains a relatively constant current with changes in voltage (arc length).*

> *When the current selector switch on the power source is set to DCEN, electrons flow from the electrode to the work. When the current selector switch is set to DCEP, electrons flow from the work to the electrode. When the current selector switch is set to AC, the flow of electrons alternates between electrode positive and electrode negative.*

GTAW CURRENT SELECTION			
Metal	**AC Current** with **High Frequency Stabilization**	**DC Current**	
		Electrode Negative	**Electrode Positive**
Magnesium up to 1/8″ thick	1	NR	2
Magnesium over 1/8″ thick	1	NR	NR
Magnesium Castings	1	NR	NR
Aluminum	1	2	NR
Aluminum Castings	1	2	NR
Stainless Steel	2	1	NR
Silver	2	1	NR
Nickel Alloys	2	1	NR
Hardfacing	2	1	NR
Cast Iron	2	1	NR
Low-Carbon Steel	2	1	NR
High-Carbon Steel	2	1	NR
Low-Alloy Steel	2	1	NR
Copper/Copper Alloys	2	1	NR

Key:
1. Excellent operation–best recommendation
2. Good operation–second recommendation
NR–not recommended

Figure 16-3. *The choice of AC or DC depends on the metal to be welded.*

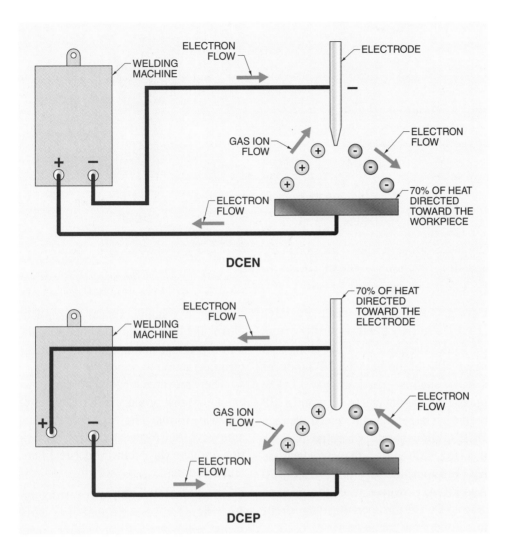

Figure 16-4. *With DCEN, electron flow from the electrode to the workpiece creates heat at the workpiece. With DCEP, electron flow from the workpiece to the electrode creates heat at the electrode.*

Direct Current Electrode Negative (DCEN). In DCEN, electrons flow from the electrode to the workpiece. Approximately 70% of the heat is directed toward the base metal. Carbon steels, stainless steels, and other ferrous metals are typically welded with DCEN. DCEN produces narrower, deeper-penetrating welds than DCEP or AC. It also allows for faster travel speed and less distortion of the base metal. Smaller diameter electrodes can be used for DCEN as compared to DCEP because most of the heat is directed toward the base metal. **See Figure 16-5.**

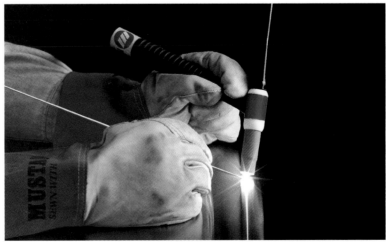

Current selection for GTAW is based on the type of metal to be welded.

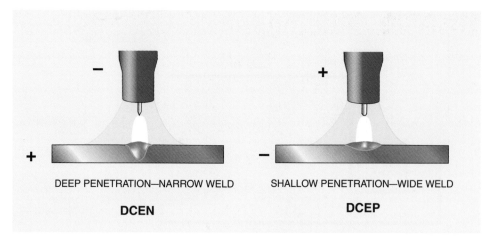

DEEP PENETRATION—NARROW WELD

DCEN

SHALLOW PENETRATION—WIDE WELD

DCEP

Figure 16-5. *The polarity of DC affects the shape of the weld and weld penetration. DCEN produces a narrow, deep-penetrating weld, whereas DCEP produces a wide weld with shallow penetration.*

Direct Current Electrode Positive (DCEP). In DCEP, electrons flow from the workpiece to the electrode along with about 70% of the heat. DCEP produces a wide, shallow-penetrating weld. The intense heat at the electrode with DCEP requires a larger diameter electrode than DCEN. For example, when the machine is set to DCEN, a 1/16″ diameter electrode has a recommended operating range of 70 A to 150 A. However, when the machine is set to DCEP, the operating range drops to an operating range of 10 A to 20 A. DCEP is rarely used in GTAW because it produces shallow penetration and requires a large diameter electrode.

Alternating Current

Alternating current (AC) is an electrical current that has alternating positive and negative values. AC is used for welding nonferrous metals such as aluminum and magnesium. Aluminum and magnesium have tough oxide coatings that melt at temperatures much higher than the metals they protect. These oxides must be removed before the metal can be welded. During the positive portion of the AC cycle, the base metal is bombarded with gas ions that produce a cleaning action. *Cleaning action* is the removal of the oxide coating on the base metal by bombarding it with gas ions. Before welding, joints should be prepared by brushing

them with a stainless steel brush. This cleans the base metal and breaks up the oxide coating, making it easier for the cleaning action to remove it.

There are three types of welding power sources that produce different types of AC waveforms. They are conventional sine wave transformer-rectifiers, conventional square wave transformer-rectifiers, and inverters.

Conventional Sine Wave Transformer-Rectifiers. A *conventional sine wave transformer-rectifier* is a welding power source that produces a sinusoidal waveform. A *sinusoidal waveform* is a 60 cycle per second (60 Hz) sine wave. A sinusoidal waveform is considered balanced because it has equal amounts of electrode positive (EP) current and electrode negative (EN) current. Sinusoidal waveforms pass through zero 120 times per second on their way to peak EP and peak EN. The arc extinguishes each time the current passes through zero, producing an unstable arc. The HF generator on the machine is set to "continuous" to allow arc starting without touching the tungsten to the work. It also stabilizes the arc through the changes from EP to EN. **See Figure 16-6.** A sinusoidal waveform is inefficient compared to a square waveform because very little time is spent at peak cleaning during the EP half-cycle or at peak penetration during the EN half-cycle.

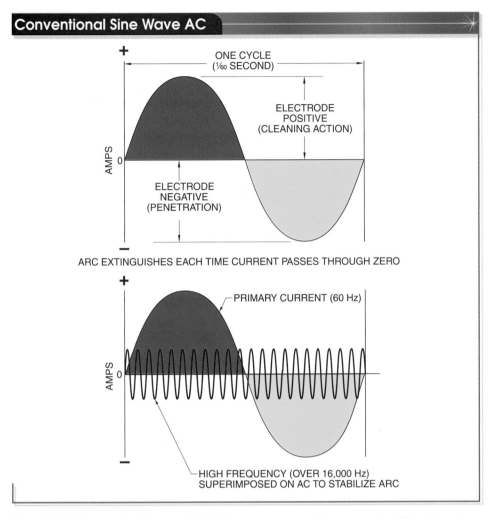

Figure 16-6. *With conventional sine wave AC, half of one complete cycle is electrode positive (EP), which produces cleaning action, while the other half of the cycle is electrode negative (EN), which produces deep penetration. High frequency (HF) is superimposed over the AC sine wave to stabilize the arc through changes from EP to EN.*

Conventional Square Wave Transformer-Rectifiers. A *conventional square wave transformer-rectifier* is a welding power source that produces a relatively square 60 Hz waveform. It provides three main advantages over a balanced, sinusoidal waveform. First, the change from positive to negative is immediate, resulting in a smoother arc. Second, a conventional square waveform is more efficient than a sine wave, because the arc spends more time at peak EP for cleaning and peak EN for penetration. Third, a conventional square wave has an adjustable balance. This means that the ratio of cycle time can be adjusted from approximately 30% cleaning action and 70% penetration to 55% cleaning action and 45% penetration. Continuous HF is still required to maintain a stable arc. **See Figure 16-7.**

Inverters. An *inverter* is a welding power source that produces a true square AC waveform. Inverters are smaller, lighter, and more efficient than transformer-rectifiers. Most inverters produce such a quick transition between EP and EN that continuous HF is not required to maintain a stable arc. Instead, HF is used for arc starting only. Inverters allow more precise control of the AC waveform. They provide expanded AC balance control, the ability to adjust the frequency of the AC cycle, and the ability to change the amplitude (height) of the EP and EN portions of the AC cycle independently.

Inverters allow for much finer balance control than transformer-rectifiers. The duration of the EN portion of the AC cycle can be adjusted from about 30% to over 90% depending on the amount of penetration required. Likewise, the EP portion of the AC cycle can be adjusted from about 1% to over 70% depending on the amount of cleaning required.

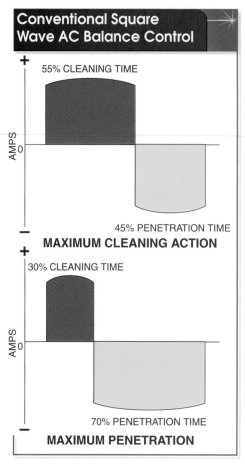

Figure 16-7. *Conventional square wave AC produces a smoother, more efficient arc than sinusoidal AC. A conventional square wave is unbalanced to allow for more cleaning action or more penetration depending on the application.*

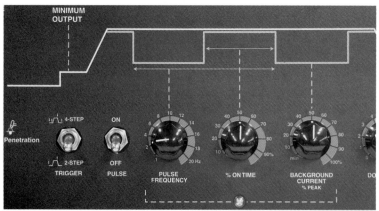

The Lincoln Electric Company

Controls on a GTAW power source can be used to change current and frequency of a waveform.

A lower percentage of EN increases the width of the cleaning zone on aluminum and decreases penetration. It also requires a larger diameter tungsten electrode to handle the increased heat. A higher percentage of EN focuses the arc, increases penetration, and allows a smaller diameter tungsten electrode to be used for a given current because less heat is directed toward the electrode. More EN also reduces the width of the cleaning zone adjacent to the weld. A maximum cleaning zone width is sometimes specified on high-quality welds for aerospace. **See Figure 16-8.**

Inverters are capable of producing AC welding current output ranging from 20 Hz to as high as 200 Hz. Lower frequencies produce a soft arc, a wide cleaning zone, and a wide bead with shallow penetration. Frequencies below 60 Hz can help to reduce porosity on oil soaked aluminum repair welds. Above 60 Hz, the arc becomes tighter and more narrowly focused. Higher frequencies are used for making very small welds on thin aluminum. **See Figure 16-9.**

Some inverters have amplitude controls. *Amplitude* is the height of the EP and EN portions of the AC cycle above or below zero. Amplitude controls make it possible to adjust the amount of cleaning action and the amount of penetration independently. Waveforms with adjustable amplitudes are referred to as asymmetric. Depending on the application, inverters can provide adequate cleaning action with as little as 15% EP. Reducing the amount of EP reduces the amount of heat directed toward the electrode. **See Figure 16-10.**

GTAW EQUIPMENT

In addition to a constant current power source, GTAW requires additional equipment to produce quality welds. GTAW equipment typically includes a torch, tungsten electrode, shielding gas, regulator and flowmeter assembly, and remote current control. **See Figure 16-11.**

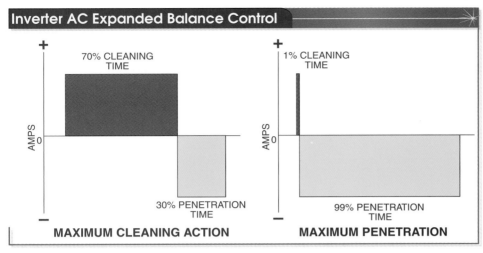

Figure 16-8. *Inverter AC produces a true square wave with fine balance control. A lower percentage of EN produces a wide cleaning zone and shallow penetration. More EN produces a narrow cleaning zone and deep penetration.*

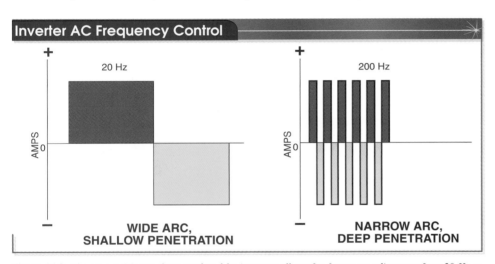

Figure 16-9. *The square AC waveform produced by inverters allows for frequency adjustment from 20 Hz to 200 Hz. Frequencies below 60 Hz produce a soft arc, wide cleaning zone, and shallow penetration. High frequencies produce narrow, deep-penetrating welds on thin aluminum.*

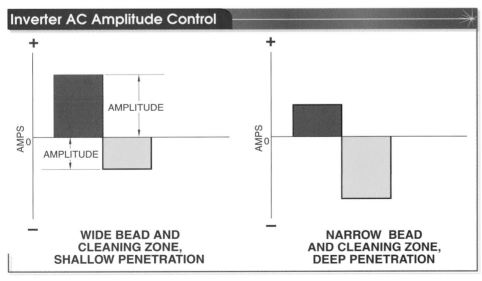

Figure 16-10. *Some inverters have asymmetric waveforms that allow amplitude of the EP and EN portions of the AC cycle to be adjusted independently. This makes it possible to reduce cleaning action to the minimum required, which reduces the heat directed toward the electrode.*

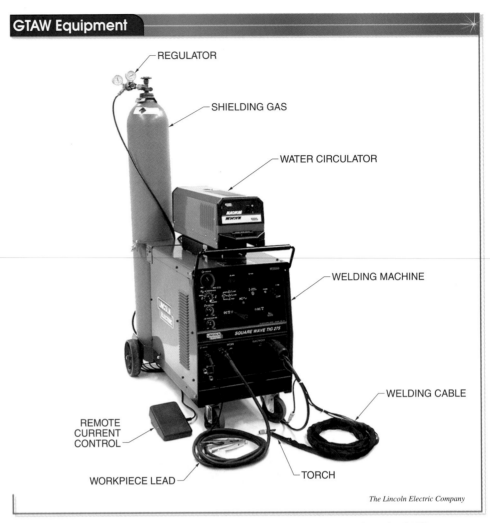

REGULATOR

SHIELDING GAS

WATER CIRCULATOR

WELDING MACHINE

WELDING CABLE

REMOTE
CURRENT
CONTROL

WORKPIECE LEAD

TORCH

The Lincoln Electric Company

Figure 16-11. *Equipment required for GTAW typically includes a torch, tungsten electrode, shielding gas, water circulator, regulator, and remote amperage control, in addition to the welding machine.*

Torches

GTAW torches hold and energize the tungsten electrode, direct shielding gas to protect the electrode and the weld zone, and insulate the electrode and electrical connections from the welder. Torches are selected for specific applications and amperage ranges. There are torches designed to work in tight spaces, torches with flexible necks, variable angle torches that can be rotated 360°, and general purpose torches with fixed heads. Torches range in capacity from under 50 A for delicate work, up to 300 A for heavy duty welding. Torches can be air-cooled or water-cooled. **See Figure 16-12.**

Air-cooled torches are designed for welding at low duty cycles on thin-gauge metals when low current values are used. They are generally used for welding up

to 200 A. The torch cable for air-cooled torches delivers welding current and shielding gas to the torch. In addition to protecting the tip of the electrode and weld zone from the atmosphere, the shielding gas is used to cool the electrode and the internal components of the torch. For this reason, air-cooled torches are also referred to as gas-cooled torches.

Torches rated above 200 A are typically water-cooled. A water-cooled torch uses a stream of water from a water circulator to maintain a safe torch operating temperature. The water circulator consists of a tank, pump, supply line, and return line. Water is delivered to the torch through the supply line in the torch cable. The water return line is enclosed in the same conduit as the power cable.

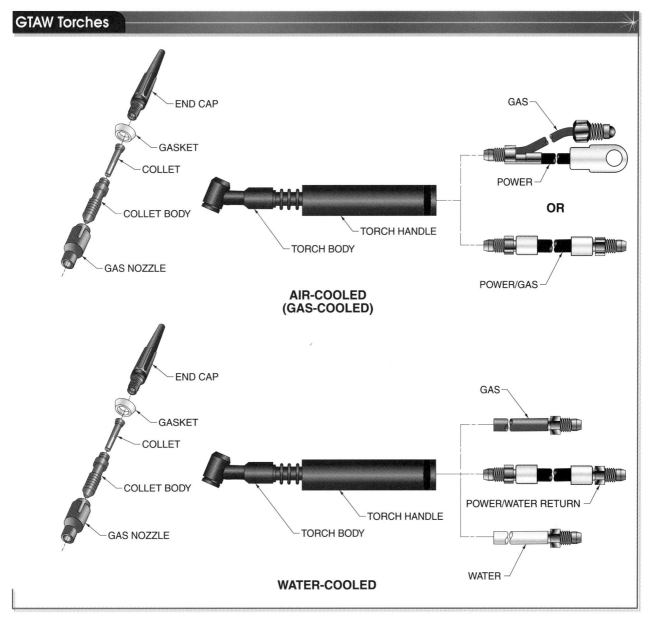

Figure 16-12. *An air-cooled torch is used for welding light-gauge metals. A water-cooled torch prevents overheating when welding requires current above 200 A.*

A GTAW torch consists of a collet, collet body, gas nozzle, and end cap (tail). The collet and collet body are made of copper. The collet body screws into the torch body and delivers shielding gas toward the weld zone during welding. The collet slides into the collet body and holds the electrode. The collet conducts welding current to the electrode. The collet and collet body must be matched to the diameter of the electrode. The end cap screws into the back of the torch body and protects the electrode. The end cap is loosened to adjust the electrode and tightened to hold it in place.

The gas nozzle directs the flow of shielding gas around the electrode and over the weld zone to protect them from the atmosphere. Nozzles are available in a variety of sizes (orifice diameters) and are interchangeable to accommodate a variety of flow rates and coverage requirements. They can be made from heat-resistant ceramic material, metal, or glass. The most commonly used nozzles are made of heat-resistant ceramic. Nozzles are sized in 1/16″ increments. For example, a number 4 nozzle has an orifice diameter of 1/4″. Gas nozzle diameter should be a minimum of three times the electrode diameter.

✓ **Point**

Gas nozzles that are too small for the welding task may overheat, crack, or deteriorate rapidly.

However, the larger the nozzle orifice diameter, the better the gas coverage. Chipped, cracked, or dirty nozzles should be replaced because they can cause shielding problems. **See Figure 16-13.**

Some nozzles are equipped with gas lenses. Gas lenses have several layers of fine stainless steel wire screen that reduce or eliminate turbulence in the gas stream. Turbulence can pull air into the weld zone, causing weld contamination, or diminish shielding of the weld. **See Figure 16-14.**

GAS NOZZLE SIZES		
Metal Thickness	Tungsten Electrode Diameter*	Gas Nozzle Orifice Diameter*
1/16	1/16	1/4 – 3/8
1/8	3/32	3/8 – 7/16
3/16	3/32	7/16 – 1/2
1/4	3/32 or 1/8	1/2 – 3/4

* in in.

Figure 16-13. *The required orifice diameter for a gas nozzle is a minimum of three times the diameter of the electrode.*

Tungsten Electrodes

Electrodes for GTAW are nonconsumable and are made from tungsten. The electrode serves as an electrical conductor to maintain the arc. It does not melt and become part of the weld like electrodes for SMAW, GMAW, and FCAW. Tungsten is ideal for electrodes because at 6100°F, it has the highest melting point of all metals. Tungsten also has a low electrical resistance, which means that it conducts electricity efficiently. The best quality tungsten electrodes have a fine-grained structure. The finer the grain structure, the better they conduct current.

Tungsten electrodes are classified by alloy content as thoriated, lanthanated, ceriated, zirconiated, and pure. Thoriated electrodes are alloyed with thorium oxide (thoria), lanthanted electrodes are alloyed with lanthanum oxide (lanthana), while ceriated and zirconiated electrodes are alloyed with cerium oxide (ceria) and zirconium oxide (zirconia) respectively. Different oxides produce slightly different operating characteristics, but they typically make arc starting easier, improve arc stability, and prolong service life by increasing the operating current of the electrode.

Tungsten Electrode Classification and Identification. All electrodes for GTAW start with the letters "EW". The "E" stands for electrode, and "W" stands for tungsten (from wolfram, which is German for tungsten).

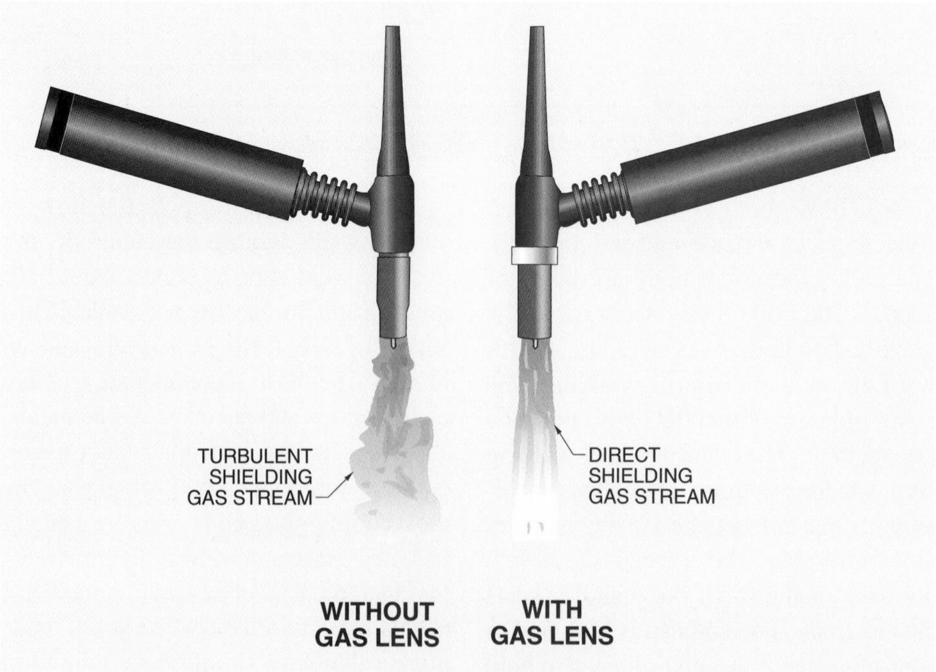

TURBULENT SHIELDING GAS STREAM

DIRECT SHIELDING GAS STREAM

WITHOUT GAS LENS

WITH GAS LENS

Figure 16-14. *A gas lens in a nozzle eliminates turbulence in the shielding gas stream.*

Electrodes are further identified by the element with which they are alloyed. For example, the letters "Th" stand for thoriated, "La" stands for lanthanated, "Ce" stands for ceriated, "Zr" stands for zirconiated, and "P" stands for pure. A number following a dash indicates the percentage of alloy contained in the electrode by weight. For example, the electrode designation EWTh-2 indicates a tungsten electrode that contains 2% thoria by weight, while the designation EWLa-1.5 indicates an electrode with 1.5% lanthana by weight. Electrodes are color-coded for easy identification. **See Figure 16-15.** AWS A5.12/A5.12M *Specification for Tungsten and Tungsten Alloy Electrodes for Arc Welding and Cutting* provides requirements and other data for GTAW electrodes.

ELECTRODES FOR GTAW				
AWS Class	**Electrode Type**	**Color**	**Current**	**Operating Characteristics**
EWTh-1	1% Thoriated	Yellow	DCEN, INVERTER AC	• Easy arc starting, excellent arc performance, long service life • Withstands high amperage • EWTh-2 is most commonly used electrode and recommended for inverters, along with EWLa-2, and EWCe-2 *Note:* Thoria is mildly radioactive. If alternatives (EWLa-1.5 and EWCe-2) are not technically feasible, proper ventilation is required when grinding.
EWTh-2	2% Thoriated	Red		
EWLa-1	1% Lanthanated	Black	DCEN, AC	• Easy arc starting and arc stability • EWLa-1.5 is best non-radioactive alternative to EWTh-2 electrodes • EWLa-2 is better suited to short weld cycles and numerous re-ignitions associated with automated applications than EWTh-2, EWCe-2, and EWLa-1.5 • Primarily used for DCEN, but shows good results in some AC applications • Good service life
EWLa-1.5	1.5% Lanthanated	Gold		
EWLa-2	2% Lanthanated	Blue		
EWCe-2	2% Ceriated	Orange	DCEN, AC	• Easy arc starting and arc stability • Non-radioactive alternative to thoriated tungsten • Good all purpose electrode for AC and DC applications • Good service life
EWZr-1	1% Zirconiated	Brown	AC	• Easier arc starting and better current capabilities than pure tungsten • Recommended over EWP for radiographic-quality welds on aluminum and magnesium • Balls up easily in AC applications • Longer service life than pure tungsten • Creates a clear and stable weld pool for easy visibility
EWP	Pure	Green	AC	• Less expensive than other electrodes • Tends to form balled tip causing arc instability, and difficulty starting arc • Prone to tungsten spitting at higher current difficulty • Creates a clear and stable weld pool for easy visibility • Short service life
EWG	General	Gray	Manufacturer specified	• Vary according to alloy content

Figure 16-15. *Tungsten electrodes are classified by alloy content and are color-coded for identification.*

Media Clip

Tungsten Electrode Preparation

Electrode Size and Current Range. Tungsten electrodes range in length from 3″ to 24″. The most common length is 7″. Electrodes range in diameter from 0.010″ to ¼″. Current-carrying capabilities of electrodes vary with electrode diameter and type of current. **See Figure 16-16.**

Electrode Selection and Preparation. Electrode selection and electrode preparation are important factors affecting weld quality in GTAW. Tip preparation affects arc starting, penetration, the size and shape of the weld bead, and the service life of the electrode.

Thoriated, lanthanated, and ceriated electrodes for use with DCEN are ground to a taper with a blunted tip. The taper is referred to as the included angle and each edge of the taper is used to form the angle. The taper should

be about 2½ to 3 times the electrode diameter. The tip should be blunted to approximately ¹⁄₆₄″. A properly blunted tip produces a narrower arc with better penetration characteristics, reduces the risk of tip erosion, which can cause tungsten inclusions in the weld, and increases the service life of the electrode.

Electrodes should be ground lengthwise so grind marks run parallel with the working direction. Never grind tungsten electrodes crosswise. Grinding across the grain creates ridges on the taper that may melt off, causing tungsten spitting. Tungsten spitting produces tungsten inclusions in the weld. Ridges also interfere with current flow, causing the arc to form before the tip of the electrode. This creates a wide arc with a tendency to wander. **See Figure 16-17.**

CURRENT RANGE FOR TUNGSTEN ELECTRODES*							
Electrode Diameter		DC (in Amperes)		AC (in Amperes)			
		DCEN	DCEP	Unbalanced		Balanced	
in.	mm	EWCe-2 EWLa-X EWTh-X	EWCe-2 EWLa-X EWTh-X	EWP	EWCe-2 EWLa-X EWTh-X EWZr-1	EWP	EWCe-2 EWLa-X EWTh-X EWZr-1
0.010	0.30	up to 15	—	up to 15	up to 15	up to 15	up to 15
0.020	0.50	5-20	—	5-15	5-20	10-20	5-20
0.040	1.00	15-80	—	10-60	15-80	20-30	20-60
¹⁄₁₆	1.60	70-150	10-20	50-100	70-150	30-80	60-120
³⁄₃₂	2.40	150-250	15-30	100-160	140-235	60-130	100-180
⅛	3.20	250-400	25-40	150-210	225-325	100-180	160-250
⁵⁄₃₂	4.00	400-500	40-55	200-275	300-400	160-240	200-320
³⁄₁₆	4.80	500-750	55-80	250-350	400-500	190-300	290-390
¼	6.40	750-1000	80-125	325-450	500-630	250-400	340-525

* Based on the use of argon for shielding

Figure 16-16. *All current values are based on argon shielding gas. Current-carrying capabilities of electrodes vary with electrode diameter and type of current.*

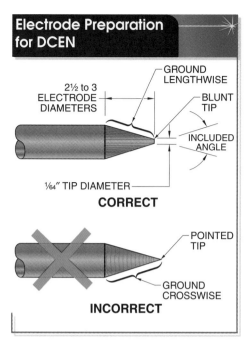

Figure 16-17. *Tapered electrodes for DCEN should be ground lengthwise with the grain, and the tip should be blunted. Grinding across the grain creates ridges that produce a wide arc and may cause tungsten spitting.*

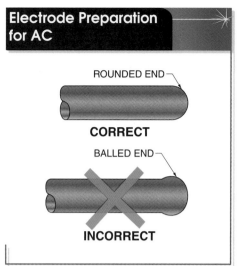

Figure 16-18. *Zirconiated and pure tungsten electrodes for sinusoidal and conventional square wave AC should have rounded tips equal to the diameter of the electrode. A balled tip larger than the diameter of the electrode causes arc instability.*

A fine-grit wheel should be used for grinding. The wheel should be reserved for grinding tungsten only. Grinding tungsten on wheels used for grinding other metals can contaminate the electrode. This can cause problems with arc stability and produce inclusions in the weld. For best results, bench-type grinders or portable tungsten grinders with diamond wheels should be used for accurate, smooth tapers and proper tip preparation.

Zirconiated and pure tungsten electrodes with rounded (hemispherical) tips are used for welding nonferrous metals like aluminum and magnesium with AC. The rounded tip should be equal to the diameter of the electrode. If the rounded tip forms a ball larger than the diameter of the electrode, the arc can wander around the ball, creating arc instability. **See Figure 16-18.**

Zirconiated electrodes are recommended for X-ray quality welds because they have a higher current-carrying capacity than pure tungsten electrodes. Pure tungsten electrodes are prone to tungsten spitting, especially when the current is set too high for the electrode diameter. Tungsten spitting may cause tungsten inclusions in the weld.

Tapered electrodes can be used to weld nonferrous metals with inverters that have amplitude controls for the EP and EN portions of the AC cycle. When the machine is adjusted correctly, the arc forms a ball at the tip of the electrode suitable for the application. **See Figure 16-19.** The prepared end should be opposite the end with the color marking to allow for future identification of the electrode.

Electrode Extension and Stickout. *Electrode extension* on a GTAW torch is the distance from the end of the collet to the tip of the electrode. *Stickout,* in GTAW, is the distance from the end of the nozzle to the tip of the electrode. **See Figure 16-20.** Stickout should be adjusted to allow the arc close enough to the joint to provide sufficient heat for penetration and adequate shielding for the weld area. Stickout is primarily determined by the type of weld joint. In general, stickout for groove welds in butt joints on thin-gauge metal should be ⅛″ to 3/16″. Stickout for fillet welds in T-joints should be 3/16″ to ¼″. In addition, the type of joint affects the retention of shielding gas in the weld area. For example, a T-joint retains shielding gas in the weld area more efficiently than a butt joint. Excessive stickout can result in inadequate shielding of the weld area.

⚠ WARNING

Thorium is mildly radioactive, and appropriate safety precautions should be taken when using thoriated electrodes. The primary health concern associated with thoriated electrodes is the inhalation of grinding dust.

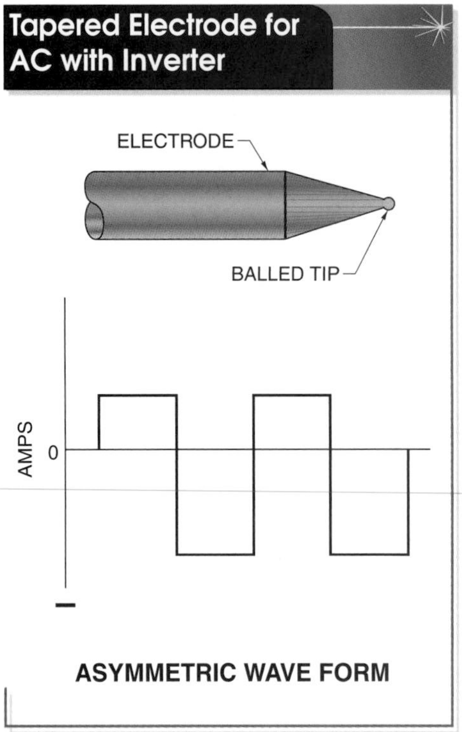

Tapered Electrode for AC with Inverter

ELECTRODE

BALLED TIP

AMPS

0

ASYMMETRIC WAVE FORM

Figure 16-19. *Tapered electrodes can be used for AC applications on inverters with asymmetric waveforms because EP amplitude can be adjusted to direct less heat toward the electrode.*

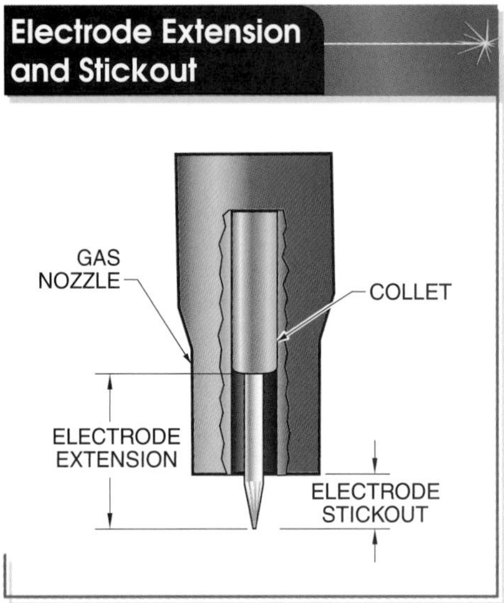

Electrode Extension and Stickout

GAS NOZZLE

COLLET

ELECTRODE EXTENSION

ELECTRODE STICKOUT

Figure 16-20. *Electrode extension is the distance from the end of the collet to the tip of the electrode. Stickout is the distance from the end of the nozzle to the tip of the electrode. Stickout is determined by the type of joint.*

An inert gas is a gas that does not readily combine with other elements. Argon is the most commonly used inert shielding gas.

Shielding Gases

GTAW requires an inert shielding gas to protect the weld pool, the tungsten electrode, and the heated end of the filler metal from oxygen and nitrogen in the atmosphere. An *inert gas* is a gas that does not readily combine with other elements. Argon, helium, or a combination of argon and helium can be used for shielding. The type of shielding gas used and the gas flow rate are determined by the welding current, type of joint, base metal, and welding position. Shielding gas can be supplied from gas cylinders or from bulk storage systems. Typically, cylinders of argon and helium contain approximately 330 cu ft and 290 cu ft respectively, at a pressure of 2000 psi. However, cylinder pressure varies with temperature. **See Figure 16-21.**

Argon. Argon has several advantages over helium, which makes it the most commonly used shielding gas for GTAW. The advantages argon has include the following:

- smoother arc at lower amperage and voltage settings, which is good for welding thin metals and for better weld pool control in the vertical and overhead positions
- easy arc starting because argon has a lower ionization potential (the amount of voltage required to turn a shielding gas into an electrical conductor)
- heavier than air, which means lower gas-flow rates are required, except for overhead position welds; performs better in cross-drafts
- better cleaning action than helium, which is desirable for welding aluminum and magnesium
- less expensive than helium

Helium. Helium is lighter than air, which means that it requires higher flow rates to provide adequate coverage. It has a higher ionization potential than argon, which means arc starting is harder. It is also more expensive than argon. Heliarc welding, an older term used to describe the GTAW process, derives its name from helium used as a shielding gas.

GTAW SHIELDING GASES			
Material	**Type**	**Preferred Gas**	**Remarks**
Aluminum	Manual Welding	Argon	Better arc starting, cleaning action, and weld quality; lower gas consumption
		Helium	Higher welding speed possible
	Machine Welding	Argon-Helium	Better weld quality, lower gas flow rates than straight helium
Magnesium	To 1/16″	Helium	Controlled penetration
	Over 1/16″	Argon	Excellent cleaning, ease of manipulation, low gas flow
Carbon Steel	To 1/8″	Argon	Ease of manipulation, freedom from overheating
	Over 1/8″	Argon	Produces high quality welds
	Spot Welding	Argon	Generally preferred for longer electrode life Better weld nugget contour
		Argon-Helium	Ease of starting, lower gas flow Helium addition improves penetration on thick material
	Manual Welding	Argon	Better weld pool control, especially for out of position welding
Stainless Steel	Machine Welding	Argon	Permits controlled penetration on thin material (up to 14 gauge)
		Argon-Helium	Higher heat input, higher welding speed possible on thicker metal
		Argon-Hydrogen (95%-5%)	Minimizes the risk of undercutting, produces desirable weld contour at low current level, requires lower gas flow rate
		Helium	Provides highest heat input and deepest penetration
Copper & Nickel and their Alloys		Argon	Better weld pool control, good bead contour on thin material
		Argon-Helium	Higher heat input to offset high thermal conductivity
		Helium	Highest heat input for high welding speed on thick material
Titanium		Argon	Low gas flow rate minimizes turbulence and air contamination of weld, improved HAZ
		Helium	Better penetration for manual welding of thick sections (inert gas backing required to shield back of weld against contamination)
Silicon Bronze		Argon	Reduces cracking tendency on cooling ("hot shortness")
Aluminum Bronze		Argon	Shallow penetration of base metal

Figure 16-21. *Argon, helium, or a combination of argon and helium may be used as shielding gases for GTAW. In most cases, backshielding (purging) is required for welding stainless steels, nickel-based alloys, and titanium alloys.*

The advantages helium has for certain applications include the following:

- hotter arc and deeper penetration, which makes it good for welding thicker material
- suitable for metals with high thermal conductivity (the rate at which metal transmits heat)
- more fluid puddle, which allows faster travel
- better coverage in overhead position because helium is lighter than air

Argon-Helium Mixtures. Argon and helium are sometimes combined to produce the increased heat input and deeper penetration of helium along with the easier arc starting, arc stability, and cleaning action of argon.

Postflow. Welding power sources for GTAW are equipped with postflow (post purge) timers. *Postflow* is the flow of shielding gas after the arc is extinguished. Postflow ensures that the weld zone has cooled enough to prevent oxidation from exposure to the atmosphere. The torch should be held in position until the postflow times out. As a general rule, allow 1 sec of postflow for every 10 A of current.

Some machines are also equipped with preflow timers. Preflow ensures that shielding gas is present for arc starting. However, preflow should be used sparingly. If the preflow timer is set too high, it wastes gas and reduces productivity because the arc will not start until preflow times out.

Backing Gas, Trailing Shields, and Atmospheric Chambers. Some metals such as stainless steels, nickel-based alloys, and titanium alloys are extremely sensitive to atmospheric contamination when welded. These reactive metals require special measures to protect the entire weld zone from contamination. Open root welds and seam welds should be protected on the underside by backup bars. The joint or seam is placed over a grooved, copper backup bar. Backing gas is shielding gas provided through a series of ports in the bottom of the backup bar to the back of the joint. **See Figure 16-22.**

Backup Bars and Trailing Shields

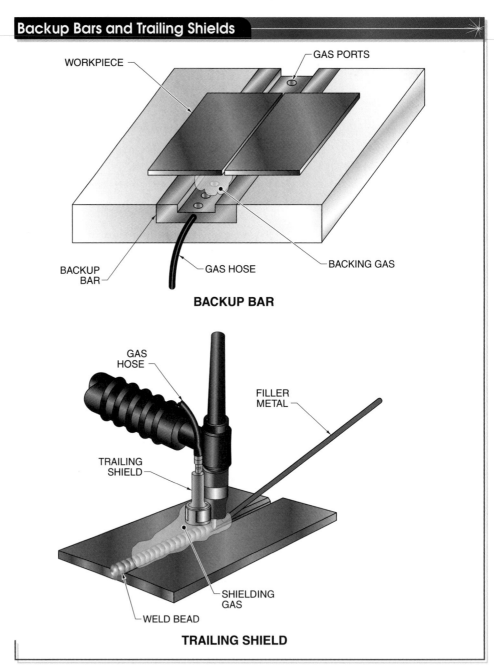

BACKUP BAR

TRAILING SHIELD

Figure 16-22. *Backup bars provide shielding gas (backing gas) to protect the root side of welds on metals sensitive to atmospheric contamination. Trailing shields provide additional coverage for the surface of the weld as it cools.*

A trailing shield attaches to the torch. As the torch is moved along the joint, the trailing shield provides additional coverage for the surface of the weld as it cools.

Atmospheric chambers provide the best protection against atmospheric contamination. The weldment, torch, and filler metal are placed inside the chamber. The chamber is then filled with shielding gas. There are a variety of atmospheric chambers available, including small portable units. **See Figure 16-23.**

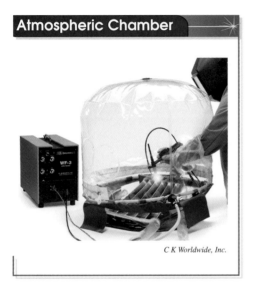

Atmospheric Chamber

C K Worldwide, Inc.

Figure 16-23. *Atmospheric chambers provide the best protection against atmospheric contamination because they surround the entire weldment with shielding gas.*

Regulator Assemblies

Regulators for GTAW are equipped with a pressure gauge for measuring cylinder pressure and a flowmeter that controls gas flow to the torch. Some regulators are also equipped with a purge flowmeter for backing gas and trailing shields. Flowmeters are calibrated to indicate gas flow in cubic feet per hour (cfh) or liters per minute (lpm). A ball inside the glass flowmeter indicates gas flow. The ball rises as the adjustment screw is turned counterclockwise to increase flow. The bottom of the ball indicates the

flow rate. The flow rate required varies depending on the type of shielding gas, joint configuration, and welding position. **See Figure 16-24.**

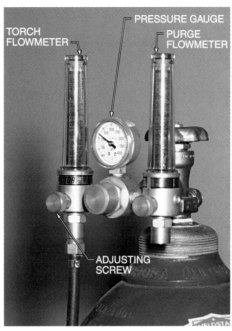

Figure 16-24. *A pressure gauge and flowmeter control the flow of shielding gas to the torch. Some regulators include a second flowmeter to control the flow of shielding gas to backup bars and trailing shields.*

Remote Current Controls

Most power sources for GTAW have a remote current control. Remote current controls can be foot operated, or they may be located on the torch and operated by finger or hand. The remote control starts the flow of electricity and shielding gas. It also initiates high frequency for arc starting. A remote control foot pedal operates like the gas pedal on a car. The farther the pedal is depressed, the more current is delivered to the arc. Heat input is reduced by easing back on the pedal. Current can be increased when starting a weld to aid in penetration and decreased at the end of a weld to fill the crater completely. **See Figure 16-25.**

Media Clip

Vertical Uphill Welding

GTAW FILLER METALS

Certain types of joints on thin-gauge metals, such as lap joints and outside corner joints, can be welded without filler metal.

Figure 16-25. *Remote amperage controls can be foot operated, or they may be located on the torch and operated by finger or hand.*

Filler metal with the similar chemical composition and mechanical properties as the base metal is generally used for GTAW. Thus, low-carbon steel filler metal is used for low-carbon steels, low-alloy steel filler metal for low-alloy steels, aluminum filler metal for aluminum, and stainless steel filler metal for stainless steels.

Filler metals for GTAW contain deoxidizers such as silicon and manganese that produce sound welds. For thin materials, the diameter of the filler metal should be about the same as the thickness of the base metal to be welded, or slightly smaller. Filler metals used for oxyacetylene are not recommended for GTAW because they lack sufficient deoxidizers to produce a quality weld.

These welds are called autogenous welds. An *autogenous weld* is a weld made without the addition of filler metal. Other types of joints, such as T-joints and butt joints, require filler metal to reinforce the joint or to reduce weld cracking. Filler metal is also required on thicker metals to completely fill the joint. Filler metal for GTAW uses the same classification system as filler metal for gas metal arc welding (GMAW). **See Figure 16-26.**

> When welding austenitic stainless steel using GTAW, hydrogen can be added to the shielding gas to reduce oxide formation. Nitrogen can be added to the shielding gas to increase mechanical properties and reduce pitting in super-austenitic and duplex stainless steels.

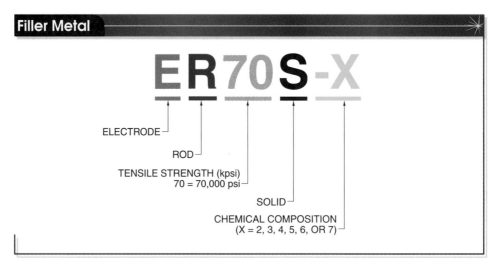

Figure 16-26. *Filler metal for GTAW uses the same classification system as filler metal for gas metal arc welding (GMAW).*

 Refer to Quick Quiz® on CD-ROM

- High frequency (HF) is used for arc starting with DC, and to start and maintain the arc with AC.
- In GTAW, DCEN is used for welding most ferrous metals because it produces deep penetration into the metal.
- DCEP is rarely used in GTAW because it produces shallow penetration and requires a large diameter electrode.
- A water-cooled torch is recommended when using currents over 200 A. Ensure cooling water is flowing before welding.
- Gas nozzles that are too small for the welding task may overheat, crack, or deteriorate rapidly.
- Before starting to weld, ensure that the tungsten electrode has the proper stickout beyond the end of the gas nozzle.
- The diameter of an electrode selected for a welding operation is determined by the required welding current.
- Argon is the most commonly used shielding gas for GTAW applications.

Questions for Study and Discussion

1. In GTAW, what type of welding machines may be used?
2. Why does AC require high frequency?
3. What type of current is commonly used to weld ferrous metals?
4. In GTAW, what results can be expected with respect to heat distribution between the base metal and the electrode with DCEN as compared to DCEP?
5. What determines whether an air-cooled or water-cooled torch is used for a particular welding application?
6. What precautions should be observed when using a water-cooled torch?
7. Why is it important to use the correct size gas nozzle?
8. What determines the size of the tungsten electrode to be used for welding?
9. What is the recommended shape of the tungsten electrode for DCEN and for AC welding with a transformer-rectifier welding machine?
10. What is the function of a flowmeter in a gas regulator assembly?
11. When using filler metal, how should it be manipulated?
12. When is filler metal used in GTAW?

Refer to Chapter 16 in the *Welding Skills Workbook* for additional exercises.

GTAW–Procedures

GTAW can be performed in all positions and produces a minimum of weld spatter. Weld spatter is greatly reduced or eliminated because no metal passes through the arc. Since GTAW produces a smooth weld surface and little or no metal finishing is required, there can be a savings in production cost. In addition, there is less distortion of the metal near the weld. However, production cost savings may be offset by low deposition rates and training for the additional skills necessary to perform GTAW.

GTAW CONSIDERATIONS

When performing GTAW, the base metal, type of weld joint, and welding position must all be considered. Adjustments required for GTAW operations include selecting current type and current setting, selecting and preparing the tungsten electrode, checking the cooling-water flow, selecting the shielding gas, adjusting the shielding gas flow rate, and adjusting the electrode stickout.

GTAW can be performed through one of four basic processes: manual, semiautomatic, mechanized, and automatic. In the manual process, welding is done by hand. GTAW, similar to OFW, is usually performed in the forehand direction with push travel. In the semiautomatic process, the operator controls the speed and direction of travel, while the filler metal is automatically fed into the weld pool.

In the mechanized process, the filler metal feed, weld size, weld length, rate of travel, and starting and stopping are controlled by equipment under the observation and control of the welding operator. In the automatic process, all welding operations are performed without constant observation or adjustment of the controls by the operator.

In GTAW, a shield of inert gas displaces air from the welding area to prevent oxidation of the filler metal, weld pool, electrode, and heat affected zone (HAZ). When GTAW is properly performed, an even ripple pattern is produced. Since GTAW produces no smoke or visible fumes, the welder can clearly observe the weld as it is being made. Additionally, the completed weld requires very little cleaning because GTAW welds are slag-free.

Joint Preparation

Regardless of the type of joint used, proper cleaning of the base metal is essential. All oxidation, scale, oil, grease, dirt, and other foreign matter must be removed by physical or chemical means before welding. Filler metal for GTAW

> ✓ **Point**
> *GTAW can be used for joining many metals and alloys, thicknesses, and joints.*

> ⊕ *AWS A5.12/A5.12M, Specification for Tungsten and Tungsten Alloy Electrodes for Arc Welding and Cutting, specifies which tungsten electrodes may be used with GTAW. Commonly used tungsten electrodes for welding ferrous metals are EWTh-2, EWLa-1.5, and EWCe-2.*

contains deoxidizers, but does not have the fluxing agents that are included in SMAW electrodes to eliminate contaminants. Ideal joint preparation is obtained using cutting tools such as a lathe for round or cylindrical joints, or a milling machine for longitudinal preparations.

Many problems that arise during GTAW are the result of improper joint preparation. Many of these problems are caused by improper grinding. Grinding wheels designed for specific metal types should be used to ensure proper metal preparation prior to welding. Small abrasive particles from grinding wheels can contaminate soft metals such as aluminum, resulting in excessive porosity. Grinding wheels must be thoroughly cleaned on a regular basis. Common joint designs used with GTAW include the butt joint, lap joint, T-joint, corner joint, and edge joint. **See Figure 17-1.**

Butt Joints. For thin metals up to ⅛″, the square butt joint is the easiest to prepare. To ensure complete penetration, a root opening between 1⁄32″ and 1⁄16″, and the use of filler metal is recommended. If a full penetration weld is to be made without filler metal, the joint edges should be properly aligned with a small gap to allow for expansion. Extreme care must be taken to prevent excessive melt-through.

To ensure complete joint penetration, the single-V groove butt joint is used on ferrous metal ranging in thickness from ⅛″ and ¼″ thick. The groove angle should be approximately 60° to 80°, with a root opening of 1⁄32″ to 1⁄16″, and a root face of about 1⁄16″ to 3⁄32″.

When the thickness of the metal exceeds ¼″ and the joint design is such that the weld can be made on both sides, a double-V groove butt joint with a root face of 1⁄16″ is used. A double-V groove butt joint ensures complete penetration, and welding on both sides of the joint helps to control distortion forces. On nonferrous metals, single- and double-U groove joints with 1⁄16″ to 3⁄32″ root faces and no root openings are recommended.

Lap Joints. The only special requirement for making a good fillet weld in a lap joint is that the members should be in close contact along the entire length of the joint. Lap joints on metals ⅛″ or less can be made with or without filler metal. The top member should fuse completely to the lower member. Filler metal is required on lap joints with members between ⅛″ and ¼″ thick. As a rule, lap joints are not recommended on metals thicker than ¼″.

T-Joints. Filler metal must be used to weld T-joints regardless of the thickness of the metal. Generally, continuous fillet welds are made on both sides of a T-joint for added strength and to balance distortion forces. The number of passes required depends on the thickness of the metal and the size of the weld. If complete penetration is required on thicknesses greater than ¼″, the vertical member should be beveled on both sides to produce a double-bevel groove. However, depending on service conditions and the type of metal, the job may call for intermittent fillet welds to minimize heat input and distortion.

Corner Joints. Joint preparation for corner joints depends on the thickness and type of base metal. When welding a corner joint on thin metals up to ⅛″ thick, no filler metal is required. On thicker metal between ⅛″ and ¼″, filler metal should be used. If the metal exceeds ¼″, one edge of the joint should be prepared with a bevel or J-groove. The root face on the prepared edge should be between 3⁄32″ and ⅛″ to allow complete penetration without excessive melt-through. The number of passes required for a corner joint depends on the size of the groove angle and the thickness of the metal.

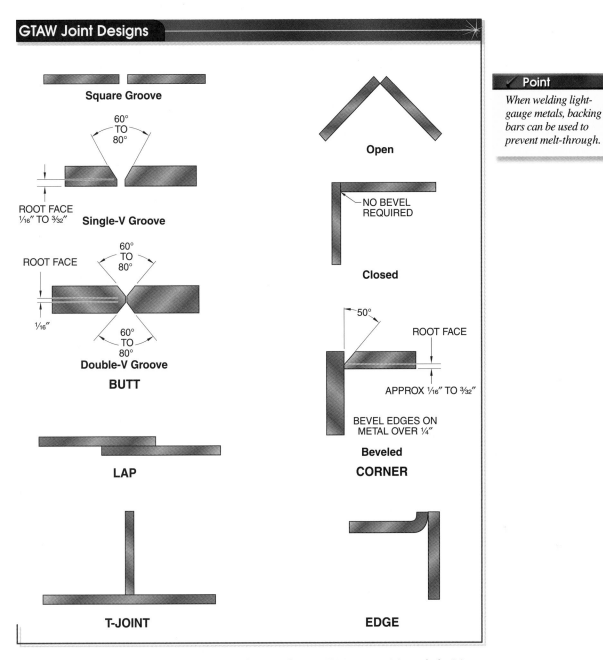

Figure 17-1. *Joint designs used with GTAW include the butt joint, lap joint, T-joint, corner joint, and edge joint.*

Edge Joints. An edge joint is suitable only on thin-gauge metal. No filler metal is needed to weld an edge joint because the two members fuse together to form the weld. Edge joints are used to join parallel, or nearly parallel members.

Weld Backing

Many welding jobs require the use of suitable backing. For butt joints on thin-gauge metals sensitive to atmospheric contamination, backing bars are used to protect the root side of the weld and prevent excessive melt-through. On thick metal, backing bars act as heat sinks. A *heat sink* is a piece of metal that draws some of the heat generated by the arc away from the weld zone.

The type of metal used as a backing bar depends on the metal to be welded. Copper bars are suitable for most applications. A backing bar should be positioned so it does not touch the weld zone. **See Figure 17-2.** Many weld backings are consumed into the weld. Consumable backings (inserts) are the same as those used for pipe welding. Consumable backings should be composed of the same material as the base metal.

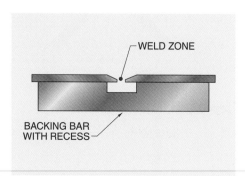

Figure 17-2. *A backing bar should be positioned so that it does not touch the weld zone.*

GTAW PROCEDURES

The procedures for GTAW are similar to those used for OFW. The torch is manipulated to distribute the heat evenly in the weld area. Filler metal, if required, is added to the weld pool using the dip technique or the lay-wire technique.

In the dip technique, the filler metal is dipped into the leading edge of the weld pool and withdrawn. The heated tip of the filler metal must be kept in the flow of shielding gas to prevent oxidation. In the lay-wire technique, the filler metal is held in the joint with a slight downward pressure as the torch is moved with a steady push or with a crescent motion. The travel angle used depends on the type of joint and the welding position. The electrode must be reground to remove contamination if it touches the weld pool or if it comes into contact with the filler metal. The setup procedures for GTAW include the following:

1. Check all electrical connections to make sure they are tight and check the torch and workpiece leads for damage.

2. Connect the remote current control to the welding machine.
3. Ensure the shielding gas cylinder is chained in place, crack the cylinder valve to blow out any debris, and install the regulator and flowmeter assembly.
4. Select the electrode type based on the type of metal to be welded and the type of welding machine (transformer-rectifier or inverter) used.
5. Select the electrode diameter based on the current requirements of the job and prepare the electrode tip.
6. Match the collet and collet body to the diameter of the electrode.
7. Install the gas nozzle. Make sure it is a minimum of three times the diameter of the electrode. Check for dirt or cracks in the nozzle that could interfere with gas flow (follow manufacturer recommendations).
8. Install the electrode and adjust the stickout based on the type of joint. Stickout should be $\frac{1}{8}''$ to $\frac{3}{16}''$ for butt joints and corner joints, and $\frac{3}{16}''$ to $\frac{1}{4}''$ for T-joints. Then tighten the end cap to secure the electrode (but do not overtighten it). **See Figure 17-3.**
9. Set the welding machine to the correct current (DCEN, DCEP, or AC).
10. Turn the welding machine ON, and set the current to the middle of the range required for the job.
11. If the torch is water-cooled, turn the water circulator ON.
12. Open the valve on the shielding gas cylinder to the fully open position, and adjust the gas flow rate.
13. Set the postflow timer. Allow one second for every 10 A. Typically, 15 seconds provides sufficient postflow.
14. Run test welds on scrap material to fine tune machine settings.

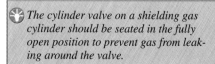

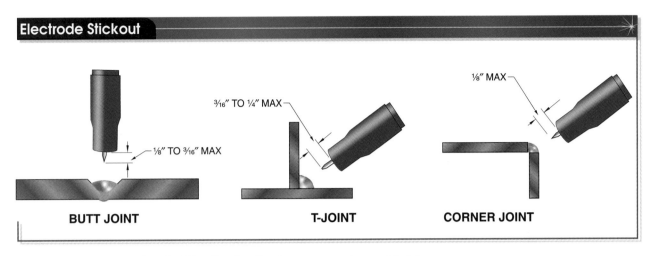

Figure 17-3. *The electrode stickout should be adjusted to allow the arc to access the root of the joint.*

Starting the Arc

There are several methods for starting the GTAW arc. The four most typical starting methods are high frequency, capacitor discharge (CD), scratch, and tap start.

High Frequency (HF) Start. *High frequency (HF) start* is an arc starting method that uses high frequency to create a path for the arc between the electrode and base metal. This makes it possible to start the arc without touching the electrode to the work. The electrode is held about ⅛″ above the joint. When the remote current control is pressed, it starts the flow of shielding gas and initiates the high frequency. The high frequency ionizes the shielding gas and creates a path for the arc to jump the gap between the electrode and the work.

For DC and inverter AC applications, HF can be set to discontinue once the arc is established. For AC applications with transformer-rectifier welding power sources, HF is set to run continuously to stabilize the arc as it transitions between electrode positive and electrode negative. **See Figure 17-4.**

Capacitor Discharge (CD) Start. *Capacitor discharge (CD) start* is an arc starting method that uses a burst of high voltage from a bank of capacitors in the welding power supply. If the power source is equipped with a remote current control, the arc can be started without touching the electrode to the work. The electrode is held in position approximately ⅛″ above the work. When the remote current control is activated, it discharges the capacitors. The burst of high voltage ionizes the shielding gas and allows the arc to jump the gap between the electrode and the work.

If the power source is not equipped with a remote current control, the electrode is touched to the work, creating a dead short. When the system senses a dead short, it charges the capacitors. When the electrode is lifted from the surface of the work, the capacitors discharge, allowing the arc to jump the gap between the electrode and the work.

CD start is typically available on small portable inverters without AC, as well as on DC welding power sources for machine and automatic applications. It is important to set the current to the required setting for the job and to adjust the gas flow rate before starting the arc.

Scratch Start. *Scratch start* is an arc starting method that uses the high open-circuit voltage of the constant-current (CC) welding power source to start the arc when the electrode comes into contact with the base metal. With the current set and the shielding gas adjusted to the proper flow rate, position the electrode approximately 1″ above the work. Lower the torch and touch the electrode to the base metal with a light scratching motion, and quickly withdraw it to a normal arc length. Start the arc in the joint area to avoid leaving an arc strike on the base metal. An *arc strike* is a discontinuity on the surface of the base metal caused by the welding arc.

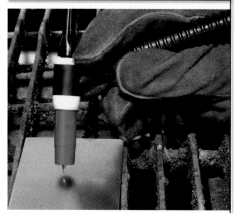

When starting an arc using DC current, set high frequency (HF) to start only, use a capacitor discharge start for inverters without HF, or use a scratch start or tap start if the welding machine is a constant-current machine designed for SMAW.

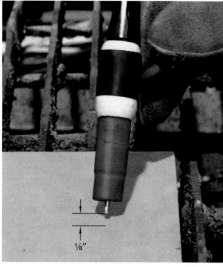

⅛″

When starting the arc using AC, set HF to continuous for transformer-rectifiers or to start only for inverters.

Figure 17-4. *Depending on the type of welding machine, the arc can be started by using a high frequency start, a capacitor discharge start, a scratch start, or tap start.*

Scratch start is used when performing GTAW with a CC welding power source designed for SMAW. It is not considered an appropriate arc starting method for most applications because it can cause tungsten inclusions in the weld.

Tap Start. *Tap start* is an arc starting method that requires the electrode to touch the work. The technique is similar to the scratch start, except the torch is lowered until the electrode touches the

joint, and then withdrawn to a normal arc length. Like the scratch start, it has a high risk of causing tungsten inclusions.

To stop the arc on a welding machine with a remote current control, gradually decrease the current at the end of the weld to fill the crater completely. When the arc extinguishes, hold the torch in place until the postflow timer stops the flow of shielding gas. To stop the arc on welding machines without a remote current control, swing the torch to the horizontal position without touching the electrode to the work.

Procedures for Various Types of Joints

Welding procedures vary for different types of joints. The three typical welding joints encountered are butt, lap, and T-joints.

Butt Joints. When depositing a groove weld in a butt joint, position the torch with a 90° work angle and a 20° push angle. Electrode stickout is ⅛″ to ³⁄₁₆″. Preheat the starting point of the weld by applying enough current to develop a weld pool. As soon as the weld pool becomes fluid, move the torch slowly and steadily along the joint to form a uniform bead. **See Figure 17-5.** To add filler metal, apply the following procedure:

1. With the arc at the rear of the weld pool, add filler metal to the leading edge of the weld pool while maintaining a 15° to 20° angle between the filler metal and the surface of the work.
2. Withdraw the filler metal from the weld pool but keep the end in the flow of shielding gas.
3. Advance the torch to the leading edge of the weld pool.
4. Repeat steps 1 through 3 for the entire length of the weld.
5. At the end of the weld, gradually reduce the current and fill the crater.
6. When the arc extinguishes, hold the torch in position until the postflow times out.

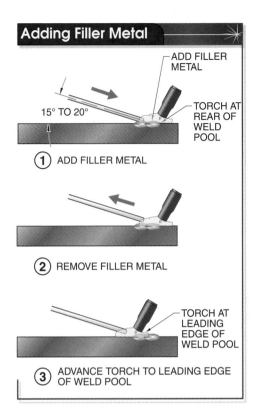

① ADD FILLER METAL

② REMOVE FILLER METAL

③ ADVANCE TORCH TO LEADING EDGE OF WELD POOL

Figure 17-5. *Filler metal is added to the leading edge of the weld pool. The filler metal should not be withdrawn from the flow shielding gas.*

Lap Joints. When depositing a fillet weld in a lap joint, the overlapping members must be in close contact. On metal ⅛″ thick or less, the weld can be made with or without filler metal. GTAW is typically not used to weld lap joints on metal thicker than ¼″. **See Figure 17-6.** For lap joints welded without filler metal, apply the following procedure:

1. Form a weld pool on the bottom member. When the weld pool forms, shorten the arc to about one electrode diameter, then rotate the torch directly over the edge of the top member until the two members fuse together. After welding is started, no further torch rotation is necessary.
2. Use a work angle of 70° to 80° with the electrode positioned over the top edge of the top plate, and a push travel angle of 10° to 15°.
3. Move the torch along the edge of the top member with a steady push. At the end of the joint, reduce the current and hold the torch in place until the postflow times out.

For lap joints welding with filler metal, apply the following procedure:

1. Start the weld pool using the technique described above.
2. Angle the filler metal about 15° to 20° above the bottom plate, and point the tip of the filler metal toward the joint with a 10° to 15° side angle. Use the same work and travel angles as the weld without filler metal.
3. Dip the filler metal into the weld pool when the molten metal forms a V-shaped notch.
4. Withdraw the filler metal from the weld pool and keep it in the flow of shielding gas to protect it from contamination.
5. As the torch moves forward along the joint, dip the filler metal into the weld pool each time the V-shaped notch forms.
6. Travel at a speed that keeps the arc at the back of the notch. Dip the filler metal into the leading edge of the puddle about every ¼″ of travel. Fill the notch completely along the length of the joint to ensure complete fusion and penetration, as well as to produce a uniform bead shape. Do not advance the torch ahead of the notch.
7. Reduce current at the end of the weld and fill the crater.
8. Hold the torch in place until the postflow times out.

> ✓ **Point**
>
> *When using GTAW in overhead position, reduce the current 5% to 10% from what is used for flat position.*

Miller Electric Manufacturing Company
Since GTAW produces no smoke or visible fumes, the welder can clearly observe the weld as it is being made.

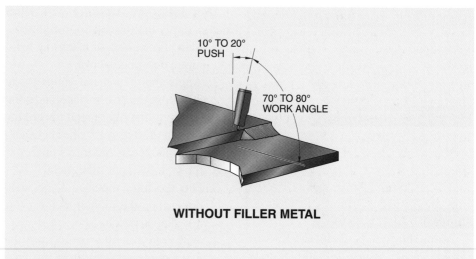

10° TO 20°
PUSH

70° TO 80°
WORK ANGLE

WITHOUT FILLER METAL

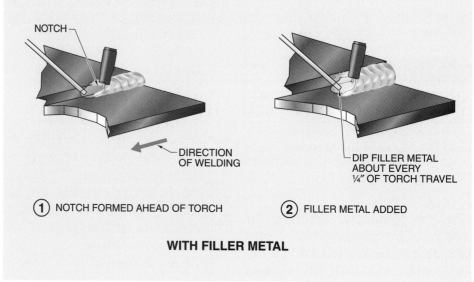

NOTCH

DIRECTION
OF WELDING

DIP FILLER METAL
ABOUT EVERY
¼″ OF TORCH TRAVEL

① NOTCH FORMED AHEAD OF TORCH

② FILLER METAL ADDED

WITH FILLER METAL

Figure 17-6. *When welding without filler, use a steady push and melt the edge of the upper plate to form the weld. When welding with filler metal, advance the torch so that the notch in the weld bead continues to form ahead of the torch. Dip filler metal into the leading edge of the weld pool.*

T-Joints. Filler metal is required when depositing fillet welds in T-joints, regardless of the thickness of the metal. To deposit a fillet weld in a T-joint, apply the following procedure:

1. Use a work angle of 45° from the bottom plate, with a 10° to 20° push angle.
2. Ensure that the electrode stickout is ³⁄₁₆″ to ¼″ to access the root of the joint.
3. Point the filler metal toward the root of the joint, with a 15° to 20° angle above the bottom plate and a 15° to 20° side angle.
4. Use the welding technique used to weld lap joints with filler metal.

Fillet welds in T-joints can also be deposited using the lay-wire technique. Excessive heat or traveling too slow can cause concavity. *Concavity* is an indentation on the side of the joint opposite the weld. This is also referred to as suck-back.

Flat and Horizontal Position Procedure

For flat (1G) and horizontal (2G) groove welds in butt joints with GTAW, start the arc about ½″ from the edge of the joint. Once the arc is started, move the torch to the edge of the joint and begin welding. Hold the torch at a work angle of 90° for 1G welds, and 85° to 90° for 2G welds, with a push angle of 10° to 15°. Arc length should be approximately one electrode diameter. The filler metal should be angled about 15° to 20° above the joint. Dip the filler metal into the front of the weld pool. For horizontal groove welds, dip the filler metal on the high side of the weld pool as the torch is advanced along the joint to help prevent undercutting. **See Figure 17-7.**

Watch the weld pool closely. Reduce current as necessary with the remote current control to prevent the weld pool from sagging on 2G welds and to prevent excessive melt-through on 1G welds. At the end of the weld, reduce the current and fill the crater. Hold the torch in position until the postflow times out. If the welding machine is not equipped with a remote current control, withdraw the torch slightly if the weld pool becomes too fluid.

Vertical Welding Procedure

Uphill welding is preferred for GTAW since deeper penetration can be achieved. Uphill welding generally requires filler metal. However, lap joints and outside corner joints on metals up to ⅛″ can be welded without filler metal.

One problem associated with uphill travel is heat buildup in the weld zone. The weld pool has a tendency to become too fluid and roll out of the joint. In order to prevent this, the welder must watch the weld pool carefully and reduce the current as necessary to maintain a consistent weld bead.

Another common problem with uphill travel is using too large a travel angle. This directs too much heat to the end of the filler metal, causing it to melt prematurely. The proper work angle and travel angle must be maintained for the entire length of the joint. **See Figure 17-8.**

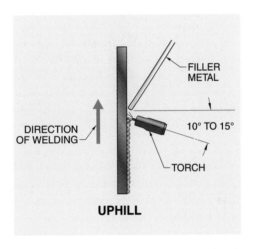

Figure 17-8. *The proper work angle and travel angle must be maintained when performing uphill welding.*

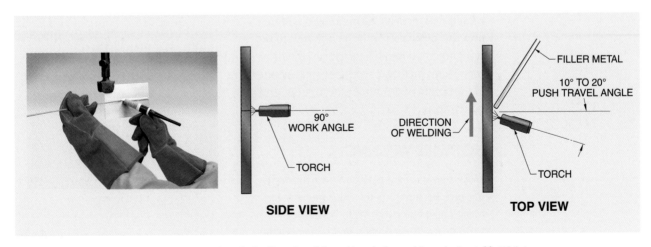

Figure 17-7. *Dip the filler metal into the high side at the leading edge of the weld pool when welding a horizontal butt joint.*

Overhead Welding Procedure

When welding with GTAW in the overhead position, the current should be reduced 5% to 10% from what is normally used for flat position. A reduced current provides better control of the weld pool. Both the torch and filler metal should be held as in flat position welding. A work angle of 90° and a push angle of 5° to 10° should be maintained. **See Figure 17-9.**

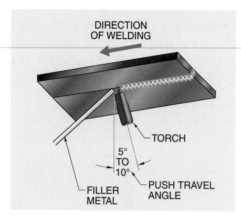

Figure 17-9. *The torch and filler rod angles in the overhead position are similar to their angles in flat position.*

Dip the filler metal in and out of the weld pool as in other welding positions. Like uphill travel in the vertical position, heat buildup in the weld zone can be a problem with overhead welding. If the weld pool becomes too large, gravity can cause it to roll out of the joint, causing incomplete penetration. Therefore, the current should be reduced as necessary to maintain a consistent weld bead. A small weld bead is advisable since it is less affected by gravity. Correct torch angles are also critical. Too large a travel angle will direct too much heat toward the filler metal, and incorrect work angles can create problems with undercut, fusion, and penetration.

HOT WIRE WELDING

Hot wire welding is a gas tungsten arc welding process in which the filler metal is preheated as it enters the weld pool. Hot wire welding produces quality welds at about the same speed as GMAW. In hot wire welding

systems, the filler metal is automatically fed from a wire feeder that runs to a hot wire torch mounted behind the GTAW torch. A *wire feeder* is a welding machine accessory that holds a filler metal spool and allows it to be fed to the hot wire torch as welding progresses. Filler metal is resistance-heated by an AC current that passes from an AC welding machine through the filler metal. The welding machine is regulated so the filler metal reaches its melting point as it enters the weld pool.

By attaching the hot wire torch behind the GTAW welding torch, the operator is given an unobstructed view of the weld. By preheating the filler metal, weld porosity is eliminated. Welds are made with greater quality and speed. Hot wire welding is a rapid and efficient welding process in many fabrication situations.

PULSED GTAW (GTAW-P)

Pulsed GTAW (GTAW-P) is a gas tungsten arc welding variation in which direct current is pulsed between a peak current (high pulse) and a background current (low pulse). Welding is done on the high pulse. The low pulse maintains the welding arc and allows the weld pool to cool slightly between pulses. The main advantage of pulsed direct current is that peak current can be set higher than continuous direct current. This produces deep penetration with less overall heat input, making it easier to control the weld pool. Pulsed GTAW is ideal for thin-gauge metals and for welding out of position. The resulting weld looks like a series of overlapping spot welds. **See Figure 17-10.**

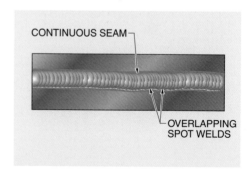

Figure 17-10. *Pulsed GTAW (GTAW-P) produces overlapping spot welds that form a continuous seam.*

Pulsed GTAW can be manual or automatic and can be used with or without filler metal depending on joint type. The process can be used for welding very thin metals where critical control of metallurgical factors is necessary, such as austenitic stainless steel. A pulsed current permits more tolerance of edge misalignment, better root penetration, and less distortion.

Pulsed GTAW is well suited to welding in the vertical and overhead positions because it allows such precise control of the weld pool. When welding curved seams or pipes, pulsed GTAW allows continuous welding without having to vary travel speed, voltage, or current. The four basic controls for setting pulsed current parameters are peak current, background current, pulses per second, and percent ON time. **See Figure 17-11.**

Peak current is the high current setting for the pulse cycle. It is usually set higher than the current for non-pulsed GTAW welding. *Background current* is the current setting for the low pulse, which maintains the arc. *Pulses per second* is the number of times per second that the current achieves peak current. Higher pulsing rates, in the hundreds of pulses per second, are possible with inverter GTAW welding machines. These higher pulsing rates can increase weld pool fluidity, allowing increased travel speeds on fusion welds, improved ability to handle surface contaminates, and improved joint fit-up tolerance. *Percent ON time* is the length of time that peak current is maintained before it drops to background current. Percent ON time is also referred to as peak time.

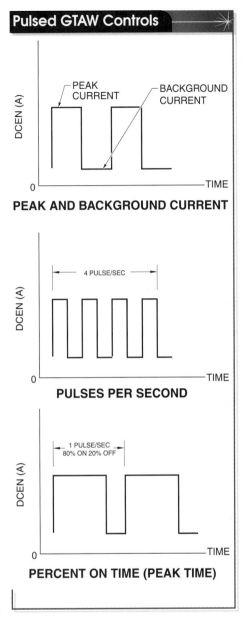

Figure 17-11. *A pulse controller has separate adjustments for peak current, background current, pulses per second, and percent ON time (peak time).*

Refer to Quick Quiz® on CD-ROM

- GTAW can be used for joining many metals and alloys in various thicknesses and for various types of joints.
- When welding light-gauge metals, backing bars can be used to prevent excessive melt-through.
- When using a water-cooled torch, ensure that the water is ON before welding.
- When using GTAW in overhead position, reduce the current 5% to 10% from what is used for flat position.

Exercises

Depositing Beads on Low-Carbon Steel in Flat Position

exercise **1**

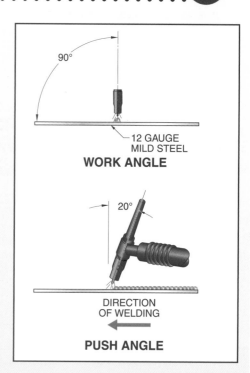

WORK ANGLE

12 GAUGE MILD STEEL

90°

PUSH ANGLE

20°

DIRECTION OF WELDING

1. Obtain a ³⁄₃₂″, EWTh-2, EWLa-1.5, or EWCe-2 electrode and prepare a tapered tip with a blunted end.

2. Insert the electrode in the torch and adjust the stickout ⅛″ to ³⁄₁₆″ beyond the end of the gas nozzle.

3. Set the welding machine output to DCEN. If the welding machine is equipped with high-frequency, set it for start only. Connect the remote current control and turn the power switch on the welding machine to ON.

4. Set the shielding gas (argon) at 15 cfh to 20 cfh (cubic feet per hour) with a postflow time of 10 sec to 15 sec.

5. Set the current at 80 A to 95 A.

6. Obtain a piece of 12-gauge low-carbon steel, 4″ wide and 6″ long, and place it in the flat position.

7. Position the electrode ⅛″ from the workpiece with a 90° work angle and a 20° push angle.

8. Press the foot pedal or use the hand-operated remote current control on the torch to start the flow of shielding gas and initiate the arc.

9. Form a weld pool about ⅛″ wide. Maintain a consistent arc length, and travel with a steady push to maintain a consistent bead across the workpiece. Adjust current as necessary to control heat input.

10. Reduce current near the end of the weld. Release the foot/hand control to extinguish the arc, and hold the torch in place until the postflow times out.

11. Deposit a series of straight, consistent beads on the workpiece approximately ³⁄₈″ apart.

Depositing Beads with Filler Metal on Low-Carbon Steel in Flat Position

exercise 2

1. Complete equipment setup and adjustment from steps 1 through 5 in Exercise 1.

2. Obtain a piece of 12-gauge low-carbon steel, 4″ wide and 6″ long, and place it in the flat position.

3. Obtain the recommended filler metal for low-carbon steel.

4. Position the electrode about ⅛″ from the workpiece with a 90° work angle and a 20° push angle. The filler metal is angled about 15° to 20° above the workpiece with the tip pointing toward the electrode.

5. Press the foot pedal or use the hand-operated remote current control on the torch to start the flow of shielding gas and initiate the arc.

6. Form a weld pool and dip the filler metal into the leading edge of the pool using an in-and-out motion. Do not touch the filler metal to the tungsten electrode and do not remove the tip of the filler metal from the flow of shielding gas.

7. Use a small circular motion with the torch. Form a weld approximately ³⁄₁₆″ wide, and maintain consistent bead width along the length of the workpiece.

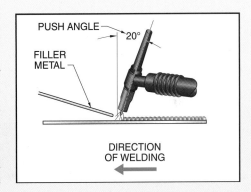

8. Reduce current near the end of the weld and add filler metal to fill the crater at the end of the bead. Release the foot/hand control to extinguish the arc. Hold the torch in place until the postflow times out.

9. Deposit a series of straight, consistent beads approximately ⅜″ apart. Allow the plate to cool briefly between welds to avoid overheating, or use two plates and alternate between them.

Welding a Butt Joint on Low-Carbon Steel in Flat Position...

exercise 3

1. Complete equipment setup and adjustment from steps 1 through 5 in Exercise 1.

2. Obtain two pieces of 12-gauge low-carbon steel, 1½″ wide and 6″ long.

3. Form a butt joint with a ³⁄₃₂″ root opening and tack together.

4. Position the workpiece so the weld joint is in flat position.

5. Obtain the recommended filler metal for low-carbon steel.

6. Position the electrode about ⅛″ from the workpiece with a 90° work angle and a 20° push angle.

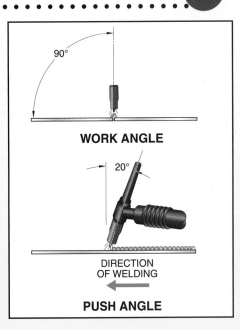

WORK ANGLE

DIRECTION OF WELDING

PUSH ANGLE

7. Use the foot/hand current control to start the flow of shielding gas and initiate the arc, and establish a weld pool.

8. Dip the filler metal into the leading edge of the weld pool using an in-and-out motion. Do not touch the filler metal to the tungsten electrode and do not remove the tip of the filler metal from the flow of shielding gas.

9. Use a small circular motion with the torch. Form a consistent bead, approximately ³⁄₁₆″ wide, and deposit the weld along the entire length of the joint.

10. Reduce current near the end of the weld and add filler metal to fill the crater at the end of the bead. Release the foot/hand control to extinguish the arc and hold the torch in place until the postflow times out.

11. The resulting weld should have complete penetration through to the root side, with a ³⁄₁₆″ bead width.

12. Prepare additional joints to practice as necessary.of the weld.

Welding a Lap Joint on Low-Carbon Steel in Horizontal Position...

exercise **4**

1. Complete equipment setup and adjustment from steps 1 through 5 in Exercise 1.

2. Obtain two pieces of 12-gauge low-carbon steel, 1½″ wide and 6″ long.

3. Form a lap joint and tack together.

4. Position the workpiece so the weld joint is in horizontal position.

5. Obtain the recommended filler metal for low-carbon steel.

6. Hold the torch at a 70° to 80° work angle and a 15° to 20° push angle. Aim the electrode into the root of the joint. Position the filler metal at a 20° angle above the bottom plate, with a 10° to 15° side angle.

7. Start the arc and form a weld pool by melting the top edge of upper plate. Add filler metal each time a V-shaped notch forms in the weld pool, withdraw the filler metal, and move the torch to the leading edge of the weld pool. Do not touch the filler metal to the tungsten electrode, and do not remove the tip of the filler metal from the flow of shielding gas.

8. Maintain a consistent bead along the length of the joint.

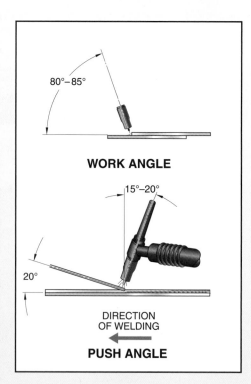

80°–85°

WORK ANGLE

15°–20°

20°

DIRECTION
OF WELDING

PUSH ANGLE

9. Reduce current near the end of the weld and add filler metal to fill the crater at the end of the bead. Release the foot/hand control to extinguish the arc and hold the torch in place until the postflow times out. Weld the joint on the other side of the workpiece.

10. Repeat the exercise on another lap joint using the lay-wire technique. Use the same torch and filler metal angles. Hold the filler metal in the root of the joint with a slight downward pressure.

11. Repeat the exercise on another lap joint without filler metal. Use the same torch angles. Position the electrode over the top edge of the upper plate and use steady push travel to produce a consistent fillet weld.

12. Prepare additional joints to practice as necessary.

Welding a T-Joint on Low-Carbon Steel in Horizontal Position

exercise **5**

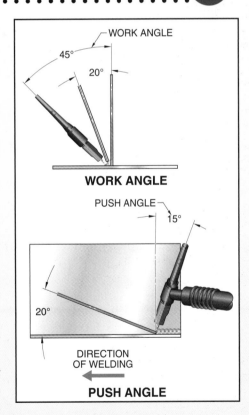

WORK ANGLE

WORK ANGLE
45°
20°

PUSH ANGLE
15°

20°

DIRECTION
OF WELDING

PUSH ANGLE

1. Complete equipment setup and adjustment from steps 1 through 5 in Exercise 1.

2. Obtain two pieces of 12-gauge low-carbon steel, 2″ wide and 6″ long.

3. Form a T-joint with the pieces at a 90° angle and tack together.

4. Position the workpiece so the weld joint is in horizontal position.

5. Obtain the recommended filler metal for low-carbon steel.

6. Hold the torch at a 45° work angle and a 15° push angle. Position the filler metal at a 20° angle from the bottom plate and a 15° to 20° side angle.

7. Establish a weld pool. Weave the torch slightly and add filler metal to the leading edge of the weld pool using an in-and-out motion.

8. Reduce current as necessary to prevent excessive heat buildup on the vertical member.

9. Reduce current near the end of the weld and add filler metal to fill the crater at the end of the bead. Release the foot/hand control to extinguish the arc and hold the torch in place until the postflow times out. Deposit a fillet weld on the other side of the joint.

10. Repeat the exercise on another T-joint using the lay-wire technique. Use the same torch and filler metal angles. Hold the filler metal in the root of the joint with a slight downward pressure.

11. Prepare additional joints to practice as necessary.ing an in-and-out motion, add filler metal to the leading edge of the weld pool.

1. Complete equipment setup and adjustment from steps 1 through 5 in Exercise 1.

2. Obtain two pieces of 12-gauge low-carbon steel, 2″ wide and 6″ long.

3. Form a T-joint with the pieces at a 90° angle and tack together.

4. Position the workpiece so the weld joint is in vertical position.

5. Obtain the recommended filler metal for low-carbon steel.

6. Weld uphill. Start at the bottom of the joint. Hold the torch at a 45° work angle and a 20° push angle. Center the filler metal on the root of the joint with a 20° angle above the plate and a 15° to 20° side angle. Add filler metal to the leading edge of the weld pool using an in-and-out motion.

DIRECTION OF WELDING

7. Reduce current as necessary to prevent excessive heat buildup in the joint.

8. Reduce current near the end of the weld and add filler metal to fill the crater at the end of the bead. Release the foot/hand control to extinguish the arc and hold the torch in place until the postflow times out. Deposit a fillet weld on the other side of the joint.

9. Prepare additional joints to practice as necessary.

1. Obtain a ⅛″ EWP or a ³⁄₃₂″ EWZr-2 tungsten electrode and insert it into the torch. Adjust the stickout ⅛″ to ³⁄₁₆″ and prepare a spherical tip. To shape the tip, set the welding machine to DCEP. Position the torch at a 90° angle and strike an arc on a piece of copper. Release the foot/hand current control when the sphere at the tip is equal to the diameter of the electrode.

2. Turn the power switch OFF, and set the welding machine output to AC. High frequency should be set to continuous.

3. Turn the power switch ON, and set the shielding gas (argon) to 20 cfh with a postflow time of 15 sec.

4. Set the current at 120 A to 150 A.

5. Obtain a piece of ⅛″ aluminum, 4″ wide and 6″ long, and place it in the flat position.

6. Brush the plate with a clean stainless steel brush to break up the oxide coating and to remove any surface contamination.

7. Position the electrode about ⅛″ above the plate, with the torch at a 90° work angle and a 20° push angle.

8. Start the arc using the foot/hand current control, and form a weld pool about ¼″ wide.

9. Travel with a steady push, and add filler metal to the leading edge of the weld pool with an in-and-out motion. Adjust the current as necessary with the foot/hand control to maintain a consistent bead.

10. Fill the crater at the end of the weld by reducing current with the foot/hand control and continuing to add filler metal. Hold the torch in place until the postflow times out.

11. Deposit a series of straight, consistent beads on the workpiece approximately ⅜″ apart. Allow time for the plate to cool between welds, or use two plates and alternate between them.

Welding Joints on Aluminum in Flat Position

1. Complete equipment setup and adjustment in Exercise 7.

2. Obtain six pieces of ⅛″ aluminum.

3. Tack up a butt joint with a ³⁄₃₂″ root opening, a lap joint, and a T-joint.

4. Position the butt joint on the work surface so the joint is in the flat (1G) position. To position the lap and T-joints in the flat (1F) position, fixture the workpieces at a 45° angle.

5. For the butt joint, use a 90° work angle and a 20° push angle. Angle the filler metal about 15° to 20° above the joint. Position the electrode about ⅛″ from the workpiece.

6. For the lap joint, use a 70° to 80° work angle from the bottom plate and a 15° to 20° push angle. Aim the electrode into the root of the joint. Position the filler metal at a 20° angle above the bottom plate, with a 10° to 15° side angle.

7. For the T-joint, use a 45° work angle and a 15° push angle. Position the filler metal at a 20° angle from the bottom plate, and a 15° to 20° side angle.

8. Brush each joint thoroughly with a clean stainless steel wire brush immediately before welding.

9. Prepare additional joints to practice as necessary. Practice depositing fillet welds in lap joints with and without filler metal.

Welding Lap and T-Joints on Aluminum in Horizontal Position

1. Complete equipment setup and adjustment in Exercise 7.

2. Obtain four pieces of ⅛″ aluminum.

3. Tack up a lap joint and a T-joint.

4. Place the workpieces flat on the work surface so the joints are in the horizontal (2F) position.

5. The torch and filler metal angles are the same as the angles for the flat position.

6. Brush each joint thoroughly with a clean stainless steel wire brush immediately before welding.

7. Prepare additional joints to practice as necessary. Practice depositing fillet welds in lap joints with and without filler metal.

Welding Lap Joints on Stainless Steel in Horizontal Position

1. Complete equipment setup and adjustment from steps 1 through 4 in Exercise 1.

2. Set the current at 40 A to 50 A.

3. Obtain several pieces of 16-gauge 308 austenitic stainless steel, 1″ wide and 4″ long, and thoroughly clean them.

4. Tack up lap joints and place them flat on the welding surface so the joints are in the horizontal position.

5. Weld the first joint without filler metal. Position the electrode over the top edge of the upper plate, with a 70° to 80° work angle, and a 10° to 15° push angle.

6. Start the arc and use a steady push travel angle to produce a consistent weld. Hold the torch in position at the end of the weld until the postflow times out. Turn the workpiece over, and weld the other joint.

7. Weld a new workpiece with filler metal. Use the same torch angles. Angle the filler metal 20° above the bottom plate, with a 10° to 15° side angle. Be careful not to withdraw the filler metal from the flow of shielding gas.

8. Weld another workpiece without filler metal and with pulsation. Move the electrode to the leading edge of the weld pool during the low pulse, and hold the torch in position for the high pulse. *Note:* Consult manufacturer recommendations for pulse controller settings.

9. Weld a workpiece with filler metal and with pulsation. Use the same torch technique. Dip the filler metal into the leading edge of the weld pool during the high pulse.

10. Prepare additional joints to practice as necessary. Practice autogenous welds with and without pulsation. Also, practice welds with filler metal using the dip and lay-wire techniques with and without pulsation.

Welding T-Joints on Stainless Steel in Horizontal, Vertical, Flat Position

1. Complete equipment setup and adjustment from steps 1 through 4 in Exercise 1.

2. Set the current at 40 A to 50 A.

3. Obtain several pieces of 16-gauge 308 austenitic stainless steel, 1″ wide and 4″ long, thoroughly clean them, and tack them up to form T-joints.

4. For horizontal welding, place a workpiece on the work surface so the joint is in the horizontal (2F) position. Use a 45° work angle and a 15° push angle. Position the filler metal at a 20° angle from the bottom plate, and a 15° to 20° side angle.

5. For vertical uphill welding, fixture the workpiece so the joint is in the vertical (3F) position. Use the same torch and filler metal angles relative to the joint.

6. For flat welding, fixture the workpiece at a 45° angle so the joint is in the flat (1F) position.

7. The welding technique is the same for all positions, however heat buildup can be a problem in the vertical position. Electrode stickout should be adjusted to ensure that the arc can access the root of the joint.

8. Practice using the dip and lay-wire techniques with and without pulsations.

9. Prepare additional joints to practice as necessary.

1. Complete equipment setup and adjustment from steps 1 through 4 in Exercise 1.

2. Set the current at 35 A to 45 A.

3. Obtain several pieces of 16-gauge 308 austenitic stainless steel, 1″ wide and 4″ long, thoroughly clean them, and tack them up to form butt joints with 1/32″ root openings.

4. Obtain a backup bar to provide shielding gas to the root side of the weld. Connect the gas hose and adjust the gas flow rate on the purge flowmeter between 5 cfh and 10 cfh.

5. For the flat (1G) position, place the backup bar on the work surface and insert a workpiece. Use a 90° work angle and a 20° push angle. Angle the filler metal about 15° to 20° above the joint.

6. Start the arc and form a weld pool. Dip the filler metal into the leading edge of the weld pool. Stainless steel is very susceptible to atmospheric contamination, so it is important to hold the torch in position at the end of the weld until the postflow times out.

7. Practice using the dip and lay-wire techniques with and without pulsation.

8. For the horizontal (2G) position, fixture the backup bar so that the joint is horizontal. Use the same torch and filler metal angles as the flat position.

9. Prepare additional joints to practice as necessary.

Questions for Study and Discussion

1. What does GTAW stand for?
2. What are some of the advantages of GTAW compared to other welding processes?
3. How is the arc started and stopped in GTAW?
4. What is the proper torch angle for welding a butt joint?
5. What is hot wire welding?
6. How are welds produced by the pulsed GTAW process?
7. What kinds of metals can be welded with the GTAW process?
8. Why is joint cleanliness more important with GTAW as compared to SMAW?

Refer to Chapter 17 in the *Welding Skills Workbook* for additional exercises.

GTAW–Applications

Chapter

GTAW is used where accurate control of weld penetration and weld purity are critical and a spatter-free weld is required. Common applications of GTAW are on carbon steels, stainless steels, aluminum, magnesium, copper and copper alloys, and other metals that cannot be welded satisfactorily using other welding processes.

GTAW is commonly used for joining metals in the aerospace and aircraft industries. The high degree of control of the GTAW arc permits welding on very thin metal with minimal distortion and/or alteration of base metal properties. GTAW is also used when welding pressure vessels and critical piping systems, such as systems in nuclear power plants, because of its weld penetration and purity.

CARBON STEEL

Carbon steel can be welded using a variety of welding processes. GTAW is used for welding low- and medium-carbon and low-alloy steels when greater protection of the weld from atmospheric contamination is required. When GTAW is used on carbon steels without filler metal, there may be some pitting (porosity) in the weld. When filler metal is used, it should contain deoxidizers to prevent porosity and should meet suggested filler metal recommendations. Filler metal for GTAW is the same as the filler metal used for gas metal arc welding (GMAW).

Medium- and high-carbon steels are weldable, but preheat, special welding techniques, and postheating are required. These requirements become increasingly important as the carbon content of the steel increases. Unless these precautions are taken, the welded area loses toughness and ductility and fails by cracking. **See Figure 18-1.**

> ✓ **Point**
>
> *Filler metal containing deoxidizers should be used when welding with GTAW to prevent porosity in the weld.*

> ✓ **Point**
>
> *Medium- and high-carbon steels require preheat and post-heating to avoid loss of toughness and ductility.*

Metal Thickness*	DCEN†	Argon Flow‡		Filler Metal Diameter*
		lpm	cfh	
1/16	100 – 140	4 – 5	20	1/16
1/8	120 – 200	4 – 5	20	1/16
1/4	150 – 250	4 – 5	25	1/8
1/2	150 – 300	4 – 5	25	1/8

GTAW—CARBON STEEL

* in in.
† amps
‡ 20 cfh

Figure 18-1. *Welding parameters should be set based on carbon steel thickness.*

STAINLESS STEEL

Stainless steels, especially those in the 300 austenitic series, are easy to weld with GTAW using DCEN. GTAW is particularly adaptable for welding thin-gauge stainless steel and high-pressure stainless steel piping.

The procedure for welding all types of stainless steels is the same. Filler metals for welding stainless steels are alloyed to prevent cracking problems. When welding without filler metals, care must be taken to prevent cracking. Welding parameters such as proper current, electrode diameter, shielding gas flow rate, and filler metal diameter, should be set based on the thickness of the stainless steel. **See Figure 18-2.**

ALUMINUM

Aluminum is a nonferrous metal. A *nonferrous metal* is a metal that contains no iron. Many types of pure and alloyed aluminums are available. Each type of aluminum has specific properties for specific uses.

Nonheat-treatable wrought aluminum alloys in the 1000, 3000, 4000, and 5000 series are readily weldable. Heat-treatable alloys in the 2000, 6000, and 7000 series can be welded, but higher welding temperatures and welding speeds are required. Weld cracking in alloys can be minimized by using filler metal that has a higher alloy content than the base metal. However, heat-treatable aluminum alloys above 7039 are not weldable.

Welding can be performed in any position. However, welding is easier if it can be done in the flat position. Copper backing bars should be used as heat sinks whenever possible to minimize distortion, especially on thin-gauge metal 1/8″ thick or less. In most cases, the torch should be moved in a straight line without a weaving motion. Best results are obtained by using AC current with argon as a shielding gas. Welding parameters such as current, electrode diameter, argon flow rate, and filler metal

diameter are based on the thickness of the aluminum and the welding position. **See Figure 18-3.**

Aluminum forms a thin oxide layer when exposed to the atmosphere. The oxide layer that forms on aluminum has a much higher melting point than the aluminum to be welded and must be removed before fusion can occur. AC produces a cleaning action that removes the oxide layer. During the electrode positive portion of the AC cycle, gas ions bombard the surface of the aluminum with enough force to remove the oxide layer. However, the joint should be brushed with a stainless steel wire brush to break up the oxide before welding. The oxide layer can also be removed with a chemical cleaner or by filing.

MAGNESIUM

The welding characteristics of magnesium are comparable to those of aluminum. Both have high thermal conductivity, low melting points, high thermal expansion, and rapid oxidization. One difference is that magnesium must be preheated before welding, and slow cooled after welding to prevent cracking.

With GTAW, several current variations are possible. AC used with helium, argon, or an argon/helium mixture can join metals from approximately 0.20″ to 1/2″ thick. Both DCEP and AC current provide excellent cleaning action of the base metal surface. Using DCEP with helium as a shielding gas produces wide weld deposits, higher heat, a large heat-affected zone (HAZ), and shallow penetration for thin material. Using DCEN with helium as a shielding gas produces a deep penetrating arc but no surface cleaning. DCEN with helium is used for mechanized welding of square-groove butt joints up to 1/4″ thick. Welding parameters such as proper current, electrode diameter, shielding gas flow rate, and backing requirements should be set based on the thickness of the magnesium and the welding position. **See Figure 18-4.**

GTAW—STAINLESS STEEL

Metal Thickness*	Joint Type	DC Current†			Electrode* Diameter	Argon Flow‡		Filler Metal Diameter*
		Flat	Horizontal & Vertical	Overhead		lpm	cfh	
1/16	Butt	80 – 100	70 – 90	70 – 90	1/16	5	15	1/16
	Lap	100 – 120	80 – 100	80 – 100	1/16	5	15	1/16
	Corner	80 – 100	70 – 90	70 – 90	1/16	5	15	1/16
	T	90 – 110	80 – 100	80 – 100	1/16	5	15	1/16
3/32	Butt	100 – 120	90 – 110	90 – 110	1/16	5	15	1/16
	Lap	110 – 130	100 – 120	100 – 120	1/16	5	15	1/16
	Corner	100 – 120	90 – 110	90 – 110	1/16	5	15	1/16
	T	110 – 130	100 – 120	100 – 120	1/16	5	15	1/16
1/8	Butt	120 – 140	110 – 130	105 – 125	3/32	5	15	3/32
	Lap	130 – 150	120 – 140	120 – 140	3/32	5	15	3/32
	Corner	120 – 140	110 – 130	115 – 135	3/32	5	15	3/32
	T	130 – 150	115 – 135	120 – 140	3/32	5	15	3/32
3/16	Butt	200 – 250	150 – 200	150 – 200	3/32	6	20	1/8
	Lap	225 – 275	175 – 225	175 – 225	3/32	6	20	1/8
	Corner	200 – 250	150 – 200	150 – 200	3/32	6	20	1/8
	T	225 – 275	175 – 225	175 – 225	3/32	6	20	1/8
1/4	Butt	275 – 350	200 – 250	200 – 250	1/8	6	20	3/16
	Lap	300 – 375	225 – 275	225 – 275	1/8	6	20	3/16
	Corner	275 – 350	200 – 250	200 – 250	1/8	6	20	3/16
	T	300 – 375	225 – 275	225 – 275	1/8	6	20	3/16

* in in.
† amps
‡ 20 cfh argon on thinner materials, helium on thicker material

Figure 18-2. *Welding parameters should be set based on stainless steel thickness.*

GTAW—ALUMINUM

Metal Thickness*	Joint Type	AC Current†			Electrode* Diameter	Argon Flow‡		Filler Metal Diameter*
		Flat	Horizontal & Vertical	Overhead		lpm	cfh	
1/16	Butt	60 – 80	60 – 80	60 – 80	3/32	7	15 – 20	3/32
	Lap	70 – 90	55 – 75	60 – 80	3/32	7	15 – 20	3/32
	Corner	60 – 80	60 – 80	60 – 80	3/32	7	15 – 20	3/32
	T	70 – 90	70 – 90	70 – 90	3/32	7	15 – 20	3/32
1/8	Butt	125 – 145	115 – 135	120 – 140	1/8	8	17 – 20	1/8
	Lap	120 – 130	125 – 145	120 – 130	1/8	8	17 – 20	1/8
	Corner	125 – 145	115 – 135	120 – 130	1/8	8	17 – 20	1/8
	T	120 – 130	115 – 135	120 – 130	1/8	8	17 – 20	1/8
3/16	Butt	160 – 180	160 – 180	160 – 170	5/32	10	20	5/32
	Lap	170 – 180	160 – 180	160 – 170	5/32	10	20	5/32
	Corner	160 – 180	160 – 180	160 – 170	5/32	10	20	5/32
	T	170 – 180	160 – 180	160 – 170	5/32	10	20	5/32
1/4	Butt	220 – 240	210 – 230	200 – 220	3/16	12	25	3/16
	Lap	230 – 250	210 – 230	200 – 220	3/16	12	25	3/16
	Corner	230 – 250	210 – 230	200 – 220	3/16	12	25	3/16
	T	230 – 250	210 – 230	200 – 220	3/16	12	25	3/16

* in in.
† amps
‡ 20 cfh

Figure 18-3. *Welding parameters should be set based on aluminum thickness.*

GTAW—MAGNESIUM								
Metal Thickness*	Joint Type	AC Current† Flat Position	Tungsten Electrode Diameter*	Argon Flow‡ lpm	cfh	Welding Metal Diameter*	Remarks	
.040	Butt	45	1/16	6	13	3/32	Backing bar	
	Butt	25	1/16	6	13	3/32	No backing	
	Lap, T	45	1/16	6	13	3/32		
.064	Butt	60	1/16	6	13	3/32	Backing bar	
	Butt, Corner	35	1/16	6	13	3/32	No backing	
	Lap, T	60	1/16	6	13	3/32		
.081	Butt	80	1/16	6	13	3/32	Backing bar	
	Butt, Corner, Edge	50	1/16	6	13	3/32	No backing	
	Lap, T	80	1/16	6	13	3/32		
.102	Butt	100	3/32	9	19	1/8	Backing bar	
	Butt, Corner, Edge	70	3/32	9	19	1/8	No backing	
	Lap, T	100	3/32	9	19	1/8		
.128	Butt	115	3/32	9	19	1/8	Backing bar	
	Butt, Corner, Edge	85	3/32	9	19	1/8	No backing	
	Lap, T	115	3/32	9	19	1/8		
3/16	Butt	120	1/8	9	19	1/8	1 pass	
	Butt	75	1/8	9	19	1/8	2 passes	
1/4	Butt	130	1/8	9	19	3/16	1 pass	
	Butt	85	1/8	9	19	3/16	2 passes	

* in in.
† amps (non-derated current levels)
‡ 20 cfh

Figure 18-4. *Welding parameters should be set based on magnesium thickness.*

A ventilation system removes toxic fumes, smoke, and dust caused by welding.

COPPER AND COPPER ALLOYS

Deoxidized copper is the type of copper most widely used in GTAW. Copper alloys such as brass and bronze, and copper alloys of nickel, aluminum, silicon, and beryllium are readily welded with GTAW. DCEN is generally used for welding these metals. However, AC or DCEP is often recommended for beryllium copper or for copper alloys less than 0.040″ thick.

Metal more than 1/4″ thick should be preheated to approximately 300°F to 500°F prior to welding. A forehand welding technique usually produces the best results. Welding parameters such as proper current, electrode diameter, shielding gas flow rate, and filler metal diameter should be set based on the thickness of the copper or copper alloy and the welding position. **See Figure 18-5** and **Figure 18-6.** A high-velocity ventilating system should be used when welding copper or copper alloys because the fumes from copper alloys can be toxic.

GTAW—DEOXIDIZED COPPER						
Metal Thickness*	Joint Type	DCEN† Flat Position	Electrode* Diameter	Argon Flow‡ lpm	cfh	Filler Metal Diameter*
1/16	Butt	110 – 140	1/16	7	15	1/16
	Lap	130 – 150	1/16	7	15	1/16
	Corner	110 – 140	1/16	7	15	1/16
	T	130 – 150	1/16	7	15	1/16
1/8	Butt	175 – 225	3/32	7	15	3/32
	Lap	200 – 250	3/32	7	15	3/32
	Corner	175 – 225	3/32	7	15	3/32
	T	200 – 250	3/32	7	15	3/32
3/16	Butt	250 – 300	1/8	7	15	1/8
	Lap	275 – 325	1/8	7	15	1/8
	Corner	250 – 300	1/8	7	15	1/8
	T	275 – 325	1/8	7	15	1/8
1/4	Butt	300 – 350	1/8	7	15	1/8
	Lap	325 – 375	1/8	7	15	1/8
	Corner	300 – 350	1/8	7	15	1/8
	T	325 – 375	1/8	7	15	1/8

* in in.
† amps
‡ 20 cfh

Figure 18-5. *Welding parameters should be set based on copper thickness.*

GTAW—COPPER ALLOYS								
Metal Thickness*	Joint Type	DCEN† Flat	Horizontal & Vertical	Overhead	Electrode* Diameter	Argon Flow‡ lpm	cfh	Filler Metal Diameter*
1/16	Butt	100 – 120	90 – 110	90 – 110	1/16	6	13	1/16
	Lap	110 – 130	100 – 120	100 – 120	1/16	6	13	1/16
	Corner	100 – 130	90 – 110	90 – 110	1/16	6	13	1/16
	T	110 – 130	100 – 120	100 – 120	1/16	6	13	1/16
1/8	Butt	130 – 150	120 – 140	120 – 140	1/16	7	15	3/32
	Lap	140 – 160	130 – 150	130 – 150	1/16, 3/32	7	15	3/32
	Corner	130 – 150	120 – 140	120 – 140	1/16	7	15	3/32
	T	140 – 160	130 – 150	130 – 150	1/16, 3/32	7	15	3/32
3/16	Butt	150 – 200	–	–	3/32	8	17	1/8
	Lap	175 – 225	–	–	3/32	8	17	1/8
	Corner	150 – 200	–	–	3/32	8	17	1/8
	T	175 – 225	–	–	3/32	8	17	1/8
1/4	Butt	150 – 200	–	–	3/32	9	19	1/8, 3/16
	Lap	250 – 300	–	–	1/8	9	19	1/8, 3/16
	Corner	175 – 225	–	–	3/32	9	19	1/8, 3/16
	T	175 – 225	–	–	3/32	9	19	1/8, 3/16

* in in.
† amps
‡ 20 cfh

Figure 18-6. *Welding parameters should be set based on copper alloy thickness.*

Points to Remember

- Filler metal containing deoxidizers should be used when welding with GTAW to prevent porosity in the weld.
- Medium- and high-carbon steels require preheat and post-heating to avoid loss of toughness and ductility.
- Ensure that there is good ventilation when welding. Fumes from some metals can be highly toxic.
- When welding light-gauge metals, a copper backing bar is usually required.

Questions for Study and Discussion

1. How can weld cracking in alloys be eliminated?
2. How can the danger of cracking when welding stainless steels be reduced?
3. What preheat temperature should be used on copper workpieces more than ¼″ thick?
4. What are some of the properties of magnesium?
5. What are the benefits of DCEP an AC when welding magnesium?
6. What defect might occur when GTAW is used on carbon steels without filler metal?

 Refer to Chapter 18 in the *Welding Skills Workbook* for additional exercises.

GMAW–Equipment

Chapter

The gas metal arc welding (GMAW) process was first conceived in the 1920s, but it was not available for commercial use until the late 1940s. Initially, the process was called metal inert gas (MIG) welding, because it used inert gas for shielding. The primary application for MIG was the welding of aluminum.

During the early 1950s, it was discovered that reactive gases, like carbon dioxide (CO_2), and mixtures of inert and reactive gases could be used to join a wider range of materials with increased weld purity and production efficiency. The ability to use both inert and reactive shielding gases resulted in the process being renamed gas metal arc welding (GMAW). The development of more versatile continuous consumable wire electrodes (welding wires) increased the popularity of GMAW. Today, GMAW is used extensively in the automotive industry and in a variety of other manufacturing and fabrication environments.

GAS METAL ARC WELDING

Gas metal arc welding (GMAW) is an arc welding process that uses an arc between a continuous wire electrode and the weld pool. GMAW equipment consists of a welding power source, a welding gun cable and gun assembly, electrode wire (welding wire), a wire feeder, shielding gas, and a workpiece lead with a workpiece connection.

The continuous wire electrode (welding wire) is fed through the wire feeder and the welding gun cable and gun assembly at a preset speed (the wire feed speed, or wfs). The molten welding wire transfers across the arc where it fuses with the base metal to form the weld. A shielding gas or a combination of shielding gases supplied from an external source is also fed through the welding gun cable and gun assembly. The shielding gas completely covers and protects the weld pool.

Metal transfer is the manner in which molten metal transfers from the end of the electrode across the welding arc to the weld pool. GMAW is capable of producing three modes of metal transfer: short circuiting transfer, globular transfer, and spray transfer, as well as a variation of spray transfer called pulsed spray transfer. The type of metal transfer used is determined by wire feed speed (which controls the amperage or current), arc voltage, welding wire diameter and composition, and type of shielding gas.

A GMAW weld can be applied by the semiautomatic, mechanized, or automatic methods. When semiautomatic welding is used, the wire feed speed, voltage setting, and gas flow rate are preset, but the welding gun is manually operated. The welder directs the welding gun along the weld joint to complete the weld.

> *A constant-voltage welding machine with direct current electrode positive is most commonly used when welding with GMAW.*

In mechanized GMAW, the welding operator sets the welding parameters and monitors the welding operation while a mechanical device controls the welding gun along the joint. In automatic GMAW, the welding parameters and welding gun movements are programmed into a computer, and all aspects of the process are controlled by the equipment, such as in a robotic cell in a manufacturing environment.

Media Clip

GMAW-Welding Machine

GMAW produces high-quality welds on a wide variety of ferrous and nonferrous metals at relatively low cost. It can be used to join a wide range of material thicknesses. The process allows higher deposition rates, faster travel speeds, less electrode waste, and it is easier to use than manual welding processes like SMAW, GTAW and OAW. There is very little if any post-weld cleaning required with GMAW because welds are slag-free. The process can also be easily adapted to mechanized and robotic applications.

However, the process is less portable than SMAW because it requires a supply of external shielding gas. Another disadvantage is that GMAW is not well-suited to outdoor applications, because wind can blow shielding gas away from the weld zone, exposing the molten weld pool to the atmosphere.

GMAW CURRENT SELECTION

The most common current selected for GMAW is direct current electrode positive (DCEP). DCEP is the most efficient current because it produces deep penetration. DCEP also provides greater surface cleaning, which is important for removing the oxide layer on metals.

A wide range of current values can be used for GMAW. In GMAW, current is a function of wire feed speed. Current is limited by metal thickness, welding wire diameter, and the type of shielding gas. The correct wire feed speed for a particular joint must often be determined by trial and error. The wire feed speed should be high enough to allow the desired penetration, but low enough to prevent undercut, overlap, or excessive melt-through. Once the wire feed speed is set, it remains constant.

DCEN should not be used for GMAW because weld penetration is shallow and wide, there is excessive spatter, and no surface cleaning occurs. DCEN is also ineffective because metal transfer is erratic and globular. AC current should not be used with GMAW since electrode melting rates (burn-off rates) are unequal for each half-cycle.

✓ Point

DCEP provides deep penetration and excellent cleaning action.

GMAW WELDING POWER SOURCES

GMAW uses a direct current (DC) welding power source capable of producing constant voltage (CV). CV welding power sources are also referred to as constant-potential (CP) machines. Unlike constant-current (CC) power sources that have a steep volt-ampere (V-A) curve, CV power sources have a slightly sloping V-A curve. **See Figure 19-1.**

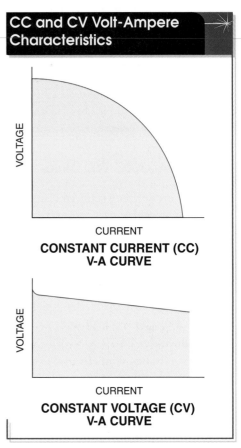

CC and CV Volt-Ampere Characteristics

CONSTANT CURRENT (CC) V-A CURVE

CONSTANT VOLTAGE (CV) V-A CURVE

Figure 19-1. *Constant-current (CC) power sources have a steep volt-ampere (V-A) curve, while constant-voltage (CV) power sources have a slightly sloping V-A curve.*

CV welding power sources for GMAW can be transformer-rectifiers or inverters. They range in size from small, compact units for light-duty work, to heavy-duty machine for high-volume production and fabrication work. *Duty cycle* is the percentage of time during a 10 min period that a power source can operate at rated power. Welding power sources rated at 350 A with 100% duty cycles are adequate for most commercial applications.

With CV power sources, welding current provides the energy necessary to melt the electrode wire. The higher the wire feed speed, the higher the welding current. An increase in current produces an increase in depth of penetration, weld metal deposition rate, and weld bead size.

Voltage is set in relation to wire feed speed and is affected by the type and diameter of the welding wire, the type of base metal, and the type of shielding gas. Voltage is directly related to arc length. The higher the welding current, the higher the voltage necessary to maintain a particular arc length. When the wire feed speed and voltage are set correctly in relation to each other, the arc length will remain constant.

From a particular voltage setting, an increase in voltage will produce a flatter, wider weld bead. If the voltage is set too high in relation to wire feed speed, the arc length becomes too long and may cause porosity, increased spatter, and undercut. Excessive voltage can cause problems with burnback. If the voltage is set too low in relation to wire feed speed, the weld bead becomes narrow with a high crown and deeper penetration. If the voltage is excessively low, it may cause the electrode to stub into the base metal.

Arc Length and Electrode Extension

CV power sources are designed to maintain selected arc voltage. They will maintain this preset voltage even with changes in electrode extension. *Electrode extension* is the distance from the contact tip to the end of the welding wire. An increase in electrode extension results in a slight increase in voltage. The increase in the electrode extension also increases the resistance in the wire, which results in a decrease in the welding current.

Similarly, a decrease in electrode extension decreases voltage slightly. The decrease in electrode extension reduces the resistance in the wire and results in an increase in the current. In either case, the arc length and voltage remain constant with changes in the electrode extension. **See Figure 19-2.**

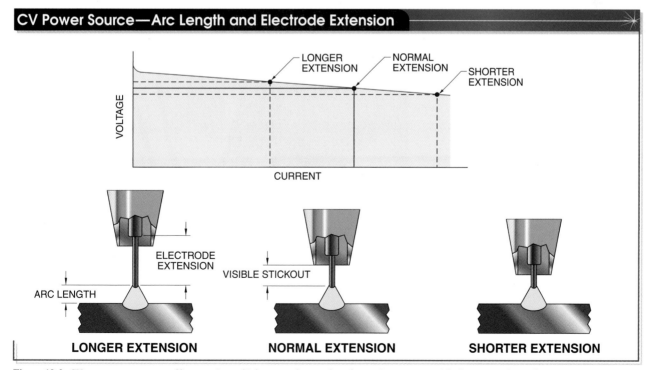

CV Power Source—Arc Length and Electrode Extension

Figure 19-2. *CV power sources are self-correcting, which means that arc length remains constant with changes in electrode extension.*

In mechanized and automatic applications, the term contact tip-to-work distance is used in place of electrode extension. *Contact tip-to-work distance (CTWD) is* the distance from the end of the contact tip to the work and includes arc length.

GMAW EQUIPMENT

In addition to a DC/CV welding power source, GMAW equipment includes a welding gun cable and gun assembly, a wire feeder, shielding gas, and a workpiece lead with a workpiece connection. **See Figure 19-3.** Additional equipment may be added to automate the system.

Welding Gun Cable and Gun Assembly

A welding gun cable conducts the welding wire, shielding gas, and welding current to the welding gun. The welding gun cable contains a separate gas line for shielding gas, a current conductor to energize the electrode, and a gun cable liner that serves as a conduit for the welding wire. If the torch is water cooled, the gun cable will also contain a water line to the gun and a return line to the water circulator. The welding gun cable must not become kinked or damaged, as restricted flow of welding wire or shielding gas may occur.

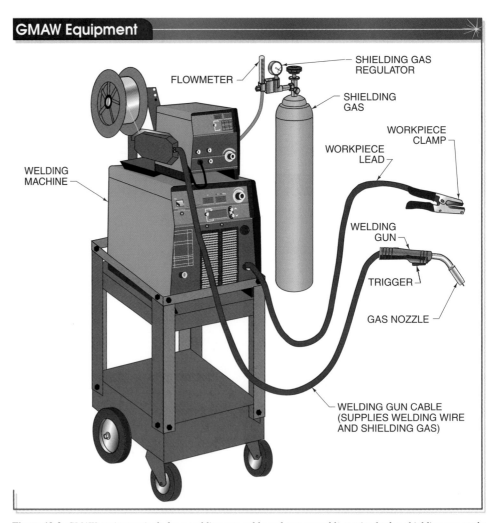

GMAW Equipment

FLOWMETER

SHIELDING GAS REGULATOR

SHIELDING GAS

WORKPIECE CLAMP

WORKPIECE LEAD

WELDING MACHINE

WELDING GUN

TRIGGER

GAS NOZZLE

WELDING GUN CABLE (SUPPLIES WELDING WIRE AND SHIELDING GAS)

Figure 19-3. *GMAW equipment includes a welding gun cable and gun assembly, a wire feeder, shielding gas, and a workpiece lead with a workpiece connection.*

Welding gun components include a handle with a conductor tube and trigger, a contact tip, a gas nozzle, a gas diffuser, and an insulator. **See Figure 19-4.** The handle and conductor tube allow easy positioning of the gun by the operator. Welding guns are available with curved or straight conductor tubes. Curved conductor tubes are typically used for semiautomatic applications. Straight conductor tubes are used for mechanized and automatic applications. **See Figure 19-5.**

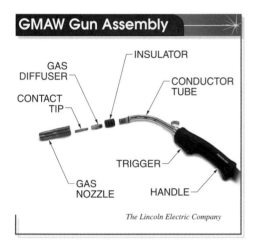

Figure 19-4. *Welding gun components include a handle with a conductor tube and trigger, a contact tip, a gas nozzle, a gas diffuser, and an insulator.*

Figure 19-5. *Welding guns are available with curved or straight conductor tubes.*

The contact tip conducts current from the welding gun to the welding wire. Contact tips are available with different hole sizes. The diameter of the hole in the contact tip should match the diameter of the welding wire.

The gas diffuser distributes shielding gas evenly around the welding wire and contact tip, and the gas nozzle directs the flow of shielding gas to the molten weld. The insulator prevents the gas nozzle from becoming energized. It also reduces heat transfer from the gas nozzle to the conductor tube. Gas nozzle size and shape may vary.

Spatter buildup on the gas nozzle, contact tip, and gas diffuser can restrict the flow of shielding gas. These components should be checked and cleaned regularly. Damaged gas nozzles should be replaced because they can cause turbulence in the flow of shielding gas. Contact tips should be checked regularly for wear and replaced if necessary. A worn contact tip may cause problems with arc starting because of poor electrical contact with the welding wire. Gas diffusers should be replaced if the gas ports become blocked with spatter.

The hand-operated trigger on the welding gun energizes the welding wire, starts the flow of shielding gas, and activates the wire feeder. Cooling of the welding gun is required to prevent overheating. Cooling is provided by the shielding gas or by shielding gas and water circulating through the gun.

The welding gun cable should be kept as straight as possible to prevent kinking or flattening of the liner, which could impede the welding wire. A damaged liner can also cause problems with burnback or bird nesting. *Burnback* is a condition that occurs when welding wire is restricted, and fuses to the end of a contact tip. *Bird nesting* is a tangle of wire that forms in a wire feeder when welding wire is restricted in the liner or by a burnback condition.

A dirty gun cable liner can also restrict welding wire, causing problems with burnback and bird nesting. Residue from the welding wire can combine with dust to create a buildup of dirt in the liner. To keep the gun liner clean, shop air should be blown through it each time a new spool of welding wire is installed.

There are two types of gun cable liners, helical steel and Teflon® lined. Helical steel gun cable liners are used with steel welding wired. Teflon gun cable liners are used for soft welding wires. They keep soft welding wire like aluminum feeding smoothly.

There are several welding guns available for GMAW, but they can be categorized into two groups: welding guns designed for semiautomatic applications and welding guns designed for automatic applications.

Semiautomatic Welding Guns. A semiautomatic welding gun allows the welder to manually control and direct welding wire to the joint. Semiautomatic welding guns are manufactured in many shapes and sizes. A variety of factors, including current requirements, determine the correct semiautomatic welding gun to use for a particular welding task. A welding gun and welding gun cable must be capable of providing sufficient current for the welding task. Semiautomatic welding guns are rated to operate between 100 A and 750 A.

A semiautomatic welding gun generally has a curved conductor tube. The curved conductor tube is used for most welding positions and provides easy access to intricate joints and difficult-to-weld patterns. Gas nozzles are commonly made of copper because copper can conduct away the intense heat that builds up near the arc. Gas nozzles are available with orifice diameters from ⅜″ to ⅞″. Gas nozzle size depends on the size of the weld pool, the gas shielding required, and the weld joint design.

A semiautomatic welding gun attaches to the welding gun cable, which contains the power cable, gun cable liner, and hose for shielding gas. Semiautomatic welding guns can be air cooled or water cooled. Air-cooled welding guns use shielding gas for cooling. A water-cooled welding

gun has two additional connections for "Water In" and "Water Out" to control water flow. Water lines that run through the welding gun cable provide water to the welding gun for cooling.

The trigger on a semiautomatic welding gun starts the wire feeder, the arc, and the flow of shielding gas. When the trigger is released, the wire feeder, arc, and shielding gas stop immediately. Preflow and postflow timers are included on some equipment to permit shielding gas to flow before and after welding to better protect the weld zone.

Automatic Welding Guns. Automatic welding guns have a design similar to semiautomatic welding guns, but the gun is usually mounted to a fixture directly below the wire feeder. The fixture may move the welding gun, the worktable, or both. An automatic welding gun does not usually have a trigger. The welding gun is energized from a control panel or a remote power control. Automatic welding guns may be rated up to 1200 A.

An air-cooled welding gun is used for welding at low currents, while a water-cooled welding gun is used for welding at high currents. Automatic welding guns are typically water-cooled because of the high currents and duty cycles at which they operate.

Wire Feeders

A wire feeder automatically advances the welding wire from the wire spool, through the welding gun cable liner and welding gun, to the arc. The wire feeder must be selected to match the power source used for the GMAW application. Constant-speed wire feeders are typically used with CV welding machines. The wire feeder may be portable, mounted on the welding machine, or mounted elsewhere to facilitate welding in a large area. **See Figure 19-6.** Wire feeders are designed for use with a wide range of solid and metal-cored welding wire from 0.023″ to ¹⁄₁₆″. The wire feed speed control on the wire feeder can be adjusted to vary the wire feed speed.

WIRE SPOOL

Miller Electric Manufacturing Company

Figure 19-6. *The wire feeder may be portable, mounted on the welding machine, or mounted elsewhere to facilitate welding in a large area.*

The wire feeder does more than feed welding wire. It also supplies welding current to the welding gun cable and contains a solenoid that activates the flow of shielding gas. Most wire feeders have a speed control, a voltage control, a purge button, and a jog button. Wire feeders designed for high wire feed rates may also have a burnback control.

Wire Feed Speed Control. The wire feed speed control adjusts the speed with which welding wire feeds into the arc. Wire feed speed determines the welding current. Wire feed speed is measured in inches per minute (ipm). Most wire feeders can be adjusted from 70 ipm to 800 ipm. A welding procedure specification (WPS) frequently provides the wire feed speed, as well as the voltage for the specified welding wire and shielding gas.

Voltage Control. The voltage control allows the welder to set the optimum arc length. The arc voltage required for a particular application depends on the type and diameter of welding wire, the wire feed speed, and the type of shielding gas used.

Purge Button. The purge button allows the welder to set the shielding gas flow rate without using the gun trigger. Using the gun trigger instead of the purge button is not recommended, because the gun trigger not only starts the flow of gas, it also starts the wire feeder and energizes the welding wire. This not only wastes welding wire, it can also cause the wire to arc if it touches a metal surface. The purge button is also used to purge the gas line in the welding gun cable before welding. The line should be purged if the welding machine has been idle for an extended period.

Jog Button. The jog button advances the welding wire without using the gun trigger. The jog button is used to advance the welding wire through the liner when installing a new spool. It can also be used to feed wire without energizing the contact tip.

Burnback Control. Burnback control prevents the welding wire from freezing to the base metal by maintaining the arc briefly after the trigger is released. At low feed rates, the welding arc and the welding wire stop when the gun trigger is released. However, at high feed rates, the arc stops when the trigger is released, but the momentum of the wire can cause it to stub into the weld pool. The burnback control maintains the arc until the wire stops to keep the wire from sticking to the work. The burnback control should not be confused with the type of burnback that occurs when the wire freezes to the contact tip due to a blockage in the liner, or to the voltage being set too high in relation to wire feed speed.

The wire feed mechanism that drives the welding wire consists of a variable speed electric motor connected to drive rolls. The drive rolls push the welding wire through the gun liner and gun. Wire feeders for light-duty applications have two drive rolls. Wire feeders for industrial applications typically have four drive rolls.

Drive rolls for GMAW are grooved, and the tension on the drive rolls can be adjusted to grip the welding wire without slipping. Grooves can be U-shaped, V-shaped, or knurled depending on the type of welding wire. U-shaped grooves are used with soft wires, V-shaped grooves are used with steel wires, and knurled grooves are used with metal-cored and flux-cored wires. **See Figure 19-7.**

Media Clip

Wire Feeder Drive Roll

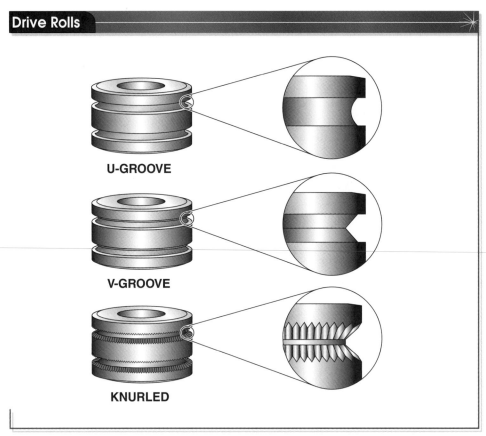

Drive Rolls

U-GROOVE

V-GROOVE

KNURLED

Figure 19-7. *Drive roll grooves can be U-shaped, V-shaped, or knurled depending on the type of welding wire.*

The drive rolls and the welding gun liner must be properly sized to match the diameter of the welding wire. The wire outlet guide must be aligned closely with the groove in the drive rolls without touching it. **See Figure 19-8.** Misalignment of the liner and the drive rolls can impede the welding wire, causing problems with burnback and bird nesting.

The wire feeder can be a push type, a pull type, or a push-pull type, depending on the location of the drive rolls. Spool gun wire feeders are also available for small spools of welding wire.

Push Type Wire Feeder. The most common wire feeder for steel welding wire is the push type wire feeder. A push type wire feeder has drive rolls that push the welding wire through the gun cable liner to the welding gun. The gun cable liner can be up to 15′ for steel wire or 6′ for aluminum wire. The push type wire feeder can handle large-diameter welding wire for ferrous metals in welding conditions where current is over 250 A.

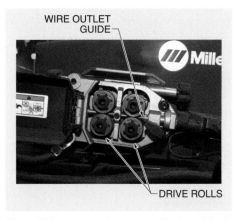

WIRE OUTLET GUIDE

DRIVE ROLLS

Figure 19-8. *The wire outlet guide must be aligned closely with the groove in the drive rolls, without touching it.*

Pull Type Wire Feeder. In the pull type wire feeder, the welding wire is fed through the liner and pulled by drive rolls located on the welding gun. A pull type wire feeder is often used for mechanized and automatic welding. The drive rolls are built into the welding gun and pull the welding wire from the wire feeder. A pull type wire feeder works best with soft welding wire and small-diameter steel welding wire up to about .045″ in diameter. A pull type wire feeder can be used with any semiautomatic welding gun.

Push-Pull Type Wire Feeder. The push-pull type wire feeder is used for driving welding wire long distances and for low-strength welding wires. The push-pull wire feeder has synchronous drive motors in the wire feeder and the welding gun. The drive rolls in the wire feeder push the welding wire from the wire feeder through the liner, while the drive rolls in the gun pull it.

Spool Gun Type. Spool gun wire feeders are designed for small spools of welding wire between 1 lb and 2 lb. The spool is mounted at the back of the welding gun. A spool gun includes a drive motor, drive rolls, and a wire feed speed control. Typically, spool guns are used for small-diameter aluminum welding wire.

The type of wire feeder used is determined by the characteristics of the welding wire. Small-diameter, soft aluminum welding wire must be pulled through the gun cable or fed from a spool gun. Large-diameter electrodes often require a push-pull wire feeder for a consistent flow of wire.

Shielding Gas

Shielding gas affects the properties of the weld deposit. The air in the weld area is displaced by the shielding gas to prevent it from contacting the weld pool. The arc is then started under a blanket of shielding gas and welding can be performed. Since the weld pool is exposed only to the shielding gas, it is not contaminated, and strong, dense weld deposits are obtained. The nozzle-to-work distance of the welding gun must be maintained to ensure an adequate shielding gas cover.

Air is made up of 21% oxygen, 78% nitrogen, 0.94% argon, and 0.04% other gases (primarily carbon dioxide). The atmosphere will also contain a certain amount of water in the form of hydrogen depending on its humidity. The elements of air that cause difficulties when welding are oxygen, nitrogen, and hydrogen. The effects of oxygen, nitrogen, and hydrogen on the weld make it essential that they be excluded from the weld area during welding.

Oxygen (O_2) is a highly reactive gas and readily combines with other elements in molten metal to form oxides and gases. The oxide-forming characteristic of oxygen can be overcome by using deoxidizers in the filler metal. Deoxidizers, such as manganese and silicon, combine with oxygen and float to the top of the weld pool, forming deposits called silica islands on the surface of the finished weld or along the toes of the weld. If deoxidizers are not provided in the filler metal, oxygen combines with the molten metal, causing porosity and other problems that affect the mechanical properties of the metal.

Nitrogen (N) causes serious problems when welding steel. When the molten weld pool is exposed to nitrogen, it forms nitrides as it cools. Nitrides increase hardness and decrease ductility and impact resistance. The loss of ductility often leads to cracking in the weld and in the HAZ. In excessive amounts, nitrogen can also cause porosity in the weld.

Hydrogen (H) is harmful because even small amounts in the weld pool can cause an erratic arc. Hydrogen can also become trapped in the solidifying metal and HAZ, causing small cracks in the weld as well as underbead cracking.

Atmospheric gases can be excluded by using an inert gas for shielding. Inert gases do not react readily with other materials, making them useful as shielding gases for arc welding. Argon and helium are inert gases that are commonly used for shielding. **See Figure 19-9.**

A mixture of 75% Ar and 25% CO_2 produces shallower penetration when welding sheet metal and less spatter for autobody applications as compared to 100% CO_2.

GMAW SHIELDING GASES		
Material	**Gas**	**Remarks**
Mild Carbon Steel and Low-Alloy Steel	100% CO_2	Produces deep, broad penetration and excessive spatter; unsuitable for sheet metal and autobody applications due to deep penetration and spatter
	75% Argon (Ar) + 25% CO_2	Produces shallower penetration with narrower penetration finger and less spatter than 100% CO_2; good for sheet metal and autobody
	98% Ar + 2% O_2	Removes oxidation; gas mixtures containing over 80% Ar produce spray transfer; welds are high quality and low spatter, but limited to the flat position; oxygen minimizes undercut
Stainless Steels	90% Helium (He) + 7.5% Ar + 2.5% CO_2	Popular blend for welding stainless steels with short circuiting transfer; high thermal conductivity of He produces flat bead with excellent fusion; promotes high travel speeds
	99% Ar + 1% O_2	Minimizes problems with undercut
	95% Ar + 5% O_2	Oxygen improves arc stability
Aluminum Alloys	100% Ar	Good cleaning action for removal of aluminum oxide coating
Aluminum Bronze	100% Ar	Less penetration of base metal; commonly used as a surfacing material
Magnesium	100% Ar	Good cleaning action for removal of magnesium oxide coating
Nickel	100% Ar	Good wetting; decreases fluidity of weld pool
Monel	100% Ar	
Inconel	100% Ar	
Titanium	100% Ar	Reduces heat-affected zone (HAZ); improves metal transfer
Silicon Bronze	100% Ar	Reduces crack sensitivity
Magnesium Aluminum Alloys	75% He + 25% Ar	Higher heat input reduces risk of porosity; good cleaning action to remove oxide coating
Copper (deoxidized)	75% He + 25% Ar	Good wetting; increased heat input to counteract high thermal conductivity

Figure 19-9. *Inert gases do not react readily with other materials, making them useful as shielding gases for arc welding.*

Although it is not inert, carbon dioxide (CO_2) can also be used for shielding the weld area if compensation is made for its oxidizing tendencies. CO_2, argon, or helium can be used alone or mixed for different applications. All shielding gases should be welding grade. Welding-grade shielding gases are over 99% pure. They provide the best protection and produce the best results. Different shielding gases produce different weld bead contour and penetration characteristics. **See Figure 19-10.**

Carbon Dioxide (CO_2). Although it may be used in other shielding-gas mixtures, CO_2 is used primarily for welding mild steel. At normal temperatures, CO_2 is essentially an inert gas. However, when subjected to high temperatures,

CO_2 dissociates into carbon monoxide and oxygen. Because of the oxidizing characteristic of CO_2 gas, the welding wire used with CO_2 must contain deoxidizing elements. The deoxidizing elements readily combine with oxygen, preventing it from causing porosity and other problems in the weld. The most common deoxidizers used in welding wire are manganese, silicon, aluminum, titanium, and vanadium.

CO_2 produces a wide, deep-penetrating weld bead. Bead contour with CO_2 is good and there is less tendency toward undercutting. Another advantage is its relatively low cost compared to other shielding gases.

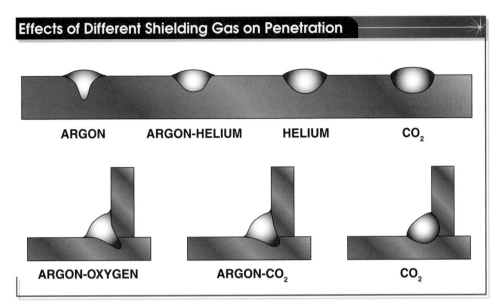

Effects of Different Shielding Gas on Penetration

ARGON   ARGON-HELIUM   HELIUM CO_2

ARGON-OXYGEN   ARGON-CO_2 CO_2

Figure 19-10. *Different shielding gases produce different weld bead contour and penetration characteristics.*

While manufacturers generally use color codes to identify gas cylinders, colors may not be consistent between suppliers. Always check the cylinder for contents before attaching and using a gas cylinder.

A drawback of using CO_2 for shielding is that it produces a somewhat violent arc. This can cause excessive spatter. For many applications, spatter is not a major problem, and the excellent penetration characteristics of CO_2 outweigh its disadvantages. Also, antispatter sprays are available to prevent spatter from sticking to the base metal, gas nozzle, contact tip, and gas diffuser.

However, 100% CO_2 produces too much penetration for sheet metal welding and too much spatter for autobody applications. A mixture of 75% Ar and 25% CO_2 reduces penetration sufficiently for sheet metal welding. An argon-CO_2 mixture also minimizes spatter for autobody work.

Many GMAW welding guns may be used at 100% duty cycle with CO_2 as the shielding gas at a particular current setting; however, using the same welding gun with argon as the shielding gas, a lower current setting must typically be used for a 100% duty cycle.

Argon (Ar). Argon is the most commonly used inert gas. Argon has a relatively low ionization potential. This means that the welding arc is easier to start, tends to be more stable, and produces little or no spatter. Since argon has a low ionization potential, the arc voltage is reduced when an argon mixture is used as a shielding gas. This results in lower power in the arc and shallower penetration. The combination of shallower penetration and reduced spatter makes the use of an argon/CO_2 mixture desirable when welding sheet metal.

Straight argon is seldom used as a shielding gas except when welding metals such as aluminum, copper, nickel, and titanium. When welding steel, the use of straight argon leads to undercutting and poor bead contour. Additionally, penetration with straight argon is shallow at the bead edges and deep at the center of the weld, which can lead to lack of fusion at the root of the weld. Argon is often mixed with other gases to improve their stability.

Helium (He). Helium is lighter than air and requires high flow rates to produce adequate coverage. Helium has a higher ionization potential than argon, which allows for higher arc voltage. Ionization potential refers to the amount of voltage required to ionize the gas column so it can conduct the arc. A shielding gas with a high ionization potential makes the arc harder to start. Helium produces a deep, broad, parabolic weld with a low bead profile.

Because of its high cost, helium is used primarily for special welding tasks and for nonferrous metals such as aluminum, magnesium, and copper. It is most commonly used in combination with other shielding gases. Argon-helium mixtures are commonly used for welding aluminum greater that 1″ thick. Argon-helium mixtures are also used on stainless steel instead of argon-CO_2 mixtures because CO_2 can adversely affect the mechanical properties in the weld and in the HAZ.

Argon-Oxygen. Oxygen is added to argon when welding mild steel to improve bead contour and penetration. A small amount of oxygen improves penetration by broadening the deep penetration finger at the center of the weld bead. It also improves bead contour and eliminates the undercutting at the edge of the weld that occurs with pure argon. Normally, oxygen is added in amounts of 1%, 2%, or 5%. Adding oxygen in amounts greater than 5% may lead to porosity in the weld. Argon-oxygen mixtures are common when welding alloy steel, carbon steel, and stainless steel.

Argon-CO_2. For some mild steel welding applications like sheet metal and autobody, welding-grade CO_2 does not provide the required arc characteristics. This is usually evident in the form of excessive penetration or excessive spatter. Using an argon-CO_2 mixture minimizes these problems. A mixture of 75% Ar and 25% CO_2 is commonly used for welding mild and low-alloy steels with short circuiting transfer because it produces shallower penetration, faster travel speeds, a smoother and more focused arc, and less spatter than 100% CO_2.

Argon-Helium-CO_2. An argon-helium-CO_2 shielding-gas mixture is used for welding austenitic, martensitic, and ferritic stainless steels. This combination of gases provides a unique characteristic to the weld. With it, it is possible to make a weld with very little buildup of the top bead profile. An argon-helium-CO_2 mixture is used for applications where a high-crowned weld is detrimental.

Gas Flow Rates. A regulator and flowmeter assembly deliver a steady preset flow of shielding gas to the weld area. The regulator reduces cylinder pressure to working pressure. The flowmeter indicates the rate of flow of shielding gas to the weld in cubic feet per hour (cfh). The amount of shielding gas required is determined by the type of welding gun, weld joint design, base metal, type of metal transfer, and conditions in the weld area. For example, welding outdoors in breezy conditions requires higher shielding-gas flow rates to provide adequate coverage than when welding in the shop.

For most welding, the gas flow rate is approximately 20 cfh to 35 cfh. A WPS typically specifies the type of shielding gas along with recommended gas flow rates. Final adjustments must often be made on a trial-and-error basis. The correct gas flow rate is determined by the type and thickness of metal to be welded, the type and diameter of welding wire, the welding position and type of joint, the shielding gas used, and the type of metal transfer used.

Information on a gas cylinder label typically includes the type of gas or mixture of gas contained, and the manufacturer or supplier name.

Proper gas shielding usually results in a rapid, crackling or sizzling sound for short circuiting transfer, and a hissing sound for spray transfer. Inadequate gas shielding produces a popping sound and results in porosity, spatter, and a discolored weld.

Gas drift may occur with high travel speeds or in drafty conditions around the weld area. Gas drift commonly results in inadequate gas shielding. The welding area should be protected with windbreaks to eliminate breezes. It may also be necessary to increase the gas flow to provide better coverage. Some welding guns also have adjustable nozzles that can be positioned to improve gas coverage. **See Figure 19-11.**

The distance from the work to the gas nozzle is determined by the nature of the weld. The gas nozzle is usually placed up to ½″ from the work. Too much space between the gas nozzle and the work reduces the effectiveness of the gas shield, while too little space may result in excessive weld spatter, which collects on the gas nozzle, contact tip and gas diffuser and shortens component life.

GMAW WELDING WIRE

Welding wire for GMAW should be similar in composition to the base metal. The copper coating on GMAW electrode wire aids in arc starting, improves wire feeding, and prolongs contact tip life. Welding wire for GMAW is classified by a letter and number system. **See Figure 19-12.**

Metal-cored welding wire is used extensively in high-production applications because it has 30% faster travel speeds than solid wires. It handles contaminants well, bridges gaps due to poor fit-up without excessive melt-through, and can be used to weld a variety of metal thicknesses.

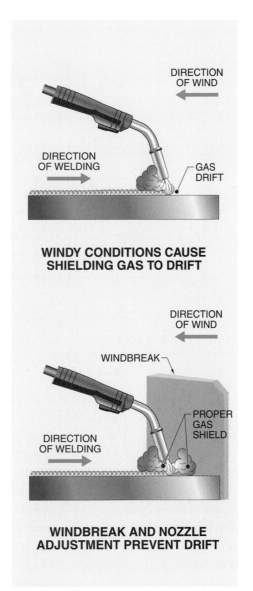

WINDY CONDITIONS CAUSE SHIELDING GAS TO DRIFT

WINDBREAK AND NOZZLE ADJUSTMENT PREVENT DRIFT

Figure 19-11. *The welding area should be protected with windbreaks to prevent gas drifts.*

Miller Electric Manufacturing Company

Welding wire is selected to match the composition of the metal to be welded. Welding wire designations are based on AWS classifications.

✓ **Point**

The correct diameter wire must be used to ensure a quality weld. Check the wire manufacturer recommendations for correct wire diameters.

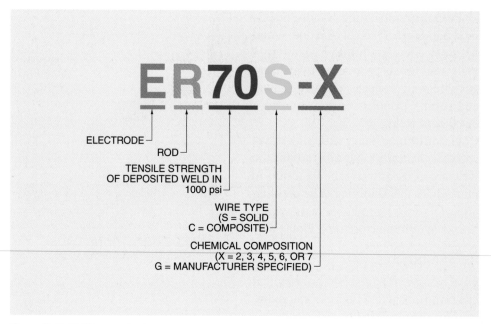

Figure 19-12. *Welding wire for GMAW is classified by a letter and number system that reflects the ideal conditions under which the welding wire should be used.*

The E identifies the device as an electrode. The R stands for rod and indicates that it can be used as a non-current-carrying filler rod, as in GTAW. The number specifies the tensile strength of the deposited weld in thousands of pounds per square inch, for example, 70 = 70,000 psi. The S indicates a solid wire. The letter C in this position would indicate composite (metal-cored) wire. The digit following the dash indicates the chemical composition of the welding wire. Using this system enables a welder to choose the correct steel welding wire base on AWS specifications. **See Figure 19-13.**

Basic welding wire diameters include 0.023″, 0.030″, 0.035″, 0.045″, 0.052″, 1/16″, and 1/8″. Generally, welding wire of 0.023″, 0.030″, or 0.035″ is best for welding thin metal, although it can be used to weld low- and medium-carbon steel and medium-thickness, high-strength/low-alloy (HSLA) steel. Medium-thickness metal normally requires 0.045″ or 1/16″ diameter welding wire. **See Figure 19-14.**

Welding position is a factor that must be considered when selecting welding wire. For vertical and overhead welding, small-diameter wire is preferred because it produces a smaller, more manageable weld pool.

Metal-cored welding wires are often preferred to solid wires for high-production GMAW applications. Metal-cored welding wires are composite electrode wires consisting of a metal sheath (tube) filled with metallic powders. Like solid GMAW wire, metal-cored wires produce slag-free welds that require little or no cleanup. Metal-cored welding wires produce a wide arc with a broader penetration profile and better fusion into the toes of the weld than solid welding wires. However, metal-cored welding wires are limited to the flat position and to horizontal fillet welds because of the fluidity of the weld pool.

Metal-cored wires handle contaminants such as rust and mill scale (a surface layer of ferrous oxide) well and bridge gaps due to poor fit-up without excessive melt-through. They can also be used to weld single and multiple-pass welds. Metal-cored welding wires can weld a variety of metal thicknesses, including thin-gauge metal used in automobile exhaust systems.

WELDING WIRE FOR GMAW	
AWS Classification	**Remarks**
AWS A5.18/A5.18M	**Specification for Carbon Steel Electrodes and Rods for Gas Shielded Arc Welding**
ER70S-2	High in silicon and manganese deoxidizers; also contains aluminum, titanium, and zirconium deoxidizers; can be used for single and multiple pass welds, as well as root passes on carbon steel pipe; deoxidizer content makes it suitable for use on steels with moderate levels of mill scale
ER70S-3	Most commonly used welding wire; contains medium levels of silicon and manganese deoxidizers; can be used for single or multiple pass welds, with 100% CO_2, Ar/CO_2 mixtures, or Ar/O_2 mixtures
ER70S-4	Higher in silicon and manganese that ER70S-3; can be used for single- or multiple-pass welds
ER70S-5	Contains aluminum; can be used for single- or multiple-pass welds on rimmed, semi killed, or killed mild steel; suitable for welding on rusty or dirty surfaces; normally used with CO_2
ER70S-6	Higher in silicon and manganese that ER70S-4; can be used for single or multiple pass welds; good for welding rusty base metal or metal with moderate to high mill scale; hvvigh deoxidizer content produces a fluid weld pool with excellent fusion into the toes and a flat bead contour
ER70S-7	Higher in manganese that ER70S-6, but lower in silicon; can be used with Ar/CO_2 mixtures; produces hardness levels between ER70S-3 and ER70S-6
ER70C-6M	Composite (metal-cored) wire high in silicon and manganese deoxidizers; better than solid wire for welding on base metal high in mill scale; 30% faster travel speeds than solid wire; commonly used for mechanized and automatic applications; excellent penetration and fusion characteristics on a wide range of thicknesses; limited to flat position and horizontal fillet welds due to fluidity of weld pool
AWS A5.28/A5.28M	**Specification for Low-Alloy Steel Electrodes and Rods for Gas Shielded Arc Welding**
ER80S-Ni1	High in silicon and manganese with 1% nickel (corrosion resistance); also contains small amounts of chromium, molybdenum, and vanadium for higher strength and impact resistance
ER80S-D2	Higher than ER80S-Ni1 in silicon and manganese; contains 0.50% molybdenum for improved strength and toughness of weld metal
AWS A5.10/A5.10M	**Specification for Bare Aluminum and Aluminum-Alloy Welding Electrodes and Rods**
ER1100	Weld aluminum with similar composition welding wire
ER4043	
ER5183	
ER5554, 5556	
ER5654	
AWS A5.9/A5.9M	**Specification for Bare Stainless Steel Welding Electrodes and Rods**
ER308L	For welding types 304, 308, 321, and 347
ER308L-Si	For welding types 301 and 304
ER309	For welding types 309 and straight chromium grades when heat treatment is not possible; also for welding types 304-clad
ER310	For welding types 310, 304-clad, and hardenable steels
ER316	For welding types 316
ER347	For welding types 321 and 347 where maximum corrosion resistance is required
AWS A5.7/A5.7M	**Specification for Copper and Copper-Alloy Bare Wire Welding Rods and Electrodes**
ECuSi (Silicon Bronze)	Special wires for welding copper and copper-based alloys
ECuAl-A1 (Aluminum Bronze)	
ECu (Deoxidized Copper)	
ECuAl-A2 (Aluminum Bronze)	
ECuAl-B (Aluminum Bronze)	

Figure 19-13. *The American Welding Society (AWS) has developed a classification system for mild steel welding wire.*

WELDING WIRE DIAMETERS					
Metal Thickness*	Wire Size*	Welding Conditions (DCEP)		Gas Flow[†]	Travel Speed[‡]
		(arc volts)	(amperes)		
.025	.030	15 – 17	30 – 50	15 – 20	15 – 20
.031	.030	15 – 17	40 – 60	15 – 20	18 – 22
.037	.035	15 – 17	65 – 85	15 – 20	35 – 40
.050	.035	17 – 19	80 – 100	15 – 20	35 – 40
.062	.035	17 – 19	90 – 110	20 – 25	30 – 35
.078	.035	18 – 20	110 – 130	20 – 25	25 – 30
.125	.035	19 – 21	140 – 160	20 – 25	20 – 25
.125	.045	20 – 23	180 – 200	20 – 25	27 – 32
.187	.035	19 – 21	140 – 160	25 – 30	14 – 19
.187	.045	20 – 23	180 – 200	25 – 30	18 – 22
.250	.045	19 – 21	140 – 160	30 – 35	10 – 15
.250	.052	20 – 23	180 – 200	30 – 35	12 – 18

NOTE: Gas flow rates will vary from values shown based on the type of metal welded
Shielding gas CO_2, welding grade
Wire stickout—¼″ to ⅜″
* in in.
[†] in cubic feet per hour (cfh)
[‡] in in./min

Figure 19-14. *Welding position is a factor that must be considered when selecting welding wire.*

Metal-cored welding wires are used extensively in high-production manufacturing and fabrication applications because they allow travel speeds 30% faster than solid wire and require little or no post-weld cleaning. The high travel speeds associated with metal-cored wires make them well-suited to mechanized and automatic applications.

Metal-cored welding wires have replaced flux-cored wires in some environments because they produce less smoke and fumes. They are also less sensitive to changes in welding gun angles and in welding wire extension. They also reduce top-toe undercut on horizontal fillet welds. This makes them extremely well-suited to semiautomatic applications as well.

 Refer to Quick Quiz® on CD-ROM

- DCEP provides deep penetration and excellent cleaning action.
- For GMAW, a constant-voltage welding machine with a nearly flat volt-ampere characteristic maintains a constant, preset voltage level during welding.
- Ensure that the wire feed speed is set for the current that is to be used for welding.
- Visible stickout is the distance the welding wire projects from the end of the nozzle of the welding gun.
- The proper nozzle-to-work distance must be maintained to ensure an adequate shielding gas cover.
- The use of CO_2 as a shielding gas is most effective and least expensive when welding steel.
- Argon produces the most effective results when welding aluminum.
- For most welding, the gas flow rate is approximately 20 cfh to 35 cfh.
- The correct diameter wire must be used to ensure a quality weld. Check the wire manufacturer recommendations for correct wire diameters.

1. What are some of the specific advantages of GMAW?
2. Why is DCEP current used for GMAW?
3. What results can be expected if DCEN current is used?
4. How does a constant-voltage (CV) welding machine differ from a constant-current (CC) welding machine?
5. What is the advantage of using a CV welding machine for GMAW?
6. What are the elements that make up air?
7. Why is oxygen generally a harmful element in welding?
8. Why does nitrogen cause the most serious problems in welding?
9. When is argon or an argon-oxygen mixture considered the ideal gas for shielding?
10. When is CO_2 better for shielding than an inert gas?
11. How is it possible to determine the proper gas flow for shielding?
12. What happens if the gas flow is allowed to drift from the weld area?
13. What factors must be taken into consideration in selecting the correct diameter welding wire?
14. How is the welding wire fed to the welding gun?
15. What determines the rate at which the wire feed should be set?
16. Why is the correct electrode extension important?

Refer to Chapter 19 in the *Welding Skills Workbook* for additional exercises.

GMAW–Procedures

GMAW is a relatively fast welding process with higher deposition rates than SMAW. Many welding applications that were once only performed with SMAW are now being completed with the GMAW process. Pipelines, railroad cars, automobiles, and heavy equipment manufacturing are industries that use GMAW for many welding jobs.

Since GMAW uses smaller diameter electrode wire (welding wire), up to $\frac{1}{16}''$, combined with good penetration characteristics, narrower beveled joint designs can be used. When performing GMAW welding outdoors, wind protection may be needed to protect the shielding gases from being blown away from the weld area.

GMAW PROCEDURES

GMAW was developed to increase the speed at which weld metal could be deposited. Although GMAW can be mechanized or fully automated, it is most often semiautomatic. When GMAW is used semiautomatically, the wire feed speed (wfs), voltage setting, and gas flow rate are preset, but the welding gun is manually operated. The welder controls travel speed, maintains the correct visible wire stickout, and guides the welding gun along the joint with the proper work and travel angles. GMAW advantages over other welding processes include the following:

- No flux or slag and little spatter are produced, minimizing clean-up time and resulting in a savings in total welding cost.
- Less time is required to train an operator. Welders who are proficient in other welding processes can easily master GMAW. The primary duty is to monitor the angle of the welding gun, the travel speed, and the visible wire stickout.

- There is no starting and stopping to change electrodes. Starting and stopping reduces welding time and minimizes problems with overlap and crater cracking that are associated with frequent restarts.
- Less heat input due to faster travel speeds results in reduced distortion and a narrower heat-affected zone (HAZ) with improved metallurgical properties.
- GMAW is more economical for welding thin-gauge metal when short circuiting transfer is used.
- Narrower bevel angles can be used on joints because of the smaller diameter of the welding wire, longer stickout, and deeper penetrating capabilities of GMAW. This results in reduced weld size.

The GMAW process was refined during World War II, when a cost-effective and efficient method of welding thick metals, such as found on ships and tanks, was needed.

✓ **Point**

GMAW is a faster welding process than SMAW and is easy to learn.

Miller Electric Manufacturing Company

GMAW is a versatile welding process that allows fast deposition on different metal thicknesses.

much smaller in diameter than SMAW electrodes. This means that groove angles can be smaller, root faces can be thinner, and root openings can be narrower than joints designed for SMAW. **See Figure 20-1.**

Butt Joints. A butt joint typically requires more welding skill than other joints. When making butt joints, distortion and residual stress must be prevented by using the proper fit-up and joint edge preparation. Butt joints have very good mechanical strength if properly prepared.

Lap Joints. A lap joint is commonly used for many welding applications. In a lap joint, the surfaces of the members to be joined overlap one another. The amount of overlap is determined by the thickness of the metal. Lap joints are welded with fillet welds, which results in good mechanical properties, especially when welded on both sides.

T-Joints. T-joints generally require little, if any, edge preparation. Edges of a T-joint may be left square. However, depending on the thickness of the material and the service conditions, the vertical member can be prepared with a single or double bevel to produce a stronger joint. T-joints are welded with a fillet weld.

Joint Preparation

Joint preparation is recommended to aid in weld penetration and to control weld reinforcement. Beveling of the joint edges is recommended for butt joints thicker than ¼″ if complete root penetration is desired. A square-groove butt joint can be used on metals ¼″ or less.

Generally, the joint design recommended for other arc welding processes can be used for GMAW. However, joints can be designed more efficiently to take advantage of the operating characteristics of GMAW. GMAW welding wires are

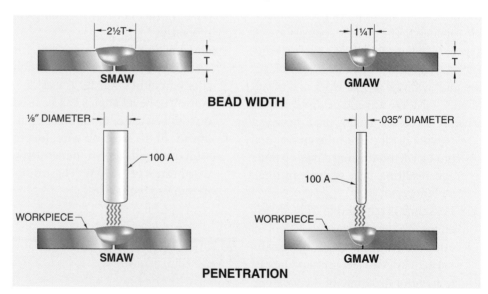

Figure 20-1. *GMAW produces a narrower weld bead and deeper penetration than SMAW, requiring a smaller root face and root opening.*

Edge Joints. Edge joints are commonly used when the finished weld will not be exposed to excess loads or heavy impact. The edges of the metal to be welded may be left square, or beveled by grinding or machining. The grooves created by beveling allow proper penetration of the weld metal.

Corner Joints. Corner joints also require little, if any, edge preparation. After a corner joint is welded, the edges can be ground smooth to improve the appearance of the finished weld.

Weld Backing

When performing GMAW, weld backing helps to obtain a sound weld at the root. Backing forms a mold or dam to retain the molten metal in the joint. Backing also conducts heat away from the joint. There are several types of material used for backing, including steel or copper blocks, strips, and bars, as well as ceramic or fired clay. The materials most commonly used for backing with GMAW are copper or steel.

Positioning Work and Welding Wire

Proper positioning of the welding gun and the work is necessary to achieve a quality weld. The flat position is preferred for most joints because it improves weld pool control, bead contour, and shielding gas protection. On thin-gauge metal, it is sometimes necessary to weld with the work inclined 10° to 20°. When the work is inclined, welding is performed downhill. Downhill welding tends to flatten the bead, increase the travel speed, and reduce penetration.

The welding wire must be properly aligned in relation to the joint. The welding wire should be on the centerline of the joint for most butt joints if the members to be joined are of equal thickness. If the workpieces are unequal in thickness, the welding wire may be angled toward the thicker metal.

Correct work angles and travel angles ensure correct weld bead formation. **See Figure 20-2.** The travel angle may be a push angle or a drag angle depending upon the position of the welding gun. If the welding gun is angled back toward the beginning of the weld, the travel angle is called a drag angle. A *drag angle* is a travel angle where the electrode wire points away from the direction of travel. If the welding gun is pointing away from the weld toward the end of the joint, the travel angle is called a push angle. A *push angle* is a travel angle where the electrode wire points in the direction of travel.

> ✓ **Point**
>
> *Keep the welding gun properly positioned to ensure a uniform weld with proper penetration.*

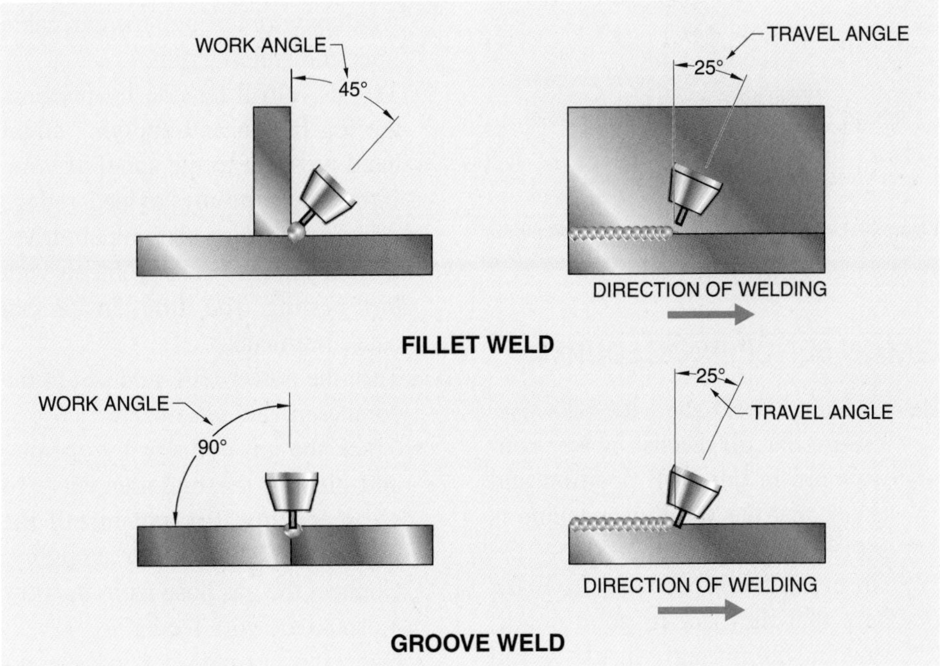

Figure 20-2. *The correct work angle and travel angle are required for correct weld bead formation.*

When the welding gun is positioned ahead of the weld with the welding wire at a drag angle, the technique is referred to as pulling (dragging) the weld metal. If the welding gun is behind the weld with the electrode wire at a push angle, it is said to be pushing the weld metal. Generally, pulling produces deeper penetration than push travel. Pulling also makes it easier to see the weld pool. The push technique allows faster travel speeds and produces less penetration and wider welds. Pushing is used on thin-gauge metals to prevent excessive melt-through. **See Figure 20-3.**

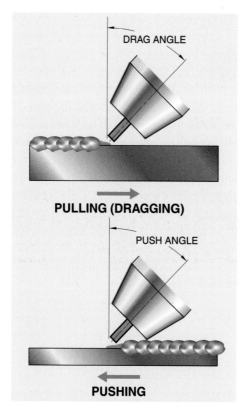

DRAG ANGLE

PULLING (DRAGGING)

PUSH ANGLE

PUSHING

Figure 20-3. *Pulling (dragging) the weld is preferred for welding thick metals, while pushing the weld is preferred for light-gauge metals.*

GMAW SET-UP PROCEDURES

Before starting to weld, check the following:
- Ensure that all electric power controls are in the OFF position and check that the welding machine is set to DCEP.
- Check all electrical cables to make sure that they are in good condition and are properly insulated and secured. Check that there are no

splices in the cables within 10 of the welding area.
- Ensure the gun cable liner, contact tip, gas nozzle, and wire feeder drive rolls are the proper size for the diameter of the welding wire.
- Blow out the welding gun cable liner with shop air before installing a new spool of welding wire. Without regular cleaning, the gun cable liner may become clogged with dust and residue from the welding wire. This can restrict the welding wire in the gun cable liner causing problems with burnback and bird nesting.
- Check the O-rings on the outlet guide of the welding gun cable for damage.
- Install the welding gun cable in the wire feeder and ensure the hole in the outlet guide is aligned with the grooves in the drive rolls.
- Install the spool of welding wire. Feed the welding wire through the inlet guide at the back of the wire feeder, across the drive rolls, and into the outlet guide in the welding gun cable.
- Remove the gas nozzle and contact tip, turn the welding machine ON, and use the jog button to feed the welding wire through the gun cable liner and welding gun.
- Test drive-roll tension by pressing the jog button and applying slight hand pressure to the spool of wire. If the wire continues to feed, reduce drive-roll tension. Too much drive-roll tension can cause problems with bird nesting. Too little tension can cause burnback.
- Turn the power OFF and install the contact tip and gas nozzle.
- Crack the gas cylinder by opening and closing the cylinder valve to clear out any dirt and install the regulator and flow meter assembly.
- Connect the gas hose from the flow meter to the wire feeder.
- Open the gas cylinder valve completely. Turn the welding machine

ON, press the "purge" button on the machine, and adjust the gas flow rate. Then turn the power OFF and check for gas leaks.

- Check the coolant level and water line connections if the welding gun is water cooled.
- Set the wire feed speed (current) and voltage to the middle of the range specified for the job.
- Run a few test welds on a piece of scrap metal to fine-tune the settings.

During any welding operation, certain welding conditions may have to be changed. Welders should be familiar with common welding variables and the required changes that must be made during welding. **See Figure 20-4.**

Starting the Arc

Make sure the workpiece connection is attached to the workpiece. Position the welding gun over the joint with the proper gun angles and visible electrode stickout. If the wire feed speed and voltage are set correctly, the arc should start when the gun trigger is pulled. To help ensure a good arc start, the welding wire can make electrical contact with the work. The welding wire must exert sufficient force on the workpiece to penetrate any surface impurities. **See Figure 20-5.**

Starting the arc becomes increasingly difficult as electrode extension increases. A reasonable balance between voltage and wire feed speed must be maintained to ensure the proper arc and to deposit the weld metal at the best wire-melting rate. Once the arc is started, the welding gun is held at the correct work and travel angles and moved at a uniform travel speed to keep the arc on the leading edge of the weld pool.

WELDING VARIABLES								
Change Required		Arc Voltage	Welding Current*	Travel Speed	Travel Angle	Electrode Extension/ Stickout	Wire Size	Gas Type
Deeper Penetration			Increase (1)		Drag max. 25°	Decrease (2)	Smaller† (5)	CO$_2$ (4)
Shallow Penetration			Decrease (1)		Push (3)	Increase (2)	Larger† (5)	A+CO$_2$
Bead Height and Bead Width	Larger Bead		Increase (1)	Decrease (2)		Increase† (3)		
	Smaller Bead		Decrease (1)	Increase (2)		Decrease† (3)		
	Higher, Narrower Bead	Decrease (1)			Drag (2)	Increase (3)		
	Flatter, Wider Bead	Increase (1)			90° or Push (2)	Decrease (3)		
Faster Deposition rate			Increase (1)			Increase† (2)	Smaller (5)	
Slower Deposition rate			Decrease (1)			Decrease† (2)	Larger (5)	

Key: (1) First choice, (2) Second choice, (3) Third choice, (4) Fourth choice, (5) Fifth choice
* Same adjustment is required for wire feed speed
† When these variables are changed, the wire feed speed must be adjusted so that the welding current remains constant

Figure 20-4. *Welding conditions may change during welding, requiring adjustments to welding variables.*

Figure 20-5. *Electrical contact is required to start an arc.*

Media Clip

GMAW-Short Circuiting Transfer

Welding a Joint

In general, the GMAW welding procedure follows a definite sequence regardless of the type of welding that is being done. For welding a joint with GMAW, use the following procedure:

1. Set the voltage, wire feed speed, and shielding gas flow to the standard conditions for the required type of welding.
2. Adjust the welding wire to the proper electrode extension stickout.
3. Start the arc and move the welding gun at a uniform speed, maintaining the proper gun angles. If the arc is not started properly, the welding wire may stick to the work. If the welding wire sticks, release the trigger to prevent bird nesting, shut the machine OFF, clip the wire with a pair of wire cutters, and remove the welding wire from the joint.
4. Move the welding gun along the joint using the pushing or pulling technique. Keep the welding wire at the leading edge of the weld pool. Be sure the welding wire is centered in the shielding gas to ensure adequate shielding. A slight weaving motion is helpful to ensure fusion in the toes of the joint when welding with short circuiting transfer.
5. Release the trigger at the end of the weld to stop the arc, wire feeder, and shielding gas. If the welding machine is equipped with a postflow timer and the postflow is set, hold the welding gun over the weld until the postflow times out.
6. Properly shut down the welding machine when welding is completed:
   a. Turn OFF wire feeder.
   b. Close the shielding gas cylinder valve.
   c. Squeeze welding gun trigger to bleed the gas lines.
   d. Shut OFF the welding machine.
   e. Hang up the welding gun.

METAL TRANSFER MODES

Metal transfer is the manner in which molten metal travels from the end of the welding wire to the base metal. The three modes of metal transfer in GMAW are short circuiting transfer, globular transfer, and spray transfer. The factors affecting the type of metal transfer are wire feed speed, arc voltage, type of shielding gas, and welding wire diameter size.

Short Circuiting Transfer

Short circuiting transfer (GMAW-S) is a metal transfer mode in which molten metal from consumable welding wire is deposited during repeated short circuits. The number of short circuits can occur up to 200 times a second depending on welding power-source controls. **See Figure 20-6.** GMAW-S occurs at current levels below 200 A, with smaller welding wire diameters of 0.045″ or less. The low heat input associated with GMAW-S make it suitable for welding thin-gauge base metals up to ¼″. GMAW-S can be used in all welding positions because it produces a small, fast-freezing weld pool.

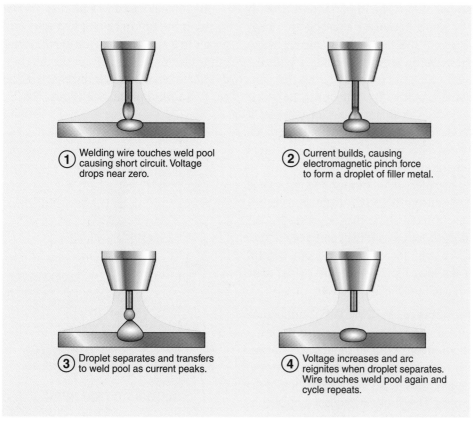

Figure 20-6. *Short circuiting transfer is practical for all welding positions, especially where control of the weld pool is difficult.*

GMAW-S works well on joints with poor fit-up without producing excessive melt-through and is commonly used to weld root passes on pipe and open root joints on plate. GMAW-S produces faster travel speeds, higher electrode efficiencies (in the range of 93%), and less distortion because of its lower heat input. GMAW-S is easy to use, which give it high operator appeal. However, the low heat input associated with GMAW-S can result in incomplete fusion. A mixture of 75% Ar and 25% CO_2 is commonly used for GMAW-S.

A mixture of 75% Ar and 25% CO_2 produces faster travel speeds, a smooth and focused arc with good penetration characteristics, and less spatter than 100% CO_2. Straight CO_2 can be used where good penetration is essential but bead contour and appearance are not particularly important. Typical electrode extensions for GMAW-S are 3/8″ to 1/2″ with the contact tip even with the gas nozzle. The transfer of a single molten droplet welding wire to the weld pool occurs as follows:

1. At the start of the cycle, the welding wire touches the molten weld pool causing a short circuit. The arc voltage drops to near zero, and the current increases.

2. As current increase, it applies electromagnetic force evenly around the electrode. This force is called the electromagnetic pinch force. The pinch force squeezes the electrode, causing a droplet of filler metal to form at the tip. The voltage also begins to rise.

3. The droplet separates from the wire and transfers to the weld pool as current peaks.

4. When the droplet separates, voltage increases and the arc reignites. The wire shorts out against the weld pool, and the cycle repeats.

Two important variables that affect droplet transfer in GMAW-S are slope and inductance.

Slope. Slope is an important function of GMAW-S because it controls the magnitude of the short-circuit current when the welding wire is in contact with the work. Slope is represented by the volt-ampere curve. By adjusting the slope to reduce peak short-circuit current, it produces a smaller electromagnetic pinch effect. This in turn causes a less violent separation of the droplet, resulting in less spatter and a more stable arc. Some constant-voltage (CV) power sources have a fixed slope that is preset for the most common welding conditions. Other CV power sources have a slope control that allows the short-circuit current to be adjusted for the application.

Inductance. In GMAW-S, inductance affects the rate of current rise during a short circuit. Without inductance, current rises quickly, and the pinch effect is applied rapidly, causing a violent separation of the droplet. This results in frequent short circuits per second, small low-energy droplets with shallow penetration, and excessive spatter. As inductance increases, current rises more slowly, and the pinch effect takes longer, causing a less violent separation of the droplet. This results in fewer short circuits per second, large high-energy droplets capable of deep penetration, and reduced spatter.

Fewer short circuits increase the amount of arc-on time. More arc-on time produces a more fluid weld pool resulting in a flatter bead contour. The goal of inductance control is to produce the smallest possible droplet, with enough energy for good fusion, and the least amount of spatter. Depending on the design of the power source, inductance can be fixed or adjustable.

Globular Transfer

Media Clip
GMAW-Globular Transfer

When the wire feed speed and voltage are increased above the upper range for GMAW-S, the transfer mode changes to globular transfer. *Globular transfer* is the transfer of molten metal in large droplets from the welding wire across an arc to the workpiece. Globular transfer occurs at the rate of a few droplets per second.

In globular transfer, a molten droplet forms at the tip of the welding wire and grows to a size about twice the diameter of the electrode. Gravity causes the large droplet to separate and transfer across the arc to the workpiece. **See Figure 20-7.**

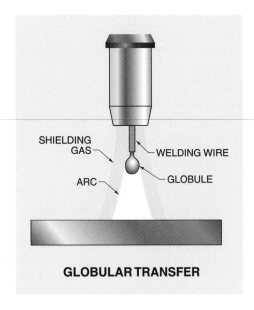

GLOBULAR TRANSFER

Figure 20-7. *In globular transfer, the molten drop grows to two or three times the diameter of the welding wire before separating and transferring to the workpiece.*

As the globule moves across the arc, it assumes an irregular shape and a rotary motion because of the physical forces of the arc. Metal transfer also occurs as the result of occasional short circuits. The result is poor arc stability, poor penetration, and excessive spatter. Globular transfer can occur at high travel speeds, but high spatter levels reduce electrode efficiency and require extensive post-weld clean up. Globular transfer is also prone to fusion defects such as overlap.

Globular transfer is limited to the flat position and horizontal fillet welds because of its large droplet size and its dependence on gravity. As a result, globular transfer is not very effective for most GMAW operations. Typical electrode extension for globular transfer is ¾″ to 1″ with the contact tip recessed about ⅛″ in the gas nozzle. Gas flow rates are in the range of 35 cfh to 50 cfh.

Spray Transfer

Spray transfer is a metal transfer mode in which molten welding wire is propelled axially across the arc. The arc is a steady, quiet column with a well-defined, narrow, cone-shaped core in which metal transfer takes place. Very fine droplets of welding wire are rapidly projected through the arc to the workpiece in the direction in which the welding gun is pointed. The droplets are equal to or smaller than the diameter of the welding wire and produce a constant spray. The welding wire may be solid or composite (metal-cored). **See Figure 20-8.**

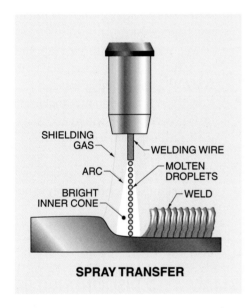

Figure 20-8. *Spray transfer occurs when very fine droplets of welding wire are projected through the arc to the workpiece.*

Spray transfer is performed at current levels above the transition current from globular to spray transfer in an argon-rich shielding gas. Argon produces a pinching effect on the molten tip of the electrode, permitting only small droplets to form and transfer during the welding process. The transition current varies depending on the type and diameter of the welding wire and the composition of the shielding gas. Spray transfer is particularly useful for welding thick sections of ferrous and nonferrous metal. It is not practical for welding thin-gauge metal because it results in excessive melt-through.

For ferrous metals, argon with mixtures of 1% to 5% oxygen are commonly used. A small amount of oxygen improves the uniformity of the weld by making the weld pool more fluid. Oxygen also increases the rate of droplet transfer and limits droplet size. However, oxygen content is limited to 5% or less to avoid porosity and other defects and discontinuities such as undercut. Argon/oxygen mixtures produce a deep finger-like penetration profile.

Argon-CO_2 mixtures are also used on ferrous metals. However, CO_2 levels cannot exceed 18%. A common gas mixture for carbon and low-alloy steels is 90% Ar and 10% CO_2, and for stainless steels is 98% Ar and 2% CO_2. An argon-CO_2 mixture produces a more rounded penetration profile compared to the finger-like penetration of argon/oxygen. Metal-cored welding wires perform well in a 90% Ar and 10% CO_2 gas shield.

Spray transfer for nonferrous metals requires 100% argon shielding gas, or a mixture of argon and helium. Even small amounts of oxygen are unsuitable for nonferrous metals because oxygen is a reactive gas. It interacts with the molten metal to form porosity and other defects.

Using a longer electrode extension with spray transfer allows for higher deposition rates. The welding wire has a longer preheat time before entering the arc, so there is less amperage needed to melt the wire and faster travel speeds are possible. If electrode extension is excessive, reduced penetration may occur. Travel speed and penetration rates must be monitored to ensure that proper penetration is taking place. Typical electrode extension for spray transfer is ¾″ to 1″, with the contact tip recessed about ⅛″ in the gas nozzle. Gas flow rates are in the range of 35 cfh to 50 cfh.

Spray transfer produces high deposition rates with high welding-wire efficiencies in the range of 98%. It can be used with a variety of welding wire types and diameters, from 0.030″ to ¹⁄₁₆″. Spray transfer produces an excellent, spatter-free bead appearance and good fusion and penetration. It requires little or no post-weld cleaning, and can be used in semiautomatic, mechanized, and automatic applications.

However, spray transfer is limited to the flat position and horizontal fillet welds due to the fluidity of the weld pool. It also produces high heat and arc radiation as well as higher fume generation, requiring added safety precautions.

Pulsed Spray Transfer (GMAW-P)

Pulsed spray transfer (GMAW-P) is a variation of spray transfer in which current is pulsed from a low background level to a peak level above the spray transfer transition current. During peak current, a single molten droplet is pinched from the end of the welding wire and transfers across the arc. At the end of the peak current portion of the cycle, current drops to the background level. Background current is set high enough to maintain the arc and heat the welding wire, but low enough that no metal transfer occurs. The pulsing cycle from peak current to background current occurs up to several hundred times per second. Pulsing frequency increases with wire feed speed. **See Figure 20-9.**

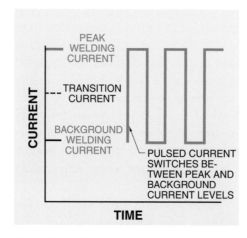

Figure 20-9. *Pulsed spray transfer (GMAW-P) occurs as the current is pulsed.*

GMAW-P requires an inverter welding machine equipped with a pulsing unit. A pulsing unit has separate controls for peak current and background current. Because welding current pulses between peak and background levels, the overall heat input is lower than continuous spray transfer. This makes it possible to weld a wide variety of metal thicknesses, including thin-gauge sheet metal. The lower overall heat input associated with GMAW-P and the small droplet size make it possible to weld in all positions. Other advantages of GMAW-P include the following:

- Spatter-free welds with minimal post-weld cleaning and excellent bead appearance
- Fewer problems with overlap and other fusion defects associated with GMAW-S and globular transfer
- Low overall heat input compensating for poor fit-up without excessive melt-through and reducing distortion problems
- Capable of high travel speeds and high welding-wire efficiencies in the range of 98%
- Produces uniform root penetration without the use of backing

GMAW WELD DISCONTINUITIES

GMAW, like any other form of welding, must be controlled properly to consistently produce high-quality welds. Welds should be analyzed to prevent repeated weld defects. Common discontinuities, such as incomplete fusion, porosity, crater cracks, insufficient penetration, excessive penetration, and whiskers, may be encountered when performing GMAW.

Porosity can be caused by setting the shielding-gas flow rate too low or too high. If the flow rate is too low, it will not properly shield the weld pool from the atmosphere. If the flow rate is too high, it causes turbulence, which draws air into the weld zone.

Incomplete Fusion (Overlap)

Incomplete fusion (overlap) is the protrusion of weld metal beyond the weld toe or weld root. It usually occurs when the arc does not melt the base metal sufficiently, causing the weld pool to flow onto unwelded base metal without fusing. **See Figure 20-10.** Often, incomplete fusion occurs when the weld pool is allowed to become too large. For proper fusion, the arc should be kept at the leading edge of the weld pool. Proper arc placement prevents the weld pool from becoming too large and flowing ahead of the welding arc. To prevent incomplete fusion, the size of the weld pool can be reduced by increasing the travel speed or reducing the wire feed speed. Also, heat input may be increased.

Porosity

Generally, surface porosity is the direct result of atmospheric contamination. Atmospheric contamination occurs if the shielding gas level is set either too low or too high. If the shielding gas level is too low, the air in the arc area is not fully displaced. If the shielding gas flow is too high, it causes turbulence, which draws air into the weld zone.

On occasion, porosity occurs if welding is performed in a windy area. Without a protective wind shield, the shielding gas envelope may be blown away, exposing the molten weld pool to the contaminating effects of the air. Subsurface porosity is caused by moisture in the shielding gas, an excessive tip-to-work distance, or surface contamination such as rust, paint, dirt, or oil on the base metal. **See Figure 20-11.**

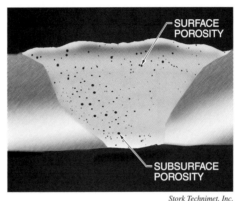

Stork Technimet, Inc.

Figure 20-11. *Porosity can occur throughout the weld area and is categorized as surface or subsurface.*

Crater Cracks

Possible causes of crater cracks include moisture in the shielding gas, excessive tip-to-work distance, improperly filled concave craters, and contaminants such as rust, paint, dirt, or oil on the base metal. **See Figure 20-12.**

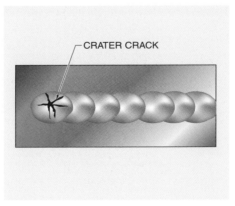

Figure 20-12. *A common cause of crater cracks is improperly filled concave craters.*

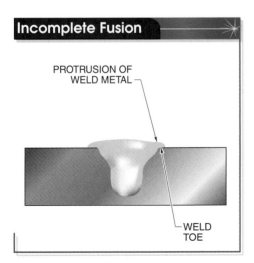

Figure 20-10. *Incomplete fusion (overlap) occurs when the arc does not melt the base metal sufficiently, causing the weld pool to flow onto unwelded base metal without fusing.*

Insufficient Penetration

Insufficient penetration (lack of penetration) is due to low heat input in the weld area, failure to keep the arc properly located on the leading edge of the weld pool, or traveling too fast. If the heat input is too low, increase the wire feed speed to increase the welding current. **See Figure 20-13.**

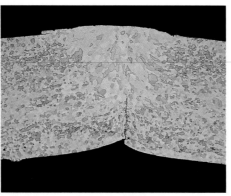

Stork-Technimet, Inc.

Figure 20-13. *Lack of penetration can result from low heat input to the weld area or a failure to keep the arc located properly on the leading edge of the weld pool.*

Excessive Penetration

Excessive penetration (excessive melt-through) is caused by excessive heat in the weld zone or too wide a root opening. Excessive penetration results in a weld root bead that protrudes below the bottom of the joint. **See Figure 20-14.**

Problems with excessive penetration can be minimized by reducing the wire feed speed, which lowers the current, or by increasing the travel speed. Too wide a root opening can be compensated for by increasing electrode extension and depositing the root pass with a weaving motion.

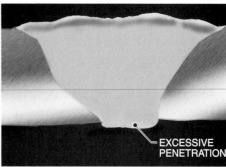

EXCESSIVE PENETRATION

Stork Technimet, Inc.

Figure 20-14. *Reducing the wire feed speed and increasing the travel speed can prevent excessive penetration.*

Whiskers

Whiskers are short lengths of electrode wire sticking through the weld joint. Whiskers are caused by pushing the wire past the leading edge of the weld pool. A small section of wire protrudes inside the joint and becomes welded to the deposited metal. To remedy this defect, reduce the travel speed, increase the contact tip-to-work distance slightly, or reduce the wire feed speed.

- GMAW is a faster welding process than SMAW and is easy to learn.
- Groove joints used with GMAW have thinner root faces, narrower root openings, and smaller groove angles. This reduces the joint area, requiring less weld metal to fill the joint.
- Keep the welding gun properly positioned to ensure a uniform weld with proper penetration.
- Ensure that the contact conductor tube and gas diffuser orifices are clean to prevent clogging, which restricts wire feed and shielding gas flow.
- If the wire feeder is equipped with a postflow timer, do not remove the welding gun from the weld area until the gas flow stops. Postflow and proper crater filling help to prevent crater cracks from developing.
- Short circuiting transfer with Ar/CO_2 shielding gas is best for welding thin-gauge carbon steels.
- A combination of high current and an argon-rich shielding gas produces spray transfer. Spray transfer has a steady, quiet arc with a well-defined core in which metal transfer takes place.
- Incomplete fusion (overlap) occurs when weld metal extends beyond the weld toe without fusing into the base metal.
- Check the weld for surface porosity. Surface porosity is usually caused by improper gas shielding.
- Insufficient and excessive penetration are the result of failure to control heat input, wire feed speed, and travel speed.

 Refer to Quick Quiz® on CD-ROM

Depositing Beads on Low-Carbon Steel in Flat Position

exercise 1

1. Obtain a 0.035″, E-70S-3 welding wire.

2. Insert the welding wire in the welding gun and set wire stickout to ¼″ to ⅜″.

3. Set the welding machine output for DCEP.

4. Set the current at 100 A to 120 A; set voltage at 19 V to 21 V.

5. Set the shielding gas (carbon dioxide) at 20 cfh.

6. Obtain a piece of low-carbon steel ³⁄₁₆″ to ¼″ thick and 4″ to 6″ long.

7. Position the workpiece in flat position.

8. Set the wire feed control so that the ammeter reads between 100 A and 120 A. To obtain the correct reading, have another person observe the current while welding is being performed.

9. Set the voltage to 26 V to 28 V using the same procedure.

10. Position the welding gun at a 90° work angle and a 10° to 15° drag angle.

11. Adjust the voltage until wire feeds properly and the bead is ⁵⁄₁₆″ wide and ⅛″ high.

12. Deposit a series of straight, consistent beads approximately ⅜″ apart.

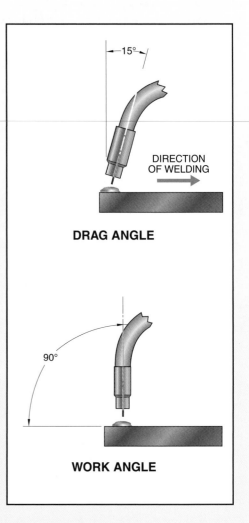

DRAG ANGLE

WORK ANGLE

Depositing Buildup on Low-Carbon Steel in Flat Position

exercise 2

1. Complete equipment setup and adjustment as in Exercise 1.

2. Obtain a piece of low-carbon steel ¼″ thick and 4″ to 6″ long.

3. Position the workpiece in flat position.

4. Position the welding gun at a 90° work angle and a 10° to 15° drag angle to deposit a bead ¼″ from the edge of the workpiece. The bead should be 5⁄16″ wide and ⅛″ high.

5. Deposit a second bead overlapping the first by half. Use an 80° to 90° work angle and a 10° to 15° drag angle.

6. Deposit consistent, overlapping beads until the workpiece is covered.

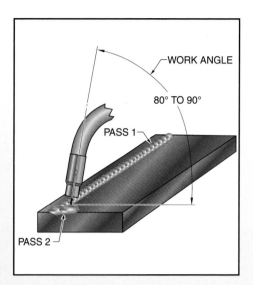

WORK ANGLE

80° TO 90°

PASS 1

PASS 2

Welding a Butt Joint on Low-Carbon Steel in Flat Position

exercise 3

1. Complete equipment setup and adjustment as in Exercise 1.

2. Obtain two pieces of low-carbon steel 3⁄16″ to ¼″ thick, 1½″ wide, and 6″ long.

3. Form a butt joint, with a 3⁄32″ to ⅛″ root opening, and tack weld.

4. Position the workpiece so the weld joint is in flat position.

5. Position the welding gun at a 90° work angle and a 10° drag angle.

6. Use a slight weaving motion to control the weld pool and a travel speed that allows for complete penetration.

7. Maintain the electrode on the leading edge of the weld pool to prevent whiskers.

8. If excessive penetration occurs, lengthen the electrode extension, decrease the current, and/or adjust the voltage for a smooth arc.

1. Complete equipment setup and adjustment as in Exercise 1.

2. Obtain two pieces of low-carbon steel ³⁄₁₆″ to ¼″ thick, 1½″ wide, and 6″ long.

3. Form a lap joint and tack together.

4. Position the workpiece so the weld joint is in flat position.

5. Position the welding gun at a 45° work angle and a 10° to 15° drag angle. Use a slight weaving motion.

6. The bead face should be flat to slightly convex.

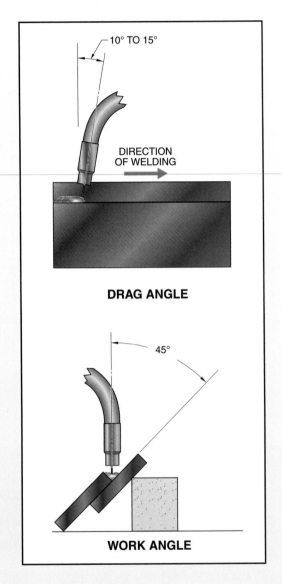

10° TO 15°

DIRECTION
OF WELDING

DRAG ANGLE

45°

WORK ANGLE

1. Complete equipment setup and adjustment as in Exercise 1.

2. Obtain two pieces of low-carbon steel ³⁄₁₆″ to ¼″ thick, 1½″ wide, and 6″ long.

3. Form a T-joint with the pieces at a 90° angle and tack together.

4. Position the workpiece so the weld joint is in horizontal position.

5. Position the welding gun at a 45° work angle and a 10° to 15° drag angle. Deposit the first pass on both sides of the T-joint.

6. Position the welding gun at a 55° work angle and a 10° to 15° drag angle. Deposit the second pass on both sides of the T-joint overlapping the first pass by half.

7. Position the welding gun at a 35° work angle and a 10° to 15° drag angle. Deposit the third pass on both sides of the T-joint overlapping the first and second passes.

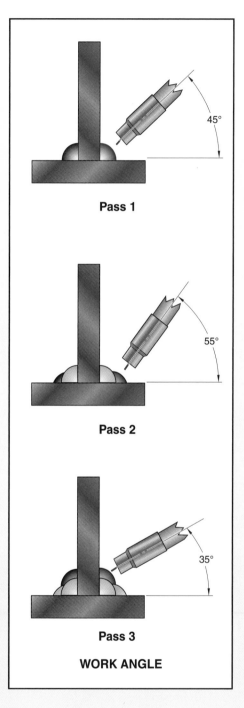

Pass 1

Pass 2

Pass 3

WORK ANGLE

1. Complete equipment setup and adjustment as in Exercise 1.

2. Obtain two pieces of low-carbon steel ³⁄₁₆″ to ¼″ thick, 2″ wide, and 6″ long.

3. Form a T-joint with the pieces at a 90° angle and tack together.

4. Position the workpiece so the weld joint is in vertical position.

5. Position the welding gun at a 45° work angle and a 10° to 20° drag angle.

6. Weld downhill using a slight weaving motion. Pause at the toes of the weld to prevent undercutting.

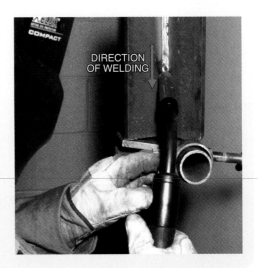

DIRECTION OF WELDING

1. Complete equipment setup and adjustment as in Exercise 1.

2. Obtain two pieces of low-carbon steel ³⁄₁₆″ to ¼″ thick, 2″ wide, and 6″ long.

3. Form a T-joint with the pieces at a 90° angle and tack together.

4. Position the workpiece so the weld joint is in overhead position.

5. Position the welding gun at a 45° work angle and a 5° to 10° drag angle. Deposit the first pass on both sides of the T-joint.

6. Position the welding gun at a 50° work angle and a 5° to 10° drag angle. Deposit the second pass on both sides of the T-joint using a slight weaving motion.

7. Position the welding gun at a 40° work angle and a 5° to 10° drag angle. Deposit the third pass on both sides of the T-joint using a slight weaving motion.

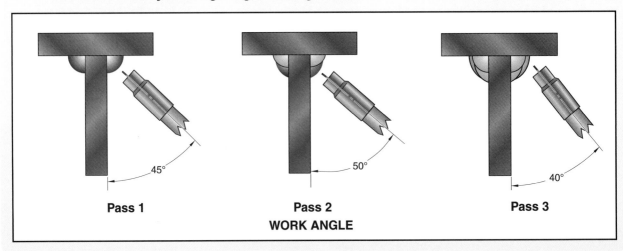

Pass 1   45°

Pass 2   50°

Pass 3   40°

WORK ANGLE

1. Obtain two pieces of aluminum, ³⁄₁₆″ to ¼″ thick, 2″ wide, and 6″ long.

2. Form the proper joint and tack together.

3. Position the workpiece so the weld joint is in proper position.

4. Set voltage, wire feed speed, and shielding gas flow.

5. Angle the welding gun to a 15° drag angle with a wire stickout of ½″ to ¾″.

6. Use the pushing technique. Use the procedures for welding mild steel to complete a butt joint, a lap joint, and a T-joint on aluminum.

? Questions for Study and Discussion

1. How does GMAW differ from GTAW?
2. At what work angle should the welding gun be held for horizontal fillet welding?
3. At what work angle should the welding gun be held for flat fillet welding?
4. What determines whether a pulling or pushing technique should be used?
5. What is the difference between spray transfer and globular transfer?
6. Why is globular transfer ineffective for welding heavy-gauge metals?
7. What is meant by short circuiting transfer? For what type of welding is this most effective?
8. What is the probable cause for the formation of overlap in a weld?
9. What should be done to prevent surface porosity in a weld?
10. How can crater porosity or cracks be prevented?
11. What should be done if weld penetration is insufficient?

Refer to Chapter 20 in the *Welding Skills Workbook* for additional exercises.

GMAW–Applications

GMAW has become an accepted process for joining all types of metals. Production welding can be easily mechanized with GMAW, substantially reducing manufacturing costs. Generally, the same type of equipment and welding techniques apply to all metals when performing GMAW.

GMAW can be used to weld carbon steel, aluminum, stainless steel, and copper. Welding parameters such as edge preparation, electrode diameter, shielding gas flow rate, proper current, and electrode feed speed are set based on weld requirements.

CARBON STEEL

Carbon steels can be welded using short circuiting transfer, spray transfer, and pulsed spray transfer. Short circuiting transfer is excellent for welding thin-gauge steel in all positions. Spray transfer is excellent for production welding of thicker steels in the flat position and for horizontal position fillet welds. Pulsed spray overcomes the limitations associated with short circuiting transfer and spray transfer. With more heat input than short circuiting transfer, it can eliminate overlap (incomplete fusion) on thicker steel and still be used in all positions. With less heat input than spray transfer, it can minimize distortion, better handle poor fit-up, and minimize problems with undercut and overlap associated with spray transfer. Edge preparation and joint design requirements vary depending on the thickness of the carbon steel.

Steel from .035″ to ⅛″ thick can be butt welded with no edge preparation. When welding a square butt joint on thin steel, a root opening of 1/16″ or less is recommended. For wider openings, short circuiting transfer should be used

because relatively large gaps at the root opening are more easily bridged without excessive penetration.

A butt joint with no edge preparation and a 1/16″ to 3/32″ root opening may be used for carbon steel 3/16″ to ¼″ thick. Two passes are generally required to ensure complete penetration and fill the joint.

Beveling is required on steel more than ¼″ thick. A single-V or double-V bevel with a 50° to 60° groove angle is used for carbon steel up to 1″ thick. A U-groove with a root opening of 1/32″ to 3/32″ is necessary on carbon steel thicker than 1″. **See Figure 21-1.** The sequence of bead deposits for multiple pass welds is similar to SMAW.

Shielding gases used with GMAW are straight CO_2 or an argon-CO_2 mixture. For short circuiting transfer on carbon steels and low-alloy steels, a 75% Ar and 25% CO_2 mixture is preferred. The argon-CO_2 mixture improves arc stability and minimizes spatter. A mixture of argon and 5% to 10% CO_2 may also be used and will result in deeper penetration with faster welding speeds.

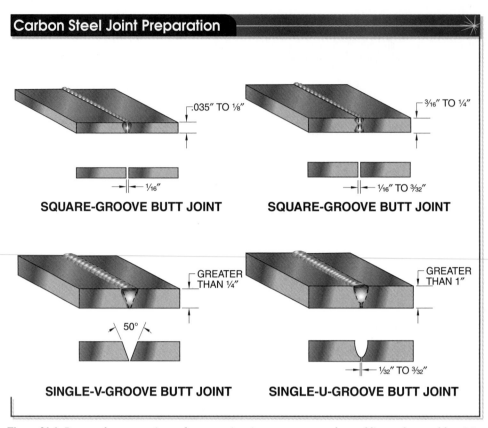

SQUARE-GROOVE BUTT JOINT
.035″ TO ⅛″
1/16″

SQUARE-GROOVE BUTT JOINT
³⁄₁₆″ TO ¼″
1/16″ TO ³⁄₃₂″

SINGLE-V-GROOVE BUTT JOINT
GREATER THAN ¼″
50°

SINGLE-U-GROOVE BUTT JOINT
GREATER THAN 1″
1/32″ TO ³⁄₃₂″

Figure 21-1. *Proper edge preparation and root opening size are necessary when welding carbon steel butt joints using GMAW.*

Carbon steel welding with spray transfer can be performed using a 95% to 98% Ar and 2% to 5% O_2 mixture. Adding oxygen to the shielding gas mixture provides a more stable arc, minimizes undercutting, and permits faster travel speeds. Straight CO_2 may be used for high-speed production welding. However, straight CO_2 produces globular transfer with excessive spatter. Pulsed spray transfer (GMAW-P) can be performed with 90% Ar and 10% CO_2. Welding parameters such as electrode diameter, proper wire feed speed (current) and voltage, and shielding gas flow are set based on the carbon steel thickness. **See Figure 21-2.**

ALUMINUM

Joint design for aluminum is similar to that for steel. However, a narrower root opening and lower welding current are recommended due to the higher fluidity of the metal. Generally, aluminum ⅛″ thick or more can be welded using GMAW. However, some welders may be able to weld thinner sections.

Argon gas is the preferred shielding gas for GMAW on aluminum up to 1″ thick because it provides better metal transfer and arc stability with less spatter. When welding aluminum between 1″ and 2″ thick, a mixture of 90% Ar and 10% He provides the higher heat input possible with helium and the good cleaning action obtained with argon. Neither oxygen, hydrogen, or CO_2 should ever be used for welding aluminum with GMAW, not even trace amounts.

Using short circuiting transfer on aluminum produces a colder arc than is produced with spray transfer, permitting the weld pool to solidify rapidly. This action is especially useful for vertical, overhead, and horizontal welding and for welding thin aluminum. When using GMAW in vertical position, a downhill technique is preferred. Welding parameters such as edge preparation, electrode diameter, argon flow, proper current and voltage, and electrode feed speed for short circuiting transfer should be set based on aluminum thickness. **See Figure 21-3.**

✓ **Point**

Welding parameters are set based on the thickness of the metal used.

GMAW — CARBON STEEL (SHORT CIRCUITING TRANSFER)

Metal Thickness*	Electrode Diameter*	DCEP Welding Current†	DCEP Arc Voltage‡	Wire Feed Speed§	Travel Speed§	Shielding Gas Flow‖
.025	.030	30 – 50	15 – 17	85 – 100	12 – 20	15 – 20
.031	.030	40 – 60	15 – 17	90 – 130	18 – 22	15 – 20
.037	.035	55 – 85	15 – 17	70 – 120	35 – 40	15 – 20
.050	.035	70 – 100	16 – 19	100 – 160	35 – 40	15 – 20
.063	.035	80 – 110	17 – 20	120 – 180	30 – 35	20 – 25
.078	.035	100 – 130	18 – 20	160 – 220	25 – 30	20 – 25
.125	.035	120 – 160	19 – 22	210 – 290	20 – 25	20 – 25
.125	.045	180 – 200	20 – 24	210 – 240	27 – 32	20 – 25
.187	.035	140 – 160	19 – 22	210 – 290	14 – 19	20 – 25
.187	.045	180 – 205	20 – 24	210 – 245	18 – 22	20 – 25
.250	.035	140 – 160	19 – 22	240 – 290	11 – 15	20 – 25
.250	.045	180 – 225	20 – 24	210 – 290	12 – 18	20 – 25

* in in.
† in amps
‡ in volts
§ in inches per minute (ipm)
‖ in cfh

Figure 21-2. *Welding parameters should be set based on carbon steel thickness.*

GMAW — ALUMINUM (SHORT CIRCUITING TRANSFER)

Metal Thickness*	Edge Preparation	Electrode Diameter*	Argon Flow†	DCEP Current‡	Voltage§	Electrode Feed Speed‖
.040	Lap, T, or tight Butt	.030	30	40	15	240
.050	Lap, T, or tight Butt	.030	15	50	15	290
.063	Lap, T, or tight Butt	.030	15	60	15	340
.093	Lap, T, or tight Butt	.030	15	90	15	410

* in in.
† in cfh
‡ in amps
§ in volts
‖ in ipm (approximate)

Figure 21-3. *Welding parameters for short circuiting transfer on thin aluminum should be set based on aluminum thickness.*

Point

Argon gas is the preferred shielding gas for GMAW on aluminum up to 1″ thick because it provides better metal transfer and arc stability with less spatter.

Spray transfer on aluminum is especially suitable for thick sections. With spray transfer, more heat is produced to melt the electrode and the base metal. Vertical, horizontal, and overhead welds are typically more difficult with spray transfer than with short circuiting transfer. Welding parameters such as edge preparation, electrode diameter, argon flow, proper current and voltage, and electrode feed speed for spray transfer should be set based on aluminum thickness. **See Figure 21-4.**

Pulsed spray transfer (GMAW-P) is capable of welding thin and thick aluminum. On thicker materials, GMAW-P can reduce heat input and improve out-of-position welding. The forehand, or push technique, is preferred for aluminum welding.

STAINLESS STEEL

Stainless steel was initially developed to prevent the rusting and corrosion that occured with carbon steel. Stainless steel is produced at a higher quality level than carbon steels and has fewer impurities, making it a reliable material for welding. On stainless steel 1/4″ thick or more, the welding gun should be moved back and forth with a slight side-to-side movement. Thin stainless steel is best welded with a slight back-and-forth motion along the joint. **See Figure 21-5.** The forehand technique is generally used for welding stainless steel.

GMAW — ALUMINUM (SPRAY TRANSFER)

Metal Thickness*	Edge Preparation	Electrode Diameter*	Argon Flow†	DCEP Current‡	Voltage§
.250	Single-V-groove butt (60° groove angle) sharp root face use backing bar	3/64	35	180	24
	Square-groove butt with backing bar	3/64	40	250	26
	Square butt with no backing bar	3/64	35	220	24
.375	Single-V-groove butt (60° groove angle) sharp root face use backing bar	1/16	40	280	27
	Double-V-groove butt (75° groove angle 1/16″ root face). No backing bar. Back chip after root pass	1/16	40	260	26
	Square-groove butt with no backing bar	1/16	50	270	26
.500	Single-V-groove butt (60° groove angle) sharp root face use backing bar	1/16	50	310	27
	Double-V-groove butt (75° groove angle 1/16″ root face). No backing bar. Back chip after root pass	1/16	50	300	27

* in in.
† in cfh
‡ in amps
§ in volts

Figure 21-4. *Welding parameters for spray transfer on thick aluminum should be set based on aluminum thickness.*

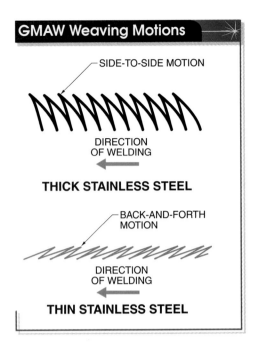

GMAW Weaving Motions

SIDE-TO-SIDE MOTION

DIRECTION
OF WELDING

THICK STAINLESS STEEL

BACK-AND-FORTH
MOTION

DIRECTION
OF WELDING

THIN STAINLESS STEEL

Figure 21-5. When using GMAW to weld stainless steel plates, a slight weaving motion is used.

Many of the characteristics of stainless steel, such as its corrosion resistance, sensitivity to heat, and low thermal and electrical conductivity, can be controlled once a welder understands how welding affects these characteristics. Properly identifying the type of stainless steel and its particular characteristics is necessary to determine which characteristics to control during welding. When using GMAW on stainless steels, proper ventilation is necessary due to the fumes given off by the metal.

Short circuiting transfer can be used on thin stainless steel in overhead or vertical position. Welding parameters such as edge preparation, electrode diameter, shielding gas flow, proper current and voltage, electrode feed speed, welding speed, and welding passes for short circuiting transfer should be set based on stainless steel thickness. **See Figure 21-6.**

Quality welds can be produced on stainless steel using the spray transfer process with a $1/16''$ diameter electrode and high current. DCEP with argon and 1% to 2% O_2 may be used for spray transfer on stainless steel. Welding parameters such as edge preparation, electrode diameter, shielding gas flow, proper current, electrode feed speed, welding speed, and welding passes for spray transfer should be set based on stainless steel thickness. **See Figure 21-7.**

GMAW — STAINLESS STEEL (SHORT CIRCUITING TRANSFER)								
Metal Thickness*	Edge Preparation	Electrode Diameter*	Gas Flow†	DCEP Current‡	Voltage§	Electrode Feed Speed‖	Welding Speed‖	Welding Passes
.063	Non-positioned T or lap	.030	12 – 30	85	15	184	18	1
.063	Square butt	.030	12 – 30	85	15	184	20	1
.078	Non-positioned T or lap	.030	12 – 30	90	15	192	14	1
.078	Square butt	.030	12 – 30	90	15	192	12	1
.093	Non-positioned T or lap	.030	12 – 30	105	17	232	15	1
.125	Non-positioned T or lap	.030	12 – 30	125	17	280	16	1

* in in.
† CO_2 in cfh
‡ in amps
§ in volts
‖ in ipm

Figure 21-6. Welding parameters for short circuiting transfer on stainless steel should be set based on stainless steel thickness.

GMAW — STAINLESS STEEL (SPRAY TRANSFER)								
Metal Thickness*	Edge Preparation	Electrode Diameter*	Gas Flow	DCEP Current‡	Voltage§	Electrode Feed Speed‖	Welding Speed‖	Welding Passes
.125	Square-groove butt with backing	1/16	35†	200 – 250	24	110 – 150	20	1
.250	Single-V-groove butt (60° groove angle) no root face	1/16	35†	250 – 300	25 – 26	150 – 200	15	2
.375	Single-V-groove butt (60° groove angle) 1/16″ root face	1/16	(1% O_2)	275 – 325	25 – 26	225 – 250	20	2
.500	Single-V-groove butt (60° groove angle) 1/16″ root face	3/32	(1% O_2)	300 – 350	26 – 27	75 – 85	5	3 – 4
.750	Single-V-groove butt (90° groove angle) 1/16″ root face	3/32	(1% O_2)	350 – 375	25 – 27	85 – 95	4	5 – 6
1.000	Single-V-groove butt (90° groove angle) 1/16″ root face	3/32	(1% O_2)	350 – 375	25 – 27	85 – 95	2	7 – 8

* in in.
† in cfh
‡ in amps
§ in volts
‖ ipm

Figure 21-7. *Welding parameters for spray transfer on stainless steel should be set based on stainless steel thickness.*

✓ **Point**

Steel backing bars are required for welding copper 1/8″ thick or less.

Copper backing bars should be used when welding stainless steel up to 1/16″ thick. Precautions must be taken to prevent air from reaching the underside of the weld while the weld pool is solidifying because oxygen and nitrogen in the air will embrittle the weld. To prevent air from contacting the underside of the weld, an argon back-up gas is often used.

COPPER

Using GMAW on copper is usually restricted to the deoxidized types of copper. Welding electrolytic copper or tough pictch copper is not advisable because of the potential for embrittlement exhibited by such welds. Argon is preferred as the shielding gas for thin copper. For copper 1″ thick or more, a mixture of 65% He and 35% Ar is recommended.

✓ **Point**

Preheating copper at 400°F is advisable on sections 3/8″ thick or more.

Steel backing bars are required for welding copper 1/8″ thick or less. Preheating at this thickness is not necessary. Preheating at 400°F is advisable on sections 3/8″ thick or more. Welding parameters such as edge preparation, electrode diameter, proper current and voltage, electrode feed speed, and welding speed should be set based on copper thickness. **See Figure 21-8.**

Filler metals that can be used to weld copper are specified in ANSI/AWS A5.6, Specifications for Covered Copper and Copper Alloy Arc Welding Electrodes.

GMAW — COPPER						
Metal Thickness*	Edge Preparation	Electrode Diameter*	DCEP Current†	Voltage‡	Electrode Feed Speed§	Welding Speed§
1/8	Square-groove butt, with steel backing bar	1/16	310	27	200	30
1/4	Square-groove butt	3/32	460	26	135	20
1/4	Square-groove butt	3/32	500	26	150	20
3/8	Double-bevel-groove, 90° groove angle, 3/16″ root face	3/32	500	27	150	14
3/8	Double-bevel-groove, 90° groove angle, 3/16″ root face	3/32	550	27	170	14
1/2	Double-bevel-groove, 90° groove angle, 1/4″ root face	3/32	540	27	165	12
1/2	Double-bevel-groove, 90° groove angle, 1/4″ root face	3/32	600	27	180	10

* in in.
† in amps
‡ in volts
§ in ipm

Figure 21-8. *Welding parameters should be set based on copper thickness.*

Refer to Quick Quiz® on CD-ROM

Points to Remember

- Welding parameters are set based on the thickness of the metal used.
- Argon gas is the preferred shielding gas for GMAW on aluminum up to 1″ thick because it provides better metal transfer and arc stability with less spatter.
- When using GMAW on stainless steels, proper ventilation is necessary to remove the fumes emitted.
- Steel backing bars are required for welding copper 1/8″ thick or less.
- Preheating copper at 400°F is advisable on sections 3/8″ thick or more.

1. When welding carbon steels, what thickness range may be butt welded with no edge preparation?
2. What type of joint is required for carbon steel greater than 1″ thick?
3. Which shielding gas mixture is recommended for welding carbon steels?
4. Why is spray transfer preferred for welding thick sections of aluminum?
5. Which technique should be used for GMAW in vertical position?
6. What type of backing is required when welding stainless steel?
7. What type of backing is required when welding copper?
8. When should preheating be used on copper?

Refer to Chapter 21 in the *Welding Skills Workbook* for additional exercises.

Flux-Cored Arc Welding (FCAW)

Chapter

The flux-cored arc welding (FCAW) process was developed in the 1950s. It is an arc welding process similar to GMAW in that it uses a continuously fed electrode. FCAW has become more commonly used as a result of developments and improvements in welding machines, wire feed systems, and fluxes. Welding guns equipped with fume extractors have also improved FCAW welding conditions. FCAW can be used to weld carbon steels, low-alloy steels, various stainless steels, and high-strength quenched and tempered steels.

Self-shielded flux-cored arc welding (FCAW-S) is a variation of FCAW in which the shielding gas is provided solely by the flux material within the electrode. FCAW-S is commonly used on medium thicknesses of metal and can be used for all-position welding. Gas-shielded flux-cored arc welding (FCAW-G) is an FCAW variation that requires external shielding gas. FCAW-G produces high-quality welds at a lower cost and with less effort than SMAW. FCAW-G is generally limited to thicker base metals because it produces a deeper-penetrating weld than FCAW-S.

FLUX-CORED ARC WELDING (FCAW)

Flux-cored arc welding (FCAW) is an arc welding process that uses a tubular electrode with flux in its core. FCAW is very similar to GMAW with respect to operation and the type of equipment used. In FCAW, weld metal is transferred as in GMAW globular or spray transfer. FCAW is capable of high weld metal deposition rates and deep penetration.

The flux-cored arc welding process was developed in the 1950s with the development of an electrode that contained a core of powdered flux material. However, even with the flux-cored electrode, an external shielding gas was required. In 1959, a flux-cored electrode was developed that did not require an external shielding gas. Shielding gas could be generated solely by the flux contained in the core of the electrode as it was being consumed during the welding process. This reduced the cost of the welding process by eliminating the need for additional shielding gas and its accompanying equipment.

The two variations of FCAW are self-shielded flux-cored arc welding (FCAW-S)

and gas-shielded flux-cored arc welding (FCAW-G). *Self-shielded flux-cored arc welding (FCAW-S)* is an FCAW variation in which shielding gas is provided exclusively by the flux core within the electrode. The heat of the welding arc causes the flux to melt, creating a gaseous shield around the arc and the weld pool. **See Figure 22-1.** FCAW-S is also called Innershield® in the field. Innershield® is a flux-cored arc welding process developed by the Lincoln Electric Company.

Gas-shielded flux-cored arc welding (FCAW-G) is an FCAW variation in which auxiliary shielding gas is supplied from an external source through the welding gun cable and gas diffuser, as well as from the flux core of the electrode. The shielding gas can be 100% CO_2, a mixture of argon-CO_2 or argon-oxygen, and other gases depending on the base metal. FCAW-G is commonly performed in the flat position and on horizontal position fillet welds. However, vertical or overhead welding is possible with small-diameter electrodes.

Both FCAW-S and FCAW-G produce a slag coating over the weld to protect it from the atmosphere. The slag produced by FCAW electrode wires is easily removed.

✓ **Point**

Self-shielded flux cored arc welding (FCAW-S) is an FCAW variation in which shielding gas is provided exclusively by the flux within the electrode.

Media Clip

Flux Cored Arc Welding

✓ **Point**

Gas-shielded flux cored arc welding (FCAW-G) is an FCAW variation in which the shielding is obtained from shielding gas flowing from the gas nozzle and from the flux core of the electrode.

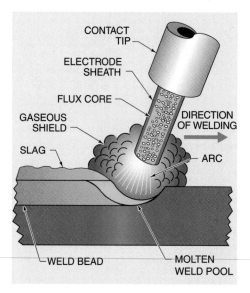

CONTACT TIP
ELECTRODE SHEATH
FLUX CORE
GASEOUS SHIELD
DIRECTION OF WELDING
SLAG
ARC
WELD BEAD
MOLTEN WELD POOL

Figure 22-1. *In FCAW-S, a tubular electrode containing flux ingredients is used to produce a gaseous shield around the weld pool.*

FCAW-G, shielding gas and a shielding gas supply system are required.

The welding equipment can be designed for semiautomatic or mechanized operation. With semiautomatic equipment, the welder moves a hand-held welding gun along the weld joint. With mechanized equipment, the operator makes equipment adjustments as required while observing the welding operation. **See Figure 22-2.**

FCAW electrode wires produce smoke and fumes. Some base metals can also produce toxic fumes when welded. Fume extraction equipment should be used to keep smoke and fumes from entering under the welder's helmet and to protect others who are working in the area. A fume extractor with a flexible hose arm can be positioned close to the fume source. **See Figure 22-3.** It is good safety practice to use a welding helmet equipped with a respirator to protect the face, eyes, nose, and lungs from smoke and fumes.

Specially designed welding guns are available with built-in fume extraction systems to evacuate smoke and fumes from the weld area while providing maximum visibility. FCAW-S may be used in the field under conditions where the air will remove the smoke and fumes from the area.

Advantages of FCAW

The FCAW process combines the best qualities of SMAW, SAW, and GMAW. FCAW uses fluxing agents that remove detrimental materials from the weld pool and improve the chemical and mechanical properties of the weld. The FCAW process enables the welder to weld continuously for long periods. FCAW produces a quality weld with less effort than SMAW and is more flexible than SAW. Some additional benefits of FCAW include the following:

- requires less precleaning of base metals than GMAW
- produces less distortion than SMAW
- produces smooth, uniform beads with an excellent weld appearance
- has a high deposition rate
- is capable of relatively high travel speeds
- welds a variety of steels and a wide range of metal thicknesses

FCAW EQUIPMENT

Equipment for FCAW is similar to that used for GMAW. A welding machine, welding gun cable and gun assembly, wire feeder, flux-cored electrode wire, and workpiece lead with a workpiece connection are required. Additionally, for

Flux cored arc welding (FCAW) is used for many of the same welding applications that use gas metal arc welding (GMAW) or shielded metal arc welding (SMAW). With FCAW, higher deposition rates are possible, there is no stub loss, and less time is wasted switching electrodes.

Welding Power Sources for FCAW

FCAW uses the same direct current, CV welding power sources used for GMAW. However, welding power sources for FCAW must be capable of the higher currents and voltages than GMAW in order to handle large-diameter FCAW welding wire. Large-diameter flux-cored welding wire can require up to 650 A. **See Figure 22-4.**

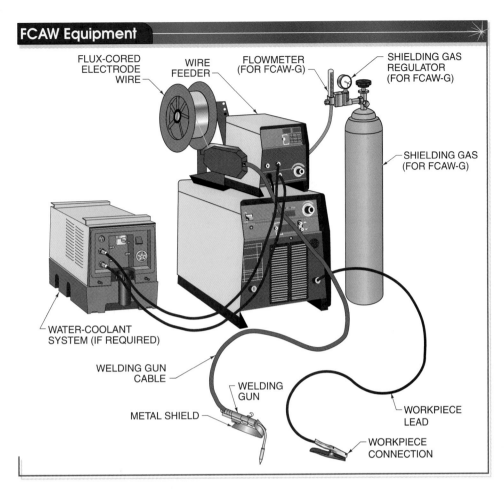

FLUX-CORED
ELECTRODE
WIRE

WIRE
FEEDER

FLOWMETER
(FOR FCAW-G)

SHIELDING GAS
REGULATOR
(FOR FCAW-G)

SHIELDING GAS
(FOR FCAW-G)

WATER-COOLANT
SYSTEM (IF REQUIRED)

WELDING GUN
CABLE

METAL SHIELD

WELDING
GUN

WORKPIECE
LEAD

WORKPIECE
CONNECTION

Figure 22-2. *FCAW equipment consists of a DC welding machine, a wire feeder, welding cables, and a welding gun. Additionally, a flowmeter, shielding gas regulator, and shielding gas are required for FCAW-G.*

The Lincoln Electric Company

Figure 22-3. *Portable fume extractors are commonly used to protect workers and to remove smoke from the work area.*

Miller Electric Manufacturing Company

Figure 22-4. *The welding machine used for flux-cored arc welding is typically a CV welding machine similar to that used for GMAW.*

Welding Gun

The type and size of welding gun for a particular application depends on the type of electrode wire (self-shielded or gas-shielded) and the maximum current levels required for the job. The types of welding guns available include pistol grip, air-cooled, and water-cooled.

Pistol grip welding guns provide for straighter feeding of large-diameter electrode wire than other types of welding guns. Air-cooled welding guns are used with lower welding currents, and water-cooled guns are designed for higher welding currents. Generally, welding applications above 600 A require a water-cooled gun to prevent overheating.

The welding gun used for FCAW-S, which has self-shielded welding wire, is equipped with a metal shield to protect the welder from heat and spatter. FCAW-S welding guns do not have a shielding gas nozzle, allowing greater access to the weld joint. **See Figure 22-5.**

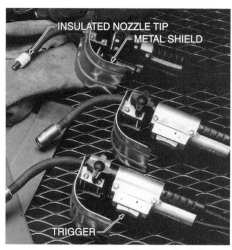

INSULATED NOZZLE TIP
METAL SHIELD

TRIGGER

The Lincoln Electric Company

Figure 22-5. *An FCAW-S welding gun has an insulated nozzle tip, a metal shield to protect the welder from slag and spatter, and a trigger to start and stop welding.*

Some FCAW welding guns are equipped with built-in fume extractors to remove smoke and fumes from the welding area. Built-in fume extractors improve visibility of the weld but add weight and bulk to the welding gun. The vacuum on the fume extractor must be set so that it does not remove shielding gas from the weld area.

Flux-Cored Electrodes

Flux-cored electrodes are composite tubular electrodes consisting of a metal sheath and a core of various powdered materials that produce an extensive slag covering on the face of a weld bead. The

AWS has developed a classification system for carbon steel electrodes for flux-cored arc welding. **See Figure 22-6.**

The E stands for electrode. The digit next to the E indicates the tensile strength of the deposited weld in ten thousand pounds per square inch, for example, 7 = 70,000 psi. The digit to the left of the T specifies welding position. A 1 means the electrode can be used in all positions. A 0 means the electrode is limited to the flat position and horizontal fillet welds. The T itself means that the electrode is tubular. A number 1 through 14 following the dash specifies usability. This includes the type of welding current (DCEN or DCEP), the composition of the powdered core, whether the electrode is gas shielded or self-shielded, and whether it is suitable for single or multiple passes.

Single-pass electrodes are high in deoxidizers for welding rusty metal and metal with mill scale. They should not be used for multiple-pass welds because a buildup of deoxidizers in the weld, such as manganese, can alter the chemical and mechanical properties of the weld metal. This can cause reduced ductility and centerline cracks.

The AWS classification may contain additional letters and numbers. For example, the letters M or C following the usability digits indicates the type of shielding gas. An M stands for mixed gas, and a C indicates 100% CO_2. The letter J in the classification number indicates that the electrode meets the minimum impact requirements of the improved toughness test (20 ft-lbf at –40°F). An H followed by a 4, 8, or 16 indicates a low-hydrogen electrode with a maximum hydrogen limit of 4, 8, or 16 ml per 100 grams of deposited weld metal. Electrodes with low hydrogen limits are designed for welding crack-sensitive steels.

Flux-cored electrodes for carbon steels are commonly identified by their letter-number suffix (T-1 through T-14). The one or two digit suffix indicates the usability characteristics of the electrode. **See Figure 22-7.**

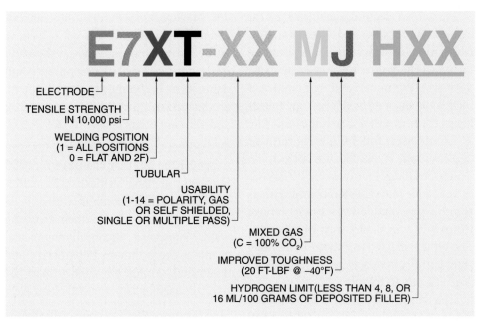

Figure 22-6. *The American Welding Society (AWS) classification system for carbon steel electrodes for FCAW consists of a series of letters and numbers.*

COMMON CARBON STEEL FLUX-CORED ELECTRODES			
AWS Classification	**Welding Current**	**Shielding**	**Single or Multiple Pass**
EXXT-1C EXXT-1M	DCEP	CO_2 75–80% Ar/bal CO_2	Multiple
EXXT-2C EXXT-2M	DCEP	CO_2 75–80% Ar/bal CO_2	Single
EXXT-3	DCEP	None	Single
EXXT-4	DCEP	None	Multiple
EXXT-5C EXXT-5M	DCEP or DCEN[†]	CO_2 75–80% Ar/bal CO_2	Multiple
EXXT-6	DCEP	None	Multiple
EXXT-7	DCEN	None	Multiple
EXXT-8	DCEN	None	Multiple
EXXT-9C EXXT-9M	DCEN	CO_2 75–80% Ar/bal CO_2	Multiple
EXXT-10	DCEN	None	Single
EXXT-11	DCEN	None	Multiple
EXXT-12C EXXT-12M	DCEN	CO_2 75–80% Ar/bal CO_2	Multiple
EXXT-13	DCEN	None	Single
EXXT-14	DCEN	None	Single
EXXT-G	Not specified*	Not specified*	Multiple
EXXT-GS	Not specified*	Not specified*	Single

[†]DCEN may be recommended for some EXXT-5C (T-5M) electrodes for out-of-position welding
*Agreed between purchaser and supplier

Figure 22-7. *Common carbon steel flux-cored electrodes are typically identified by their letter-number suffix (T1, T2, T3, etc.).*

Flux-cored electrodes are typically designed for high current densities and deposition rates.

⚠ CAUTION

Electrodes selected for FCAW must be compatible with the shielding gas used.

Flux-cored electrodes are designed for high current densities and deposition rates, which result in sharply increased production rates. The electrode size and base metal thickness determine welding parameters such as current, wire feed speed, and shielding gas flow required. **See Figure 22-8.**

Electrodes for FCAW-S require higher-current levels and longer electrode extensions than electrodes for FCAW-G. For example, the typical electrode extension for gas-shielded electrodes ranges from ¾″ to 1½″ (19 mm to 38 mm). The typical electrode extension for self-shielded electrodes ranges from ¾″ to 3½″ (19 mm to 89 mm).

The longer extension for self-shielded electrodes provides the necessary preheating to activate the shielding-gas formers and arc stabilizers in the flux. Longer electrode extension also produces deeper penetration and increases deposition rates. If the electrode extension is too short for the application, the gas formers and arc stabilizers will not burn efficiently in the arc, causing problems with arc stability and porosity in the weld.

The powdered elements in the electrode core consist of deoxidizers and denitrifiers (scavengers), slag formers, arc stabilizers, gas formers (gasifiers), and alloying elements.

Deoxidizers and Denitrifiers. Oxygen trapped in the weld can cause porosity, and nitrogen can cause brittleness. Deoxidizers and denitrifiers combine with oxygen and nitrogen and remove them from the weld. Manganese, silicon, aluminum, and titanium are common deoxidizers and denitrifiers. Aluminum is the primary deoxidizer used in self-shielded electrodes.

Slag Formers. Slag is a nonmetallic covering over the weld created by slag-forming elements in the flux. Slag protects the weld and weld pool from the atmosphere, helps to control the cooling rate of the weld, and aids in shaping the weld bead. Common slag-forming elements are calcium, potassium, silicon, sodium, and titanium.

FLUX-CORED ARC WELDING CONDITIONS

	Material Thickness*[1]	Current DCEP[†]	Arc Voltage	Wire Feed[‡]	Shielding Gas Flow[§][2]	Travel Speed[‡]	No. of Passes	Electrode Extension
Flux cored arc welding of steel using ³⁄₃₂″ diameter electrode (flat and horizontal positions)	⅛	300 – 350	24 – 26	100 – 120	35 – 40	25 – 30	1	¾ to 1½
	³⁄₁₆	350 – 400	24 – 28	120 – 150	35 – 40	25 – 35	1	¾ to 1½
	¼	350 – 400	24 – 28	120 – 150	35 – 40	20 – 30	1	¾ to 1½
	⅜	475 – 500	28 – 30	180 – 210	35 – 40	15 – 20	1	¾ to 1½
	½	400 – 450	25 – 28	150 – 170	35 – 40	18 – 20	2 – 3	¾ to 1½
	⅝	400 – 450	25 – 28	150 – 170	35 – 40	14 – 18	2 – 3	¾ to 1½
	¾	400 – 450	25 – 28	150 – 170	35 – 40	14 – 18	5 – 6	¾ to 1½

Electrode Size*	Flat Position[3]		Horizontal Position[3]		Vertical Position[3]	
	(current)[4]	(voltage)[5]	(current)[4]	(voltage)[5]	(current)[4]	(voltage)[5]
.045	150 – 225	22 – 27	150 – 225	22 – 26	125 – 200	22 – 25
¹⁄₁₆	175 – 300	24 – 29	175 – 275	25 – 28	150 – 200	24 – 27
.068	175 – 325	24 – 26	175 – 325	24 – 26	—	—
⁵⁄₆₄	200 – 400	25 – 30	200 – 375	26 – 30	175 – 225	25 – 29
.072	225 – 350	23 – 25	175 – 315	22 – 24	175 – 225	23 – 25
³⁄₃₂	300 – 500	25 – 32	300 – 450	25 – 30	—	—
⁷⁄₆₄	400 – 525	26 – 33	—	—	—	—
⅛	450 – 650	28 – 34	—	—	—	—

* in in.
† in amps
‡ in ipm
§ for FCAW-G, in cubic feet per hour (cfh)
[1] for groove and fillet welds. Material thickness also indicates fillet size. Use V groove for ¼″ and thicker double-V for ½″ and thicker
[2] welding grade CO_2
[3] applies to groove, bead, or fillet welds in position shown
[4] current range can be expanded. Higher currents can be used, especially with automatic travel
[5] voltage range can be expanded. It will increase when higher electrode–to–work distance is used

Figure 22-8. *FCAW conditions must be properly maintained and are determined by the electrode size and the material thickness.*

Slag can be classified into acid slag systems and basic slag systems. Acid slag systems promote spray transfer with low spatter and fast-freezing slag. This makes acid slag systems good for welding in all positions. Basic slag systems produce a globular-type transfer and slow-freezing slag. They are useful for welding metal coated with rust, mill scale, or contaminants because the slow-freezing slag absorbs porosity-forming gases from the weld pool. Basic slag systems produce welds with excellent mechanical properties but are typically limited to the flat position and horizontal fillet welds.

Arc Stabilizers. Arc stabilizers are added to flux to produce a smooth arc and to reduce weld spatter. Common elements used as arc stabilizers are potassium and sodium.

Gas Formers. Gas formers are added to flux to produce shielding gas that protects the electrode, molten weld pool, and HAZ from the atmosphere. Both self-shielded and gas-shielded electrodes contain gas formers. Common elements used to produce shielding gas are fluorspar and limestone.

Alloying Elements. Alloying elements are added to flux to produce desired chemical and mechanical properties in the deposited weld metal, such as hardness, strength, toughness, ductility, and corrosion resistance. Common alloying elements are carbon, chromium, manganese, molybdenum, nickel, and vanadium. The ability to manufacture FCAW electrode wires with different combinations of alloying elements for specific applications makes FCAW an extremely versatile process.

Wire Feeders

The wire feeder used with FCAW is matched to the size of the welding machine and the application. Wire feeders range in size from small units integrated into the welding machine to large industrial units capable of driving two spools of welding wire. The wire feeder provides a constant, preset feed speed of welding wire into the welding arc.

As with GMAW, the wire feed speed determines the welding current that the CV welding machine supplies. Increasing or decreasing the wire feed speed on the wire feeder changes the welding current. For some applications, a CC welding machine may be used with a voltage-sensing wire feeder that varies the wire feed speed depending upon the arc length.

A push type wire feeder is most commonly used for FCAW because of the rigidity of the flux-cored electrode. Wire feeders may be equipped with two or four drive rolls. Four-drive-roll wire feeders are used with large-diameter electrode wires. A four-drive-roll wire feeder helps to straighten the wire before feeding it into the welding gun cable. Too much drive roll tension can collapse tubular electrodes, so FCAW uses knurled drive rolls. Knurled drive rolls provide the necessary grip to push the electrode through the liner with less tension than smooth rollers.

The Lincoln Electric Company

FCAW can be used to fabricate products in all positions.

Shielding Gas

Carbon dioxide is used as a shielding gas for FCAW, and many electrodes are manufactured specifically for use with CO_2. CO_2 is commonly used as a shielding gas because it yields deep penetration, has good impact properties, and produces less smoke and fumes than other gases. CO_2 is also one of the least expensive shielding gases available, making it a cost-effective gas for FCAW-G.

Gas mixtures such as an argon-CO_2 may be used for FCAW. A common mixture is 75% Ar and 25% CO_2. This mixture may be used for out-of-position welding and when high tensile and yield strengths are required. It provides better arc characteristics for out-of-position welding.

When welding stainless steels using FCAW, a 98% Ar and 2% O_2 shielding-gas mixture may be used. The externally supplied shielding-gas mixture works with gas produced by gas formers in the flux-cored electrode to shield the arc.

Shielding-gas cylinders, a regulator and flowmeter assembly, and gas hoses deliver the shielding gas to the weld area. Gas flow rates vary depending on the electrode type, electrode extension, joint design, air movement around the weld, etc. Gas flow rates can range from 30 cfh to 45 cfh.

When using 100% CO_2, a nonmetallic washer should be inserted in the regulator connection to prevent icing. High-volume regulators or heater-equipped regulators help prevent icing as well. Icing of the regulator may allow moisture to enter the weld area, causing porosity.

Electrodes designed for 100% CO_2 should never be used with mixed gas because excessive amounts of deoxidizers from the electrode will remain in the weld. This will alter the chemical and mechanical properties of the weld deposit.

Always check for shielding gas leaks prior to welding and use proper ventilation equipment, especially when welding in confined spaces. Shielding gas is odorless and can displace oxygen in enclosed spaces without the welder's knowledge.

FCAW APPLICATIONS

FCAW combines the production efficiency of GMAW and the penetration and deposition rates of SMAW. In addition, FCAW is useful when shielding gas is unavailable. The most common application of FCAW is structural fabrication. High deposition rates achieved in a single pass make FCAW popular in the railroad, shipbuilding, and automotive industries.

FCAW can be used in all positions with the proper electrode and shielding gas. A steady push or a Z-weave technique can be used for welding in the vertical position with uphill travel. When using the weave technique, move up the joint in small increments, and stop momentarily at each toe to prevent undercut. FCAW can be used to weld carbon steels, low-alloy steels, various stainless steels, and high strength quenched and tempered steels.

A fume extractor may be needed for FCAW.

Points to Remember

- Self-shielded flux-cored arc welding (FCAW-S) is an FCAW variation in which shielding gas is provided exclusively by the flux within the electrode.
- Gas-shielded flux-cored arc welding (FCAW-G) is an FCAW variation in which the shielding is obtained from both the CO_2 gas flowing from the gas nozzle and from material contained within the flux core of the electrode.
- FCAW produces a quality weld at lower cost with less effort than SMAW and is more flexible than SAW.
- The welding gun selected is determined by the type of FCAW process used, and the highest current required for welding.
- Care must be taken to prevent buildup of deoxidizing agents in the weld. Buildup of deoxidizing agents can result in lower ductility of the weld.
- The flux in a flux-cored electrode includes ionizers to stabilize the arc, deoxidizers to purge the deposits of gas and slag, and other metals to produce high strength, ductility, and toughness in weld deposits.
- Flux-cored electrodes are typically for high current densities and deposition rates.
- When straight CO_2 is not used as a shielding gas for FCAW, a common gas mixture is 75% Ar and 25% CO_2.

Depositing Beads on Low-Carbon Steel in Flat Position

exercise   **1**

1. Obtain a ³⁄₃₂″, E70T-1 flux cored electrode.

2. Feed the electrode through the wire feeder to the welding gun and set the electrode extension between 1″ and 1½″.

3. Set the welding machine output for DCEP.

4. Set the shielding gas (carbon dioxide) at 40 cfh.

5. Set the wire feed control so that the ammeter reads between 390 A and 410 A. To obtain the correct reading, have another person observe the current while welding is being performed.

6. Set the voltage to 26 V to 28 V using the same procedure as in step 5.

7. Obtain a piece of low-carbon steel ½″ to 1″ thick, 4″ wide, and 6″ long.

8. Position the workpiece in flat position.

9. Position the welding gun at a 90° work angle and a 20° to 30° drag angle.

10. Maintain a bead that is approximately ¾″ wide and ⅛″ to ¼″ high.

11. Deposit a series of straight, consistent beads approximately ⅜″ apart.

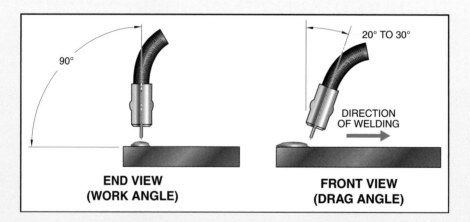

90°

END VIEW
(WORK ANGLE)

20° TO 30°

DIRECTION
OF WELDING

FRONT VIEW
(DRAG ANGLE)

Welding a Multiple-Pass Lap Joint on Low-Carbon Steel in Flat Position

1. Complete equipment setup and adjustment as in Exercise 1.

2. Obtain two pieces of low-carbon steel ¾″ to 1″ thick, 2″ wide, and 6″ long.

3. Form a lap joint and tack together.

4. Position the workpiece so the weld joint is in flat position.

5. Position the welding gun at a 45° work angle and a 20° drag angle. Deposit the first pass on both sides of the lap joint.

6. Use a weaving motion and deposit the second pass on both sides of the lap joint. Pause at the toes of the weld to prevent undercutting.

7. Deposit the third pass on both sides of the joint using the same procedure as for the second pass.

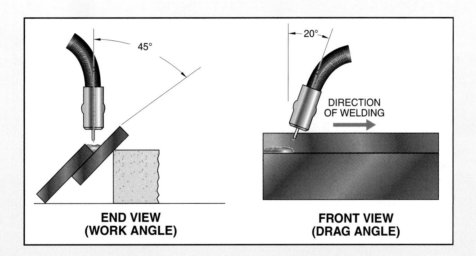

END VIEW (WORK ANGLE)   **FRONT VIEW (DRAG ANGLE)**

45°

20°

DIRECTION OF WELDING

Welding a Multiple-Pass T-Joint in Horizontal Position

exercise   **3**

1. Complete equipment setup and adjustment as in Exercise 1.

2. Obtain two pieces of low-carbon steel ¾″ to 1″ thick, 2″ wide, and 6″ long.

3. Form a T-joint with the workpieces at a 90° angle and tack together.

4. Position the workpiece so the weld joint is in horizontal position.

5. Position the welding gun at a 45° work angle and a 20° drag angle and deposit the first pass on both sides of the T-joint.

6. Position the welding gun one electrode diameter below the bottom toe of the root bead and deposit the second pass on both sides of the T-joint using a 50° to 60° work angle and a 20° drag angle.

7. Deposit the third pass on both sides of the T-joint using a 30° to 40° work angle and a 20° drag angle.

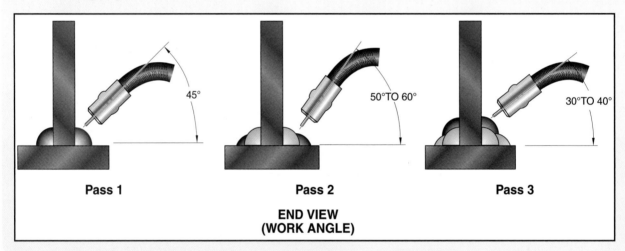

| Pass 1 | Pass 2 | Pass 3 |

**END VIEW
(WORK ANGLE)**

❓ Questions for Study and Discussion

1. What is the basic difference between FCAW and GMAW?
2. What type of welding machine is most commonly used for FCAW?
3. How does flux protect the weld metal from contaminants?
4. Why is the push type wire feeder used for FCAW?
5. What equipment is required for the FCAW-S process?
6. Why must buildup be prevented when using flux-cored electrodes?
7. What equipment is required for the FCAW-G process?
8. Why is CO_2 preferred as a shielding gas for FCAW-G?
9. How are electrodes selected for FCAW-G?

Refer to Chapter 22 in the *Welding Skills Workbook* for additional exercises.

Brazing, Braze Welding, and Soldering

Brazing and soldering differ from welding in that joining occurs when filler metal is added at temperatures below the melting point of the metals joined. Soldering also uses nonferrous filler metals with melting temperatures below 840°F.

Braze welding is slightly different from conventional brazing. In braze welding, filler metal is deposited in standard weld joints. Capillary action is not a factor in distribution of the filler metal. Braze welding is adaptable for joining or repairing metals such as cast iron, malleable iron, copper, and brass. Braze welding can also be used to join dissimilar metals such as cast iron and steel.

BRAZING

Brazing (B) is a group of joining processes that produce a coalescence of metals using nonferrous filler metals that have a melting point below that of the base metal. Filler metals suitable for brazing are those that begin to melt, or change to a liquid state, above 840°F. During brazing, the joined metals remain in a solid state. Filler metal is distributed between the closely fitted surfaces of the joint by capillary action. *Capillary action* is the force that distributes liquid filler metal through surface tension between the faying surfaces of the joint. The faying surface is the point of contact between two members to be joined.

Most metals can be brazed, including copper and copper alloys, stainless steels, magnesium alloys, aluminum alloys, carbon and low-alloy steels, cast irons, titanium and titanium alloys, and zirconium and zirconium alloys. Brazing is also used for joining dissimilar metals. One exception is that copper and copper alloys cannot be brazed directly to aluminum or aluminum alloys.

Most brazed joints have a relatively high tensile strength, but they do not possess the full strength properties produced by other welding techniques.

A characteristic of brazing is that the properties of the HAZ are not impaired during brazing because lower bonding temperatures are used than in welding. For sound brazed joints the following requirements must be met:

- Use proper joint design to allow capillary action of the filler metal and adequate surface area.
- Use proper surface preparation to ensure wetting of surfaces by the filler metal.
- Use correct fluxes for a controlled atmosphere and to prevent surface oxidation.
- Use correct filler metal, which should meet AWS standards when possible.
- Use proper heating equipment to provide specified brazing temperature and heat distribution.

Additionally, for brazing, the following criteria are necessary:

- Parts are joined without melting of the base metal.
- Filler metal begins to melt above 840°F.
- Filler metal wets the base metal and is drawn into, or held in, the joint by capillary action.

Brazing filler metal must be molten before it flows into a joint. The melting temperature of filler metals varies depending on the type of filler metal. Filler metal must have a liquidus temperature lower than the solidus temperature of the base metal. *Liquidus temperature* is the temperature at which a metal is completely molten. *Solidus temperature* is the highest temperature that a metal can reach and remain in a solid state. The lowest effective brazing temperatures possible should be used to minimize the effects of heat on the base metal. Excess heat on the base metal can cause grain growth, warpage, and hardness reduction.

Joint Design

Joint design is based on adhesive qualities of the filler metal. Two joints used for brazing are the lap joint and butt joint. A lap joint is commonly used because it offers a large surface area for the greatest strength. For maximum efficiency, the overlap should equal or exceed three times the thickness of the thinnest member. The main drawback of a lap joint is that metal thickness at the joint is increased. For joint design purposes, T-joints and corner joints are treated as butt joints.

A butt joint does not provide the same strength as a lap joint because its cross-sectional area is equal only to the cross-sectional area of the thinnest member. Higher strengths can be achieved by scarfing the edges; however, greater care is required to prepare the joint and keep the pieces aligned. The strength of a butt joint can be improved using a sleeve. **See Figure 23-1.**

Joint design is also based on joint clearance. Joint clearance has a major effect on the mechanical properties of a brazed joint. Surfaces that fit too tightly together hinder the flow of molten filler metal.

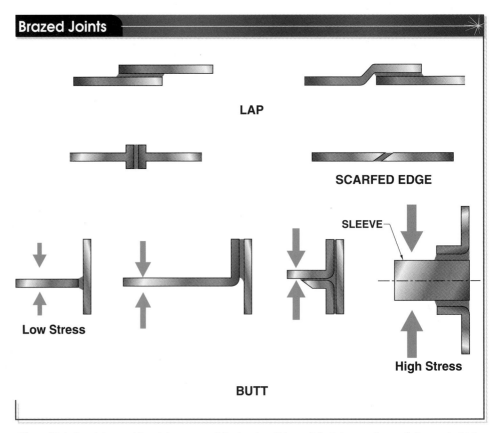

Brazed Joints

LAP

SCARFED EDGE

SLEEVE

Low Stress

High Stress

BUTT

Figure 23-1. *Lap joints and butt joints are used for brazing. Edges of the joint may be scarfed to attain higher joint strength.*

Surfaces that fit too loosely at the joint prevent the full effects of capillary action, leaving voids and poor distribution of filler metal. Adequate joint clearance is in the range between 0.001″ and 0.010″. Recommended joint clearances vary with the type of filler metal used. **See Figure 23-2.**

When welding dissimilar metals, particular attention must be paid to joint design, as all metals have different expansion rates. Varying expansion rates must be considered if parts are to be clamped, fitted together, or restrained in a jig.

Surface Preparation

Clean, oxide-free surfaces are necessary to make sound brazed joints. Uniform capillary action is only possible when surfaces are completely free of foreign substances such as dirt, oil, grease, and oxide. Foreign substances can be removed by immersing a part in a commercial-cleaning solvent or salt bath; by pickling in acid (sulfuric, nitric, or hydrochloric); or by using a vapor-degreasing unit. Surface oxide can be eliminated by sanding, grinding, filing, machining, blast cleaning, or wire brushing. The method used depends on the contaminants, the joint design, and type of metal to be brazed.

When cleaning the surface, prevent wearing the faying surfaces too smooth.

If the faying surfaces are too smooth, filler metal will not be able to effectively wet the joint. Smooth surfaces can be roughened by rubbing with a 30-grit (coarse) or 40-grit emery cloth. Brazing should be performed as soon as the metal is cleaned to prevent contamination from atmospheric exposure or handling.

Flux and Stopoffs

Metal surfaces are easily contaminated from the atmosphere after they are cleaned. Some metals are more susceptible to contamination than others. Any chemical reaction resulting from air exposure is accelerated as the temperature is raised during the brazing process; therefore, a flux is needed to dissolve and remove oxides that may form during brazing. Flux may contain boric acid, borates, fluorides, fluoroborates, chlorides, and/or wetting agents. The purpose of a flux is to prevent or inhibit the formation of oxide during the brazing process. Flux is not intended to remove contamination that is already present on metal, such as dirt, grease, and oil. **See Figure 23-3.** A *stopoff* is a material used to outline areas that are not to be brazed. Stopoffs consist of various compounds made into slurries that effectively prevent the ingress of filler metal.

> ✓ **Point**
> *Surfaces to be brazed must be completely free of oil, grease, dirt, and oxide.*

BRAZING JOINT CLEARANCE		
Filler Metal Group	**Joint Clearance***	**Brazing Joint Clearance†**
BAlSi	.002 – .008	For lap length less than ¼″
	.008 – .010	For lap length greater than ¼″
BCuP	.001 – .005	No flux and flux brazing
BAg	.002 – .005	Flux brazing
	.001 – .002‡	Gas phase (atmosphere) brazing
BAu	.002 – .005	Flux brazing
	.000 – .002‡	Gas phase (atmosphere) brazing
BCu	.000 – .002‡	Gas phase (atmosphere) brazing
BCuZn	.002 – .005	Flux brazing
BMg	.004 – .010	Flux brazing
BNi	.002 – .005	Flux or gas phase (atmosphere) brazing
	.000 – .002	Free flowing or gas phase (atmosphere) brazing

* in in.
† joint clearance on the radius when rings, plugs, or tubular members are used. Use recommended clearance on the diameter to prevent excessive clearance when all the clearance is on one side. Excessive clearance produces voids especially in gas phase brazing
‡ for maximum strength, use a press fit of 0.001 in./in. of diameter

Figure 23-2. *An accurate joint clearance is necessary for optimum strength of brazed joints.*

✓ Point

Always use an appropriate filler metal and flux that is recommended for the metal to be brazed.

The flux used for brazing must readily promote the fluidity of the filler metal. Equally important is its surface tension, since this affects the wettability of the base metal and its flow in the joint. Finally, a flux must last long enough to counteract any reactive effects developed during brazing. Some brazing filler metals are precoated with a flux.

Flux is available in powder, paste, or liquid form. Fluxes must be selected to suit a particular metal. Paste flux and powder flux are commonly used for brazing. Paste flux can be applied to a joint before brazing and provides good adherence. Powder flux is sprinkled on the joint or applied to the heated end of the filler metal by dipping the filler metal into the flux container. **See Figure 23-4.** A liquid flux is used mostly for torch brazing. The fuel gas is passed through the liquid flux, which carries the

flux along and deposits it wherever the flame is applied.

Controlled Atmosphere. A controlled atmosphere may also be used to prevent the formation of oxides during brazing. In a controlled atmosphere, a gas is continuously supplied to a furnace and circulated at slightly higher than atmospheric pressures. The gas used may consist of high-purity hydrogen, carbon dioxide, carbon monoxide, nitrogen, argon, ammonia, or some form of combusted fuel gas.

Brazing uses a higher temperature than soldering to join metal together. In brazing, the metal parts are not melted together. Brazing is best suited for joining dissimilar or thinner metals and metals that are difficult to weld or solder. Brazing should not be used on metal that will be subjected to high temperatures in service.

BRAZING FLUX					
AWS Brazing Flux*	Base Metals	Filler Metals†	Useful Range‡	Flux Agent	Available as:
1	All brazeable aluminum alloys	BAlSi	700 – 1190	Chlorides Fluorides	Powder
2	All brazeable magnesium alloys	BMg	900– 1200	Chlorides Fluorides	Powder
3A	All except those listed under 1, 2, and 4	BCuP, BAg	1050 – 1600	Boric Acid Borates Fluorides Fluoroborates	Powder Paste Liquid
3B	All except those listed under 1, 2, and 4	BCu, BCuP, BAg, BAu, RBCuZn, BNi	1350 – 2100	Wetting Agent Boric Acid Borates Fluorides Fluoroborates	Powder Paste Liquid
4	Aluminum bronze, aluminum brass and iron or nickel base alloys containing minor amounts of Al and/or Ti	BAg (all) BCuP (Copper based alloys only)	1050 – 1600	Wetting Agent Chlorides Fluorides Borates Wetting Agent	Powder Paste
5	All except those listed under 1, 2, and 4	Same as 3B (excluding BAg-1 through-7)	1400 – 2200	Borax Boric Acid Borates Wetting Agent	Powder Paste Liquid

* flux type No.
† recommended
‡ °F

Figure 23-3. *Flux prevents the formation of oxide or other undesirable substances during brazing, but does not remove contamination that is already present on the metal.*

Figure 23-4. *To apply a powder flux, heat the filler metal and dip it into the flux, making sure the filler metal is thoroughly coated.*

Filler Metals (Brazes)

Some filler metals for brazing are manufactured with a flux coating. Brazing filler metals are available in wire, rod, strip, and powder forms. Filler metals are designed to braze different metals. The AWS specifies that brazing filler metals have the following characteristics:

- be able to wet the base metal and form a strong bond between the base metal and filler metal
- have a melting temperature that permits adequate distribution by capillary action

- have sufficient homogeneity and stability to minimize separation by liquation (separation of the solid and liquid portion) and not be excessively volatile
- be capable of producing a brazed joint to meet service requirements such as strength and corrosion resistance

Filler metals may be designated by commercial names or AWS classification symbols. The AWS classification consists of the letter B, which identifies it as brazing filler metal, followed by the chemical symbols of the metallic elements included in the filler metal. **See Figure 23-5.** Digits following a dash are shown after the chemical symbols to designate specific filler metals within the group.

Filler Metal Application. Brazing filler metal and flux can be applied manually after the work is heated, or pre-placed in a suitable position before the work is heated. Rod and wire are generally used for manual face-feeding. Pre-placed filler metals are usually in the form of rings, washers, formed wire, shims, and powder, and are located near the joint to ensure a uniform flow of filler metal into the joining surfaces. Although pre-placed filler metals can be used in manual brazing, they are more commonly used for production work in furnace, induction, or dip brazing.

BRAZING FILLER METALS	
AWS Classification of Brazing Filler Metals	**Types of Metals to be Brazed**
BAlSi (aluminum-silicon)	Aluminum, aluminum alloys
BCuP (copper-phosphorus)	Copper, copper alloys
BAg (silver)	Ferrous and nonferrous metals except aluminum and magnesium
BAu (precious metals)	Iron, nickel, and cobalt base metals
BCu (copper)	Ferrous and nonferrous metals
BCuZn (copper-zinc)	Ferrous and nonferrous metals
BNi (nickel)	Stainless steels, carbon steels, low-alloy steels, copper
BMg (magnesium)	Magnesium, magnesium alloys

Figure 23-5. *The AWS classifies filler metals by the symbol of the metallic elements that are included in the filler metal.*

Manual Brazing

The heat required for manual brazing methods is typically applied using a gas torch. The gas mixture can be oxyacety-lene, air-gas, gas-oxygen, oxyhydrogen, or MAPP-oxygen. The gas mixture used depends on the thermal conductivity, type, and thickness of the metal to be brazed. **See Figure 23-6.**

Smith Equipment

Figure 23-6. *A gas torch is commonly used for manual brazing.*

Oxyacetylene and MAPP-oxygen are generally more versatile because their heat can be controlled over a wide temperature range. With either of these gas mixtures, a slightly reducing flame is required. Only the outer envelope of the flame and not the inner cone should be applied to the work.

The air-gas torch provides the lowest heat and has greater application in brazing light-gauge metals. Air-gas mixtures may use air at atmospheric pressure and city gas or an air-acetylene mixture.

A gas-oxygen mixture uses oxygen and city gas, bottled gas, propane, or butane. The gas-oxygen mixture produces a high flame temperature and is effective where higher brazing heat is required.

An oxyhydrogen mixture, due to the low heat it produces, is used for brazing aluminum and other nonferrous metals. The low temperature prevents overheating and the hydrogen provides a cleaning action and shielding during the brazing process. For brazing applications on most metals, follow the procedure:

1. Determine the most suitable joint for the work to be brazed.
2. Review safety practices per ANSI Z49.1, *Safety in Welding and Cutting.* This includes assessing the need for personal protective equipment, assessing ventilation requirements, reviewing relevant MSDSs, and assessing potential hazards such as fires.
3. Remove dirt, grease, oil, and oxides from surfaces to be brazed.
4. Select the correct flux and apply it to both the workpieces and the filler metal by brushing, dipping, sprinkling, or spraying.
5. Assemble the workpieces and keep them in alignment using clamps, fixtures, or jigs. Do not apply excessive pressure because enough clearance between the faying surfaces must exist to allow a free flow of filler metal.
6. Preheat the entire work area to a uniform brazing temperature by playing a torch over the workpiece surface.
7. Once the flux is completely fluid, touch the filler metal to the joint. Keep applying filler metal until it flows completely through the joint. Use a slightly reducing flame and do not apply the inner cone of the flame directly to the filler metal or the workpiece.
8. Clean the brazed joint to remove flux residue or debris.
9. Visually inspect the brazed joint. The joint should be free from grease, paint, oil, oxide film, and stopoff. The part should retain dimensional conformance and there should be no visible interruption to the flow of filler metal. There should be no cracks or porosity. Visual inspection cannot detect internal discontinuities. The procedure specification will indicate specific nondestructive procedures that must be performed.

Brazing using silver filler metals can be used for high stress applications that may be subjected to system vibration, and expansion and contraction that occurs on heating and cooling. **See Figure 23-7.**

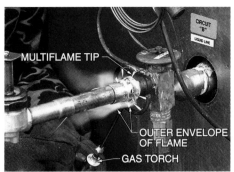

Smith Equipment

Figure 23-7. *When brazing, only the outer envelope of the flame should be applied to the work.*

Production Brazing

Although torch brazing can be mechanized for production purposes, higher production rates are usually accomplished using furnace heating, induction brazing, resistance brazing, or dip-brazing techniques. Production brazing methods ensure accurate heat control and high-quality brazed joints.

With furnace heating, parts to be brazed are positioned on trays, which are then placed in a gas, electric, or oil-fired furnace. Flux is generally used on the parts, unless the furnace atmosphere performs the function of a flux, or if cleaning of the brazed surfaces is not possible due to design complexities. The correct atmosphere must be used in a furnace and is determined by the type of base metal and filler metal used. **See Figure 23-8.**

With induction brazing, the workpiece is placed near an induction coil. As current flows through the coil, the resistance of the coil to the flow of current causes instant heating to occur. The parts are placed in an AC current field, but do not become part of the circuit. Induction brazing is commonly used for high-volume manufacturing applications. Induction brazing provides rapid heating; however, it is difficult to obtain a uniform heating rate. **See Figure 23-9.**

✓ Point

Remove all flux residue after brazing is completed.

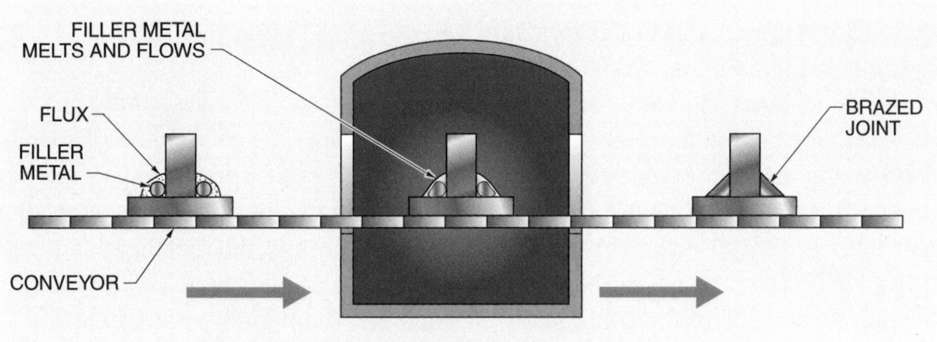

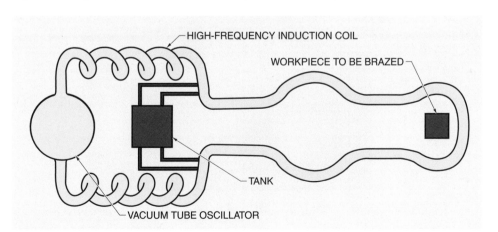

Figure 23-8. *Pre-placed filler metals are generally used for production brazing in a furnace.*

Figure 23-9. *In induction brazing, current flows through an induction coil. Resistance of the coil to the flow of current creates the necessary heat.*

Resistance brazing is similar to spot welding where heat is generated by the passage of low-voltage current through carbon electrodes that are clamped around the work. In resistance brazing, current flows through the parts being brazed and the parts become part of the electrical circuit. Resistance brazing is used with pre-placed flux and for low-volume production applications. **See Figure 23-10.**

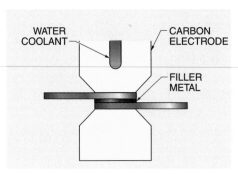

Figure 23-10. *In resistance brazing, current passes through carbon electrodes clamped around the work.*

One dip-brazing method consists of immersing parts in a bath of molten brazing metal. The brazing metal is contained in an externally heated crucible. **See Figure 23-11.** Dip-brazing is limited to small assemblies such as wire connections or metal strips that can easily be held in fixtures. A second dip-brazing method involves the placement of parts in a molten salt bath. The salt bath is heated either by passing electrical current through the bath or by heating the outside of the container.

Flux Removal

Once brazing is completed, flux residue must be removed to prevent corrosion from developing in the brazed joint. Flux residue has a glass-like surface appearance. Flux residue can be removed by washing the part in hot water. In some instances, the joint can be immersed in cold water before it has completely cooled from the brazing temperature. The thermal shock of the cold water will usually crack off the flux residue. For heavy residue, a chemical dip is sometimes used. Wire or fiber brushing, steam jet cleaning, or blast cleaning are also effective means of removing heavy residues or of removing flux residue from large objects. On some soft metals such as aluminum, residue must be removed mechanically and then cleaned with fluid to ensure removal of small flux particles that may have become embedded in the surface.

BRAZE WELDING

Braze welding (BW) is a joining process that produces a coalescence of metals with filler metals that begin to melt at temperatures above 840°F, below the melting point of the metals joined, and in which the filler metal is not distributed into the joint by capillary action.

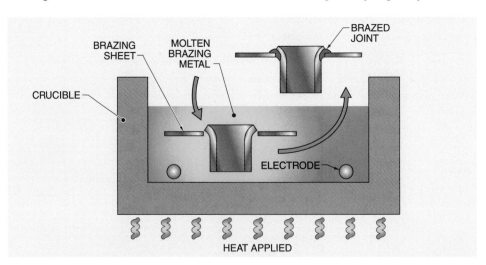

Figure 23-11. *In dip-brazing, parts are immersed in molten brazing metal inside an externally heated crucible.*

The braze welding procedure usually must be qualified. Eight basic steps are required to perform braze welding. For braze welding, follow the procedure:

1. Clean the surfaces to be brazed thoroughly with a stiff wire brush. Remove all scale, dirt, or grease; otherwise the braze will not stick. If a surface has oil or grease on it, remove these substances by heating the area to a bright red to burn them off.

2. On thick sections, especially when repairing castings, bevel the edges to form a 90° single-V. Edges can be beveled by chipping, machining, filing, or grinding.

3. Arrange the work in flat position.

4. Adjust the torch to a neutral flame then gently heat the surfaces of the weld area. The surfaces should not be melted, but only heated to a dull red (tinning temperature).

5. Heat the brazing filler metal and dip it in the flux. (This step is not necessary if the filler metal has been prefluxed.) When heating filler metal, do not apply the inner cone of the flame directly to the rod.

6. At the start, concentrate the flame on the base metal until the base metal begins to turn red. Melt a small amount of brazing filler metal onto the surface and allow it to spread along the entire seam. The flow of this thin film of filler metal is known as tinning. Unless the surfaces are tinned properly, braze welding cannot be carried out successfully.

If the base metal is too hot, filler metal bubbles or runs like drops of water on a warm stove. If the base metal is not hot enough, filler metal forms into balls that roll off the base metal as water would if placed on a greasy surface. When the base metal is the proper temperature, the filler metal spreads out evenly.

7. Once the base metal is tinned sufficiently, deposit the proper size beads over the joint. Use a slight circular motion with the torch and deposit the beads as in regular fusion welding with a filler metal. Continually dip the filler metal in the flux as the weld progresses forward. **See Figure 23-12.**

Smith Equipment

Figure 23-12. *Use a slight circular motion with the torch when braze welding and deposit the beads using filler metal.*

8. If the pieces to be welded are grooved, use several passes to fill the groove. On the first pass, ensure that the tinning action takes place along the entire bottom surface of the groove and about halfway up on each side. The number of tinning passes to be made depends on the depth of the groove. When depositing several passes, be sure that each pass is fused into the previous one.

When making a braze weld with the work in vertical position, first build up a slight shelf at the bottom. The shelf acts as a support for additional filler metal. As the weld is carried upward, swing the flame from side to side to maintain uniform tinning and to produce an even bead.

Cast Iron Braze Welding

Braze welding is primarily used to repair broken cast iron parts. High preheat temperatures are not usually required unless the part is very heavy or complex in geometry. A maximum preheat of 200°F is typically sufficient. The heat of the flame or the arc is

⚠ CAUTION

When brazing or braze welding, make sure that the base metal is not allowed to overheat.

✓ Point

Use a neutral flame unless otherwise specified. Use a circular torch motion.

✓ Point

Do not braze weld a metal that will be subjected to high temperatures or high stresses.

sufficient to bring the surface of the cast iron to a temperature at which the filler metal will bond to the cast iron. The filler metal ductility compensates for the brittleness of the cast iron, and the weld and adjacent area of the base metal are machinable after the weld is completed. Braze welding broken cast iron is acceptable if a color difference between the filler metal and the cast iron is not objectionable.

Filler Metal and Flux

Braze welding filler metals are usually brasses, with an approximate composition of 60% copper and 40% zinc, and which produce adequate tensile strength and ductility. In addition, filler metal contains small quantities of tin, iron, manganese, aluminum, lead, nickel, chromium, and silicon. These elements help deoxidize the weld metal, decrease the tendency to fume, and increase the free-flowing action of molten metal. **See Figure 23-13.**

A clean metal surface is essential for braze welding. For the filler metal to provide a strong bond, it should flow smoothly and evenly over the entire weld area. Adhesion of the molten filler metal to the base metal takes place only if the surface is chemically clean. Even after a metal surface has been thoroughly cleaned by mechanical means, certain oxides may still be present. These oxides can only be compensated for by using the correct flux.

Prefluxed brazing filler metal eliminates the need to apply flux while brazing. The flux may also be applied by dipping the heated filler metal into the powdered flux. The flux adheres to the surface of the filler metal and can then be transferred to the weld. Another method of applying flux is to dissolve the flux in boiling water and brush it on the filler metal before welding is started.

Braze Welding Disadvantages

One precaution that must be considered in braze welding is not to weld a metal that will be subjected to high temperatures in service. Filler metal loses its strength when exposed to high temperatures. Also, braze welding should not be used on steel parts that must withstand unusually high stresses.

SOLDERING

Soldering (S) is a group of joining processes that produce a coalescence of metal and nonferrous filler metal that has a melting point below that of the base metal. Filler metals suitable for soldering are those that are completely molten below 840°F. In soldering, the joined metals remain in a solid state and filler metal is distributed between the closely fitted surfaces of the joint by capillary action.

In both brazing and soldering, wetting and capillary action occur; however, in soldering, a small amount of alloying occurs between the base metal and the filler metal (solder). A major benefit of soldering is that low temperatures are involved, with a minimum effect on base metal properties. Many low-temperature heating methods can be used in soldering with high reliability. Soldering is the primary

COPPER-ZINC FILLER METAL FOR BRAZE WELDING

AWS Classification*	Approximate Chemical Composition†					Min Tensile Strength		Liquidus Temperature	
	Copper	Zinc	Tin	Iron	Nickel	ksi	MPa	°F	°C
RBCu Zn-A	60	39	1			40	275	1650	900
RBCu Zn-C	60	38	1	1		50	344	1630	890
RBCu Zn-D	50	40			10	60	413	1714	935

* see AWS A5.7 and A5.8
† in %

Figure 23-13. *A copper-zinc filler metal is commonly used for braze welding.*

method of making joints in electrical and electronic circuits. It is also commonly used in the sheet metal and plumbing industries. Precautions that must be followed for soldering include the following:

- Parts to be soldered must have the proper fit-up so that solder can travel by capillary action along the joint. Solder will cease to flow where gaps occur in the workpiece.
- Parts to be soldered must be clean because solder will not stick to dirt, oil, or oxide-coatings on the surface. Dirt and grease can be removed with a cleaning solvent. Steel wool or an abrasive cloth is used to eliminate the oxide. Application of a flux completes the cleaning process and keeps the metal free from oxide during heating and soldering.
- Parts must be held together during soldering so there is no movement. Movement during heating causes the pieces to be misaligned. The slightest disturbance to solder causes it to solidify without forming an optimum bond, resulting in a weak joint.
- Parts to be soldered must have a suitable joint design to withstand the necessary load imposed on the joint. A lap joint is a satisfactory joint for most purposes.
- Parts must be washed in hot water after soldering to eliminate the corrosive action of the flux.

Filler Metals (Solders)

Soldering uses filler metals composed of tin, lead, antimony, and sometimes silver, and produces joints with relatively low tensile strength. Most metals such as steel, galvanized sheet steel, tin plate, stainless steel, copper, brass, and bronze can be joined with a soft solder. Tin-lead alloy solders have a melting range from about 370°F, for a mixture of 70% tin and 30% lead, to about 590°F for a 5% tin and 95% lead mixture. **See Figure 23-14.**

The most common general-purpose solder is known as half-and-half or 50/50 solder. It contains 50% lead and 50% tin and melts at approximately 471°F. Alloys with a low tin content have higher melting points and do not flow as readily as high-tin alloys. Solders with a high tin content have better wetting properties and produce less cracking. Solders are available as bar, cake, solid wire, flux-core wire, ribbon, or paste. Flux-core wire solder has an acid or rosin flux in the center of the wire. With 50/50 solders, no additional flux is needed.

Special solders are available for welding aluminum and where special characteristics are required of the soldered joint. Tin-zinc solders are intended primarily for joining aluminum. A tin-antimony solder is designed to solder food-handling vessels where lead contamination must be prevented. Lead-silver solders are used for applications in which strength at elevated temperatures is required.

Flux

Just as in brazing, a flux is required for most soldering applications. The flux prevents the formation of oxides during soldering and increases the wetting action so the solder can flow more freely. General-purpose fluxes can be used on most metals.

Fluxes are classified as corrosive or noncorrosive. Rosin is the most common noncorrosive flux. Zinc chloride is the most frequently used corrosive flux. Although the corrosive types are most effective, they must be washed away from the metal after soldering. They should never be used for electrical or electronics work. Zinc chloride is prepared by adding small pieces of zinc to muriatic (commercial hydrochloric) acid until the zinc no longer dissolves. The cut, or killed, acid is then diluted with an equal quantity of water.

SOLDER COMPOSITIONS AND MELTING TEMPERATURES

Alloy Grade	Composition*				Melting Range†			
	Sn	Pb	Sb	Ag	Solidus		Liquidus	
					°F	°C	°F	°C
Sn96‡	96.2	.10	.12	3.4 – 3.8	430	221	430	221
Sn95‡	95.2	.10	.12	4.4 – 4.8	430	221	473	245
Sn94‡	94.2	.10	.12	5.4 – 5.8	430	221	536	280
Sn70	69.5 – 71.5	28.5 – 30.5	.50	.015	361	183	377	193
Sn63	62.5 – 63.5	36.5	.50	.015	361	183	361	183
Sn62	61.5 – 62.5	34.5	.50	1.75 – 2.25	354	179	372	189
Sn60	59.5 – 61.5	39.0	.50	.015	361	183	374	190
Sn50	49.5 – 51.5	49.0	.50	.015	361	183	421	216
Sn45	44.5 – 46.5	54.0	.50	.015	361	183	441	227
Sn40-A	39.5 – 41.5	59.0	.50	.015	361	183	460	238
Sn40-B	39.5 – 41.5	59.5	.50	.015	365	185	448	231
Sn35-A	34.5 – 36.5	64.0	1.8 – 2.4	.015	361	183	447	247
Sn35-B	34.5 – 36.5	62.7	50	.015	365	185	470	243
Sn30-A	29.5 – 31.5	69.0	1.6 – 2.0	.015	361	183	491	255
Sn30-B	29.5 – 31.5	67.9	.50	.015	365	185	482	250
Sn25-A	24.5 – 26.5	74.0	1.1 – 1.5	.015	361	183	511	266
Sn25-B	24.5 – 26.5	74.2	.50	.015	365	185	504	263
Sn20-A	19.5 – 21.5	79.0	.8 – 1.2	.015	361	183	531	277
Sn20-B	19.5 – 21.5	78.5	.50	.015	363	184	517	270
Sn15	14.5 – 16.5	84	.50	.015	437	225	554	290
Sn10-A	9.0 – 11.0	89.5	.20	.015	514	268	576	302
Sn10-B	9.0 – 11.0	87.8	.50	1.7 – 2.4	514	268	570	299
Sn5	4.5 – 5.5	94.5	.50	.015	586	308	594	312
Sn2	1.5 – 2.5	97.5	4.5 – 5.5	.015	601	316	611	322
Sb5‡	94.0 min	.20	.40	.015	450	233	464	240
Ag1.5	.75 – 1.25	97.1	.40	1.3 – 1.7	588	309	588	309
Ag2.5	.25	96.85	.40	2.3 – 2.7	580	304	580	304
Ag5.5	.25	93.85	.40	5.0 – 6.0	580	304	716	380

* limits are % max. unless shown as a range or stated otherwise
† temperatures given are approximate and for information only
‡ contains less than .2% lead (Pb)

Figure 23-14. *Solders are composed primarily of tin, lead, antimony, and silver.*

Joint Design and Clearance

The strength of a soldered joint depends on the design of the joint and the joint clearance. As with brazing, lap joints are the most common design, with sufficient overlap to provide the required strength. Proper joint clearance is required for maximum strength. If greater strength is needed, some type of mechanical joint should be made before soldering. A joint clearance of 0.003″ to 0.005″ is required for most applications. **See Figure 23-15.**

> Soldering uses a lower temperature than brazing (typically below 840°F) to join metal together.

⚠ WARNING

When zinc is dissolved in muriatic acid, harmful chlorine fumes are given off. Preparation must always be carried out in areas with adequate ventilation. Uncut or raw acid (straight) is preferred for galvanized steel, but cut acid (diluted) may be used and is safer to handle.

Heating Devices

In any soldering operation, both workpieces must be hot enough to melt the solder. A strong bond is achieved only if the molten solder spreads evenly over the surface. A number of devices—soldering coppers, electric soldering devices, and gas torches—are available for heating. The type used depends on the size and configuration of the assembly to be soldered. **See Figure 23-16.**

Soldering Coppers. A *soldering copper* is a tool that consists of a copper or steel heating tip fastened to a rod with a wooden handle. These coppers vary in size and have heads forged in several shapes. Generally, a lightweight copper is used for soldering light-gauge metal and a heavyweight copper is used for soldering heavy- gauge metal. Using a lightweight soldering copper on heavy metal does not produce enough heat to adequately heat the metal or allow the solder to flow smoothly. Soldering coppers are heated either in a furnace or with a blowtorch.

The point of a soldering copper must be covered with a thin coat of solder. Overheating or failing to keep the copper clean causes the point to become covered with oxide. The process of replacing this coat of solder is called tinning. To tin copper:

1. File each side of the point until all oxide and pits are removed.
2. Heat the soldering copper until it is hot enough to melt solder.
3. Rub the point of the soldering copper on a block of ammonium chloride (sal ammoniac) and apply solder while rubbing. Ammonium chloride helps clean the point of the soldering copper. Another method of applying solder is to dip the point of the soldering copper in a liquid or paste flux and then apply the solder.
4. Remove excess solder by wiping the soldering copper with a clean cloth.

Electric Soldering Devices. Electric soldering irons and pencils are often more convenient than soldering coppers because they maintain uniform heat. Electric soldering devices vary in size from 25 W to 550 W. Lightweight, low-voltage irons with replaceable heating elements and tips are called soldering pencils and are used for electrical and electronic work. An electric soldering gun is also very popular for electronic soldering work. Electric soldering guns produce instant heat at the tip of a long, small point when the trigger is pulled. On some soldering guns, the trigger also turns on a light, which focuses at the point.

Soldering Joint Designs

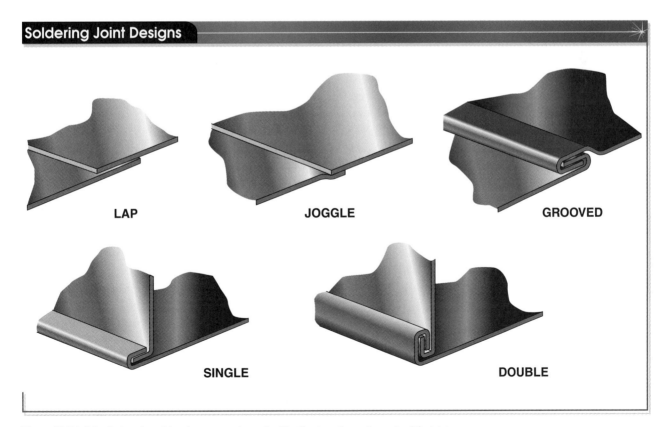

LAP   JOGGLE   GROOVED

SINGLE   DOUBLE

Figure 23-15. *Joint designs for soldered seams are determined by the strength requirements of the joint.*

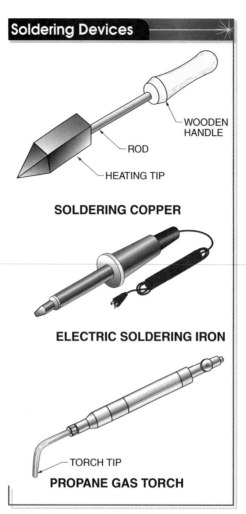

Soldering Devices

WOODEN HANDLE

ROD

HEATING TIP

SOLDERING COPPER

ELECTRIC SOLDERING IRON

TORCH TIP

PROPANE GAS TORCH

Figure 23-16. *A number of joint design devices are available to provide the necessary heat for soldering.*

Gas Torches. Some soldering operations are very difficult, or impossible, to perform with a soldering copper or iron. For such soldering tasks, a flame is used as the heat source. The flame can be produced with a gas torch. The gases used depend on the nature of the task. The most efficient, safe, and versatile gas torch is one that uses a variety of gases such as acetylene, MAPP, natural gas, propane, and compressed air.

A gas torch used for soldering is equipped with changeable tips that can produce a range of flame sizes. A gas-air torch has two needle valves; one valve controls the gas pressure and the other controls the compressed air. **See Figure 23-17.** To light a gas-air torch, the gas-needle valve is opened slightly and the gas is ignited with a sparklighter.

Then the air valve is turned on and adjusted until a neutral flame results. The length of the flame is controlled by the amount of gas and air allowed to flow to the tip.

Smith Equipment

Figure 23-17. *A gas torch can be used to solder copper pipe.*

Bottled-gas torches are also used for soldering, especially when a stationary torch is not available. The bottled-gas torch must be operated with care. Follow manufacturer instructions carefully.

Soldering Techniques

The soldering technique required is determined by the size and configuration of the joint. Common manual soldering techniques are seam soldering and sweat soldering. **See Figure 23-18.**

Seam Soldering. In seam soldering, a layer of solder is deposited along the outside edge of the joint. To solder a seam directly, place the fluxed workpieces together and tack weld the seam in several places. Tacking is done by holding the soldering copper on the metal until the flux begins to sizzle. Apply a small amount of solder directly in front of the soldering copper point. The metal should be hot enough to melt the solder. Do not apply the solder to the soldering copper. Once the workpiece is tack welded, start at one end of the seam and heat the metal. Apply solder as needed in front of the soldering copper point. If necessary, press each newly soldered section together.

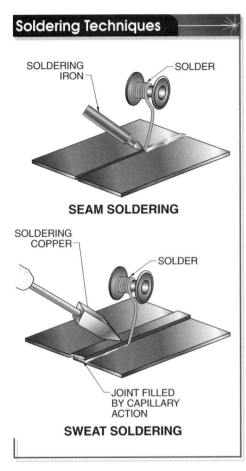

SEAM SOLDERING

SWEAT SOLDERING

SOLDERING IRON — SOLDER

SOLDERING COPPER — SOLDER

JOINT FILLED BY CAPILLARY ACTION

Figure 23-18. *In seam soldering, a layer of solder runs along the outside edge of the joint. In sweat soldering, two pieces are joined without any solder being visible.*

Sweat Soldering. *Sweat soldering* is a process whereby two surfaces are soldered together without allowing the solder to be seen. To perform sweat soldering, follow the procedure:

1. Coat the workpiece to be soldered with flux after all dirt, oil, grease, and oxide have been removed.
2. Apply a uniform coating of solder to each of the surfaces to be joined.
3. Place the surfaces together with the soldered sides in contact.

4. Place the flat side of a heated copper on one end of the seam. To avoid smearing the exposed surfaces of the metal with solder, remove any excess solder on the copper by quickly wiping the point with a damp cloth before placing it on the joint.
5. As the solder between the two surfaces begins to melt and flow out from the edges, press down on the metal with a punch. Draw the copper slowly along the seam and follow with the punch. Do not move the copper faster than the solder melts.

Inspecting Soldered Joints

Soldered joints may be visually inspected for quality as follows:

- Joint integrity. Joint should be smooth, with no porosity. A smooth transition should exist between the soldered joint and the base metals.
- Non-wetting and de-wetting. Non-wetting occurs when the solder fails to wet the metal, which retains its original color. De-wetting occurs when solder flows across the metal, but is pulled back into globules, leaving a dirty, discolored-looking surface. Both are indications of improper precleaning or flux selection.
- Overheating or underheating. Overheating is exhibited by burned fluxes and oxides on the solder joint. Underheating is exhibited by poor flow of solder into the joint. They are both indicative of poor bonding between the solder and the joint.

Refer to Quick Quiz® on CD-ROM

- Use the lowest effective brazing temperatures to minimize grain growth, warpage, and hardness reduction.
- Joint design for brazing is based on the adhesive qualities of the filler metal and on joint clearance.
- Surfaces to be brazed must be completely free of oil, grease, dirt, and oxide.
- Always use an appropriate filler metal and flux that is recommended for the metal to be brazed.
- When using oxyacetylene or MAPP-oxygen gas mixtures, heat the surfaces with the outer envelope of the flame and not the inner cone.
- Remove all flux residue after brazing is completed.
- Use a qualified procedure for braze welding.
- Clean surfaces thoroughly before applying filler metal.
- Be sure the surfaces are properly tinned before depositing beads.
- Use a neutral flame unless otherwise specified. Use a circular torch motion.
- Do not braze weld a metal that will be subjected to high temperatures or high stresses.
- Parts to be soldered must be clean and their surfaces should fit closely together.
- Do not allow the parts to move during soldering while the solder is molten.
- Wash the soldered work in hot water to eliminate the corrosive action of the flux.
- Be sure the soldering heat is adequate for the soldering job to be done.

❓ Questions for Study and Discussion

1. Why is a lap joint better than a butt joint for brazing?
2. Why is joint clearance an important factor in brazing?
3. What procedure should be used in cleaning surfaces to be brazed?
4. Why is a flux needed for brazing?
5. Why should all flux residue be removed after brazing is completed?
6. What do the AWS classification symbols for brazing filler metal represent?
7. How should the torch flame be applied to the work to carry out a brazing operation?
8. What is meant by liquidus and solidus temperatures?
9. What is the difference between braze welding and brazing?
10. What are some of the advantages of braze welding?
11. When should braze welding not be used?
12. What kind of filler metal is needed for braze welding?
13. How should flux be applied?
14. When is a surface hot enough for braze welding?
15. During the soldering process, why should parts be held firmly in place?
16. What is meant by tinning copper?

Refer to Chapter 23 in the *Welding Skills Workbook* for additional exercises.

Surfacing and Hardfacing

Chapter

The surfacing process applies a new layer of material to surfaces or edges of parts. A variety of welding processes may be used for surfacing. Hardfacing is a branch of surfacing that deals with applying a hard, wear-resistant layer, which is an economical method of extending the life of worn-out machine parts, earthmoving tools, and construction equipment. Many types of wear can be corrected with hardfacing. Surfacing is also used to restore original dimensions or apply corrosion-resistant layers to a metal.

SURFACING

Surfacing is the application of one or more layers of additional material to a surface to obtain desired properties or dimensions. The most common type of surfacing is hardfacing. *Hardfacing* is the application of a surfacing layer with hard, wear-resistant properties. Hardfacing techniques retain the ductility of the base metal while providing a surface resistant to wear. **See Figure 24-1.** This combination is particularly useful for certain machine parts or heavy equipment.

Hardfacing is only applied to surfaces or edges that may wear excessively if not protected. Contact surfaces, screw flight edges, journal bearings, seal-wiped areas, hammer tips, and shear edges are some examples. For example, shovel or bucket teeth on earthmoving equipment are subject to heavy wear, and hardfacing is used to protect and repair the teeth. In fact, hardfacing can also be applied in a pattern that traps abrasive material, which becomes a wear surface itself. **See Figure 24-2.** Different patterns are better suited for the different types of wear

expected from different materials. Areas of a component that do not wear do not require hardfacing.

The Lincoln Electric Company

Figure 24-1. *Hardfacing is used on heavy equipment to retain the ductility of the base metal, while providing a surface resistant to abrasive wear.*

> ✓ **Point**
>
> *With surfacing welds, the surfacing material creates a metallurgical bond with the base metal. With thermal spray coating, the bond is mechanical.*

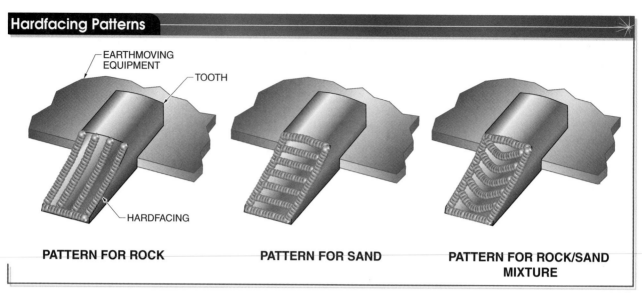

PATTERN FOR ROCK

PATTERN FOR SAND

PATTERN FOR ROCK/SAND MIXTURE

Figure 24-2. *The hardfacing pattern can be changed to accommodate the type of wear expected.*

Surfacing is also used to add layers with corrosion resistance or other desirable material properties. Surfacing may also be used to restore original or desired dimensions. In this case, surfacing is used to build up the size of the component, which is then machined or ground down to the proper dimensions and surface finish.

Surfacing is used on both new components and for repairs. For example, the initial fabrication of a new component may include surfacing processes if the expected wear is concentrated to a small area that can be welded. Or, if wear has caused a loss of material, surfacing can be used as a repair technique.

Surfacing can be applied by welding or thermal spraying. When surfacing is applied by welding, it is also called weld overlay. *Weld overlay* is the application of surfacing using a welding process that creates a metallurgical bond with the base metal through melting of the surfacing metal. A *metallurgical bond* is the joining of two components by fusion. Weld overlay processes include OAW, SMAW, GTAW, GMAW, SAW, and PAW. Depending on the weldability of the base metal, preheating, interpass temperature control, and postheating may be required.

Thermal spraying methods apply overlays that are mechanically bonded to the surface, though not fused. These methods include plasma spraying, flame spraying, high-velocity oxyfuel (HVOF) flame spraying, spray and fuse, and arc spraying.

The area and thickness of the applied surfacing must be minimized to reduce distortion. With high hardness deposits, it is usually not possible to apply more than two layers without cracking. If the desired thickness of the hard material is inadequate, a soft metal buildup is used to minimize cracking before the final hard deposit is applied.

WEAR TYPES

The wear experienced by metal components while in service is the primary reason for applying surfacing layers. *Wear* is mechanical abrasion that results in gradual loss of material from a component surface. There are many types of wear. Most wear can be repaired by surfacing. However, not every type of surfacing process may be applied to every type of wear. The specific wear type must be determined before specifying a surfacing method.

Wear types include erosion, gouging, solid particle impingement, liquid impingement, cavitation, slurry erosion, fretting, adhesive wear and galling, pitting (spalling), impact damage, and brinelling. **See Figure 24-3.**

WEAR TYPES AND SURFACING METHODS

Wear Type	Description	Examples	Surfacing Methods
Erosion (Low-Stress Abrasion)	Abrasive forces result in scratching of the surface; low force, does not crush abrasives	Particles sliding in chutes; packing cartons that run on shafting; sandy soil being plowed; abrasive material being cut	Welding; thermal spraying
Gouging (High-Stress Abrasion)	Abrasive forces result in deep scratches; surface has insufficient compressive strength to resist damage; plastic deformation; chip removal after repeated compressive loading	Rollers running on dirty tracks; ball mills for grinding minerals; farm implements in hard soil; heavily loaded metal sliding systems in dirty environments; gyratory crusher parts; hammer mill hammers; jaw crushers	Welding
Solid Particle Impingement	Wearing of surface caused by repeated impact of solid particles; forms small craters; removes chips of material	Abrasive blasting; aircraft operating in sand or dirt; cyclone separators	Welding; thermal spraying
Liquid Impingement	Progressive material removal caused by the striking action of liquid	Steam turbine vanes; fans exhausting liquid droplets	Welding; thermal spraying; rubber lining
Cavitation	Progressive loss of material caused by air bubbles of a liquid collapsing near surface	Ship propellers; pump impellers and casings; ultrasonic cleaners	Welding
Slurry Erosion	Progressive loss of material caused by a slurry	Slurry pipelines and pumps; oil-well downhole equipment; mud pumps; well pumps; agitators	Welding;* thermal spraying*
Fretting	Oscillatory movement with little displacement; produces oxide debris; leads to pitting and fatigue failure	Gears and sheaves held on shafts with setscrews; bearings loose-fitting on shafts; drive coupling components; metal parts in vibrating contact; bolted components subjected to repetitive stress	Welding; thermal spraying
Adhesive Wear and Galling	Localized damage in solid-state welding between sliding surfaces leading to material transfer between surfaces	Face seals; gears; bushings; drive chains; actuators	Welding†
		Heavily loaded sliding members; austenitic stainless steel gate valves; plug valves; threaded fastener assemblies	Welding
Pitting (Spalling)	Removal or displacement of a surface caused by repetitive sliding or rolling surface stresses; leads to subsurface cracking	Cam paths; gear teeth; rolling element raceways; sprockets	Welding
Impact Damage	Removal of material from a surface caused by repetitive impact collisions of two surfaces	Hammerheads; riveting tools; pneumatic drills	Welding
Brinnelling	Localized plastic deformation or surface denting caused by repeated local impact or overload	Wheels or rails; rolling element bearings and cams	Welding

* when slurry is corrosive, must have adequate corrosion resistance
† determine proper consumable by trial and error

Figure 24-3. *Parts in service are commonly subjected to many types of wear, which can often be repaired with overlays applied by welding and thermal spraying.*

Erosion

Erosion is a type of wear in which an abrasive impacts a surface with little force, but over time, causes the gradual removal of surface material. Erosion is a form of low-stress abrasion. Erosion can occur in moving liquids containing abrasive particles. If the liquid is corrosive, the form of damage is called erosion-corrosion.

Areas in which erosion occurs include coal and ore chutes and slurry pipelines. Hardfacing by welding or thermal spraying may be used to prevent or repair erosion.

Gouging

Gouging is a severe type of wear in which an abrasive impacts a surface with enough force to macroscopically gouge, groove, or deeply scratch the surface. Gouging is a form of high-stress abrasion.

An example of gouging is the action of backhoe teeth against a surface. Hardfacing by welding may be used to prevent or repair gouging. Thermal spraying should not be used because the abrasive forces are typically too strong for thermal spray coating to withstand.

Solid Particle Impingement

Solid particle impingement is a type of wear caused by repeated impact from solid particles. Solid particle impingement forms small craters and removes tiny chips from the surface.

Solid particle impingement occurs in abrasive blasting or cyclone separators. Hardfacing by welding and thermal spraying may be used to prevent or repair solid particle impingement erosion. The angle of impact of the particle and its hardness affects which process should be used for surfacing.

Liquid Impingement

Liquid impingement is a type of wear caused by the striking action of a liquid. The removal of material may be aggravated by corrosive liquids. Liquid impingement occurs in steam turbine vanes and fans that exhaust liquid droplets.

Hardfacing by welding and thermal spraying may be used to prevent or repair liquid impingement. The corrosiveness of the liquid may influence the surfacing process used. When liquid impingement is caused by liquid droplets, a rubber lining may be used because it provides better protection from repeated impact without damage.

Cavitation

Cavitation is a type of wear caused by vapor bubbles collapsing in a flowing liquid. The vapor bubbles form because of changes in flow velocity and/or direction, or because of a reduction in the cross section of the flow passage. An increase in pressure causes the bubbles to collapse. The collapsing bubbles form shock waves or tiny explosions that cause contact stresses on the metal surface. Repetitive shock waves or explosions lead to spalling and pitting of the surface.

Cavitation is common in pumps and engine cylinders and can occur in ship propellers, pump impellers, and casings. Hardfacing by welding may be used to prevent or repair cavitation. Thermal spraying should not be used.

Slurry Erosion

Slurry erosion is a type of wear caused by slurry moving over the surface. A *slurry* is a mixture of solid particles in a liquid. If the slurry is corrosive, erosion of the base metal is accelerated.

Areas in which slurry erosion can occur include slurry pipelines and pumps and oil-well downhole equipment. Hardfacing by welding and thermal spraying may be used to prevent or repair some types of slurry erosion. When slurry is corrosive, the surfacing material must provide corrosion resistance.

Fretting

Fretting is a type of wear caused by oscillatory movement between the surfaces of two materials, usually metal. Fretting produces oxide debris and leads to pitting and, eventually, fatigue failure.

Fretting commonly occurs on bolted components subjected to repetitive stresses and can occur in loose-fitting bearings, metal parts in vibrating contact, and gears and sheaves at the setscrews. Hardfacing by welding and thermal spraying may be used to prevent or repair fretting.

Adhesive Wear and Galling

Adhesive wear is a type of wear caused by the welding together and subsequent shearing of minute areas of two surfaces that slide across each other under pressure. In advanced stages, adhesive wear leads to galling.

Adhesive wear may occur in drive chains, gears, and bushings. Hardfacing by welding may be used to prevent or repair adhesive wear. The appropriate weld overlay material is selected based on experience or trial-and-error methods. Thermal spraying should not be used.

Galling is a condition that occurs when excessive friction, caused by rubbing of high spots on the surface, results in localized welding with subsequent pitting (formation of surface slivers) and further roughening of the rubbing surfaces. Galling is a result of an improper mating combination between components and not a failure of any one component. Galling may result in seizure of a component.

Examples of components that often experience galling include valve trim, engine camshafts, and threaded connections. Galling may be prevented or repaired using welding hardfacing and heat treatment procedures. Thermal spraying should not be used.

Pitting (Spalling)

Pitting (spalling) is a type of wear characterized by the forming of localized cavities in metal from corrosion, repetitive sliding or rolling surface stresses, or poor electroplating. Pitting leads to subsurface fatigue cracking. Pitting appears on the surface as cavities, depressions, or flakes.

Pitting can occur in cam paths, gear teeth, rolling element raceways, and sprockets. Hardfacing by welding may be used to prevent or repair pitting, but the type of material used must be carefully selected. Thermal spraying should not be used.

Impact Damage

Impact damage is a type of wear caused by repetitive collisions or impact between two surfaces. Impact damage can occur in hammerheads, riveting tools, and pneumatic drills. Hardfacing by welding may be used to minimize impact damage. Thermal spraying should not be used.

Brinelling

Brinelling is a type of wear characterized by localized plastic deformation or surface denting due to repeated local impact or overload. Brinelling occurs in wheels or rails, rolling element bearings, and cams. Hardfacing by welding may be used to prevent or repair brinelling. Thermal spraying should not be used.

> **✓ Point**
>
> *Adhesive wear is the removal of metal from a surface by welding together and subsequent shearing of minute areas of two surfaces that slide across each other under pressure.*

> **✓ Point**
>
> *Pitting (spalling) is the forming of localized cavities in metal resulting from corrosion, repetitive sliding or rolling surface stresses, or poor electroplating.*

ASI Robicon

Conveyor systems are exposed to many types of wear including fretting, which results from repetitive stresses, and adhesive wear, which results from parts sliding across each other.

SURFACING AND HARDFACING BY WELDING

Most welding processes may be used for surfacing and hardfacing, though each has characteristics that may affect the resulting surface. **See Figure 24-4.** Base metal preparation for weld overlay depends on the required quality of the finished surface. For dirty work, such as guide plates, coke chutes, or power shovels, where some degree of surface porosity or inclusions may be tolerated, loose scale, dirt, or other foreign substances are removed by wire brushing, grinding, or sandblasting.

For critical work such as valve seats, pump shafts, or coating rolls, where no porosity or inclusions are permitted, the base metal must be prepared by machining or grinding to bright metal. Otherwise, surface irregularities can lead to gas voids and inclusions. All foreign matter such as grease, oxides, or dirt must be removed completely. Any metal remaining from a previous surfacing operation must also be removed. The surface may also be degreased by scrubbing with methanol. Handling of the component after preparation should be minimized because even fingerprints can interfere with good wetting action during surfacing.

When applicable, worn metal is rebuilt to within 3/16″ to 3/8″ of the original surface dimension, depending on how much wear may be tolerated in the part. Although many alloys can be used for weld overlay, most surfacing is done with one of a few filler metal types. These are not of the same composition as filler metals used for joining metals. Weld overlay can only be used on metal combinations that can be joined by welding. The effect of dilution is an important factor to consider in the application of weld overlays.

Thermal stresses may develop in the overlay as it contracts during cooling, creating residual stresses. Since overlays have high hardness and limited ductility, they tend to relieve the stresses by cracking. Usually a fine cracking pattern is acceptable unless the overlay is to be used for sealing purposes.

For some base metals, heat treatment after surfacing may prevent cracking of the surfacing or base metal. Heat treatment methods usually depend on the type of the base metals. For example, steels may require heat treatment before and after surfacing to ensure they remain tough and serviceable.

SURFACING PROCESSES				
Process	**Application**			**Surfacing Metal Forms**
	Manual	**Semiautomatic**	**Automatic**	
OFW	✓	✓		Powder and bar cast and tubular rods
SMAW	✓			Covered electrodes, solid cast electrodes, and tubular electrodes
FCAW		✓	✓	Composite electrode of metallic sheath and powder core
GMAW		✓	✓	Bare solid and tubular electrodes
SAW			✓	Bare solid and tubular wire and strip
GTAW	✓		✓	Powder, bare solid and tubular wires, and base cast rods
PAW			✓	Powder and bare solid and tubular wires
Plasma arc spraying		✓	✓	Powder
Electric arc spraying		✓	✓	Bare solid and tubular wires
Flame spraying		✓	✓	Powder and bare and tubular wires
High-velocity oxyfuel spraying		✓	✓	Powder

Figure 24-4. *Most welding processes can be used to apply surfacing overlays, though each has characteristics that may make it more suitable for certain types of surfacing.*

Dilution

Dilution is a change in the composition of welding filler metal in the weld deposit caused by the addition of melted base metal. **See Figure 24-5.** As melted base metal is incorporated into the melted filler metal, it changes the composition of the deposited metal. Dilution is an important factor in surfacing. The first layer of overlay material is diluted by the base metal, affecting the desired surface properties. Therefore, it may require two or more layers to achieve the intended wear properties. The amount of dilution is reduced for successive layers, since each layer is diluted only by the layer before it. A one-layer surfacing deposit may be possible, but only if it is carefully applied and the welding process causes little melting of the base metal (such as with OAW). Welding variables that can be manipulated to control dilution include the following:

- Polarity—DCEN provides less penetration and therefore less dilution than DCEP. Dilution with AC is between that of DCEN and DCEP.
- Current—Decreasing the current decreases the temperature of the arc, which decreases dilution.
- Electrode diameter—A smaller electrode diameter requires less current and therefore produces less dilution. However, for GMAW, if larger diameters result in globular transfer, dilution is less when compared to spray transfer.
- Electrode extension (stickout)—A long electrode extension decreases dilution for consumable electrode processes such as GMAW and FCAW. A worn contact tip with GMAW can lead to longer electrode extension and cause the welder to increase the current to compensate, leading to greater dilution.
- Bead spacing—Greater overlap between beads reduces dilution because more of the previously deposited bead is remelted and less base metal is melted and incorporated into the weld pool.
- Welding technique—A weave bead results in lower dilution than a stringer bead because a weave bead allows more liquid metal between the arc and the base metal to act as a cushion, absorbing arc energy that would otherwise cause deeper penetration into the base metal. A greater oscillation width results in less dilution. Consequently, a stringer bead results in maximum dilution.

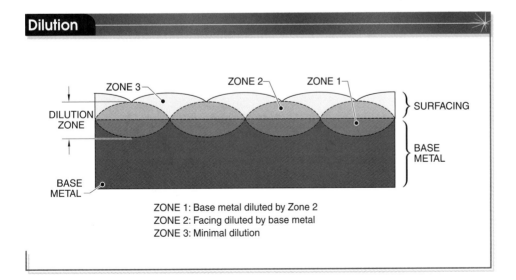

Figure 24-5. *Dilution is the mixing of the base metal with the overlay metal during welding such that the composition of the weld metal becomes a combination of the two. Subsequent overlays are then diluted by the weld metal below them.*

- Travel speed—Reducing travel speed decreases the amount of base metal melted and increases the amount of weld consumable added, which decreases dilution. The decrease is the result of the changing shape and thickness of the bead and because the arc force is expended on the weld pool rather than in the base metal.
- Welding position—Depending on the welding position and work inclination, the pool of molten metal runs ahead of, remains under, or runs behind the arc. If the pool stays ahead of or under the arc, the penetration and dilution is reduced. The pool acts like a cushion, absorbing some of the energy of the arc before it affects the base metal. The pool cannot be too far ahead of the arc, however, because then it will insufficiently melt the base metal and fusion will not occur. **See Figure 24-6.** Surfacing and hardfacing overlays are usually applied in the horizontal position. Flat positions result in the least dilution.

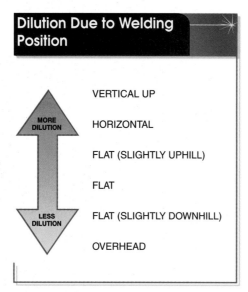

Figure 24-6. *Welding position affects the amount of dilution of a surfacing weld.*

- Arc shielding—The shielding medium, gas, or flux influences the fluidity and surface tension of the weld pool, which determine how much the weld metal wets the base metal and blends in along the edges of the weld bead. The shielding also has a significant influence on the type of welding current that can be used. Gas mixes with helium or carbon dioxide and granular fluxes without alloy addition generally increase dilution. Gas mixes with argon and granular fluxes with alloy addition generally decrease dilution.
- Auxiliary surfacing metal—The addition of auxiliary surfacing metal (material other than in the electrode) to the weld pool can greatly reduce dilution. It may be added as powder, strip, or wire, with or without flux. Some of the arc energy is forced to melt the auxiliary metal rather than the base metal, reducing penetration and dilution.

Weld Overlay Processes

Weld overlay may be applied using OAW, SMAW, GTAW, GMAW, SAW, and PAW processes. The amount of dilution varies depending on the welding process. **See Figure 24-7.**

OAW Overlays. OAW overlays are widely used on steels where maximum hardness (due to minimum dilution) and minimum crack susceptibility are required. An OAW weld overlay can be applied to most materials, copper alloys being one of the exceptions. This is because there is a greater loss of aluminum or silicon in copper alloy overlays by oxidation as compared to arc welding processes, resulting in a softer deposit.

The base metal surface must be clean of dirt and oxides and preheated to produce a sweating condition on the surface. During preheating, the tip of the surfacing filler metal is held on the fringe of the flame until the metal has been sufficiently heated. The filler metal is then moved into the center of the flame and melted. Filler metal is deposited using a regular forehand welding technique with a slight weaving motion. Generally, a slightly reducing flame is recommended, as this adds carbon to the deposit. Filler metal used for OAW weld overlay should be composed of low-melting-point, high-carbon metal.

Dilution Due to Welding Process

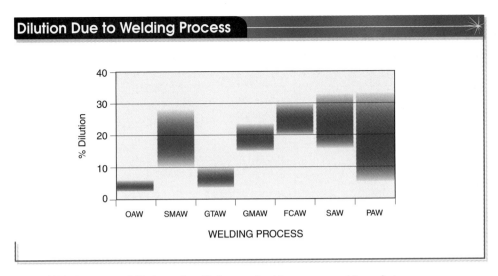

Figure 24-7. *The amount of dilution varies with the type of welding process used for surfacing.*

The deposition rate with OAW is not as high as with other processes. However, OAW minimizes fusion of the base metal, resulting in less dilution, so one layer may be sufficient. The lack of steep thermal gradients in OAW also reduces cracking or spalling of the weld overlay due to thermal stresses. OAW is useful for depositing weld overlays on small or thin parts such as engine valves, plowshares, and tools.

SMAW Overlays. SMAW is commonly used to apply overlays because of the high deposition rate and relatively low dilution. However, many weld variables affect the surfacing characteristics. **See Figure 24-8.** SMAW is also widely used for surfacing large areas or for heavy parts that would normally require excessive time to heat with an oxyacetylene flame. SMAW weld overlays are especially suitable for manganese steel and other steel alloys where heat buildup must be restricted.

The surface of the base metal must be thoroughly cleaned before surfacing, although cleanliness is not as stringent a requirement in SMAW as in other processes because the flux helps with the cleaning process. Although some porosity and cracking may be present, such discontinuities are usually acceptable in the types of applications for which SMAW is used, such as thick overlays on earthmoving equipment and mining

equipment. It is generally necessary to apply several layers of surfacing to achieve the intended surface hardness.

Either AC or DC can be used to produce a satisfactory weld overlay. Proper application of a weld overlay using SMAW involves the following procedure:

1. Remove all rust, scale, and other foreign matter from the surface.
2. Set current just high enough to provide sufficient heat to maintain the arc yet prevent dilution.
3. If possible, position the workpiece so the section to be surfaced is in flat position. Most surfacing electrodes are designed for use in flat position only.
4. Maintain a medium arc length and do not allow the electrode to touch the base metal. When making the deposit, use a straight or weaving motion. A weaving or whipping motion should be used on thin metals. A weaving motion is preferred when only a thin bead deposit is required. The width of the weaved bead should not be more than ¼″. A whipping action is often used when surfacing an area along a thin edge. The arc is held over the heavy portion and then whipped out to the thin edge. In this manner, a shallow deposit is made before the heat builds up enough in the base metal to burn through.

> ✓ **Point**
>
> *Use a minimum amount of heat when surfacing using SMAW.*

> ✓ **Point**
>
> *When applying surfacing with SMAW, maintain a medium arc length and do not allow the electrode coating to contact the base metal.*

CHAPTER 24—Surfacing and Hardfacing

CHAPTER 24—Surfacing and Hardfacing **295**

EFFECT OF SMAW VARIABLES OF SURFACING CHARACTERISTICS*				
Variable		Dilution	Deposition Rate	Deposit Thickness
Polarity	AC	Moderate	Moderate	Moderate
	DCEP	High	Low	Thin
	DCEN	Low	High	Thick
Current	High	High	High	Thick
	Low	Low	Low	Thin
Electrode Diameter	Small	High	High	Thick
	Large	Low	Low	Thin
Arc Length	Long	Low	No effect	Thin
	Short	High	No effect	Thick
Bead Spacing	Narrow	Low	No effect	Thick
	Wide	High	No effect	Thin
Welding Technique	Stringer	High	No effect	Thick
	Weave	Low	No effect	Thin
Travel Speed	Fast	High	No effect	Thin
	Slow	Low	No effect	Thick
Position	Flat	High	No effect	Thin
	Uphill	Moderate	No effect	Moderate
	Downhill	High	No effect	Thin
	Horizontal	Moderate	No effect	Thickest
	Vertical Up (Forehand)	Highest	No effect	Thinnest
	Vertical Up (Backhand)	Lowest	No effect	Thick

* assuming only one variable at a time is changed

Figure 24-8. *Variables in the SMAW process may affect the characteristics of surfacing overlays.*

5. Remove slag from the surface after each pass.
6. Manipulate the electrode carefully to secure adequate penetration into previous passes. Hold the electrode over the deposited bead momentarily to allow heat to build up in the adjoining beads. This procedure also minimizes undercutting.

GTAW Overlays. GTAW uses wrought, tubular, or cast welding rod, or it uses powder and a nonconsumable tungsten electrode shielded with argon, helium, or both. Surfacing materials are easily deposited to form a smooth, uniform, porosity-free weld overlay. The shielding gas required by GTAW prevents oxidation and the loss of alloying ingredients. Surfacing with GTAW is somewhat slower than with other arc welding processes, but the resulting weld overlay is of slightly higher quality. **See Figure 24-9.**

Surfacing with GTAW ordinarily requires very little preheating. However, the high heat input of the process causes dilution and steep thermal gradients that increase cracking susceptibility from high thermal stresses. GTAW is often used where thin overlays are required. It is particularly effective in applying cobalt-base alloys. Other materials that can be deposited include high-alloy steels, chromium stainless steels, nickel and nickel alloys, copper and copper alloys, and cobalt and cobalt alloys.

A variety of special filler metals are available for practically every surfacing operation. Composite filler metal is typically used. Composite filler metals, such as flux-cored electrodes, consist of a tubular steel shell with metallic powders or fine particles of hard compounds incorporated into the center or into the coating.

GMAW Overlays. GMAW is not as widely used for surfacing as the other arc welding processes, although it allows for many types of base metals to be surfaced. With its continuous wire feed, GMAW is faster (has higher deposition rates) than GTAW and produces excellent weld overlays with low dilution. **See Figure 24-10.** The electrode may be solid or tubular and shielding gases include argon, carbon dioxide, helium, or mixtures with a small amount of oxygen. An auxiliary wire may also be fed into the weld pool, which reduces dilution by absorbing some of the heat energy.

GMAW modes of metal transfer include spray transfer, short circuiting transfer, and pulsed arc transfer. Spray transfer results in relatively high dilution compared with the other modes. Short circuiting mode may be used in some out-of-position configurations, and produces slightly higher deposition rates than covered electrode processes while minimizing dilution and distortion. Pulsed arc may also be used for out-of-position surfacing for metals that are very fluid when molten, with less dilution than spray transfer but lower deposition rates.

FCAW Overlays. FCAW is similar to GMAW and can be used to surface the same base metals, though FCAW results in higher deposition rates and more dilution. However, many types of flux-cored electrodes are available and alloy powders may be added, increasing versatility. Depending on their specification, flux-cored electrodes may or may not require auxiliary shielding gas. The welding equipment is similar to GMAW except that the electrode feed rolls are modified for the softer-cored wires.

EFFECT OF GTAW VARIABLES OF SURFACING CHARACTERISTICS*				
Variable		**Dilution**	**Deposition Rate**	**Deposit Thickness**
Polarity	AC	Moderate	Moderate	Moderate
	DCEP	Low	Low	Thin
	DCEN	High	High	Thick
Current	High	High	High	Thick
	Low	Low	Low	Thin
Voltage	High	Low	No effect	Thin
	Low	High	No effect	Thick
Electrode Diameter	Small	High	Low	Thin
	Large	Low	High	Thick
Electrode Extension	Short	No effect	No effect	No effect
	Long	No effect	No effect	No effect
Bead Spacing	Narrow	Low	No effect	Thick
	Wide	High	No effect	Thin
Welding Technique	Stringer	High	No effect	Thick
	Weave	Low	No effect	Thin
Travel Speed	Fast	High	No effect	Thin
	Slow	Low	No effect	Thick
Position	Flat	Low	No effect	Thin
	Uphill	Moderate	No effect	Moderate
	Downhill	Low	No effect	Thin
	Horizontal	Moderate	No effect	Thickest
	Vertical Up (Forehand)	Highest	No effect	Thinnest
	Vertical Up (Backhand)	Lowest	No effect	Thick
Shielding Gas	Argon	Lowest	Lowest	Thinnest
	Helium	Highest	Highest	Thickest
Auxiliary Wires		Low	High	Thicker

* assuming only one variable at a time is changed

Figure 24-9. *GTAW variables can be manipulated to affect a surfacing overlay.*

EFFECT OF GMAW VARIABLES OF SURFACING CHARACTERISTICS*		Dilution	Deposition Rate	Deposit Thickness
Variable				
Polarity	DCEP	High	Low	Thin
	DCEN	Low	High	Thick
Current	High	High	High	Thick
	Low	Low	Low	Thin
Voltage	High	Low	No effect	Thin
	Low	High	No effect	Thick
Arc Transfer	Spray	Highest	Highest	Thickest
	Globular	Low	Low	Thin
	Short Circuit	Lowest	Lowest	Thinnest
Electrode Diameter	Pulsed	High	High	Thick
	Small	High	High	Thick
Electrode Extension	Large	Low	Low	Thin
	Short	High	Low	Thin
Bead Spacing	Long	Low	High	Thick
	Narrow	Low	No effect	Thick
Welding Technique	Wide	High	No effect	Thin
	Stringer	High	No effect	Thick
Trave Speed	Weave	Low	No effect	Thin
	Fast	High	No effect	Thin
	Slow	Low	No effect	Thick
Position	Flat	Moderate	No effect	Thin
	Uphill	High	No effect	Moderate
	Downhill	Low	No effect	Thin
	Horizontal	Moderate	No effect	Thickest
	Vertical Up (Forehand)	Highest	No effect	Thinnest
	Vertical Up (Backhand)	Lowest	No effect	Thick
Shielding Gas	Argon	Lowest	Lowest	Thinnest
	Helium	Highest	Highest	Thickest
	Carbon Dioxide	Moderate	Moderate	Moderate
Auxiliary Wires		Low	High	Thicker

* assuming only one variable at a time is changed

Figure 24-10. *Surfacing overlays can be applied with the GMAW process.*

Important characteristics of FCAW that affect dilution are the production of slag that must be removed before the next layer is deposited, greater sensitivity to electrode extension, and limitations of the feed radius of the cored electrode wire.

SAW Overlays. SAW overlay is the most widely used method of automated surfacing. The SAW process is used when surfacing an extensive area and on parts that require heavy deposits of surfacing. Since SAW uses a high welding current, it has a high deposition rate and results in high-quality deposits with no spatter. **See Figure 24-11.**

The electrode is a continuous solid or tubular wire, or a solid or composite strip. Additional surfacing metal powder may also be fed by metering into the work to create alloy compositions that cannot easily be created in wire or strip form. The arc is protected by a blanket of granular flux. Electrode compositions are formulated to be compatible with a specific flux, which is tailored to the application in order to influence dilution, deposition rate, and deposit thickness. Dilution is generally higher with SAW and two or more layers are required to compensate for this.

EFFECT OF SAW VARIABLES OF SURFACING CHARACTERISTICS*				
Variable		**Dilution**	**Deposition Rate**	**Deposit Thickness**
Polarity	AC	Moderate	Moderate	Moderate
	DCEP	High	Low	Thin
	DCEN	Low	High	Thick
Current	High	High	High	Thick
	Low	Low	Low	Thin
Voltage	High	Low	No effect	Thin
	Low	High	No effect	Thick
Electrode Diameter	Small	High	High	Thick
	Large	Low	Low	Thin
Electrode Extension	Short	High	Low	Thin
	Long	Low	High	Thick
Bead Spacing	Narrow	Low	No effect	Thick
	Wide	High	No effect	Thin
Welding Technique	Stringer	High	No effect	Thick
	Weave	Low	No effect	Thin
Travel Speed	Fast	High	No effect	Thin
	Slow	Low	No effect	Thick
Position	Flat	Moderate	No effect	Moderate
	Uphill	High	No effect	Thick
	Downhill	Low	No effect	Thin
Process Variations	1 electrode	High	Lowest	Thinnest
	1 electrode and surfacing wire	Moderate	Lowest	Thin
	1 electrode and hot surfacing wire	Low	Low	Thin
	2-wire series	Moderate	Low	Thin
	2-wire series and cold wire	Low	Moderate	Moderate
	Multiple wire	High	High	Thick
	Strip electrode	Highest	High	Moderate
	Hot and cold strip	Lowest	Highest	Thickest
	Powder addition	Low	Moderate	Moderate

* assuming only one variable at a time is changed

Figure 24-11. *The SAW process is widely used for automated surfacing.*

PAW Overlays. PAW overlays provide a mechanized process that uses a metal powder as the surfacing material and a constricted arc between a nonconsumable tungsten electrode and the weld pool. The metal powder is carried from a hopper to the electrode holder in an argon gas stream. From the torch, the powder moves into the arc stream where it is melted and then fused to the base metal. A variety of cobalt-, nickel-, and iron-based surfacing powders is available that have varying degrees of impact resistance, abrasion resistance, and corrosion resistance.

The power source for PAW is a conventional DCEN power supply unit. A second DC unit is connected between the tungsten electrode and the arc-constricting orifice to support a nontransferred arc. The second power supply serves as a pilot arc to start the transferred arc and supplements the heat of the transferred arc. Argon gas is used to form the plasma as well as for shielding. The arc is struck between the electrode and the workpiece while surfacing material is supplied in powder form. The metallurgical structure of the deposit is similar to a GTAW overlay. Dilution and distortion are relatively low compared with other processes, but equipment cost is relatively high.

Plasma arc hot-wire surfacing is a variation of PAW overlay in which the surfacing material is heated to almost melting temperature and the deposited metal is melted on the surface of the workpiece. The temperature is held to the minimum required to fuse the surfacing to the base metal, resulting in minimal dilution. This process is used for overlaying inside of vessels and equipment.

Surfacing Consumables

Consumables used for weld overlays are formulated to provide wear resistance, corrosion resistance, or some other desirable property to the surface. They are not the same formulations as filler metals used for joining because surfacing requires different properties. Weld overlay materials may be available in several forms: bare metal or wire, coated electrode, flux-cored electrode, metal powder, metal-cermet self-fluxing powder, or composite.

AWS specifications identify a limited number of surfacing-consumable families. **See Figure 24-12.** The AWS designation uses symbols to describe the main elements, such as RNiCr-A. The R prefix stands for bare wire or rod, which does not normally conduct current. An E prefix would stand for electrode, which can conduct current. The A, B, and C suffixes identify specific alloys within the group. However, most materials used for surfacing are commercially identified by manufacturer's trade names and have no corresponding AWS designation, making classification difficult.

High-Speed Steels. High-speed steels and tool steels may be used to restore dimensions to tool steel parts that have lost material. Where a surfacing consumable matching the tool steel is available, it should be used.

Austenitic Manganese Steels. Austenitic manganese steels are alloys of iron that contain sufficient manganese to cause them to harden considerably with cold work, such as by repeated impact, with minimal distortion. Although their as-deposited hardness is a low 20 HRC, cold work can raise their hardness to 50 HRC. These overlay materials are used for battering type applications, such as rock crushers, earth loader bucket teeth, and railroad switches.

High Chromium-Iron Alloys. High chromium-iron alloys are essentially cast irons with sufficient chromium to produce a large percentage of hard chromium carbides in the microstructure. They may be applied by SMAW, GTAW, FCAW, or SAW. The high chromium content also provides resistance to elevated temperatures, with retention of hardness up to 1000°F. However, they have relatively poor corrosion resistance. Their low cost and overall wear performance makes them suitable for many applications such as metal-metal wear, abrasion, and erosion.

Nickel Alloys. Nickel alloys contain cobalt, chromium, and boron as the principal alloying elements. Wear resistance is achieved by the formation of hard chromium boride constituents. They are available as coated rods for arc welding and bare cast rods for arc and gas welding. Some proprietary versions are available as pastes that can be brushed on and fused with OAW. They are suitable for abrasive wear and metal-to-metal wear in applications such as extruder screws, pump and fan impellers, and machine ways.

Cobalt Alloys. Cobalt alloys contain chromium as a principal alloying element, plus some nickel and tungsten. They were first developed under the name Stellite®, but this should not be used as a generic name because many other compositions are available from other manufacturers. Cobalt alloys have good corrosion and oxidation resistance in addition to wear resistance, specifically metal-metal wear and low-stress abrasion. Surfacing with cobalt alloys has many uses, including sleeves.

Composite Materials. Composite materials used for consumables in surfacing are composed of steel tubes filled with wear-resistant particles, usually carbides. The tubes can be coated with flux for SMAW or used bare for OAW or GTAW.

WELD OVERLAY FILLER METALS			
Material	Designation	Rockwell Hardness C Scale (HRC)	Application
High-Speed Steel	R Fe 5-B OR E Fe 5-B	60	• repairs on tool steels
High Chromium Iron Alloy	R Fe Cr-Al OR E Fe Cr-Al	58	• high-stress abrasion resistance for – heavily-loaded metal – metal sliding systems in dirty environments
Nickel Alloy	R Ni Cr-C OR ENi Cr-C	35–56	• low-stress abrasion resistance • metal-to-metal wear • deposits that must be machined – shafts – running in packing – ash handling equipment
Cobalt Alloy	RCo Cr-A OR ECo Cr-A	38–47	• metal-to-metal wear • low-stress abrasion resistance • elevated temperatures • corrosive environments
Composite material	RWC 20/30 OR EWC 20/30	60	• high-stress and gouging abrasion resistance for – crushers – earth-moving equipment

Figure 24-12. *A variety of filler metals can be used for surfacing, although the AWS identifies only a few specific groups.*

As deposited, the hardness of the carbides in a matrix of steel is extremely high at 60 HRC. Excessive welding currents should not be used or the carbides will dissolve, reducing hardness. Since they cannot be machined, they are used in applications where the as-deposited surface is acceptable, like bucket teeth, sand augers, and coal loading devices.

SURFACING AND HARDFACING BY THERMAL SPRAYING

Thermal spraying (*THSP*) is a group of processes that apply a coating of molten or semimolten particles by impact onto a base metal. Thermal sprayed coatings generally have poor strength, ductility, and impact properties when compared with weld overlays. Coatings are generally used for wear or corrosion resistance, not

strength. Fatigue strength of the base metal is decreased. Since the base metal temperature usually does not exceed 300°F, distortion is significantly less a problem than with weld-applied surfacing.

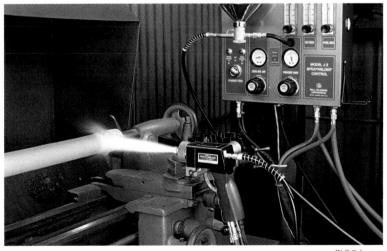

Wall Colmonoy

Thermal spraying is commonly used to build up shafts.

THERMAL SPRAY METHODS

Process	Method	Bond Strength PSI	Coating Material (Max. Thickness*)	Porosity %	Cost
Flame Spray (metallizing)	Consumables in the form of rod, wire, or powder are heated in an oxygen and acetylene flame and propelled towards the workpiece	1500–5000	400 series SST (.125) 300 series SST (0.30)	10–20	Lowest
Spray and Fuse† (only on newly fabricated equipment)	As above, but a powdered nickel or cobalt alloy is used that is fused to the workpiece by torch or furnace heating after spraying, creating a nonporous surface	Metallurgical Bond	Nickel base HRC 40 (.125) Nickel base HRC 50 (.100) Nickel base HRC 60 (.080) Cobalt base (.080)	0	Mid-range
Twin Wire Electric Arc Spray	Two consumable wires of coating materials are melted by an electric arc, atomized, and propelled towards the workpiece with compressed air	5000–9000	400 series SST (.125) 300 series SST (.030)	5–15	Mid-range
Plasma Spray	Powder is fed into a plasma created by striking an arc in an inert gas and propelled towards the workpiece with compressed air	4000–6000	Metals (.060) Ceramics (.030) Carbides (.015)	2–5	Mid-range
High Velocity Oxy Fuel (HVOF)	Hot gas from fuel combustion melts powder that is directed towards the workpiece at supersonic speed	>10,000	Conventional HVOF (Jet Kote, Diamond Jet) Metals (.030) Carbides (.015) High Pressure HVOF (JP-5000) Metals (.060) Carbides (.030)	0–2	High
Detonation Gun (D-gun)	Acetylene and oxygen are ignited by a spark plug and powder is propelled towards the workpiece by rapid detonations	>10,000	Carbides (.010)	<.5	Highest

*in in.
† If heated treated condition is not critical

Figure 24-14. *Classification of thermal spraying processes is by the method required to generate the heat, either electric arc or gas combustion.*

Plasma Arc Spraying

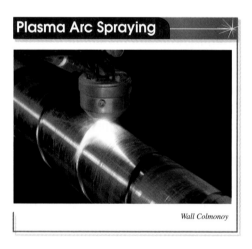

Wall Colmonoy

Figure 24-15. *In plasma spraying, an argon gas stream carries the metal-powder surfacing material from a hopper to the electrode holder. The powder moves into the arc stream, is melted, and fuses to the base metal.*

Since both inert gas and high gas temperatures are used, the mechanical and metallurgical properties of the resulting coatings are generally superior to those applied by either type of flame spraying. Plasma spray coatings are also denser, but more costly. Since the temperature for plasma spraying is so high, coating materials with high melting points can be applied. Most inorganic materials that melt without decomposition can also be used.

Electric Arc Spraying. *Electric arc spraying (twin wire arc spraying)* is a thermal spraying process that uses an electric arc to heat and atomize the coating material, which is propelled toward the workpiece with compressed air. The coating material is fed into the arc in the form of a pair of wires, which provides a high deposition rate. **See Figure 24-16.** The arc reaches a temperature of approximately 7000°F, which allows the highly heated particles to bond strongly with the surface. The resulting porosity is similar to that produced by flame spraying.

Flame Spraying. *Flame spraying* is a thermal spraying process that uses an oxyfuel gas flame as a source of heat to melt the coating material. Wire, rod, or powder coating material is introduced into a stream of oxygen and fuel gas that atomizes the material, allowing it to be propelled by a stream of air to the

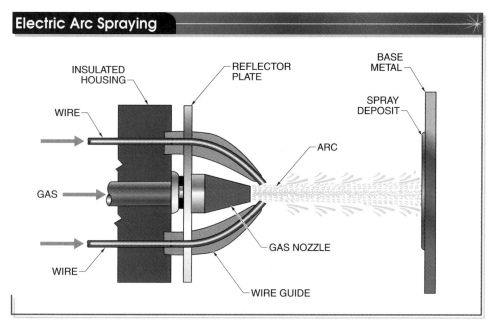

Figure 24-16. *Electric arc spraying equipment produces coatings with greater bond strength and lower oxide content than coatings from oxyacetylene spraying guns.*

surface. **See Figure 24-17.** Compressed air is used for atomizing and propelling the material to the workpiece. The special spray gun torch is set with the proper flame, and the trigger is pressed to propel the material to the surface.

Wall Colmonoy

Figure 24-17. *An oxyfuel torch is used in flame spraying. When the trigger is pressed, the coating material is propelled to the base metal surface.*

The most commonly used gas for the oxyfuel flame is acetylene, which is capable of producing temperatures exceeding 5600°F. Hydrogen or propane may be used for materials that melt at a lower temperature.

A flame spray gun consists of two major parts: the power unit and the gas head. The power unit feeds the coating material into the nozzle of the gun through a center orifice. Around the orifice are a number of gas jets that provide the flame and the airstream. As the coating material comes through the orifice, it is melted and atomized by the flame. The gas head controls the flow of oxygen, fuel gas, and compressed air. The fine molten particles are picked up by the airstream and projected toward the workpiece.

Flame spraying can be applied manually or automatically. Flame spraying allows hard, thin coatings to be deposited quickly and uniformly. Deposits range from 0.01″ to 0.08″ thick. Flame spray guns can deposit about 4 lb to 12 lb of metal per hour. The coatings are porous and usually brittle, as such, they do not resist excessive mechanical abuse.

There are two variations of flame spraying. One uses coating material in wire or rod form and is sometimes referred to as metallizing. The other uses coating material in powder form. In both variations, the coating material is fed through the spray gun and melted in the oxyfuel gas flame at the nozzle.

Wire spray coating materials are metals that can be made into flexible wire that melts in an oxyacetylene flame. Wire spray materials are commonly zinc, aluminum, carbon steel, 300 series stainless steel, bronze, or molybdenum. Flame spraying with wire is used to coat metals for rust protection, heavy rebuilding, or restoring dimensions. It is not used on parts that are subject to rigorous service conditions.

Rod spray materials are usually ceramics such as aluminum oxide, chromium oxide, and zirconium. However, other thermal spraying methods are often more appropriate for applying ceramics because they usually provide a better coating.

For powdered materials, a hopper mounted on or near the torch body feeds the coating material (metal alloy) into the gas stream while the operator controls the flow of the powdered alloy. **See Figure 24-18.** The alloy particles become molten as they are sprayed through the flame and onto the workpiece. Powdered spray materials include carbides, high-alloy steels, stainless steel, cobalt alloys, and ceramics. Powdered thermal spraying machines are usually more complex than other flame spraying equipment and are used for more sophisticated work.

Spray and Fuse (Spraywelding). *Spray and fuse (spraywelding)* is a two-step thermal spray process in which a coating is first deposited and then subsequently fused by heating with a torch or by placing the part in a furnace. Spray and fuse is a variation of flame spraying in which the coating material is fused after application. The spray and fuse process contains characteristics of both weld overlay and thermal spraying and provides better corrosion resistance with less effort.

Spray and fuse coating materials are usually made of nickel or cobalt self-fluxing alloys that contain silicon or boron. The material is then fused to the base metal with a torch or furnace at a temperature between 1875°F and 2000°F. The silicon and boron react with oxide films on the surface and with powder particles, allowing them to wet and diffuse with the base metal.

The spray and fuse process creates a hard, nonporous welded surface that can be ground and lapped to a low surface roughness. Tungsten carbide particles are added for increased wear resistance. The high fusing temperature means that metals that have been heat treated for high strength may lose this mechanical property and may need to be heat treated again after coating.

High-Velocity Oxyfuel (HVOF) Spraying. *High-velocity oxyfuel (HVOF) spraying* is a thermal spraying process that uses repeated blasts of molten powder to apply a surfacing coating. A mixture of oxygen and a combustible gas, such as acetylene, is fed into the barrel of a spray gun with a charge of surfacing powder. The mixture is ignited and the detonation wave accelerates the powder to the workpiece while heating it close to or above its melting point. The cycle is repeated many times a second. The noise level is extremely high, so the process must be performed in a soundproof room.

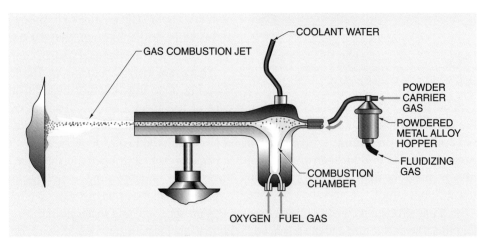

Figure 24-18. *An oxyacetylene metal spray torch has a hopper that feeds powdered metal alloy into the gas stream. The metal particles melt as they are sprayed through the flame and are fused to the workpiece.*

HVOF flame spraying is most successful in applying dense and hard carbide and oxide coatings to critical areas of precision components. Since the base metal surface is seldom heated above 300°F, the component can be fabricated and fully heat-treated prior to coating.

Detonation Gun Spraying. Detonation gun spraying is very similar to HVOF spraying. It is a proprietary process done at a limited number of centers where the equipment and technology are available. Powder is fed into the gun under low gas pressure. Valves are opened to introduce a prescribed mixture of oxygen and acetylene into the gun's combustion chamber. The explosive mixture is detonated by a spark plug. The temperature of the detonated fuel is about 7000°F and the powder reaches velocities up to 2400 f/s. Detonations are repeated four to eight times per second and nitrogen gas flushes the combustion chamber after each detonation. Each detonation produces a coating thickness of a few microns.

Spraying must be done in a sound-proof room because of the detonations, so the equipment is mechanized. Coating density and bond strength results are the highest of all the gas combustion and arc processes with the exception of spray and fuse.

 Refer to Quick Quiz® on CD-ROM

- With surfacing welds, the surfacing material creates a metallurgical bond with the base metal. With thermal spray coating, the bond is mechanical.
- Erosion (low-stress abrasion) is a form of abrasive wear in which the force of an abrasive against the surface causes the removal of surface material.
- Adhesive wear is the removal of metal from a surface by welding together and subsequent shearing of minute areas of two surfaces that slide across each other under pressure.
- Weld overlay is the application of surfacing material using a welding process that creates a metallurgical bond with the base metal through melting of the surfacing metal.
- Pitting (spalling) is the forming of localized cavities in metal resulting from corrosion, repetitive sliding or rolling surface stresses, or poor electroplating.
- Weld overlay may be applied using the OAW, SMAW, GTAW, GMAW, SAW, or PAW processes.
- Use a minimum amount of heat when surfacing using SMAW.
- When applying surfacing with SMAW, maintain a medium arc length and do not allow the electrode coating to contact the base metal.
- When depositing surfacing welds, remove slag after each pass.
- Consumables may be bare filler metal or wire, coated electrode, flux-cored electrode, metal powder, metal-cermet self-fluxing powder, or composite materials.
- When thermal spraying flat surfaces, the surfacing gun is moved back and forth to allow a full, uniform deposit. Thermal spraying should begin beyond one edge of the area to be covered and continue beyond the opposite end.
- Properly cleaning and roughing the part's surface ensures that a thermal spray coating can be successfully applied.

? **Questions for Study and Discussion**

1. What is surfacing?
2. What benefits does surfacing provide when used for repair work?
3. What types of wear do parts encounter in service?
4. What is solid particle impingement?
5. What is pitting?
6. What are the two common surfacing overlay methods?
7. What is dilution?
8. What surface defects can occur on critical work if the surface is not properly prepared?
9. Why should surfacing be done in flat position?
10. How should the torch be manipulated when surfacing large objects with SMAW where a high deposition rate is required?
11. What is the most commonly used form of thermal spraying?
12. When is high-velocity oxyfuel (HVOF) flame spraying commonly used?

Refer to Chapter 24 in the *Welding Skills Workbook* for additional exercises.

Cutting Operations

Cutting operations are methods of rough or final preparation of shapes and edges of metals for welding. Gouging is related to cutting and refers to excavation of metal from the surface in preparation for welding. Safety considerations are an integral part of any cutting operation.

Cutting may be controlled manually or with mechanized equipment. In manual cutting, a torch is manipulated over the area to be cut. In mechanized cutting, the torch is guided entirely by automatic controls. The cutting process used depends largely on the type of base metal, the type of cut to be performed, and the cost of the operation. Common cutting processes used are oxyfuel gas cutting (OFC), plasma arc cutting (PAC), and air carbon arc cutting (CAC-A).

OXYFUEL CUTTING (OFC)

Oxyfuel cutting (OFC) is a group of cutting processes that use heat generated by an oxyfuel gas flame. The fuel/oxygen mixture accelerates the chemical reaction between oxygen and the base metal, removing the metal. Cutting metal using a flame is common in many industrial, construction, and fabrication applications. The cutting is done by means of a hand cutting torch or by an automatically controlled cutting machine (track burner). **See Figure 25-1.**

Victor, a division of Thermadyne Industries, Inc.

Figure 25-1. *OFC is commonly performed with a hand cutting torch.*

While the flame produced by OFC is not as bright as the intense light produced by arc welding and other cutting processes, it can still cause eye damage. Therefore, goggles equipped with the appropriate filter-plate shade should be worn while performing OFC.

The cutting of metal occurs when ferrous metals are subjected to rapid oxidation. When a piece of steel is left exposed to the atmosphere, a chemical reaction (rusting) takes place. Rust is the result of oxygen in the air uniting with the metal, causing it to oxidize. Occurring naturally, the rusting process is very slow. But if metal is heated to its ignition temperature it oxidizes and rusts much faster. The intense heat causes the mixture of oxides and metal to melt. The mixture is swept away by the flow of oxygen, resulting in a cutting action. The width of the resulting cut is called the kerf.

The oxygen used for cutting must be 99% pure. Efficiency and cutting speed are reduced with lesser oxygen purity. Just as in oxyfuel welding, a neutral flame with an approximately 1:1 mixture of oxygen and fuel is used for cutting. Too much fuel in relation to oxygen produces a carburizing flame. The area to be cut should be preheated with a neutral flame to ensure complete melt-through. However, too much preheat when cutting with OFC can cause the top edge of the metal to melt away.

Oxyacetylene cutting is commonly performed on the job site since oxygen and acetylene are readily available and easy to transport.   Victor, a division of Thermadyne Industries, Inc.

OFC is only suitable for carbon steel applications. Cast iron is not easily cut because of its high carbon content. Stainless steels cannot be cut because chromium oxide is formed on the surface, which resists melting and shields the metal surface. Copper and aluminum form similar high-melting-point oxides and in addition possess high thermal conductivity, making it difficult for them to be heated sufficiently.

Oxyfuel Cutting Gases

Gases mixed with oxygen and used for OFC include acetylene, natural gas, propane, methylacetylene-propadiene stabilized (MAPP), and proprietary gases. Procedures and equipment used in the OFC process do not vary much regardless of which gas is used. The heat of a neutral oxyfuel flame brings the base metal up to melting temperature and a jet of pure oxygen is introduced to create the rapid oxidation of the steel.

Natural gas has a low-temperature, low-heat flame, making it inadequate for many welding operations. However, natural gas is commonly used for OFC

because it works well for preheating and cutting materials. Propane, also called LPG or liquefied petroleum gas, may also be used for OFC and for preheating or postheating. Natural gas and propane are commonly available in many shops, making them inexpensive options for cutting operations. However, both natural gas and propane draw an excessive amount of oxygen during heating, which may offset their initial savings. Acetylene is the gas most commonly used for OFC. Because oxygen and acetylene are the most common gases used in OFC, oxyacetylene equipment and procedures are depicted. Fuel lines are typically red and oxygen lines are green. Fuel fittings have left-hand threads to prevent them from being attached to the wrong supply.

Torches

For the rapid cutting of metal to be possible, it is necessary to use a cutting torch that will heat the iron or steel to a certain temperature with a neutral flame, and then direct a jet of oxygen onto the heated section to perform the cutting action and blow the molten metal clear of the kerf. Therefore, the oxygen working pressure is higher than the fuel pressure when cutting jets are open. **See Figure 25-2.**

Victor, a division of Thermadyne Industries, Inc.

Figure 25-2. *OFC is used to rapidly cut metal by subjecting a heated section of metal to a blast of oxygen that produces the cutting action.*

The cutting torch has conventional oxygen and acetylene needle valves. These are used to control the flow of oxygen and acetylene when heating the metal. Some cutting torches have two oxygen needle valves for fine adjustment of the neutral flame. The cutting tip is composed of an orifice in the center surrounded by several smaller orifices (preheat holes). The center orifice permits the flow of the cutting oxygen and the smaller holes are for the preheating flame. The cutting tip should be selected based on the type of fuel used. **See Figure 25-3.**

A cutting torch differs from a regular welding torch in that it has an additional lever to control the oxygen discharged through the center orifice.

A number of different tip sizes are provided for cutting metals of varying thicknesses. In addition, special tips are made for other purposes, such as for cleaning metal; cutting rusty, scaly, or painted surfaces; rivet washing; etc. It is possible to convert a welding torch into a cutting torch by replacing the mixing head with a cutting attachment.

Oxygen and Acetylene Pressures. A fully charged oxygen cylinder contains approximately 2200 psi at 70°F. A fully charged acetylene cylinder contains approximately 225 psi at 70°F. The maximum safe working pressure of acetylene is 15 psi. The correct oxygen and acetylene pressures for a particular application depend upon the tip size used, the type of cutting to be performed, and the thickness of the metal to be cut. **See Figure 25-4.**

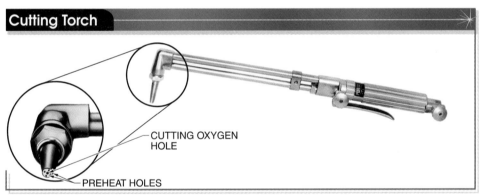

Cutting Torch

CUTTING OXYGEN HOLE

PREHEAT HOLES

Figure 25-3. *An oxygen cutting torch has one oxygen needle valve and one acetylene needle valve. The torch tip includes the cutting oxygen hole and several preheat holes.*

CUTTING PRESSURE FOR METALS			
Tip No.	Metal Thickness*	Oxygen Pressure†	Acetylene Pressure†
0	1/4	30	3
1	3/8	30	
	1/2	40	
2	3/4	40	
	1	50	
3	1 1/2	45	
4	2	50	
5	3	45	4
	4	60	
6	5	50	5
	6	55	
7	8	60	6
	10	70	

* in in.
† in psi

Figure 25-4. *The correct oxygen and acetylene pressure must be used when cutting metals; correct pressures are determined by the tip size and the thickness of the metal to be cut.*

Always consult manufacture recommendations as to the proper oxygen and acetylene pressure settings for a particular torch and tip. The given oxygen pressure cannot always be strictly followed because cutting conditions are not the same for every metal.

Piercing Holes

For steel up to ½″ thick, hold the torch over the area where the hole is to be cut until the flame has heated a small, round spot. Gradually press down the oxygen lever and at the same time raise the tip slightly. A small, round hole is quickly pierced through the metal. **See Figure 25-5.** For steel more than ½″ thick, move the torch slowly in a circular motion as the oxygen lever is depressed to pierce the metal.

Piercing Holes

(1) Heat a small round spot

(2) Depress oxygen lever and raise tip slightly

(3) Pierce small round hole

Figure 25-5. *A cutting torch can be used to pierce a hole through metal.*

When larger holes and circular shapes are required, trace the shapes with a soapstone. If the holes are located away from the edge of the workpiece, first pierce a small hole near the desired area, and then start the cut from the hole, gradually working to the drawn line and continuing around the outline. **See Figure 25-6.**

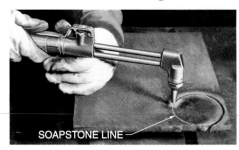

SOAPSTONE LINE

LARGE HOLE

CIRCULAR SHAPE

Figure 25-6. *The cutting tool must be held steady when cutting circles and large curves.*

Beveling

To make a bevel cut on steel, incline the head of the torch to the desired angle rather than holding it vertically. An even bevel may be made by resting the edge of the torch tip on the workpiece as a support, or by clamping a piece of angle iron across the workpiece. Travel speed is important when cutting, especially with OFC. Traveling too slow (or excessive preheat) can cause the top edge to melt away. Traveling too fast prevents complete melt-through. A cutting machine (track burner) can be set to automatically cut the proper beveled edge. **See Figure 25-7.**

Keep the preheating cones burning with a neutral flame. Hold the torch with the inner cone of the heating flame about ¹⁄₁₆″ above the metal until a spot is heated to a bright red. Move the cutting torch just fast enough to make a fast but continuous cut. If the cut does not go through the metal, start the cutting process over again.

Figure 25-7. *The torch must be positioned at an angle to make a beveled edge.*

Cutting Round Stock

To cut round stock, start the cut about 90° from the top edge. Keep the torch in a vertical position (perpendicular to the cutting line) and gradually lift it to follow the circular outline of the bar. Maintain the position of the torch while ascending as well as descending on the opposite side. **See Figure 25-8.** Depending on the thickness of the round stock, preheat may be necessary before cutting.

Cutting Cast Iron

When cutting cast iron, the chemical composition of the iron must be considered. Since cast iron has such a wide range of uses, a vast difference in quality and chemical composition can be expected. The better grades of castings are more easily cut. Do not start a cut in cast iron or heavy steel unless it can be completed without stopping.

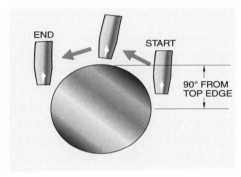

Figure 25-8. *When cutting round stock, start 90° from the top edge. Then follow around the contour of the bar.*

Random grades of scrap, such as counterweights, grate bars, and floor plates, present greater difficulty in cutting and require more fuel, a wider kerf, and correspondingly, a slower cutting speed. If the cut is stopped on a heavy section of cast iron or heavy steel, it is extremely difficult to start again.

The oxygen pressure and acetylene pressure needed for cutting cast iron depend upon the tip size used and the thickness of the cast iron. Always consult manufacturer recommendations for the proper oxygen and acetylene pressure settings for cutting cast iron. **See Figure 25-9.**

CUTTING PRESSURE FOR CAST IRON			
Tip Size	Cast Iron Thickness*	Oxygen Pressure†	Acetylene Pressure†
L-3	½	40	7 to 8
	¾	45	
	1	50	
	1½	60	
	2	70	
L-4	3	80	8 to 10
	4	90	
	6	110	
	8	120	
	10	150	
	12	170	

* in in.
† in lb

Figure 25-9. *The oxygen and acetylene pressure settings required for cutting cast iron are determined by the tip size and the thickness of the cast iron.*

Excess heat, sparks, and slag are generated when cutting cast iron. Proper personal protective equipment is required when cutting. Welding gloves are essential, and a firebrick or suitable torch rest is desirable.

PLASMA ARC CUTTING (PAC)

Plasma arc cutting (PAC) is a cutting process that uses a constricted arc to remove molten metal with a high-velocity jet of plasma (ionized gas). The high-velocity jet of plasma issues from a constricting orifice. Plasma is superheated gas capable of conducting an electric arc. The orifice directs the superheated plasma stream toward the base metal. As the arc melts the base metal, the high-velocity jet removes the molten metal to form a narrow kerf. The molten residue from cutting is called dross.

PAC is one of the best processes for high-speed cutting of a variety of metals including carbon steel, stainless steel, and aluminum. It cuts carbon steel up to 10 times faster than any oxyfuel mixture, with equal quality and at a lower cost. **See Figure 25-10.**

Media Clip

Plasma Arc Cutting

✓ Point

In plasma arc cutting, set the polarity to direct current electrode negative.

PAC uses the same type of direct current power sources as SMAW and GTAW. PAC is similar to GTAW because it uses tungsten electrodes and requires shielding gas to protect the electrode and the molten metal.

When cutting aluminum and stainless steel, best results are obtained with an argon-hydrogen or nitrogen-hydrogen gas mixture. Air has proven to be the most efficient gas for use with plasma cutting; however, oxygen can also be used. Carbon steels require an oxidizing gas. In order for PAC to operate efficiently, the air must be completely dry, and there should be no moisture on or around the equipment.

Several metals produce toxic gases when they are cut. For example, stainless steel produces chromium fumes, and galvanized steel produces zinc fumes. Therefore, proper ventilation should be used at all times.

Manual PAC

In a plasma arc cutting torch, the tip of the electrode is located within the constricting nozzle, which has a relatively small opening constricting the arc. The gas must flow through the arc where it is heated to the plasma temperature range. Since the gas cannot expand due to the construction of the nozzle, it is forced through the opening and emerges at an extremely high velocity and hotter than any flame (30,000°F to 43,000°F). This heat melts any known metal and its velocity blasts the molten metal through the plate creating a kerf. PAC produces a narrower kerf than OFC. **See Figure 25-11.**

Because the maximum transfer of heat to work is essential in cutting, plasma arc torches use a transferred arc (the workpiece itself becomes an electrode in the electrical circuit). The workpiece is subjected to both plasma heat and arc heat. Precise control of the plasma jet can be obtained by controlling the variables—current, voltage, type of gas, gas velocity, and gas flow (cfh).

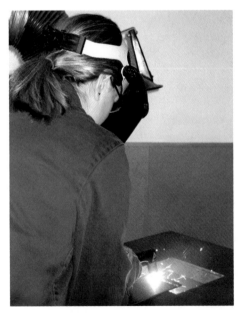

Figure 25-10. *The plasma arc cutting process is one of the best high-speed cutting processes for nonferrous metals and stainless steels.*

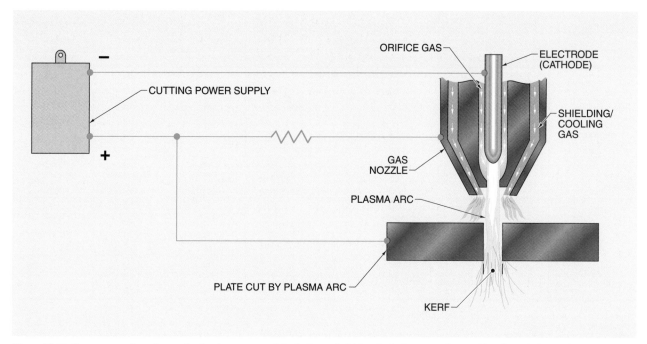

Figure 25-11. *Gases emerge from the nozzle of a plasma arc torch in the form of a high-velocity jet stream that can blast through the metal, creating a kerf.*

The power supply for PAC is a special rectifier-type with an open-circuit rating of 400 V. DCEN is also used. A control unit automatically controls the sequence of operations—pilot arc, gas flow, and carriage travel. A water pressure input of 60 psi to 80 psi for gas cutting and 100 psi for air cutting is necessary to keep the torch cool.

PAC produces intense heat and arc radiation. Proper PPE, including a welding helmet with a filter plate matched to the current, should always be worn when working with a plasma arc cutting torch.

Mechanized PAC

To make a proper plasma arc cut, the power supply and the gas flow must be adjusted to the appropriate settings. **See Figure 25-12.** When the operator pushes the START button on the remote control panel, the control unit performs all ON-OFF and sequencing functions. The cooling water must also be turned ON or the water-flow interlock will block the starting circuit.

To make a mechanized cut, the operator locates the center of the torch about ¼″ above the surface of the workpiece to be cut and pushes the START button.

Current flows from the high-frequency generator to establish the pilot arc between the workpiece and the cathode in the nozzle. Gas starts to flow, and welding current flows from the power supply. The pilot arc sets up an ionized path for the cutting arc. As soon as the cutting arc is established, the high-frequency pilot arc is shut OFF, and the carriage starts to move. **See Figure 25-13.**

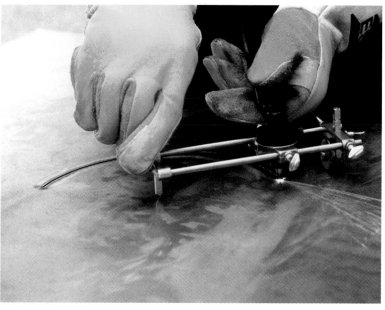

ESAB Welding and Cutting Products
A plasma cutting tool is commonly used for accurately cutting circles and large curves.

PLASMA ARC CUTTING CONDITIONS						
Type of Metal	Thickness*	Speed†	Orifice Type	Insert‡	Power§	Gas Flow‖
Stainless steel	½	25	4 × 8	⅛	45	130 N_2
	½	70	4 × 8	⅛	60	130 N_2
	1½	25	5 × 10	5/32	85	10 H_2 175 N_2
	2½	18	8 × 16	¼	150	15 H_2 175 N_2
	4	8	8 × 16	¼	160	15 H_2
	½	25	4 × 8	⅛	50	100#
	½	200	4 × 8	⅛	55	100#
Aluminum	1½	30	5 × 10	3/32	75	100#
	2½	20	5 × 10	5/32	80	150#
	4	12	6 × 12	3/16	90	200#
	¼	200	4 ×	⅛	55	250
Carbon steel	1	50	5 ×	5/32	70	300
	1½	35	6 ×	3/16	100	350
	2	25	6 ×	3/16	100	350

* in in.
† ipm
‡ diameter
§ kW
‖ cfh
\# 65% argon, 35% hydrogen mixture
** multiport orifice

Figure 25-12. *The operator must adjust the power supply and gas flow to the appropriate settings for a particular PAC operation.*

Welding Tool Corp.

Figure 25-13. *A semiautomatic plasma arc cutting unit is commonly used to ensure an even cut to the metal.*

When the cutting operation is completed, the arc goes out automatically because there is no ground to sustain it and the control unit stops the carriage, opens the main contactor, and shuts OFF the gas flow.

AIR CARBON ARC CUTTING (CAC-A)

Air carbon arc cutting (CAC-A) is a cutting process in which the cutting of metals is accomplished by melting with the heat of an arc between a carbon electrode and the base metal. A compressed-air line is attached directly to the torch. When the torch is in operation, the jet orifices must be positioned under the electrode. As the metal melts during cutting, a jet of compressed air is directed at the arc to blow the molten metal away from the cutting area. The jet airstream is controlled by depressing the pushbutton on the electrode holder.

Power can be supplied with either an AC or DC welding machine. However, the power requirements for a given diameter carbon electrode are higher than those for a comparable diameter SMAW electrode. Air is supplied by an ordinary compressor. In general, the required air pressure range is from 40 psi to 80 psi. Heavy-duty CAC-A systems require air pressure between 80 psi and 100 psi. **See Figure 25-14.**

Electrodes used for air carbon arc cutting are plain or copper-clad carbon-graphite electrodes. Plain carbon-graphite electrodes are less expensive, but copper-clad carbon-graphite electrodes last longer, carry higher currents, and produce more uniform cuts. Electrode holders are specially designed for air carbon arc cutting. **See Figure 25-15.**

AIR CARBON ARC CUTTING CONDITIONS								
Electrode Diameter		DCEP				AC	Air Pressure	
Inches	mm	DC Electrode		AC Electrode		AC Electrode		psi	psi
		Min A	Max A	Min A	Max A	Min A	Max A		
5/32	4	90	150	–	–	–	–	40 / 80	280 / 550
3/16	4.8	150	200	150	180	150	200	40 / 80	280 / 550
1/4	6.4	200	400	200	250	200	300	40 / 80	280 / 550
5/16	7.9	250	450	–	–	–	–	80	550
3/8	9.5	350	600	300	400	300	500	80	550
1/2	12.7	600	1000	400	500	400	600	80	550
5/8	15.9	800	1200	–	–	–	–	80	550
3/4	19.1	1200	1600	–	–	–	–	80	550

Figure 25-14. *The cutting air pressure and power settings are determined by the size electrode used for air carbon arc cutting.*

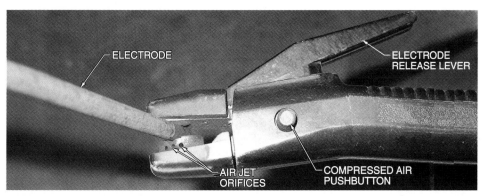

Figure 25-15. *The carbon-graphite electrode must be held in a special electrode holder designed for air carbon arc cutting.*

Air carbon arc cutting is used to cut metal, to gouge out cracks, to remove risers and pads from castings, to remove inferior welds, and to backgouge and prepare grooves for welding. Air carbon arc cutting is used when slightly ragged edges are not objectionable. The cut area is small, and since metal is melted and removed quickly, the surrounding area does not reach high temperatures. This reduces the tendency toward distortion and cracking.

Air carbon arc cutting may be used for alumimum alloys, copper alloys, carbon steels, cast irons, nickel alloys, and stainless steels. It is not recommended for titanium or zirconium. After air carbon arc cutting, but before welding, grinding must be used to remove the surface that has picked up carbon.

CAC-A produces intense arc radiation and high noise levels. Proper PPE, including appropriate hearing protection and a welding helmet with a filter plate matched to the current, should always be worn when working with CAC-A equipment. As with any welding or cutting process, fumes always pose a potential hazard, so proper ventilation is required.

The air carbon arc cutting process must be properly performed when gouging, cutting, washing, or beveling metals to prevent carburized molten metal from remaining on the surface.

Gouging

Gouging is a cutting process that removes metal by melting or burning off a portion of the base metal to form a bevel or groove. The depth and contour of the groove are controlled by the electrode angle and travel speed. For a narrow, deep groove, a steep electrode angle and slow speed are used. A flat electrode

✓ **Point**
Use plain or copper-clad carbon-graphite rods when cutting metals with the air carbon arc process.

angle and fast speed produce a wide, shallow groove. The width of the groove is also influenced by the diameter of the electrode. During all gouging operations, using the proper travel speed produces a smooth, hissing sound.

The electrode holder should be gripped so that a maximum of 6″ of electrode extends from the electrode holder to the work. For aluminum alloys the distance should be reduced to 4″. Hold the electrode holder so the electrode slopes back from the direction of travel. The jet air stream should be behind the electrode. Maintain a short arc and travel fast enough to keep up with metal removal. The arc must provide sufficient clearance so the compressed air blast can sweep beneath the electrode and remove all molten metal.

Gouging in flat position is typically performed toward the left (as the work is viewed). The electrode holder should be held perpendicular to the direction of travel with the electrode pointing to the left. The air jet orifices should be under the electrode and should follow the electrode. **See Figure 25-16.**

Flat-Position Gouging

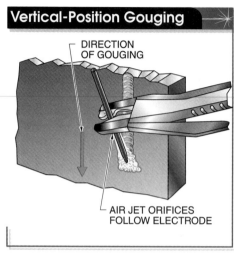

Figure 25-16. *When gouging in flat position, the electrode holder is held so that the electrode slopes back from the direction of travel.*

For gouging in vertical position, hold the electrode holder perpendicular to the workpiece and move downward. The air jet orifices should follow the electrode so that gravity assists in removing the molten metal. **See Figure 25-17.**

Gouging in horizontal position can be done by moving the electrode to either the right or the left. When traveling to the right, hold the electrode holder perpendicular to the direction of travel, with the electrode pointing toward the

right, the release lever in the down position, and the air jet orifices following the electrode. When traveling to the left, reverse the position of the electrode holder so the air jet orifices are under the electrode, the release lever is on top, and the electrode faces toward the left. **See Figure 25-18.**

Vertical-Position Gouging

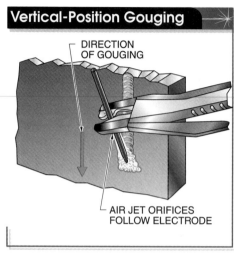

Figure 25-17. *In vertical position gouging, the electrode holder is held perpendicular to the workpiece and the air jet orifices follow the electrode, permitting gravity to remove the molten metal.*

Horizontal-Position Gouging

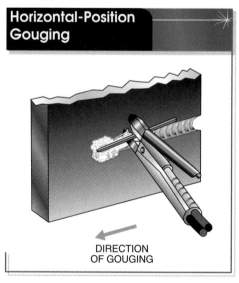

Figure 25-18. *Gouging in horizontal position can be performed from either the left or the right. The air jet orifices should always follow the electrode.*

Cutting

The cutting technique is the same as gouging except that the electrode is held at a steeper angle and is directed at a point that permits the tip of the electrode to pierce the metal being cut.

For cutting thick, nonferrous metals, hold the electrode in vertical position with a push angle of 45° and, with the air jet above it, move the arc up and down through the metal with a sawing motion.

Washing

Washing is a process of removing metal from large areas, such as removal of surfacing and of riser pads on castings. When using air carbon arc cutting for washing, weave the electrode from side to side in a forward direction to the depth desired. A push angle of 55° is recommended, with the air jet orifice following the electrode. The steadiness of the operator determines the smoothness of the surface produced. **See Figure 25-19.**

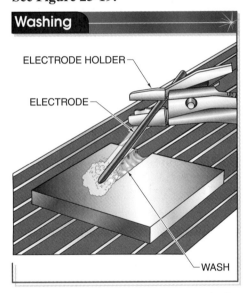

Figure 25-19. *Weave the electrode from side to side when washing with air carbon arc.*

Beveling

For beveling, hold the electrode at approximately a 45° angle, with the oxygen blast between the electrode and the metal surface. Draw the electrode smoothly along the edge being beveled. **See Figure 25-20.**

SAFETY PRECAUTIONS

In any cutting operation, a large amount of molten metal (dross) always falls to the floor. Turn pant cuffs down over the shoes to prevent dross from lodging inside the cuffs or shoes.

Figure 25-20. *When beveling a plate, hold the electrode at a 45° angle.*

Be sure there are no combustible materials near the work area when performing cutting operations. When an excessive amount of cutting is to be done, in an area with a concrete floor, the concrete should be protected with a layer of sand. Hot dross can cause air pockets in the concrete to expand, resulting in particles of concrete flying upward. Another alternative is to cut over a workbench tray partially filled with sand. If the bench lacks a tray, a sand-filled pan can be placed on the floor. A helmet with the appropriate filter-plate shade number should be worn to protect the eyes and face from arc radiation, sparks, and dross.

Fumes are a potential health hazard, especially when working in confined spaces. Cutting processes, such as plasma arc cutting and air carbon arc cutting, can produce a high volume of dust, smoke and fumes. Also, several metals produce toxic gases when they are welded or cut. For example, stainless steel produces hexavalent chromium fumes, and galvanized steel produces zinc fumes. Therefore, proper ventilation should be used at all times. In some cases, additional protection, such as a welding helmet equipped with a ventilator may be required.

CAC-A generates excessive noise levels that can damage hearing. Therefore, proper hearing protection should be worn with CAC-A and when working around loud noises in general.

! Points to Remember

- The correct oxygen and acetylene pressures to be used depend upon the tip size used, the type of cutting to be performed, and the thickness of the metal to be cut.
- When cutting cast iron, adjust the preheating flame so it is slightly carburizing.
- For high-speed cutting of nonferrous metals, plasma arc cutting is the most effective.
- In plasma arc cutting, set the polarity to direct current electrode negative.
- Use plain or copper-clad carbon-graphite rods when cutting metals with the air carbon arc process.

EXERCISES

Cutting Steel Using Oxyfuel Cutting...  exercise 1

1. Obtain a piece of mild steel.

2. Use a soapstone to draw a line on the workpiece about ¾″ from one edge.

3. Position the workpiece so the line clears the edge of the welding bench.

   When an exceptionally straight cut is desired, clamp a bar across the workpiece alongside the cutting line to act as a guide for the torch.

4. Turn ON the acetylene needle valve and light the gas with a sparklighter as for welding. Turn ON the oxygen valve and adjust it for a neutral flame.

   The neutral flame is used to bring the metal to a kindling temperature. In the case of plain carbon steel, for example, the kindling temperature is between 1400°F and 1600°F.

5. Observe the nature of the cutting flame by pressing down the oxygen control lever. When the oxygen pressure lever is depressed, additional adjustments may be needed to keep the preheating cone burning with a neutral flame.

6. Grasp the torch handle in such a way as to permit instant access to the oxygen control lever. The valve is usually operated with either the thumb or forefinger.

   Hold the torch steady to ensure making a clean, straight cut. If the tip is allowed to waver from side to side, a wide kerf is formed, which results in a rough cut, slower cutting speed, and greater oxygen consumption. To help keep the torch steady, support the elbow or forearm.

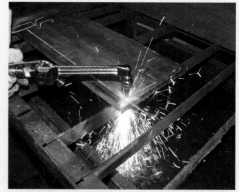

Smith Equipment

7. Start the cut at the edge of the workpiece. Hold the torch with the tip vertical to the surface of the metal and the inner cone of the heating flame approximately 1/16″ above the line. Hold the torch steady until a spot in the metal has been heated to a bright red.

8. Gradually press down the oxygen pressure lever and move the torch forward slowly along the line.

   The torch should be moved just rapidly enough to ensure a fast but continuous cut. A shower of sparks falling from the underside of the cut indicates that penetration is complete and the cut is proceeding correctly.

9. If the cut does not seem to penetrate the metal, close the oxygen pressure lever and reheat the metal until it is a bright red again. If the edges of the cut appear to melt and have a very ragged appearance, the metal is not burning through and the torch is being moved too slowly.

10. Initially, the workpieces may stick together, even when the cut has penetrated through. This is due to the slag produced by the cutting flowing across the workpiece. Slag is not a serious problem because it is quite brittle, and a slight blow with the hammer will separate cut sections.

11. It may occasionally be necessary to start the cut in from the edge of the plate. If so, hold the preheating flame slightly longer on the metal; then raise the cutting nozzle about 1/2″ and depress the oxygen lever. When a hole is cut through, lower the torch to its normal position and proceed with the cut in the usual manner.

1. Obtain a piece of cast iron.

2. Use a soapstone to draw a line on the workpiece about ¾″ from one edge.

3. Position the workpiece so the line clears the edge of the welding bench.

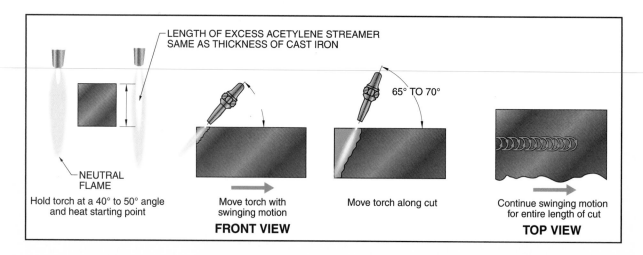

LENGTH OF EXCESS ACETYLENE STREAMER SAME AS THICKNESS OF CAST IRON

65° TO 70°

NEUTRAL FLAME

Hold torch at a 40° to 50° angle and heat starting point

Move torch with swinging motion

FRONT VIEW

Move torch along cut

Continue swinging motion for entire length of cut

TOP VIEW

4. Light the torch and adjust the preheating flame so that it shows an excess of acetylene.

5. The excess acetylene, as indicated by the length of the white cone, must be varied to best suit the grade and thickness of the cast iron to be cut. Experience is the best guide; however, it generally varies from little or no excess of acetylene for extremely thin workpieces to an excess of a 1″ to 2″ white cone for thick workpieces.

6. Bring the tip of the torch to the starting point. Hold the torch at an angle of approximately 40° to 50° and heat a spot about ½″ in diameter to a molten condition.

7. With the end of the preheating cone about ³⁄₁₆″ from the metal, start to move the torch and open the high-pressure cutting valve. A swinging motion may be required for thick metals.

   If adjustments to the flame are needed, they should be made with the high-pressure valve wide open to avoid any change in the character of the flame during the cutting operation.

8. Gradually bring the torch along the line of the cut, continuing the swinging motion. As the cut progresses, gradually straighten the torch to an angle of 65° to 70° to ensure thorough penetration. Continue the swinging motion along the entire length of the cut.

9. On thick workpieces, ensure that there is sufficient heat to allow the cut to proceed without interruption. On thin workpieces, it is easy to lose the cut as the surface of the metal cools too rapidly and only a slight groove is made with the flame.

   Restart a cut by heating a small circle as previously described. Gradually raise the torch and incline it to cut away the lower portion of the workpiece. Proceed as before, with the exposed side of the cutting groove appearing bright. Continue to cut until finished.

1. What causes metal to rust?
2. What principle makes possible the cutting of metal by OFC?
3. How does a cutting torch differ from a welding torch?
4. What determines the oxygen and acetylene pressure that must be used for cutting?
5. What aids may be used to facilitate an even cut?
6. How can it be determined that the cut is penetrating through the metal?
7. What is the position of the torch when cutting round stock?
8. How is it possible to make a bevel cut with a cutting torch?
9. Describe the operation for piercing small holes with a cutting torch.
10. What type of flame is used for cutting cast iron, assuming a good grade of iron?
11. How is the torch held when cutting cast iron?
12. What torch motion is used for cutting cast iron?
13. What is meant by PAC?
14. What types of metals can be cut by PAC?
15. What type of electrode is used in the CAC-A process?
16. What causes the removal of molten metal when using CAC-A?
17. What does the term washing mean when using CAC-A?
18. What are some of the precautions that should be observed before engaging in any cutting operation?

Refer to Chapter 25 in the *Welding Skills Workbook* **for additional exercises.**

Repair Welding

Repair welding is a method of restoring components that have failed or have lost their ability to perform as designed. All repair options, usually mechanical repair and weld repair, must be evaluated after identifying the cause of failure. Distortion, flammability, and related safety concerns may prevent the use of repair welding. When repair welding is appropriate, a repair welding plan must be prepared that complies with relevant repair codes and safety requirements.

EVALUATING REPAIR METHODS

Repair welding is used only if it is economical or if a replacement part is not available. If a piece of equipment fails within the warranty period, the manufacturer of the equipment is contacted to determine replacement options before developing a repair plan.

Certain repairs may require the approval of authorized personnel, use of a qualified procedure, and/or preparation of supporting documentation. Repairs that are regulated by applicable codes and standards, such as repairs to aircraft, pressure vessels, and transportation containers, require documentation that repairs were made appropriately.

An understanding of how a part failed is necessary before considering repair methods. Some repair methods may not be effective on certain failures. For example, a leaking stainless steel tank that has failed by chloride stress corrosion cracking cannot be repair welded. The heat created by grinding to remove the cracks or by welding over the cracks actually accelerates the spreading of the crack, making it worse. For an effective weld repair, the leaking area must be cut out and an insert plate welded flush with the tank wall. **See Figure 26-1.**

Failure Analysis

Failure analysis provides an accurate explanation of the cause of a failure or loss of performance. Failure analysis techniques consist of failure modes and effects analysis, physical failure analysis, and root cause failure analysis.

Failure modes and effects analysis is a failure analysis process that provides a diagnosis of the technical cause of failure using experience gained from previous failures. *Physical failure analysis* is a failure analysis process that provides a diagnosis of the technical cause of failure using various analytical methods. *Root cause failure analysis* is a failure analysis process that determines how to prevent a failure from recurring by understanding how the actions of humans or systems may have led to the technical cause of the failure. Root cause failure analysis seeks to eliminate defects so that the failure does not recur.

> ✓ **Point**
>
> *An understanding of how a part failed is necessary before considering repair methods. Some repair methods may not be effective on certain failures.*

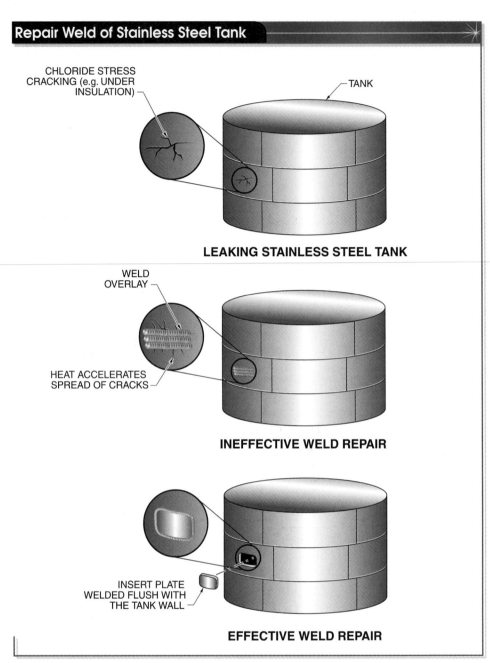

CHLORIDE STRESS
CRACKING (e.g. UNDER
INSULATION)

TANK

LEAKING STAINLESS STEEL TANK

WELD
OVERLAY

HEAT ACCELERATES
SPREAD OF CRACKS

INEFFECTIVE WELD REPAIR

INSERT PLATE
WELDED FLUSH WITH
THE TANK WALL

EFFECTIVE WELD REPAIR

Figure 26-1. *An understanding of how a part failed is necessary when selecting an effective repair method.*

✓ **Point**

Root cause failure analysis identifies and links the three levels of deficiency that lead to failures: technical, human, and system.

Root cause failure analysis uses failure modes and effects analysis or physical failure analysis to determine the root cause of failure. Overall, root cause failure analysis identifies and links three levels of deficiency that lead to failures:

- technical causes that lead to equipment unreliability
- human causes that lead to technical causes
- system operations causes that lead to human causes

Once the technical cause is identified, human behavior and system operations causes that contribute to the root cause of failure can be determined. Once the root cause of the failure is determined, a repair plan can be established. Mechanical repair and weld repair are two options for conducting repair welding.

An understanding of how a part failed is necessary before considering repair methods. Some repair methods may not be effective on certain failures.

MECHANICAL REPAIR METHODS

Mechanical repair is a repair weld process that consists of methods that do not create a metallurgical bond between the restored parts or at the restored surface. Mechanical repair methods produce a physical joining or resurfacing of parts without metallurgical bonding. Mechanical repair does not involve a significant heat input, which reduces the potential for distortion and residual stresses that may occur in weld repair.

Some mechanical repair methods may be performed in the field where the failure occurred; however, the failed part is generally taken to a shop for repair. Mechanical repair methods include adhesive bonding, cold mechanical repair, electroplating, and blend grinding.

Adhesive Bonding

Adhesive bonding is the joining of parts with an adhesive placed between the faying (mating) surfaces, which produces an adhesive bond. A satisfactory adhesive bond requires close contact between the surfaces to be joined.

Adhesive bonded parts normally have a high resistance to shear and tension stresses because the entire surface area of the joint contributes to the strength of the bond. On the other hand, adhesive bonded parts exhibit relatively low resistance to cleavage and peeling. Thus, if the load is concentrated at the end of the bond, the joint may start to fail from the loaded end, leading to incremental separation into the body of the joint ("unzipping").

To minimize or eliminate the negative effects of peeling or cleavage, adhesive bonding may be combined with an additional mechanical fastening method such as riveting. Riveting coupled with adhesive bonding also increases the fatigue strength. **See Figure 26-2.**

> ✪ An adhesive's mode of attachment varies. Depending on the type of adhesive, bonds may be made by mechanical or by chemical means.

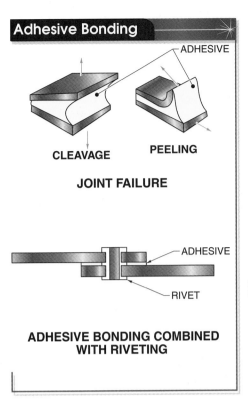

Adhesive Bonding

JOINT FAILURE

CLEAVAGE   PEELING

ADHESIVE BONDING COMBINED WITH RIVETING

Figure 26-2. *Adhesive bonded joints may resist bonding, resulting in cleavage or peeling. However, when combined with riveting, fatigue strength is increased.*

> **✓ Point**
> *Mechanical repair methods do not create a metallurgical bond with the surface that is being repaired.*

Surface Preparation. A clean, dry surface is necessary for a quality adhesive bond. Joint failure commonly occurs because of inadequate cleaning. Surface cleanliness may be evaluated by pouring a small quantity of water over the cleaned surface. If the water breaks into individual droplets, some contamination is present. If the water uniformly covers the surface in a thin layer, the surface is clean. Surface preparation methods for adhesive bonding are abrasive cleaning, solvent cleaning, and chemical conversion.

Abrasive cleaning includes sandblasting, sanding, and wire brushing. Abrasive cleaning is used to remove heavy layers of rust or other deposits. Solvent cleaning includes hot alkaline washing, solvent wiping, and vapor degreasing. When using solvent cleaning, the solvents must be properly disposed of once cleaning is finished. Chemical conversion includes anodizing and phosphating.

> **✓ Point**
> *Adhesive bonding is the joining of parts with an adhesive that is placed between the faying (mating) surfaces, producing an adhesive bond.*

Adhesive Application. Adhesives are selected based on the service requirements of the bonded part and the permitted application methods. **See Figure 26-3.** Epoxy phenolic adhesives form strong bonds and have good moisture retention, making them suitable for joining some metals, glass, and phenolic resins. Polyacrylate esters are not suitable for structural joints but may be used as pressure-sensitive tape.

An adhesive supplier is a good source of technical advice on adhesive selection. To make any recommendation, the adhesive supplier must be provided the following information:

- how the part is expected to perform mechanically
- thermal history of the materials being joined
- expected service temperature, including any temperature cycling

ADHESIVE APPLICATION		
Adhesive Type	**Characteristics**	**Curing Requirements**
Acrylic	• variable impact resistance • moderate strength • aluminum, copper, and steel	Room temperature up to 130°F (55°C) for 10 min to 20 min
Cyanoacrylate	• strong but brittle • jewelry, electronic components	30 sec to 5 min, room temperature
Epoxy: amine, amide, and anhydride cured	• high tensile strength, low peel strength, and moisture- and chemical-resistant • widely used for metals, ceramics, and rigid plastics	Varies from slow to fast, room temperature to 350°F (175°C)
Epoxy-phenolic	• strength retention from 300°F (150°C) to 500°F (250°C) • some metals, aluminum, steel, glass, and phenolic resins	1 hr at 350°F (175°C)
Nitrile-phenolic or neoprenephenolic	• flexible with impact resistance • metals and some plastics	Up to 12 hr, 250°F (120°C) to 300°F (150°C)
Polyamide	• good room temperature strength and toughness • aluminum and copper	Applied as hot melt and cures by cooling
Polyester	• develops strong bond • fiberglass, sometimes for metal	Min to hr, room temperature
Polyhydroxyether	• moderate strength, flexible, good adhesion • nickel and copper	Applied as hot melt and cures by cooling
Rubber-containing (e.g. neoprene, natural rubber)	• flexible • limited load bearing ability, but high impact strength and moisture resistance • most types of materials	Varies, but mostly pressure sensitive

Figure 26-3. *Various adhesives may be used depending on the service requirements of the part to be repaired.*

Manual adhesive application (using brushes, rollers, and squeeze bottles) is the simplest method of applying adhesives. Adequate ventilation is required, and personal protective equipment must be used to prevent skin contact with the adhesive.

After the adhesive is applied, the parts are brought into contact with each other and the adhesive is allowed to cure. *Curing* is a process that converts the adhesive from its applied condition to its final solid state. Curing occurs by solvent evaporation or by chemical reaction between two or more chemical components. For example, contact cements cure by solvent evaporation, whereas epoxy and urethane adhesives cure by chemical reaction.

Heat and/or pressure may be used to assist curing. The adhesive supplier should set the limits of using heat and pressure for a particular product. Excessive heat and/or pressure may result in an unacceptable bond. Excessive pressure causes adhesive to be squeezed out of the bond area, leading to a starved joint. A *starved joint* is a joint that contains insufficient adhesive to create an optimum bond.

However, incomplete contact or insufficient contact pressure leads to an excessively thick adhesive bond line, increasing the probability of a major flaw developing within the bond, which can eventually cause joint separation.

Cold Mechanical Repair

Cold mechanical repair (metal stitching) is a repair method that consists of locks and stitching pins installed into the surface of a cracked metal part. A *lock* is a precision, high-strength steel member with a multi-lobed outer contour. When installed into precision-drilled hole patterns they add strength across a crack by pulling and holding the two sides together.

Stitching pins are installed along the crack in a continuous, overlapping pattern. The special threads on the stitching pins grip the side walls of the drilled and tapped holes and draw the sides together rather than spread them apart as normal threads do.

Cold mechanical repair is appropriate for repairing metals that are difficult to weld, such as cast iron. Welding on cast iron at temperatures above 1000°F often causes more problems than it solves. Cold mechanical repair prevents those problems. Cold mechanical repair can also be used on fabricated metal parts that may be difficult to replace or where welding may cause distortion.

Cold mechanical repair is commonly used on metals such as gray iron, ductile iron, aluminum, bronze, steel, and fabricated steel sections. Some applications include engine blocks and heads, pumps, compressors, machine tools, and gear boxes. However, the repair cannot typically be used to seal vessels that hold pressure. Locks, stitching pins, and tooling are manufactured by companies that specialize in cold mechanical repair. Some repairs can be done with just the stitching pins or the locks. **See Figure 26-4.** For repairs requiring both locks and stitching pins, follow the procedure:

1. Determine the extent of cracking by liquid penetrant (PT) or magnetic particle (MT) examination.
2. Drill the hole patterns transverse to the crack along its length using the precision-drill fixtures.
3. Drive the locks into the hole pattern to lock the opposite sides of the crack together.
4. Drill, tap, and install the stitching pins in an overlapping pattern along the crack.
5. Grind or machine the repair flush with the surface.

Metal stitching is not appropriate for all repair applications, but is ideal for certain situations. For example, cast iron is particularly difficult to weld, but can often be repaired quickly, easily, and with good finish characteristics by metal stitching. This technique is often used on engine blocks and cylinder heads.

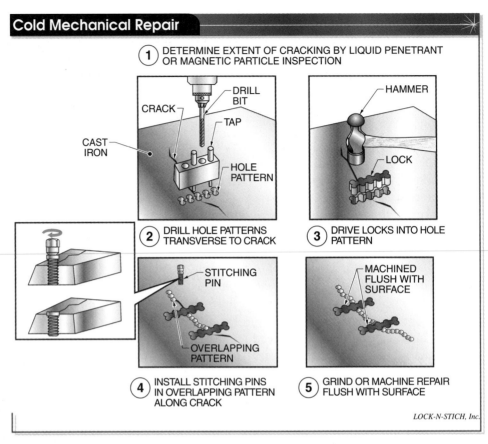

1. DETERMINE EXTENT OF CRACKING BY LIQUID PENETRANT OR MAGNETIC PARTICLE INSPECTION

DRILL BIT

CRACK

TAP

CAST IRON

HOLE PATTERN

HAMMER

LOCK

2. DRILL HOLE PATTERNS TRANSVERSE TO CRACK

3. DRIVE LOCKS INTO HOLE PATTERN

STITCHING PIN

OVERLAPPING PATTERN

MACHINED FLUSH WITH SURFACE

4. INSTALL STITCHING PINS IN OVERLAPPING PATTERN ALONG CRACK

5. GRIND OR MACHINE REPAIR FLUSH WITH SURFACE

LOCK-N-STICH, Inc.

Figure 26-4. *Cold mechanical repair is used on large or complex castings and forgings that are difficult to replace.*

Electroplating

Electroplating is the application of a thin, hard, chrome coating to repair minor damage. The coating is typically between 5 mil and 10 mil thick. The part is masked with a nonconducting compound such as wax to screen areas not to be plated. The part is then placed in an electroplating tank that contains electroplating solution. An electric current is applied, with the part as the negative electrode in the circuit, so that metal (plating) is deposited on the part. Heavier plating thickness can be achieved by applying a copper flash plate, creating the bulk of the buildup (up to 1/16″) with electroplated nickel, and then applying the final thin coating with the required properties. When electroplating is intended to produce wear resistance, a thin layer of electroplate is used.

Electroplating reduces the fatigue strength of rotating or reciprocating equipment such as shafts. Fatigue strength can be restored by peening the part before electroplating, and baking the part after electroplating. Baking should be performed for 4 hr at 350°F or higher.

Electroplating may also be used for minor repairs by means of selective plating. *Selective plating* is a form of electroplating used for touch-up repairs on worn or damaged parts. Selective plating can be performed in the field or in the shop to repair nicks, scratches, or dings in rolls, bearing journals, wear surfaces, or oil seal surfaces. **See Figure 26-5.**

An anode saturated with special plating solution is used for selective plating. A rectified AC power supply is connected to the workpiece and the plating anode. Selective plating is accomplished by the relative motion of the solution-soaked anode and the workpiece. A variety of plating types is available; however, hard nickel and nickel alloy plating are most common.

Figure 26-5. *Selective plating may be used in the field or repair shop to repair nicks, scratches, or dings on wear surfaces.*

Blend Grinding

Blend grinding is a mechanical repair method in which a thinned, pitted, or cracked region of a part is ground away to create a gradual transition with the unaffected surface. The smooth transition reduces stress that might lead to failure in service. Two specific conditions must be addressed before blend grinding is used: design thickness and the corrosion allowance of the structure.

Design thickness is the thickness of metal required to support the load on a part. Most parts are built with a thickness in excess of their design thickness because the product forms used to fabricate them, such as plate and pipe, are available in standard thicknesses. The designer selects the nearest available thickness above the design thickness.

Corrosion allowance is an additional thickness of metal above the design thickness that allows for metal loss from corrosion or wear without reducing the design thickness. The corrosion allowance is based on the anticipated severity of the environment in which the part is to operate. For example, a part having a ⅛″ corrosion allowance may withstand a general corrosion rate of 20 mil/yr for approximately 6 yr before repair or replacement may be necessary.

Blend grinding may be applied to restore the wall of a storage tank that exhibits a sharp ring of corrosion. Blend grinding eliminates the need for weld repair, provided the metal thickness in the corroded ring is above the minimum required for the service loads. **See Figure 26-6.** If it does not reduce the metal wall to below its design thickness and corrosion is not an issue, blend grinding may be preferable to weld repair.

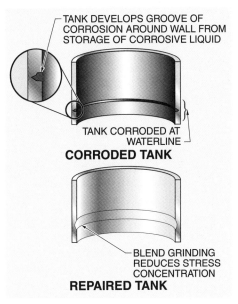

Figure 26-6. *Blend grinding can be used to repair tanks in which corrosion has developed, and is an adequate repair if the remaining tank wall has sufficient structural strength. Blend grinding helps to prevent weld repair buildup on the surface.*

Wall thickness checks are performed using ultrasonic thickness measurements to measure the extent of the area to be ground. There must be an adequate wall thickness in the tank to ensure a successful blend grinding operation.

Blend grinding is useful when weld repair is too difficult or unsafe, as in a storage tank that contains flammable vapors. Precautions must be taken to prevent sparks during grinding and to ensure that confined space entry procedures are followed. A wall thickness check is performed using ultrasonic thickness measurements to map and measure the extent of the area to be repaired, and the available wall thickness in the area.

✓ **Point**

Blend grinding may be used to avoid weld repair when the remaining thickness of a structure provides adequate strength.

WELD REPAIR METHODS

Weld repair is a repair weld process that consists of methods that join failed parts or restore their surface using a welding process. The heat and residual stresses created by weld repair methods must be anticipated and allowed for in the repair plan. Therefore, weld repair is not always an option. The heat generated by welding may cause significant problems, such as reduced corrosion resistance, weld discontinuities, and residual stress. Weld repair methods consist of structural weld repair, surfacing weld repair, sheet lining (wallpapering), and sleeving.

Structural Weld Repair

Structural weld repair is restoration of a load-bearing structure by welding to meet performance requirements. Examples of structural weld repairs are restoration of a worn or broken rotating shaft, or rebuilding a storage tank wall that has corroded to less than the design thickness.

Before structural weld repair is performed, confirm that the residual stresses introduced by the weld will not worsen the failure. Welding must be done in a region away from the critical high stress region where the failure occurred to prevent continued failure of the part.

When fatigue stresses are a factor, the location of the weld repair must be in a region away from a change in section thickness, where the stress concentration is highest. **See Figure 26-7.**

Preheat, postheat, and distortion control requirements must be detailed in the repair plan. The structural weld repair technique must minimize distortion and prevent the introduction of excessive residual stresses. If the repair is greater than ½″ thick or the joint is highly restrained, low-hydrogen electrodes should be used. A fillet weld joint is commonly a highly restrained joint and the toe should be undercut (gently grooved) when a fillet weld joint is to be used in fatigue or high stress applications.

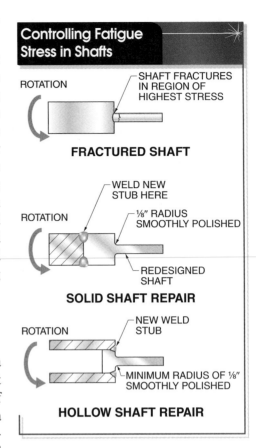

Controlling Fatigue Stress in Shafts

FRACTURED SHAFT

ROTATION — SHAFT FRACTURES IN REGION OF HIGHEST STRESS

SOLID SHAFT REPAIR

ROTATION — WELD NEW STUB HERE — ⅛″ RADIUS SMOOTHLY POLISHED — REDESIGNED SHAFT

HOLLOW SHAFT REPAIR

ROTATION — NEW WELD STUB — MINIMUM RADIUS OF ⅛″ SMOOTHLY POLISHED

Figure 26-7. *When making a structural weld repair, ensure the weld is made away from the region of highest stress.*

Surfacing Weld Repair

Surfacing weld repair is the application of a layer, or layers, of specially formulated weld metal to restore worn or corroded components to extend their useful life. Surfacing weld repair can be used for many applications. The compatibility of the base metal and the surfacing material determines which surfacing weld repair technique to use. As with structural weld repair, preheat postheat, and distortion control must be detailed in the repair plan.

Surfacing weld repair may be done in the shop or in the field and may be performed automatically or manually. Automatic surfacing weld repair is typically performed on large corroded areas that must be rebuilt. Automatic surfacing welding machines have one or two GMAW heads and deposit metal on the vertical inside surface. The welding heads are mounted on a boom that rotates around a centerline and makes the metal deposits on the inside diameter of the corroded surface. Many automatic surfacing weld repair machines are portable enough to be used in the field as well. For manual surfacing repair in the field, SMAW is preferred, however, OFW processes can also be used because the equipment is portable.

The thickness of the surfacing weld repair should not be greater than twice the amount of wear. For hard deposits, each layer should be as thin as possible to prevent cracking, and no more than two passes should be made. A single pass is adequate if dilution between the surfacing and the component can be minimized. Surfacing weld repair should not be applied over an existing deposit that has partially worn away. A worn deposit must be completely removed by grinding.

SMAW or FCAW may be used after a worn deposit is removed by grinding and for applications in which the same component is resurfaced on a regular basis. GMAW is used with automatic or semiautomatic processes. GTAW and PAW are typically only used on small components because they are more expensive, take longer to apply, and there is less availability of wires.

OAW is used for bronze bearing surfacing weld repair. Only one pass of weld metal is required, which helps minimize distortion. However, if the repair is being done to improve corrosion resistance, the level of dilution by the base metal must be determined to ensure that one pass is adequate. If dilution of the surfacing weld could be a problem, a second pass may be required.

Peening is used to minimize distortion and crosschecking in surfacing deposits. *Crosschecking* is a series of parallel cracks about ½″ apart that occur in brittle deposits (with hardness greater than HRC 50) as they undergo stress relief. Peening is done by battering the surface with a blunt-nosed hammer while the temperature exceeds 1000°F (540°C). Peening compresses the surface and reduces residual tensile stresses. For manual repair weld processes, the welder deposits no more than 6″ of surfacing and then peens the surface.

To prevent hardening of some steels during surfacing weld repair, the base metal must be preheated to the appropriate temperature. Preheat temperature must be maintained throughout the procedure, and the component must be blanket-cooled after surfacing weld repair is completed.

Sheet Lining (Wallpapering)

Sheet lining (wallpapering) is a weld repair method that uses thin, usually ¹⁄₁₆″, sheets of corrosion-resistant material that are welded to a corroded surface to cover a large area of damage. Storage tanks, scrubbers, and other large vessels subject to corrosion or pitting may be sheet lined. The corrosion-resistant sheets are usually made of nickel alloy. GMAW is commonly used for sheet lining using short circuiting transfer or pulsed spray transfer. An intermittent fillet weld is used with adjacent sheets overlapping one another. A continuous fillet weld is made between the new sheet and the previously installed sheet. **See Figure 26-8.** Continuous fillet welds are required around the entire outer edge of the corrosion-resistant sheet so that no leakage occurs. If sheets larger than 1 sq ft are used, spot welds or plug welds should be made in the middle of each sheet to provide additional reinforcement.

However, if a subsequent leak occurs in a vessel that was repaired by sheet lining, the space behind the lining may become contaminated. This may make further repair difficult.

Corrosion-resistant alloys are relatively expensive, so sheet lining may be used to clad selected areas of cheaper alloys.

✓ Point

The compatibility of the base metal and the surfacing material determines which surfacing weld repair technique to use.

✓ Point

Sheet lining (wallpapering) and sleeving are specialized repair welding processes and are less commonly used than other repair welding processes.

Haynes International, Inc.
Figure 26-8. *Corrosion-resistant sheets are overlapped, with a fillet weld made between new and previously installed sheets, to repair a corroded surface.*

Sleeving

Sleeving is a weld repair method that applies surfacing to badly worn shafts by welding snug-fitting semicircular forms to cover the shaft surface. Half sleeves are usually made of a wear-resistant cobalt alloy. Transverse shrinkage tends to pull the half sleeves tightly down on the shaft when sleeves are longitudinally welded to one another. Longitudinal relief grooves are cut into the shaft to prevent heat buildup that might lead to cracking where the half sleeves are welded together.

WELD REPAIR PLANS

When developing a weld repair plan, all factors that lead to a successful repair must be considered. Factors include determining necessity of repairs, repair codes, identifying base metal, joint profile, distortion control, and repair welding procedures. **See Figure 26-9.**

Determining Necessity of Repairs

Before any repairs can take place, it must be determined whether a weld repair is the best course of action for the part. Some components can be repaired, while others are normally replaced. Many factors determine the type of repair to be made, or whether any repair at all should be made. Based on the effects of distortion and residual stress in the per-

formance of the component, mechanical repair methods may offer better options. Mechanical repair methods may provide better results because if an incorrect welding procedure is used, there is an increased chance of failure.

The component or equipment name must be documented. A fabrication drawing number is assigned to the component to be designed. The fabrication drawing number for the component contains essential data, such as materials of construction and heat treatment requirements, when applicable. For shafts, the fabrication drawing number contains other essential information necessary to make an effective repair. Essential information such as shaft finish; radii, if stepped locations; special fits; existing surface treatments; and run out are included.

The technical cause of failure must be known and understood. Using an inappropriate weld repair method may cause the repair to fail rapidly. Inappropriate weld repair methods are a common problem when performing weld repair on fatigue failures.

Repair Codes

Weld repair may be governed by a code, in which case it is necessary to follow the applicable requirements, or risk penalties for violation of the code. Some codes address repair welding requirements as a specific subject and others require welding qualifications that apply to both new and repair welding. Codes dealing specifically with repairs are sometimes called in-service inspection and repair codes. All repair codes require qualified welding procedures, qualified welders, and proof that welders are qualified to perform the required procedures. Most codes rely on ASME Section IX and AWS D1.1 to describe the requirements for qualified welding procedures and welders. Nondestructive examination must be performed by qualified examiners, and the results interpreted by inspectors qualified in the applicable code. The following specific types of equipment are covered by repair codes:

WELD REPAIR PLAN	
Step	**Description**
Determining Necessity of Repairs	
Part/equipment	
Fabrication drawing numbers	
Failure mode	
Requirements for stress analysis	
Repair options	
Repair Codes	
Name of code	
Base Metal	
Base metal identification	
Effect of welding on mechanical properties	
Repair Profile	
Access of welder and filler metal to repair area	
Method of defect/crack excavation	
Nondestructive examination	
Method of checking for total defect/crack removal	
Surface preparation before repair	
Cleaning requirements before welding	
Special disassembly procedures	
Distortion Control	
Effect of welding on distortion	
Special alignment methods	
Special repair weld sequence	
Special support/bracing	
Welding Personnel and Equipment	
Method of qualifying welders	
Welding equipment	
Support personnel	
Safety requirements	
Repair Procedures	
Welding procedure	
Welding process	
Gas shielding	
Filler metal type and diameter	
Tack welds or buttering	
Number of passes (if surfacing)	
Welding position	
Welding technique	
Preheat/interpass temperature control	
Peening	
Postheating or cooling method	
Repair Wrap-up	
Inspection	
Testing of repair	
Final machining	
Cleanup	
Reassembly	

Figure 26-9. *A weld repair plan details all required steps to successfully complete a repair weld.*

- Bridges, steel-frame buildings, and ships may only be repaired with special authorization. Structural steel and bridges are built in accordance with American Welding Society Codes. The repair work must be designed and approved; welders must be qualified according to the code used; and the work must also be inspected. Written welding procedures are required.

- Transportation equipment and containers, such as railroad locomotives and railroad car wheels, high-strength, low-alloy steel truck frames, and compressed gas containers are not usually weld repaired. Welding is only permitted with special permission and approval.

- Aircraft may be repaired by welding, under stringent controls. The welder performing repairs on aircraft should be certified in accordance with MIL-T-5012D, *Tests; Aircraft and Missile Operators Certification*. The welder must be qualified for the type of metal being welded, the welding process to be used, and the category of parts involved.

Furthermore, the welder should be qualified in accordance with the requirements of the Federal Aviation Administration (FAA) guidelines contained in two applicable documents: *Acceptable Methods, Techniques and Practices— Aircraft Inspection and Repair*, and the *Air Frame and Power Plant Mechanics Air Frame Handbook*. These documents provide precautionary information, techniques, practices, and methods that may be used in repair welding. Alternative techniques must be approved by the FAA. The latest version of any standard or code should always be used.

Strict regulation of repair welding for aircraft is required because many aircraft parts are made of materials heat-treated to obtain high strength. Welding repair may compromise their mechanical properties.

- Rotating equipment such as turbines, generators, and large engines are generally covered by casualty insurance.

Weld repair is performed only after approval of a written welding procedure by the insurance company.

- Fired boilers and pressure vessels are covered by various repair codes. Boilers and pressure vessels are covered by ANSI/NB-23, *National Board Inspection Code*. Additionally, pressure vessels are covered by API 510, *Pressure Vessel Inspection Code; Materials, Inspection, Rating; Repair; and Alteration*. An authorized inspector (AI) must be involved or give approval during repairs and alterations.

The repair firm contacts the jurisdictional authority, the insurer, and the owner of the boiler or pressure vessel to ensure that the method and extent of repair is approved before making the repair.

- Storage tank repairs are covered by API 653, *Tank Inspection, Repair, Alteration, and Reconstruction*, and piping repairs by API 570, *Piping Inspection, Repair, Alteration, and Rerating*.

All welding procedures and welders must be qualified in accordance with ASME Section IX, *Welding and Brazing Qualifications*.

Some companies specialize in the repair and alteration of boilers and pressure vessels. They are authorized by the appropriate jurisdictional authority; possess a current ASME code symbol stamp covering the scope of the repair work; or have a current National Board "R" repair code symbol stamp.

Analysis of operating stresses may be required to ensure that adequate weld metal is applied.

Point

The base metal must be identified before attempting a weld repair to determine the weldability of the metal and the type of repair that should be performed.

If a component to be repaired was originally a welded fabrication, then information on the original process is important to a successful repair. If access to this information is not practical, then an analysis of the base metal, including previous weld deposits, is mandatory. Many construction codes require that a written procedure be prepared prior to any repair welding.

Identifying Base Metal

The base metal to be welded must be identified, including its heat-treated and/or mechanically worked condition to determine the weldability of the metal and the proper repair performed. The integrity of the base metal is influenced by heat treatment and mechanical work.

Original documentation and drawings are helpful in determining the specifications or description of the base metal. Without documentation, the base metal can be identified by using a spark test, chemical analysis, or X-ray fluorescence analysis. A spark test can determine the approximate base metal chemistry for carbon steels. Chemical analysis requires that drillings be taken from the item and provides an accurate composition of the base metal. X-ray fluorescence analysis can identify many types of alloys.

Even when the chemical analysis or material type is known, the hardness may not be indicated. Hardness is a key indicator of mechanical properties for carbon and alloy steels. Heat treatment may be required after weld repair to restore the part to its optimum mechanical condition, particularly with machinery components, which are usually quenched and tempered to high strength levels.

Distortion Control

The effect of heat and residual stresses on tolerances and other critical dimensions must be estimated before weld repair. During weld repair, special alignment methods may be required to monitor distortion, coupled with special supports or bracing and special repair sequences to reduce distortion.

For repair of mechanical equipment, alignment markers may be used. An *alignment marker* is a center punch mark made across the joint in various locations. Alignment markers are useful in precise repair work to maintain dimensional control or alignment during welding. When repairing shafts, a dial indicator may be used to measure distortion.

Intermittent welding and back-step welding are used to reduce distortion. Intermittent welding and back-step welding help balance out stresses and the effects of heat in the repair.

Support or bracing during weld repair is required on complex or large jobs to ensure the parts under repair do not move and are free of unnecessary forces. The supports or braces are necessary to align the parts and should not interfere while the weld repair is being made. It may be necessary to temporarily tack weld the supports or braces to the structural members to support the load.

> *Per AWS, D1.1, Structural Welding Code—Steel, the old weld is removed by grinding, machining, chipping, or gouging. Additionally before weld repair methods are performed, the surface of the structure to be repaired must be thoroughly cleaned of foreign matter, including paint to at least 2″ from the root of the weld.*

Joint Profiles

The repair area must be accessible by the welder and the welding electrode or filler metal. Otherwise, making the repair is difficult and the chance of successfully repairing the area is low. Access to the repair area must allow the most comfortable welding position.

Root opening and backgouging requirements must be specified to achieve a full-penetration weld. If the backside of the weld cannot be backgouged, a backing bar may be required. The groove angle should be the minimum possible to reduce the amount of weld metal added, but should be sufficient to allow room to manipulate the electrode at the root of the repair.

When weld repairing casting defects, the groove angle depends on the alloy. The groove angle is opened up with cast austenitic stainless steels compared with cast steels because the former are more prone to hot cracking. Buttering the sides of the joint may be necessary with cast

> **⚠ WARNING**
> Never perform weld repair on an unidentified base metal.

austenitic stainless steels to overcome susceptibility to hot cracking. Low heat input is also beneficial. Nickel alloys are even more susceptible to hot cracking than austenitic stainless steels, and the groove angle may be opened up even further.

Evaluating Defects and Cracks. Crack or defect removal is usually performed with oxyfuel gas cutting or with air carbon arc gouging. Special gouging tips should be selected based on joint preparation. For some metals, air carbon arc gouging can introduce carbon into the surface and the HAZ. By closely inspecting the joint surface, it is possible to see if cracks are spreading. Metals that require preheating in welding require the same preheat conditions during cutting or gouging.

Testing for Defect and Crack Removal. Liquid penetrant examination is the most common method of checking for cracks. All liquid penetrant residues must be removed after examination by wiping with a rag soaked in a suitable solvent for solvent removable penetrant or with water for water washable penetrant.

Final surface preparation before repair is required when the surface produced by cutting or gouging is not as smooth as desired, or has been contaminated. Unless the resulting groove is smooth without undercutting or contamination, grinding or machining may be required. Grinding must achieve bright metal without excessive heat buildup. If machining is used, all fluids must be cleaned off the surface. After grinding or machining, the surface is carefully inspected for cracks and oxide particles to ensure they are removed. If magnetic particle or liquid penetrant examination is required, the surface must be cleaned one final time before testing.

Weld Repair Procedures

Weld repair procedures must take into consideration cleaning methods, disassembly, preheating, welding process, postheating, weld repair equipment, welding support personnel, safety requirements, inspecting weld repairs, and cleanup.

Cleaning Methods. Surface or subsurface contaminants can lead to cracking, porosity, or lack of fusion in a weld repair. Contaminants penetrate pits, cracks, patches, plating, and pinholes, and are difficult to remove. Metals respond in different ways to contaminants introduced from the heat of welding. Nickel alloys crack when exposed to sulfur compounds such as grease or oil. Stainless steels crack when exposed to zinc, such as from contaminated grinding wheels.

The immediate work area must be cleaned of all dirt, grease, paint, galvanizing, or any other coating. The method of cleaning depends on the material to be removed. For most construction and production equipment, steam cleaning is used. If steam cleaning is inadequate, solvent cleaning may be used, provided proper disposal conditions are established. For small components, acid or solvent dipping may be advisable. Acids must be completely removed from the base metal after dipping to prevent excessive corrosion.

Mechanical cleaning methods include grinding with discs or wheels, power wire brushing, and blast cleaning with abrasives. Blast cleaning with abrasives is very effective, but the abrasive must not be recycled or contamination can return. Blast cleaning is often the only way of removing zinc contamination because grinding and power wire brushing only smear the zinc over the surface.

Where zinc is a specific contaminant, the dithizone test for residual zinc must be carried out. It is important to perform the dithizone test to check for cleanliness when performing repair welding on carbon steels, stainless steels, nickel alloys, or heat-resistant castings.

Disassembly. Components that are sensitive to the heat of welding must be protected or disassembled. Instrument tubing, wiring, lubrication lines, and critical surfaces must not interfere with the repair and must not be exposed to damage by heat, sparks, or weld spatter. Disassembly may require skilled mechanics experienced with the equipment. Sheet metal baffles may be used to protect adjacent machinery and fireproof cloth may be used to protect critical surfaces.

Preheating. The preheating rate depends on the amount of metal involved. For large sections of metal, the temperature rise in the component should not be greater than 100°F/hr and the entire section thickness of the area to be repaired must be at the preheat temperature for ½ hr before starting the repair. Based on the service conditions to which the component is exposed, it may be necessary to perform a bake-out. A bake-out is a temperature-control process used on a casting to remove hydrogen and other contaminants that could cause cracking during welding. A typical bake-out would be performed at 600°F to 800°F for 4 hr to 8 hr.

Peening is done by battering the surface with a blunt-nosed hammer, while the temperature is greater than 1000°F, to compress the surface and reduce residual tensile stresses. When manual welding, the welder deposits 6″ of weld metal and then peens the welded area.

Welding Process. The welding process is selected to achieve a sound repair consistent with the conditions at the time of repair. For these reasons, OAW and SMAW are often preferred for field weld repairs. Stringer (straight) beads are preferred over weave beads to reduce heat input. Whenever possible, the joint should be welded in the flat (1G) position to produce the most effective weld quality.

The filler metal must be selected for optimum weldability. A smaller diameter is preferred to reduce heat input, which is beneficial in reducing distortion.

Tack welds may be required to maintain alignment of the joints. Tack welds should be performed to the same qualified procedures as the main repair weld. If not, they must be ground away and the ground area inspected for cracks using liquid penetrant examination. Buttering may be used to avoid the need for preheat or postheat.

When performing a surfacing repair, the minimum number of passes necessary to meet dilution requirements is preferable.

Postheating. Postheating may be required after weld repair to restore mechanical properties. Postheating is required to stress-relieve the repair weld and reduce distortion. In some cases, slow cooling under a blanket may serve to stress-relieve the part so that a complete postheating cycle is not required. The problem with a complete postheating cycle is the possibility of further distortion.

Weld Repair Equipment. Welding equipment required for weld repairs must be readily available to prevent delays to the work. Standby equipment is also required. Equipment required for weld repair includes electrode holders, grinders, wire feeders, and welding cables. Sufficient power sources must be available to power all necessary equipment. If the job runs around the clock, provisions for lighting and personnel comfort (such as windbreaks or covers) should be provided. Wind and rain are two conditions that adversely affect field welding. When GMAW and GTAW processes are used outside, they are restricted to fully sheltered locations since it takes very little wind to disturb them.

Miller Electric Manufacturing Company

Weld repairs are commonly performed in the field. Welding equipment must be available to complete weld repairs with a minimum of downtime for the equipment.

Welding materials must also be readily available for the entire job. Welding materials that are needed include filler metals, inserts, reinforcement, fuel for preheat and interpass temperature control, shielding gases, and fuel for engine-powered welding machines.

Welding Support Personnel. Trained welders and assistants should be capable of performing the entire job. There should be a sufficient number of welders available to perform the weld repair and, if necessary, they should be rotated to maintain quality output. For code repairs, the welder must be formally qualified to the applicable welding procedures. For non-code work, welders should be qualified to a mock-up. A *mock-up* is a simulation of the repair area on which the welder performs work in the expected position of the repair.

Safety Requirements. Special safety requirements must be met when performing weld repair. Safety requirements include confined space entry procedures, proper grounding, and correctly sized welding cables.

Confined spaces can contain life-threatening atmospheres such as oxygen deficiency, combustible gases, and/or toxic gases, and can cause entrapment. Oxygen deficiency is caused by the displacement of oxygen as welding takes place, the combustion or oxidation process, oxygen being absorbed by the vessel by corrosion, and/or oxygen being consumed by bacterial action. Oxygen-deficient air can result in injury or death. **See Figure 26-10.**

Before entering a permit-required confined space, an entry permit must be posted at the entrance or otherwise made available to entrants. The permit must be signed by the entry supervisor. A signed entry permit verifies that pre-entry preparations have been completed and that the space is safe to enter. **See Figure 26-11.**

POTENTIAL EFFECTS OF OXYGEN-DEFICIENT ATMOSPHERES*	
Oxygen Content†	Effects and Symptoms‡
19.5	Minimum permissible oxygen level
15–19.5	Decreased ability to work strenuously. May impair condition and induce early symptoms in persons with coronary, pulmonary, or circulatory problems
12–14	Respiration exertion and pulse increases. Impaired coordination, perception, and judgment
10–11	Respiration further increases in rate and depth, poor judgment, lips turn blue
8–9	Mental failure, fainting, unconsciousness, ashen face, blue lips, nausea, and vomiting
6–7	8 min, 100% fatal; 6 min, 50% fatal; 4 min–5 min, recovery with treatment
4–5	Coma in 40 sec, convulsions, respiration ceases, death

* values are approximate and vary with state of health and physical activities
† % by volume
‡ at atmospheric pressure

Figure 26-10. *Oxygen-deficient atmospheres in confined spaces can cause life-threatening conditions.*

☑ Confined Space   ☑ Hazardous Area   Permit No. 4672 555 11

ALL COPIES OF PERMIT WILL BE POSTED AT JOB SITE UNTIL JOB IS COMPLETED. PERMIT GOOD ONLY ON DATE(S) INDICATED.

SITE LOCATION and DESCRIPTION Bunker Water Tank #2	**PERMIT SPACE HAZARDS** (indicate specific hazards with initials)

PERMIT SPACE HAZARDS (indicate specific hazards with initials)

_____ Oxygen deficiency (less than 19.5%)
_____ Oxygen enrichment (greater than 23.5%)
_____ Flammable gases or vapors (greater than 10% of LEL)
_____ Airborne combustible dust (meets or exceeds LEL)
_____ Toxic gases or vapors (greater than PEL)
_____ Mechanical Hazards
_____ Electrical Shock
_____ Materials harmful to skin
_____ Engulfment
_____ Other: _____

SITE LOCATION and DESCRIPTION
Bunker Water Tank #2

PURPOSE OF ENTRY
Repair Crack in Tank

SUPERVISOR(S) in charge of crew(s). (Type of Crew-Phone #)
Michael Green   Maintenance Shift II - x5924

AUTHORIZED DURATION OF PERMIT

DATE: 10/2 to 10/4

TIME: 7:00 AM to 3:00 PM

---

* BOLD DENOTES MINIMUM REQUIREMENTS TO BE COMPLETED AND REVIEWED PRIOR TO ENTRY*

REQUIREMENTS COMPLETED	DATE	TIME	REQUIREMENTS COMPLETED	DATE	TIME
Lock Out/De-energize/Tag-out	10/2	09:00	**Full Body Harness w/"D" ring**	10/4	08:00
Line(s) Broken-Capped-Blanked	10/2	11:00	**Emergency Escape Retrieval Equip**	10/4	08:00
Purge-Flush and Vent	10/3	09:00	**Lifelines**	10/4	08:00
Ventilation	10/3	10:00	**Fire Extinguishers**	10/4	08:00
Secure Area (Post and Flag)	10/2	08:00	**Lighting (Explosiveproof)**	10/4	08:00
Breathing Apparatus	10/4	08:00	**Protective Clothing**	10/4	08:00
Resuscitator-Inhalator	10/4	08:00	**Respirator(s) (Air Purifying)**	10/4	08:00
Standby Safety Personnel	10/4	08:00	**Burning and Welding Permit**	10/4	08:00

Note: Items that do not apply enter N/A in the blank.

** RECORD CONTINUOUS MONITORING RESULTS EVERY 2 HOURS

CONTINUOUS MONITORING** TEST(S) TO BE TAKEN	Permissible Entry Level		10/4							
PERCENT OF OXYGEN	**19.5% to 23.5%**		20.5	20.6	20.7	20.5	20.5	___	___	___
LOWER FLAMMABLE LIMIT	**Under 10%**		5	5	5	5	6	___	___	___
CARBON MONOXIDE	**35 PPM+**		0	0	0	0	0	___	___	___
Aromatic Hydrocarbon	1 PPM+	5 PPM*	2	1	2	1	1	___	___	___
Hydrogen Cyanide	(Skin)	4 PPM*	N/A							
Hydrogen Sulfide	10 PPM+	15 PPM*	N/A							
Sulfur Dioxide	2 PPM+	5 PPM*	3	2	2	2	2	___	___	___
Ammonia		35 PPM*	N/A							

* Short-term exposure limit: Employee can work in area up to 15 min.
+ 8 hr time weighted avg: Employee can work in area 8 hr (longer with appropriate respiratory protection).

REMARKS: _____

GAS TESTER NAME & CHECK #	INSTRUMENT(S) USED	MODEL &/OR TYPE	SERIAL &/OR UNIT #
Marty James	Combination Gas Meter	Industrial Scientific	15A

SAFETY STANDBY PERSON IS REQUIRED FOR ALL CONFINED SPACE WORK

SAFETY STANDBY PERSON(S)	CHECK #	NAME OF SAFETY STANDBY PERSON(S)	CHECK #
Kate Washington	3312		
Tony Linder	3318		

SUPERVISOR AUTHORIZING ENTRY
ALL ABOVE CONDITIONS SATISFIED _Michael Green_

AMBULANCE 2800
Safety 4901

FIRE 2900
Gas Coordinator 4529/5387

Figure 26-11. *A confined space entry permit form documents preparations, procedures, and required equipment.*

The workpiece connection must be connected to the workpiece with good electrical contact. The workpiece lead should make a firm, positive connection with the welding power source. The placement of the workpiece connection determines arc characteristics and prevents or minimizes arc blow.

Welding cables must be sized correctly for the job. A hot cable indicates the cable is too small, or the connections are inadequate. In all cases, welding cables must not be used if the insulation becomes damaged or the connections become hot.

> *Many repairs of in-place storage tanks are performed by companies specializing in such repairs. Repair methods must be approved by the authority having jurisdiction and the insurer of the storage tank. When repairs are made to a pressure vessel, a Form R-1, Report of Welded Repair or Alteration must be signed by the inspector who authorized the repairs and by the contractor performing the repairs. Copies must be sent to the proper state authorities, and must be retained by the vessel owner and the authorized inspector.*

Inspecting Weld Repairs. Inspection should be done informally during the repair and formally at the end of the repair. During welding, the quality of the repair should be continually checked to catch and correct problems before the formal inspection. The final welds should be smooth and without notches, and reinforcing, if used, should blend smoothly into the existing structure. Grinding may be necessary to maintain smooth contours and to perform surface repairs on seals. A formal inspection brief should be prepared with the types of nondestructive examinations required. Examiners must be qualified if required by specific codes; the acceptance criteria of the governing code must be used.

Testing of a repair weld may involve a pressure test such as a hydrostatic test. A pressure test is not mandatory unless specifically required by the applicable code. Other procedures may be substituted if approved by the authorized inspector or code. Examples of alternate tests include 100% radiography of repair welds; liquid penetrant examination or magnetic particle examination of all welds not radiographed; or a sensitive leak test such as a vacuum box test.

Cleanup. Cleanup includes removal of strongbacks or other clamps, and smooth grinding of the locations where such items were attached. The ground area should be inspected with liquid penetrant examination or magnetic particle examination to ensure it contains no cracks or other defects. Ground areas that require weld repairs must follow exactly the same weld procedure as the repair itself, including preheat and interpass temperature control if necessary. All weld stubs, weld spatter, slag, and other residues must be removed. Grinding dust must be removed since it is abrasive and may infiltrate working joints and bearings, creating future problems.

Reassembly is required for pieces of machinery taken apart for repair welding. Particular attention must be paid to proper fit-up. If necessary, remachining or grinding may be necessary to restore proper fit-up or alignment. All other items such as lubrication lines, cable, and conduit should be reassembled. Once all wrap-up procedures have been completed, the repaired machinery should be given an operational test before being returned to service.

Points to Remember

- An understanding of how a part failed is necessary before considering repair methods. Some repair methods may not be effective on certain failures.
- Root cause failure analysis identifies and links the three levels of deficiency that lead to failures: technical, human, and system.
- Mechanical repair methods do not create a metallurgical bond with the surface that is being repaired.
- Adhesive bonding is the joining of parts with an adhesive that is placed between the faying (mating) surfaces, producing an adhesive bond.
- Cold mechanical repair is a repair method that consists of locks and stitching pins installed into the surface of a cracked metal part to add strength across the crack.
- Electroplating is the application of a thin, hard, chrome coating to repair minor damage.
- Selective plating is a form of electroplating used for touch-up repairs that can be performed in the field or in the shop.
- Blend grinding may be used to avoid weld repair when the remaining thickness of a structure provides adequate strength.
- The compatibility of the base metal and the surfacing material determines which surfacing weld repair technique to use.
- Sheet lining (wallpapering) and sleeving are specialized repair welding processes and are less commonly used than other repair welding processes.
- A documented repair welding plan must be created before every job.
- The base metal must be identified before attempting a weld repair to determine the weldability of the metal and the type of repair that should be performed.
- After the crack is removed, grinding is required to smooth the surface. Nondestructive examination must be used to ensure cracks have been removed.
- During welding, the quality of the repair should be continually checked to prevent problems during the formal inspection at the completion of the job.

1. What are the three levels of deficiency analyzed by root cause failure analysis?
2. When are mechanical repairs required rather than repair welding?
3. What is adhesive bonding?
4. Why must an adhesive bonded joint be allowed to cure?
5. On what type of metal is cold mechanical repair primarily used?
6. What is electroplating?
7. What type of bond is produced between thermal spray coating and the base metal?
8. What conditions must be present in order to use blend grinding as a repair option?
9. What are the two major types of weld repair?
10. What is the benefit of peening a surface?
11. What welding process is commonly used for wallpapering?
12. Why is it important to know the type of base metal before attempting a weld repair?
13. Why is cleaning of the surface so important to repair welding?
14. What common types of inspection are used to inspect weld repairs?

Refer to Chapter 26 in the *Welding Skills Workbook* for additional exercises.

Pipe Welding

Chapter

Pipe is used to transport oil, gas, and water in a system. Pipe is also used to transport chemicals (nitrogen, air) or utilities. Pipe is used extensively for piping systems in buildings, refineries, and industrial plants. The use of pipe has gained acceptance in construction and often takes the place of beams, channels, or angle iron. Pipe is commercially available in a wide range of diameters, wall thicknesses, and lengths.

Welding is the easiest, most common method of joining sections of pipe. Pipe welding eliminates or reduces mechanical joint designs that may leak, permits free flow of liquids, and reduces installation costs. Welding is also a practical and effective cost-cutting technique in joining noncritical low-pressure piping for refrigeration or HVAC systems. Welded joints are not designed to be disassembled. Repair or replacement requires removal of a section by cutting.

PIPE CLASSIFICATION

Pipe for most applications is made from stainless steel or low-carbon steel. Special applications may use chrome-moly steel, nickel steel, low-alloy steel, copper, aluminum, brass piping, or nickel alloy. Pipe is selected based on the working load pressure and the material to be controlled in the pipe. For example, steam lines in a nuclear power plant must be strong enough to withstand high pressures, high temperatures, and corrosion.

Pipe dimension is determined using the nominal pipe size (NPS). For pipe 14″ and larger in diameter, the NPS is the same as the outside diameter. Pipe wall thickness is specified using one of two standards: ANSI or ASTM/ASME. ANSI classifies pipe thicknesses using schedule numbers (Schedule 40, 60, 80, etc.). ASTM and ASME classify pipe thickness using nominal inside diameter as required by load requirements. The nominal inside diameter is determined using standard weight pipe. The three standard pipe weights are standard (STD), extra-strong (XS), and double extra-strong (XXS). As the wall thickness of extra-strong pipe and double extra-strong pipe is increased, the inside diameter is reduced. The outside diameter remains constant in pipe classifications. **See Appendix.**

For example, pipe with an NPS of 3″ has an outside diameter of 3.5″. Standard pipe has an inside diameter of 3.068″. Extra-strong pipe has an inside diameter of 2.9″. Double extra-strong pipe has an inside diameter of 2.3″. **See Figure 27-1.**

Pipe can be classified as thin-wall or thick-wall. Thin-wall pipe has a wall thickness from ⅛″ to ⁵⁄₁₆″. Thick-wall pipe has a wall thickness greater than ⁵⁄₁₆″.

> **⚠ CAUTION**
>
> Determine certification requirements for welders before proceeding with any pipe welding application.

PIPE DIAMETERS							
NOMINAL PIPE SIZE*	OUTSIDE DIAMETER†	INSIDE DIAMETER*			NOMINAL WALL THICKNESS*		
		STANDARD	EXTRA-STRONG	DOUBLE EXTRA-STRONG	SCHEDULE 40	SCHEDULE 60	SCHEDULE 80
⅛	.405	.269	.215		.068	.095	
¼	.540	.364	.312		.088	.119	
⅜	.675	.493	.423		.091	.126	
½	.840	.622	.546	.252	.109	.147	.294
¾	1.050	.824	.742	.434	.113	.154	.308
1	1.315	1.049	.957	.599	.133	.179	.358
1¼	1.660	1.380	1.278	.896	.140	.191	.382
1½	1.900	1.610	1.500	1.100	.145	.200	.400
2	2.375	2.067	1.939	1.503	.154	.218	.436
2½	2.875	2.469	2.323	1.771	.203	.276	.552
3	3.500	3.068	2.900	2.300	.216	.300	.600
3½	4.000	3.548	3.364	2.728	.226	.318	
4	4.500	4.026	3.826	3.152	.237	.337	.674
5	5.563	5.047	4.813	4.063	.258	.375	.750
6	6.625	6.065	5.761	4.897	.280	.432	.864
8	8.625	7.981	7.625	6.875	.322	.500	.875
10	10.750	10.020	9.750	8.750	.365	.500	
12	12.750	12.000	11.750	10.750	.406	.500	

* in in.
† bw gauge

Figure 27-1. *The inside diameter of pipe changes as the wall thickness increases.*

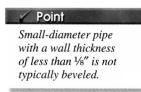

✓ **Point**

Small-diameter pipe with a wall thickness of less than ⅛" is not typically beveled.

The wall thickness of the pipe to be welded determines the joint preparation required. Thin-wall pipe with a wall thickness of ⅛" or less does not commonly require edge preparation or beveling. Pipe with a wall thickness greater than ⅛" usually requires edge preparation. **See Figure 27-2.**

PIPE CONNECTIONS

Welding is the most common method of joining large-diameter pipe. Welded pipe connections cause less restriction to the flow of materials in the pipe. When properly welded, there is no gap between pipe sections and joint strength is

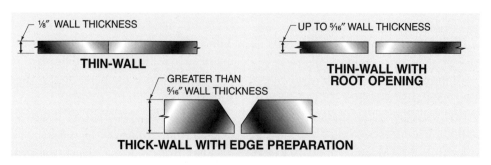

Figure 27-2. *Thick-wall pipe has a wall thickness greater than ⁵⁄₁₆". Wall thicknesses over ⅛" require some edge preparation.*

consistent with the surrounding sections of pipe. Welded joints may be made with butt-welded fittings or socket fittings. **See Figure 27-3.**

Butt-welded fittings require edge preparation of the pipe with a maximum ³⁄₁₆″ root opening. The groove angle should be between 70° and 80°. A backing ring is recommended for butt-welded pipe with a wall thickness over ¾″. Socket fittings join pipe and fittings using sleeves that are welded, brazed, or soldered. A socket fitting does not require edge preparation. Socket fittings are used on pipe with less than 2″ outside diameter (OD).

PIPE JOINT PREPARATION

Pipe joints must be properly prepared before welding. Common joint preparations used with pipe include the single-V-groove and single-U-groove. The groove angle, root face, and root opening vary based on pipe diameter. Pipe weld specifications commonly used on ⁵⁄₁₆″ thick-wall pipe call for a 75° groove angle (37.5° bevel angle). The root opening is approximately ³⁄₃₂″ to ⅛″. The root face is approximately ³⁄₃₂″ to ⅛″. The root opening and groove angle increase as required for pipe with greater wall thicknesses. **See Figure 27-4.**

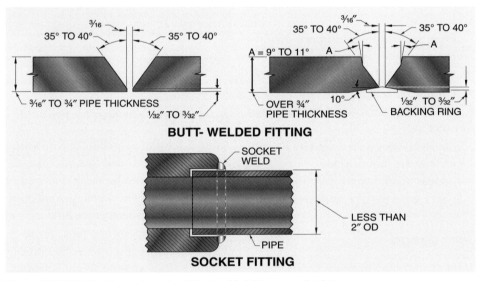

Figure 27-3. *Welded joints may be made with butt-welded fittings or socket fittings.*

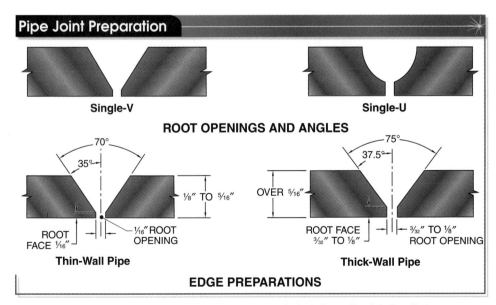

Figure 27-4. *Root openings, angles, and joint preparations vary for both thin-wall and thick-wall pipe.*

In addition, various joint preparations may be required to ensure proper penetration during welding. The groove face can be altered to allow access and to limit the amount of filler metal required without compromising weld strength. Joint preparation is determined by pipe wall thickness, pipe composition, and the welding process used.

Small-diameter pipes with wall thicknesses less than ⅛″ are commonly welded without any joint preparation. The ends are simply butted together with a small separation to ensure complete fusion. Most pipe with wall thicknesses over ⅛″ require some joint preparation.

Pipe welding techniques are affected by pipe dimensions, location, requirements of the pipe and the weld, and welding equipment available. The following steps are used to prepare pipe for welding:

1. Select proper joint design for the job.
2. Clean the joint surface.
3. Align and fit-up the pipe joints.
4. Tack weld the pipe sections together.

Joint Design

Joint design specifications vary depending on the size and composition of the pipe and the thickness of the pipe wall. However, a single-V groove is used for most thick-wall pipe welding. The joint edges must be smooth and free of defects and contaminants. Edges should be worked with a wire brush, if necessary, to remove defects or discontinuities and contaminants.

Pipe typically arrives from the supplier already prepared and beveled to standard specifications. A bevel can be cut or ground in the field; however, this method is time-consuming and is less accurate than machine-beveling.

Whether performed at the supplier's or at the shop, beveling of the joint is usually done with an oxyacetylene beveling machine or pipe machine. A cutting torch can be used to bevel the edges if

joint preparation must take place at a job site where a beveling machine is not available. **See Figure 27-5.**

Thermadyne Industries, Inc.

Figure 27-5. *A plasma arc beveling machine can be used to bevel pipe.*

Joint Alignment and Fit-Up

After the joint is prepared, it must be accurately aligned and spaced. Surfaces to be welded must be clean and free of foreign matter before welding. Joint fit-up must be as consistent as possible around the circumference of the pipe. *Fit-up* is the positioning of pipe with other pipe or fittings before welding. Weld specifications and details indicate fit-up requirements. Line-up clamps are used to hold pipes or pipe and fittings securely and to ensure proper alignment during welding. The pipe may be aligned with consumable inserts, spacers, backing rings, or pipe jigs. **See Figure 27-6.**

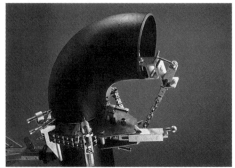

Mathey Dearman

Figure 27-6. *Line-up clamps hold pipe sections securely in position while tack welds are made.*

A consumable insert provides the proper opening of the weld joint and becomes part of the weld. Consumable insert rings are used to ensure an accurate root opening before welding. Consumable insert rings are placed when the pipe is tack welded. When the root pass is deposited, the insert ring is consumed into, and becomes part of, the completed weld. Five classes and compositions of inserts are available for use as required by specific jobs. The classification numbers refer to AWS classes of consumable inserts. Class 1 is A-shaped, or inverted-T, and extends beneath both pipe on the opposite side. Class 2 is J-shaped and extends under one pipe on the opposite side. Classes 3 and 5 are rectangular. Class 4 is Y-shaped, extending to both sides on the welding side. **See Figure 27-7.**

Spacers may be used to provide a gap between the joint until the joint is tack welded. The spacers are removed before welding. Backing rings are commonly used in the GTAW process. Backing rings have spacers attached to a ring which fits in the pipe before welding.

Liners or backing rings can be fitted into the pipe before welding. Liners and backing rings assist in securing penetration without burning through the surface. Liners and backing rings also prevent spatter and slag from entering the pipe at the joint. Backing rings are useful in maintaining pipe alignment and preventing metal slag and spatter from entering the pipe at the joint. **See Figure 27-8.**

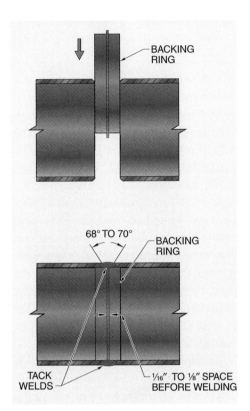

Figure 27-8. *Backing rings are fitted inside the pipe before welding and keep the sections of pipe aligned, preventing slag and spatter penetration.*

> The AWS categorizes consumable inserts into five classes, which are detailed in AWS A5.30, *Specifications for Consumable Inserts.*

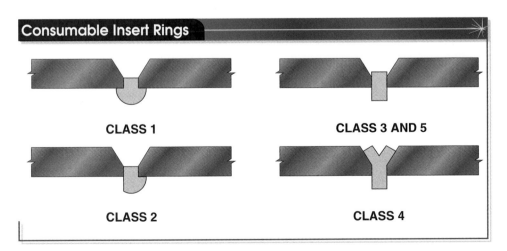

Figure 27-7. *Consumable insert rings are categorized by class and may be used to maintain an accurate root opening before welding.*

Tack Welding

A tack weld keeps the joint members from moving out of their required positions during welding. The pipes to be joined must be properly fitted and can be held in position with a pipe jig. A *pipe jig* is a device that holds sections of pipe or fittings before tack welding. Once the pipe joints are properly aligned, they are tack welded to hold the alignment during welding. For most pipe welding, four evenly spaced tack welds are made around the pipe. Tack welds should typically be about ½″ to ¾″ long. Tack welds should penetrate to the root of the groove, since they will become part of the root bead.

To make a tack weld, the electrode is inclined 10° to 15°. The arc is struck in the joint slightly ahead of where the tack weld is to be made. The arc is then quickly lengthened to stabilize it and give it time to form a protective gas shield. A sliding motion is used after the electrode has been lightly pushed into the joint. When the tack weld is completed, the electrode is pulled away. This procedure produces a strong and fully penetrating tack weld. **See Figure 27-9.**

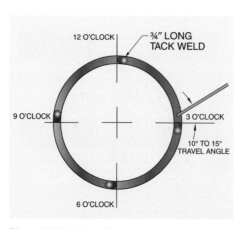

Figure 27-9. *Tack welds are used to hold properly aligned pipes in position.*

WELDING PASSES

Welding passes are used to fill the groove to the specified depth. Welding codes and standards publications specify the required depth of welds. Welding codes and specifications should be followed closely.

Most gas tungsten arc pipe welding standards require complete root penetration with uniform welds and allow for few, if any, defects. The passes used for pipe welding are the root pass, intermediate weld pass(es), and the cover pass.

Root Pass

A *root pass* is the initial weld pass that provides complete penetration through the thickness of the joint member. The current is set to provide maximum penetration without excessive weld metal deposited on the inside of the pipe.

The root pass deposits weld metal in the root of the weld as a "keyhole". The keyhole is formed by the penetration of the root pass. A properly deposited root pass (root bead) should penetrate to the root and leave a solid bead below the surface with a slight crown that does not exceed approximately ¹⁄₁₆″ or the maximum allowed by the governing code. An improper root bead is cause for the entire weld to be rejected.

The success of a pipe weld depends on the correct penetration of the root bead because it forms the base upon which successive layers are made. Subsequent weld passes cannot compensate for a defective root bead.

Some undercutting may occur on the face of the groove, but undercutting is not objectionable since it will be eliminated by successive passes. **See Figure 27-10.**

There may be times when the root opening will vary due to poor fit-up. If the root opening is narrow, the speed of travel and electrode angle should be reduced. Where a widened root opening exists, the travel speed should be increased.

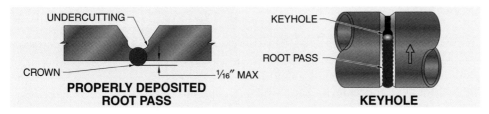

Figure 27-10. *A properly deposited root pass forms a crown on the inside of the pipe. A keyhole is formed if the root pass is stopped and must be filled when welding starts again.*

Intermediate Weld Pass(es)

An *intermediate weld pass* is a single progression of welding subsequent to the root pass and before the cover pass. In pipe welding, the first intermediate weld pass (also called the hot pass) is designed to fill the undercut caused by the root pass. It burns out particles of slag that may remain in the groove and ensures complete fusion of the base metal and the root bead. The first intermediate weld pass also eliminates the possibility of slag inclusions or porosity left from the root bead. **See Figure 27-11.**

The number of intermediate weld passes (or fill passes) required depends on the wall thickness, the groove angle, the size of the electrode, and the welding process used. Intermediate weld passes are deposited with large diameter electrodes and are intended to fill the weld joint. **See Figure 27-12.** Each intermediate weld pass penetrates completely into the previous weld bead. When intermediate weld passes are made using SMAW, the slag produced must be entirely removed after each pass.

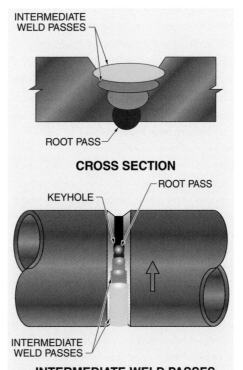

Figure 27-12. *Intermediate weld passes are used to build up the weld and fill the joint, creating a strong, sound weld.*

⊕ *The term intermediate weld pass replaces the formerly used terms hot pass and fill pass. Intermediate weld pass identifies each pass deposited after the root pass and before the cover pass.*

✓ **Point**

Intermediate weld passes are used to fill the weld joint.

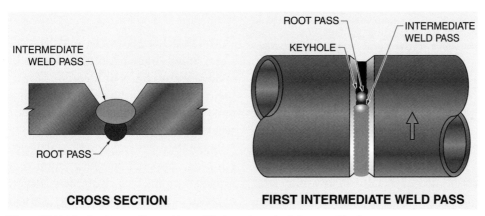

Figure 27-11. *The first intermediate weld pass fills the undercut keyhole created by the root pass.*

Cover Pass

A *cover pass* is the final weld pass deposited. The cover pass provides maximum reinforcement to the weld joint and gives the weld a neat appearance. The cover pass should have a slight crown extending about ¹⁄₁₆″ above the surface of the pipe. The cover pass is usually made using a weaving motion to provide a complete cover for and a neat appearance to the weld. A slant or semicircular motion can be used; however, it must be wide enough to cover the entire weld joint. The cover pass also provides the weld reinforcement required for strength and protection. **See Figure 27-13.**

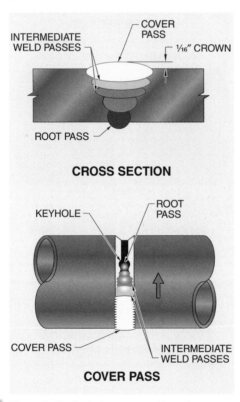

CROSS SECTION

COVER PASS

Figure 27-13. *The final cover pass adds reinforcement to the weld and provides a neat appearance. The cover pass should form a slight crown above the surface of the pipe.*

Electrodes

Most shielded metal arc pipe welding for carbon steel is done with E-6010 or E-6011 electrodes, except where high strength welds are required. When high strengths are needed, especially on low-alloy steel pipe, electrodes in the E-70XX series are used. **See Figure 27-14.**

RECOMMENDED ELECTRODE SIZES FOR PIPE WELD PASSES

Pass	Electrode	Current*†
Root Pass‡	⁵⁄₃₂	140 – 165
Intermediate Weld Pass(es)	⁵⁄₃₂ – ³⁄₁₆	170 – 200
Cover Pass	³⁄₁₆	160 – 180

* the ideal current is within the ranges shown. The best quality is obtained at the lower end of each range
† in amps
‡ Weld root pass at 24 to 26 arc volts and 10 to 16 ipm arc speed

Figure 27-14. *Electrode size is recommended based on the weld pass and required weld strength.*

PIPE WELDING TECHNIQUES

Pipe welding is recognized as a specialty within the welding trade. Although many pipe welding skills and practices are similar to other types of welding, pipe welders must develop certain techniques that are characteristic to pipe welding alone. Pipe welders have to pass certain tests to be certified because public health, environmental restrictions, and safety concerns are involved (especially in welding cross-country transmission pipelines and high-pressure lines that convey steam, oil, air, or corrosive materials). **See Figure 27-15.**

Pipe welding techniques vary depending on the welding conditions and the type of pipe being used. Pipe welders should be proficient in welding techniques such as, downhill welding, uphill welding, roll welding, and position welding.

Downhill Welding

Downhill welding is used to weld thin-wall pipe. Small-diameter pipe is typically welded with GTAW or GMAW. Downhill welding is preferred for welding cross-country pipelines because it is a fast welding technique.

After the pipes are securely tack welded, a root pass is deposited completely around the joint. The electrode is held in approximately the same position as when making the tack welds. The arc is struck

The Lincoln Electric Company

Figure 27-15. *Pipe welders must be certified in specific pipe weld joint positions.*

slightly ahead of the weld to preheat the area where the weld bead will be started. After the arc has stabilized, the electrode is lowered into the root opening and moved along the groove. Intermediate weld passes are usually made with a side-to-side (weaving) motion and consist only of a light bead deposit. The electrode should pause at the end of each stroke to ensure good fusion at each edge of the weld. As the electrode reaches the bottom of the weld a semicircular or horseshoe weave is used. **See Figure 27-16.**

Intermediate weld passes are made with the same diameter electrode used for the root bead but with slightly higher current.

For downhill welding, follow the procedure:

1. Deposit four tack welds to hold the pipe in alignment.
2. Start welding the root pass in the 12 o'clock position.
3. Carry the root pass weld downward to the 6 o'clock position.
4. Follow same procedure on the other side of the pipe. **See Figure 27-17.**

If the electrode sticks and fails to glide smoothly because of built-up heat, a slight side-to-side oscillating motion will usually correct the problem. After the root pass is completed, additional weld passes are deposited. The number of passes depends on the thickness of the pipe.

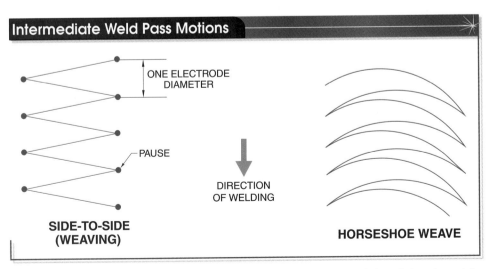

Figure 27-16. *Two motions used to make intermediate weld passes are the side-to-side (weaving) motion and the horseshoe weave.*

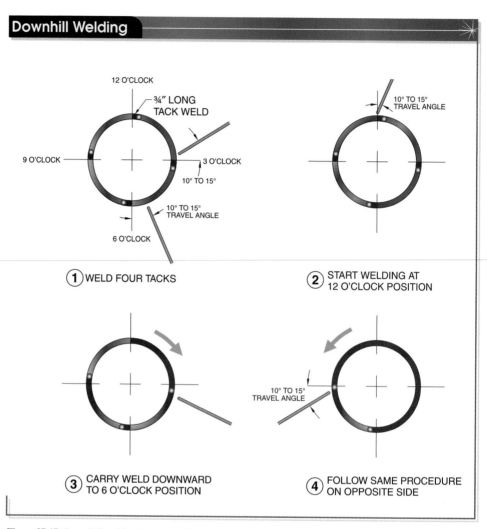

12 O'CLOCK

¾" LONG
TACK WELD

9 O'CLOCK

3 O'CLOCK

10° TO 15°

10° TO 15°
TRAVEL ANGLE

6 O'CLOCK

① WELD FOUR TACKS

10° TO 15°
TRAVEL ANGLE

② START WELDING AT
12 O'CLOCK POSITION

③ CARRY WELD DOWNWARD
TO 6 O'CLOCK POSITION

10° TO 15°
TRAVEL ANGLE

④ FOLLOW SAME PROCEDURE
ON OPPOSITE SIDE

Figure 27-17. *Downhill welding is commonly used to weld thin-wall pipe.*

> ⚠ **CAUTION**
>
> When restarting the arc, completely tie together the welded section with the next section.

> ✓ **Point**
>
> *The ends of the weld must always be tied together.*

One problem encountered in downhill welding is controlling the heat input. Lack of heat input control is especially a problem when welding small-diameter pipe where heat does not dissipate fast enough and excessive heat builds up in the weld zone. Heat input can usually be regulated using a smaller diameter electrode and reducing the current setting.

Another problem in downhill welding is maintaining proper control of the weld pool. The molten metal tends to flow downward in the same direction the arc is moving. If the flow is not controlled, penetration cannot be achieved and slag becomes entrapped in the molten metal, producing slag inclusions in the weld. Slag inclusion is only a problem when welding with SMAW or FCAW. Control of the molten weld pool is accomplished

using a fast travel speed and a high-current setting to keep the arc ahead of the weld pool.

Starting and Stopping. There is a certain amount of starting and stopping during welding due to the need to change electrodes or weld position. When welding must be stopped and then restarted, the ends of each weld bead must be tied together. To restart a weld, the arc is struck about ½" ahead of the crater and then moved back into the crater with a long arc. As soon as the arc is stabilized, the arc is shortened and the electrode is momentarily held in the crater of the last bead to regenerate the molten weld pool. The electrode is then raised slightly and the weld continued. When the weld approaches the end and must be tied into the other deposited bead, the electrode is moved up the sloping side of the previous bead, and the direction of travel is briefly reversed after the molten pool blends smoothly between the two beads. The arc is then withdrawn quickly by flicking the electrode downward and away from the center.

Uphill Welding

Uphill welding is used for welding thick-wall pipe. Welding progresses upward on one side of the pipe and then upward on the opposite side. **See Figure 27-18.** For uphill welding, follow the procedure:

1. Weld four tacks to hold the pipe in alignment.
2. Start welding the root pass in the 6 o'clock position.
3. Carry the weld upward to the 12 o'clock position.
4. Follow the same procedure on the opposite side.

As in downhill welding, tack welds are used to maintain alignment of the pipes. The root pass is deposited just back of the bottom, or 6 o'clock, position. The arc is struck ahead of the 6 o'clock position and a long arc is maintained for a short period to preheat the surface; then it is brought back to the weld area and welding is begun.

> **✓ Point**
> *Uphill welding should be used on thick-wall pipe.*

> **✓ Point**
> *In uphill welding, the root pass should be started just back of the 6 o'clock position.*

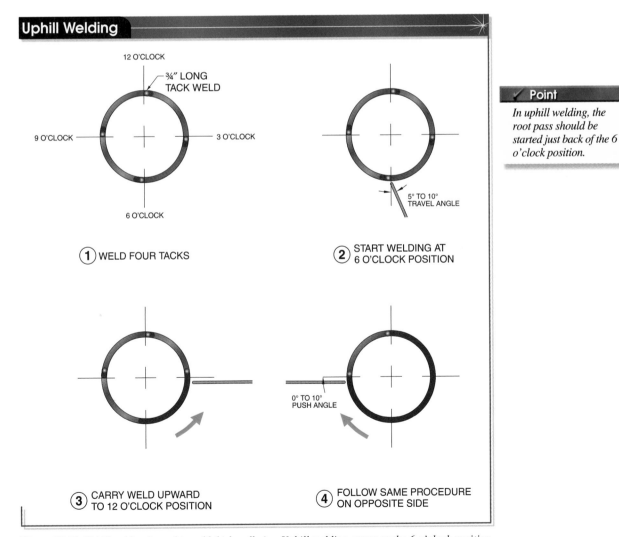

Figure 27-18. *Uphill welding is used to weld thick-wall pipe. Uphill welding starts at the 6 o'clock position and works upward on both sides of the pipe.*

While the root pass is being deposited, no electrode weaving motion is necessary. The electrode is simply advanced at an angle of 5° to 10° with a slow and uniform movement along the joint. As the electrode approaches the upper part of the pipe, the molten metal begins to flow downward at a faster rate. A slight whipping motion helps to control the weld pool and prevent metal flow.

After the root pass is completed, one or more intermediate weld pass layers are deposited followed by the final cover pass.

Roll Welding

Roll welding is a welding procedure that reduces time required and improves quality of welding. Roll welding is usually performed with GMAW using a hand-held welding gun. The roll welding method requires that two or more sections of pipe be lined up and tack welded. Special pipe clamps hold the pipe in alignment until they are tacked. **See Figure 27-19.** The weld is then completed in flat position while helpers rotate the pipe. After the short pipe sections are welded, they are placed in line with the existing or previously installed pipe and welded in a stationary position.

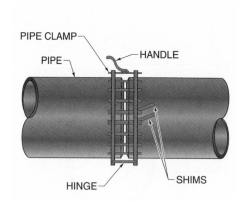

PIPE CLAMP
PIPE
HANDLE
HINGE
SHIMS

Figure 27-19. *Pipe clamps are used to hold pipe in alignment until tack welds are made.*

Position Welding

Position welding consists of lining up each section, length by length, and welding each joint while the pipe remains stationary.

Since the pipe is not rotated, the welding has to be done in various positions—flat, horizontal, vertical, and overhead. **See Figure 27-20.**

The Lincoln Electric Company

Figure 27-20. *Position welding requires that the pipe be welded in various positions around stationary pipe.*

PIPE WELDING STANDARDS

Standards ensure pipe welding quality. Pipe welding standards have been established by the American Petroleum Institute (API), the American Society of Mechanical Engineers (ASME), and the American Welding Society (AWS) for specifying material requirements, preparation, welder proficiency, and weld testing. Other agencies may adopt these standards for specific applications. For example, the U.S. Department of Defense has adopted several standards published by the AWS.

Welder Certification

Certification of welders is based on the proficiency of the welder making welds in specific positions. Pipe weld joint positions are identified as test positions. Because pipe welds are usually groove welds, they are identified by the letter G, for groove weld. Test positions are 1G, 2G, 5G, 6G, and 6GR. There is no 3G or 4G test position in pipe welding. The axis of the pipe may vary ±15° for the 1G, 2G, and 5G test positions, but only ±5° for the 6G and 6GR positions. **See Figure 27-21.**

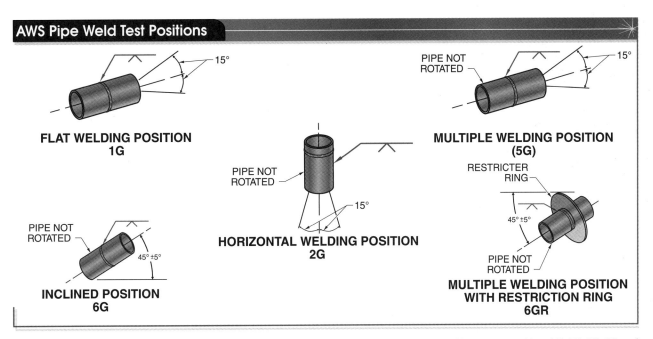

Figure 27-21. *The American Welding Society (AWS) has identified weld positions for pipe and tubing welding as test positions 1G, 2G, 5G, 6G, and 6GR.*

1G Position. Test position 1G is the flat welding test position. The axis of the pipe is in horizontal position. The axis of the weld is in flat position. The weld is completed in flat position as the pipe is rotated. The axis of the pipe should be within 15° above or below the horizontal. Test position 1G qualifies a welder to weld in flat position.

2G Position. Test position 2G is the horizontal position. The axis of the pipe is in vertical position and the axis of the weld is in horizontal position. The weld is completed in vertical position. The axis should be within 15° on any side of vertical. Test position 2G qualifies a welder to weld in flat and horizontal positions.

5G Position. Test position 5G is the multiple welding test position. The axis of the pipe is in horizontal position. The axis of the weld is in vertical position, but the pipe is not turned or rolled during welding. The weld is completed in flat, vertical, or overhead fixed positions. The axis should be within 15° above or below the horizontal. Test position 5G qualifies a welder to weld in flat, vertical, and overhead positions.

6G Position. Test position 6G is the inclined position. The pipe is fixed in position and is not rotated during welding. The weld is completed with the axis of the pipe at a 45° angle ±5°. The axis of the pipe is set and the pipe is not rotated while welding.

6GR Position. Test position 6GR is multiple position welding with a restriction ring. Restricted accessibility is often added by placing a restriction ring near the weld. Test position 6GR requires the axis of the pipe to be positioned at a 45° angle, ±5°. The pipe is fixed in position and is not rotated during welding.

WELDING METHODS

Welding methods used for pipe are the same as are used for other welding processes. The method used depends on the pipe material, diameter, and function of the piping system. The composition of the pipe determines the filler metal and welding process used. For example, welding stainless steel pipe with a ⅜″ wall thickness requires deep penetration. Pipe in a critical application may be purged with shielding gas. GTAW is used to ensure weld purity.

> ✓ **Point**
>
> *Most pipe welding jobs require that the welder be certified.*

Most pipe welding is done with either SMAW or GMAW. The advantage of GMAW over SMAW is that with GMAW, no slag occurs in the weld. Also, the gas protection shield over the weld area prevents atmospheric contamination of the weld. Since GMAW requires no slag removal, less welding time is required. There is no significant difference in welding techniques and procedures between SMAW and GMAW. General descriptions of pipe welding processes apply to both SMAW and GMAW. GMAW-P, FCAW-S, and FCAW-G are used to improve productivity.

GTAW may be used when shop welding small-diameter pipes. GTAW is also sometimes used to deposit the root bead of large-diameter pipe jobs. Pipe welding is commonly performed using manual, semiautomatic, mechanized, or automatic welding.

Manual Welding

Manual welding is welding with an OAW torch, GTAW torch, or SMAW electrode holder, held and manipulated by hand. Manual welding using OAW was commonly used for many years to weld pipe. It worked well for thin-wall pipe, but thick-wall pipe required too much time and was difficult to weld. Although not commonly used, some thin-wall pipe is still welded with OAW.

SMAW is a manual welding process commonly used for welding pipe because of the flexibility and mobility of the equipment, and the accessibility to the weld area. GTAW is used for critical root pass and complete welds where quality standards are extremely high.

Semiautomatic Welding

Semiautomatic welding is manual welding with equipment that controls one or more welding conditions automatically. Semiautomatic welding requires a welder to manually weld while equipment controls one or more welding condition(s).

A constant voltage welding machine provides the power and a wire feeder delivers the electrode to the weld pool while the welder controls the welding gun manually. **See Figure 27-22.** Common semiautomatic welding processes are GMAW and FCAW.

Figure 27-22. *Gas metal arc welding is often used to join small-diameter pipe while the welder controls the welding gun.*

Mechanized Welding

Mechanized welding is a welding process in which the welding is automatic, but the operator must make process adjustments manually. In mechanized welding, the machine controls the welding gun. The welding gun moves along the weld at a set height. If the seam is not flat or straight, the operator must adjust the equipment. The operator must observe the progress of the welding gun or electrode holder and make adjustments as necessary.

Automatic Welding

Automatic welding is a welding process that requires minimal observation by the operator and no manual adjustment of the controls. The welding equipment automatically controls one or more welding condition(s). The automatic welding system monitors the arc voltage and adjusts the height of the welding gun from the base metal to maintain a consistent distance and a quality weld. Most large diameter (24″ and over) pipe is welded using an automatic GMAW process.

Automatic GMAW machines speed up the welding process and produce welds without slag inclusions, which can be a problem with SMAW. Unlike conventional pipe welding procedures where the root bead is deposited externally, with some automatic welding the root pass is deposited inside the pipe. A special bevel is made on the pipe for this purpose. **See Figure 27-23.** Usually, four welding heads mounted on an internal line-up clamp are used to make the internal root bead in a single pass.

The internal welding unit is self-propelled along the inside of the pipe and held in place at the weld site by clamp shoes. Welding heads are positioned precisely over the joints by means of special aligner blocks. Once the unit is correctly positioned, the next section of pipe is slipped over the reach rod of the unit. The joint is properly spaced and another set of clamp shoes is actuated to hold the joint in place for welding. A control box mounted on the handle of the reach rod controls the starting and stopping of welding. Each welding head welds a 90° arc. All welds are made downhill with two heads moving clockwise and the other two moving counterclockwise. Shielding gas for internal welding consists of 75% argon and 25% CO_2.

External Welding. The external welding process includes a root pass, intermediate weld pass, and cover pass. Passes are made with special welding units that travel around the external perimeter of the pipe on pre-positioned circumferential pipe bands. Two welding machines, sometimes referred to as bugs, move simultaneously on the pipe. **See Figure 27-24.** One bug starts at the 12 o'clock position and travels downward to 6 o'clock. The other bug starts at the 3 o'clock position and stops at the 9 o'clock position. The bug is then moved to the 12 o'clock position to complete the pass at the horizontal. External welds are made with 100% CO_2 because it has a higher deposition rate and better penetrating qualities.

Weld Tooling Corp.

Figure 27-24. *External welding bugs are used to make intermediate and cover passes while positioned outside the pipe.*

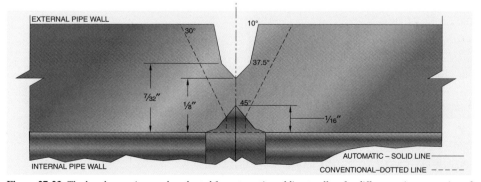

Figure 27-23. *The bevel on a pipe can be adapted for automatic welding to allow for differences in penetration of the welds.*

PIPE WELD TESTING

Pipe weld testing can be conducted using nondestructive examination and destructive testing. Testing methods for pipe are similar to those used for other types of metals.

Destructive Testing

Destructive testing is used primarily in the qualification of welding procedures and are often used to test welder performance. In destructive testing, a test specimen is analyzed using the tensile test or the guided bend test. A test specimen is a section of welded metal that includes the weld area. In the tensile test, the test specimen is subjected to force in opposite directions. The tensile strength achieved is compared with weld strength requirements. **See Figure 27-25.** In the guided bend test, a test specimen is used in a guided bend tester to identify points of failure when the test specimen is subjected to a bending force. The guided bend test requires two test specimens, a face bend specimen and a root bend specimen.

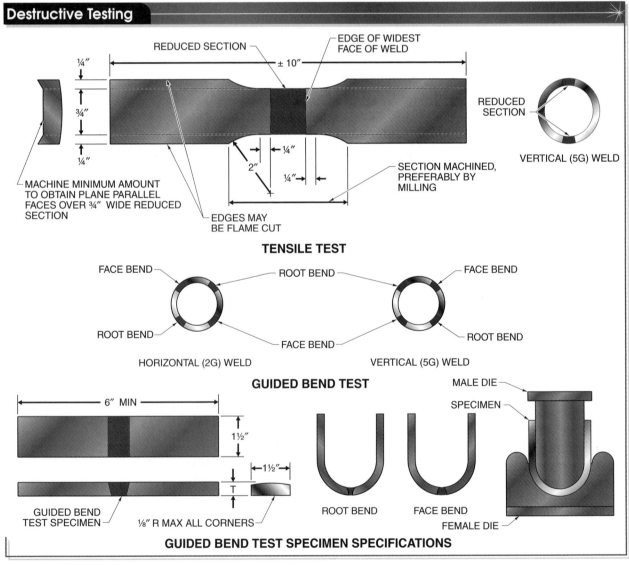

Destructive Testing

Figure 27-25. *Tensile and guided bend tests are used for destructive testing of pipe welds.*

The face-bend specimen is checked for incomplete penetration, porosity, inclusions, or other defects. The specimen is placed in the guided bend tester with the face side down. The root-bend specimen is tested for complete penetration. The specimen is placed in the guided bend tester with the root side down. After bending, the test specimen is inspected for cracks, inclusions, incomplete penetration, porosity, and other defects.

Nondestructive Examination

Nondestructive examination (NDE) is the development and application of methods to examine materials or components in ways that do not impair their usefulness or serviceability.

Nondestructive examination is used to determine weld quality without affecting performance of the weld. Nondestructive examination methods include liquid penetrant, radiographic, ultrasonic, and visual examination. **See Figure 27-26.**

The Lincoln Electric Company

A welder must be able to weld pipe from various positions around the pipe since most pipe cannot be rotated.

NONDESTRUCTIVE EXAMINATION	
Method	**Letter Designation**
Acoustic emission	AET
Electromagnetic	ET
Leak	LT
Magnetic particle	MT
Neutron radiographic	NRT
Liquid penetrant*	PT
Proof*	PRT
Radiographic*	RT
Ultrasonic*	UT
Visual*	VT

*methods used for testing pipe welds

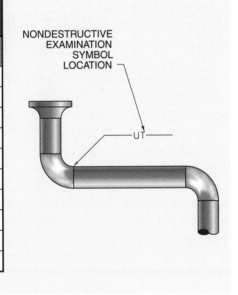

Figure 27-26. *Nondestructive examination does not adversely affect the performance of the weld.*

Refer to Quick Quiz® on CD-ROM

- Small-diameter pipe with a wall thickness of less than ⅛″ is not typically beveled.
- Tack welded pipes must be properly aligned before welding.
- The root pass should completely penetrate into the root of the joint.
- Intermediate weld passes are used to fill the weld joint.
- A pipe weld should be finished with a final cover pass.
- Downhill welding should be used to weld large diameter thin-wall pipe.
- The ends of the weld must always be tied together.
- Uphill welding should be used on thick-wall pipe.
- In uphill welding, the root pass should be started just back of the 6 o'clock position.
- Most pipe welding jobs require that the welder be certified.

Questions for Study and Discussion

1. How is thin-wall pipe distinguished from thick-wall pipe?
2. As a rule, how many tack welds are made on pipe?
3. What is the function of a backing ring?
4. If the electrode sticks in the groove when making a tack weld, what should be done?
5. Why is a proper root opening very important in pipe welding?
6. What is a root pass?
7. What is the function of the first intermediate weld pass?
8. What is the function of the cover pass?
9. Why is a whipping action of the electrode sometimes used?
10. Why should each layer start and stop at different points?
11. What electrode motions are used to make intermediate weld passes?
12. The external welding process includes what passes?
13. Which electrodes are used for most shielded metal arc pipe welding?
14. How is welding performed when started at the 6 o'clock position?
15. What is the difference between uphill and downhill welding?
16. What are some of the problems that may be encountered in downhill welding?
17. Downhill welding is used for welding what kind of pipe?
18. How are the ends of a weld tied together?
19. At what angle should the electrode be held for downhill welding?
20. What is the starting position for thin-wall pipe welded with the downhill technique?
21. Why do pipe welders usually have to meet certification requirements?

Refer to Chapter 27 in the *Welding Skills Workbook* for additional exercises.

Production Welding

Production welding refers to welding techniques used in the fabrication of goods in mass production. Industries involved in manufacturing use welding processes that allow the joining of metal rapidly and automatically. Since production techniques depend on the nature of the goods made, the welding process and equipment used vary from one industry to another.

Special welding machines are often designed for a particular industry. An aircraft company may need a spotwelder designed to join certain types of aluminum structures, while an automotive manufacturer may require a resistance-type seam welder specially made to weld structural steel. Other applications may require a stud-welding gun to fasten studs on metal components.

Welding processes used for production welding include resistance welding (RW), gas tungsten arc spot welding (GTAW), gas metal arc welding (GMAW), stud welding, electron beam welding (EBW), friction welding (FRW), laser beam welding (LBW), plasma arc welding (PAW), submerged arc welding (SAW), ultrasonic welding (USW), electrogas welding (EGW), and adhesive bonding (AB). Other welding processes that may be used for production welding are explosion welding (EXW), forge welding (FOW), roll welding (ROW), and cold welding (CW).

RESISTANCE WELDING

Resistance welding (RW) is the most commonly used of the production welding processes. *Resistance welding (RW) is a group of welding processes in which fusion occurs from the heat obtained by resistance to the flow of welding current through the metals joined.* All RW processes are based upon the following fundamental principles:

- Heat is generated by the resistance of the workpieces to be joined to the passage of a heavy electrical current.
- The heat at the juncture of the workpieces changes the metal to a plastic state.
- When the workpieces, heated to a plastic state, are combined with the correct amount of pressure, with a combination of melting and plastic deformation, fusion takes place.

RW machinery is similar whether the machine is a simple or complex design. The main difference is the type of jaws or electrodes that hold the object to be welded. **See Figure 28-1.** A standard resistance welder has four principal elements:

- The frame is the main body of the machine, which differs in size and shape for stationary and portable types.
- The electrical circuit, which includes a step-down transformer that reduces voltage and proportionally increases current to provide the necessary heat at the point of welding.
- The electrodes include the mechanism for making and holding contact at the weld area.

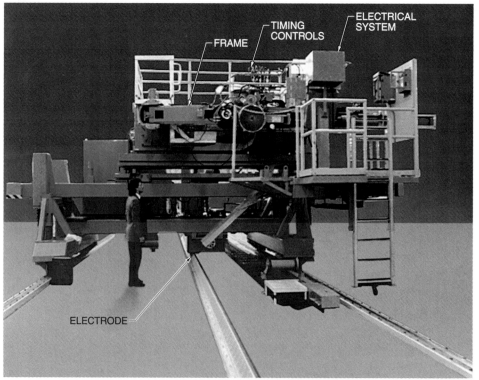

Media Clip

Resistance Spot Welding

• The timing controls use switches that regulate the amount of current, current duration, and the contact period.

The most common types of RW are spot welding, seam welding, projection welding, multiple-impulse welding, flash welding, and upset welding.

Spot Welding

Spot welding is the most commonly used RW process. The material to be joined is placed between two electrodes, pressure is applied, and a charge of electricity is sent from one electrode, through the material, to the other electrode.

There are three stages in making a spot weld. First, the electrodes are brought together against the metal and pressure is applied before the current is turned ON. Next, the current is turned ON for a short time. The third step is the hold time, in which the current is turned OFF but the pressure continues. The hold time fuses the metal while it is cooling.

Spot welding usually leaves slight depressions on the metal that are undesirable on the "show side" of the finished product. Depressions can be minimized using large-diameter electrode tips on the "show side".

Spotwelders are made for both direct current and alternating current. The amount of current must be controlled. Too little current produces a light tack and provides insufficient strength to the weld. Too much current causes excessive heat.

To dissipate the heat and cool the weld as quickly as possible to prevent overheating, the electrodes that conduct the current and apply the pressure for spot welding are water-cooled. The electrodes are made of low-resistance copper alloy and are usually hollow to facilitate water-cooling. Electrodes must be kept clean and in the correct shape to produce good results. For example, if a ¼″ dia. electrode face is allowed to increase to ⅜″ by wear or mushrooming, the contact area is doubled and, correspondingly, the current density decreases. A current density decrease results in weak welds unless the decrease is

Pandjiris, Inc.

Figure 28-1. *RW machinery is similar whether the machine is a simple or complex design.*

compensated for by an increase in current setting. Additional factors that cause poor welds are misalignment of electrodes, improper electrode pressure, and convex or concave electrode surfaces.

Two basic types of spotwelders are single-spot and multiple-spot. A single-spot has two long horizontal horns, each holding a single electrode, with the upper arm providing the moving action.

Multiple-spot spotwelders have a series of hydraulic- or air-operated welding guns mounted in a framework or header but use a common (or bar) mandrel for the lower electrode. The guns are connected by flexible bands to individual transformers or to a common busbar attached to the transformer. Two or four guns can be attached to a transformer.

Although many spotwelders are stationary, portable spotwelders are becoming more popular. A portable spotwelder, or spot-welding gun, consists of a welding head connected to the transformer by flexible cables. The jaws of the welding head can be operated manually, pneumatically, or hydraulically.

The self-contained portable spotwelder contains a built-in timer, electrode contactors, and transformer, and requires only a 115 V power connection. **See Figure 28-2.** With this apparatus, spot welds can be made on irregularly shaped objects. A self-contained portable spotwelder is especially suitable for sheet metal and auto body welding.

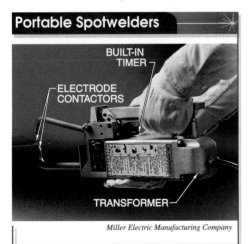

Portable Spotwelders

BUILT-IN TIMER

ELECTRODE CONTACTORS

TRANSFORMER

Miller Electric Manufacturing Company

Figure 28-2. *A self-contained portable spotwelder contains a built-in timer, electrode contactors, and a transformer.*

Spotwelders are used extensively for welding steel. When equipped with an electronic timer, spotwelders can be used for other commercial metals such as aluminum, copper, and stainless steel. They are also very effective for welding galvanized metal.

Seam Welding

Seam welding is similar to spot welding except that the welds overlap, making a continuous weld seam. In seam welding, the metal pieces pass between roller-type electrodes. As the electrodes revolve, the current is automatically turned ON and OFF at intervals corresponding to the speed at which the parts move. With proper control, it is possible to obtain airtight seams suitable for containers such as barrels, water heaters, and fuel tanks. When an intermittent current is used and the spot welds are not overlapped long enough to produce a continuous weld, the process is referred to as roller spot welding. **See Figure 28-3.**

Seam welding is an effective welding method because of its short current cycle. The rollers may be cooled to prevent overheating, with consequent wheel dressing and replacement problems reduced to a minimum. Cooling is accomplished by internally circulating water or by an external spray of water over the electrode rollers. **See Figure 28-4.**

Since the heat input is low, very little of the welded area is hardened, and the yield point is not materially affected. Very little grain growth takes place during seam welding, which makes seam welding applicable to corrosion-resistant alloys such as alloys and other alloys whose mechanical properties are modified by grain growth.

> *Seam welding is often used to fabricate cylindrical, leak-tight vessels such as pipes and tanks. Seam welding can be used either along the side of a cylinder or around its circumference.*

> ✓ **Point**
> *Spotwelders are available to produce single-spot welds or multiple-spot welds.*

> ✓ **Point**
> *Seam welding produces a series of overlapping spot welds, thereby making a continuous-weld seam.*

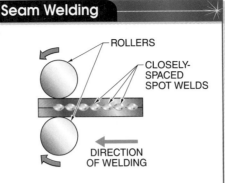

CONTINUOUS CURRENT

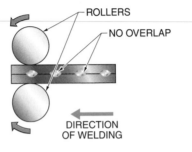

INTERMITTENT CURRENT

Figure 28-3. *In a continuous spot weld, welds must be closely spaced to provide an airtight seam. In intermittent spot welding, a seam weld is produced in which the spot welds are spaced apart.*

Pandjiris, Inc.

Figure 28-4. *Seam welding is an advantageous welding method because of its short current cycle. The rollers may be cooled to prevent overheating.*

Projection Welding

Projection welding (PW) is a welding process that produces a weld using heat obtained from resistance of the workpiece to the welding current. The PW process is similar to spot welding. The point where welding is to be performed has one or more projections that have been formed by embossing, stamping, casting, or machining. The PW process consists of placing the projections in contact with the workpiece and aligning them between the electrodes. **See Figure 28-5.** The projections serve to concentrate the welding heat at the weld area and cause fusion without requiring high current. Single or multiple projections can be welded simultaneously.

There are many variables involved in PW such as metal thickness, type of metal, and number of projections, which make it impossible to predetermine the correct current setting and pressure required. Only by trial runs followed by careful inspection can proper control settings be established.

Projection Welding

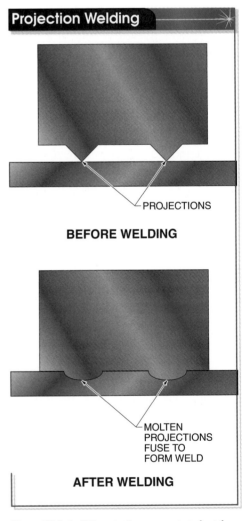

Figure 28-5. *In PW, projections concentrate heat from the resistance to the welding current.*

Not all metals can be welded with PW. Brass and copper do not lend themselves to PW because brass and copper projections usually collapse under pressure. Aluminum PW is limited to extruded parts (shapes formed by forcing metal through a die). However, galvanized iron and tin plate, as well as most other light-gauge steels, can be successfully welded with PW. PW is also widely used for attaching fasteners to structural members.

Multiple-Impulse Welding

Multiple-impulse welding is a form of resistance welding in which welds are made with repeated electrical impulses. In regular spot welding, interruption of the flow of welding current is controlled manually; with multiple-impulse welding, the current is regulated electronically to cycle on and off at regular intervals during the making of one weld. **See Figure 28-6.** Multiple-impulse welding permits thicker materials to be spot-welded. The interrupted current helps keep electrodes cooler, minimizing electrode distortion and reducing the tendency of the weld to spark, as well as increasing the life of the electrode.

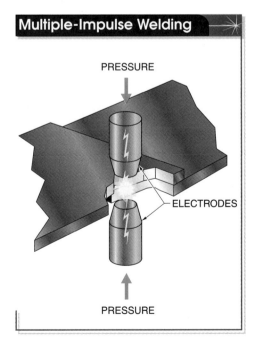

Multiple-Impulse Welding

PRESSURE

ELECTRODES

PRESSURE

Figure 28-6. *In multiple-impulse welding, the current is regulated by precise electronic control.*

Flash Welding

Flash welding (FW) is a resistance welding process that produces a weld at the faying surfaces of a joint by the intense heat of an arc that occurs when the workpieces are contacted and by the application of pressure after heating. The weld is completed by a rapid upsetting of the workpieces.

The workpieces to be joined are clamped in copper alloy dies shaped to fit each piece that conducts the electric current to the work. The ends of the two workpieces are moved together until an arc is established. The flashing action across the gap, caused by very high current densities at small contact points between the workpieces, melts the metal, forcibly expels material from the joint, and creates fusion as the two molten ends are moved together. **See Figure 28-7.** The current is turned OFF as soon as the fusing action is completed. For some operations the dies are water-cooled to dissipate the heat from the welded area. Parts to be flash welded must be precisely aligned. Misalignment results in a poor joint and produces uneven heat and telescoping of one piece over another.

FW is used to butt- or miter-weld sheet, bar, rod, tubing, and extruded sections. It has almost unlimited application for both ferrous and nonferrous metals, but it is not generally recommended for welding cast iron, lead, or zinc alloys.

A problem in FW is the bulge (flash) or increased size that results at the weld. If the profile of the finish area of the weld must be smooth, then the flash must be removed by grinding or machining after welding.

Upset Welding

Upset welding (UW) is a resistance welding process that produces a weld on the faying surfaces by the heat obtained from resistance to the flow of current through the surface contact areas while under constant pressure. The metals to be welded are brought into contact under pressure, an electric current is passed through them, and the edges are softened and fused together.

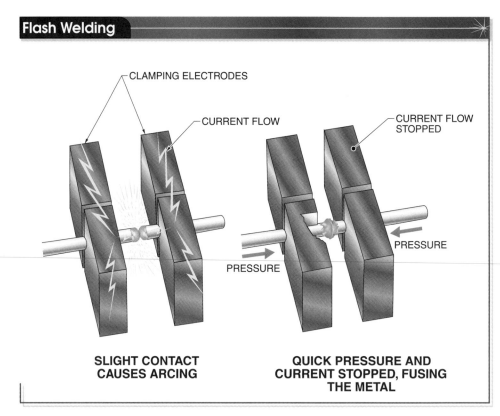

CLAMPING ELECTRODES

CURRENT FLOW

CURRENT FLOW STOPPED

PRESSURE

PRESSURE

SLIGHT CONTACT CAUSES ARCING

QUICK PRESSURE AND CURRENT STOPPED, FUSING THE METAL

Figure 28-7. *In FW, an intense arcing—caused by the electrical current flowing through the two workpieces being brought together—melts the metal and creates fusion.*

UW differs from FW in that constant pressure is applied during the heating process, which eliminates flashing. The heat generated at the point of contact results entirely from resistance. Although the operation and control of the UW process is almost identical to FW, the basic difference is that less current is used and more time must be allowed for the weld to be completed. **See Figure 28-8.**

GAS TUNGSTEN ARC SPOT WELDING

Gas tungsten arc spot welding is an arc welding process that produces localized fusion similar to resistance spot welding but does not require accessibility to both sides of the joint. The gas tungsten arc spot welding process has many applications in fabricating sheet-metal products with joints that cannot be welded using RW because of the location of the weld, the size of the parts, or where welding can be made from only one side. Gas tungsten arc spot welding provides a deeper, more localized penetration compared to conventional RW. **See Figure 28-9.** Heat is generated from resistance of the work to the flow of electrical current in a circuit of which the work is a part.

Lincoln Electric Company

Welding is used to join sections of metal together in shipbuilding.

With gas tungsten arc spot welding (GTAW), a melted spot on the bottom of a workpiece is indicative of a good weld.

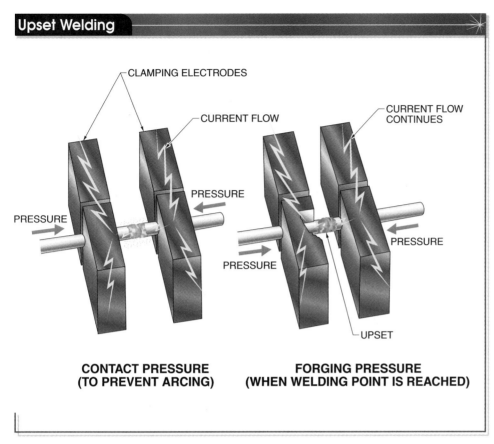

CONTACT PRESSURE
(TO PREVENT ARCING)

FORGING PRESSURE
(WHEN WELDING POINT IS REACHED)

Figure 28-8. *UW involves passing high current through the workpieces while continuous pressure is applied.*

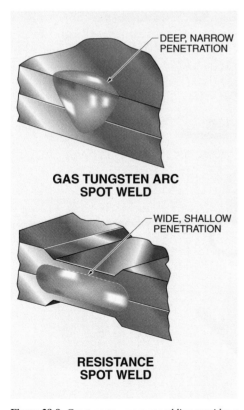

**GAS TUNGSTEN ARC
SPOT WELD**

**RESISTANCE
SPOT WELD**

Figure 28-9. *Gas tungsten arc spot welding provides a deeper and more localized penetration compared to that obtained by conventional resistance welding.*

Gas Tungsten Arc Spot Welding Equipment

Any DC welding machine that provides up to 250 A with a minimum open circuit voltage of 55 V can be used for gas tungsten arc spot welding.

The tungsten arc welding gun has a nozzle that holds a tungsten electrode. Various shapes of nozzles are available to meet particular job requirements. **See Figure 28-10.** The standard nozzle can also be machined to permit access in tight corners or its diameter reduced to weld on items such as small holding clips.

For most operations, a ⅛″ diameter electrode is used. The end of the electrode should normally be flat and of the same diameter as the electrode. However, when working at low current settings (100 A or less), better results are obtained if the end of the electrode is tapered slightly to provide a blunt point approximately one-half the diameter of the electrode. The blunt point helps to prevent the

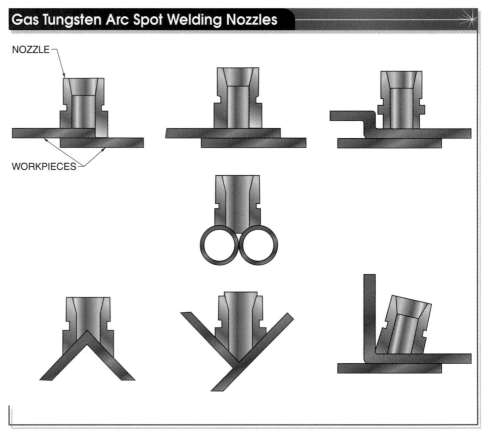

arc from wandering. If the end of the electrode balls excessively after only a few welds have been made, it is usually an indication of excessive current, dirty material, or insufficient shielding gas. Helium used as a shielding gas produces greater penetration than argon, although argon produces a larger weld diameter. Gas flow should be set at approximately 6 cubic feet per hour (cfh).

Gas Tungsten Arc Spot Welding Procedure

To make a spot weld, the end of the welding gun is placed against the workpiece and the trigger is pulled. Squeezing the trigger starts the flow of cooling water and shielding gas and advances the electrode to touch the workpiece. As soon as the electrode touches the workpiece, it automatically retracts, establishing an arc. The arc is extinguished at the end of a preset length of time. The welding gun is usually preset at the factory to provide an arc length of ¹⁄₁₆″, which is satisfactory for most welding applications.

The current required for welding is determined by the thickness of the metal to be welded. The major effect of increasing the current when both workpieces are approximately the same thickness is to increase penetration. However, it also tends to increase the weld diameter. Increasing the current when the bottom workpiece is considerably heavier than the top workpiece results in an increase in weld diameter with little or no increase in penetration. **See Figure 28-11.**

> **Point**
>
> *When gas tungsten arc spot welding, set the current based on the thickness of the metal to be spot welded.*

> *Manual gas tungsten arc spot welding can be performed using automatic sequencing controls to set the gas and water flow rates, control arc starting and intervals, and provide necessary postweld shielding gas and water flow.*

Gas Tungsten Arc Spot Welding Nozzles

NOZZLE

WORKPIECES

Figure 28-10. *The gas tungsten arc spot welding nozzle can be shaped or machined for a variety of welding jobs.*

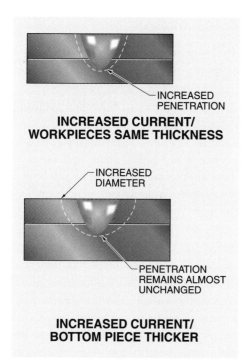

Figure 28-11. *The thickness of the workpieces being welded has an effect on weld diameter and penetration when the current is increased.*

Figure 28-12. *Good surface contact is important to making a sound spot weld.*

Weld time is set on the dial of the control panel. The dial is calibrated in 60ths of a second and is adjustable from 0 sec to 6 sec. The effect of increasing the weld time is to increase the weld diameter. However, by increasing the weld diameter, penetration is also increased.

Mill scale, oil, grease, dirt, paint, and other foreign materials on or between the contacting surfaces prevent good contact and reduce weld strength. The space between the two contacting surfaces resulting from these surface conditions or poor fit-up acts as a barrier to heat transfer and prevents the weld from penetrating into the bottom workpiece. Consequently, good surface contact is important for sound welds. **See Figure 28-12.**

Stress relieving is used with GMAW to prevent distortion that occurs through localized heating. Stress relieving methods include electric resistance heating blankets, induction coils, and special furnaces.

Gas tungsten arc spot welding can be done from one side only so the bottom workpiece must have sufficient rigidity to permit the two workpieces to be brought into contact with pressure applied by the welding gun. If the thickness, size, or shape of the bottom workpiece is such that it does not provide enough rigidity, then some form of backing is required. Backing may be either steel or copper.

GAS METAL ARC WELDING IN PRODUCTION WELDING

GMAW is an economical and effective method of joining light-gauge, hard-to-weld metals such as nickel, stainless steel, aluminum, brass, copper, titanium, columbium, molybdenum, Inconel®, Monel®, and silver, as well as structural plates and beams. GMAW may be performed semiautomatically, mechanically, or automatically.

Semiautomatic GMAW uses welding equipment that controls only the filler metal feed. An operator controls the welding speed. When the trigger of the welding gun is pressed, the shielding gas, current, and filler metal automatically begin to flow. **See Figure 28-13.** The operator simply keeps the weld concentrated in the designated area of the workpiece and maintains the proper travel speed.

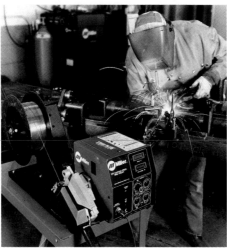

Miller Electric Manufacturing Company

Figure 28-13. *With a portable GMAW welding gun, the gas, current, and wire feed automatically begin to flow when the trigger is pressed.*

Mechanized GMAW uses welding equipment that performs the welding operation under the constant observation and control of the operator. The welding head is stationary rather than portable. The welding head is either mounted on a carriage that travels over the workpiece or it is in a fixed position and the workpieces to be welded are moved beneath the unit.

Automatic GMAW uses welding equipment that performs the entire welding operation without constant observation and adjustment of the controls by the operator. The controls are set to the specified welding schedule and the machine performs the entire operation.

STUD WELDING

Stud welding is a form of electric arc welding. Stud welding is a term used for the process of joining a metal stud, or similar part, to a workpiece. Two methods of stud welding have been developed, the Nelson method and the Graham method, each with a different principle of operation.

Nelson Method

The Nelson method uses a flux and a ceramic guide or ferrule. Welding equipment consists of a welding gun, a timing device that controls the DC welding current, specially designed studs, and ceramic ferrules. Studs are available in a variety of shapes, sizes, and types to meet many applications. The studs have a recess in the welding end, which contains the flux. The flux acts as an arc stabilizer and a deoxidizing agent. An individual porcelain ferrule is used with each stud when welding. The ferrule is the most important part of the operation because it concentrates the heat, acts (with the flux) to prevent air from contacting the molten metal, confines the molten metal to the weld area, shields the glare of the arc, and prevents charring of the workpiece through which the stud is being welded.

A stud is loaded into the chuck of the welding gun and a ferrule is positioned over the stud. When the trigger is depressed, the current energizes a solenoid coil, which lifts the stud away from the workpiece, causing an arc that melts the end of the stud and the area on the workpiece. A timing device shuts the current OFF at the proper time. The solenoid releases the stud, a spring action plunges the stud into the weld pool, and the weld is made. **See Figure 28-14.**

Graham Method

The Graham method uses a small cylindrical tip on the joining face of the stud. The diameter and length of the tip vary with the diameter of the stud and the workpiece. The Graham method operates on AC current and requires an air source that can supply about 85 lb of air pressure.

The welding gun is air-operated with a collet (to hold the stud) attached to the end of a piston rod. Constant air pressure holds the stud away from the workpiece until it is sufficiently heated; then air pressure drives the stud against the workpiece. When the small tip touches the workpiece, a high-current, low-voltage discharge results, creating an arc that melts the entire area of the stud and the corresponding area of the work. Arcing time is about one millisecond (0.001 sec); thus a weld is completed with little heat penetration, no distortion, and practically no fillet. The stud is driven at a velocity of about 31″ per second and the explosive action as it meets the workpiece cleans the area to be welded. A minimum workpiece thickness of 0.02″ is preferred, particularly if no marking on the reverse side is required. **See Figure 28-15.**

Both methods of stud welding are adaptable for welding most ferrous and nonferrous metals, their alloys, or combinations of them that are metallurgically compatible.

In stud welding, ferrules must be positioned exactly as required. Locating fixtures and equipment are used to accurately place ferrules in the proper location. A template can locate a ferrule to within ± ¹⁄₃₂″ of the specified location. A spacer on the template ensures accurate spacing between ferrules.

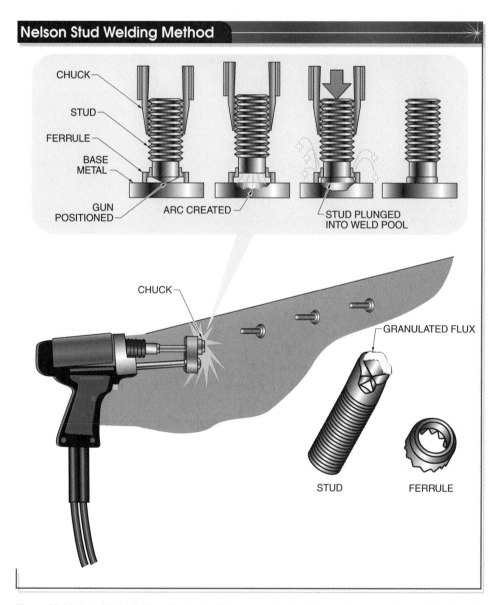

Nelson Stud Welding Method

CHUCK
STUD
FERRULE
BASE METAL
GUN POSITIONED
ARC CREATED
STUD PLUNGED INTO WELD POOL

CHUCK
GRANULATED FLUX
STUD
FERRULE

Figure 28-14. *A stud is loaded into the chuck of the gun and a ferrule is positioned over the stud. As the stud contacts the workpiece, an arc is started that melts the end of the stud and an area on the workpiece to which the stud is welded.*

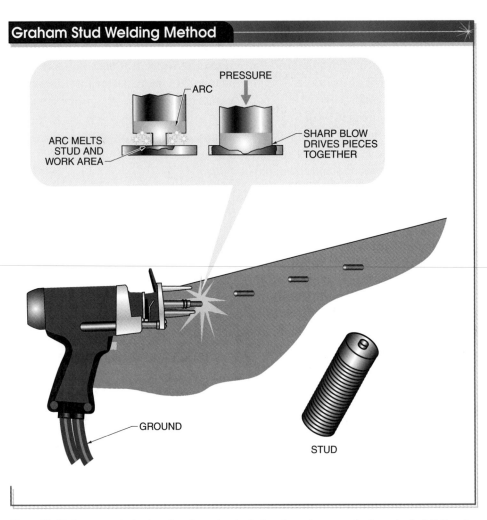

PRESSURE

ARC

ARC MELTS
STUD AND
WORK AREA

SHARP BLOW
DRIVES PIECES
TOGETHER

GROUND

STUD

Figure 28-15. *On contact, ionization takes place, cleaning both surfaces. An arc results that melts the full diameter of the stud and a corresponding area of the workpiece. A sharp blow drives the two together, completing the weld.*

ELECTRON BEAM WELDING

Electron beam welding (EBW) is a welding process that produces coalescence with a concentrated beam, composed primarily of high-velocity electrons, impinging on the joint. EBW is performed without shielding gas and without exerting pressure on the weld. EBW is a fusion welding process. Fusion is achieved by focusing a high-power-density beam of electrons on the area to be joined. Upon striking the metal, the kinetic energy of the high-velocity electrons changes to thermal energy, causing the workpiece to melt and fuse.

Electrons are emitted from a tungsten filament heated to approximately 3630°F (2000°C). Since the filament would quickly oxidize at this temperature if it were exposed to normal atmosphere, welding must be done in a vacuum chamber. The entire electron beam generation system, consisting of the gun to generate electrons and the electron optics to control and focus the beam, must be contained in a vacuum. Otherwise, the electrons would collide with molecules of air and this would cause them to lose focus and energy.

EBW can be used to join metals that range from thin foil to 2″ thick. It is particularly adaptable for welding refractory metals such as tungsten, molybdenum, columbium, and tantalum, and metals that oxidize readily, such as titanium, beryllium, and zirconium. It can also be used to join dissimilar metals, aluminum, standard steels, and ceramics.

> **✓ Point**
>
> *EBW is a fusion process where a high-power-density beam of electrons is focused on the area to be joined.*

EBW Processes

Electron beam welding is done using either of two processes: the vacuum chamber process or the beam-in-air process.

The vacuum chamber process uses a controlled vacuum environment where the welding gun and the workpieces are enclosed. **See Figure 28-16.** Because the vacuum chamber is free from atmospheric contaminants, the vacuum chamber process produces a cleaner weld without a shielding gas. The weld is more precise because the beam is much narrower in the vacuum chamber.

Sciaky, Inc.

Figure 28-16. *The welding gun and workpieces are enclosed in the vacuum chamber of an electron beam welding machine.*

The beam-in-air process uses a gun that has a vacuum chamber that surrounds the area where electrons exit from the welding gun; welding is done in the open atmosphere. To shield the weld area from atmospheric contaminants, argon is used as a shielding gas. The welds produced by the beam-in-air process are similar to welds made using GTAW.

EBW has several advantages over other processes. It welds with a low total-energy input. Workpiece distortion and effects on the properties of the workpiece are reduced. The weld size and location can be controlled relative to the energy input, but it is relatively narrow. EBW is chemically clean and facilitates welding without contamination of the work-piece. Using the beam-in-air process allows greater welding speed than GTAW.

EBW is often associated with the joining of difficult-to-weld metals. It is used in aerospace fabrication where new metals require more exacting joining characteristics; however, adaptation of the process to commercial applications is increasing. There is every indication that this growth will continue.

One of the major limitations of EBW using the vacuum chamber process is that the piece must be small enough to fit into the vacuum chamber. This limitation is being reduced to some extent because large chambers are now manufactured to accommodate a variety of product sizes. Another limitation is that when the workpiece is in the chamber in a vacuum it becomes inaccessible. It must be manipulated using special controls.

EBW Equipment

An electron gun consists of a filament, cathode, anode, and focusing coil. The electrons emitted from the heated filament carry a negative charge and are repelled by the cathode and attracted by the anode. The electrons pass through the anode and then through a magnetic field generated by the electromagnetic focusing coil. An optical viewing or numerical control system determines the path of the electron beam centerline to the weld area. **See Figure 28-17.**

By varying the current to the focusing coil, the operator can focus the beam for gun-to-work distances ranging from ½″ to 25″. The electron beam can be controlled with a focusing coil to produce a spot diameter of less than 0.005″.

> *A feature of electron beam welding (EBW) is the high depth-to-width ratio of the weld, which leads to a narrower weld zone compared to other welding techniques and lower distortion. This makes it possible to weld much thicker workpieces than with other welding processes. Also, since the electron beam is tightly focused, the heat-affected zone is small, which minimizes distortion.*

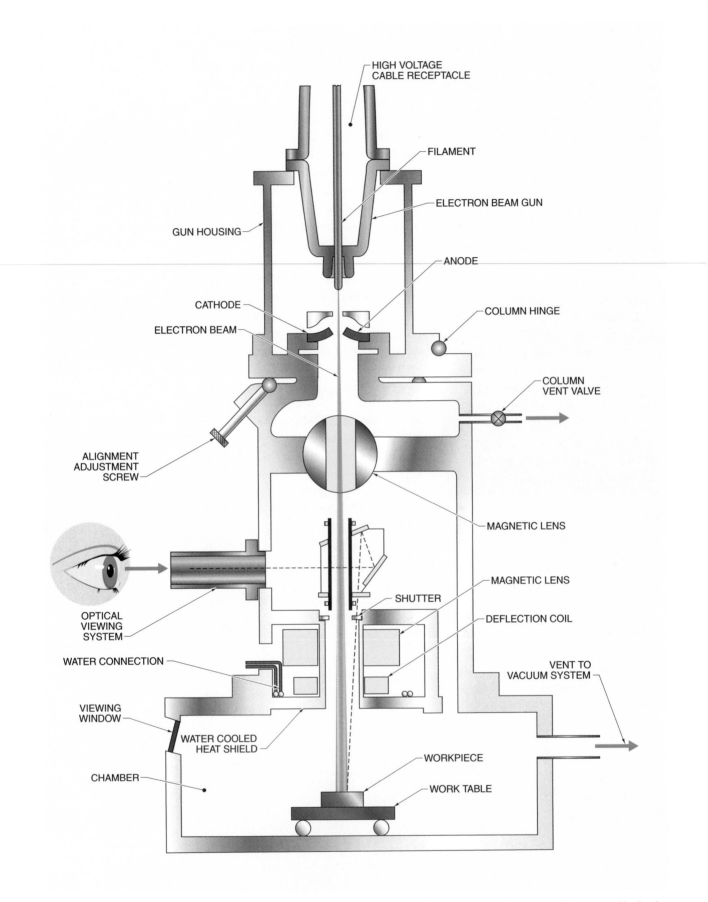

Figure 28-17. *In the electron beam column, the electrons pass through an aperture in the anode and then through a magnetic field generated by the electromagnetic focusing coil. An optical viewing system provides a line of sight down the path of the electron beam centerline to the weld area.*

A vacuum chamber has heavy glass windows to permit viewing the work. A work table in the chamber is arranged so it can be operated either manually or electrically along the x- and y-axes. T-slots are provided on the table to attach fixtures or workpieces for welding.

A vacuum pumping system is designed to clean and dry the vacuum chamber in a relatively short time. The capacity of the pump required is determined by the volume and area of the vacuum chamber and the time required to evacuate the chamber. The pumping equipment is usually completely automatic once setup is completed.

Necessary electrical controls include setup controls and operating controls. Setup controls include instruments required for the initial setup of the welding operation, such as meters for beam voltage, beam current, focusing current, and filament current. Operating controls consist of stop-and-start sequence, high-voltage adjustment, focusing adjustment, filament activation, and work table motion.

EBW Procedure

The workpiece is positioned on the work carriage in the vacuum chamber. The electron gun and the work-to-gun distance are aligned manually and visually using the optical system. Work travel or welding gun travel, depending on the type of welding facility used, is aligned. The vacuum chamber is then closed. Vacuum controls are started and the chamber is pumped down to the required vacuum, which is prescribed in the weld schedule.

Beam voltage, beam current, filament current, and focusing current controls are set based on the weld schedule. The weld schedule is usually determined by a welding engineer. Once the control settings have been checked, the beam current is switched ON for an instant and OFF again for a weld spot alignment check. The weld or weld area is viewed by opening the shutter only when the beam current is turned OFF. If the shutter is opened when the beam current is ON, the optical system can be severely damaged.

After all controls and settings have been checked and all switching made operative, welding is begun by turning the sequence start switch to the ON position. The weld is made automatically.

FRICTION WELDING

Friction welding (FRW) is a welding process that joins two metal parts that rotate or are in relative motion with respect to one another when they are brought into contact and pressure is applied between them. Friction, or inertia, welding is a process where stored kinetic energy is used to generate the required heat for fusion. The two workpieces to be joined are aligned end to end. One is held stationary by a chuck or a fixture, and the other is clamped in a rotating spindle.

The rotating workpiece is brought up to a certain speed to develop sufficient energy. The drive source is disconnected and the pieces are brought into contact under a computed thrust load. At this point, the kinetic energy contained in the rotating mass converts to frictional heat. The metal at and immediately behind the interface is softened, permitting fusion to take place between the workpieces.

FRW has several advantages over conventional fusion welding (FW) or UW. It produces improved welds at higher speed and lower cost, less electrical current is required, and costly copper fixtures are eliminated. With FRW there is less shortening of the components, which often occurs with FW or UW. Additionally, the HAZ near the weld is confined to a very narrow band. **See Figure 28-18.**

FRW can be used to weld dissimilar as well as similar metals. Weld strength is normally equal to that of the original metals.

Friction welding (FRW) is used with metals and thermoplastics in a wide variety of aviation and automotive applications. An advantage of friction welding is that it allows dissimilar metals to be joined because localized melting is not involved.

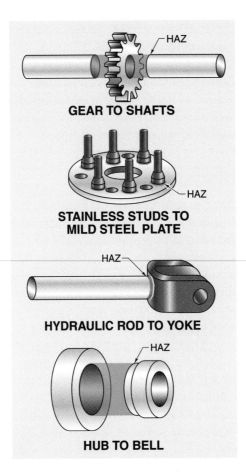

GEAR TO SHAFTS

STAINLESS STUDS TO MILD STEEL PLATE

HYDRAULIC ROD TO YOKE

HUB TO BELL

Figure 28-18. *On friction-welded workpieces, the HAZ is confined to a very narrow band.*

LASER BEAM WELDING

Laser beam welding (LBW) is a welding process that produces coalescence with the heat from a laser beam impinging on the joint. Laser beam welding is used without shielding gas and without exerting pressure on the weld. Fusion is achieved by directing a highly concentrated beam to a spot about the diameter of a human hair. The highly concentrated beam generates a power intensity of 1 billion or more watts per square centimeter at its point of focus. Because of its excellent heat input control, LBW can be used near glass or varnish-coated wires without damaging the glass or the insulating properties of the varnish.

Since the heat input to the workpiece is extremely small in comparison to other welding processes, the size of the HAZ and the thermal damage to the adjacent parts of the weld are minimized. It is possible to weld heat-treated alloys without affecting their heat-treated condition, and the weld can be held in the hand immediately after welding is completed.

LBW can be used to join dissimilar metals such as copper, nickel, tungsten, aluminum, stainless steel, titanium, and columbium. Additionally, the laser beam can pass through transparent substances without affecting them, making it possible to weld metals that are sealed in glass or plastic. Because the heat source is a light beam, atmospheric contamination of the weld joint is not a problem.

LBW is used in the aerospace and electronic industries where extreme control of the weld is required. A major limitation of LBW is its shallow penetration.

The duration of the beam is usually about 0.002 sec, with a pulse rate of one to 10 pulses per second. As each point of the beam hits the workpiece, a spot is melted that resolidifies in microseconds. The weld line consists of a series of round, solid, overlapping weld pools. The workpiece is moved beneath the beam or the energy source is moved across the weld. The beam is focused onto the workpiece using an optical system and the welding energy is controlled by a switch.

Laser Beam Theory

Atoms have been made to give off energy by exciting them in such common devices as fluorescent lights and television tubes. *Fluorescence* is the ability of certain atoms to emit light when they are exposed to external radiation of shorter wavelengths.

In LBW, the atoms that are excited to produce the laser light beam are produced in a synthetic ruby rod ⅜″ in diameter. The ruby rod is identical to a natural ruby but has a more perfect crystalline structure. About 0.05% of its weight is chromium oxide. The chromium atoms give the ruby its red color because they absorb green light from external light sources. When the atoms absorb this light energy, some of their electrons are excited. Thus, green light is said to pump the chromium atoms to a higher energy state.

The excited atoms eventually return to their original state. As they do, a portion of the extra energy they previously absorbed (as green light) is given off in the form of red fluorescent light. When the red light emitted by one excited atom hits another excited atom, the second atom gives off additional red light. The additional red light is in phase with the colliding red light wave, increasing the intensity of the light. In other words, the red light from the first atom is amplified because more red light exactly like it is produced.

By using a very intense green light to excite the chromium atoms in the ruby rod, a larger number of its atoms can be excited and the chances of collisions are increased. To further enhance this effect, the parallel ends and the sides of the rod are mirrored to bounce the red light back and forth within the rod. When a certain critical intensity of pumping is reached (the threshold energy), the chain reaction collisions become numerous enough to cause a burst of red light. The lens at the front end of the rod is only a partial reflector, allowing the burst of light to escape through it. **See Figure 28-19.**

PLASMA ARC WELDING

Plasma arc welding (PAW) is an arc welding process that uses a constricted arc between a nonconsumable tungsten electrode and the weld pool (transferred arc), or between the electrode and constricting nozzle (non-transferred arc).

The electrode and part are shielded by ionized gas (plasma) issuing from the torch, which may be supplemented by an auxiliary shielding gas. PAW uses a central plasma core of extreme temperature surrounded by a sheath of cool gas. **See Figure 28-20.** The required heat for fusion is generated by an electric arc that has been highly intensified by the injection of a gas into the arc stream. The superheated arc column is concentrated into a narrow stream, and when directed onto a workpiece, can make a groove weld ½″ thick or more in a single pass without filler metal or edge preparation.

In some respects, PAW may be considered an extension of GTAW. The main difference is that in PAW, the arc column is constricted. This constriction produces a much higher heat transfer rate.

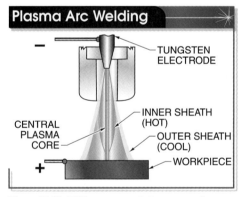

Figure 28-20. *PAW uses a central plasma core of extreme temperature surrounded by a sheath of cool gas.*

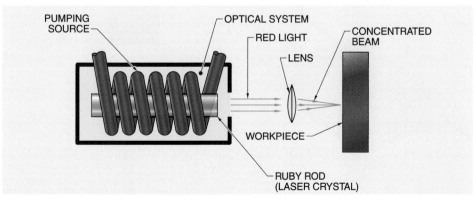

Figure 28-19. *The LBW machine has a concentrated beam that is focused on the workpiece with an optical system.*

The arc plasma actually becomes a jet of high current density. The plasma gas, upon striking the workpiece, cuts or keyholes, entirely through the workpiece, producing a small hole that is carried along the weld. During this cutting action, the melted metal in front of the arc flows around the arc column, then is drawn together immediately behind the hole by surface tension forces, reforming in a weld bead.

Plasma arc welding can be used to weld stainless steels, carbon steels, Monel®, Inconel®, titanium, aluminum, copper, and brass alloys. Filler metal is typically not needed; however, a continuous filler wire can be added.

PAW Equipment

A regular heavy-duty DC rectifier is used as the power source for PAW. A special control console is required to provide the necessary operating controls. A water-cooling pump is usually needed to ensure a controlled flow of cooling water to the torch at a regulated pressure. Proper cooling prolongs the life of the electrode and the nozzle. **See Figure 28-21.**

Torches specially designed for PAW can be hand-held or mounted for stationary or mechanized applications. **See Figure 28-22.** The shielding gas supply should be either argon or helium. In some applications, argon is used as the plasma gas and helium as the shielding gas. However, in many operations argon is used for both shielding and generating the plasma arc.

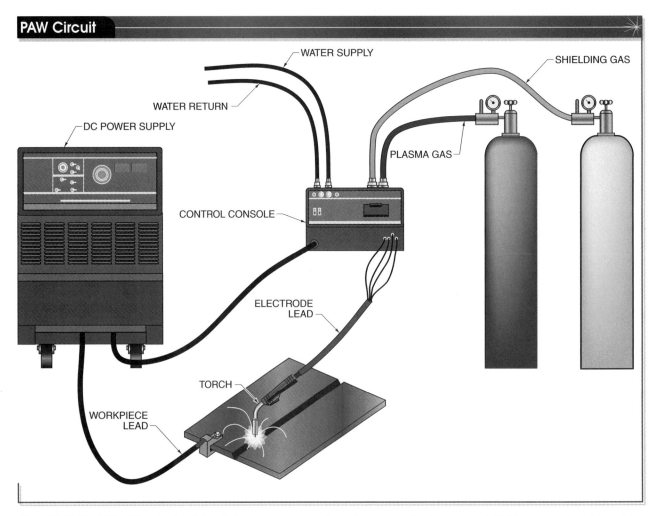

PAW Circuit

WATER SUPPLY

SHIELDING GAS

WATER RETURN

DC POWER SUPPLY

PLASMA GAS

CONTROL CONSOLE

ELECTRODE LEAD

TORCH

WORKPIECE LEAD

Figure 28-21. *The PAW welding circuit includes a DC power source, control console, water supply, plasma gas and shielding gas supply, welding cables, and torch.*

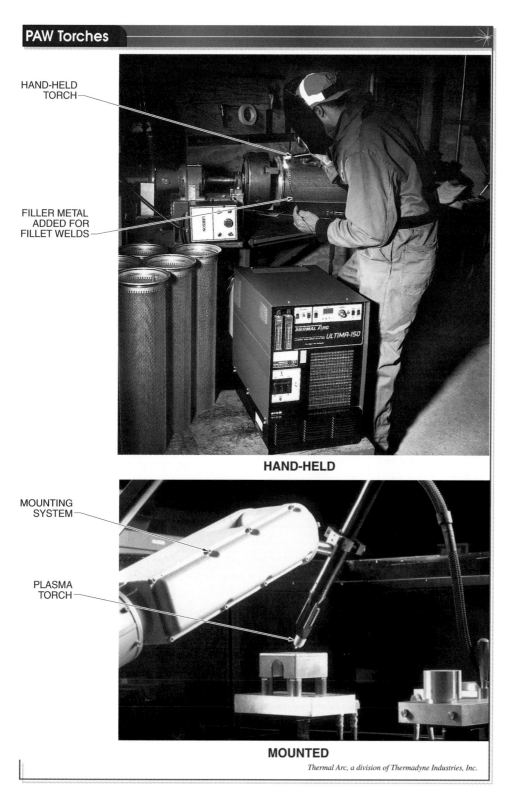

HAND-HELD TORCH

FILLER METAL ADDED FOR FILLET WELDS

HAND-HELD

MOUNTING SYSTEM

PLASMA TORCH

MOUNTED

Thermal Arc, a division of Thermadyne Industries, Inc.

Figure 28-22. *Torches specially designed for PAW can be hand-held or mounted for stationary or mechanized applications.*

SUBMERGED ARC WELDING

Submerged arc welding (SAW) is an arc welding process that uses an arc between a bare metal electrode and the weld pool. A blanket of granular flux, supplied from the electrode, forms a layer of slag that protects and shields the arc and weld pool from contamination. The granulated flux shields the welding action and covers the

Media Clip

Submerged Arc Welding

molten metal. The weld is submerged beneath the flux and slag. Pressure is not used on the weld during SAW. **See Figure 28-23.**

SAW can be either semiautomatic or automatic. The welding unit used with the automatic process is set up to move over the weld area at a controlled speed. **See Figure 28-24.** On some machines, the welding head moves and the work remains stationary. In others, the head is stationary and the work moves. Semiautomatic SAW requires the use of a special welding gun.

SAW can be used for metals from $1/16''$ thick. It is usually used for welding thick metals and where deep penetration is required. For example, it is possible to weld $3''$ plate in a single pass. However, caution is necessary as impurities in the weld collect toward the center of the weld, developing a weak area. Very little edge preparation is necessary on metal less than $1/2''$ thick. Generally, backing is essential when welding thick steel. Welding positions are limited because of the large amount of fluid molten metal.

The difference between SAW and GMAW is that no inert shielding gas is required. The welding gun is pointed over the weld area and the trigger is pressed. As soon as the trigger is pressed, the welding wire is energized and the arc is started. At the same time, flux begins to flow. Welding is then carried out in the same manner as GMAW.

Miller Electric Manufacturing Company

Figure 28-24. *The welding unit used with the automatic process is set up to move over the weld area at a controlled speed.*

Submerged arc welding (SAW) is widely used in heavy industrial applications because of higher weld quality, high deposition rates, and the ability to join relatively thick metals.

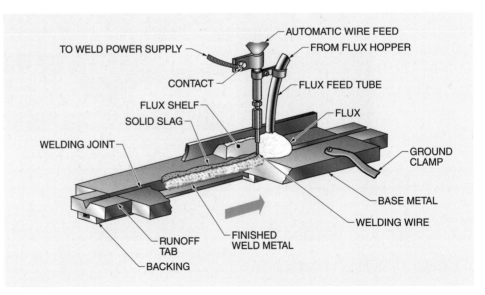

Figure 28-23. *A cutaway view of an SAW machine shows how the granulated flux shields the welding action and covers the molten metal.*

SAW Equipment

The welding equipment required for SAW includes a power source, wire feed and drive assembly, welding gun, and flux delivery system. Any regular GMAW DC welding machine can be adapted for SAW. Since SAW is usually automated, the power source must be capable of an output and a duty cycle that can match the operation being performed. The metal thickness dictates the required current. Light-gauge metal requires as little as 300 A, while thick metals may require 1000 A or more.

A constant-voltage power source sets the voltage and holds it relatively constant. Current is determined by the feed speed of the electrode wire. As the wire feed speed is increased, more current is required to burn off the wire. Conversely, when the wire feed speed is decreased, less current is required.

With a constant-current power source, a voltage-sensing wire feeder may be used. A voltage-sensing wire feeder increases the speed of the wire feed motor when the arc voltage increases and reduces the speed of the wire feed motor when the voltage decreases, maintaining a fairly constant arc voltage and length. However, it does not provide a consistent deposition rate.

Wire feed systems used for GMAW or FCAW can be used for SAW, provided they can feed the required wire size at the proper speed. For semiautomatic SAW, a standard wire feeder is normally used. When using a constant-current power source, special wire feeders that change feed rates in response to arc voltage changes are sometimes used. Burnback controls may be used for both semiautomatic and automatic SAW to prevent the electrode wire from sticking to the weld pool at the end of the weld.

The welding wire should be clipped to a sharp point as close to the flux cone as possible. Once the voltage and current are set, the welding gun is positioned over the joint. As the welding wire is fed into the weld zone, the welding gun deposits the granulated flux over the weld pool and completely shields the welding action. The arc is not visible since it is buried in the flux, thus there is no flash or spatter. The portion of the granular flux immediately around the arc fuses and covers the molten metal, but after it has solidified, it can be tapped off easily with a chipping hammer.

Flux can be delivered to the weld pool by either the gravity feed or forced-air feed method. The gravity feed method is designed for short-duration welds that are easily accessible. It is limited by the amount of flux that the operator can handle in the flux canister.

The forced-air flux feeding method is commonly used for semiautomatic welding. A conventional wire-feeding unit feeds the welding wire to the weld pool. A pressurized storage tank that holds approximately 100 lb of flux and a hand-held welding gun with a high-pressure air feed attachment are also used. An air supply is attached to the flux storage tank. The tank's regulator adjusts the pressure that feeds the flux through the tubing to the welding gun and the weld pool.

The Lincoln Electric Company

An automatic submerged arc welding machine is designed to move over the weld area while depositing the weld in flat position.

ULTRASONIC WELDING

Ultrasonic welding (USW) is a welding process that produces a weld by applying high-frequency vibratory energy to workpieces that are held together under pressure. Theoretically, if two workpieces with perfectly smooth surfaces are brought into close contact, the metal atoms of one workpiece will unite with the atoms of the other piece to form a permanent bond. However, regardless of how smooth such surfaces are a sound metallic bond normally does not occur because it is impossible to prepare surfaces that are absolutely smooth.

Whatever method is used to smooth surfaces, they still possess peaks and valleys (as seen by a microscope). As a result, only the peaks of the workpieces that come into close contact unite, producing no bond in the valleys. Also, smooth surfaces are never completely clean. Oxygen molecules from the atmosphere react with the metal to form oxides. These oxides attract water vapor, forming a film of moisture on the oxidized metal surface. Both the moisture and oxide film also act as barriers to prevent close contact.

In USW, to overcome the barriers to fusion, the interface between the workpieces is plastically deformed. This is done by means of vibratory energy, which disperses moisture and oxide and levels an irregular surface to bring the surfaces of both workpieces into close contact and form a solid bond. Vibratory energy is generated by a transducer. **See Figure 28-25.**

USW Equipment

The welding equipment used for USW consists of two units: a power source or frequency converter, which converts 60 Hz line power into high-frequency electrical power; and a transducer. The components to be joined are simply clamped between a welding tip and supporting anvil with just enough pressure to hold them in close contact.

USW Procedure

High-frequency vibratory energy is transmitted to the joint for a required period. Bonding is accomplished without applying external heat or adding filler metal. Welding variables such as power, clamping force, weld time for spot welds or welding rate for continuous-seam welds can be preset and the cycle completed automatically. A switch lowers the welding head, applies the clamping force, and starts the flow of ultrasonic energy.

Ultrasonic Welding

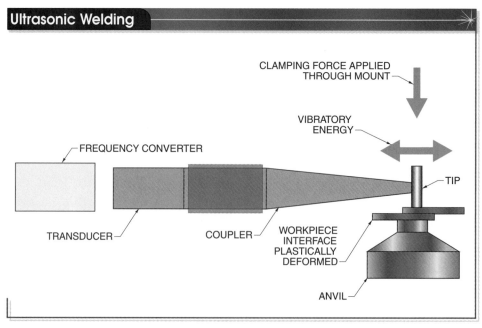

Figure 28-25. *An ultrasonic continuous-seam welder is often used for complete sealing of components used in electronics.*

Successful USW depends on the proper relationship between welding variables, which is usually determined experimentally for each application. Clamping force may vary from a few grams for very thin metals to several thousand pounds for thick metals. Weld time may range from 0.005 sec to 1 sec for spot welding and a few feet per minute (fpm) to 400 fpm for continuous-seam welding. The high-frequency electrical input to the transducer may vary from a fraction of a watt to several kilowatts.

USW is particularly adaptable for joining electrical and electronic components, hermetic sealing of materials and devices, splicing metallic foil, welding aluminum wire and sheet, and fabricating nuclear fuel elements. Spot welds or continuous-seam welds can be made on a variety of metals ranging in thickness from 0.00017″ (aluminum foil) to 0.10″. Thick sheet and plate can be welded if the machine is specifically designed for them. High-strength bonds are possible on similar and dissimilar metal combinations.

ELECTROGAS WELDING

Electrogas welding (EGW) is a welding process that uses an arc between a filler metal electrode and the weld pool, using approximately vertical welding and a backing bar to control the weld metal. EGW can be used with or without shielding gas and without exerting pressure on the weld. EGW uses a gas-shielded metal arc and is designed for single-pass welding of vertical joints on steel ranging in thickness from 3/8″ to 1½″.

The welding head is suspended from an elevator mechanism that provides automatic control of the vertical travel speed during welding. This mechanism raises the welding head automatically at the same rate as the advancing weld metal. The welding head is self-aligning and can follow any alignment irregularity in the metal or in the joint.

Once the equipment is positioned on the joint, welding is completely automatic.

The wire feed speed and the current levels remain constant. At the end of the weld, the process stops automatically. The EGW technique is especially adaptable for shipbuilding and fabrication of storage tanks and large-diameter pipes.

OTHER WELDING PROCESSES

Other welding processes approved by the AWS may be used for particular applications. These processes include explosion welding, forge welding, roll welding, and cold welding.

Explosion Welding

Explosion welding (EXW) is a welding process that produces a weld by extreme impact of the metals through controlled detonation. Coalescence occurs from the explosive force of the impact on the heated surface. EXW forms a strong bond between many metals, including dissimilar metals that cannot be joined by arc welding. EXW is commonly used for cladding steel with thinner metals.

Forge Welding

Forge welding (FOW) is a welding process that produces a weld by heating the metals to welding temperature and applying forceful blows to cause deformation at the faying surfaces. FOW is one of the oldest welding procedures, commonly used by blacksmiths for joining metals. The metals are heated to a red-hot temperature and a hammer and anvil are used to deform the surface. Flux is often applied to aid in bonding the joint.

Roll Welding

Roll welding (ROW) produces a weld by applying heat and pressure using rollers to cause deformation at the faying surfaces. ROW is similar to forge welding except that the weld is formed by rollers rather than a hammer. ROW is commonly used for welding pipe and for cladding mild- or low-alloy steel with high-alloy steel.

Cold Welding

Cold welding (CW) is a welding process in which a weld is produced using substantial pressure at room temperature to cause deformation at the joint. Coalescence occurs because of the pressure that is applied. Surface oxides and contaminants must be removed before CW takes place. Power brushing is the best method to clean the surface. Metals that do not work harden rapidly are most suited for cold welding. Since there is no localized melting, many dissimilar metal combinations, such as aluminum and copper, can be joined using cold welding.

  Refer to Quick Quiz® on CD-ROM

Points to Remember

- Spot welding is a form of RW with wide application in industry.
- Spotwelders are available to produce single spot welds or multiple spot welds.
- Seam welding produces a series of overlapping spot welds, thereby making a continuous-weld seam.
- PW is widely used in attaching fasteners to structural members.
- In multiple-impulse welding, the current is regulated to cycle on and off at regular intervals.
- When gas tungsten arc spot welding, set the current based on the thickness of the metal to be spot welded.
- EBW is a fusion process where a high-power-density beam of electrons is focused on the area to be joined.
- In FRW, heat resulting from the parts being rotated together plus the application of pressure is used to fuse the pieces.
- Laser beam welding (LBW) is a welding process that produces coalescence with the heat from a laser beam impinging on the joint.
- PAW uses an electric arc that is highly intensified by the injection of gas into the arc stream, which results in a jet of high current density.
- In SAW, the electric arc is completely hidden beneath a flux.
- No inert shielding gas is required for SAW since the flux completely surrounds the electric arc.
- USW is a process where vibratory energy disperses the moisture, oxide, and surface irregularities between the workpieces, thereby bringing the surfaces into close contact to form a permanent bond.

1. What is the basic principle of resistance welding?
2. What is projection welding?
3. What is meant by multiple-impulse welding?
4. How does upset welding differ from flash welding?
5. What is the advantage of gas tungsten arc spot welding over conventional resistance spot welding?
6. What are some advantages and limitations of electron beam welding?
7. What is the principle of friction welding?
8. In laser beam welding, how is the high-intensity laser light beam generated?
9. How does PAW differ from regular GTAW?
10. What is SAW and what are some of its advantages?
11. How is fusion of metal accomplished in ultrasonic welding?
12. What are some advantages of adhesive bonding?

Refer to Chapter 28 in the *Welding Skills Workbook* for additional exercises.

Automation and Robotic Welding

Chapter

Automation in production welding offers greater efficiency and weld quality control to manufacturers and fabricators. Automation requires that some or all of the steps of an operation be performed in sequence by electronic or mechanical means. Many welding processes can be automated for production welding that requires consistent, rapidly repeated welds. Automatic welding is most commonly used in automation systems.

Automation in production welding can be broadly classified as fixed automation and flexible automation. Fixed automation uses mechanically directed movements of the torch and workpiece. Flexible automation uses programmable movements of a robotic torch and the workpiece. Automated welding equipment is used to achieve the accuracy and speed needed in a production environment.

AUTOMATION IN PRODUCTION WELDING

Automation in production welding offers greater efficiency and weld quality control to manufacturers and fabricators. Production welding processes have evolved as new welding technology has been developed. Production welding processes used for automation in welding are mechanized, semiautomatic, and automatic welding processes. Automatic welding is most commonly used in automation systems. Automation in production welding can be broadly classified as fixed automation and flexible automation.

Fixed Automation Systems

A *fixed automation system* is a system that uses machines designed for a specific production function. Fixed automation systems are primarily used for simple production path welds such as circles, linear seams, or radial seams. A fixed automation system is generally used in production facilities demanding high volume and repeated welds.

Fixed automation equipment uses mechanical and electrical means to guide the torch and the workpiece. Fixed automation equipment provides more arc-on time, better accuracy and speed, and lower cost than manual welding processes. The torch may be fixed and the workpiece moved about the torch, such as on a pipe weld; or the workpiece may be fixed and the torch moved, such as on a seamer. **See Figure 29-1.**

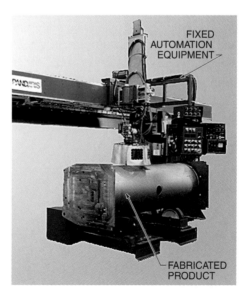

Pandjiris, Inc.

Figure 29-1. *Fixed automation equipment is designed for a specific production function.*

Since the welding equipment is mechanical, the operator must adjust the mechanical path to make changes to the torch movement. Adjustments require time to retool the equipment for the next weld. The single-purpose design of the equipment makes intricate welding applications prohibitive if not impossible. Fixed automation welding systems equipment commonly includes operator controls, a torch positioner/holder, and a workpiece positioner/holder.

Operator Controls. Operator controls are used to start and stop the welding cycle. The operator controls may be connected to a programmable logic controller (PLC), which in turn controls the positioners and the welding equipment. The PLC sequences through the weld cycle, controlling when to move the torch, start the arc, feed the wire, and turn on the shielding gas, as well as other welding sequences. Some welding equipment uses internal controls rather than PLCs to control the welding equipment.

Torch Positioner/Holder. Fixed automation equipment such as seamers and orbital welders use a torch positioner, or holder. A *torch positioner* is a fixed-path mechanical apparatus that moves the torch in a specified path. A seamer is designed to weld linear seams in rolled tubes or flat plates. When a torch positioner is used on a seamer, it keeps the torch on a linear path along the joint and maintains a constant rate of speed. **See Figure 29-2.** A PLC or some other automated controller is typically used to direct the weld sequence.

✓ **Point**

A torch positioner is a fixed-path mechanical apparatus that moves the torch in a specified path.

✓ **Point**

A robot is a programmed path device used to position the torch and at times the workpiece.

Workpiece Positioner/Holder. The design of a workpiece positioner/holder needs to be quite sophisticated and elaborate to ensure that the simple path required by the torch can be maintained. The workpiece positioner must hold the workpieces without interfering with the path of the torch.

Workpiece positioners can be controlled manually, pneumatically, or hydraulically. The workpiece positioner may rotate or tilt by means of an electric motor that allows easier access to the weld seam.

> ✦ *The purpose of automation is to reduce manufacturing costs by increasing productivity and quality.*

Flexible Automation Systems

A *flexible automation system* is a system that uses programmable movements of the torch and sometimes the workpiece. In flexible automation, programmed equipment guides the torch. The most common type of programmed device is the robot. A *robot* is a programmed path device used to position the torch and at times the workpiece. A robot can perform complex movements in order to follow a complex path. The robot can provide the fabricator with extended arc-on time. With the advances in electrical motors and motor control circuits, robots' speeds have almost matched those of fixed automation equipment, and the air cut time of the robot systems has been reduced. *Air cut time* is the time that a piece of equipment spends in the nonproductive activity of moving from one weld to another. In the past, fixed automation equipment provided much faster movements than flexible automation equipment. The biggest advantage that flexible automation has over fixed automation is the reprogrammability of the robot movement, allowing for varied movement of the robot. This feature makes it easier and quicker to change weld settings, locations, and workpiece positions. The robot is capable

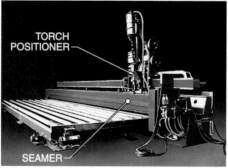

TORCH POSITIONER

SEAMER

Pandjiris, Inc.

Figure 29-2. *A torch positioner used on a seamer keeps the torch on a linear path along the joint and maintains a constant speed.*

of storing weld programs, which can quickly be changed, permitting a variety of parts and welds to be efficiently made by one robot.

In the past, flexible automation equipment was much more expensive than fixed automation equipment. However, with the increased variety and availability of flexible automation systems, the costs have become comparable. A flexible automation system typically incurs additional costs for the fixture designs and tooling associated with flexible automation systems.

An additional cost to the users of flexible automation systems is the cost associated with training. Operators, programmers, and maintenance personnel all must be trained in the proper use of the robot. In each case, the most efficient and beneficial training is to train process experts — welders — on how to operate and program the robot.

The components of a robot welding system (robot cell) used for flexible automation consist of a robot controller, robot manipulator, teach pendant, operator controls, and workpiece positioner.

Robot Controller. The robot controller provides the control for the servomotors and communicates with the welding equipment and other equipment in the system. A *servomotor* is an AC or DC motor with encoder feedback to indicate how far the motor has rotated. AC servomotors provide higher speeds and torques than DC servomotors and are the preferred method of control for robotic systems. With encoder feedback technology, AC servomotors provide faster and more accurate movement than the stepper motors used in the past. Stepper motors would rotate 360° in stepped increments, with an accuracy of 0.5°. AC servomotors can be controlled with an accuracy of 0.1°, without the need for the complex driving circuits associated with stepper motors.

The robot controller directs the starting and stopping of the servomotors as well as the rate of speed and acceleration

of each servomotor. **See Figure 29-3.** The robot controller not only controls movement of the manipulators from point to point but also controls the path of the torch from point to point. Controlled movement paths may be linear (straight line) or circular (curved line).

> Robotic welding is an application of industrial robotics that is very similar to several other processes, such as CNC machining. Both use multi-axis, computer-controlled robots to perform a precise sequence of motions encoded in a program file. However, while a CNC machining center controls a cutting bit at the end of the robot, a welding robot controls a welding torch or electrode.

✓ **Point**

The robot controller controls the movement of the manipulator from point to point and the path of the torch from point to point.

Robot Controller

The Lincoln Electric Company

Figure 29-3. *A robot controller is used to direct the starting, stopping, speed, and acceleration of the servomotors.*

🎞 **Media Clip**

Robotic Welding

A robot controller may control more than the servomotor in the manipulator, it may also control servomotors in the workpiece positioner and other equipment in the robot cell. Some robot controllers can control as many as 16 servomotors simultaneously for a synchronized motion. The robot controller must accelerate and decelerate each servomotor individually to maintain the controlled path at the tool center point (TCP). Advanced software and digital hardware are typically required to continually adjust the servomotors to the correct speed, acceleration, and deceleration.

Robot Manipulator. The robot manipulator is the robot arm. A servomotor moves the robot manipulator from one point to another. The robot manipulator consists of a base and several links and joints. The base provides the mounting for the robot manipulator, much like a human torso. The links are the arm structures, similar to a human upper arm and forearm. The joints slide and rotate to allow the movements of the links, just like the human shoulder, elbow, or wrist. The joints of the robot manipulator are referred to as axis joints. Thus, a six-axis robot will have six axis joints.

✓ Point

The robot manipulator is the robot arm and consists of a base and several links and joints (or axes).

The robot manipulator is the most important part of the robot cell design. Manipulators can be found in many different configurations, sizes, and speeds. Early robot manipulators, such as the rectilinear robot manipulator, were designed for easy control. These robot manipulators were large and slow, but easy to design and control. Rectilinear robot manipulators are still used for some spot welding applications, but are limited in their access to various welding positions.

✓ Point

A teach pendant is the input method that the robot programmer uses to create robot movement programs.

The arrangement and number of axes on a robot determine its ability to perform complex motions and reach into hard-to-reach spaces. Each axis is also known as a degree of freedom. Robots designed for permanent, well-defined tasks may utilize fewer degrees of freedom, but those that may need to be adapted or reprogrammed for different welding requirements may require six or more.

The most common configuration for a welding robot is a six-axis articulated robot manipulator driven by AC servomotors. **See Figure 29-4.** The servomotor provides the speed and repeatability needed for the welding operation. An articulated configuration allows the arm link and wrist joints to be small and compact. A small and compact joint can be easily maneuvered into tight areas.

Robot Manipulator

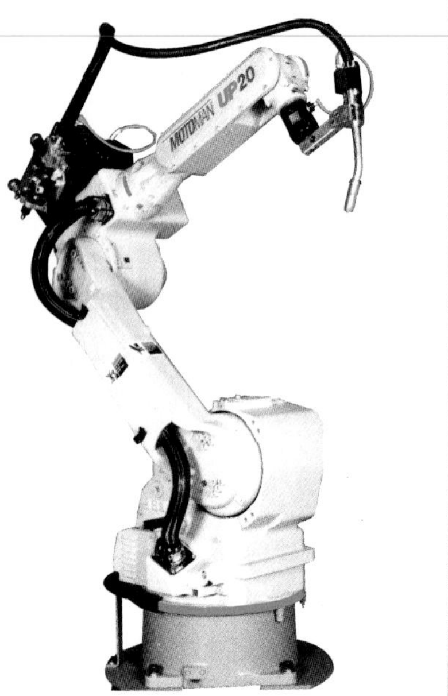

Motoman, Inc.

Figure 29-4. *A common configuration for a welding robot is a six-axis articulated robot manipulator.*

Welding manipulators are usually only required to lift small, light loads, but they must be able to move the loads quickly and with high repeatability. Most manipulators can return to a programmed point within approximately 0.004″, which gives the manipulator a 0.004″ repeatability factor.

Teach Pendant. The teach pendant and the robot controller are the brains of the robot welding system. A *teach pendant* is the device that the robot programmer uses to create robot movement programs. **See Figure 29-5.** The programmer uses the teach pendant to move the manipulator in either the axis plane or the Cartesian coordinate plane. **See Figure 29-6.** Axis motion is created by each axis servomotor individually creating movement to position the TCP at the point programmed. Points in the Cartesian plane are found using the Cartesian coordinate system. The *Cartesian coordinate system* is a system of locating points in space defined by perpendicular planes. The Cartesian coordinate system uses a three-dimensional box with a horizontal X direction, a vertical Y direction, and a depth Z direction. The robot controller controls all of the axis servomotors simultaneously in order to maintain a straight-line X, Y, or Z direction when moving the manipulator. This type of motion allows the programmer to easily position the TCP to the point programmed because the movement is similar to that of a human welder.

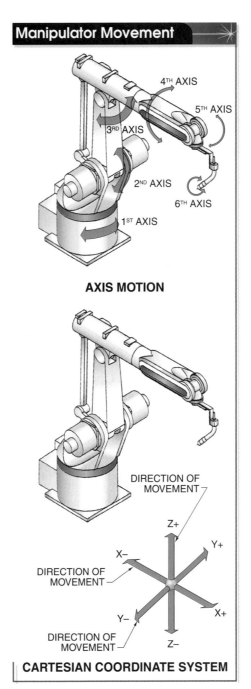

AXIS MOTION

CARTESIAN COORDINATE SYSTEM

Figure 29-6. *Robot manipulator movement is produced by axis motion specified using the Cartesian coordinate system.*

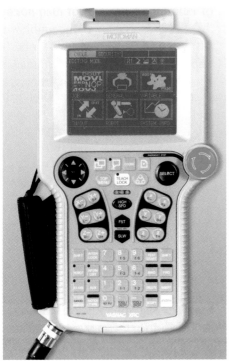

Motoman, Inc.

Figure 29-5. *The teach pendant is the input method that the robot programmer uses to move the robot and create robot programs.*

Three methods of programming, or teaching, a robot are the lead through method, off-line programming, and the walk through method. The lead through method is used for teaching most robots. Off-line programming requires an experienced programmer. Walk through programming is rarely used any longer.

Newer robot systems use off-line programming software to create robot programs. Off-line programming software is run on a PC, which simulates the robot work cell. **See Figure 29-7.** When the program is complete, the programmer downloads the program to the robot to be verified and run. This type of programming reduces the amount of downtime needed to program or modify a robot.

programmer. While the weave pattern allows the robot to relearn the program and alter its trajectory to maintain a quality weld, the weaving action slows the welding process and may cause undercut to the joint. Laser seam trackers use a laser beam to sweep across the weld path looking for weld seam deviations. Laser seam trackers do not need to weave the welding wire along the weld joint, which allows for an increased welding speed and reduced chance of undercut from the weaving action. The laser systems require more care and maintenance to prevent damage to the optical laser lens and receiver.

Robotic GTAW and PAW systems may also include motorized torch adjustments. These systems monitor the arc voltage and adjust the height of the torch by moving the torch up and down on a slide mounted between the torch mount and the torch. These systems usually add bulk to the torch and thus reduce the ability to access certain weld joints.

AUTOMATIC WELDING EQUIPMENT

Automatic welding equipment used for automatic welding and robotics is generally semiautomatic equipment outfitted to perform automatic operations. For example, for the GMAW process, automatic welding equipment includes a welding power source, wire feeder, torch, and shielding gas system.

The power source used is typically the same type of power source used for semiautomatic welding. The wire feeder is modified to accommodate the sequence commands needed for the automated welding process. Robotic systems use torches and a shielding gas system specifically designed to be mounted and operated on a robot manipulator.

Torch mounting includes breakaway plates or crash detection mounts so the torch breaks away or the robot stops if the torch hits something while moving. Torch mounts protect the robot manipulator from severe damage.

Many robot controllers provide the signals for the gas solenoid, wire feed speed, and welding voltage through arc start and arc end parameters programmed into the controller. These signals need to be communicated to the welding equipment. Some welding systems integrate the power source, wire feeder, and shielding gas system into one piece of equipment. For example, submerged arc flux delivery and recovery systems are added to the robotic system to provide solid flux for SAW. Extra equipment, such as additional torches or multiple shielding gas systems, may be required for some automatic welding processes.

The American Welding Society maintains standard AWS/NEMA D16.2/D16.2M, *Guide for Components of Robotic and Automatic Arc Welding Installations,* which details the components within robotic and automatic welding systems.

Welding Process Parameters

The majority of arc welding robots are designed for the GMAW process. The GMAW process provides arc control and filler wire control simultaneously. **See Figure 29-9.** The weld programmer only needs to control the placement of the wire in the weld joint and ensure the use of the correct welding parameters.

Miller Electric Mfg. Co.
Seam trackers help identify any deviations of the weld seam while a robot is welding.

Fanuc Robotics North America

Figure 29-9. *The GMAW process provides arc control and filler wire control to the robot simultaneously.*

The GMAW process in automated welding is slightly different from the GMAW process in manual welding. The goals in automatic welding are high travel speed with maximum weld penetration and minimum weld spatter. The main welding parameter that must be controlled is the weld travel speed. In manual welding, weld travel speeds rarely exceed 15 inches per minute (ipm). In automatic welding, manufacturers strive for 30 ipm to 40 ipm and with special GMAW processes may attain 50 ipm to 60 ipm.

To achieve these goals in a practice setting, some of the GMAW welding parameters must be adjusted to optimize the weld. The angle of the torch must be adjusted to a 15° push angle to allow for maximum penetration at maximum weld travel speeds. The wire feed speed and the welding voltage must be adjusted for the material to be welded and the weld joint to produce a GMAW spray transfer arc at manual welding travel speeds. After a good arc at manual welding travel speeds is achieved, the weld travel speed can be increased to production speed levels. These weld parameters offer the speed and deposition rate of the GMAW spray transfer process with the penetration and control of GMAW short circuiting transfer. The arc produces a distinctive sound that some have termed a "GMAW production spray."

WORK AREA AND SAFETY

The robot work area poses potential dangers for maintenance personnel, operators, and programmers. Robot operators, programmers, installers, and manufacturers must be aware of potential dangers. A primary hazard posed by the robot is through mechanical movement of the robot. The robot may hit, trap, or crush a person. A safeguarded space is established to protect personnel from hazards. Protection is usually provided by perimeter guarding devices such as fencing and safety gates to prevent access to the safeguarded space without conscious action. **See Figure 29-10.** Additional presence-sensing devices, such as safety mats, should be installed within the perimeter guards to ensure that no personnel enter the safeguarded space during operation of the robot. Protection from welding is provided by screens.

All personnel require protection from the potential hazards that may occur as a result of interactions with the robot. The Robotic Industries Association (RIA), in conjunction with ANSI, maintains ANSI/RIA R15.06, *Industrial Robots and Robot Systems —Safety Requirements.* All companies involved with robotics should follow these safety requirements.

The ANSI/RIA standard separates the robot work area into the operating space of the robot, the restricted space, and the safeguarded space. The operating space is the space where the robot runs the processes associated with the robot. In the case of welding, this work area includes the workpiece and tooling. The restricted space is the space that the robot operates in with limiting devices attached to the system. Without limiting devices, the robot would move in its maximum space, or work envelope, which is the space encompassing the maximum movement of the robot, the end-effector, the workpiece, and the attachments. The safeguarded space confines the mechanical hazards from all personnel not operating or teaching the robot. The safeguarded space cannot be smaller than the restricted space.

The ANSI/RIA standard calls for proper training of all personnel. Training includes safeguard training, teacher training, operator training, maintenance training, and installer training. The end user of the robot and robot system is responsible for training their employees and maintaining training documentation. With proper training and protection, robot system hazards can be significantly reduced or eliminated.

Guidelines for the levels of qualification and safety and health considerations may be found in AWS D16.4, *Specification for the Qualification of Robotic Arc Welding Personnel.*

The American Welding Society also publishes AWS D16.3M/D16.3, *Risk Assessment Guide for Robotic Arc Welding,* which includes guidelines for risk assessment, robot classification, and potential hazards primarily associated with arc welding robots and robotic arc welding systems.

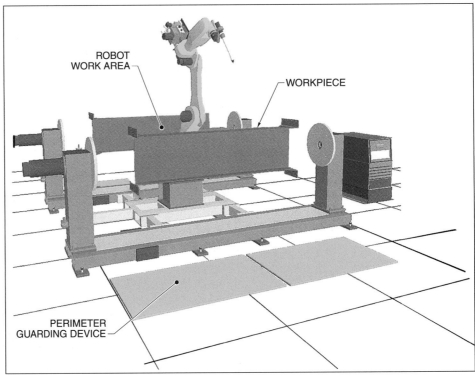

The Lincoln Electric Company

Figure 29-10. *Protection from the robot work area is provided by perimeter guarding devices to prevent access to the safeguarded space without conscious action.*

Refer to Quick Quiz® on CD-ROM

- A torch positioner is a fixed-path mechanical apparatus that moves the torch in a specified path.
- A robot is a programmed path device used to position the torch, and at times the workpiece.
- The robot controller controls the movement of the manipulator from point to point and the path of the torch from point to point.
- The robot manipulator is the robot arm and consists of a base and several links and joints (or axes).
- A teach pendant is the input method that the robot programmer uses to create robot movement programs.
- The robot interface provides the communication path for the robot controller to the welding power source, gas solenoid, and wire feeder.
- The majority of arc welding robots are designed for the GMAW process, which provides arc control and filler wire control simultaneously.
- Automatic welding equipment includes a welding power source, wire feeder, torch, and shielding gas system.
- Robotic welding equipment can be dangerous. Always follow the safety requirements found in the Robotic Industries Association (RIA) standard ANSI/RIA R15.06, *Industrial Robots and Robot Systems—Safety Requirements.*

Questions for Study and Discussion

1. What are the two categories of automation?
2. What manual welding process is most commonly adapted for robotic welding?
3. What is the purpose of torch mounts?
4. Why is fixed automation preferable to manual welding processes?
5. What are the components of a robotic welding system?
6. What are the two torch motion patterns?
7. Why is a robot interface added to a robotic welding system?
8. How does the mechanical movement of the robot pose hazards to operators, programmers, and other personnel?
9. How are personnel protected from the hazards posed by the robot?

 Refer to Chapter 29 in the *Welding Skills Workbook* for additional exercises.

Other Joining Techniques

Chapter

Welding is often used to fasten parts in the fabrication of plastic products. Welding can be used for assembling such products as storage tanks, boxes, and other containers. Installation of plastic pipe and ductwork with welding is also common. The manufacture of many custom plastic products is made possible by plastic welding techniques. Joining dissimilar metals, plastics, and composites in manufacturing and repair operations can be accomplished with adhesive bonding.

TYPES OF PLASTICS

Most plastics are identified by trade names or by the principal compound from which they are made. Plastics are broadly grouped as thermosetting plastics and thermoplastics.

Thermosetting plastics soften only once when exposed to heat. Once thermosetting plastics have been molded into a particular shape and cured (hardened), no subsequent heating can soften them again. Thermosetting plastics are not weldable. They are joined by mechanical methods, principally adhesive bonding. Typical thermosetting plastics are ureas, phenolics, melamines, polyesters, silicones, epoxies, and urethanes.

Thermoplastics can repeatedly soften when heat is applied. These plastics can easily be welded. There are many kinds of thermoplastics, such as acrylics, polystyrenes, polyamides, polyfluorides, and vinyls. Generally the more common thermoplastics used where welding is involved are polyethylene, polyvinyl chloride (PVC), and polypropylene. Welding these types of plastics produces seams that are as strong or stronger than the materials being bonded. Compressed air is best for welding PVC and several other types of

plastics. Both the gas and compressed air are controlled by regulators to provide the correct pressure flow. **See Figure 30-1.**

THERMOPLASTIC WELDING CHART		
Thermo-plastic Material	Welding Temperature*	Welding Gas
PVC	525	Air
Polyethylene	550	Nitrogen
Polyethylene	575	Nitrogen
Penton	600	Air
ABS	500	Nitrogen
Plexiglas™	575	Air

* °F

Figure 30-1. *The welding gas used is determined by the type of thermoplastic material to be welded.*

PLASTIC WELDING TECHNIQUES

Plastic welding is similar to metal welding in that localized heat is used to produce fusion. Joint preparation requirements such as proper fit-up and root opening, joint design, and beveling are required in plastic welding as in metal welding, with one significant difference. In metal welding, a sharply defined melting point develops and the base material and filler material melt and flow together to form the weld joint. However, plastics are poor heat conductors, and consequently they

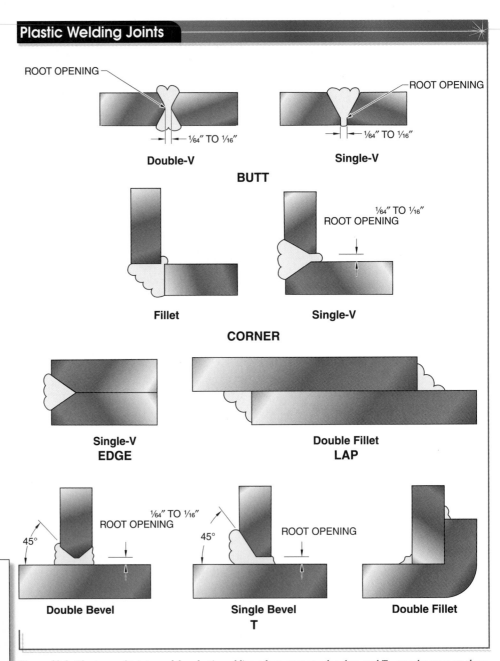

do not readily melt and flow. To achieve a permanent bond, the filler material and base materials must be heated to a point at which the materials will fuse together, but not so high that the plastic decomposes.

Joint Preparation

The types of joints used in plastic welding are the same as those used in metal welding—butt, corner, edge, lap, and T. The edges of the joints are beveled to provide a sufficient area on which to form a good bond. The beveled edges should have a groove angle of 60° with a root opening between ¹⁄₆₄″ and ¹⁄₁₆″, although the root opening may be deeper if a larger filler material is required. **See Figure 30-2.**

> **Point**
>
> *Bevel all edges to secure a proper weld joint. Interlocking the corners produces the best results on corners.*

> *Plastic welding involves many of the same techniques and considerations as metal welding, such as joint design, filler selection, and heat control.*

Plastic Welding Joints

ROOT OPENING

ROOT OPENING

¹⁄₆₄″ TO ¹⁄₁₆″

¹⁄₆₄″ TO ¹⁄₁₆″

Double-V

Single-V

BUTT

Fillet

¹⁄₆₄″ TO ¹⁄₁₆″
ROOT OPENING

Single-V

CORNER

Single-V
EDGE

Double Fillet
LAP

45°

¹⁄₆₄″ TO ¹⁄₁₆″
ROOT OPENING

45°

ROOT OPENING

Double Bevel

Single Bevel

T

Double Fillet

> **⚠ WARNING**
>
> Some plastic materials, such as vinyl, produce HCl gas or obnoxious odors. Polyvinyl chloride produces poisonous fumes. Precautions must be taken to avoid inhaling these fumes. If necessary, a respirator should be used.

Figure 30-2. *The types of joints used for plastic welding—butt, corner, edge, lap, and T—are the same as those used for welding metal.*

Welding Procedure

Select the correct shape tip and insert in the gun. Guns should be able to supply a temperature varying from 400°F to 600°F or more (up to 925°F). Different materials and different plastic thicknesses have differing heat requirements.

Set the air or gas pressure according to the plastic manufacturer's recommendations. Although the wattage of the heating element determines the range of heat, the air or gas pressure determines the actual amount of heat at the tip. **See Figure 30-3.**

AIR PRESSURE SETTINGS

Element*	Air Pressure†	Temperature‡
320	2–3	400
340*	2–3	410
350	2½–3½	430
450	3–4	540
460*	3–4	600
550	4–5	700
650	4½–5½	800
750	5–6	860
800*	5–6	900

* Note: Three-heat unit with a rotary heat selector switch:
 (in W) Low – 340 W, Medium – 460 W, High – 800 W
† psi
‡ °F ³⁄₁₆″ from tip

Figure 30-3. *The air or gas pressure setting determines the amount of heat at the tip during welding.*

During the welding cycle, 3 psi to 5 psi of pressure should be applied to ensure weld integrity. Exerting excessive pressure on the filler material may cause excessive stretching, particularly when welding vinyl. The length of the filler used should be the same as the length of the weld. Equally important is to avoid overheating the weld area as the filler material and the base material can char and discolor, resulting in an unacceptable weld. Underheating is also objectionable since it produces a cold weld that has poor tensile strength.

Check the weld by bending a test weld 90°. If the weld is made properly, the weld beads will not separate from the base material nor will it be possible to pry the filler material out of the weld when cooled. A cross-section of the test weld also reveals whether complete penetration has occurred. **See Figure 30-4.**

Plastic welding should always be done in a well-ventilated area. Follow the manufacturer's recommendations for safe practices when welding specific types of plastics. The basic welding processes used for plastic welding are hot gas, heated-tool, and induction. With some restrictions, friction welding can also be used.

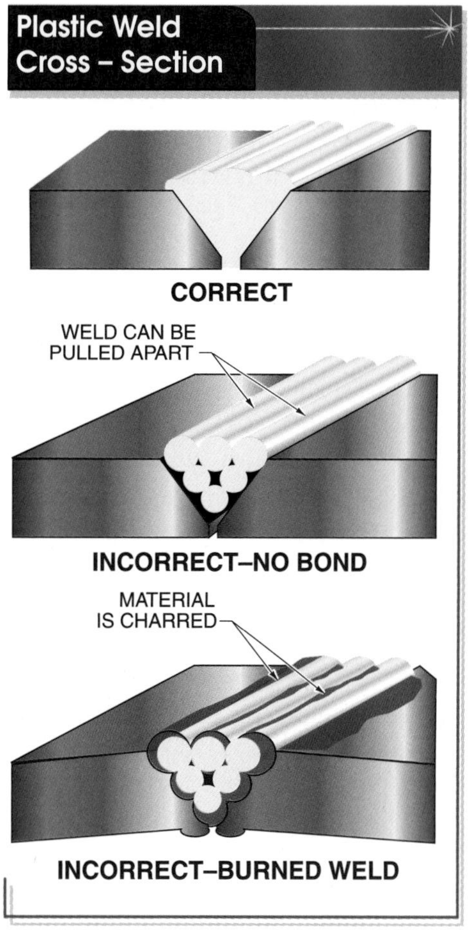

Plastic Weld Cross – Section

CORRECT

WELD CAN BE PULLED APART

INCORRECT–NO BOND

MATERIAL IS CHARRED

INCORRECT–BURNED WELD

Figure 30-4. *A cross-section cut through a test weld shows the amount of penetration that has occurred.*

HOT GAS WELDING

Hot gas welding is accomplished with a specially designed gun containing an electrical heating unit. A stream of compressed air or inert gas (nitrogen) is directed over the heated element, which then flows out of the nozzle and onto the surface of the material being bonded. The gun permits the use of several different tips for different welding operations. The type of tip used depends on the plastic welding application. [Four types of tips are designed for hand welding: tacker, round, flat, and V-shaped.] High-speed tips may also be used for plastic welding. **See Figure 30-5.** The increased speed of a high-speed tip is achieved by the design of the tip, which holds the filler material and applies the needed pressure as the weld is made. A tacker tip is used for tack welding.

Hand Feed Welding

The technique for hand feed welding of plastic is similar to oxyacetylene welding of metals. The gun is held in one hand and the filler material is held in the other. The correct filler material is selected and cut. The filler material should be of the same basic composition as the base material. Either flat, round, or triangular strips may be used. Triangular strips are particularly advantageous in V or fillet welds since the area can be filled with one pass. One-pass welding reduces welding time and minimizes the chances of lack of fusion, which may occur with multiple passes of round strips. **See Figure 30-6.** The hand feed operation for welding plastic is as follows:

1. Hold the tip of the gun about 3/16″ to 1/2″ away from the start of the weld and begin a fanning motion. Place the filler material in a vertical position so the heat from the gun is directed on both the filler material and the base material.

2. When both the base material and the filler material become tacky, press the filler firmly into the joint and bend it back at a slant with the point away from the direction of welding.

3. As the gun is moved along the seam, continue to exert pressure on the filler material to force it into the groove. Maintain a constant fanning motion at a 45° angle so both the filler material and the joint area are heated equally. When welding heavy-gauge plastic with filler material, most of the heat should be directed on the joint.

> *Most plastic welding requires that the base materials and filler material be the same type of plastic in order to form an adequate bond.*

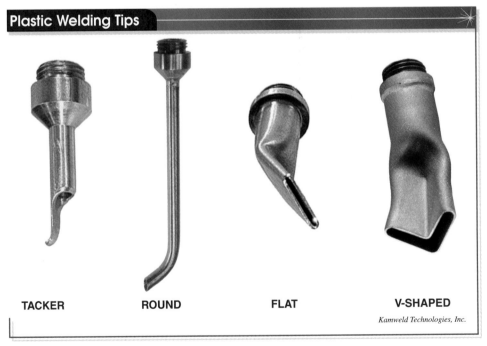

Plastic Welding Tips

| TACKER | ROUND | FLAT | V-SHAPED |

Kamweld Technologies, Inc.

Figure 30-5. *Several types of tips are available for hot gas welding of plastic.*

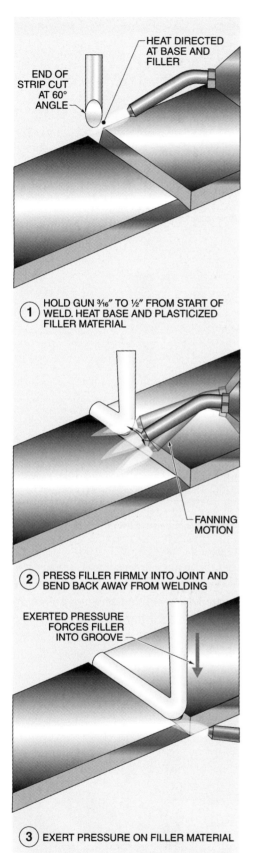

HEAT DIRECTED AT BASE AND FILLER

END OF STRIP CUT AT 60° ANGLE

① HOLD GUN ³⁄₁₆″ TO ½″ FROM START OF WELD. HEAT BASE AND PLASTICIZED FILLER MATERIAL

FANNING MOTION

② PRESS FILLER FIRMLY INTO JOINT AND BEND BACK AWAY FROM WELDING

EXERTED PRESSURE FORCES FILLER INTO GROOVE

③ EXERT PRESSURE ON FILLER MATERIAL

Figure 30-6. *When the base material and the filler material become tacky, press the filler firmly into the joint, continuing to heat the area and exert pressure as the welding progresses.*

High-Speed Welding

The speed of making welds can be substantially increased using the high-speed welding process. As-welded, round, and triangular filler materials are often used for high-speed welding. Filler material must be cut into the required lengths, with one or two inches allowed for trimming. **See Figure 30-7.** The high-speed welding procedure is as follows:

1. Insert the filler material into the high-speed tip. Start the weld by holding the tool at a 90° angle and tamping the broad shoe of the tip on the surface until the first inch of the filler adheres firmly to the base material. Hold the high-speed welding tool at a 45° angle to the work and press the end of the filler into the weld. Feed the filler material manually until the weld bead has been sufficiently started.

2. Maintain an angle of 45° while moving forward along the seam. Once the welding operation is under way, a firm downward pressure of 3 lb to 5 lb is placed on the gun to automatically feed the filler material into the preheated tube.

3. Keep the gun moving at a sufficient speed. Correct speed can be observed by the formation of flow lines on both sides of the filler material. Insufficient speed causes the filler material to stretch because of built-up excessive heat. This condition can be corrected with a quick tamping motion of the shoe as used in starting the weld.

Kamweld Technologies, Inc.

Parts to be joined by high-speed welding should be firmly clamped to prevent movement out of position.

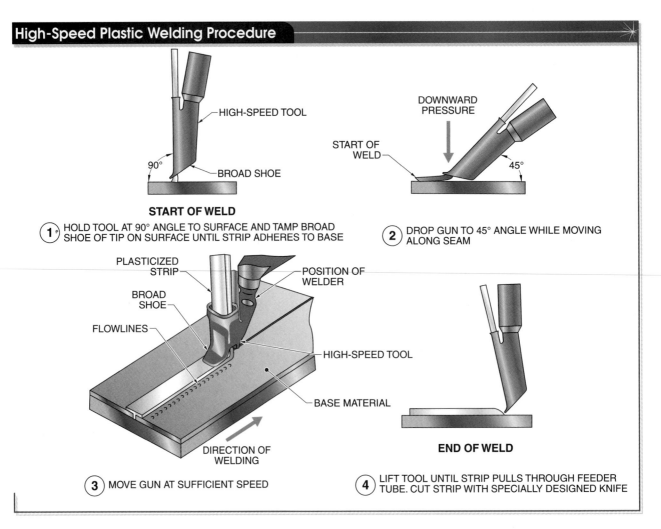

START OF WELD

HIGH-SPEED TOOL

90°

BROAD SHOE

① HOLD TOOL AT 90° ANGLE TO SURFACE AND TAMP BROAD SHOE OF TIP ON SURFACE UNTIL STRIP ADHERES TO BASE

DOWNWARD PRESSURE

START OF WELD

45°

② DROP GUN TO 45° ANGLE WHILE MOVING ALONG SEAM

PLASTICIZED STRIP

BROAD SHOE

FLOWLINES

POSITION OF WELDER

HIGH-SPEED TOOL

BASE MATERIAL

DIRECTION OF WELDING

③ MOVE GUN AT SUFFICIENT SPEED

END OF WELD

④ LIFT TOOL UNTIL STRIP PULLS THROUGH FEEDER TUBE. CUT STRIP WITH SPECIALLY DESIGNED KNIFE

Figure 30-7. *Special tips can be used for high-speed welding to hold the filler material in the correct position; however, the proper procedure must also be used to ensure sufficient penetration and strength.*

Tack Welding

Tack welding is used to fuse materials together prior to welding in order to eliminate the use of clamps or fixtures. A tacker tip is used for tack welding and is used with all types of joints to be welded. **See Figure 30-8.**

HEATED-TOOL WELDING

In the heated-tool (heated surface) welding process, heat for welding is generated in a hot tool. The edges to be joined are heated to the proper temperature, then brought into contact and allowed to cool under pressure. The edges of the plastic sheet are softened with some heat-producing unit such as an electrical strip or bar heater, hot plate, or resistance-coil heater. The heater should be aluminum or nickel since hot steel and copper have a tendency to decompose plastic. The heated-tool welding technique is commonly used for joining sections of pipe and tubing and in the assembly of many molded articles.

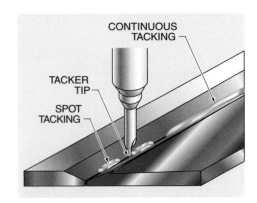

CONTINUOUS TACKING

TACKER TIP

SPOT TACKING

Figure 30-8. *Common tack welding techniques used to eliminate the need for clamps and fixtures during welding include continuous tacking and spot tacking.*

Heated-Tool Welding Procedure

The heated-tool welding procedure is a machine process in which heat is applied by holding the edges in contact with the heating unit until the surface is softened. When the material has reached a molten state, it is removed from the heater and the edges are quickly pressed together. The pressure on the pieces should be enough to force out air bubbles and form a solid contact. Normally, pressures of 5 lb to 15 lb produce good bonded joints. Pressure can be applied by hand or, in production work, with jigs. The pressure must be maintained until the weld has cooled. The most important factor in securing sound welds by the heated-tool technique, outside of proper softening of materials and firm contact, is the elapsed time between removing the pieces from the heating unit and joining them together. The elapsed time interval should be as short as possible to prevent any degree of solidification before the edges come in contact.

FRICTION WELDING

Friction welding, or spin-welding, consists of rubbing the surfaces of the parts to be joined until sufficient heat is developed to bring them to a fusing temperature. Pressure is then applied and maintained until the unit is cooled. In friction welding, one piece is held in a fixed, locked position and the other is rotated. When sufficient melt occurs, the spinning is stopped and the pressure is increased to squeeze out air bubbles and distribute the softened plastic uniformly between the surfaces.

The principal advantages of friction welding are the speed and simplicity of the process. However, friction welding is limited to circular areas and small items. Sometimes friction welding produces a flashing out of soft material beyond the weld area, but usually the excess flashing can be directed to the interior of the part if the weld is properly designed.

As a rule, welds made by the induction process are not as strong as those obtained by other heating methods. Excess flashing can also be avoided by preventing the parts from overheating and by maintaining the proper pressure.

A tack weld can be used in plastic welding to eliminate the need for clamps and fixtures.

ADHESIVE BONDING

Adhesive bonding (AB) is used to join parts with an adhesive placed between the faying (mating) surfaces. AB is useful for joining dissimilar metals, plastics, and composites in manufacturing and repair operations. AB can be used to reduce the number of fasteners required and to strengthen joints prone to failure from vibration. **See Figure 30-9.**

Thin metals subject to heat distortion can be joined with adhesives. For example, auto body panels joined with adhesives do not have depressions caused by resistance welding heat. Workpiece joint dimensions do not affect bonding strength.

Thin metals can be joined with thick metals. Adhesives fill the voids between workpieces without breaking surface contours. The flexibility of adhesives also allows distortion without failure. Joint types for AB require large contact areas for adhesion, as in brazing and soldering.

Fel-Pro Chemical Products

Figure 30-9. *Adhesive bonding is used to join dissimilar materials and strengthen joints prone to failure from vibration.*

AB requires proper surface preparation, application, and curing procedures. The faying surfaces must be clean and free of foreign matter. Adhesives are selected by the material and application of the parts to be joined. **See Figure 30-10.** Adhesive application processes can be manual, semiautomatic, mechanized, automatic, and robotic, depending on the equipment available. Equipment required for adhesive bonding varies depending on application and curing methods. Adhesives are cured by chemical action using catalyst cure (two parts), evaporation, ultraviolet (UV) light, heat, pressure, or both heat and pressure.

Adhesives are available in various viscosities. *Viscosity* is the resistance of a substance to flow in a fluid or semi-fluid state. Low-viscosity adhesives are liquid in form, and flow readily into small spaces. High-viscosity adhesives range from gels to plastic-like forms. In some applications, an adhesive functions as a sealant. A *sealant* is a product used to seal, fill voids, and waterproof parts. Adhesive selection is based on the material and application of the parts to be joined.

Adhesive Types

Adhesives can be broadly classified by chemical content or base as acrylic, anaerobic, cyanoacrylate, epoxy, hot melt, polyurethane, polysulfide, silicone, solvent-base, or water-base adhesives.

ADHESIVE BONDING

Adhesive	Components	Cure Time	Viscosity	Void-Filling	Flexibility	Heat Resistance	Cold Resistance	Thermal Resistance	Water Resistance	Metal Bonding
Acrylic	Two-part One-part (UV or Heat cure)	Medium to Fast	Medium	Good	Good	Good	Good	Good	Good	Good
Anaerobic	One-part	Medium	Low	Poor to Fair	Good	Good	Good	Good	Good	Fair
Cyanoacrylate	One-part	Fast	Low	Poor to Fair	Poor to Fair	Fair	Fair	Good	Fair	Good
Epoxy	Two-part One-part (heat cure)	Slow to Medium	Medium to High	Excellent	Fair	Good	Fair	Good	Good	Good
Hot Melt	One-part	Fast	High	Excellent	Fair to Good	Poor to Fair	Fair	Fair	Good	Fair
Polyurethane	One-part Two-part	Medium	Medium	Good	Good	Fair	Good	Good	Fair	Good
Polysulfide	One-part Two-part	Medium	High	Excellent	Good	Good	Good	Excellent	Good	Good
Silicone	One-part Two-part	Medium	High	Excellent	Excellent	Excellent	Excellent	Excellent	Excellent	Fair
Solvent-base	One-part	Medium	Low to Medium	Poor to Fair	Good	Good	Good	Good	Good	Good
Water-base	One-part	Medium	Low to Medium	Poor to Fair	Poor to Fair	Fair	Fair	Poor	Poor	Poor to Fair

Figure 30-10. *Adhesives are selected based on the material and application of the parts to be joined.*

An *acrylic* is a one-part UV (heat cure), or a two-part adhesive that can be used on a variety of materials. It has a fast setting time and excellent flexibility. An *anaerobic adhesive* is a one-part adhesive or sealant that cures due to the absence of air which has been displaced between mated parts. Low-viscosity anaerobic adhesives are commonly used for locking metal parts together such as screws, nuts, and other fasteners. High-viscosity anaerobic adhesives are used for joining parts that have large gaps between faying surfaces.

A *cyanoacrylate adhesive* is a one-part adhesive that cures instantly by reacting to trace surface moisture to bond mated parts. Cyanoacrylate adhesives have common names such as instant glue or super glue and have a low resistance to high temperatures, moisture, vibration, and shock. *Epoxy* is a two-part adhesive that cures when resin and hardener are combined. Some epoxies are heat-cured.

A *hot melt adhesive* is thermoplastic material that is applied in a molten state and cures to a solid state when cooled. A hot melt adhesive is not as strong as epoxy but is very fast setting. *Polyurethane* is a one- or two-part adhesive with excellent flexibility that cures by evaporation, catalyst, or heat. A *polysulfide adhesive* is a one- or two-part adhesive or sealant that cures by evaporation or catalyst. It is commonly used in the aerospace and building materials industry.

Silicone is a one- or two-part adhesive or sealant that cures by evaporation or catalyst. It has high temperature resistance and excellent sealing characteristics. A *solvent-base adhesive* is a one-part adhesive with a rubber or plastic base that cures by solvent evaporation. It is commonly used as contact cement for bonding large surface areas and lamination applications. A *water-base adhesive* is a one-part adhesive that cures by water evaporation. A water-base adhesive is low in flexibility and is primarily used for wood and paper products.

 Refer to Quick Quiz® on CD-ROM

 Points to Remember

- Bevel all edges to secure a proper weld joint. Interlocking the corners produces the best results on corners.
- Use a filler material of the same composition as the base material.
- Weld plastics only in a well-ventilated area. If a ventilation system is not in place, portable ventilating equipment should be used to ensure adequate ventilation during welding.
- Do not allow the surface to char or discolor.
- Use a fanning motion to ensure uniform heat distribution over the filler material and the edges of the joint.

1. What is the main difference between plastic welding and metal welding?
2. Why are thermosetting plastics not weldable?
3. At what range of temperatures are plastics generally welded?
4. What governs the degree of heat that is to be used in plastic welding?
5. What is the particular advantage of using triangular filler material over round?
6. How far from the surface should the gun be held when welding plastics?
7. Why is a fanning motion necessary in manipulating the gun over the weld joint?
8. Why should excessive pressure on the filler material be avoided?
9. What happens if insufficient heat is used when a welder is making a plastic weld?
10. What test can determine if a weld is made properly?
11. What precautions should be taken when welding plastics?
12. How does the high-speed plastic welding technique differ from the regular hot gas welding technique?
13. When using high-speed welding, why should the filler material not be allowed to remain in the feeder tube?
14. How is the heated-tool welding technique accomplished?
15. What is one of the main limitations of induction plastic welding?
16. How are plastic joints bonded by friction welding?

Refer to Chapter 30 in the *Welding Skills Workbook* for additional exercises.

Destructive Testing

Chapter

Destructive testing involves taking sample portions of a welded structure and subjecting them to loads until they fail. The nature of the test is dictated by the service requirements of the finished product. Destructive testing is performed on welds to qualify both welders and welding procedures, to develop manufacturing quality control acceptance specifications, and to determine if electrodes and filler metals meet the requirements of the specifications. Destructive testing is also used to measure residual stresses associated with welds. Several types of standardized destructive tests are used. Destructive test types and the location(s) of specimens in the weld joint are indicated in the controlling fabrication code or standard. Specimen preparation techniques are necessary for reliable test results.

DESTRUCTIVE TEST TYPES

Destructive tests are used to measure quality, strength, ductility, toughness, and hardness of welded joints. Destructive tests are relatively expensive since they involve preparing materials, making welds, cutting and often machining, and testing of specimens to failure, followed by interpretation of tests by qualified personnel.

Tests are used in specific applications to qualify welding procedures and welders. To qualify a welding procedure or welder, welds are made to welding procedure specification (WPS) parameters, cut into standardized sizes and shapes, and tested to destruction. The welding process, filler metals, and welding technique are selected to make the weld in the position required on the base metal used. Welding joint details and material thicknesses may not be exactly as used

in making the production weld. Requirements vary between fabrication codes and standards so that test specimens are not always the same, nor are they taken from the same locations in a test weld. It is essential that the current edition of the controlling fabrication code or standard be followed when making test welds and test specimens, and when conducting destructive tests. Destructive tests consist of tensile, shear, bend, hardness, toughness, and break tests.

Tensile Test

A *tensile test* is a destructive test that measures the effects of a tensile force on a material. Tensile testing involves the placement of a weld specimen in a universal testing machine and pulling the piece until it breaks. Tensile force occurs when a mechanical load is applied axially (parallel to the axis) to stretch a test specimen. **See Figure 31-1.**

> ✓ **Point**
>
> *The current edition of the controlling fabrication code or standard must be followed when making test welds and test specimens, and when conducting destructive tests.*

411

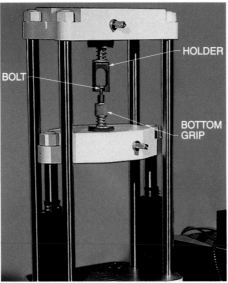

Tinius Olsen Testing Machine Co., Inc.

Figure 31-1. *A universal testing machine is used to perform a tensile test on weld specimens, such as a bolt, to determine the tensile strength of the welds.*

The specimen is cut either from an all-weld area or from a welded butt joint for plate and pipe. The specimen for an all-weld area should conform to specific dimensions and it should be cut from the welded section so its reduced area contains only weld metal. **See Figure 31-2.**

The transition from the ends of the tensile specimen to the reduced section is either shouldered or made with a fillet. Shoulders and fillets minimize stress concentrations. This is particularly important for brittle materials because they are more likely to fail catastrophically at a region of high stress concentration. The longitudinal axis of the specimen and the specimen grips are symmetrical along the longitudinal axis of support to avoid the introduction of bending loads during

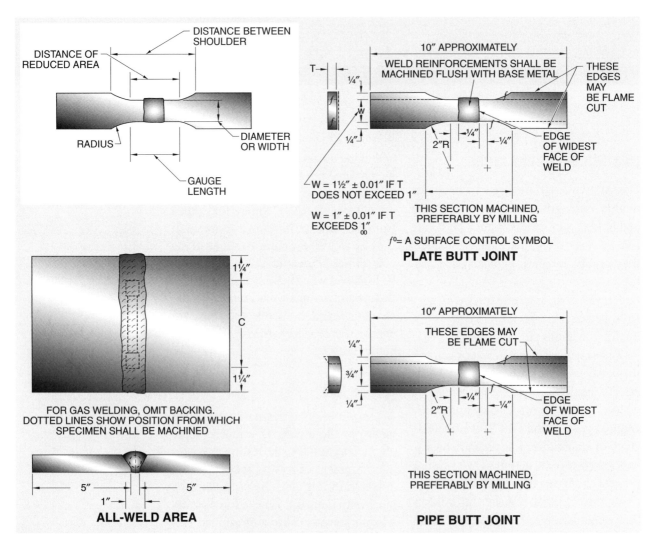

Figure 31-2. *A radius is used on the tensile test specimen to remove stress concentrations. Specimens for all-weld areas must conform to specific dimensions. Butt welds are used for tensile tests on plate and pipe.*

the test. The gauge length (distance over which the elongation measurement is made) is always less than the distance between the shoulders. The most common gauge lengths are 2″ and 8″. The gauge length is normally marked on the specimen using a pair of center punch marks spaced a prescribed distance apart. The gauge marks are always an equal distance from the center of the length of the reduced section. The weld in a welded tensile specimen is always located at the center. Before the specimen is placed in the tensile machine, an accurate measurement should be taken of the gauge length.

Data recorded from weld tensile tests on a welding procedure qualification record (PQR) are maximum load, tensile strength, and failure location. In certain cases, percent elongation and percent reduction of area are also reported.

In addition to qualifying welding procedures and welders, the tensile test also provides information on the load-bearing capacities, joint efficiencies, strain-hardening properties, and ductility of welded joints. Tensile test results provide quantitative data that can be compared or analyzed and used in the design of welded structures. Fracture surface appearance at the failure location provides information on the presence and effects of discontinuities such as incomplete fusion, incomplete joint penetration, porosity, inclusions, and cracking.

A *tensile test machine* is a testing machine composed of two major components that are the means of applying the load to the specimen and the means of measuring the applied load. Some machines are designed for one type of testing only, for example machines that test chain and wire. Universal testing machines apply loads to test specimens in tension or compression.

In a universal testing machine, the load is applied mechanically to the specimen by means of a screw and gears, or it is applied hydraulically. The applied load is measured by a dynamometer (load cell) for mechanically driven machines and by a Bourdon tube for hydraulically driven machines. A *load cell* is a device that uses the elastic deformation of a spring or diaphragm that is calibrated to indicate the mechanical load applied to the specimen. A *Bourdon tube* is a coiled fluid-containing tube that straightens out as the internal pressure on the fluid is increased. The motion of the tube is used to rotate a pointer over a scale that is calibrated to read the hydraulic load applied to the specimen.

Tensile specimens are usually dog-bone shaped in that the central portion of the specimen is reduced in cross section compared with the two ends. This shape causes the test specimen to fail in the narrower central portion rather than at the ends, where the gripping devices affect the stress configuration.

Tensile specimens have a round cross section (round specimen) or a rectangular cross section (rectangular specimen). In general, tensile specimens obtained from welded joints are rectangular, unless taken from locations where it is not possible to obtain a specimen with a rectangular cross section, such as when testing filler metals.

The shape of the ends of the specimen is determined by the specimen gripping device that is used. The ends of round specimens are either plain, shouldered, or threaded. Rectangular specimens are generally made with plain ends, but occasionally pin ends are used. A *pin end* is a rectangular specimen that contains a hole for a pin bearing. **See Figure 31-3.**

Destructive testing of weld mock-ups is a powerful method of evaluating their integrity, or qualifying welders and welding procedures. The tests have evolved to meet the requirements of various industries or equipment types, but they are all based on common principles.

The tensile test procedure consists of fixing the specimen firmly in the grips of the testing machine. An extensometer, which is a device for measuring the extension or elongation of the test specimen, is fitted to the specimen across its gauge length. **See Figure 31-4.**

The specimen is stretched to failure at some steady rate. Unless otherwise specified, the rate of straining is between 0.05″ and 0.5″ per min. Differences in the rate of straining could result in testing inconsistencies. Tensile test procedure is described in ASTM E 8, *Tensile Testing of Metallic Materials.*

As the test specimen is stretched, a load-extension (stress-strain) curve is plotted; the extensometer is removed before the specimen breaks. **See Figure 31-5.**

The load-extension curve shows load and extension limits for metals. Point A is the proportional limit.

The *proportional limit* is the maximum stress at which stress is directly proportional to strain. Between points A and B, the line starts to curve. Up to point B, the tensile specimen will return to its original length if the load is removed.

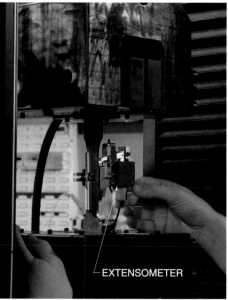

EXTENSOMETER

Tinius Olsen Testing Machine Co., Inc.

Figure 31-4. *An extensometer measures the extension or elongation of the tension test specimen.*

The tensile strength of metals is typically high, with tensile strengths of 60,000 psi or 70,000 psi common. A large machine would be needed to test a full-size part; so small test specimens are tested instead. A section is cut and tested, and the result is multiplied by a size ratio to find the tensile strength equivalent.

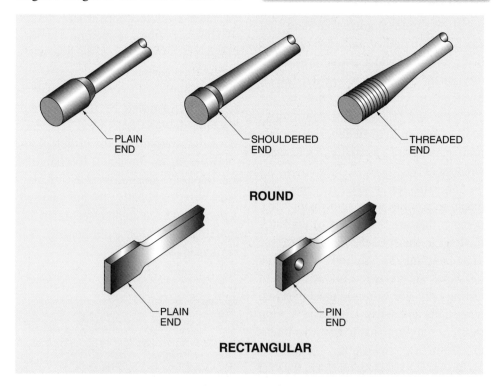

PLAIN END   SHOULDERED END   THREADED END

ROUND

PLAIN END   PIN END

RECTANGULAR

Figure 31-3. *A variety of tensile test specimen ends may be used to ensure that the testing machine securely and uniformly grips the test sample.*

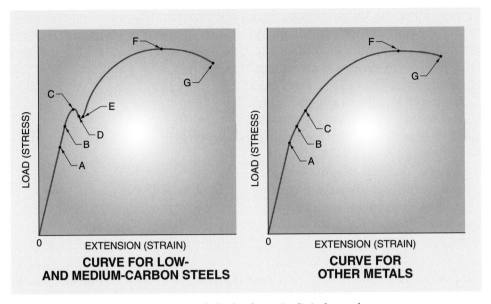

Figure 31-5. *The load-extension curve shows the load and extension limits for metals.*

Point B is the elastic limit. *Elastic limit* is the maximum stress to which a material is subjected without any permanent strain remaining after stress is completely removed. Beyond point B, strain is permanent, or the strain in the specimen is plastic. *Plastic strain* is strain that remains permanent after the stress is removed. Beyond point B, the shape of the curve varies for different metals.

Low- and medium-carbon steels show a jog in their curve, which peaks at point C, or the yield point. *Yield point* is the location on the stress-strain curve where an increase in strain occurs without an increase in stress. Yield point behavior leads to Luders bands (ripples) on the test specimen. Stretcher strains (elongated markings) are observed in low-carbon steel pressings when deformed to the yield point. Yield point behavior is only exhibited by low- and medium-carbon steels.

Between points C and D the curve falls, indicating a plastic strain. The curve continues down to point E, the lower yield point. The curve eventually regains its upward movement and peaks at point F. Point F is the ultimate tensile strength. *Ultimate tensile strength* is a measure of the maximum stress (load) that a metal can withstand. Between points F and G, the specimen begins to neck down or develop a pronounced waist. Point G is the point of failure, the point at which fracture occurs. With all materials, the slope of the load-extension curve decreases and peaks at point F, with failure occurring at point G. With brittle metals, fracture may occur while the load is increasing toward point F.

When the tensile test is completed, the broken specimen is removed from the testing machine. The percent of elongation can be found by fitting the broken ends of the two pieces together and measuring the new gauge length. The new, increased gauge length and the reduced diameter at the narrowest point are measured. Measurement can be made on either side of the break. **See Figure 31-6.**

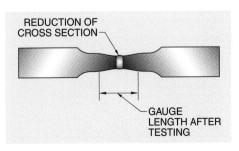

Figure 31-6. *The increased gauge length and reduced diameter at the narrowest point are measured and used to calculate the percent of elongation and the percent of reduction in area.*

Tensile Strength Measurement. The actual tensile strength is found by dividing the maximum load needed to break the piece by the cross-sectional area of the specimen. The cross-sectional area is determined by multiplying the width of the bar by its thickness. Tensile strength is an imprecise value because the original cross-sectional area is not the same as the reduced cross-sectional area that actually exists at the maximum load. Tensile strength is measured in thousands of pounds per square inch (ksi) or megapascals (MPa).

The standard ½″ diameter tensile specimen is referred to as a "505" because a diameter of 0.505″ has an area of 0.2″. To permit easy calculation of stress from loads, specimen diameters of 0.505″, 0.357″, 0.252″, 0.160″, and 0.13″ are convenient because computing of stress or strength may be done using the multiplying factors 5, 10, 20, 50, and 100, respectively. For rectangular specimens the cross-sectional area is calculated from the product of the width and thickness of the specimen.

> *The manufacturer of a base metal typically performs tensile testing and reports the results.*

Percent Elongation and Percent Reduction Measurement. Percent elongation and percent reduction of area are measures of the ductility of a tensile specimen. Measurements are used to calculate the percentage of elongation and reduction in area of a material. They indicate the amount of plastic deformation prior to fracture of the test specimen. To find percent elongaton of a tensile test specimen, apply the formula:

$$\% E = \frac{L_f - L_g}{L_g} \times 100$$

where

$\% E$ = percent elongation

L_f = final length

L_g = gauge length

100 = constant

For example, what is the percent elongation of a tensile specimen that has an inita gauge length of 2″ and a final length of 2.45″?

$$\% E = \frac{2.45 - 2}{2} \times 100$$

$$\% E = \frac{0.45}{2} \times 100$$

$$\% E = 0.225 \times 100 = 22.5$$

$$\% E = \mathbf{22.5\%}$$

Percent elongation is calculated from the gauge length. The longer the gauge length, the less the effect necking down of the specimen has on final length, resulting in lower a percent elongation. When the gauge length is made equal to k√A, where k is a constant equal to 4.47 and A is equal to the cross-sectional area of the specimen, the percent elongation value remains practically constant for different gauge lengths. The most common gauge length in tensile testing is 2″. To find percent reducion of area of a tensile specimen, apply the formula:

$$\% RA = \frac{D_o - D_f}{D_o} \times 100$$

where

$\% RA$ = percent reduction of area

D_o = original diameter

D_f = final diameter

100 = constant

For example, what is the percent reduction of area of a tensile specimen with an original imeter of 0.505″ and a reduced diameter of 0.350″?

$$\% RA = \frac{0.505 - 0.350}{0.505} \times 100$$

$$\% RA = \frac{0.155}{0.505} \times 100$$

$$\% RA = 0.307 \times 100$$

$$\% RA = \mathbf{30.7\%}$$

Round tensile specimens must be used to calculate percent reduction of area. Rectangular specimens have significant rounding of their corners during the test, making measurement of the cross-sectional area less accurate.

Failure location is the region of the specimen at which final failure occurs. Failure location is categorized as base metal, heat-affected zone (HAZ), or weld. Failure location is recorded on the PQR. Fabrication codes and standards usually require that failure location be in the base metal and not in the HAZ or in the weld. Percent elongation and percent reduction in area values are not usually provided for routine weld testing because the bend test is most often used to indicate ductility.

Shear Tests

A shear test is used to determine the shear strength of fillet welds, brazed joints, and spot welds. Shear occurs when some force causes a material to separate, parallel to the load. An acceptable shear strength is usually at least 60% of the minimum specified tensile strength of the base metal. A shear test specimen is prepared to prevent it from rotating during the test. If the specimen were to rotate, interference from other types of stresses would be introduced into the test.

A *fillet weld shear test* is a shear test in which a tensile load is placed on a fillet weld specimen so that the load shears the fillet weld in a longitudinal or a transverse direction. The longitudinal test measures the longitudinal shear strength of the specimen for loads parallel to the axis of the weld. The transverse test measures the transverse shear strength of the specimen for loads normal to the axis of the weld. To prevent rotation and bending stresses during testing, transverse shear specimens are tested as double lap joints. The two shear test types are tension shear test and peel test.

Refer to AWS C1.1 and AWS C1.1M, Recommended Practices for Resistance Welding, for shear test specimen dimensions, test fixtures, and evaluation methods.

Tension Shear Test. A *tension shear test* is a shear test in which a prepared specimen is pulled to failure in a tensile testing machine. Specimens can be pulled from fillet welded, brazed, or spot-welded assemblies. The shear strength of the material is calculated from the load at failure.

To check the shear strength of a transverse weld, a specimen is prepared, placed in a tensile testing machine, and pulled until it breaks. Dividing the maximum load in pounds by twice the width of the specimen will indicate the shearing strength in pounds per linear inch. If the shearing strength in pounds per square inch (psi) is desired, the shearing strength in pounds per linear inch is divided by the throat dimension of the weld.

Tension shear tests are necessary if service failure is caused by two overlapping members being forced apart by a load that is parallel to them. The types of tension shear test are the brazed joint tension shear test and the spot-weld tension shear test.

The *brazed joint tension shear test* is a shear test that determines the strength of filler metal in a brazed joint. The specimen is composed of two single ⅛″ thick sheets joined by brazing with a filler metal. The parts should be fixtured during brazing to maintain accurate specimen alignment. The shear strength of the filler metal is calculated from the tensile load at failure divided by the brazed area.

The *spot-weld tension shear test* is a shear test that determines the strength of arc welds and resistance spot welds. The specimen is made by overlapping materials of suitable size and creating an arc or resistance spot weld in the center of the overlapped area. The load on the weld causes bending and rotation of the weld, resulting in failure around the edges of sheet thicknesses less than about 0.040″. On thicker sheets, the base metal resists bending and the spot will fail at or near the weld. With specimen thicknesses of 0.19″ or larger, the grips of the test machine are offset to reduce loading on the weld.

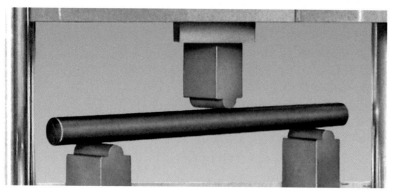

Tinius Olsen Testing Machine Co., Inc.

Bend test specimens allow inspection of all sides of a weld joint to determine the ductility and plastic deformation capabilities of a weld.

specify the maximum allowable size for discontinuities in welding procedure qualification or welder performance qualification bend tests. The most common bend test used for groove welds is the guided bend test.

Peel Test. A *peel test* is a shear test in which a specimen is gripped in a vise and then bent and peeled apart with pincers to reveal the weld. The peel test is an inexpensive alternative to the spot-weld tension shear test. The weld size is measured and compared to that required for the joint. If the weld size is equal to or exceeds the standard size for the design, the production weld is acceptable. The peel test may not be suitable for high-strength base metal or for thick sheets of metal. **See Figure 31-7.**

Bend Tests

Media Clip

Bend Test

A *bend test* is a destructive test used to determine the ductility of a weld by bending a welded specimen around a standardized mandrel. Bend tests provide qualitative information for a specific welding procedure qualification record (WPQR) on the acceptability of a weld. Bend tests are also used for welder performance qualification (WPQ).

Bend tests provide information on the plastic deformation capability of a welded joint. The plastic deformation capability is shown through the ability of the weld to resist tearing. The weld orientation and the bend location must be specified for bend tests. Each welding code (AWS D1.1, API 1104, and ASME Section IX) specifies the requirements for bend test specimens.

A bend test may also reveal discontinuities on the surface and can be used to expose incomplete fusion and delamination. Fabrication codes and standards

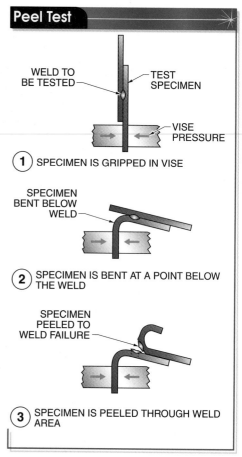

Peel Test

WELD TO BE TESTED — TEST SPECIMEN

— VISE PRESSURE

① SPECIMEN IS GRIPPED IN VISE

SPECIMEN BENT BELOW WELD

② SPECIMEN IS BENT AT A POINT BELOW THE WELD

SPECIMEN PEELED TO WELD FAILURE

③ SPECIMEN IS PEELED THROUGH WELD AREA

Figure 31-7. *A peel test is an inexpensive alternative production control tool for testing shear strength in a weld joint.*

Guided Bend Test. A *guided bend test* is a bend test in which a rectangular piece of welded metal is bent around a U-shaped die and forced into a U shape. The weld and the HAZ must be completely within the bent portion of the specimen after testing. The guided bend test is the most commonly used ductility test for groove welds, surfacing welds, and fillet welds.

Guided bend test fixtures can be bottom guided or bottom ejecting. The bottom guided bend fixture is designed to support the specimen in the die as it is bent. The bottom ejecting guided bend fixture allows the specimen to be ejected from the die after it is bent. Both types of testing are described in ASTM E 190, *Method for Guided Bend Test for Ductility of Welds.* **See Figure 31-8.**

Two specimens must be used for the guided bend test. A face-bend specimen is used to check the quality of fusion, or whether the weld is free of defects such as porosity and inclusions. A root-bend specimen is used to check the degree of weld penetration. **See Figure 31-9.**

To perform a face-bend test, the test specimen is placed in a test jig with the weld face down and forced by a plunger into a U-shaped die. The specimen is substantially bent through 180°, but when it is removed from the die the specimen will spring back slightly and no longer exhibit a perfect 180° bend. The specimen is removed and evaluated. If upon examination cracks greater than

⅛″ appear in any direction, the weld is considered to have failed. The localized overstrain on the convex side of the U-shaped bend reveals the presence of weld defects such as lack of fusion. The convex side of the specimen is inspected for slag inclusions, porosity, and cracks. If these exceed the requirements of the applicable fabrication code or standard, the weld must be rejected.

In a root-bend test, the test specimen is placed in a jig with the root down, or in the reverse position of the face-bend piece. To be an acceptable weld, the specimen must show no cracks.

When preparing guided bend test coupons, grinding and polishing may make it hard to distinguish the location of the weld. An acid etch may be needed to locate the weld area.

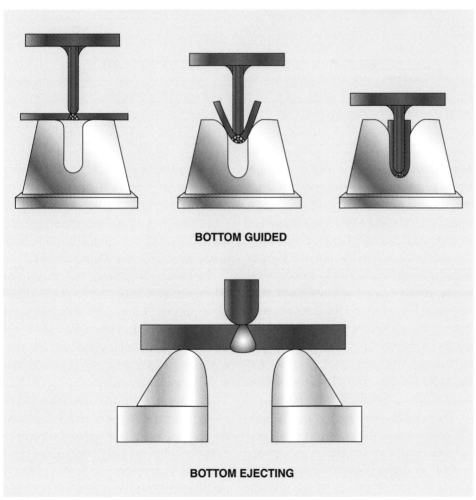

BOTTOM GUIDED

BOTTOM EJECTING

Figure 31-8. *Guided bend test fixtures may be bottom guided or bottom ejecting.*

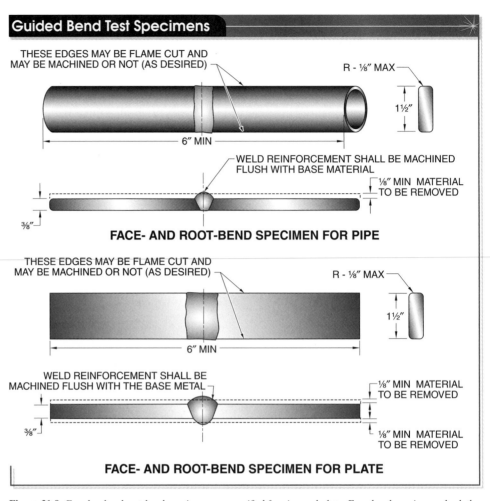

THESE EDGES MAY BE FLAME CUT AND
MAY BE MACHINED OR NOT (AS DESIRED)

R - ⅛″ MAX

1½″

6″ MIN

WELD REINFORCEMENT SHALL BE MACHINED
FLUSH WITH BASE MATERIAL

⅛″ MIN MATERIAL
TO BE REMOVED

⅜″

FACE- AND ROOT-BEND SPECIMEN FOR PIPE

THESE EDGES MAY BE FLAME CUT AND
MAY BE MACHINED OR NOT (AS DESIRED)

R - ⅛″ MAX

1½″

6″ MIN

WELD REINFORCEMENT SHALL BE
MACHINED FLUSH WITH THE BASE METAL

⅛″ MIN MATERIAL
TO BE REMOVED

⅜″

⅛″ MIN MATERIAL
TO BE REMOVED

FACE- AND ROOT-BEND SPECIMEN FOR PLATE

Figure 31-9. *Face-bend and root-bend specimens are specified for pipe and plate. Face-bend specimens check the quality of fusion and root-bend specimens check the degree of weld penetration.*

Wraparound Guided Bend Test. A *wraparound guided bend test* is a bend test in which a specimen is bent around a stationary mandrel a specified amount to expose weld discontinuities. One end of the specimen is fixed to prevent it from sliding during bending and a roller is used to force the specimen to bend around the mandrel. The weld and the HAZ must be completely within the bent portion of the specimen after testing. The test specimen is removed from the bend fixture when the outer roll has moved 180° from the starting point. **See Figure 31-10.**

The bend location may be on the face, the root, or the side. A face-bend test is made with the weld face in tension. A root-bend test is made with the weld root in tension. A side-bend test is made with the weld cross section in tension.

Side-bend tests are useful for exposing discontinuities near the mid-thickness of the weld that might not be seen in face- or root-bend tests. Side-bend specimens are normally used for relatively thick sections (over ⅜″). **See Figure 31-11.**

Transverse face-bend specimens have a longitudinal axis that is perpendicular to the weld and bent with the weld face in tension. Longitudinal face-bend specimens have a longitudinal axis that is parallel to the weld and bent with the weld face in tension.

Transverse root-bend specimens have a longitudinal axis that is perpendicular to the weld and bent with the root surface of the weld in tension. Longitudinal root-bend specimens have a longitudinal axis that is parallel to the weld and bent with the root surface of the weld in tension.

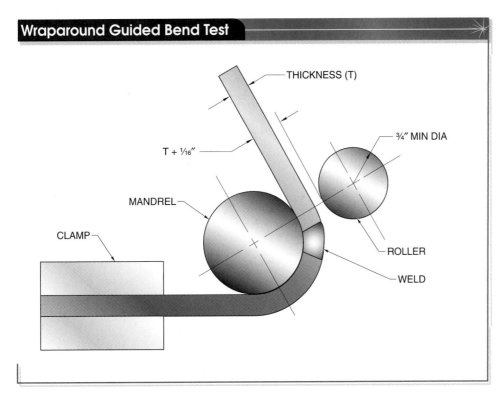

THICKNESS (T)

T + 1/16″

MANDREL

CLAMP

¾″ MIN DIA

ROLLER

WELD

Figure 31-10. *A wraparound guided bend test uses a roller to ensure that the specimen bends to the correct radius around the mandrel.*

Transverse side-bend specimens have a longitudinal axis that is perpendicular to the weld and is bent with the surface that shows the most significant discontinuities in tension. Transverse side-bend specimens are used for plate or pipe that is too thick for a face-bend or root-bend specimen and are recommended for welds with a narrow fusion zone. If the thickness of single- or double-groove joints is more than 1½″, the specimen may be cut into equal strips between ¾″ and 1½″ wide, which are then bent to the required radius determined in the bend test.

The length of a surfacing weld specimen is perpendicular to the weld direction for transverse bend specimens and parallel to the weld direction for longitudinal bend specimens. If the surface is not relatively smooth, fracture might initiate at small regions of local stress concentration and provide misleading information. A minimal amount of surfacing weld is removed from the face-bend specimen surface to obtain a smooth surface. The minimum thickness of surfacing after finishing is ⅛″.

Bend Locations

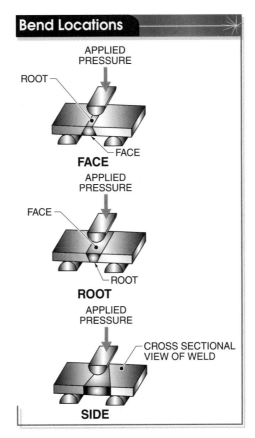

APPLIED PRESSURE

ROOT

FACE

FACE

APPLIED PRESSURE

FACE

ROOT

ROOT

APPLIED PRESSURE

CROSS SECTIONAL VIEW OF WELD

SIDE

Figure 31-11. *The bend location in a guided bend test may be on the face, the root, or the side of the specimen.*

The specimen thickness and bend radius are chosen according to the ductility of the metal being tested. Most qualification tests of low-carbon steel require that the specimen be bent around a mandrel having a diameter four times the thickness of the specimen.

Calculation of strain in the guided bend test or wraparound bend test on the outside surface of a bend specimen is given, approximately, by the following formula:

$$e = \frac{100t}{2r+t}$$

where

e = strain, in percent

t = bend test specimen thickness, in in.

r = inside bend radius, in in.

For example, what is the precentage strain when a ⅜″ (0.375) specimen is bent around a 1.5″ diameter mandrel using a guided bend test?

$$e = \frac{100(0.375)}{2(1.5) \ + \ 0.375}$$

$$e = \frac{37.5}{2(1.5) \ + \ 0.375}$$

$$e = \frac{37.5}{3 \ + \ 0.375}$$

$$e = \frac{37.5}{3.375}$$

$$e = 11\%$$

Low-carbon steel welds can easily achieve the 11% strain value. However, if weld defects are present, the bend test specimens will consistently fail.

When the deposited weld metal is stronger than the base metal, bending will begin in the base metal, resulting in more bending there and little, if any, bending in the weld metal. In this situation the severest test region is the fusion zone between the weld and the base metal.

When the deposited weld metal is weaker than the base metal, bending begins in the weld, resulting in more bending in the weld than in the base metal. A more severe test of the weld results, and failure may occur because the weld metal ductility is exceeded and not because of a defect in the weld. See Figure 31-12.

When testing welds in dissimilar strength metals, such as medium-carbon steel or low-carbon steel, the unequal strength capabilities of the metals may cause the specimen to slide sideways in the guided bend test fixture.

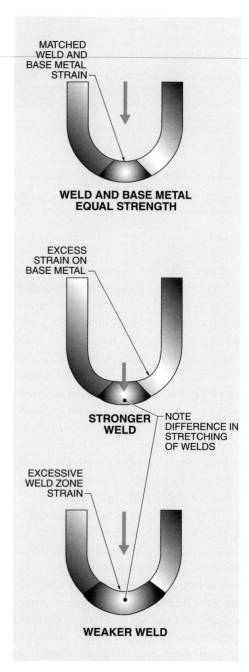

MATCHED WELD AND BASE METAL STRAIN

WELD AND BASE METAL EQUAL STRENGTH

EXCESS STRAIN ON BASE METAL

STRONGER WELD

NOTE DIFFERENCE IN STRETCHING OF WELDS

EXCESSIVE WELD ZONE STRAIN

WEAKER WELD

Figure 31-12. *Problems associated with the guided bend test relate to the relative strengths of the weld and base metal, and the applied loads during testing.*

Hardness Tests

The hardness of a material is its resistance to deformation (particularly permanent deformation), indentation, or scratching. A *hardness test* is a destructive test used to determine the relative hardness of the weld area as compared with the base metal. **See Figure 31-13.**

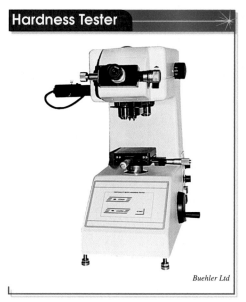

Hardness Tester

Buehler Ltd

Figure 31-13. *A hardness test is performed using a hardness tester, such as a microhardness tester, to determine the relative hardness of the weld area as compared with the base metal.*

Hardness testing is inexpensive and well established, so that experience from years of testing in many different situations may be applied to predict service behavior. Hardness tests are a widely used quality control tool in metal processing operations such as heat treatment because they are sensitive, rapid, and relatively nondestructive. Hardness tests are less commonly used for welds because the critical area for hardness testing, the HAZ, requires special preparation. Additionally, hardness testing does not provide adequate information on the physical quality of the weld compared with other tests such as the guided bend test. Hardness is indicated by values obtained from various hardness testing machines.

Hardness testing can provide information on metallurgical changes caused by welding. In alloy steels or medium-carbon steels, a high hardness value in the HAZ might indicate insufficient preheating or postheating. Welding may significantly reduce the HAZ hardness in cold-worked or age-hardened alloys by annealing or over-aging, respectively, which reduces the overall strength of the joint. Indentation hardness testing is most often applied to weld testing and uses the surface impression produced by a standardized-shape indenter and standardized load to determine hardness. The depth or size of the impression is measured to obtain the hardness value for the test specimen.

Indentation hardness tests for welds consist of Brinell, Rockwell, and Vickers tests, which provide information on the bulk properties of the metal, and microhardness tests, which provide information on the weld and the HAZ in the metal. Converting hardness numbers between different tests must be done carefully.

Brinell Hardness Test. The *Brinell hardness test* is an indentation hardness test that uses a machine to press a 10 mm diameter, hardened steel ball into the surface of a test specimen. The Brinell hardness test is used to determine base metal hardness. The load must remain on the specimen 15 sec for ferrous materials and 30 sec for nonferrous materials. Sufficient time is required for adequate flow of the material being tested; otherwise the readings will be in error. **See Figure 31-14.**

Hardness is calculated by dividing the load by the area of the curved surface of the indentation. The Brinell hardness number is found by measuring the diameter of the indentation and then finding the corresponding hardness number on a calibrated chart. The test is described in ASTM E 10, *Brinell Hardness Testing of Metallic Materials.* Portable Brinell testers are sometimes used in the field where a part cannot be moved to the laboratory for examination. These units are less accurate.

✓ **Point**

Hardness testing, although considered destructive, does not necessarily require that the specimen be cut into pieces, and is thus convenient and relatively rapid.

Brinell Hardness Test

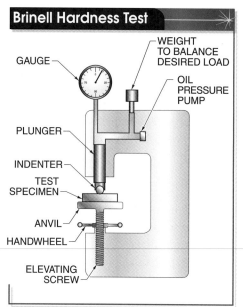

Figure 31-14. *A Brinell hardness test applies a load for a specific period and causes an indentation in the metal that is used to calculate hardness.*

The Brinell ball makes the deepest and widest indentation of any hardness test, so that it indicates an average hardness value over many grains of the metal. Consequently, the Brinell hardness test is the least affected by surface irregularity or inhomogeneity. Sometimes it is necessary to grind a flat spot on the surface to improve the diametrical measurement. The Brinell test is not suitable for very thin, case-hardened, or hard-faced components.

Brinell Test Indentations

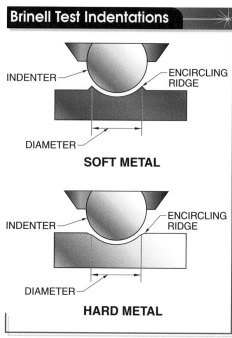

Figure 31-15. *Careful measurement must be taken after a Brinell hardness test to accurately determine the size of indentations in soft and hard metals.*

The Brinell hardness number followed by the abbreviation HB indicates a hardness value made under standard conditions using a 10 mm diameter hardened steel ball, a 3000 kg load, and an indentation time of 15 sec to 30 sec. However, the load applied to the steel ball depends on the type of metal under test. A 500 kg steel ball is used for aluminum castings and a 3000 kg steel ball is used for ferrous metals. The diameter of the indentation is measured to 0.05 mm using a low-magnification portable microscope. Care must be taken to measure the exact diameter of the indentation and not the apparent diameter caused by edge effects that result in a ridge or depression encircling the true indentation. **See Figure 31-15.**

A code is used when other test conditions are required. For example, 75 HB 10/500/30 indicates a Brinell hardness number of 75 obtained in a test using a 10 mm diameter hardened steel ball with a 500 kg load applied for 30 seconds. For extremely hard metals, a tungsten carbide ball is substituted for the steel ball, allowing readings as high as 650 HB.

Rockwell Hardness Test. The *Rockwell hardness test* is an indentation hardness test that uses two loads, supplied sequentially, to form an indentation on a metal test specimen to determine hardness. The Rockwell hardness test is the most commonly used and versatile hardness test. The Rockwell hardness test is commonly used for weld and base metal measurement. The Rockwell testing machine has a variety of attachments that enable it to measure the hardness of a wide range of materials.

A ¹⁄₁₆″ diameter steel ball and a 120-diamond cone are the two types of indenters. A minor load of 10 kg is applied that helps seat the indenter and removes the effect of surface irregularities. A major load, which varies from 60 kg to 150 kg, is then applied. **See Figure 31-16.**

The amount of the major load determines the type of indenter used. For example, a steel ball is used with the 60 kg load and a diamond cone with the 150 kg load. The difference in depth of indentation between the major and minor loads provides the Rockwell hardness number. This number is taken directly from the dial on the machine. The Rockwell hardness test is described in ASTM E 18, *Rockwell Hardness and Rockwell Superficial Hardness of Metallic Materials.*

Several types of Rockwell hardness scales are used for measuring hardness. The designation system has a hardness number followed by HR, followed by another letter that indicates the specific Rockwell scale. The two most common scales are Rockwell B (HRB) and Rockwell C (HRC).

The Rockwell B scale uses a ¹⁄₁₆″ diameter steel ball and a 100 kg load for relatively soft materials. It is used on annealed low-carbon steel, which may exhibit a hardness of approximately 85 HRB.

The Rockwell C scale uses a diamond cone and a 150 kg load for relatively hard materials. For example, a quenched and tempered medium-carbon, low-alloy steel usually exhibits hardness between 30 HRC and 45 HRC, depending on the tempering temperature. **See Appendix.**

For Rockwell hardness testing, both sides of the test specimen must be clean, scale-free, dry, and parallel. Special jigs help support round or oversize test specimens to ensure immobility during the test.

Vickers Hardness Test. The *Vickers hardness test* is an indentation hardness test that uses an indenter with a 136° square-base diamond cone, and that may be used to test hardness in the base metal and weld metal. The applied load varies from 1 kg to 120 kg. The Vickers hardness number is determined from the load divided by the surface area of the indentation.

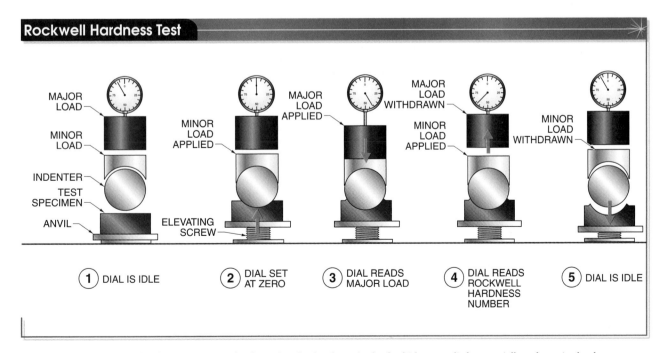

Rockwell Hardness Test

① DIAL IS IDLE ② DIAL SET AT ZERO ③ DIAL READS MAJOR LOAD ④ DIAL READS ROCKWELL HARDNESS NUMBER ⑤ DIAL IS IDLE

Figure 31-16. *The Rockwell hardness test uses two loads, a minor load and a major load, which are applied sequentially to determine hardness.*

To conduct a Vickers hardness test, the specimen is placed on an anvil and raised by a screw until it is close to the point of the indenter. The starting lever is tripped, allowing the load to be slowly applied to the indenter. The load is released, the anvil lowered, and a filar microscope is used to measure the diagonals of the square indentation to ±.001 mm. Diagonal measurements are averaged to obtain the Vickers hardness number, which is followed by the letters HV. The Vickers hardness test is described in ASTM E 92, *Vickers Hardness Testing of Metallic Materials*.

The Vickers hardness test allows extremely accurate readings to be taken. Additionally, one type of indenter covers all types of metals and surface treatments. However, test specimen preparation is critical because a poor surface finish makes the measurement of the diagonals extremely difficult. A fine emery finish is the coarsest face allowable.

Microhardness Test. A *microhardness test* is a type of indentation hardness test that uses light loads of less than 200 g. Microhardness tests are at the opposite end of the scale to the Brinell or Rockwell hardness tests. A polished surface, coupled with the light loads, allows the hardness of individual grains of metal or other microconstituents to be measured.

To conduct a microhardness test, the test specimen is placed under the microscope of the microhardness tester. The area of interest is positioned at the intersection of the cross wires. The indenter is swung into place and the load applied for a set period. The load is then removed, the microscope swung back, and the length of the diagonals of the indentation measured. The microhardness reading is obtained from the measurements and from a chart. Microhardness testing is described in ASTM E 384, *Test Method for Microhardness of Metals*.

Microhardness testing of welds is usually done on ground and polished, or ground, polished, and etched cross sections of a weld. Measurements can be made in any specific area, but they are most frequently made as a series of regularly spaced indentations across the base metal, HAZ, and weld metal for single- or multiple-pass welds. The space between readings is usually between 1/16″ and 1/8″. **See Figure 31-17.**

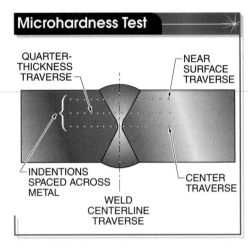

Figure 31-17. *Microhardness measurements are taken at regular intervals across a ground, polished, and etched cross section of a weld.*

Regular conversion between different hardness scales should be avoided unless there is a large amount of experience and data available to justify making such correlations. Indentation hardness readings are based on a combination of properties such as friction, elasticity, and viscosity of the indenter and the specimen. These vary with the type of specimen and test. The distribution of plastic strain in the test specimen, which is caused by the particular type of indenter, is also an important factor.

Separate conversion tables are required for different alloy families. ASTM E 140, *Standard Hardness Conversion Tables for Metals (Relationship between Brinell Hardness, Vickers Hardness, Rockwell Hardness, Rockwell Superficial Hardness, and Knoop Hardness),* contains hardness conversion tables for several major families of alloys. Pocket-size conversion charts supplied by vendors are usually an extract from the steels portion of ASTM E 140. **See Appendix.**

Toughness Tests

Toughness tests measure the ability of materials to absorb energy at high strain rates and deform plastically rather than fracture in a brittle manner, particularly in the presence of stress raisers such as cracks and notches. A *toughness test* is a dynamic test in which a specimen is broken by a single blow and the energy absorbed in breaking the piece is measured in foot-pounds (ft-lb). The purpose of the test is to compare the toughness of the weld metal with the base metal. It is especially significant in determining whether any of the mechanical properties of the base metal have been destroyed due to welding. Toughness of welds is an important property because structural metals must be able to deform and give warning of impending failure.

The mechanical properties of a metal are strongly affected by the rate of straining. A metal tested at a low strain rate may break with a large amount of strain (elongation), but a metal tested at a high strain rate may break with little or no elongation. The metal is tough and ductile at the low strain rate and is brittle at the high strain rate. **See Figure 31-18.**

Toughness is also affected by the test temperature and presence of stress raisers in the specimen. The toughness of certain metals decreases significantly below a characteristic temperature. Stress raisers in welds, such as a sharp change in weld profile at the surface or internal inclusions, may decrease toughness.

Toughness tests include the Charpy V-notch test, plane-strain fracture toughness test, and nil ductility transition temperature test.

Charpy V-Notch Test. The *Charpy V-notch test* is a toughness test that uses the energy produced by a dynamic load, and measures the energy needed to break a small machine-notched test specimen. The Charpy specimen is a square-shaped bar containing a machined V-shaped notch. The purpose of the notch in the test specimen is to facilitate fracture in a controlled location. The resulting measurement is an indicator of toughness.

A Charpy V-notch test is performed in a universal pendulum impact tester. **See Figure 31-19.** The specimen is placed horizontally against the two supports at the bottom of the tester. The pendulum is raised to a standard height, giving it a potential energy of 240 ft-lb [325 joules (J)]. The pendulum is released and the specimen is struck and broken as the pendulum swings through its arc. The swing of the pendulum after it strikes the specimen indicates the energy absorbed on impact and is measured in foot-pounds or joules. When struck by the pendulum, tough materials absorb a significant amount of energy and brittle materials fracture with relatively little energy absorbed. Tough materials cause the pendulum to travel a shorter distance after striking the test specimen. With brittle materials, the pendulum travels a longer distance after impact.

✓ **Point**

Toughness testing requirements depend on the specific applicable fabrication code or standard.

✓ **Point**

The Charpy V-notch test uses the energy produced by a dynamic load, and measures the energy needed to break a small machine-notched test specimen.

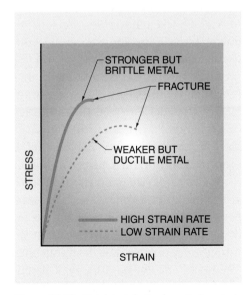

Figure 31-18. *Metal tested at a low strain rate is ductile compared with the same metal tested at a high strain rate.*

ASTM 23, Notched Bar Impact Testing of Metallic Materials, gives requirements for the Charpy V-notch test, which is the most common impact test used.

Charpy V-Notch Test

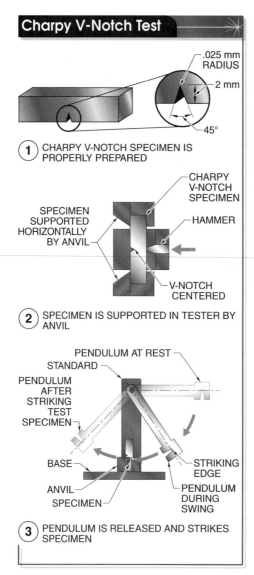

① CHARPY V-NOTCH SPECIMEN IS PROPERLY PREPARED

.025 mm RADIUS

2 mm

45°

SPECIMEN SUPPORTED HORIZONTALLY BY ANVIL

CHARPY V-NOTCH SPECIMEN

HAMMER

V-NOTCH CENTERED

② SPECIMEN IS SUPPORTED IN TESTER BY ANVIL

PENDULUM AT REST STANDARD

PENDULUM AFTER STRIKING TEST SPECIMEN

BASE

ANVIL

SPECIMEN

STRIKING EDGE

PENDULUM DURING SWING

③ PENDULUM IS RELEASED AND STRIKES SPECIMEN

Figure 31-19. *The Charpy V-notch test requires very small specimens, allowing for multiple orientations of a test to be performed on a part. The swing of the pendulum after striking the specimen indicates the energy absorbed on impact.*

The Charpy V-notch test is widely used because it requires a small specimen size and it permits correlating the results of a large body of tests with service experience. The simple method of specimen support allows testing to be performed over a range of test temperatures. Specimens are heated in a furnace or cooled in a refrigerator to the test temperature, removed, and then rapidly tested with minimal temperature change.

The small specimen size required for the Charpy V-notch test is also convenient because specimens may be cut at various orientations or locations within a part. Since the properties of metals may vary according to orientation or location, in quality control programs it is often necessary to check for properties in orientations that would exhibit the lowest toughness. For example, with plate products, a test specimen with a transverse orientation usually exhibits lower quality, or lower mechanical properties. With welds, specimens that have notch locations in the weld metal, HAZ, or base metal may exhibit significantly different notch toughness values. **See Figure 31-20.**

Charpy V-Notch Test Specimens

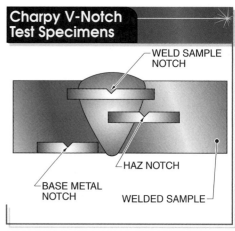

WELD SAMPLE NOTCH

HAZ NOTCH

BASE METAL NOTCH

WELDED SAMPLE

Figure 31-20. *Charpy V-notch test specimens machined from different locations in a weld may exhibit different notch toughness values.*

The behavior of metals in Charpy V-notch testing is dependent on the rate of loading, test temperature, and type of notch. These variables make it difficult to translate the absorbed energy values into design criteria. Nevertheless, the long history of Charpy V-notch testing allows acceptance or rejection limits to be placed on large quantities of materials. For example, some specifications require a minimum Charpy V-notch requirement for steel products of 15 ft-lb at the minimum expected service temperature. However, this does not mean that a test specimen exhibiting 60 ft-lb is four times tougher than the minimum. The main value of the Charpy V-notch test is as a criterion of acceptance of material when reliable service behavior has been established.

Notch toughness testing requirements depend on the specific fabrication code or standard. When applicable in the ASME Boiler and Pressure Vessel Code, they are known as supplementary essential variables. For example, notch toughness testing may be required for carbon and low-alloy steel equipment subject to cooling in service, such as through operation upsets or auto-refrigeration. *Auto-refrigeration* is cooling that occurs when gas expands, as in the sudden release of gas from a pipe or piece of equipment. Materials whose properties are enhanced by heat treatment may also require notch toughness testing.

Notch toughness values may be altered with an increase in heat input during welding. Conditions that may contribute to higher heat input include higher welding heat input; higher maximum interpass temperature; longer postheat time at temperature; reduction in base metal thickness; change to an uphill progression in vertical welding; change from stringer bead to weave bead welding; and the physical location of specimens taken from some pipe test samples. A welding procedure specification must be established that accounts for these variables to ensure that the notch toughness properties of the weld metal and HAZ are not reduced.

Plane-Strain Fracture Toughness Test. The *plane-strain fracture toughness test* is a toughness test that measures the resistance of metals to brittle fracture propagation in the presence of stress raisers such as weld defects. High stress concentrations may occur at the tips of internal discontinuities (such as lack of fusion) in some metals and produce a running (brittle) crack.

The fracture toughness of a metal at a given temperature is proportional to the stress level, measured in thousand pounds per square inch (ksi) or megapascals (MPa), and the square root of the crack length, measured in inches or meters. The unit of measure for fracture toughness is ksi√in. (ksi root inch).

Plane-strain fracture toughness test data are used in the design of structures, as when determining the allowable internal size of a discontinuity that might lead to a catastrophic failure.

Various types and sizes of specimens are used in the plane-strain fracture toughness test. A compact tension specimen is a block containing a machined notch that is placed in a fatigue-testing machine to produce a small fatigue crack at the root of the machined notch. The tip of the fatigue crack extending from the root of the machined notch is a localized region of high stress intensity.

The test specimen is pulled to failure in a testing machine and the load is plotted against the opening of the notch. The load and crack extension at the sudden failure of the test specimen are measured and used to calculate the fracture toughness of the material. The test method is described in ASTM E 399, *Plane-Strain Fracture Toughness Testing of Metallic Materials.* Fracture toughness testing is used to determine the critical stress intensity, which is a measure of the resistance of a metal to brittle fracture propagation in the presence of flaws and cracks. Pressure vessels, storage tanks, airplanes, and ships are examples of structures that are designed and manufactured in accordance with fracture toughness principles.

Nil Ductility Transition Temperature Test. The *nil ductility transition (NDT) temperature test* is a toughness test that measures the temperature at which the fracture behavior of a metal changes from ductile to brittle in the presence of a stress raiser. This temperature is sometimes referred to as the ductile-to-brittle transition temperature (DBTT). Some metals, especially carbon and low-alloy steels, show a sharp transition in toughness when temperature decreases.

> *Catastrophic failures of pressure vessels, ships, and other structures have occurred when subject to stresses below their DBTT. In some cases, welds have caused initiation of the failure by providing a stress concentration effect.*

The change in toughness capability may be the controlling factor in determining a metal's serviceability. Carbon steels lose ductility below a certain temperature, leading to brittleness. Large steel storage tanks have failed catastrophically in cold weather because the NDT temperature of the plate material was higher than the atmospheric temperature at the time of failure.

The Charpy V-notch test and the drop weight test are used to determine the NDT temperature. The Charpy V-notch test determines the NDT temperature by testing specimens over a range of temperatures. The results are plotted as impact strength against test temperature.

The drop weight test is a more reliable method than the Charpy V-notch test for measuring NDT. The specimen is a slab or plate that is up to ⅝″ thick. A weld bead made from a brittle alloy is laid down the center of the plate. The plate is brought to the test temperature and placed in the test fixture. It is supported along both ends parallel to the weld, with the weld side facing down. A weight located vertically above the center of the plate is allowed to drop, causing the plate to bend.

Cracking of the weld bead is initiated at 3° of bend. After that point the weld bead continues to crack, which initiates a fracture in the plate. To ensure the strain induced in the plate is elastic, a stop is placed below the weld bead. The stop limits the amount of deflection of the plate to 5° of bend. **See Figure 31-21.**

If the temperature of the plate is below the NDT temperature, the crack runs and the plate breaks in two. At any temperature above the NDT temperature, the crack stops before it spreads through the plate. The NDT temperature is the lowest temperature at which the plate will not break in two. The drop weight test is described in ASTM E 208, *Drop Weight Test to Determine Nil Ductility Transition Temperature of Ferritic Steels.*

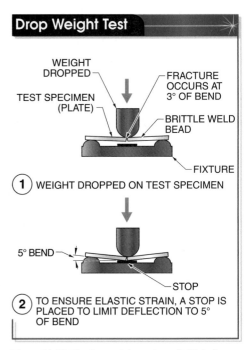

Figure 31-21. *The drop weight test is a reliable indicator for measuring NDT temperature.*

Break Tests

Break tests are a rapid and convenient method of evaluating certain types of welds and are used for welder qualification. Break tests include the nick-break test and the fillet weld break test.

Tinius Olsen Testing Machine Co., Inc.
Electromechanical tensile testing machines can be used to perform tension, compression, and flex tests on a specimen.

Nick-Break Test. The nick-break test is conducted by saw cutting a small notch in a weld assembly or specimen, followed by breaking it with hammer blows, stretching, or bending. A test specimen is prepared and placed on supporting members. A load is applied to the specimen until it breaks. The surface of the fracture is then examined for defects such as porosity, slag inclusions, overlaps, etc. **See Figure 31-22.**

For a more accurate check of the weld, the fractured pieces should be subjected to an etch test. The nick-break test is used primarily in the pipeline industry, as described in American Petroleum Institute (API) standards such as API Standard 1104, *Standard for Welding Pipeline and Related Facilities.* The nick-break test may also be used to evaluate fusion welds, flash butt welds, pressure welds, or friction welds in pipe or plate.

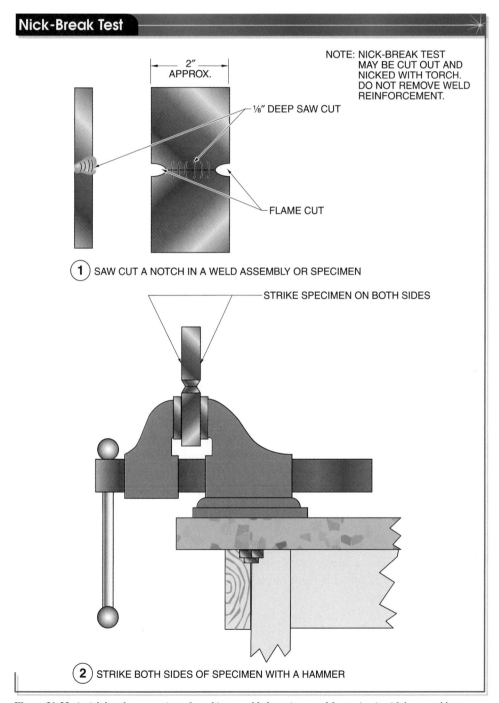

Nick-Break Test

2″ APPROX.

NOTE: NICK-BREAK TEST MAY BE CUT OUT AND NICKED WITH TORCH. DO NOT REMOVE WELD REINFORCEMENT.

⅛″ DEEP SAW CUT

FLAME CUT

① SAW CUT A NOTCH IN A WELD ASSEMBLY OR SPECIMEN

STRIKE SPECIMEN ON BOTH SIDES

② STRIKE BOTH SIDES OF SPECIMEN WITH A HAMMER

Figure 31-22. *A nick-break test consists of notching a welded specimen and fracturing it with hammer blows.*

Specimens for nick-break testing are either a full-size welded piece or a specimen cut from a full-size piece. The weld region is notched and then firmly supported at one or both ends. Once the specimen is supported, it is fractured by a hammer blow. One side is hit twice and then the specimen is turned 180° and the other side is hit twice. This procedure is continued until the specimen breaks. Alternatively, the specimen may be fractured by loading in tension or by three-point loading on a universal testing machine. The method of breaking the specimen is not significant because the sole purpose of the nick-break test is to cause failure through the weld zone to determine the presence of discontinuities or defects.

Slag inclusions on steel may have a glass-like appearance or a dark contoured appearance. The nick-break fracture will travel from the cut metal to the slag inclusion and through the center of the inclusion. The location of a slag inclusion is sometimes smooth because the slag has been dislodged by the force of the hammer blows breaking the specimen. It is useful to match the two broken specimens together and rotate in good light to identify discontinuities. Sometimes discontinuities are easier to read on the fracture surface than the other matching side.

Porosity may be spherical or cylindrical in shape and may be isolated or grouped in clusters. The key to the identification of porosity is the shape and the absence of nonmetallic solid material. Porosity has a bright white or silvery appearance on steel if it is not exposed to the atmosphere. Surface-connected porosity usually has a black oxide appearance. The sound metal surface has a gray color without voids.

The observation of incomplete fusion depends on the joint design. If the joint is a single-V groove, base metal incomplete fusion would be planar in shape, showing the area on the groove face that is not fused. In some cases the grind marks on the original groove face can be identified.

It is helpful to place the two broken nick-break specimens together and identify the location of the first weld pass and the last, as well as the weld reinforcement area. If the discontinuity is located on the groove face or between weld passes and is planar in shape, it could be incomplete fusion.

Incomplete joint penetration is easy to identify in that it is always located at the weld root and is planar in shape. Incomplete joint penetration can be detected in the nick-break specimen before it is broken.

The broken nick-break specimen shows how deep the incomplete joint penetration extends into the weld metal. On steel, incomplete joint penetration is black to bluish in color.

In steel, cracks are flat and have a silvery color if they occur after welding is completed. If the fractured surface of a crack shows a blue oxide color, the metal cracked before the final weld passes were completed and the crack surface was heated to the temper color range by subsequent weld passes.

Fillet Weld Break Test. A *fillet weld break test* is a break test in which the specimen is tested with the weld root in tension. The fillet weld break test is used for the qualification of welders and is the only test required to qualify as a tack welder in accordance with AWS D1.1, *Structural Welding Code—Steel.*

Tack welding is a vital part of many fabrications such as fabrication of pressure vessels or structures. Except for fully automatic welding operations, most construction codes or standards have qualification rules for tack welders. A high-heat-input, mechanized process may be selected for the welding application, but the tack weld may be applied manually leading to very rapid cooling and a brittle, crack-sensitive structure, commonly at the root of the weld. Subsequent weld passes with the high-heat-input process do not remove the cracks, but help them propagate further into the base metal and/or weld metal. Poorly applied tack welds are also the cause of entrapped slag, porosity, lack of penetration, and cracks.

If the WPS is qualified with preheating and postheating, the tack weld should be similarly qualified within the range specified. Most construction codes require tack welds of any length to follow a qualified WPS for the following reasons:

- tack weld is to be removed or left in place
- tack weld is attaching a component to the piece to be welded
- tack weld is incorporated into the weld as a tack in the root

To perform the fillet weld break test, a welder places a fillet weld on one side of a T-joint specimen. The specimen is placed in a press and bent to produce fracture at or near the weld. The fracture surface is examined for evidence of fusion with the root and absence of incomplete fusion or porosity larger than $\frac{3}{32}''$ in its greatest dimension. **See Figure 31-23.**

Fillet Weld Break Test

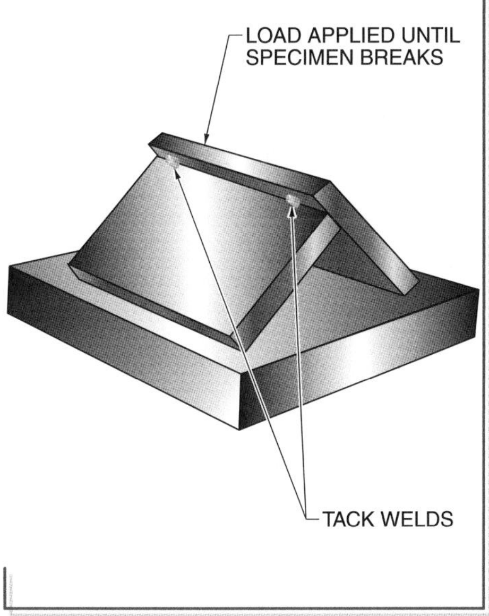

LOAD APPLIED UNTIL SPECIMEN BREAKS

TACK WELDS

Figure 31-23. *The fillet weld break test is used by an inspector to qualify tack welds.*

SPECIMEN PREPARATION

Fabrication codes and standards indicate how to obtain specimens from welds for mechanical testing. Good specimen preparation ensures that undesirable surface features are not introduced that have an undesirable effect on the test results.

Mechanical test specimen preparation is described in AWS B4.0, *Standard Methods for Mechanical Testing of Welds*. Specimen preparation may vary according to the type of weld. Safety practices must be followed when preparing specimens to prevent injury from grinding wheels, hot surfaces, or sharp edges.

Groove Welds

When using groove weld specimens, specific information must be recorded to document the results of the testing. When a double-groove weld specimen is used, identification stamps must be used to mark the side of the joint from which the test specimen was taken. Samples may be removed from specific locations in groove weld test plates and pipes to ensure representative specimens are obtained. **See Figure 31-24.** Groove weld specimens include tension specimens, root- and face-bend specimens, hardness specimens, fracture toughness specimens, and nick-break specimens.

Tension Specimens. Tension specimens for groove welds may be rectangular or round. Deep machine cuts or surface tears must be avoided during surface preparation as they can cause invalid test results. Imperfections that are present in the gauge length that are incidental to welding do not need to be removed.

Rectangular specimens may be taken from plate or from tubing greater than 2″ diameter and with wall thickness greater than $\frac{3}{8}''$. The weld orientation may be longitudinal or transverse. For tubing less than 2″ diameter, only a full-section specimen may be tested.

✓ **Point**

Specimen preparation must provide a smooth surface for testing. Nicks or sharp edges are undesirable because they introduce local stress raisers that might cause premature failure.

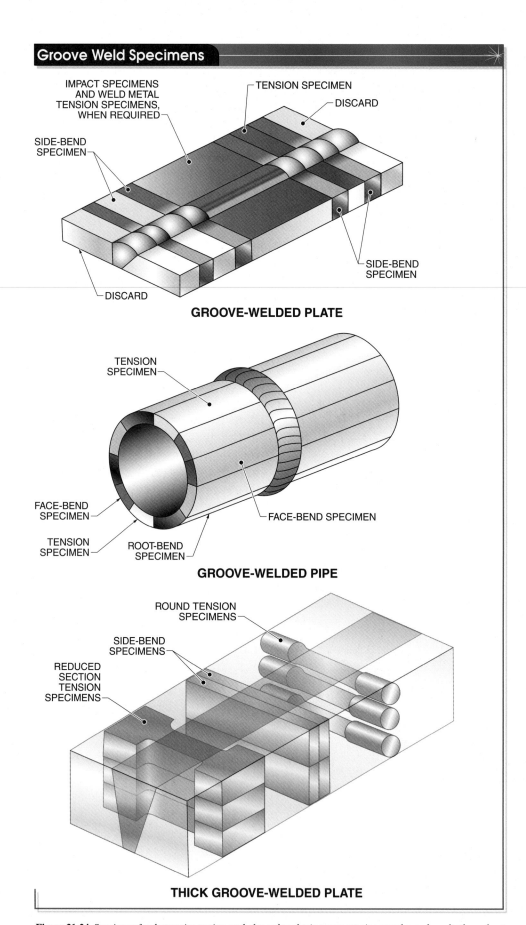

IMPACT SPECIMENS
AND WELD METAL
TENSION SPECIMENS,
WHEN REQUIRED

TENSION SPECIMEN

DISCARD

SIDE-BEND
SPECIMEN

SIDE-BEND
SPECIMEN

DISCARD

GROOVE-WELDED PLATE

TENSION
SPECIMEN

FACE-BEND
SPECIMEN

TENSION
SPECIMEN

ROOT-BEND
SPECIMEN

FACE-BEND SPECIMEN

GROOVE-WELDED PIPE

ROUND TENSION
SPECIMENS

SIDE-BEND
SPECIMENS

REDUCED
SECTION
TENSION
SPECIMENS

THICK GROOVE-WELDED PLATE

Figure 31-24. *Specimens for destructive testing are balanced to obtain representative samples and results throughout groove-welded plate, groove-welded pipe, and thick groove-welded plate.*

When the thickness of the test weld is greater than the capacity of the test equipment, the weld may be divided through its thickness into as many specimens as necessary to cover the full weld thickness and still maintain the specimen size within the equipment capacity. Usually, the results of partial-thickness specimens are averaged to determine the properties of the full-thickness joint. For specimens taken transversely to the centerline of the weld, only the ultimate tensile strength is determined because of possible material or structural inhomogeneities.

Round, all-weld metal specimens with the largest possible diameter that can be machined from a location are used. Specimens smaller than ¼″ diameter should not be used unless there is no other way to obtain the sample. Minor variations in the surface finish and test machine alignment may lead to irreproducible results due to the small size of the sample.

Bend Specimens. Bend specimens are used for welding procedure qualification and welder performance qualification tests. Similar preparation requirements are usually specific to groove weld and surfacing weld bend specimens. Bend specimens are prepared by cutting the weld metal and the base metal to form a rectangular cross-section specimen. At least ⅛″ of material must be mechanically removed from thermally cut surfaces to prevent the influence of heat on the test results. Longitudinal surfaces may be no rougher than 125 microinches (3 microns). Grinding or sanding marks should run parallel to the direction of bending to prevent them from acting as stress raisers that can lead to premature failure. Additionally, the corners of the specimen should be radiused to relieve excessive stresses.

Weld reinforcement and backing must be removed to be flush with the specimen surface. For welder performance qualification testing, undercuts may be removed, provided sufficient material remains to maintain the required specimen dimensions. When testing weld joints between base metals that have differing thicknesses, the specimen is reduced to a constant thickness using the thinner base metal.

The surfaces perpendicular to the weld axis are designated as the sides of the specimen. The other two surfaces are designated as the face or root surfaces. Transverse weld specimens may have the side, face, or root of the weld as the bend surface. Longitudinal weld specimens may have the face or the root of the weld as the bend surface.

The acceptability of a bend specimen is based on the size and/or number of defects that appear on the bend surface. The main purpose of the bend test for welding procedure qualification is to determine the ductility of a sound weld. Governing fabrication codes or specifications dictate exact acceptance or rejection criteria.

A discontinuity does not become a defect until it exceeds the limits allowed by the relevant code. A Project Engineer can ignore discontinuities that are less than the maximum, but all discontinuities must be recorded. AWS D1.1, *Structural Welding Code—Steel,* allows a total accumulation of discontinuities of ³⁄₃₂″. With the ASME Boiler and Pressure Vessel Code, bend specimens may have no open defects in the weld or the HAZ exceeding ⅛″ measured in any direction on the convex face after bending.

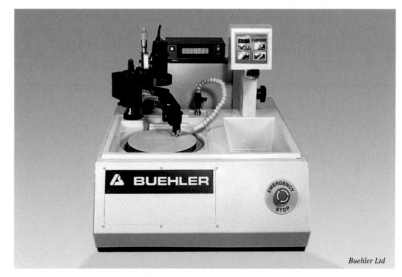

Buehler Ltd

A cross-sectioning system is used to prepare cross-sectioned specimens for welder performance qualification tests.

Etching may be required to determine whether the discontinuity is in the weld or the HAZ. Open discontinuities on the corners of specimens during testing are not considered unless there is evidence that they result from lack of fusion, slag inclusions, or other internal weld discontinuities.

Hardness Specimens. Hardness specimens for groove welds and surfacing welds are ground, machined, or polished depending on the type of hardness test to be performed. Surface preparation requirements become increasingly stringent as the size and depth of the indentation decreases. At the very minimum, it is necessary to remove rust and scale from the surface. Excessive heat must be avoided when preparing the test area of the specimen.

Weld metal hardness tests are only permitted either on weld joint cross section samples or on local areas of weld reinforcement that are ground smooth before testing. The edge of the indentation must be no closer than three times the major dimension of the indentation from the edge of the ground area of the reinforcement on welded assemblies. Specimens must be supported to prevent rocking under the tester.

It may be necessary to grind the backside of the specimen to make it flat. The indenter should be perpendicular to the specimen. With a round specimen such as bar, it is usually necessary to grind a small area flat to make a test. The specimen must be thick enough so that an anvil effect (bulge) does not appear on the opposite side when the indentation is made. For the Rockwell and Brinell hardness tests, the specimen should be at least 10 times as thick as the depth of the impression. For the Vickers hardness test, the test specimen should be one and one-half times as thick.

For evaluation of weld metal hardness, the edge of the indentation must be within the weld metal and no closer than $\frac{1}{8}''$ from the weld metal interface with the base metal. The minimum spacing between indentations depends on the type of test. If the indentations are too close together, there will be disturbed zones of metal. The minimum separation between indentations should be four diameters (4 d center to center) for the Brinell and Rockwell hardness tests and two and one-half diagonals (2½ D center to center) for the Vickers hardness test. **See Figure 31-25.**

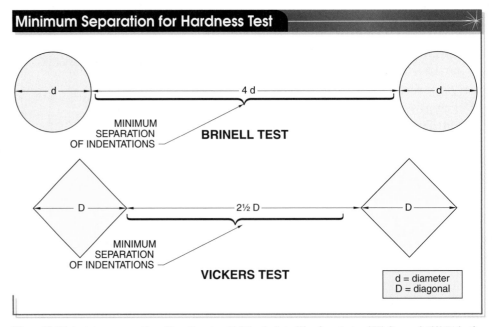

Figure 31-25. *A minimum separation of four diameters (4 d) for the Brinell hardness test and 2½ diagonals (2½ D) for the Vickers hardness test must be maintained to prevent disturbances between the base metal and weld metal zones.*

More than one reading must be taken to allow for surface irregularities and test specimen inhomogeneity. The minimum number of readings required for a specific test is determined by experience. For the Brinell test, three readings are usually taken and averaged. For the Rockwell and Vickers tests, three to five readings are usually taken and averaged.

Fracture Toughness Specimens. Fracture toughness tests may be performed to indicate the performance of the base metal, the HAZ, or the weld metal. A fracture toughness test uses a specimen that has a notch cut into it. The specimen is then tested to determine the fracture strength of the metal.

When the test is performed on the base metal or in the HAZ, the location of the notch is specified to be in the applicable region of the joint. When the test is performed on the weld metal, the width of the weld metal must be equal to or greater than the thickness of the specimen.

When specimens from double-groove welds are used, identification letters or numbers are stamped on the specimen to indicate the side of the joint from which the test specimen was taken. The location of identification stamps must not influence the failure of the specimen by creating a notch effect. Fracture toughness specimens for groove welds are made for Charpy V-notch tests, plane-strain fracture toughness tests, and drop weight tests.

The geometry and surface area of the notch are critical. Machining and finishing operations on the notch must adhere to applicable ASTM test standards. Nonstandard methods of notch preparation such as saw cutting may seem to be easier or cheaper, but they introduce variables into the test that could affect test results, and must never be used.

Nick-Break Specimens. Nick-break specimens for groove welds are prpared by machine cutting or flame cutting. The joint and base metal are cut to form a rectangular cross section. The weld is notched with a hacksaw, band saw, or thin abrasive wheel. Small weld assemblies may be tested using the complete assembly as the specimen. The notch is made at the weld edges to a depth of approximately ⅛″ and into the weld reinforcement to a depth of approximately ¹⁄₁₆″.

Fillet Welds

Fillet weld specimens include tension shear specimens, bend specimens, nick-break specimens, and hardness specimens.

Tension Shear Specimens. Tension shear specimens for fillet welds consist of longitudinal shear strength specimens and transverse shear strength specimens. Both types are sensitive to preparation procedures. The stress concentration at the root of transverse fillet welds increases with increasing root opening, and variations in root opening may lead to inconsistent test results. Both transverse and longitudinal specimens are sensitive to HAZ cracking, undercut, and bead surface contour. The longitudinal edges of transverse test specimens should be machined to eliminate crack effects and to provide smooth surfaces. Corners should be lightly rounded. These efforts will contribute to a more accurate test by focusing on the performance of the weld.

A longitudinal shear strength specimen is made using two identical welded specimens that are machined and tack welded together to prevent bending during testing. The surface contour and size of the fillet welds must meet applicable fabrication codes or standards. A transverse tension shear specimen is made by cutting from plate containing lap-welded patches on both sides. Wider plate widths may be used to obtain multiple test specimens. When multiple specimens are prepared from a single welded assembly, the results for each individual specimen are reported.

Bend Specimens. Bend specimens for fillet welds are prepared for the longitudinal guided bend test or the wraparound guided bend test. The bend specimen is prepared by making two fillet welds on a T-joint and machining the specimen to allow accommodation in the test jig. The specimen is positioned in the test jig and bent at ambient temperature. Deformation should occur in 30 sec to 2 min.

Nick-Break Specimens. Nick-break specimens for fillet welds can be prepared for pipe branch welds, pipe sleeve welds, and plate fillet welds. Pipe branch weld nick-break specimens are machine cut or flame cut samples taken from the crotch (point) area and at 90° from the crotch area. Nick-break specimens are approximately 2″ wide and 3″ long.

Pipe sleeve weld nick-break specimens can be either flame cut or machine cut. Specimens are equally spaced around the circumference of the pipe and must be at least 3″ wide and 6″ long.

Plate fillet weld specimens can be either flame cut or machine cut from the lap joint. Fillet weld specimens must be at least 3″ wide and 6″ long.

Hardness Specimens. Hardness specimens for fillet welds are prepared similarly to hardness specimens for groove welds. Specimens for fillet welds may be ground, machined, or polished, depending on the hardness test to be performed. Rust and scale must be removed from the surface. Excessive heat must be avoided when preparing the test area of the specimen. Specimens must be supported to prevent rocking during testing. If necessary, grind the backside of the specimen flat to prevent rocking.

The Mathar-Soete drilling technique and the Gunnert drilling technique are the two types of hole-drilling methods used to measure residual stresses.

Specimen Preparation Safety

Specimen preparation safety rules must be observed to prevent injury from sharp edges, hot metal, falling objects, or electrical items. In areas where grinding, burning, or welding are performed, there is a potential for toxic or flammable atmospheres that can be hazardous to the skin, eyes, and hearing. Such areas should not be entered without proper authorization.

Proper personal protective equipment must be worn, including eye and ear protection and correctly tinted glasses to observe welding in progress. Personnel should watch for tripping hazards and improper hose connections. Electrical cables and hoses that may be lying loose on a floor can be a tripping hazard. Hoses under pressure can break loose and inflict injury.

RESIDUAL STRESS MEASUREMENT

Residual stresses are locked-in stresses in materials that result from manufacturing processes such as casting, welding, forming, or heat treatment. Residual stresses can be detrimental, and when coupled with normal service stresses can be the predominant factor in fatigue and other mechanical failures. Residual stresses can also lead to stress corrosion cracking of some materials in specific corrosive environments. For example, welded carbon steel equipment and piping operating in hot caustic service must be given a stress relief heat treatment to prevent caustic stress cracking at welds, which are regions of high residual stress. The insidious aspect of residual stresses is that their presence generally goes unrecognized. Residual stresses may be measured. The most widely used technique to measure residual stresses is the hole-drilling method.

Hole-Drilling Method

The hole-drilling method is performed per ASTM E 837, *Method for Determining*

Residual Stresses by the Hole-Drilling Strain-Gauge Method. A special three-element strain gauge rosette is placed on the specimen to be tested and, using a milling guide, a ¹⁄₁₆″ or ⅛″ diameter hole is drilled on the geometric center of the strain gauge rosette to a depth equal to the hole diameter. The relieved strains measured by the three radially oriented elements of the strain gauge provide information to calculate the maximum and minimum principal residual stresses and their orientation.

The hole-drilling method requires that a blind hole be drilled into a specimen or component. However, the hole-drilling method is considered semi-destructive if, as in many cases, it does not impair the structural integrity of the component, or if the hole can be welded up without introducing detrimental residual stresses. **See Figure 31-26.**

Tinius Olsen Testing Machine Co., Inc.

Figure 31-26. *The hole-drilling method is semi-destructive if it does not impair the structural integrity of the component.*

 Refer to Quick Quiz® on CD-ROM

 Points to Remember

- The current edition of the controlling fabrication code or standard must be followed when making test welds and test specimens, and when conducting destructive tests.
- Tensile specimens obtained from welded joints are typically rectangular, unless taken from a location where it is not possible to obtain a sample of rectangular cross section.
- Bend testing is an economical way of judging weld quality to qualify a procedure or welder.
- The guided bend test is the most commonly used ductility test for groove welds, surfacing welds, and fillet welds.
- Hardness testing, although considered destructive, does not necessarily require that the specimen be cut into pieces, and is thus convenient and relatively rapid.
- Toughness testing requirements depend on the specific applicable fabrication code or standard.
- The Charpy V-notch test uses the energy produced by a dynamic load, and measures the energy needed to break a small machine-notched test specimen.
- Break tests are also rapid methods of assessing weld quality and may be called out by specific industries.
- Specimen preparation must provide a smooth surface for testing. Nicks or sharp edges are undesirable because they introduce local stress raisers that might cause premature failure.
- Proper personal protective equipment including eye and ear protection and correctly tinted goggles must be worn to observe welding.
- Residual stress measurement is a method of measuring the stress in materials produced by manufacturing processes such as welding.

1. What artificial value is created for metals that do not exhibit a yield point?
2. Which has a lower value in a tensile test, yield point or ultimate tensile strength?
3. What are the two measures of ductility obtained in a tensile test?
4. What types of welds are usually assessed in a shear test?
5. A peel test can be applied to what type of weld?
6. Is a bend test a qualitative or quantitative assessment method?
7. What is a common test used for qualifying welding procedures and welders?
8. What types of weld orientations may be specified in a bend test?
9. What types of bend locations may be specified in a bend test?
10. Why is hardness testing commonly used to measure properties of materials?
11. What is the most common method of hardness testing?
12. What are the main types of indentation hardness tests?
13. What are static and dynamic conditions during toughness testing?
14. What is the most common toughness test for welded samples?
15. Is a material with a Charpy value of 60 ft-lb four times tougher than a material with a Charpy value of 15 ft-lb?
16. What is the name used to describe the transition of a material from ductile to brittle behavior and vice versa?

Refer to Chapter 31 in the *Welding Skills Workbook* for additional exercises.

Nondestructive Examination

Chapter

Nondestructive examination is used to evaluate a part or weldment without destroying it or necessarily removing the part from service. Nondestructive examination discloses common surface and internal defects that occur with improper welding procedures or practices. A variety of testing devices are available that provide effective data about the reliability of a weldment. These devices are often more convenient to use than regular destructive testing techniques, particularly on large and costly welded units.

NONDESTRUCTIVE EXAMINATION (NDE) TERMINOLOGY

Nondestructive examination (NDE) is the development and application of technical methods to examine materials or components in ways that do not impair their future usefulness and serviceability. NDE techniques for welds are used to detect, locate, and measure discontinuities. Discontinuities in welds appear as flaws (indications). Appearance of the flaws varies depending on the NDE process. NDE results are compared with the allowable discontinuity limits in the applicable fabrication code or standard to determine acceptance or rejection of the weld.

A *flaw (indication)* is a discontinuity that can be detected through NDE techniques. Indications are categorized as relevant, nonrelevant, or false. A *relevant indication* is an NDE indication caused by a discontinuity that requires evaluation. A *nonrelevant indication* is an NDE indication caused by a discontinuity that, after evaluation, does not need to be rejected. A *false indication* is an NDE indication interpreted to be caused by a discontinuity at a location where no discontinuity actually exists. False indications are nonrelevant indications. **See Figure 32-1.**

A *defect* is one or more indications whose aggregate size, shape, orientation, or location fail to meet the acceptance criteria of the applicable fabrication code or standard. Defects are cause for rejection of the part or component.

NDE is performed by an examiner, with the results evaluated by an inspector. Qualification and certification requirements for examiners and inspectors are described in the applicable fabrication code or standard. An *examiner* is a person who is qualified, or qualified and certified, to conduct certain types of NDE processes. Examiners are qualified and certified to American Society of Nondestructive Testing (ASNT) Recommended Standard SNT-TCIA. An *inspector* is a person who is qualified, or qualified and certified, to apply the results of NDE flaw characterization to determine whether the flaws meet the acceptance criteria of the applicable fabrication code or standard. **See Appendix.**

> **✓ Point**
>
> *A flaw is not necessarily a defect. A flaw may be relevant (requiring evaluation by nondestructive testing), nonrelevant (rejection is not necessary after evaluation), or false (no discontinuity actually exists).*

> **✓ Point**
>
> *Nondestructive examination is performed by an examiner, who is a person qualified to conduct specific NDE processes.*

> **✓ Point**
>
> *An inspector is a person qualified to interpret nondestructive examination results according to the controlling code or standard for the job.*

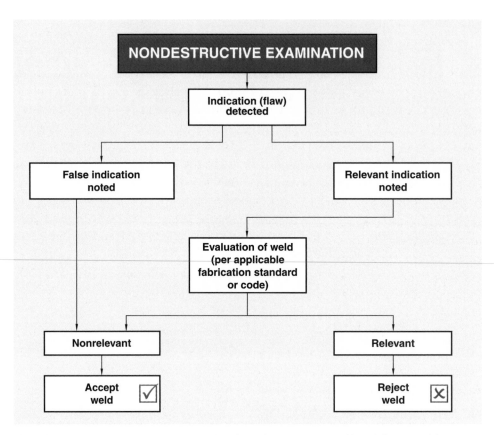

NONDESTRUCTIVE EXAMINATION

Indication (flaw) detected

False indication noted

Relevant indication noted

Evaluation of weld (per applicable fabrication standard or code)

Nonrelevant

Relevant

Accept weld ☑

Reject weld ☒

Figure 32-1. *Nondestructive examination is used to detect discontinuities in welds and determine if they are acceptable or must be rejected.*

✓ Point

Common nondestructive examination methods are visual, liquid penetrant, magnetic particle, ultrasonic, radiographic, and electromagnetic.

✓ Point

Visual examination is used to check surface condition; alignment of mating surfaces; conformance of the weld shape to a specific code or standard; and to locate leakage. Visual examination may be used before, during, or after welding.

NONDESTRUCTIVE EXAMINATION TECHNIQUES

Specific NDE techniques are selected for the detection of different types of discontinuities. NDE techniques consist of visual examination (VT), liquid penetrant examination (PT), magnetic particle examination (MT), ultrasonic examination (UT), radiographic examination (RT), electromagnetic examination (ET), and proof testing.

VISUAL EXAMINATION (VT)

Visual examination (VT) is application of the naked eye, assisted as necessary by low-power magnification and measuring devices, to monitor weld quality. A thorough examination of the weldment may disclose such surface defects as cracks, shrinkage cavities, undercuts, inadequate penetration, lack of fusion, overlaps, and crater deficiencies. VT measuring devices include rulers, calipers, straightedges, and welding gauges.

VT is generally used to determine surface condition, alignment of mating surfaces, conformance to specific shape, or to locate leakage. Direct VT requires sufficient access to place the eye within 24″ of the surface to be examined, and at an angle of not less than 30° to the surface to be examined. Mirrors are used to improve the angle of vision. Optical aids such as a magnifying glass can be used to assist in improving the quality of examinations. VT requires illumination with natural or supplemental white light at a minimum level of 50 fc (footcandles). The light source used, a verification report, and the VT technique used are documented in the examiner's report.

The limitation of visual examination is that there is no way to detect internal defects in the weld area. The weld may appear satisfactory, yet cracks, porosity, slag inclusions, or excessive grain growth may be present in the weld. VT is done before welding, during welding, and after welding.

Visual Examination Before Welding

Visual examination before welding consists of verifying the condition of materials to be welded, the conformity of partially assembled or tack welded parts, and the physical setup of the welding equipment.

Condition of Materials. The condition of the materials to be welded is verified by checking for scabs, seams, scale, and other harmful conditions on the base metal surface and for laminations in cut edges of plate. Conformance with specified dimensions is done by measurement and comparison with the specification drawing.

Conformity of Parts. Conformity of partially assembled or tack welded parts is verified after they are in position for welding. Joint dimensions, joint preparation, tack welds, and clamping must not impair the quality of the welded joint and must meet tolerances shown on the drawing. Joint dimensions include root spacing and offset. Joint preparation must ensure that rust, dirt, oil, paint, and other contaminants are removed from the weld area before welding.

Welding Equipment Setup. The physical setup of the welding equipment is verified by examining the condition of cables and connectors, how the cables are affixed to the welding machine, and how the ground cables are affixed to the work. Tack welds and clamps must maintain the root opening to ensure adequate penetration and alignment. Improper setup may lead to wasted power and erratic behavior during welding, caused by the following:

- Loose connections at the power source, work connector, or electrode holder
- Poor quality repair splices in the cable or a cable with broken strands
- Undersized cable for the required current or duty cycle
- Excessively long cables that cause an abnormal voltage drop

Visual Examination During Welding

Visual examination during welding provides details of the work while fabrication is in progress. VT during welding includes root pass examination, welding parameter verification, welding sequence monitoring, and weld bead quality checking.

Root Pass Examination. Root pass examination is done to ensure the quality of the root pass. The root pass is inspected for cracks, porosity, or blow-holes, all of which should be ground out before continuation of welding.

VT is used to check that slag deposits have been removed by chipping, grinding, or gouging before welding on the opposite side of the groove. The root opening must be examined as root pass welding progresses because it may close up from the effects of thermal expansion and lead to lack of penetration. This is especially important for branch and angle joints that are more difficult to inspect after the weld has been completed.

Welding Parameter Verification. Altering the welding parameters can affect weld quality features such as penetration or dilution. Portable meters are used to ensure compliance with specified welding current and polarity.

Compliance with preheat and interpass temperature control parameters ensures that the metal temperatures are achieved by heat soaking and not by rapid surface heating.

All welders assigned to the welding job or joint should be identified and their qualifications checked for conformance to the job requirements. If the welder does not appear to have the necessary skill for the job, the inspector can, in consultation with the supervisor, request that the welder pass requalification tests.

Welding Sequence Monitoring. Welding sequence monitoring ensures that welding is first done on the most restrained joints or, whenever possible, allowing restrained joints a small amount of movement and a measure of stress relaxation. The proper welding sequence helps prevent warpage and distortion.

Weld Bead Quality Checking. Weld bead quality checking may be done using a workmanship standard. A *workmanship standard* is a section of a joint similar to the one in manufacture in which portions of each successive weld pass are shown. Each bead of the production weld may be compared with the corresponding bead of the workmanship standard. Multiple-pass weld beads are examined for evidence of ropy, piled-up beads, or bead rollover, which could trap slag. **See Figure 32-2.** Since workmanship standards usually represent ideal conditions, there must be allowances for production tolerances.

Visual Examination After Welding

Visual examination is performed after welding or repair welding to confirm the dimensional accuracy, weld appearance, and base metal integrity of the material. VT is also used to verify application of postwelding procedures. VT for repair welding ensures that a part meets the requirements of the original fabrication.

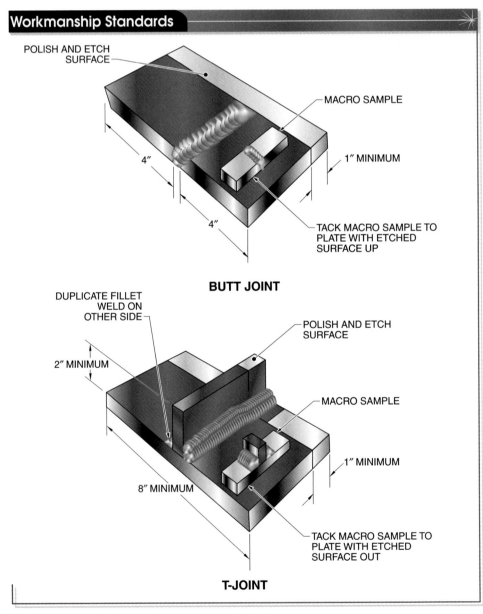

Workmanship Standards

POLISH AND ETCH SURFACE

MACRO SAMPLE

4″

4″

1″ MINIMUM

TACK MACRO SAMPLE TO PLATE WITH ETCHED SURFACE UP

BUTT JOINT

DUPLICATE FILLET WELD ON OTHER SIDE

POLISH AND ETCH SURFACE

2″ MINIMUM

MACRO SAMPLE

1″ MINIMUM

8″ MINIMUM

TACK MACRO SAMPLE TO PLATE WITH ETCHED SURFACE OUT

T-JOINT

Figure 32-2. *A workmanship standard allows assessment of the quality of intermediate passes of multiple-pass welds.*

Dimensional Accuracy. Confirmation of dimensional accuracy ensures that distortion is within acceptable limits and that all welding has been done in accordance with the drawing. Weld reinforcement in groove and fillet welds is checked to ensure that it complies with the applicable fabrication code or standard. Weld dimensions are checked with a weld gauge. A *weld gauge* is a device for measuring the size and shape of welds. There are various kinds of weld gauges. **See Figure 32-3.**

Weld Appearance. Weld appearance is examined for evidence of transverse cracks, toe cracks, crater cracks, surface porosity, incomplete root penetration, undercut, underfill, overlap, joint misalignment, incomplete joint penetration, excessive or insufficient weld reinforcement, and excessive penetration.

Some weld regions are more susceptible to discontinuities. Edges where fillet welds blend into base metal are susceptible to toe cracking and must be closely examined. Cracks are likely to be found in areas of starts and stops in the welding process and in welds with high restraint. Intermittent fillet welds are susceptible to crater cracks. Undercut that exceeds specification limits must be repaired by blend grinding, or in extreme situations, more filler metal must be added.

Base Metal Integrity. Base metal integrity must be maintained in areas where temporary attachments are welded on and subsequently removed, such as fit-up lugs, handling lugs, and machining blocks. After removal of these items, the attachment areas at the base metal must be ground smooth, and pits or tears must be filled with filler metal and ground. If indicated in the welding procedure, preheat, interpass temperature control, and postheating are required when thermal cutting or welding is done in attachment areas. Arc strikes and spatter must be removed in accordance with fabrication code or standard requirements.

Weld Gauge Measurement

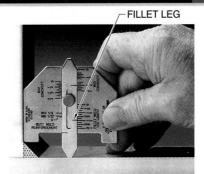

To determine size of a convex fillet weld, place gauge against toe of shortest leg of fillet and slide pointer out until it touches structure

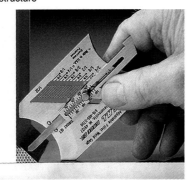

To determine size of a concave fillet weld, place gauge against structure and slide pointer out until it touches the face of fillet weld

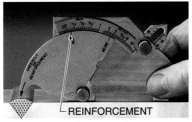

To determine reinforcement of a groove weld, place gauge so that reinforcement comes between legs of gauge and slide pointer out until it touches the face of groove weld

G.A.L. Gage Company

Figure 32-3. *A weld gauge allows the dimensions of a weld to be verified by the examiner.*

LIQUID PENETRANT EXAMINATION (PT)

Liquid penetrant examination (PT) is an NDE technique that uses dyes suspended in high-fluidity liquids to penetrate solid materials and indicate the presence of discontinuities. Application of a suitable developer brings out the dye and outlines the defect. Very small and tight discontinuities can usually be shown.

When properly applied, PT is a reliable method for detecting discontinuities open to the surface. However, it cannot be used on materials with excessively porous surfaces, such as sintered metals. Liquid penetrant examination uses the force of capillary action, which draws the liquids into all surface defects. **See Figure 32-4.**

Figure 32-5. *A portable visible-penetrant, solvent-removable PT kit is useful in determining indications when the testing needs to be done at a remote location.*

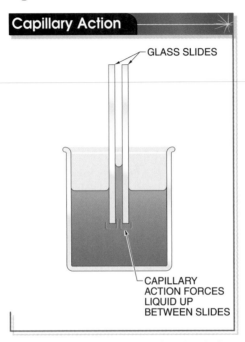

Capillary Action

GLASS SLIDES

CAPILLARY ACTION FORCES LIQUID UP BETWEEN SLIDES

Figure 32-4. *Capillary action occurs when a liquid, where it is in contact with a solid, is elevated or depressed.*

Liquid Penetrant Examination Procedure

The PT procedure consists of several steps requiring a cleaner, penetrant, and developer. **See Figure 32-5.** A cleaner is used to ensure that the surface is clean and free from dirt, oil, grease, or other materials that may adversely affect the test.

A *penetrant* is a solution or suspension of dye. Penetrants for PT have low surface tension and a high cohesive force (high capillarity). If the discontinuity is small or narrow, such as a surface crack or surface porosity, capillary action assists the penetration. When the opening is gross, such as a hot crack, the liquid may be physically trapped as it flows over the surface rather than being retained by capillary action.

✓ **Point**

Liquid penetrant examination is used to detect defects open to the surface, particularly in nonferrous metals such as aluminum, which cannot be examined by magnetic particle testing.

A *developer* is a material that is applied to the test surface to accelerate bleedout and enhance the contrast of indications. Capillary action again assists the blotting action of the developer as it draws penetrant from the discontinuity. The penetrant appears on the surface as an indication corresponding to the location of the discontinuity.

To produce the best visibility of indications, liquid penetrant contains either a colored dye easily visible in white light, or a fluorescent dye visible under black (ultraviolet) light. Liquid penetrant dyes visible in white light are available in a variety of colors, although red is most common. Some liquid penetrants have dual sensitivity, meaning they are visible in white light or black light. To perform liquid penetrant examination (PT), follow the procedure:

1. Clean the surface to be examined.
2. Dry the surface to be examined.
3. Apply penetrant to the surface. Allow sufficient time for penetrant to seep into discontinuities.
4. Remove excess penetrant from surface.
5. Apply developer to draw penetrant back to the surface.
6. Visually examine the part to locate penetrant indications that have formed in the developer coating.

Once examination is completed, the part can be cleaned to remove the penetrant and developer residue. **See Figure 32-6.**

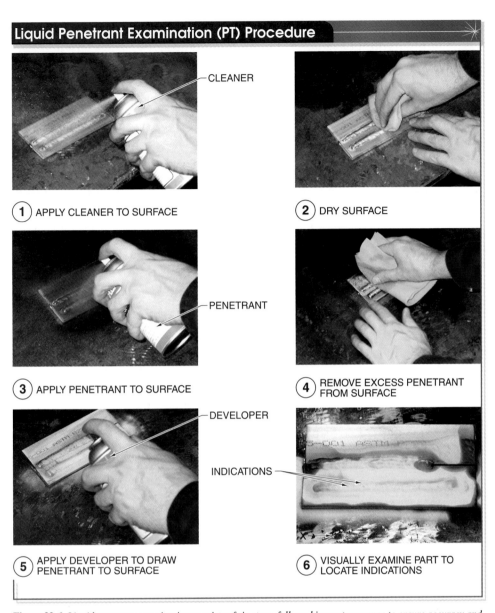

1 APPLY CLEANER TO SURFACE — CLEANER

2 DRY SURFACE

3 APPLY PENETRANT TO SURFACE — PENETRANT

4 REMOVE EXCESS PENETRANT FROM SURFACE

5 APPLY DEVELOPER TO DRAW PENETRANT TO SURFACE — DEVELOPER / INDICATIONS

6 VISUALLY EXAMINE PART TO LOCATE INDICATIONS

Figure 32-6. *Liquid penetrant examination consists of six steps, followed in a set sequence to ensure accuracy and reproducibility.*

The method of applying and developing fluorescent dyes is the same as for liquid dye penetrants; however, the fluorescent penetrant must be viewed under ultraviolet (black) light. Ultraviolet light causes the penetrants to fluoresce (glow) to a yellow-green color, which is a more clearly defined color than regular dye penetrants. Fluorescence is the emission of visible radiation by a substance as a result of, and only during, the absorption of black light radiation.

Surface Precleaning. The surface of a part must be completely clean and dry before administering liquid penetrant examination. Surface precleaning opens up surface discontinuities to penetration. Precleaning methods are detergent cleaning, vapor degreasing, steam cleaning, solvent cleaning, ultrasonic cleaning, rust and scale removal, paint removal, and etching. Precleaning methods that close up surface discontinuities must not be used.

Cleaning chemicals, such as sulfur and chlorine, must not have an adverse effect on the materials of construction. Nickel alloys may be damaged by degreasers containing sulfur; titanium alloys and stainless steels are affected by degreasers containing chlorine.

> ✓ **Point**
> *The surface of a part must be completely clean and dry before administering liquid penetrant examination.*

Cracking may result if degreasers are not completely removed from test areas that are subsequently exposed to heat or high-temperature service.

Penetrant Application. Penetrant application is done by immersion, spraying, or swabbing (brushing) on dry parts over the areas to be examined. The surface of the weldment is coated with a thin film of the penetrant, which is allowed to remain on the surface for a predetermined amount of time, known as the dwell time. *Dwell time* is the total time penetrant is in contact with the component surface, including application and drain times. **See Figure 32-7.** Dwell time is directly related to the size and shape of anticipated discontinuities since discontinuity size determines the rate of penetration. For example, tight cracks require more than 30 min for penetration if an adequate indication is to be achieved. On the other hand, gross

DWELL TIME					
Material	Form	Type of Discontinuity	Water-Washable Penetration Time*†	Post-Emulsified Penetration Time*†	Solvent-Removed Penetration Time*†
Aluminum	Castings	Porosity	5–15	5‡	3
		Cold Shuts	5–15	5‡	3
	Extrusions & Forgings	Laps	NR§	10	7
	Welds	Lack of Fusion	30	5	3
		Porosity	30	5	3
	All	Cracks	30	10	5
		Fatigue Cracks	NR§	30	5
Magnesium	Castings	Porosity	15	5‡	3
		Cold Shuts	15	5‡	3
	Extrusions & Forgings	Laps	NR§	10	7
	Welds	Lack of Fusion	30	10	5
		Porosity	30	10	5
	All	Cracks	30	10	5
		Fatigue Cracks	NR§	30	7
Steel	Castings	Porosity	30	10‡	5
		Cold Shuts	30	10‡	7
	Extrusions & Forgings	Laps	NR§	10	7
	Welds	Lack of Fusion	60	20	7
		Porosity	60	20	7
	All	Cracks	30	20	7
		Fatigue Cracks	NR§	30	10
Brass & Bronze	Castings	Porosity	10	5‡	3
		Cold Shuts	10	5‡	3
	Extrusions & Forgings	Laps	NR§	10	7
	Brazed Parts	Lack of Fusion	15	10	3
		Porosity	15	10	3
	All	Cracks	30	10	3
Plastic	All	Cracks	5–30	5	5
Glass	All	Cracks	5–30	5	5
Carbide-tipped Tools		Lack of Fusion	30	5	3
		Porosity	30	5	3
		Cracks	30	20	5
Titanium & High-temp Alloys	All		NR§	20–30	15
All Metals		Stress or Intergranular Corrosion	NR§	240	240

* for parts 60°F (16°C) to 125°F (25°C)
† in min.
‡ precision castings only
§ not recommended

Figure 32-7. *Penetrant dwell time is related to the size and shape of the discontinuities expected.*

discontinuities may be suitably penetrated in 3 min to 5 min. After allowing time for the penetrant to flow into the defects, the part is wiped clean. Only the penetrant in the defects remains.

Ambient temperature and humidity can affect dwell time. Generally, the higher the ambient temperature, the shorter the dwell time. Excessively high temperature or excessively low humidity can cause penetrant to dry too rapidly. This makes the subsequent steps of PT difficult, if not impossible. For reliable PT, the penetrant must remain wet. In some cases rewetting of the test surface is required. If penetrant has been allowed to dry, the test must be started again, beginning with surface preparation. Heating the part is not recommended. Although heating of the part accelerates penetration and shortens dwell time, it also causes evaporation of penetrant and reduces sensitivity.

> *Dwell time is determined by the type of anticipated discontinuities and the recommendation of the penetrant manufacturer.*

Penetrant Removal. Penetrant removal must ensure that all penetrant is removed from the surface without disturbing any penetrant that has entered a discontinuity. Penetrant removal is done after dwell time is complete, or after dwell time plus emulsification time. Complete penetrant removal is required to prevent the formation of false indications.

Penetrants used in water-rinsable PT have a built-in emulsifier to permit removal of the penetrant with a water rinse. Water-rinsable penetrants are sometimes called self-emulsifiable penetrants.

Post-emulsified penetrants are removed with a water rinse after completion of dwell time plus emulsification time. Light scrubbing may be required for complete penetrant removal.

Solvent-removable penetrants require a solvent designated by the penetrant manufacturer for effective penetrant removal. Solvents should not be substituted without consulting the manufacturer. Excess penetrant is first wiped from the test surface with clean, lint-free, solvent-dampened towels. Solvent is never applied directly to the surface because it might wash out or dilute the penetrant in a discontinuity.

Developer Application. After the penetrant has been sufficiently wiped clean, an absorbent material called a developer is applied to the weldment and allowed to remain until the liquid from the imperfection flows into the developer. Developer application consists of coating the test surface with a material to accelerate bleedout and enhance indication contrast. Developer acts as a blotting agent, accentuating the presence of penetrant in a discontinuity. Developer also serves as a color-contrast background for the dye. Developer causes the penetrant within a discontinuity to seep over a greater area so that the size of the indication in the developer is larger than the actual size of the discontinuity. **See Figure 32-8.** Once the developer is applied, the dye clearly outlines any defects. Developer is selected according to the manufacturer's recommendation for the type of penetrant used.

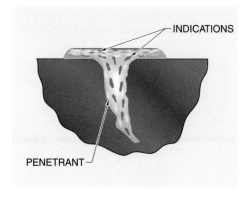

Figure 32-8. *Developer causes the penetrant to bleed within the discontinuity, causing it to seep over a greater area, making the indication appear larger than the actual discontinuity.*

PT Examination. Examination of the test surface occurs after sufficient developing time has been allowed. *Developing time* is the elapsed time between the application of the developer and the examination of the part. Insufficient developing time does not allow indications to fully develop. Excessive developing time causes indications to blur or distort. Correct developing time depends on the developer used. Generally, developing time is about half of dwell time.

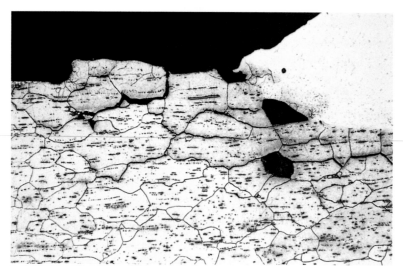

Stork Technimet, Inc.

Magnetic particle examination may be used on a magnetic stainless steel to locate hot cracks near the surface of the weld.

False or nonrelevant PT indications occur when the surface contour of the weld contains sharp depressions between weld beads that interfere with complete cleaning and complete penetrant removal. Such surfaces should be ground smooth before examination. Since smooth grinding may not be cost-effective, other NDE methods may be preferred. Diffused or weak indications appearing over a larger area are usually false indications and indicative of improper cleaning.

Nonrelevant indications are caused by surface discontinuities from the fabrication process or part geometry, which have no bearing on the service life of a component. Nonrelevant indications may appear on press fit, keyed, splined, or riveted objects, or on castings containing an adherent scale or burned-on sand.

Relevant indications are caused by discontinuities. Relevant indications are categorized as continuous line, intermittent or broken line, small dots, or round. Indications may also be categorized as faint or gross, depending on their dimensions. **See Figure 32-9.** Relevant indications must be evaluated against the requirements of the applicable fabrication code or standard. All possibilities that the indication is nonrelevant or false are first eliminated, after which the cause of the indication may be determined. It is then determined whether the indication is allowable per the applicable fabrication code or standard.

Liquid Penetrant Examination (PT)—Relevant Indications

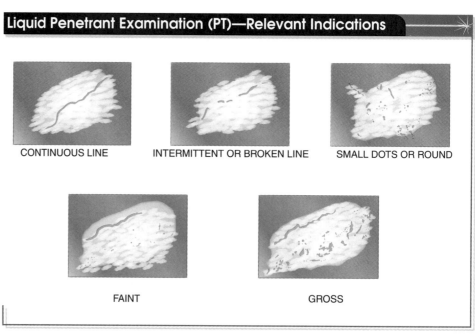

CONTINUOUS LINE

INTERMITTENT OR BROKEN LINE

SMALL DOTS OR ROUND

FAINT

GROSS

Figure 32-9. *Relevant indications fall into several categories: continuous line, intermittent or broken line, small dots, or round.*

Recording Liquid Penetrant Examination Results

PT results are recorded by the examiner in a format that records the PT method, base metal, filler metals, weld procedure identification, and location and interpretation of discontinuities.

MAGNETIC PARTICLE EXAMINATION (MT)

Magnetic particle examination (MT) is an NDE method that uses a strong magnetizing current and a finely divided powder to detect defects. Magnetic particle examination uses magnetic leakage fields and suitable indicating materials to detect surface and near-surface discontinuity indications in ferromagnetic metals.

MT consists of magnetizing the area to be examined and applying magnetic particles to the surface. However, not all materials can be magnetized.

Magnetic particles concentrate at the defect. Impurities or discontinuities in the magnetized material interrupt the lines of magnetic force, showing the size, shape, and location of defects. **See Figure 32-10.** The patterns are usually characteristic of the type of discontinuity detected.

MT detects surface discontinuities and defects resulting from very fine cracks, lack of fusion, and inclusions or internal flaws that are slightly below the surface of the weldment, including those too fine to be seen with the naked eye. All types of surface cracks can be detected using magnetic particle examination since it is one of the most reliable techniques for nondestructive examination.

Magnetic sensitivity is greatest for surface discontinuities, but diminishes rapidly for subsurface discontinuities with increasing depth. Typical discontinuities detected by MT include cracks, overlap, and laminations.

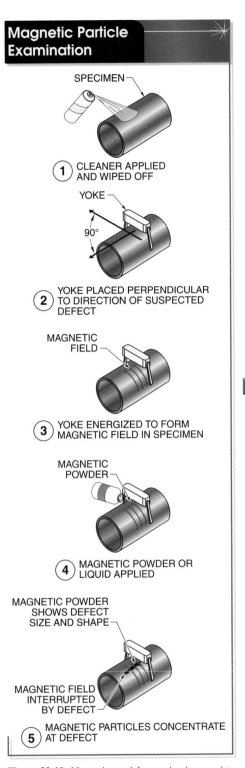

Magnetic Particle Examination

SPECIMEN

1. CLEANER APPLIED AND WIPED OFF

YOKE

90°

2. YOKE PLACED PERPENDICULAR TO DIRECTION OF SUSPECTED DEFECT

MAGNETIC FIELD

3. YOKE ENERGIZED TO FORM MAGNETIC FIELD IN SPECIMEN

MAGNETIC POWDER

4. MAGNETIC POWDER OR LIQUID APPLIED

MAGNETIC POWDER SHOWS DEFECT SIZE AND SHAPE

MAGNETIC FIELD INTERRUPTED BY DEFECT

5. MAGNETIC PARTICLES CONCENTRATE AT DEFECT

Figure 32-10. *Magnetic particle examination consists of magnetizing the area to be examined and applying magnetic particles to the surface.*

Maximum sensitivity with MT is obtained from linear discontinuities oriented perpendicular to the lines of magnetic flux. For this reason, each area should be examined twice, with the lines of magnetic flux during the second examination approximately perpendicular to the lines of flux during the first examination. **See Figure 32-11.**

Lines of Magnetic Flux

DISCONTINUITY

LINES OF MAGNETIC FLUX
PARALLEL
(MINIMUM SENSITIVITY)

**LINES OF MAGNETIC FLUX
PARALLEL TO DISCONTINUITY**

DISCONTINUITY

LINES OF MAGNETIC FLUX
PERPENDICULAR
(MAXIMUM SENSITIVITY)

**LINES OF MAGNETIC FLUX
PERPENDICULAR TO DISCONTINUITY**

Figure 32-11. *Maximum sensitivity is obtained when the lines of magnetic flux are perpendicular to the orientation of the discontinuity.*

Magnetic Particle Examination Principles

A magnetic field can be generated by the flow of electricity (magnetizing current) through a conductor. The generated magnetic field reveals the presence of discontinuities when magnetic particles are applied to the surface.

The *magnetic field* is the space within and around a magnetized part or conductor carrying current in which a magnetic force is exerted. Ferromagnetic materials are influenced by magnetic fields. A *ferromagnetic material* is a material that can be magnetized or strongly attracted by a

magnetic field. Ferromagnetic materials include carbon and low-alloy steels; martensitic and ferritic stainless steels; and tool steels.

When a magnetic field is established in a piece of ferromagnetic material containing one or more discontinuities, minute magnetic poles are set up at the discontinuities. Discontinuity sites have a stronger attraction for magnetic particles than the surrounding area of material. A *magnetic particle* is a finely divided ferromagnetic material that is capable of being individually magnetized and attracted to distortion in a magnetic field.

When a part with discontinuities is magnetized, a magnetic leakage field is produced at the discontinuities. The *magnetic leakage field* is the magnetic field that leaves or enters the surface of a part at a discontinuity or change in section configuration of a magnetic circuit. **See Figure 32-12.** Magnetic particles congregate at leakage fields and indicate the approximate shape of a discontinuity. Magnetizing current used for MT is circular or longitudinal.

Magnetic Leakage Field

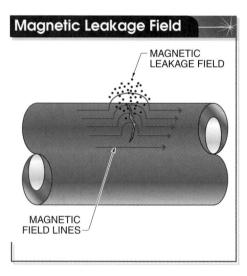

MAGNETIC
LEAKAGE FIELD

MAGNETIC
FIELD LINES

Figure 32-12. *A disruption in the magnetic field causes a magnetic leakage field as the magnetic field lines enter or leave a discontinuity, resulting in an accumulation of magnetic particles at the location of the discontinuity.*

Circular Magnetization. *Circular magnetization* is a concentric magnetic field produced by a straight conductor, such as a piece of wire, carrying an electrical current. Circular magnetization is produced by a contact head, central conductor, or prods. **See Figure 32-13.**

A *prod* is a set of hand-held electrodes used to transmit the magnetizing current from the source to the material being inspected. Prods are used where the size or location of the part does not permit the use of contact heads. The magnetic field is distorted by the interaction of the fields produced by the prods.

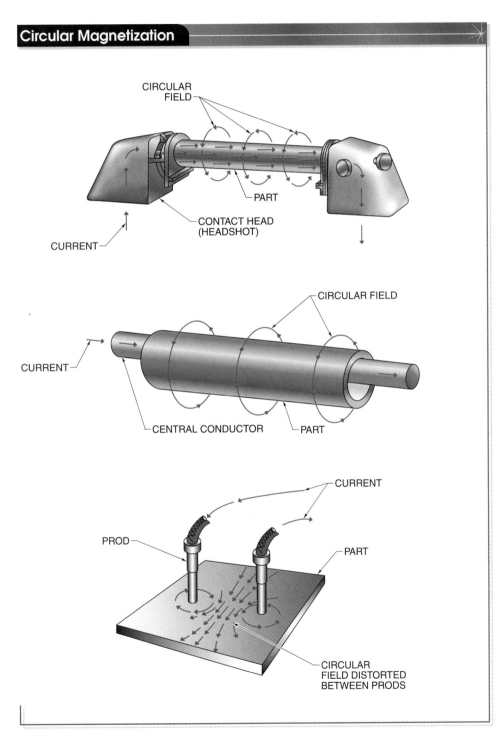

Circular Magnetization

CIRCULAR FIELD

PART

CONTACT HEAD (HEADSHOT)

CURRENT

CIRCULAR FIELD

CURRENT

CENTRAL CONDUCTOR   PART

CURRENT

PROD

PART

CIRCULAR FIELD DISTORTED BETWEEN PRODS

Figure 32-13. *Circular magnetization is produced by contact heads, prods, and central conductors.*

Longitudinal Magnetization. *Longitudinal magnetization* is a magnetic field produced when the current-carrying conductor is coiled and the magnetic field is parallel to the axis of the coil. The magnetic field strength produced within a coil increases in proportion to the number of loops within the coil. Longitudinal magnetization is achieved by coil or yoke. **See Figure 32-14.**

A coil is used when the length of the part is several times larger than its diameter. The coil is constructed by wrapping the electrical wire around the part. A *yoke* is a temporary horseshoe magnet made of soft, low-retentivity iron that is magnetized by a small wire wound around the horizontal bar. When current is passed through the wire, the magnetic flux lines flowing between the heads of the yoke in contact with the part induces a magnetic field in the part. No current flows through the part with the coil or yoke methods.

With both circular magnetization and longitudinal magnetization, the magnetic field orientation must be perpendicular or nearly perpendicular to the discontinuities to produce indications. The best results are obtained when the magnetic field is at right angles to the discontinuity and the current flow is parallel to the discontinuity.

Magnetic Particle Examination Procedure

MT procedure requirements are steps that must be followed to create an effective MT examination. MT procedure requirements include surface preparation, MT method identification, and demagnetization.

Surface Preparation. Surface preparation must ensure that the test surface is dry and free of dirt, paint, grease, lint, scale, weld flux, weld spatter, oil, and/or other extraneous matter that might interfere with the formation or interpretation of magnetic particle indications. For welds, the area to be prepared must include the weld and at least 1½″ of base metal on both sides of the weld, measured from the toe of the weld.

MT Method Identification. MT method identification determines which process to use for MT examination. The method of magnetization (continuous or residual) and the state of the magnetic particles (wet or dry) to be used determine the method.

The *continuous magnetization method* is an MT examination technique in which the magnetic particles are applied while the magnetizing force is maintained. The current continues to flow the entire time the magnetic particles are applied and excess magnetic particles removed.

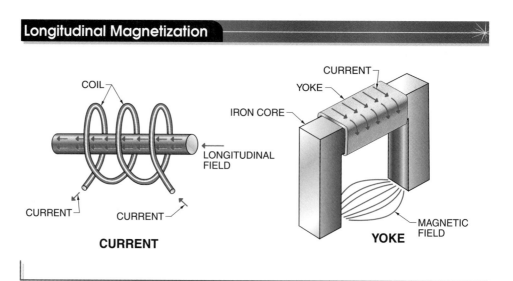

Longitudinal Magnetization

COIL

LONGITUDINAL FIELD

CURRENT CURRENT

CURRENT

CURRENT

YOKE

IRON CORE

MAGNETIC FIELD

YOKE

Figure 32-14. *Longitudinal magnetization is achieved using a coil or yoke.*

If the current is turned off before excess particles are removed, the only indications remaining will be those held by the residual magnetic field.

The *residual magnetization method* is an MT examination technique in which magnetic particles are applied after the magnetizing force has been disconnected. The residual method relies on the amount of residual magnetism retained in the test specimen. The accuracy and sensitivity of the residual method depends on the strength of the residual magnetic field. The residual method cannot be used on materials with low retentivity, such as low-alloy steel. *Retentivity* is the ability of a material to retain a portion of the applied magnetic field after the magnetizing force has been removed.

The *dry magnetization method* is an MT examination technique in which the magnetic particles are in a dry powder form. The *wet magnetization method* is an MT examination technique in which the magnetic particles are suspended in a liquid medium. Particles for the wet magnetic method are available in red or black. Red improves visibility on dark surfaces. Sensitivity of the wet method may be increased by coating the magnetic particles with a dye that fluoresces brilliantly under ultraviolet (black) light.

Demagnetization. *Demagnetization* is the elimination or reduction of residual magnetism created by MT. Demagnetization is only necessary if the residual field interferes with subsequent machining operations or arc welding, or on structures where sensitive instruments may be affected.

Demagnetization is mandatory for engine and machine parts that have been strongly magnetized. Filings, grindings, and chips resulting from operational wear are attracted to magnetized parts and interfere with performance. Demagnetization is also mandatory in aircraft construction for all steel parts in close proximity to the compass.

Selecting Magnetic Particle Examination Methods

Portable MT units are used for most weld testing. The MT method is determined by the size and shape of the workpiece and the expected discontinuities. MT methods commonly used for weld testing are dry continuous using the prod method and dry continuous using the yoke method.

Prod Method. The *prod method* is a wet or dry continuous method in which portable prod-type electrical contacts are pressed against the areas to be examined to magnetize them. Arcing may cause burns and cracking of the base metal. To prevent arcing, a remote control switch may be built into the prod handles, allowing the current to be turned on only after the prods have been properly positioned.

Wet or dry magnetic powder is applied to the surface while the magnetizing current is switched on and the prods are in contact with the surface. For efficient coverage of welds, the prods must be crisscrossed. **See Figure 32-15.**

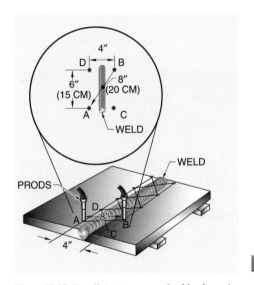

Figure 32-15. *For efficient coverage of welds when using the prod method, prods must be crisscrossed and spaced appropriately.*

Yoke Method. The *yoke method* is a dry continuous method of MT for detection of surface discontinuities. When the energized yoke is placed on the part, the flux flowing from the yoke's north pole,

✓ **Point**

Magnetic powder may be applied by the dry magnetization method or the wet magnetization method.

✓ **Point**

Demagnetization is mandatory for parts in critical service, such as engines and aircraft, that have been strongly magnetized. Filings, grindings, and chips resulting from operational wear are attracted to magnetized parts and interfere with performance.

through the part, to the south pole induces a local longitudinal field in the part. If magnetic powder is applied sparingly to the area between the poles, surface discontinuity indications are easily seen. However, the magnetic field produced by the yoke does not lie entirely within the part. An external field is present that is a deterrent to locating subsurface discontinuities.

After the test surface is prepared, the yoke is positioned on the surface and the current is turned on. Magnetic powder is lightly dusted on the surface being examined and the excess removed with a gentle air stream. The particle pattern is observed for indications. After examination is complete, the current is turned off. The examination procedure is repeated with the yoke turned at approximately a right angle to its former position. The yoke is then repositioned over the next area with sufficient overlap to ensure 100% coverage of the area to be examined. After examination and recording of discontinuities, the test surface is completely wiped clean with a cloth.

Baker Testing Services, Inc.

A cracked truck suspension is tested by magnetic particle examination. The magnetic yoke is attached to the failed part and the yellow magnetic powder is drawn to, and identifies, the crack.

Magnetic Particle Examination Indications

Magnetic particle examination indications are examined after the magnetic particles have been allowed to interact with any discontinuities. For MT examination follow the procedure:

1. Identify indications.
2. Reject false indications.
3. Interpret relevant indications according to applicable fabrication code or standard to determine if they are cause for rejection or repair.
4. Record relevant indications.

Crack types detected by MT are crater cracks, transverse cracks, and toe cracks. MT indications for crack-type discontinuities are sharply defined, tightly held, and usually heavily built up with powder. The deeper the crack, the heavier the magnetic powder buildup. Crater cracks can be a single line in any direction or star-shaped. MT indications for subsurface cracks are fuzzier and their sharpness decreases with an increase in crack depth below the surface.

The magnetic powder patterns of subsurface porosity detected by MT are fuzzy and not pronounced, yet are readily distinguished from indications of surface porosity. MT detects slag inclusions as a pattern similar to subsurface porosity when high magnetizing field strength is used.

Incomplete fusion appears as an accumulation of powder at the edge of a weld. The pattern is sharper the nearer the discontinuity is to the surface. Incomplete fusion is rarely visible at the surface and so the magnetic powder indication will not be clear and sharp. Incomplete penetration may exhibit a magnetic powder pattern similar to a subsurface crack. It will be wide and fuzzy, but the pattern should be linear.

MT may produce false indications that have no significance for weld quality. False indications are mostly attributed to physical contour effects or magnetic characteristic changes. Physical contour effects include a change in section thickness or a hole in a part. The magnetic particle patterns for physical contour effects are usually easy to identify by their location and the shape of the part at the location.

MT is not recommended for dissimilar metal welds. When two materials with differing magnetic properties are joined, such as carbon steel welded with austenitic stainless steel filler metal, an indication develops at the junction. The indication is difficult to distinguish from a crack.

Recording Magnetic Particle Examination Results

The record form for MT indications shall contain a sketch showing the geometry of the part, cable arrangement and connections, and areas of examination where adequate field strength was obtained. The information should be accompanied by the results of the examination such as a sketch or permanent record.

ULTRASONIC EXAMINATION (UT)

Ultrasonic examination (UT) is an NDE method that introduces ultrasonic waves (vibrations) into, through, or onto the surface of a part and determines various attributes of the material from its effects on the ultrasonic waves. Ultrasonic examination is very sensitive, and is capable of locating very fine surface and subsurface cracks, as well as other internal defects. High-frequency vibrations or waves are used to locate and measure defects in both ferrous and nonferrous materials. A high-frequency sound beam is directed into a part on a predictable path. The sound beam is reflected back when it encounters an interruption in the continuity of a material. The reflected

beam is detected and analyzed to define the presence and location of the discontinuity. **See Figure 32-16.**

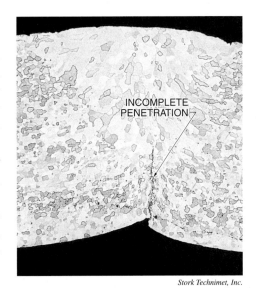

Figure 32-16. *Ultrasonic examination can be used to find subsurface discontinuities such as incomplete penetration.*

If a high-frequency vibration is sent through a sound piece of metal, a signal will travel through the metal to the other side, be reflected back, and be shown on a calibrated screen of an oscilloscope. Any discontinuities within a structure interrupt the signal and reflect it back sooner than the signal of the sound piece of material.

The reflection is shown on the oscilloscope screen and indicates the depth of the defect. Only one side of the weldment needs to be tested.

The primary purpose of UT for welds is to detect laminar discontinuities such as cracks or lack of fusion that might be more difficult to detect with other NDE techniques. A *laminar discontinuity* is a discontinuity that is relatively thin and flat. UT can also be used to detect laminations, shrinkage voids, porosity, slag inclusions, incomplete joint penetration, and other discontinuities in welds. With the proper technique, the position and depth of the discontinuity can be determined, and in some cases, the size of the discontinuity. All types of joints can be evaluated by UT and the size and location of defects can be measured.

Ultrasonic Examination Principles

The principles of ultrasonic examination are based on the ability of ultrasonic waves (vibrations) to pass through metal and to be reflected at a discontinuity. A search unit is used to send and receive the ultrasonic waves. A couplant is required to improve transmission of ultrasonic energy. Electronic components are required to generate the ultrasonic waves and record testing information.

Search Unit. A *search unit (probe)* is an electroacoustic device for transmitting or receiving ultrasonic energy, or both. A crystal (transducer) made of piezoelectric material in the search unit converts electrical energy to ultrasonic energy and vice versa.

When excited with high-frequency electrical energy, the crystal produces mechanical vibrations. The crystal also receives reflected vibrations, transforming them into low-energy electrical impulses.

Search unit configurations in weld testing are straight beam and angle beam. A *straight beam* is a vibrating pulse wave traveling perpendicular to the surface. An *angle beam* is a vibrating pulse wave traveling other than perpendicular to the surface.

Couplant. A *couplant* is a liquid substance used between the search unit and the test surface to permit or improve the transmission of ultrasonic energy. A gas interface such as air reflects almost all of the ultrasonic energy it receives. The purpose of the couplant is to exclude air between the transducer and the test surface. Couplants consist of liquids such as water, glycerin, light oil, or cellulose gum powder mixed with water. After examination, couplants must be completely removed with an acceptable solvent if heat is to be applied to the test surface at a later stage.

The weld metal or base metal must be smooth and flat to allow close contact with the search unit if required by the UT procedure. If the search unit is to be placed on the weld itself, removal of the weld reinforcement by grinding may be necessary. Weld spatter, slag, or other irregularities must be removed where the search unit might contact them.

UT Electronic Components. Electronic components required for UT include:

- An electronic signal generator to provide bursts of alternating voltage
- A sending transducer (crystal) to emit a beam of ultrasonic waves when the AC voltage is applied
- A receiving transducer to convert the sound waves to AC voltage (the receiving transducer and the sending transducer may be combined).
- An electronic device to amplify and demodulate or otherwise change the signal from the receiving transducer
- An electronic timer to control the operation
- A CRT display to characterize or record the output from the test piece. The CRT display uses A-scan presentation. **See Figure 32-17.**

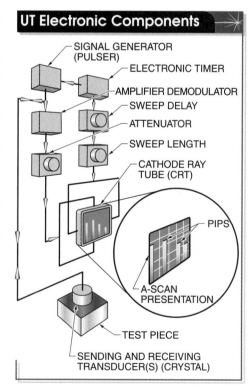

UT Electronic Components

SIGNAL GENERATOR (PULSER)
ELECTRONIC TIMER
AMPLIFIER DEMODULATOR
SWEEP DELAY
ATTENUATOR
SWEEP LENGTH
CATHODE RAY TUBE (CRT)
PIPS
A-SCAN PRESENTATION
TEST PIECE
SENDING AND RECEIVING TRANSDUCER(S) (CRYSTAL)

Figure 32-17. *The basic equipment components required for UT are a signal generator, sending and receiving transducers, an amplifier/demodulator, a CRT display, and an electronic timer.*

A-scan presentation is a method of data presentation using a horizontal base line that indicates distance or time, and a vertical deflection from the base line that indicates relative amplitude of the returning signal. The screen is graduated in both horizontal and vertical directions to facilitate measurement of pulse displays.

Ultrasonic Waves. Ultrasonic waves (vibrations) can be passed through particles that make up liquids, solids, and gases. Ultrasonic waves are above the audible range, with frequencies of about 22.5 kHz and higher. Ultrasonic waves used in weld testing are longitudinal waves and shear waves. **See Figure 32-18.**

Ultrasonic Waves

NOTE: INTERNAL ARROWS REPRESENT THE PHYSICAL MOVEMENT OF PARTICLES WITHIN THE MATERIAL

LONGITUDINAL WAVES

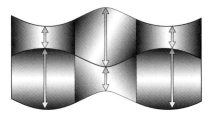

SHEAR WAVES

Figure 32-18. *Longitudinal waves and shear waves are typically used for ultrasonic weld testing.*

A *longitudinal wave* is a compression wave that represents wave motion in which the particle oscillation is in the same direction as wave propagation. Longitudinal waves can travel through solids, liquids, and gases.

A *shear wave* is a transverse wave that represents wave motion in which the particle oscillation is perpendicular to wave propagation direction. Shear waves are more easily dispersed than longitudinal waves and only travel through solids, since they cannot be propagated in liquids or gases. Shear waves have a lower velocity that allows easier electronic timing and greater sensitivity to small indications.

Shear waves are more effective than longitudinal waves at detecting weld discontinuities because they can furnish three-dimensional coordinates for discontinuity location, orientation, and characteristics. Shear wave sensitivity is about double longitudinal wave sensitivity for the same frequency and search unit size.

Longitudinal waves and shear waves complement one another in weld testing. The base metal zones adjacent to a weld are first tested with longitudinal waves to ensure that the base metal does not contain discontin-uities that would interfere with shear wave evaluation of the weld.

> *Amplifier controls include amplification control (sensitivity, gain, and uncalibrated gain), attenuation (attenuator, calibrated gain), frequency control (frequency, MHz), and display control (display control, rectified trace, unrectified trace, B-scan trace).*

Ultrasonic Examination Procedure Requirements

UT procedure requirements define how the instrumentation is set up and used for weld testing. UT procedure requirements for weld testing consist of pulse-echo mode, amplifier controls, calibration standards, and instrument calibration procedures.

Pulse-Echo Mode. *Pulse-echo mode* is a UT examination in which the presence and position of a reflector are indicated by the echo amplitude and time. The pulse-echo mode produces repeated bursts of high-frequency sound from the crystal with a time interval between bursts to receive signals from the test piece and from any discontinuities in the weld or base metal. Each pulse sets off a wave of mechanical vibrations. The initial distortion and subsequent vibrations of the crystal are fed to the amplifier and cause a pip on the CRT.

The ultrasonic unit senses reflected impulses, amplifies them, and presents them as spikes, called pips, on the CRT. The horizontal location of a reflector pip on the screen, such as from a flaw, is proportional to the distance the sound has traveled in the test piece. This makes it possible to determine the location of reflectors such as flaws by using horizontal screen graduations as a distance-measuring ruler.

Calibration Standards. Reliable information can be obtained about the specimen on the CRT by comparing signals from the specimen with those obtained from specially machined blocks, known as calibration standards. A *calibration standard* is a calibration block or a reference block.

A *calibration block* is a piece of material of specified composition, heat treatment, geometric form, and surface finish, by which ultrasonic equipment can be assessed and calibrated for the examination of material of the same general condition. A calibration block may be a simple step wedge of a particular material to allow the time base to be calibrated for accurate thickness measurement. A calibration block may also be a more complex block, allowing calibration of time base, search unit angle, resolution, index, and other features. A *reference block* is a test piece of the same material, shape, and significant dimensions as a particular object under examination, and which may contain natural or artificial discontinuities or defects.

Ultrasonic Examination Methods

Applicable standards for UT of welds are detailed in ASTM E164, *Standard Practice for Ultrasonic Contact Examination of Weldments*. The standard covers examination of specific weld configurations in wrought ferrous and aluminum alloys to detect weld discontinuities. Recommended procedures for testing butt, corner, and T-joints are given for weld test piece thicknesses from 0.5″ to 8″. Required procedures for calibrating equipment and appropriate calibration blocks are included in the standard.

UT of Base Metal. UT of the base metal is done on either side of the weld over a band that extends as far as a full skip for the shallowest angle probe, usually a 70° probe, plus half the weld reinforcement width. A *full skip* is one complete reflection of the ultrasonic beam. By checking the base metal thickness, actual thickness values are obtained for subsequent shear wave calibrations rather than the nominal thickness obtained from the prints.

Systematic scanning of base metal in the band where subsequent shear wave scans will be made allows detection of laminations, which, although they may not affect the strength of the structure, might interfere with the shear wave beam. A large lamination causes the beam to reflect up to the weld reinforcement, giving a signal that might be mistaken for a normal root bead. At the same time, the lamination can cause the beam to miss a discontinuity such as lack of penetration.

UT of Root Pass. UT of the root pass is carried out from both sides of the weld, whenever possible, using a suitable angled probe. UT of the root pass detects incomplete penetration or incomplete fusion. Scanning lines are marked at half skip distance back from the original root face on either side of the weld. A guide is then placed so that when the heel of the selected angle probe is butted against the guide, the probe index is on the scanning line. Flexible magnetic strips are useful guides for magnetic materials such as steel. **See Figure 32-19.**

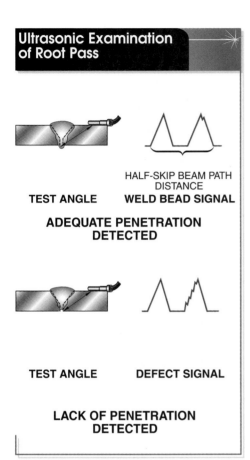

HALF-SKIP BEAM PATH DISTANCE

TEST ANGLE   **WELD BEAD SIGNAL**

ADEQUATE PENETRATION DETECTED

TEST ANGLE   **DEFECT SIGNAL**

LACK OF PENETRATION DETECTED

Figure 32-19. *UT of the weld root is carried out using an angled probe and is performed from both sides of the weld whenever possible.*

Ultrasonic Examination of Fusion Face and Weld Body. Ultrasonic examination of the fusion face and the weld body requires examining the entire weld volume. The probe is positioned to produce full skip distance to the nearest edge of the weld reinforcement. The probe index is located at a distance from the weld centerline equal to full skip distance plus one-half the full weld reinforcement width. The base metal is marked with two lines, parallel to the weld centerline, on both sides of the weld. The lines are at half skip and full skip distances and mark the boundaries of the scanning pattern. **See Figure 32-20.**

Ultrasonic Examination of Fusion Face and Weld Body

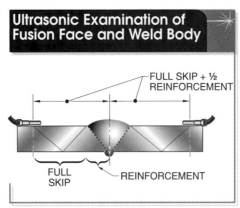

FULL SKIP + ½ REINFORCEMENT

FULL SKIP

REINFORCEMENT

Figure 32-20. *UT of the fusion face and the weld body consists of examining the entire weld volume, which is a full skip distance plus one-half the weld reinforcement width.*

The initial probe angle for the weld body scan depends upon the weld bevel angle. For maximum response, the probe angle selected should meet any sidewall lack of fusion at right angles. The required angle is calculated by dividing the weld bevel angle by 2 and subtracting from 90°. For a weld bevel angle of 60° the probe angle is 60° (90 − $^{60}\!/_2$ = 60).

Recording Ultrasonic Examination Results

Recording of UT results consists of documenting the inspection background, equipment used, equipment calibration, UT technique, and results.

RADIOGRAPHIC EXAMINATION (RT)

Radiographic examination (RT) is the use of X rays or nuclear radiation (gamma rays) to detect various types of internal and external discontinuities in material. RT images are presented on a recording medium. RT requires a source of radiation, a recording device enclosed in a light-tight holder, a qualified radiographer to produce a satisfactory exposure, and an examiner qualified to interpret radiographs. RT is used extensively to examine welds for internal discontinuities by exposing them to penetrating radiation.

Radiographic Examination Principles

X rays and gamma rays are two types of electromagnetic waves used to penetrate opaque materials. A permanent record of the internal structure is obtained by placing a sensitized film in direct contact with the back of the weldment. When the X or gamma rays pass through a weldment of uniform thickness and structure, they fall upon the sensitized film and produce a negative of uniform density. If the weldment contains gas pockets, slag inclusions, or cracks or has a lack of penetration, more rays will pass through the less dense areas and will register on the film as dark areas, clearly outlining the defects and showing their size, shape, and location. RT principles are governed by penetration and absorption, radiographic image quality, RT personnel, and radiation safety.

A radiograph is a permanent, visible image on a recording medium produced by penetrating radiation passing through a material being tested. **See Figure 32-21.**

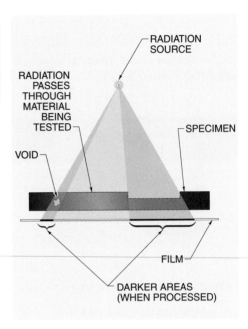

Figure 32-21. *The test material absorbs radiation, but less absorption takes place where there is a void, leading to darker areas on the processed radiograph.*

Radiographic film is placed on the opposite side of the test specimen to record the internal image of the component. The recording medium can be photographic film, sensitized paper, a fluorescent screen, or an electronic radiation detector. Photographic film is the most commonly used method.

Since more radiation passes through thin sections or locations containing voids, the corresponding areas of the film are darker. The relative positioning of the source and film in relation to the part or weld affects the sharpness, density, and contrast of the radiograph. The radiograph image quality is affected by image enlargement, image sharpness, and image distortion.

Image distortion occurs when the plane of the part and the plane of the film are not parallel. To minimize image distortion, the radiation beam must be directed in a direction perpendicular to the plane of the film. If distortion of the film image is unavoidable, the radiographer must take into consideration that all parts of the image are distorted; otherwise, the radiograph may be incorrectly interpreted. **See Figure 32-22.**

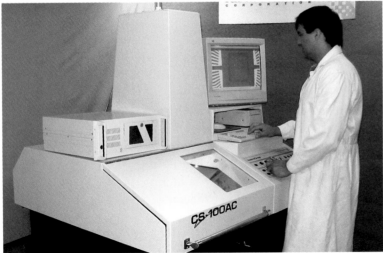

Faxitron X-ray Corporation

The radiographer must consider all parts of the image, including areas that may be unavoidably distorted, to ensure correct interpretation of the radiograph.

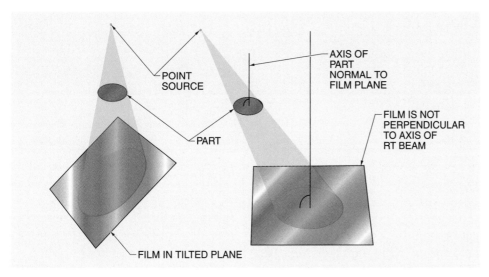

Figure 32-22. *Image distortion occurs when the plane of the film is not perpendicular to the radiation beam.*

Radiographic Examination Procedure

Radiographic examination procedure requirements are necessary to ensure the correct application of RT for weld examination. RT procedure requirements may be influenced by applicable codes and standards. RT requirements are governed by radiation source type, isotope camera, intensifying screens and filters, image quality indicator, lead identification markers, film type and film processing method.

Radiation Source Types. Radiation sources for weld inspection may be X rays from X-ray machines and gamma rays from radioactive isotopes. Both types have extremely short wavelengths, enabling them to penetrate materials that absorb or reflect light. Although the wavelength and radiation produced can be quite different, both X and gamma rays behave similarly for RT purposes.

The wavelengths of X radiation are determined by the voltage applied between the elements of an X-ray tube. Higher voltages produce X rays of shorter wavelengths and increased intensities, resulting in deeper penetration capability. The penetrating ability of X rays depends on the X-ray absorption properties of the particular metal. **See Figure 32-23.**

STEEL THICKNESS LIMITATIONS FOR X-RAY MACHINES		
Maximum Voltage*	Maximum Steel Thickness†	
	in.	mm
100	0.33	8
150	0.75	19
200	1	25
250	2	50
400	3	75
1000	5	125
2000	8	200

* in KV
† approximate

Figure 32-23. *The penetrability of X rays from the X-ray machine into the part depends on the voltage applied across the elements of the X-ray tube. Maximum voltages are established based on the thickness of the metal to be tested.*

Gamma rays are produced from portable sources and are used extensively for field-testing of welds. The gamma ray source is made as small as possible in the shape of a cylinder whose diameter and length are approximately equal. The cylindrical shape permits the use of any surface of the source as the focal spot since all surfaces, as viewed from the test specimen, are equal in area. The wavelength of the gamma rays (energy level) is determined by the nature of the source. Gamma rays have different ranges of energy and different thickness limitations for materials examined. **See Figure 32-24.**

STEEL THICKNESS LIMITATIONS FOR RADIOISOTOPES			
Radioisotope	Equivalent X-Ray Machine kV*	Maximum Steel Thickness*	
		in.	mm
Iridium-192	800	.5–2.5	12–65
Cesium-137	1000	.5–3.5	12–90
Cobalt-60	2000	2–9	50–230

* approximate

Figure 32-24. *Radioisotopes have different ranges of energy, making them suitable for different thicknesses of metals.*

Iridium-192 is equivalent to the output of an 800 kV X-ray machine. It is used for the radiography of steel. The radioisotope is supplied in the form of a capsule. The relatively low-energy radiation and high specific gravity of iridium-192 combine to make it an easily shielded, strong radiation source with a small focal spot size.

Cesium-137 is equivalent to the output of a 1000 kV X-ray machine. The radioisotope is supplied in the form of a capsule and is used on a limited scale for low-density metals such as aluminum.

Cobalt-60 is equivalent to the output of a 2000 kV X-ray machine. Cobalt-60 is used for the radiography of steel, copper, brass, and other medium-density metals. Because of its penetrating radiation, cobalt-60 requires thick shielding with resulting weight and handling difficulty.

Isotope Camera. The isotope camera consists of the equipment needed for safe handling and storage of an isotope source.

Image Quality Indicator (IQI). An *image quality indicator (IQI)* is a device or combination of devices whose demonstrated image determines radiographic quality and sensitivity. The image or images demonstrated by an IQI provide visual data, quantitative data, or both to determine the radiographic quality. An IQI is not intended for use in judging size of, or acceptable limits for, discontinuities. An IQI is also called a penetrameter, or penny. Each IQI is identified by an identification number that gives the maximum thickness of material for which the IQI is normally used. **See Figure 32-25.**

The IQI is placed on the source side of the part to provide a built-in discontinuity of known thickness containing three hole diameters. The IQI measures the ability of the RT technique to show contrast (IQI thickness) and definition (hole images).

Shim stock is sometimes used in RT of welds because the area of interest (the weld) is thicker than the part thickness. Shims are selected so that the thickness of the shim(s) equals the thickness added to the specimen by the weld in the area of interest. Shim stock is placed underneath the IQI, between it and the part. In this way, the image of the IQI is projected through a thickness of material equal to the thickness in the area of interest. The shim stock length and width are greater than those of the IQI. **See Figure 32-26.**

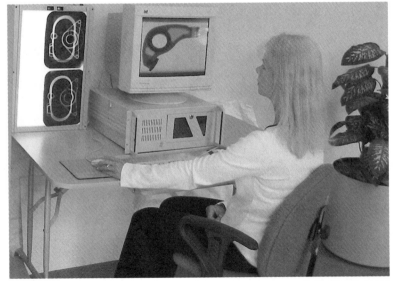

Faxitron X-ray Corporation

The results of radiographic testing are downloaded to a computer for storage and future reference.

IMAGE QUALITY INDICATOR (IQI) SIZES

Applies to Design Material Thickness* (T_m) up to and including inches[†]	ID No.	"T"[‡]	1T Hole Diameter[†]	2T Hole Diameter[†]	4T Hole Diameter[†]
¼ (0.25)	25	0.005	0.010[§]	0.020[§]	0.040[§]
⅜ (0.375)	37	0.008	0.010[§]	0.020[§]	0.040[§]
½ (0.5)	50	0.010	0.010	0.020	0.040
⅝ (0.625)	62	0.013	0.013	0.025	0.050
¾ (0.75)	75	0.015	0.015	0.030	0.060
⅞ (0.875)	87	0.018	0.018	0.035	0.070
1 (1)	1	0.020	0.020	0.040	0.080
1⅛ (1.125)	1.1	0.023	0.023	0.045	0.090
1¼ (1.25)	1.2	0.025	0.025	0.050	0.100
1½ (1.5)	1.5	0.030	0.030	0.060	0.120

* Defined as the thickness of the material (T_m) upon which the thickness of the IQI is based. For welds, T_m shall be the thickness of the strength member
† in in.
‡ IQI thickness
§ Hole size required by standard does not correspond directly to ID number or IQI thickness

NOTES:

Chart extends in ¼″ increments up to 2½″, then in ½″ increments up to 8″, and then in 1″ increments.

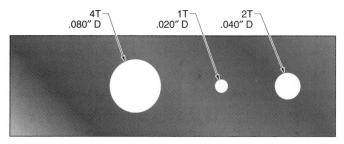

STANDARD IQI FOR 1″ MATERIAL

Figure 32-25. *An image quality indicator (IQI), or penetrameter, determines the radiographic quality level (sensitivity). The IQI thickness ("T") is 2% of the thickness of the part being radiographed.*

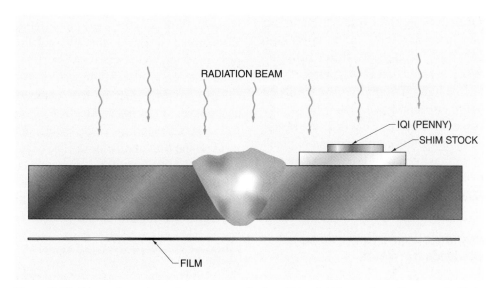

Figure 32-26. *Shim stock may be used to compensate for the additional thickness of a weld compared with the base metal.*

The Lincoln Electric Company

Nondestructive examination is often used for structures that must remain in service both during and after testing.

be set up for the type of part. Some RT methods for welds include single-wall RT for plate, pipe, or tubing; double-wall RT for pipe or tubing less than 1¼" ID (inside diameter); and double-wall RT for pipe or tubing from 1¼" to 2½" ID.

RT exposure setup conditions are based on the following factors that influence the radiographic image formation:

- Obtaining best coverage of the weld in the shortest exposure time
- Detecting image discontinuities most likely to be present
- Using multiple perpendicular exposures rather than one or more angled exposures to cover all areas of interest
- Using single- or double-wall exposures with a pipe weld
- Adhering to all radiation safety requirements

Lead Identification Markers. Lead identification markers are placed on the source side of the part to provide a clear record of the test or test location. These markers consist of a letter and numbers and must not interfere with subsequent interpretation of the radiograph by masking potential indications.

Film Type. The film type selected is based on the need for radiographs of specific contrast and definition quality. RT film consists of thin, transparent plastic sheeting coated on one or both sides with an emulsion of gelatin, approximately 0.001" thick, containing very fine crystals of silver bromide. When exposed to X rays, gamma rays, or visible light, silver bromide crystals undergo a reaction that makes them more susceptible to the chemical process of developing that converts them to black-metallic silver. The greater the amount of exposure, the greater the blackening effect on development.

Radiographic Examination Methods for Welds

Selection of RT methods for welds requires consideration of how the exposure can

Single-Wall RT for Plate and Pipe or Tubing. Single-wall RT for plate, pipe, or tubing welds is relatively simple to achieve because the critical areas of the weld are clearly defined in terms of their length, width, and thickness. **See Figure 32-27.** The film is placed in direct contact with the part on the side opposite to the source with an exposure angle of 90°. Single-wall RT should be used whenever possible for flat or circular objects. Subject contrast is small and exposure calculation is relatively simple.

Double-Wall RT for Pipe or Tubing Less than 1¼" ID. Double-wall RT for pipe or tubing welds less than 1¼" ID is done with an elliptical shot. An elliptical shot involves placing the source at an angle less than 90° to the plane of the part to view the full circumference of the weld on the film as an ellipse. The exact angle is determined by the pipe or tubing diameter. Three elliptical exposures should be made to provide sufficient coverage. **See Figure 32-28.**

Just as a photograph is an image produced by visible light, a radiograph is an image produced by radiation.

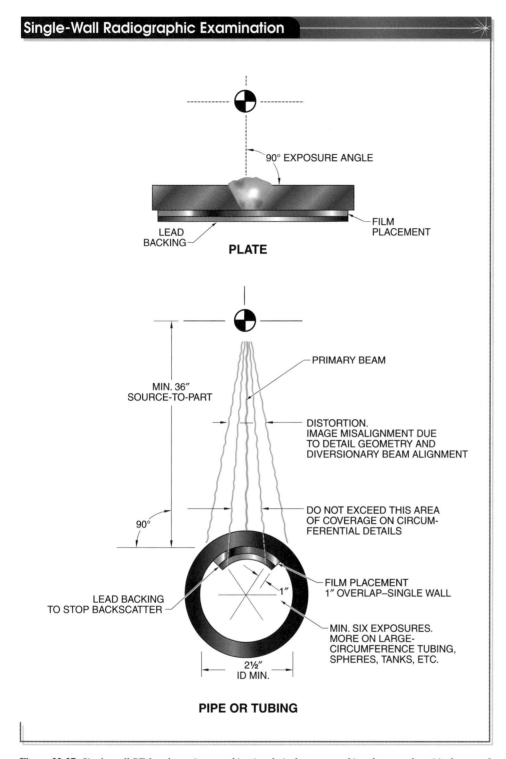

90° EXPOSURE ANGLE

FILM
PLACEMENT

LEAD
BACKING

PLATE

PRIMARY BEAM

MIN. 36″
SOURCE-TO-PART

DISTORTION.
IMAGE MISALIGNMENT DUE
TO DETAIL GEOMETRY AND
DIVERSIONARY BEAM ALIGNMENT

90°

DO NOT EXCEED THIS AREA
OF COVERAGE ON CIRCUM-
FERENTIAL DETAILS

FILM PLACEMENT
1″ OVERLAP–SINGLE WALL

LEAD BACKING
TO STOP BACKSCATTER

1″

MIN. SIX EXPOSURES.
MORE ON LARGE-
CIRCUMFERENCE TUBING,
SPHERES, TANKS, ETC.

2½″
ID MIN.

PIPE OR TUBING

Figure 32-27. *Single-wall RT for plate, pipe, or tubing is relatively easy to achieve because the critical areas of the weld are clearly defined.*

Double-Wall RT for Pipe or Tubing 1¼″ to 2½″ ID. Double-wall RT for pipe or tubing welds from 1¼″ to 2½″ ID is done with a 15° elliptical shot. As with pipe or tubing less than 1¼″ diameter, two IQIs should be used. Six elliptical shots provide sufficient coverage of the entire circumference and reveal discontinuity orientation. In addition, two 90° opposing, superimposed shots should be taken to show discontinuities in the perpendicular position. **See Figure 32-29.**

Double-Wall Radiographic Examination for Pipe and Tubing Less Than 1¼" ID

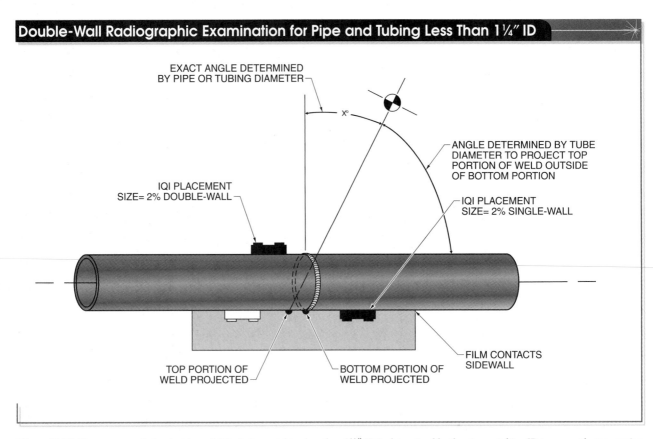

Figure 32-28. *The source angle for double-wall RT of pipe or tubing less than 1¼" ID is determined by the pipe or tubing ID to ensure the top portion of the weld projects outside of the bottom portion.*

Double-Wall Radiographic Examination for Pipe and Tubing 1¼" ID to 2½" ID

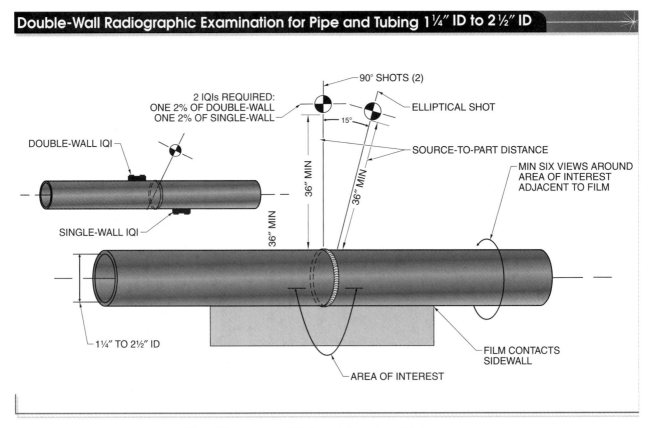

Figure 32-29. *RT of pipe or tubing from 1¼" to 2½" ID requires a 15° elliptical shot and two 90° shots.*

To evaluate a radiograph, follow the procedure:

1. Compare the identification of the radiograph against accompanying records for accuracy.
2. Determine the weld design and welding procedure used.
3. Determine the radiographic set-up procedure and the correctness of technique attributes.
4. Review film under optimum viewing conditions.
5. Identify any film artifacts (see below) and request re-radiography if necessary.
6. Identify any surface marks or unsoundness on the part not associated with the weld and verify their type and presence.
7. Evaluate and propose disposition of discontinuities revealed in the radiograph.
8. Prepare complete radiographic report.

RT for fillet welds is difficult to set up and interpret. Fillet weld RT requires a great degree of skill and in-depth knowledge of the welding conditions. Also, it is difficult to place the film ideally to obtain good resolution of discontinuities in fillet welds. Therefore, RT is not usually viable for fillet welds.

Identification of Discontinuities

RT reveals both surface and subsurface weld discontinuities including cracks and incomplete fusion; slag inclusions and tungsten inclusions; porosity and wormholes; incomplete joint penetration; undercut; excessive weld reinforcement; and insufficient weld reinforcement. RT does not reveal very narrow discontinuities that are not closely aligned (parallel) to the weld.

Cracks appear as fine dark lines of significant length, but without great width. Some crater cracks may be detected by RT if of sufficient size. Cracks may not be detected if they are small or not aligned with the beam.

Incomplete fusion has a similar appearance to cracks, but usually appears at the boundary between the weld and base metal.

Slag inclusions usually appear as irregularly shaped dark areas and have some width. Slag inclusions are generally observed at the junctions between weld passes. Tungsten inclusions appear as highly contrasted light areas (white spots).

Porosity appears as nearly round voids recognizable as dark spots whose radiographic contrast varies directly with the diameter of the pores.

Wormholes appear as dark rectangles if their long axis is perpendicular to the beam and as concentric circles if the long axis is parallel to the beam.

Incomplete joint penetration is observed as a very narrow dark line near the center of the weld.

Undercut appears as a dark zone of varying width along the edge of the fusion zone. The darkness or density of the line is an indicator of the depth of the undercut.

Excessive weld reinforcement is seen as a lighter zone along the center of the weld seam. There is a sharp change in image density where the reinforcement meets the base metal, and the edge of the reinforcement image is usually irregular.

Insufficient weld reinforcement is seen as the opposite of excessive weld reinforcement, that is, a darker zone along the center of the weld seam. The change in image density is not as pronounced as with excessive weld reinforcement.

Artifacts. An *artifact* is a nonrelevant indication that appears on a radiograph. Artifacts may occur during exposure or during handling or processing of the film, if handling or processing has been done improperly. Artifacts also may occur because of various causes including electrostatic discharge, pressure marks, and film processing defects. Artifacts must be avoided.

Electrostatic discharge during film handling exposes the film to light and causes an easily recognized pattern of sharp black lines on the radiograph. Pressure marks result from localized pressure on pre-processed film when the film is being processed.

Film processing defects lead to many kinds of artifacts. Colored stains or blisters may result from an improper acid stop bath application. Streaks may result from improper agitation during development. Fogging may be caused by overexposure of film to a safelight lamp before fixing or by using old film. Stains may be caused by improperly mixed or exhausted solutions, and water marks can result from handling partially dried film. Fingerprints are caused by improper handling of film. Scratches result from rough handling, especially during processing when the emulsion is soft. Chemical fog may be caused by overdeveloping.

Recording Radiographic Examination Results

Recording of RT results is done on a form consisting of a sketch identifying the weld locations, a description of the RT method used, and identification and interpretation of all discontinuities. RT results may be recorded in a standard format. The owner of the part tested shall retain radiographs related to the examination. In digital radiography, the film is scanned and uploaded to a digital format. The data is easy to maintain, share, and archive.

ELECTROMAGNETIC EXAMINATION (ET)

Electromagnetic examination (ET) is an NDE method that uses electromagnetic energy having frequencies less than visible light to yield information on the quality of the part being tested. ET, also called eddy current testing, uses electromagnetic energy to detect discontinuities in welds and is effective in testing both ferrous and nonferrous materials for porosity, slag inclusions, internal cracks,

external cracks, and lack of fusion. ET is applied to both magnetic and nonmagnetic materials.

Electromagnetic Examination Principles

Electromagnetic examination principles are based on the phenomenon of electromagnetic induction, meaning that an electric current flows, or is induced, in a conductor subject to a changing magnetic field. The frequency of the magnetic field varies from 50 Hz to 1 MHz, depending on the type and thickness of the materials tested. In weld testing, ET is used principally in automatic production testing of welded pipe and tube.

Electromagnetic Induction. Electromagnetic induction creates different responses in metals according to their electromagnetic properties. The part to be inspected is placed within or adjacent to an electric coil through which alternating current (the exciting current) is flowing. The exciting current induces a magnetic field and causes eddy currents to flow in the part because of electromagnetic induction. An *eddy current* is an electrical current caused to flow in a conductor by the time or space variation, or both, of an applied magnetic field. To achieve electromagnetic induction, the electric coil may be an encircling coil or an inside coil. **See Figure 32-30.**

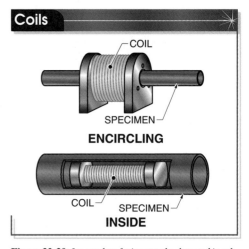

Coils

ENCIRCLING

INSIDE

Figure 32-30. *Incomplete fusion may be detected in tubing as it passes through an encircling coil or an inside coil during electromagnetic examination.*

An encircling coil is wound so that the test specimen passes through the center of the coil, causing the eddy currents to flow around the rod or tube being tested. The specimen must be centered in the coil for accurate test results. This is because the flow of eddy currents is zero at the center of the rod. An inside coil is used to test steam generator tubes. Inside coils pass through the inside of tubing and eddy currents flow around the tubing. For accurate test results, the coil and the test specimen should be close together.

The eddy current path is distorted by the presence of a discontinuity. The distortion is measured by a change in the associated electromagnetic field. Such changes have an effect on the exciting coil or other coil(s) used for sensing the electromagnetic field adjacent to the part. For example, the change in flow of eddy currents caused by incomplete fusion can be detected as the tubing passes through the coil. When fusion within the weld is complete, the eddy current flow is symmetrical. As a section containing incomplete fusion passes through the coil, the eddy current flow is impeded and changed in direction, causing a significant change in the associated electromagnetic field, which is detected on the measuring equipment.

For electromagnetic examination, the induced voltage of the exciting coil or the adjacent coil is used to monitor the condition of the part being inspected.

Electromagnetic Examination Requirements

Electromagnetic examination requirements indicate the parameters that must be controlled and documented to ensure effective, repeatable applications. ET requirements include ET inspection equipment, ET equipment calibration, and ET procedures.

ET Inspection Equipment. ET inspection equipment consists of a generator, inspection coil, amplifier, detector, and display. The generator supplies excitation current to the inspection coil and a synchronizing signal to the phase shifter, which provides switching signals to the detector. The probe may be an external coil, as used for tubing inspection.

ET Equipment Calibration. An *equipment calibration standard* is a test piece that contains typical discontinuities that demonstrate that calibration equipment is detecting the discontinuities for which the part is being inspected. Equipment calibration standards for ET contain natural or artificial discontinuities. The discontinuities in the calibration standards can accurately reproduce the exact change in the electromagnetic characteristics expected when production items containing discontinuities are tested.

Equipment calibration standards are necessary because ET does not detect discontinuities, but rather the effect they have on the electromagnetic properties of the part being inspected. It is necessary to correlate the change in electromagnetic properties with the cause of the change. Equipment calibration standards are used to facilitate the initial adjustment or calibration of the test equipment and to periodically check on the reproducibility of the measurements.

ET Procedures. ET procedures reference the type of equipment calibration standards that are required. Electromagnetic examination procedures must be standardized, often using full-scale or mock-up calibration standards with simulated discontinuities. Equipment calibration standards must meet the following requirements:

- Conform to the applicable specification.
- Be easily fabricated.
- Be reproducible in precisely graduated sizes.
- Produce an indication on the ET tester that closely resembles those produced by natural discontinuities.

✓ **Point**

Electromagnetic examination procedures must be standardized, often using full-scale or mock-up calibration standards with simulated discontinuities.

Point

Proof testing is used to demonstrate the ability of the welded structure to carry loads equal to or in excess of the anticipated service conditions.

Electromagnetic Examination Methods for Welds

Electromagnetic examination methods for welds are primarily applied to longitudinal welded pipe or tubing as a production quality control tool.

ET of Longitudinally Welded Pipe or Tubing. ET of longitudinally welded pipe or tubing is done using an encircling external energizing coil and a probe-type differential detector coil. The probe-type detector coil is located at the longitudinal center in the inner perimeter of the primary coil and is arranged so that it inspects the outside surface of the longitudinal weld.

Examination is performed by passing the pipe or tubing longitudinally through the primary energizing coil, causing the probe-type detector coil to move across the longitudinal weld from end to end. The primary coil is energized with an alternating frequency that is suitable for the part being inspected and induces eddy currents into the part. **See Figure 32-31.**

Rath Gibson, Inc.

Figure 32-31. *To inspect longitudinal weld quality in welded pipe or tubing, an energizing coil and a detector coil are required.*

Point

Adequate venting must be ensured during hydrostatic testing to prevent collapse (sucking in) of the tank.

The DC coil is energized at high current levels to magnetically saturate the pipe or tubing, improving penetration of the eddy current and canceling the effects of magnetic variables. This type of inspection is effective in detecting most types of longitudinal weld discontinuities, such as open welds, weld cracks, and hot cracks. Many discontinuities may be detected at relatively high speeds (speeds of 300 ft/min are common). The speed must be constant to within ±10%.

PROOF TESTING

Proof testing is the application of specific loads to welded structures, without failure or permanent deformation, to assess their mechanical integrity. Proof tests are usually designed to subject parts to stresses exceeding those anticipated during service, but maintained below or at the specified yield strength of the metal. Proof testing is used to demonstrate the ability of the welded structure to carry loads equal to or in excess of the anticipated service conditions. Proof tests must be designed by an engineer familiar with in-use requirements, and consist of hydrostatic testing, pneumatic testing, spin testing, leak testing, vacuum box testing, and acoustic emission testing.

Hydrostatic Testing

Hydrostatic testing (hydrotesting) is proof testing of closed containers such as vessels, tanks, and piping systems by filling them with water and applying a predetermined test pressure. Hydrostatic testing is the most common type of proof test.

Adequate venting must be ensured during hydrostatic testing to prevent collapse (sucking in) of the tank. **See Figure 32-32.** For components built to the ASME Boiler and Pressure Vessel Code, this pressure is 150% of design pressure. For other components, the test pressure may be based upon a fixed percentage of the minimum yield strength. After a fixed holding time, the container is inspected for soundness by visually checking for leakage, or by monitoring the hydrostatic test pressure for any drop.

Open containers such as storage tanks may also be hydrostatically tested by filling them with water; ships or barges may be tested by partially submerging them in water. The hydrostatic pressure exerted against any boundary is governed by the head of water.

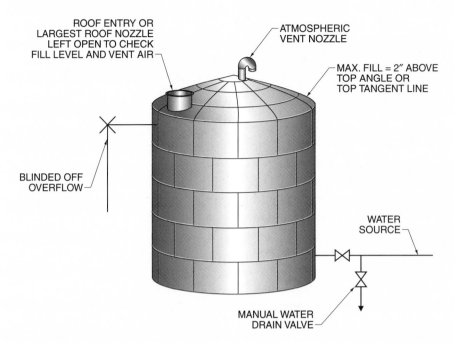

NOTE: VENT NOZZLE MUST BE LARGER THAN WATER DRAIN VALVE

ROOF ENTRY OR
LARGEST ROOF NOZZLE
LEFT OPEN TO CHECK
FILL LEVEL AND VENT AIR

ATMOSPHERIC
VENT NOZZLE

MAX. FILL = 2″ ABOVE
TOP ANGLE OR
TOP TANGENT LINE

BLINDED OFF
OVERFLOW

WATER
SOURCE

MANUAL WATER
DRAIN VALVE

Figure 32-32. *When performing hydrostatic testing on an atmospheric pressure storage tank, there must be adequate venting to prevent the tank from collapse (sucking in) when it is drained.*

Three questions to consider before using hydrostatic testing are (1) whether the foundation or support is strong enough to hold the container filled with water, (2) whether energy in the form of compressed air can build up in the container, and (3) whether there is adequate notch toughness to ensure that small leaks or discontinuities will not propagate into a catastrophic failure.

Hydrostatic testing is a relatively safe operation because water is practically noncompressible and therefore stores little energy. A small leak results in meaningful pressure drop that limits the driving force available to propagate a crack.

Several important questions must be asked before hydrostatic testing is carried out to prevent permanent equipment damage or catastrophic failure. The following safety issues must be considered:

- Are the foundations and support structure strong enough to hold the water-filled container? This is especially important if the foundation and support container were originally designed to hold a gas or light-weight liquid.
- Are there any pockets where energy can build up in the form of a compressed gas? Pockets may include high points in the system that are difficult to completely fill with water.
- Is the water temperature above the minimum temperature for hydrostatic testing carbon steel equipment? It may be necessary to warm the water slightly to assure that a relatively small leak or discontinuity will not propagate into a catastrophic brittle fracture. A minimum water temperature of 50°F is required for hydrostatic testing.
- Is the water of sufficient purity to avoid rapid localized pitting of stainless steel? Fabrication codes and standards usually limit the chloride content of hydrotest water to ensure it is not damaging to stainless steels.

⚠ WARNING

During pneumatic testing, large amounts of energy may be stored in compressed air or gas in a large volume or under high pressure, or both. A small leak or rupture can easily grow into a catastrophic failure, and can endanger life and adjacent property.

- Is the water drained and the equipment dried out completely after hydrostatic testing? This applies to stainless steel equipment where stagnant water may lead to microbiologically induced corrosion. Fabrication codes and standards usually allow a maximum time of 72 hr for water to be left in stainless steel equipment.

Pneumatic Testing

Pneumatic testing is a proof test in which air is pressurized inside a closed vessel to reveal leaks. Pneumatic testing must be used with care to prevent a catastrophic failure from release of the stored energy. Pneumatic testing is usually performed on small units that can be submerged in water during testing. The presence of air bubbles is a convenient leak indicator and immersion in water is an effective energy absorber in case the component fails.

Pneumatic testing may be applied to equipment such as equipment mounted on foundations not able to support the weight associated with hydrostatic tests, and to equipment where water or liquid may be harmful and cannot be removed, for example a plate-fin heat exchanger.

Pneumatic testing acceptance is based on freedom from leakage. Small leaks are seldom detected without some indicating devices. If a unit cannot be submerged in water, spraying it with a soap or detergent solution and checking for bubbles is an effective alternative for determining the location of leaks. This procedure is called an air-soap test.

If both pneumatic testing and hydrostatic testing are to be done, the pneumatic test should be carried out first. If done in reverse order, there is a possibility that the larger water molecules from the hydrostatic test will locate and block fine leak passages and prevent them being discovered by the smaller air molecules during the pneumatic test.

Spin Testing

Spin testing is proof testing of rotating machinery done by spinning it at speeds above design values to develop desired stresses from centrifugal forces. Visual and other nondestructive testing plus dimensional measurements are employed to determine the acceptability of the parts. Spin testing is conducted in a safe enclosure such as a specially constructed pit in case the component should rupture.

Vacuum Box Testing

Vacuum box testing is the application of a partial vacuum to one side of a structure and examining for the presence of leaks. The test involves applying soap or detergent solution to an area such as a longitudinal weld, placing a transparent box with an adequate seal over the area to be examined, and evacuating the box to achieve partial vacuum of not less than 2 psi. The area is examined for bubbles, which are the sign of a leak. Vacuum box testing is quick and convenient.

Acoustic Emission Testing

Acoustic emission testing (AE) is a proof test that consists of detecting acoustic signals produced by plastic deformation or crack formation during mechanical loading or thermal stressing of metals. Transducers strategically placed on a structure are activated by arriving acoustic signals and allow the locations of discontinuities to be identified. Once the discontinuity location is identified, it must be examined by other techniques such as RT or UT to describe and measure it.

Refer to Quick Quiz® on CD-ROM

- A flaw is not necessarily a defect. A flaw may be relevant (requiring evaluation by nondestructive testing), nonrelevant (rejection is not necessary after evaluation), or false (no discontinuity actually exists).
- Nondestructive examination is performed by an examiner, who is a person qualified to conduct specific NDE processes.
- An inspector is a person qualified to interpret nondestructive examination results according to the controlling code or standard for the job.
- Common nondestructive examination methods are visual, liquid penetrant, magnetic particle, ultrasonic, radiographic, and electromagnetic.
- Visual examination is used to check surface condition; alignment of mating surfaces; conformance of the weld shape to a specific code or standard; and to locate leakage. Visual examination may be used before, during, or after welding.
- Liquid penetrant examination is used to detect defects open to the surface, particularly in nonferrous metals such as aluminum, which cannot be examined by magnetic particle testing.
- The surface of a part must be completely clean and dry before administering liquid penetrant examination.
- Magnetic particle examination is used to detect surface or near-surface discontinuity indications in ferromagnetic metals.
- A magnetic field may be induced in the part by circular magnetization or longitudinal magnetization.
- Magnetic powder may be applied by the dry magnetization method or the wet magnetization method.
- Demagnetization is mandatory for parts in critical service, such as engines and aircraft, that have been strongly magnetized. Filings, grindings, and chips resulting from operational wear are attracted to magnetized parts and interfere with performance.
- Ultrasonic waves used in weld testing are longitudinal waves and shear waves.
- Radiographic film is placed on the opposite side of the test specimen to record the internal image of the component.
- Electromagnetic energy is used to detect surface and internal quality of welds in electromagnetic testing.
- Electromagnetic examination procedures must be standardized, often using full-scale or mock-up calibration standards with simulated discontinuities.
- Proof testing is used to demonstrate the ability of the welded structure to carry loads equal to or in excess of the anticipated service conditions.
- Adequate venting must be ensured during hydrostatic testing to prevent collapse (sucking in) of the tank.
- Pneumatic testing must be used with care to prevent a catastrophic failure caused by release of stored energy.

1. Can visual examination be used to find every type of cracking that a weld may exhibit?
2. What types of checks may be performed on metal before welding using visual examination?
3. Why is visual examination important during welding?
4. After welding, how should visual examination be applied to dimensional accuracy, weld appearance, and base metal integrity?
5. What is the difference between penetrant and developer in liquid penetrant examination?
6. Why must defects be open at the surface for liquid penetrant examination to be effective?
7. What is dwell time?
8. Why is arcing undesirable when prods are used for magnetic particle examination?
9. What is the purpose of couplant used in ultrasonic examination?
10. Why is radiographic examination commonly used to assess weld quality?
11. What types of artifacts may be present in radiographs that detract from accurate assessment of weld quality?

Refer to Chapter 32 in the *Welding Skills Workbook* for additional exercises.

Metallography

Chapter

Metallography is the visual examination of the microscopic features of metal or weld surfaces that have been specially prepared by cutting, grinding, polishing, and etching. Metallography is used in failure analysis and as a quality control tool for production. In failure analysis, metallography is used to compare the actual weld quality with the specification and to reveal contributing causes of the failure. When used as a quality control tool, the tested specimen must be representative of the overall weld. The specimen is then compared against set standards. Metallography includes microscopic examination and macroscopic examination used to analyze discontinuities, weld passes and location, and metallurgical structure of the weld.

MICROSCOPIC EXAMINATION

Microscopic examination is concerned with the microscopic features of material surfaces. The purpose of microscopic examination is to look for clues as to how a metal was made and/or how it performed under load or working conditions. Microscopic examination is conducted at high magnification. Small specimens, representative of the component, are required. The sequence of steps in microscopic examination consists of cutting and rough grinding; mounting and fine grinding; rough and final polishing; and etching and examination.

Cutting and Rough Grinding

Cutting and rough grinding are performed to obtain a representative metallographic specimen from the joint. Specimen orientation must first be determined. Special techniques, such as preventing flattening when cutting thin-wall tubing, may be

required to preserve the specimen from damage so that the essential features are not destroyed. For rough grinding, sequential cutting may be performed to obtain a suitably sized specimen. Rough grinding removes coarse material and features that result from the cutting process.

Specimen Orientation. The specimen orientation is selected to obtain a representative section of the joint. The most common specimen orientation is transverse. It can be used to investigate weld profile, weld width, weld penetration (depth of fusion), weld reinforcement, and weld area. If a transverse section is not cut exactly perpendicular to the plane of the weld, errors in weld penetration and weld area measurement may be introduced. Except in the most severe cases, errors introduced in sectioning are likely to be lower than sampling errors from variability along the length of the weld. Transverse sections may be supplemented by longitudinal sections.

> ✓ **Point**
>
> *Microscopic examination consists of cutting and rough grinding; mounting and fine grinding; rough and final polishing; and etching and examination.*

If additional details are required, other specimen orientations may be necessary. **See Figure 33-1.**

Cutting. Cutting is the most common method of obtaining specimens from a component. Large specimens must be reduced in size using flame or plasma cutting. Subsequent cutting is accomplished using a power hacksaw, band saw, abrasive cutoff wheel, or diamond-tipped cutoff wheel. Power hacksaws or band saws are used on specimens that are too large or awkward to cut using an abrasive cutoff wheel. Abrasive cutoff wheels are used to obtain specimens that are close to the final size. Diamond-tipped cutoff wheels are used on small specimens where precision cuts are

required. If a diamond-tipped cutoff wheel is used, the rough grinding steps are bypassed. **See Figure 33-2.**

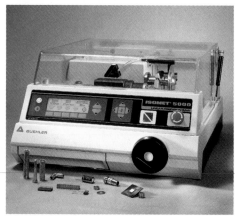

Buehler Ltd

Figure 33-2. *Diamond-tipped cutoff wheels are used on small specimens where precision cuts are required.*

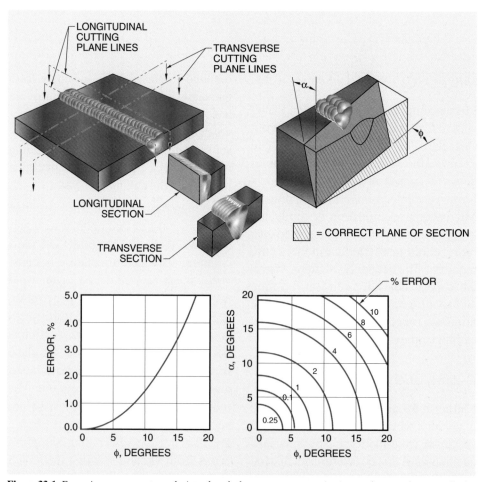

Figure 33-1. *Errors in measurement may be introduced when a transverse section is not taken exactly perpendicular to the plane of the weld.*

Overheating is microstructural damage or change caused by cutting operations. Flame or plasma cutting must be performed at a minimum distance of 1" away from the area to be examined to prevent overheating, so that final cutting can be done with less damaging techniques. Cutoff wheels and saws use a coolant at the cutting surface to prevent overheating. Materials with hardness values greater than 35 HRC may require the use of an abrasive cutoff wheel or a diamond-tipped cutoff wheel for cutting operations.

Subsurface deformation is microstructural damage or change produced by cutting and that occurs below the surface of the specimen. Coarse cutting tools and heavy applications of force increase subsurface deformation that must be removed by grinding to prevent false interpretations of the microstructure.

Rough Grinding. Rough grinding prepares specimens for mounting by removing subsurface deformation, unnecessary roughness, and flash or scale caused by cutting operations. Specimens are ground flat on a wet abrasive belt sander using an 80-grit or 150-grit belt, or they are machined flat in a milling machine. When a diamond-tipped wheel is used to make the final cut, rough grinding is usually unnecessary.

Mounting and Fine Grinding

Specimen mounting is usually permanent, meaning that the specimen is permanently encased in resin. Some mounts have a temporary clamping device that holds the specimen flat and rigid during fine grinding. Before mounting, any burrs at the edges of the specimen caused by cutting or machining are carefully removed using a smooth file or coarse abrasive paper or cloth.

Mounting prevents rounding of the edges of specimens and allows handling during the polishing and etching stages. Selection of the correct mounting resin is based on a combination of factors. Fine grinding prepares the mount for the final stages of specimen preparation.

Hot Mounting. Hot mounting is usually performed in a mounting press that encapsulates the specimen with a thermosetting resin under pressure and at an elevated temperature. **See Figure 33-3.** The specimen is placed face down in a vertical, cylindrical mold in the mounting press. A predetermined amount of thermosetting resin is poured into the mold and the mold is closed. The temperature is raised and pressure is maintained while the resin cures, making the resin hard and strong. After the mold cools, the mount is removed from the mold.

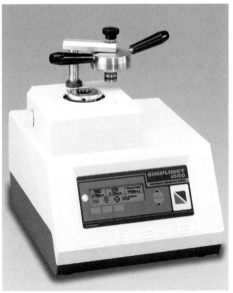

Buehler Ltd

Figure 33-3. *Mounting presses use compression and heat to encapsulate the specimen in a plastic mounting resin.*

A suitable mounting resin must cure at a temperature and pressure that does not alter the microstructure of the specimen. The mounting resin selected must resist chemical attack by the etchant, which is applied to the face of the mount to reveal microstructural features. The mounting resin must provide good adhesion to the edges of the specimen to prevent rounding of the edges and entry of lubricant or etchant during specimen preparation. Lubricant or etchant that enters the mount during preparation will flow out after final preparation and cause staining of the specimen as it dries.

The mounting resin must fill pores and crevices on the exposed face of the specimen to prevent staining. It must also be electrically conductive if electrolytic polishing or etching is to be used. If side views of the specimen are required, the mounting resin must be transparent. **See Figure 33-4.**

Cold Mounting. Cold mounting is an alternative to hot mounting and is performed when the specimen is too large for the mounting press or when the heat involved might alter the microstructure. Cold mounting is performed in a vacuum to remove air bubbles from the mount. Room temperature and atmospheric pressure must be maintained when performing cold mounting using a thermoplastic resin.

> *Mounting is used to conveniently hold the specimen, to mount multiple specimens, and to store and label specimens. Mounting also protects the edges of the specimen and provides the proper specimen orientation.*

MOUNTING RESINS

Plastic	Type	Molding Conditions			Heat-Distortion Temperature*‡	Transparency	Chemical Resistance
		Temperature*	Pressure†	Curing Time			
Phenolic molding powder	Thermosetting§	170	4000	5 min	140	Opaque	Not resistant to strong acids or alkalis
Acrylic (polymethyl methacrylate) molding powder	Thermoplastic	150	4000	none	65	Water white	Not resistant to strong acids
Epoxy casting resin	Thermosetting‖	20-40	—	24 hr	60#	Clear but light brown in color	Fair resistance to most alkalis and acids; poor resistance to nitric and glacial acetic acids
Diallyl molding compound	Thermosetting**	160	2500	6 min	150	Opaque	Not resistant to strong acids and alkalis
Formvar® (polyvinyl formal) molding compound	Thermoplastic	220	4000	none	75	Clear but light brown in color	Not resistant to strong acids
Polyvinyl chloride molding compound	Thermoplastic††	160‡‡	3000	none	60	Opaque	Highly resistant to most acids and alkalis

* °C
† in psi
‡ determined by method in *ASTM D 648-56*, at a fiber stress of 264 lb/in.²
§ wood-filled grade, preferably with low filler content
‖ liquid epoxy resin with an aliphatic amine hardener
depends on curing schedule (can be as high as 110°C with heat curing)
** diallyl phthalate polymer with a mineral filler
†† stabilized ridged PVC
‡‡ must not exceed 200°C

Figure 33-4. *Mounting resins must satisfy a variety of conditions to be acceptable.*

Fine Grinding. Fine grinding is the last stage before polishing of the mount. It consists of abrading the mount on a series of successively finer abrasive papers. Before fine grinding, any resin on the face of the specimen or any remaining burrs on the edges are removed by a 120-grit abrasive paper or cloth. During fine grinding, a series of water-lubricated papers, ranging from 240-grit to 600-grit, are used. The mount is lightly washed between abrasive papers or belts to prevent carryover of coarser abrasive material.

Two commonly used types of fine grinding are four-stage belt sanding and four-stage wheel grinding. Four-stage belt sanding uses an assembly of four strips of abrasive paper of increasing fineness. The mounted specimen is moved up and down on each grade of paper without rocking the mount. The mounted specimen is abraded backward and forward without rotation until all sanding marks from the previous coarser abrasive paper have been eliminated. **See Figure 33-5.**

Buehler Ltd

Figure 33-5. *Grinding in four-stage belt sanding starts with 240-grit and finishes with 600-grit paper.*

Four-stage wheel grinding is performed on a grinding wheel by changing the abrasive material at each stage to eliminate successively finer scratches. Ample water lubrication must be used to prevent overheating.

With either four-stage belt sanding or four-stage wheel grinding, the amount of time spent on each abrasive material is increased as finer grades of material are used. Excessive sanding with any grade of abrasive paper must be avoided as it may cause subsurface deformation that cannot be eliminated by subsequent grades of abrasive paper and that leads to artifacts. The mount is thoroughly washed and dried after fine grinding is completed.

The direction of grinding is changed 90° with each change of abrasive paper, so that complete removal of the previous grinding marks is achieved. **See Figure 33-6.**

Rough and Final Polishing

Rough and final polishing procedures are used to develop a scratch-free mirror finish on the specimen. The specimen is polished using manual, mechanical, electrolytic, or chemical techniques. The surface must be free from pits (small, sharp depressions) and subsurface deformation effects that lead to artifacts (false indications) when the specimen is etched. Pits are caused by the polishing operation that removes tiny nonmetallic particles such as carbides from the metal surface.

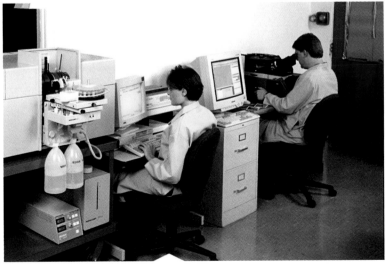

Wall Colmonoy

Microscopic examination procedures are typically performed in a lab by trained technicians.

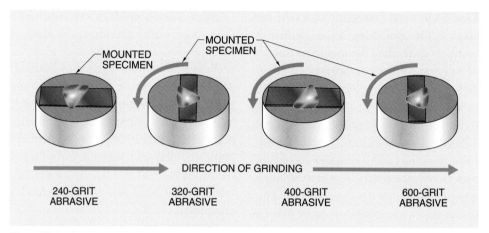

MOUNTED SPECIMEN

MOUNTED SPECIMEN

MOUNTED SPECIMEN

DIRECTION OF GRINDING

240-GRIT ABRASIVE

320-GRIT ABRASIVE

400-GRIT ABRASIVE

600-GRIT ABRASIVE

Figure 33-6. *The mount is rotated 90° and thoroughly washed between successive papers to prevent carryover of abrasive materials.*

Rough Polishing. *Rough polishing* is a polishing process that is performed on a series of rotating wheels covered with a low-nap cloth (cloth containing a small amount of fiber). Successively finer grades of diamond rouge (polishing powder) are applied to each wheel, usually starting at 45 μm size. The grades usually decrease from 30 μm to 6 μm to 1 μm. A small amount of lubricant is applied to the cloth to prevent overheating of the mount. The mount is washed with liquid soap and water, alcohol, or acetone between each polishing to prevent carryover of diamond rouge.

Final Polishing. Final polishing is similar to rough polishing, but during final polishing very light hand pressure is applied to the mount. After washing and drying in a current of warm air, the mount is examined under a metallurgical microscope for scratches. If the mount is scratch-free, it is ready for etching and examination under a metallurgical microscope.

Final polishing is done by rubbing the mount against a medium-nap cloth that has a 0.3 μm to 0.05 μm alumina slurry applied to it. If the specimen surface is to be subjected to microanalysis, alumina should not be used. The presence of alumina during microanalysis may lead to misinterpretation of the results of the microanalysis. *Microanalysis* is chemical analysis of extremely small regions of the specimen surface using tools such as energy-dispersive X-ray analysis or electron probe microanalysis.

Automatic Polishing. Automatic polishing is a process that establishes a complex motion for the mount relative to the rotation of the polishing wheel. The rough and final polishing steps are performed in an automatic polishing machine. The machine setting is determined from operator experience. Automatic polishing is used for large batches of repetitive work, for radioactive specimens, and for polishing techniques that add corrosives to the wheel. **See Figure 33-7.**

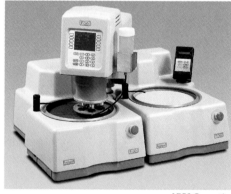

LECO Corporation

Figure 33-7. *Automatic polishing in an automatic polishing machine establishes a complex motion for the mount relative to the rotation of the polishing wheel.*

Electrolytic and Chemical Polishing. Electrolytic polishing and chemical polishing are methods of preparation that bypass the rough and final polishing stages.

Electrolytic polishing is a polishing process in which the mount is the anode (connected to the positive terminal) in an electrolytic solution and current is passed from a metal cathode (connected to the negative terminal). The current is passed through the electrolytic solution between the anode and the cathode. The current removes the rough peaks on the specimen surface. If the grain structure is homogeneous and single-phase (consisting of one crystallographic component), a mirror-polished surface is obtained. **See Figure 33-8.**

Chemical polishing is a polishing process that uses chemical reactions to remove the rough peaks on the specimen surface. The mount is immersed in a specific chemical that dissolves the high peaks on the specimen to produce a mirror-polished finish. Chemical polishing is similar to electrolytic polishing in that it removes the rough peaks on the specimen surface and produces a mirror-polished surface.

Etching and Examination

Etching, followed by examination of the mounted specimen with a metallurgical microscope, is the last stage of metallographic preparation before microstructural examination. Etching is necessary to reveal the microstructural detail of the polished specimen, assists in determining the features of the weld, and makes visible the boundary between the weld metal and the base metal. The etched specimen is examined under reflected light in a metallurgical microscope.

Etching a Specimen. Etching is the controlled selective attack on a metal surface to reveal the microstructural detail of a polished specimen. Before etching, the specimen is examined with a metallurgical microscope in the as-polished condition. Besides revealing minor scratches that must be removed, etching also makes microstructural features such as inclusions and porosity easy to observe.

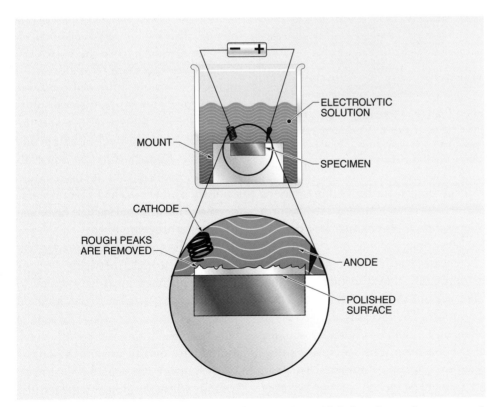

Figure 33-8. *Electrolytic polishing removes rough peaks on a specimen with the flow of current between an anode and a cathode.*

The specimen is then thoroughly degreased, dried, and prepared for etching. Etching is the last stage before examination. Etchants selectively dissolve specific microstructural components, giving the as-polished surface a relief appearance. Etchants are selected to distinguish various microstructural components to provide the best view of the microstructural features. **See Appendix.**

Etching is usually performed by immersion. The specimen is immersed with the polished face upward in a small dish of etching solution, which is gently swirled. The specimen is removed when a bloom appears. A *bloom* is a slight haze that appears on the surface of the specimen and is evidence of the first appearance of the microstructure. **See Figure 33-9.**

Etched Specimen

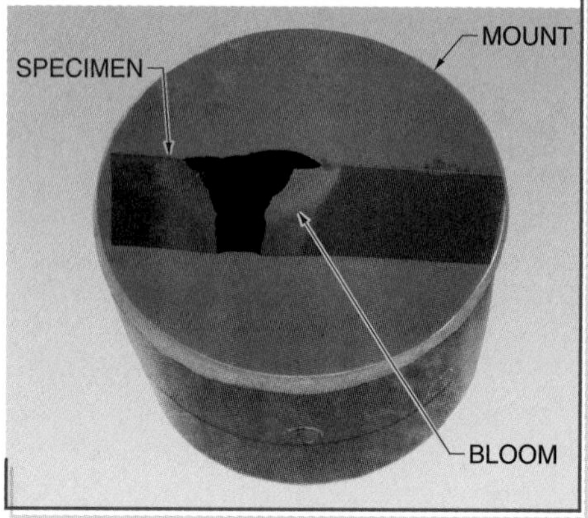

Figure 33-9. *For optimum viewing of the microstructure, the mount is etched until a bloom appears on the surface.*

If necessary, further etching may be performed after examination under a microscope to strengthen any details. However, over-etching may cause loss of contrast. After etching, the specimen is thoroughly rinsed in running water. Then acetone or alcohol is sprayed over the surface. The excess is allowed to run off against a cloth that is held at one side of the specimen. The specimen is then dried in a stream of hot air. The specimen should be etched and fine polished at least twice to remove flawed metal from the surface.

Specimen Examination. Metallurgical microscope examination uses light reflected from the specimen surface to examine microstructural details. The surface of the specimen must be widely scanned to gain a representative view of the microstructure. Details are revealed because etching attacks the grains of metal at different rates, which results in various shading effects. The proper amount of etching is required for optimum viewing of the microstructure. Improper amounts of etching lead to overetching or underetching, resulting in false effects. **See Figure 33-10.**

The etched specimen is placed in a metallurgical microscope and examined at low-power magnification of 25x or 50x to obtain an overall impression of the microstructure. It is then examined at increasing magnifications of 100x to 1000x to reveal fine detail. Higher magnifications up to 2500x cannot be achieved within the air space available between the lens and the specimen.

Higher magnifications require the use of water or oil immersion. A small amount of water or oil is daubed on the objective lens, which is lowered towards the specimen. If water or oil immersion is to be followed by lower magnification work, the water or oil is removed from the specimen and the mount may require repolishing and re-etching. Surface films on some alloys may require that specimens are repolished and re-etched several times to remove the affected surface layer and reveal their true structure.

When focusing the metallurgical microscope, contact between the lens and the specimen must be avoided to prevent surface damage to the mounted specimen. The microscope is focused in two steps. First, the microscope stage is gradually moved toward the objective lens using the coarse adjustment. Second, when the image appears, the focusing is completed using the fine adjustment.

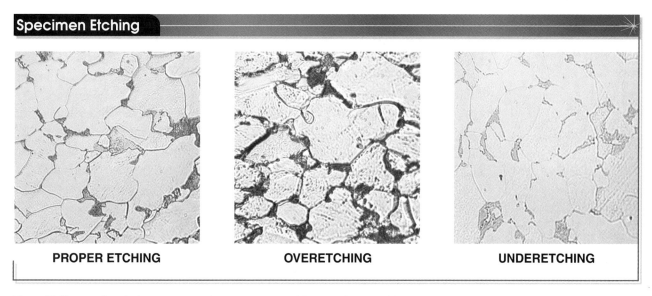

PROPER ETCHING   **OVERETCHING**   **UNDERETCHING**

Figure 33-10. *Properly etched specimens reveal true microstructural features when viewed by a metallurgical microscope.*

Metallurgical microscopes vary from small benchtop units to larger units that have their own framework. Some are equipped with a video camera and monitor that are used to view microstructures. **See Figure 33-11.** A *metallograph* is a metallurgical microscope equipped to photograph microstructures and produce photomicrographs. Photomicrographs are photographs of microstructures.

Interpretation problems such as artifacts and surface films may hinder metallurgical microscopic examination. An artifact does not correspond to the true microstructure and occurs during metallographic specimen preparation. Artifacts result from incomplete removal of a thin surface layer that has been affected by the specimen preparation process. For example, overheating during cutting may give the false impression that the specimen was heat-treated.

Metallurgical Microscopes

LECO Corporation

Figure 33-11. *The benchtop metallurgical microscope is commonly used for speciman examination.*

Illumination. Different types of illumination enhance the appearance of the microstructural characteristics of the specimens. These include brightfield illumination, darkfield illumination, polarized illumination, and Nomarski illumination. **See Figure 33-12.**

Brightfield illumination is an illumination process in which the surface features perpendicular to the optical axis of the microscope appear the brightest. Brightfield illumination is the most common form of illumination used with a metallurgical microscope. The surface of the specimen is placed perpendicular to the optical axis of the microscope and a white light is used.

Darkfield illumination is an illumination process that illuminates the specimen at sufficient obliqueness (a narrow angle to the surface) so that the contrast is completely reversed from that obtained with brightfield illumination. Those areas that are bright in brightfield will be dark in darkfield and vice versa. Darkfield illumination is useful for highlighting microstructural features (inclusions, grain boundaries, and cracks) that are dark and difficult to distinguish under brightfield illumination.

Illuminations for Micrographs

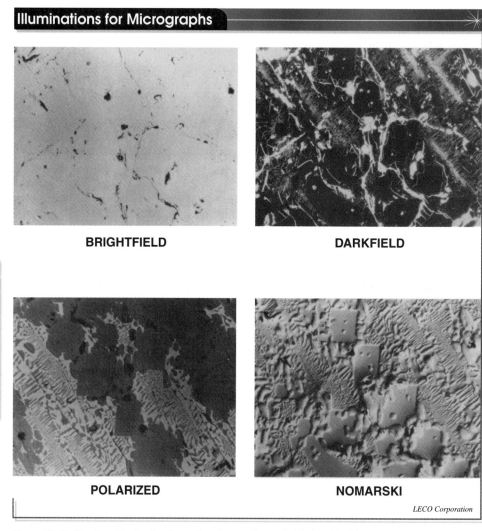

BRIGHTFIELD

DARKFIELD

POLARIZED

NOMARSKI

LECO Corporation

Figure 33-12. *The four illumination forms for micrographs are brightfield, darkfield, polarized, and Nomarski illumination.*

✓ **Point**

When examining metallographic samples under a metallurgical microscope, illumination techniques such as brightfield, dark-field, polarized, and Nomarski may be used to reveal microstructural features.

Polarized illumination is an illumination process that reveals microstructural features in metals that are optically anisotropic. Optically anisotropic describes a microstructural feature in which the microstructure has optical properties that vary with changes in the viewing direction. The light is polarized by placing a polarizer in front of the condenser lens of the microscope and placing an analyzer behind the eyepiece. A *polarizer* is a device into which normal light passes and from which polarized light emerges.

Nomarski illumination is an illumination process that illuminates the specimen using polarized light that is separated into two beams by a biprism. A *biprism* is two uniaxial, double-refracting crystals. The beams are reflected back through the biprism off of the specimen surface. The biprism combines the beams into one beam, which is run though an analyzer and viewed through an eyepiece. Images produced are three-dimensional and vary in color. This variation in dimension and color is used to identify metals and their various phases.

MACROSCOPIC EXAMINATION

Macroscopic examination is used to reveal the general structure of large areas of a specimen because they might not be revealed under the higher magnifications used in microscopic examination. Macroscopic examination is performed with the naked eye or at magnifications up to 10x using a binocular microscope. Larger specimens are used for macroscopic examination than are used for microscopic examination. A specimen for macroscopic examination is usually an entire section through a component. These specimens are used to reveal gross elements of fabrication quality, such as size of weld.

Macroscopic examination consists of specimen preparation; rough and fine grinding; and macroetching and examination. Photography may be used to document macroscopic examination.

Specimen Preparation

Specimen preparation for macroscopic examination consists of removing a slice, by flame cutting or sawing, in the plane to be examined. Fabrication codes and standards may indicate where cuts must be taken to produce acceptable specimens for macroscopic examination.

Rough and Fine Grinding

Rough and fine grinding procedures are similar to those used in microscopic examination. Fine grinding is performed to a final finish with a 240-grit abrasive. Unlike microscopic examination, the specimen is not mounted.

> *Applications of macroetching are to study the weld structure; to measure joint penetration; to detect lack of fusion; and to determine whether slag, flux, porosity, or cracks are present in the weld and the heat-affected zone.*

Macroetching and Examination

Macroetching differs from etching used for microstructural examination and requires the use of macroetchants. *Macroetchants* are deep etchants that are intended to develop gross features such as weld solidification structures. Macroetchants are designed to attack metal more deeply and more quickly than metallographic etchants. **See Figure 33-13.**

Different etchants, such as hydrochloric acid, ammonium hydroxide-peroxide, or nitric acid, are used to reveal specific types of microstructural details. After the etching process, the specimen is ready for examination in a metallurgical microscope.

A hydrochloric acid solution should contain equal parts by volume of concentrated hydrochloric (muriatic) acid and water. Immerse the weld in the boiling reagent. Hydrochloric acid will etch unpolished surfaces. It usually enlarges gas pockets and dissolves slag inclusions, enlarging the resulting cavities.

An ammonium hydroxide-peroxide solution should contain one part ammonium persulfate (solid) and nine parts water by weight. Vigorously rub the surface of the weld with cotton saturated with the ammonium persulfate reagent at room temperature.

Nitric acid etches rapidly and should only be used on polished surfaces. It will show the refined zone as well as the metal zone. Mix one part concentrated nitric acid to three parts water by volume. Either apply the reagent to the surface of the weld with a glass stirring rod at room temperature or immerse the weld in boiling reagent, provided the room is well ventilated. After etching, wash the weld immediately in clear, hot water. Remove excess water. Dip the etched surface in ethyl alcohol; then remove and dry it in a steady blast of warm air.

An iodine and potassium iodide solution can also be used as an etchant and is obtained by mixing one part powdered iodine (solid) to 12 parts of a solution of potassium iodide by weight. The potassium iodide solution should consist of one part potassium iodide to five parts water by weight. Brush the surface of the weld with this reagent at room temperature.

MACROETCHANTS		
Etching Solution	Surface Preparation*	Comments
Carbon and Low-Alloy Steels 10 g $(NH_3)_2S_2O_8$ (ammonium persulfate) + 100 mL H_2O	B	Swab; macroetch brings out fusion line, heat-affected zone, reheated zones, columnar zones
15 mL HNO_3 + 85 mL H_2O + 5 mL methanol or ethanol	A, B	Swab; macroetch brings out fusion line, heat-affected zone, reheated zones, columnar zones; scrub gently under running water to remove any black residue
8 mL HNO_3 + 2 g picric acid + 10 g $(NH_3)_2S_2O_8$ + 10 g citric acid + 10 drops (.5 mL) benzalkonium chloride + 1500 mL H_2O	B	Immerse; highlights partially transformed regions in reheat and heat-affected zones (Ref 13)
Aluminum Alloys Tucker's reagent, 45 mL HCl + 15 mL HNO_3 + 15 mL HF (48%) + 25 mL H_2O	A, B	Immerse or swab; use freshly mixed general macroetch; all alloys
Poulton's reagent, 60 mL HCl + 30 mL HNO_3 + 5 mL HF (48%) + 5 mL H_2O	A, B	Immerse or swab; general macroetch, all alloys
Copper and Copper Alloys 50 mL HNO_3 + .5 g $AgNO_3$ (silver nitrate) + 50 mL H_2O	A, B	Immerse; general macroetch, all alloys
Titanium Alloys Kroll's reagent, 10–30 mL HNO_3 + 5–15 mL HF + 50 mL H_2O	B	Immerse; general macro- and microetch; increase HNO_3 and reduce HF to bring out the fine structures in weldments

* surface preparation: A, finish grind; B, polish

Figure 33-13. *Macroetchants are deep etchants intended to develop gross features such as weld solidification structures.*

Macroetching. Macroetching is usually performed by gently daubing the sample with the macroetchant or by immersing smaller specimens in the macroetchant and gently swirling. Higher temperatures accelerate the etching rate. Prolonged etching is avoided because it leads to darkening of the specimen, which obscures detail. When the structural features are developed, the specimen is immediately rinsed in warm running water. During rinsing, the surface should be scrubbed with a soft bristle brush to remove deposits formed during macroetching. Deposits may contain residual macroetchant and may lead to localized overetching if the macroetchant is not thoroughly removed. The washed specimen is dried by squirting it with alcohol or acetone, which is allowed to drain into a cloth that is held at one side of the mount, and then drying the specimen in a current of warm air.

The surface must be preserved as quickly as possible after drying, once it has been determined that the amount of macroetching is adequate. Preservation consists of coating the surface with a clear lacquer. If the surface is not preserved it will oxidize and darken with time and lose surface features.

Examination. Examination of macroetched samples may be performed with the naked eye or under a binocular microscope. A *binocular microscope* is a light microscope that provides a low-magnification, three-dimensional view of the surface. A binocular microscope is limited to a magnification of 30x to 50x for most work. The magnification of a binocular microscope is limited by the required depth of field. *Depth of field* is the total depth of the image that can be maintained in focus within a lens. The rougher the surfaces of a weld or its macroetched face, the lower the useful magnification and the greater the depth of field required. Macroscopic examination is described in ASTM E 381, *Macroetch Testing, Inspection, and Rating of Steel Products, Comprising Bars, Billets, Blooms, and Forgings.*

Photomacrography is the documentation of macroetched samples using photography. Photomacrography is performed using an overhead digital camera. A ruler is placed alongside the specimen to indicate scale. The roughened surface attained through macroetching must be in focus to achieve adequate resolution. Resolution is controlled by the depth of field of the lens. Depth of field is controlled by three factors: the focal length of the lens, the aperture (area) of the lens, and the distance between the surface and the lens. Depth of field varies as the inverse square of the focal length. For example, if the focal length is reduced by one half, the depth of field increases by a factor of four. Depth of field doubles as the aperture setting (f-stop number of the lens) doubles. The f-stop indicates the aperture size of the lens. Depth of field is proportional to the square of the distance of the surface from the lens. For example, if the aperture size of the lens is increased by a factor of three, the depth of field increases by a factor of nine.

Lighting

Lighting has the greatest overall effect on the appearance of a surface. Proper use of lighting sources and lighting methods permits key features of the specimen to be revealed. The four types of lighting sources are spotlight, diffused light, reflected light, and flashlight. **See Figure 33-14.**

A *spotlight* is an intense lighting source that uses a single bulb in a reflector. *Diffused light* is a lighting source that uses a semi-opaque screen (such as ground glass) to diffuse the light source, reduce glare, and soften harsh details. *Reflected light* is a lighting source that bounces light off a white card, wall, or ceiling. The effect produced is similar to the effect produced by diffused light. A *flashlight* is a lighting source that provides a pulse of very intense light. A flashlight is the best light source (next to direct sunlight) for color photography of uneven surfaces.

> ✓ **Point**
>
> *Magnification, resolution, and lighting are the three most important methods of photography used for documenting macroscopic examination features.*

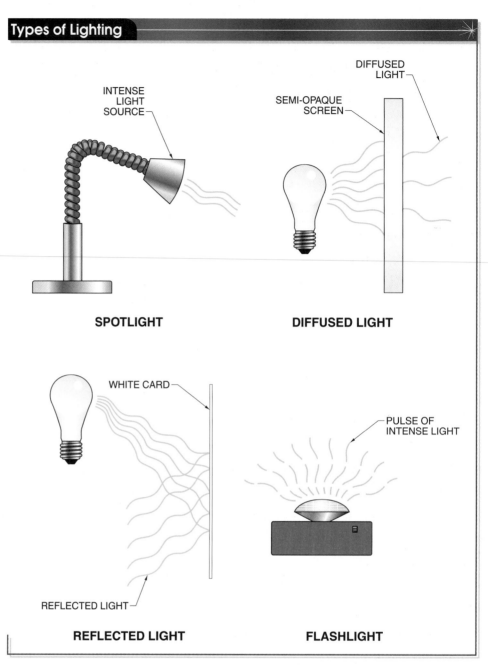

INTENSE
LIGHT
SOURCE

DIFFUSED
LIGHT

SEMI-OPAQUE
SCREEN

SPOTLIGHT

DIFFUSED LIGHT

WHITE CARD

PULSE OF
INTENSE LIGHT

REFLECTED LIGHT

REFLECTED LIGHT

FLASHLIGHT

Figure 33-14. *The four types of lighting sources are spotlight, diffused light, reflected light, and flashlight.*

The four types of lighting methods are main lighting, fill lighting, backlighting, and buildup lighting. These lighting methods use combinations of the four types of light sources to achieve the desired lighting effect.

Main lighting is a primary lighting method that uses a light source at a vertical angle of 40° to 60° to the subject. *Fill lighting* is a lighting method that uses a small region of a brighter light to increase detail on a dark area of a subject. The light source for fill lighting may be spotlight, diffused light, or reflected light. *Backlighting* is a lighting method that uses a diffused light source to eliminate or soften shadow detail. A light box (lighted ground-glass screen) behind the specimen is the most common diffused light source for backlighting. *Buildup lighting* is a lighting method that combines (adding or deleting) light sources to achieve the desired lighting effect. **See Figure 33-15.**

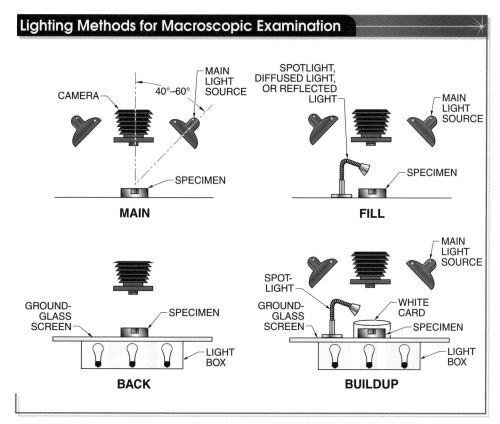

Figure 33-15. *Proper selection and use of lighting methods permits key features on a fracture surface to be revealed.*

Refer to Quick Quiz® on CD-ROM

Points to Remember

- Microscopic examination consists of cutting and rough grinding; mounting and fine grinding; rough and final polishing; and etching and examination.
- Etching often requires the use of strong acids, and all safety precautions must be observed. Always add acid to water when diluting, not vice versa.
- When focusing the metallurgical microscope, contact between the lens and the specimen must be avoided to prevent surface damage to the mounted specimen.
- When examining metallographic samples under a metallurgical microscope, illumination techniques such as brightfield, darkfield, polarized, and Nomarski may be used to reveal microstructural features.
- Macroscopic examination may be used to examine specimens with large test surface areas.
- Macroscopic examination consists of specimen preparation; rough and fine grinding; and macroetching and examination.
- Magnification, resolution, and lighting are the three most important methods of photography used for documenting macroscopic examination features.

1. What types of weld attributes may be studied using metallography?
2. Why must overheating be avoided when cutting specimens for metallographic examination?
3. What is the minimum distance the heat source must be kept from the area of interest when burning is used to remove specimens?
4. What is the purpose of rough grinding for metallographic examination?
5. What is macroscopic examination?
6. Why are successively finer stages of grinding and polishing used to prepare a specimen for metallographic examination?
7. Why must a polished specimen be etched before examination under a metallurgical microscope?
8. What is the difference between microetchants and macroetchants?
9. What type of light does a metallurgical microscope use?
10. How does the magnification range of a metallurgical microscope compare with that of a binocular microscope?
11. What is the meaning of depth of field when used in the examination of macroscopic specimens?

Refer to Chapter 33 in the *Welding Skills Workbook* for additional exercises.

Weld Discontinuities and Failures

Chapter

Weld discontinuities are interruptions in the structure of a weld and are not necessarily weld defects. Weld discontinuities are caused by poor weld design, improper welding procedures, and improper welder techniques. Weld discontinuities are grouped according to their nature.

Weld defects are weld discontinuities that fail to meet the requirements of the codes or standards by which the weld is made. A weld defect requires that the weld be rejected, or repaired and reexamined. Weld defects are not permitted by controlling codes or standards because they can lead to premature failure. Various nondestructive examination (NDE) techniques are used to detect weld defects and discontinuities and measure their size and orientation.

A welder should always perform a thorough visual inspection to look for obvious problems like undercut, overlap (cold lap), porosity, cracks, underfill, and other surface defects or discontinuities before submitting welds for further testing.

WELD DISCONTINUITIES

A *weld discontinuity* is an interruption in the typical structure of a weld. Weld discontinuities can occur as mechanical, metallurgical, or physical flaws in the weld metal, the heat-affected zone (HAZ), or the base metal. Their location varies depending upon the type of weld. A weld discontinuity is not always considered a defect. The transition point between a discontinuity and a defect depends upon the fabrication standard or code that controls the welded joint design and quality.

Discontinuities are detected by nondestructive examination (NDE). The most common NDE techniques used are visual examination (VT), magnetic particle examination (MT), liquid penetrant examination (PT), radiographic examination (RT), and ultrasonic examination (UT). The most applicable NDE technique or techniques are selected to locate and measure the size and orientation of the discontinuity. The discontinuity size and orientation are then compared with what

is allowable in the applicable fabrication standard or code to decide whether the discontinuity is a defect and whether the weld should be accepted or rejected.

Weld Defects

Weld defects result from weld discontinuities that by their nature or their accumulated effect are unable to meet the minimum acceptable requirements of the applicable fabrication standard or code. An unacceptable discontinuity under certain service conditions may be acceptable in a less demanding service or in another metal. Refer to the requirements of the fabrication code or standard that governs the quality of the welded joint under consideration. **See Figure 34-1.** A weld defect requires rejection of the part.

> ✓ **Point**
>
> *Discontinuities are classified as defects when they exceed the minimum requirements permitted by the controlling code or standard.*

> *A discontinuity is a crack, flaw, or imperfection in a base material or weld metal. Discontinuities are classified by their nature (how they alter stresses in the weld) and by their shape, which encompasses their orientation with respect to the working stress and their location with respect to the weld.*

WELD DEFECT EVALUATION GUIDE			
Types of Defect	**Pipelines (per API Std. 1104)**	**Storage Tanks (per API Std. 650)**	**Power Boilers (per ASME Section 1)**
Cracks	None allowed (except shallow crater cracks in the cover pass with maximum length of 5⁄32″)	None Allowed	None Allowed
Incomplete Penetration at root pass	• Maximum of 1″ in length in 12″ of weld, or 8% of weld length if less than 12″ • Maximum individual length of 1″	None Allowed	None Allowed
Incomplete Penetration due to high-low fit-up	• Maximum individual length of 2″ • Maximum accumulated length of 3″ in 12″ of continuous weld	None Allowed	None Allowed
Lack of Fusion at root pass	• Maximum of 1″ in length in 12″ of weld, or 8% of weld length if less than 12″ • Maximum individual length of 1″	None Allowed	None Allowed
Lack of Fusion at groove face or between beads, "cold lap"	• Maximum individual length of 2″ • Maximum accumulated length of 2″ in 12″ of continuous weld	None Allowed	None Allowed
Melt-through	**Pipe Diameter** / **Maximum Defect** / **Maximum Total** less than 2¾″ OD / ¼″ / 1″ greater than or equal to 2¾″ OD / ¼″ / ½″ in 2″	Not Covered	Not Covered
Internal Concavity	If density of radiographic image of internal concavity is less than base metal, any length is allowable. If more dense, then see burn-through above	Shall not reduce weld thickness to less than thinner material. Contour of concavity shall be smooth	Not Covered
Undercut at Root pass or cover pass	• Maximum depth 1⁄32″ or 12½% wall thickness, whichever is smaller. • Maximum 2″ length or 1⁄6 wall thickness, whichever is less, for depth of 1⁄64″ to 1⁄32″ or 6% to 12½% of wall thickness, whichever is less	• For horizontal butt joints: maximum depth 1⁄32″ • For vertical butt joints: maximum depth 1⁄64″	1⁄32″ or 10% t[†], whichever is less
Slag Inclusions elongated, except as noted	• Maximum length is 2″ and width 1⁄16″ • Maximum total length 2″ in 12″ of weld. Parallel slag lines are considered separate if width of either exceeds 1⁄32″. Isolated Slag Inclusions: • Maximum width ⅛″ and ½″ length in 12″ of weld. • No more than 4 isolated inclusions of ⅛″ maximum width.	**Material Thickness** / **Maximum Slag Length** less than or equal to ¾″ / ¼″ ¾″ to 2¼″ / ⅓t[†] greater than 2¼″ / ¾″ Maximum length of t[†] in 12t[†] length	**Material Thickness** / **Maximum Slag Length** less than or equal to ¾″ / ¼″ ¾″ to 2¼″ / ⅓t[†] greater than 2¼″ / ¾″ Maximum total length of t[†] in 12t[†] length
Porosity	Spherical: Maximum dimension ⅛″ or 25% of wall	• For aligned rounded indications, the	• For aligned rounded indications, the

* 100% X-Ray, Random X-Ray, and Spot X-Ray are quality level designations used by the ASME pressure vessel and ANSI piping codes and are also used when other NDE methods of evaluation are used
† t = weld thickness
‡ T = thinner material thickness
§ w = weld width
‖ see UHT-20 for special heat-treated ferritic steels
\# joint category A
** joint categories B, C, and D

Figure 34-1. *Fabrication standards and codes govern the acceptable quality of a welded joint, and are the determining factor in the acceptance or rejection of a weld. See Appendix.*

> **✓ Point**
> *Weld stresses may be concentrated or enhanced by the presence of discontinuities, leading to failure under load.*

Weld Stresses

Weld stresses may be increased or concentrated in a specific area by defects. Weld stresses are magnified when discontinuities reduce the cross-sectional area of the weld that is available to support the load. The average stress on a weld is in direct proportion to the reduction of the load-bearing cross-sectional area caused by the discontinuity. The lower the load-bearing cross-sectional area, the higher the stress. If the load-bearing cross-sectional area of a weld is reduced sufficiently, structural failure may occur under load. **See Figure 34-2.**

Concentrated weld stresses occur at discontinuities that create abrupt changes of geometry, resulting in a notch effect. A *notch effect* is a stress-concentrating condition caused by an abrupt change in section thickness or in continuity of the structure. The sharper the change of geometry in the notch effect, the greater the stress concentration. Tensile stresses perpendicular to the notch and shear stresses parallel to the notch are concentrated at the tip of the notch. Extremely high stress concentrations can develop at extremely sharp notches created by planar-type discontinuities such as cracks, laminations, or incomplete fusion. Such discontinuities may lead to catastrophic fracture in service. **See Figure 34-3.**

> ⊛ *Cracks are the most serious type of imperfection that can occur in welds. Cracks should always be removed because they reduce the strength of the weld.*

Discontinuities that concentrate stress can be extremely harmful. Weld discontinuity types are cracks, cavities, inclusions, incomplete fusion and incomplete penetration, incorrect shape, and miscellaneous discontinuities.

CRACKS

A *crack* is a fracture-type discontinuity characterized by a sharp tip and a high ratio of length to width, and width to opening displacement. Cracks are the most serious discontinuities in weldments and are not permitted in fabrication standards and codes. Cracks are not permitted because they create significant stress concentrations at their tips. **See Figure 34-4.** When cracking is observed during welding, it must be removed before welding continues. Weld metal that is deposited over a crack can result in extension of the crack into newly deposited weld metal.

> ✓ **Point**
> *Cracks are fracture-type discontinuities and are not permitted in fabrication standards and codes.*

> ✓ **Point**
> *The higher the carbon content of steel, the higher the risk of cracking. This is why it is critical to pay proper attention to heat treatment.*

Weld Stresses

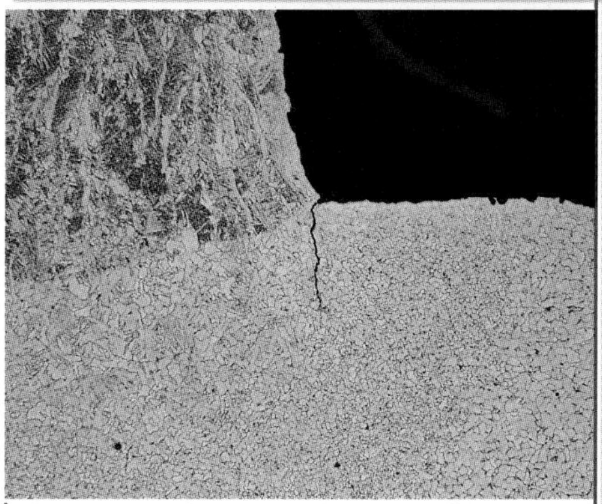

Stork Technimet, Inc.

Figure 34-2. *If the load-bearing area of a weld is sufficiently reduced, structural failure may occur when the part is placed under load.*

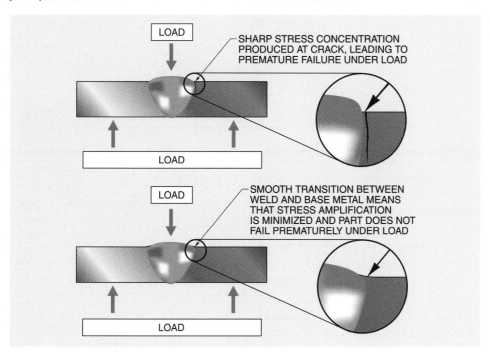

Figure 34-3. *Discontinuities that concentrate stresses are more detrimental than discontinuities that amplify stresses, and may lead to catastrophic fracture and failure in service.*

CRACK PREVENTION	
Problem	**Preventive Measure**
Weld Metal	
Highly rigid joint	Preheat
	Relieve residual stresses mechanically
	Minimize shrinkage stresses using back-step or intermittent welding sequence
Excessive dilution	Change welding current and travel speed
	Weld with covered electrode, DCEN; butter the joint faces prior to welding
Defective electrodes	Change to new electrode; properly store and maintain electrodes to prevent moisture and damage
Poor fit-up	Reduce root opening; build up edges with weld metal
Small weld bead	Increase electrode size; raise welding current; reduce travel speed
High-sulfur base metal	Use low-sulfur filler metal
Angular distortion	Change to balanced welding on both sides of joint
Crater cracking	Fill crater before extinguishing the arc; use a welding current decay device when terminating the weld bead
Heat-Affected Zone	
Hydrogen in welding atmosphere	Use low-hydrogen welding process; preheat and hold for 2 hr after welding or postheat immediately
Hot cracking	Use low heat input; deposit thin layers; change base metal
Low ductility	Use preheat; anneal base metal
High residual stresses	Redesign the weldment; change welding sequence; apply intermediate stress-relief heat treatment
High hardenability	Preheat; increase heat input; heat treat without cooling to room temperature
Brittle phases in the microstructure	Solution heat treat prior to welding

Figure 34-4. *Cracks are not permitted in metal because they create significant stress concentrations at their tips. Cracks must be removed before welding continues.*

Cracks may occur in the weld, the HAZ, or the base metal when the localized stress exceeds the ultimate strength of the metal. Cracks are classified as hot cracks or cold cracks, and may be longitudinal or transverse in their orientation. A *hot crack* is a crack formed at temperatures near the completion of solidification. A *cold crack* is a crack that develops after solidification is complete. Hot cracks propagate between the grains of metal, and cold cracks propagate both between and through the grains of the metal. A *longitudinal crack* is a crack with its major axis oriented approximately parallel to the weld axis. A *transverse crack* is a crack with its major axis oriented approximately perpendicular to the weld axis. **See Figure 34-5.**

Cracks are classified according to their location in the weld. Crack types in welds are throat cracks, crater cracks, transverse cracks, underbead cracks, lamellar tearing, toe and root cracks, fissures, and liquid metal embrittlement.

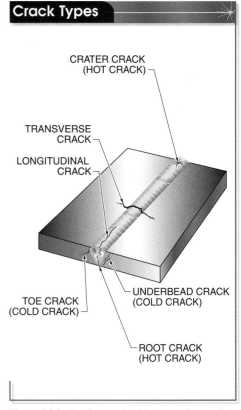

Crack Types

CRATER CRACK (HOT CRACK)

TRANSVERSE CRACK

LONGITUDINAL CRACK

TOE CRACK (COLD CRACK)

UNDERBEAD CRACK (COLD CRACK)

ROOT CRACK (HOT CRACK)

Figure 34-5. *Cracks are classified according to their location in the weld.*

Throat Cracks

Throat cracks are longitudinal cracks in the middle of the surface (throat) of a weld, extending toward the root of the weld. Throat cracks are hot cracks that are confined to the center of the weld. Throat cracks may be the extension through successive weld passes of a crack that started in the first pass (root pass). A throat crack that starts in the root pass and is not removed or completely remelted before deposition of the next pass tends to progress into it and then to each succeeding pass, until it appears at the surface. Final extension of the crack to the surface may also occur during cooling after welding has been completed. **See Figure 34-6.**

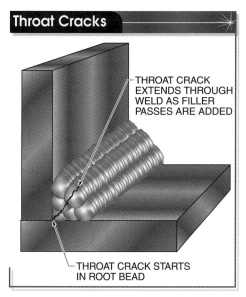

Figure 34-6. *Throat cracks are longitudinal cracks that start in the root bead and extend through the weld as filler passes are added.*

Throat cracks are detected by visual examination or liquid penetrant examination. Throat cracks appear as relatively long, straight cracks along the centerline of the weld. VT is often an adequate method of detection because throat cracks are relatively wide discontinuities.

Throat Crack Prevention. Throat cracks are prevented by using joint designs that reduce joint restraint and excessive stresses in solidifying weld metal. The weld groove dimensions must be adjusted to allow deposition of a sufficient amount of filler metal to overcome excessive joint restraint. The welding process variables must be adjusted to permit correct weld bead size for the joint thickness, sufficient heat input, and optimum travel speed to prevent excessive stresses during solidification. These may also be achieved by factors such as using a more ductile filler metal and reducing cooling rate through application of preheat.

Crater Cracks

Crater cracks are star-shaped cracks that extend from the crater (center) of the weld to the edge of the weld. Crater cracks may be the starting point for throat cracks, particularly when in the crater formed at the completion of a weld. Crater cracks are hot cracks caused by failing to back-fill the crater before terminating the arc at the end of a weld. When the crater is formed elsewhere (for instance, when an electrode is changed), the crack may be welded out when operation resumes. If not, fine star-shaped cracks are observed at various locations. Crater cracks are most often found in materials with high coefficients of expansion such as austenitic stainless steels. **See Figure 34-7.** Crater cracks are most commonly detected by VT. Crater cracks are clearly visible to the naked eye as star-shaped fissures in the small crater formed at the termination of a weld pass.

Figure 34-7. *Crater cracks are hot cracks formed from improper termination of the welding arc in the crater of the weld. Crater cracks are commonly found in materials with a high coefficient of expansion.*

Crater Crack Prevention. Crater cracks are prevented by properly terminating the weld. Methods used to prevent cracks include back-stepping the arc into previously solidified material before breaking it, using a foot pedal to allow decay of the arc; filling craters to a slightly convex shape prior to breaking the arc; or using a run-off tab. A *run-off tab* is a piece of metal of the same composition and thickness as the base metal that is tacked to the metal to allow the weld to be completed. The run-off tab is later removed by cutting it off. **See Figure 34-8.**

Media Clip

Run-off Tab

Run-off Tabs

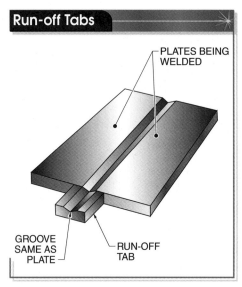

PLATES BEING WELDED

GROOVE SAME AS PLATE

RUN-OFF TAB

Figure 34-8. *A run-off tab is tack welded to the plates to be welded to allow welding to run off onto it to prevent crater cracks from forming in the weld, and to completely fill the joint.*

Transverse Cracks

Transverse cracks are cracks in a weld perpendicular to the axis of the weld and sometimes extending beyond the weld into the base metal. Transverse cracks are cold cracks resulting from high restraint acting on low ductility weld metal. Transverse cracks in steel weldments are usually related to hydrogen embrittlement. Transverse cracks are detected by VT, PT, and MT as tight, relatively straight cracks perpendicular to the weld axis.

Transverse Crack Prevention. Transverse crack prevention depends on the specific welding situation. For example,

transverse cracks may be caused by using an incorrect filler metal composition, rapid cooling, or a weld that is too small for the part being joined. Depending on the situation, transverse cracks may be eliminated by using the proper filler metal composition, higher welding current or preheat, or a larger filler metal and final weld dimension, respectively.

Underbead Cracks

Underbead cracks are cracks in the HAZ that generally do not extend to the surface of the base metal. Underbead cracks may be longitudinal or transverse, depending on the direction of the principal stresses in the weldment. Underbead cracks are cold cracks and are usually short and discontinuous. **See Figure 34-9.**

Underbead Cracks

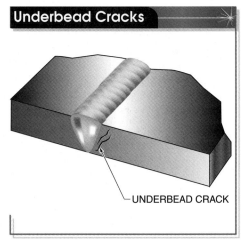

UNDERBEAD CRACK

Figure 34-9. *Underbead cracks are cold cracks that form in the heat-affected zone, and generally do not extend to the surface of the base metal.*

Underbead cracks are hydrogen cracks that occur in steels susceptible to hydrogen embrittlement during welding. Dissolved hydrogen, which is released from the electrode or from the base metal, combines with martensite formed in the HAZ during rapid cooling, creating a narrow region that is extremely brittle and sensitive to cracking from residual stresses. Underbead cracks are detected by UT or RT because the crack is usually below the surface and immediately adjacent the weld. Because of their tightness and short length, underbead cracks may be difficult to detect.

Underbead Crack Prevention. Underbead cracks are prevented by avoiding hydrogen creation in steels that are susceptible to hydrogen embrittlement. Welding conditions that encourage hydrogen creation include poor sheltering of outdoor work that permits rain, snow, or condensation to contact welded areas. Underbead crack prevention is achieved by using low-hydrogen electrodes to join susceptible steels and excluding moisture from electrodes. A drying procedure must be used to remove moisture that can absorb into the coatings on some types of electrodes when exposed to humid atmospheres. The procedure involves storing electrodes in a low-temperature oven, preheating the surface before welding to remove moisture, and postheating immediately to encourage hydrogen to escape. **See Figure 34-10.**

Figure 34-10. *Low-hydrogen electrodes can help prevent underbead cracking. Moisture is removed from electrodes before use by storing the electrodes in an oven.*

Lamellar Tearing

Lamellar tearing is a subsurface terrace and step-like crack pattern in wrought steel base metal oriented parallel to the base metal working direction. Lamellar tearing is caused by tensile stresses in the base metal from welding in a direction perpendicular to the working direction, acting upon nonmetallic inclusions in the base metal parallel to the working direction. Nonmetallic inclusions consist of metallic oxides, sulfides, and silicates that are held in steel. Nonmetallic inclusions are formed during solidification in the steelmaking process from additives to the melt or contamination from refractory in the mold. Hot or cold working elongates nonmetallic inclusions in the working direction if they are plastic at the working temperature. The net result of the elongated nonmetallic inclusions is to decrease through-thickness ductility. This results in lamellar tearing parallel to the direction of the inclusions.

Lamellar tearing is most likely to occur when welding a steel plate using groove welds, fillet welds, or combinations of them. T-joints may be especially susceptible to lamellar tearing. **See Figure 34-11.** The two members of a T-joint are located at approximately right angles to each other in the form of a T. Under these conditions, high tensile stresses can develop perpendicular to the midplane of the steel plate. The magnitude of the tensile stresses depends on the size of the weld, the welding procedure, and the amount of joint restraint imposed by the welding design. Lamellar tearing detection is difficult because it usually does not break to the surface. RT and UT are the most applicable methods for detection of lamellar tearing, which has the appearance of step-like, jagged cracking, with each step nearly parallel to the midplane of the plate.

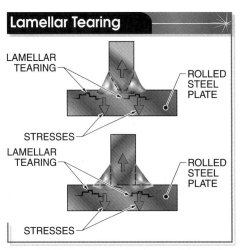

Figure 34-11. *Lamellar tearing is caused by welding stresses in the base metal perpendicular to the working direction.*

Lamellar Tearing Prevention. Lamellar tearing is prevented most reliably by the use of specially processed steel products that do not contain elongated nonmetallic inclusions. Such steel products are used in critical applications where lamellar tearing is detrimental.

Other methods of reducing lamellar tearing in regular steel products rely on reduction of the stress in the welded joint. **See Figure 34-12.** These methods include changing the location and/or design of the weld joint to minimize through-thickness strains; using a lower strength weld metal; reducing hydrogen in the weld; using preheat and interpass temperatures of at least 200°F; and peening the weld bead immediately after completion of a weld pass.

Laminations

A *lamination* is a defect or discontinuity that is aligned parallel to the worked surface of a metal. Typically, a lamination is the result of ingot steel production. As a steel ingot cools, it may develop a cavity called a "pipe." When the ingot is rolled, the pipe becomes elongated and forms a lamination in the steel.

Toe Cracks and Root Cracks

Toe cracks and root cracks have similar causes but different appearances. Toe cracks are cracks that proceed from the weld toe into the HAZ and base metal. The weld toe is the junction of the weld face and base metal. Root cracks are cracks that proceed into the base metal from the root of a fillet weld. Toe cracks and root cracks are generally cold cracks and initiate in regions of high residual stress. **See Figure 34-13.** Toe cracks are generally caused by stresses from thermal shrinkage acting on a brittle HAZ. Toe cracks are identified by VT, PT, and MT, and form their location at the weld toe. Root cracks are difficult to detect unless they have propagated through to the opposite side of the base metal.

Toe and Root Crack Prevention. Toe and root crack prevention requires welding procedures and techniques that eliminate embrittlement or excessive stresses in the HAZ of the base metal. With hardenable steels, toe and root crack prevention may be achieved by retarding the cooling rate of the base metal and HAZ with high preheat, or by stress relief after welding with postheat.

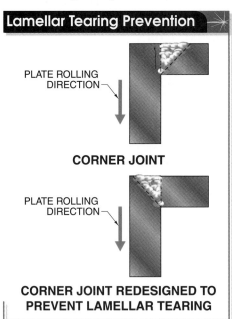

Figure 34-12. *The most effective way of preventing lamellar tearing is to redesign the weld joint to minimize stresses on the joint.*

Liquid Metal Embrittlement

Liquid metal embrittlement is intergranular penetration (cracking) of the HAZ. *Intergranular penetration* is penetration of molten metal along the grain boundaries of

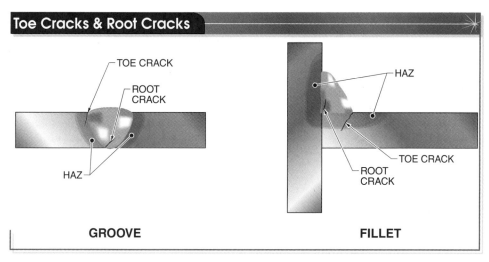

GROOVE

FILLET

Figure 34-13. *Toe cracks proceed from the weld toe into the heat-affected zone and base metal. Root cracks initiate in regions of high residual stress.*

the base metal that leads to embrittlement of the base metal. Liquid metal embrittlement can occur with specific combinations of base metals and liquid metals, usually in the presence of stress. **See Figure 34-14.**

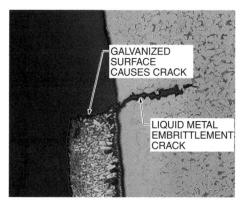

Figure 34-14. *Liquid metal embrittlement commonly occurs in certain types of metals, usually where a part is exposed to excess stress.*

Brazes are a common cause of liquid metal embrittlement in susceptible alloys. For example, many nickel alloys, when in a stressed condition, may crack from liquid metal embrittlement in contact with molten brazing metal. Liquid metal embrittlement may also occur during welding from contamination of a base metal by other metals. For example, when welding austenitic stainless steels to galvanized steels, zinc contamination may cause liquid metal embrittlement of the austenitic stainless steel base metal. The zinc contamination may be introduced by grinding dust or direct contact between the two metals, such as when welding austenitic stainless steel to galvanized carbon steel. Liquid metal embrittlement may be detected by PT as a relatively wide, jagged crack revealed under magnification.

Liquid Metal Embrittlement Prevention. Liquid metal embrittlement is prevented by avoiding susceptible braze-base metal couples or by ensuring cleanliness of the joint surfaces before welding or brazing. For example, when welding galvanized steel to austenitic stainless steel, all zinc must be removed by grit blasting a minimum of 2″ from the joint face to ensure that the zinc does not melt and mix with the austenitic stainless steel, resulting in liquid metal embrittlement. Liquid metal embrittlement susceptibility may be assessed prior to welding or brazing by testing combinations of weld metal and base metal under simulated joint restraint conditions.

CAVITIES

Cavities are weld discontinuities consisting of rounded holes of various types, either within the weld or at the surface of the weld. Two causes of cavities are gas entrapment during solidification of the weld or contraction (suckback) of the weld during solidification, which cannot be replaced by molten metal. Porosity and wormholes are cavity types formed by gases. Shrinkage voids are a cavity type formed by contraction of the weld metal during solidification. **See Figure 34-15.**

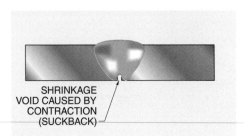

Figure 34-15. *Shrinkage voids are a cavity type formed by contraction (suckback) of the weld metal during solidification.*

Cavities are usually less serious than cracks because their rounded shape causes significantly lower stress concentration than cracks. Fabrication standards and codes allow certain types of cavities such as porosity, depending on their size, number, and orientation.

Porosity

Porosity consists of cavity-type discontinuities formed by gas entrapment during solidification. **See Figure 34-16.** Porosity may be surface porosity or subsurface porosity. Surface porosity (blowholes) consists of discrete spherical pits on the surface of the weld. Surface porosity is formed if dissolved gases cannot fully escape before the weld metal solidifies. Surface porosity may be caused by insufficient shielding gas coverage. It may also be caused by excessive gas flow rates that create turbulence that expose the molten weld to oxygen in the air. Surface porosity may be detrimental to fatigue strength if aligned in a direction perpendicular to the direction of stresses.

Subsurface porosity consists of discrete spherical holes within the body of a weld. Subsurface porosity distribution is classified as uniformly scattered, cluster, or linear. Uniformly scattered porosity exhibits a uniform distribution of pores throughout the weld metal, with size varying from almost microscopic to ⅛″ in diameter. Cluster porosity voids occur in the form of clusters separated by considerable lengths of pore-free weld metal. Cluster porosity is associated with changes in welding conditions, such as stopping or starting of the arc. Linear porosity is characterized by an accumulation of pores in a relatively straight line. The number and size of the pores and their linear distribution with respect to the axis of the weld usually define linear porosity. Linear porosity generally occurs in the root pass.

Primary causes of porosity are dirt, rust, and moisture on the surface of the base metal; on the welding consumables; and in the welding equipment.

Porosity is usually the least harmful type of weld discontinuity. Many fabrication standards and codes provide comparison charts that show the amount of porosity that may be acceptable. When porosity exceeds the amount allowable, it must be ground out and repaired. Porosity is detected by RT for internal porosity and by VT or PT for surface porosity. With RT, internal porosity has the appearance of sharply defined dark shadows of rounded contour.

Porosity

Stork Technimet, Inc.

Figure 34-16. *Porosity is formed by gas entrapment within the weld during solidification if dissolved gases cannot fully escape before the metal solidifies.*

Porosity Prevention. Porosity is prevented by improving welding housekeeping conditions that can cause the porosity. Good housekeeping includes the use of clean materials and well-maintained equipment. Also, avoiding the use of excessive current and arc lengths can prevent porosity. High currents and excessive arc lengths may cause high consumption of the deoxidizing elements in the covering of shielded metal arc electrodes, leaving insufficient quantities available to combine with the gases in the molten metal during cooling.

Specific methods of preventing porosity depend on the type of welding process. For example, changing welding conditions such as gas flow rate and gas purity for gas shielded processes compensates for improper arc length, welding current, or electrode manipulation. Reducing travel speed may also decrease porosity. **See Figure 34-17.**

Wormholes

Wormholes are elongated or tubular cavities caused by excessive entrapped gas. Wormholes are detected by RT where they have the appearance of sharply defined dark shadows of rounded or elongated contour, depending on the orientation of the wormholes.

Wormhole Prevention. Wormholes are prevented by methods that are similar to those that prevent porosity.

Shrinkage Voids

Shrinkage voids (pipe or hollow bead) are cavity-type discontinuities normally formed by shrinkage during solidification and are usually in the form of long cavities parallel to the root of the weld. Shrinkage voids are detected by RT.

Shrinkage Void Prevention. Shrinkage voids are prevented by providing sufficient heat input to maintain molten filler metal to all areas of a weld during solidification.

> ✓ **Point**
> *Inclusions consist of foreign matter in the weld metal, either from the base metal, filler metal, or nonconsumable electrode.*

POROSITY PREVENTION	
Problem	**Preventive Measure**
Excessive hydrogen, nitrogen, or oxygen in welding atmosphere	Use low-hydrogen welding process; use filler metals high in deoxidizers; increase shielding gas flow
High solidification rate	Use preheat or increase heat input
Dirty base metal	Clean joint faces and adjacent surfaces
Dirty filler wire	Use specially cleaned and packaged filler wire, and store it in clean area
Improper arc length, welding current, or electrode manipulation	Change welding conditions and techniques
Volatilization of zinc from brass	Use copper-silicon filler metal; reduce heat input
Galvanized steel	Use E6010 electrodes and manipulate the arc heat to volatilize the zinc ahead of the molten weld pool; remove zinc from weld region
Excessive moisture in electrode covering or on joint surfaces	Use recommended procedures for baking and storing electrodes Preheat the base metal
High-sulfur base metal	Use electrodes with basic slagging reactions

Figure 34-17. *Porosity prevention methods are determined by the type of welding process; corrective measures are based on the type of problem that has occurred.*

INCLUSIONS

Inclusions are entrapped foreign solid material in deposited weld metal, such as slag or flux, tungsten, or oxide. Inclusion types are slag inclusions, oxide inclusions, and tungsten inclusions. **See Appendix.**

Slag Inclusions

Slag inclusions are nonmetallic materials that are formed by slag reactions and trapped in a weld. Slag is a nonmetallic product resulting from mutual dissolution (chemical reactions) of flux and nonmetallic impurities in some welding and brazing processes. Slag inclusions can occur between passes or at the groove face. **See Figure 34-18.** Slag inclusions may occur in welds made by flux shielded welding processes such as SMAW, FCAW, and SAW. Slag inclusions have a lower specific gravity than the surrounding metal and usually rise to the surface of molten metal, unless they become entrapped in the weld metal.

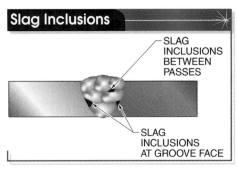

Figure 34-18. *Slag inclusions are nonmetallic materials formed by slag reactions that are trapped in a weld. Slag inclusions can occur between passes or at the groove face.*

Multiple-pass welds are more prone to slag inclusions than single-pass welds because slag from the preceding pass, if not completely removed, will become entrapped in the subsequent pass. Slag inclusions are detected by RT where they appear as dark lines, more or less interrupted, parallel to the edges of the weld. Slag inclusions are usually elongated and rounded, and run in the direction of the axis of the weld. Slag inclusions can be continuous, intermittent, or randomly spaced.

Slag Inclusion Prevention. Slag inclusions are prevented by using proper welding preparation, such as thoroughly removing slag from the weld and cleaning the weld groove between each pass of a multiple-pass weld. Failure to thoroughly remove slag between each pass increases the probability of slag entrapment and the production of a defective weld. Slag may be removed from the weld surface by chipping, wire brushing, grinding, or air arc gouging. **See Figure 34-19.**

Complete and efficient slag removal requires that each weld bead be properly contoured and blend smoothly into the adjacent bead or base metal. Small weld beads cool more rapidly than large ones, which tends to make slag removal easier from small beads. Concave or flat beads that blend smoothly into the base metal or any adjoining beads minimize undercutting and avoid a sharp notch along the edge of the bead where slag could stick.

INCLUSION PREVENTION

Problem	Preventive Measure
Slag inclusions	Clean surface and previous weld bead
Entrapment of refractory oxides	Power wire brush the previous weld bead
Tungsten in the weld metal	Avoid contact between the electrode and the work; use larger electrode
Improper joint design	Increase groove angle of joint
Oxide inclusions	Provide proper gas shielding
Slag flooding ahead of the welding arc	Reposition work to prevent loss of slag control
Poor electrode manipulative technique	Change electrode or flux to improve slag control
Entrapped pieces of electrode covering	Use undamaged electrodes

Figure 34-19. *Slag inclusion prevention can be achieved through proper cleaning of the weld groove before depositing additional weld beads. Slag may be removed from the surface by chipping, wire brushing, grinding, or air arc gouging.*

Oxide Inclusions

Oxide inclusions are particles of surface oxides on the base metal or weld filler metal that have not melted and mix with the weld metal. Oxide inclusions occur when welding metals that have tenacious surface oxide films, such as stainless steels, aluminum alloys, and titanium alloys. Oxide inclusions are detected by RT.

Oxide Inclusion Prevention. Oxide inclusion prevention is achieved by cleaning out the joint and weld area thoroughly before welding. **See Figure 34-20.** The weld area should be thoroughly cleaned after each pass using a wire brush.

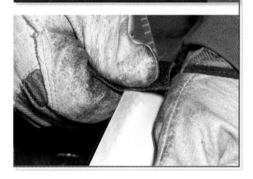

Figure 34-20. *Oxide inclusions can be prevented by cleaning out the joint and weld area thoroughly before welding.*

Tungsten Inclusions

Tungsten inclusions are particles from the nonconsumable tungsten electrode that enter the weld. **See Figure 34-21.** The occasional contact between the electrode and the work or the molten metal may transfer particles of the tungsten into the weld deposit. Improper grinding of the tungsten electrode, or using an electrode too small in relation to the amperage setting, may cause tungsten spitting that can also result in tungsten inclusions in the weld. A limited number of tungsten inclusions may be acceptable according to the applicable fabrication standard or code, but it will depend on the thickness of the part being welded. Tungsten inclusions are detected by VT or RT. VT is used for tungsten inclusions at the surface. However, as with most other types of inclusions, tungsten

inclusions are generally detected using RT, where they appear as isolated, sharp, irregular shapes.

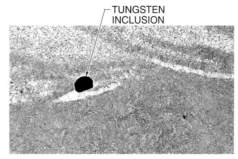

Stork Technimet, Inc.

Figure 34-21. *Tungsten inclusions are particles found in the weld metal as a result of the nonconsumable tungsten electrode coming in contact with the work or the molten metal.*

Tungsten Inclusion Prevention. Tungsten inclusions are prevented at the weld start using superimposed high-frequency current for arc starting and a copper striker plate. Tungsten inclusions may be minimized by using thoriated tungsten or zirconium-tungsten electrodes and less current or larger electrodes, and by keeping the tungsten electrode out of the molten weld pool.

INCOMPLETE FUSION AND INCOMPLETE PENETRATION

Incomplete fusion (lack of fusion) and incomplete joint penetration (lack of penetration) are similar discontinuities. They result from incomplete melting at the interface between weld passes or in the root of the joint.

Incomplete Fusion

Incomplete fusion is a lack of union (fusion) between adjacent weld passes or base metal. Incomplete fusion may be caused by failure to raise the temperature of the surface layers of base metal or previously deposited weld metal to the melting temperature. Incomplete fusion is usually elongated in the direction of welding, with either sharp or rounded edges. **See Figure 34-22.**

> **✓ Point**
>
> *Incomplete fusion and incomplete penetration are found in areas with incomplete melting between the base metal and the weld metal. Incomplete fusion is less desirable than incomplete penetration.*

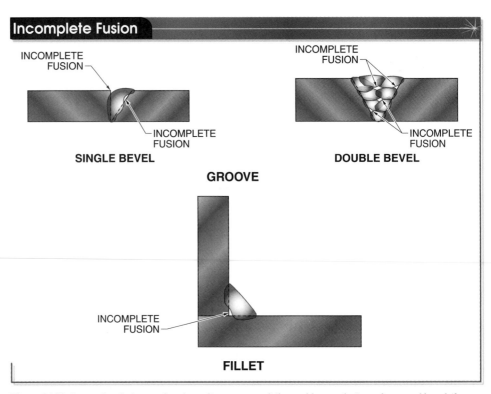

Incomplete Fusion

INCOMPLETE FUSION

INCOMPLETE FUSION

SINGLE BEVEL

INCOMPLETE FUSION

INCOMPLETE FUSION

DOUBLE BEVEL

GROOVE

INCOMPLETE FUSION

FILLET

Figure 34-22. *Incomplete fusion results when adjacent passes fail to meld properly. It can be caused by a failure to sufficiently raise the temperature of the surface layers of the base metal or deposited metal.*

Incomplete fusion occurs more commonly with some welding processes than with others. For example, the reduced heat input in the short circuiting transfer mode of GMAW results in low penetration into the base metal. This may be desirable on thin-gauge materials and for out-of-position welding, but may result in incomplete fusion, especially in the root area or along groove faces. Incomplete fusion leads to undesirable stresses and is severely restricted in most fabrication standards and codes.

Incomplete fusion can be detected by RT as a thin, dark line with sharply defined edges. Depending on the orientation of the defect with respect to the X-ray beam, the line may tend to be wavy and diffuse. However, some codes may not permit RT as a means of qualifying welders when using GMAW short circuiting transfer on test welds.

Incomplete Fusion Prevention. Incomplete fusion is prevented by ensuring an adequate surface temperature to raise the temperature of the surface layers to the melting point, which allows the deposited metal to fuse with the surface below it. This may be achieved by reducing travel speed, increasing welding current or increasing electrode diameter, using joint design to allow electrode accessibility to all surfaces within the joint, use of proper electrode angle, and reducing the effects of arc blow. **See Figure 34-23.**

Incomplete Penetration

Incomplete penetration is a condition in a groove weld in which weld metal does not extend through the joint thickness. In arc welding, the arc is established between the electrode and closest part of the base metal. All other areas of the base metal receive heat principally by conduction. If the region of base metal closest to the electrode is a considerable distance from the joint root, heat conduction may be insufficient to attain adequate temperature to achieve fusion of the root. **See Figure 34-24.**

INCOMPLETE FUSION PREVENTION	
Problem	**Preventive Measure**
Insufficient heat input	Use correct type or size of electrode; proper joint design; and proper gas shielding
Incorrect electrode position	Maintain proper electrode position
Weld metal running ahead of the arc	Reposition work; lower current; increase weld travel speed
Trapped oxides or slag on weld groove or weld face	Clean weld surface prior to welding

Figure 34-23. *Incomplete fusion prevention can be ensured using the proper welding parameters.*

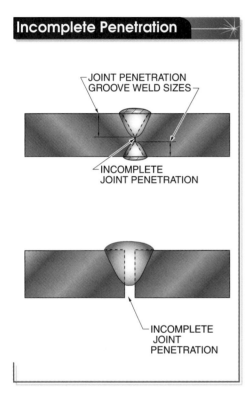

Incomplete Penetration

JOINT PENETRATION
GROOVE WELD SIZES

INCOMPLETE
JOINT PENETRATION

INCOMPLETE
JOINT
PENETRATION

Figure 34-24. *Incomplete joint penetration occurs when weld metal does not penetrate completely through the joint thickness. It can occur when the base metal is a considerable distance from the heat of the electrode.*

Incomplete penetration is not always undesirable because some weld joints are designed for partial penetration. The applicable fabrication standards and codes indicate permissible levels of incomplete penetration. Incomplete penetration is detected by RT, where it appears as a dark, continuous or intermittent line in the middle of the weld.

Incomplete penetration may occur when a groove is welded from one side only if the root face dimension is too great, if the root opening is too small, or if the groove angle of the V-groove is too small, even with an adequate root opening and a satisfactory joint design. Incomplete penetration may also be caused by electrodes that are too large or that have a tendency to bridge; or by using abnormally high rates of travel or insufficient welding current.

Incomplete Penetration Prevention. The most frequent cause of incomplete penetration is the use of an unsuitable joint design for the welding process or the conditions of the actual weld construction. Unsuitable joint designs make it difficult to reproduce qualification test results under conditions of actual production. **See Figure 34-25.**

INCORRECT SHAPE

An incorrect shape in a weld includes any weld discontinuity that produces an unacceptable weld profile or dimensional nonconformance and that adversely influences performance of the weld under load. An insufficient cross-sectional area of a weld may result in a weld that is unable to support a load, or may allow a stress-concentrating notch, leading to fracture. Incorrect shape discontinuities are undercut, overlap, excessive weld reinforcement, underfill, concave root surface, and melt-through.

> ✓ **Point**
> *Incorrect shapes, such as undercut, overlap, excess weld reinforcement, underfill, concave root surface, and melt-through, produce an unacceptable weld profile.*

INCOMPLETE PENETRATION PREVENTION	
Problem	**Preventive Measure**
Excessively thick root face or insufficient root opening	Use proper joint geometry
Insufficient heat input	Follow welding procedure
Slag flooding ahead of arc	Adjust electrode or work position
Electrode diameter too large	Use smaller electrode or increase root opening
Misalignment of second side weld	Improve visibility or backgouge weld
Failure to backgouge when specified	Backgouge to sound metal if required in welding procedure specification
Bridging of root opening	Use wider root opening or smaller electrode in root pass

Figure 34-25. *Using a proper joint design can help ensure that incomplete joint penetration does not occur in a weld.*

Undercut

Undercut is a groove melted into the base metal adjacent to the weld toe or weld root and left unfilled by weld metal. **See Figure 34-26.** Faulty electrode manipulation, excessive welding current, excessive arc length, excessive travel speed, and arc blow cause undercut. Undercut of a completed weld is undesirable because it produces a stress concentration that reduces impact strength and fatigue resistance. Undercut is detected by VT in groove or fillet welds. VT is the simplest and most effective way of detecting and measuring undercut against the particular fabrication standards or codes. RT may also detect undercut in groove welds, where it appears as a dark line, sometimes broad and diffuse, along the edge of the weld.

Undercut Prevention. Undercut is prevented by the following methods: pausing at each side of the weld bead when using the weave bead technique; using proper electrode angles; using proper welding current for electrode size and welding position; reducing arc length; reducing travel speed; and reducing the effects of arc blow. **See Appendix.**

Undercut of the sidewalls of a welding groove will in no way affect the completed weld if it is removed before the next bead is deposited in that location. A well-rounded chipping tool or grinding wheel will be required to remove the undercut. If the undercut is slight, however, it is possible for the welder to estimate how deeply the weld will penetrate and fill the undercut with the next pass.

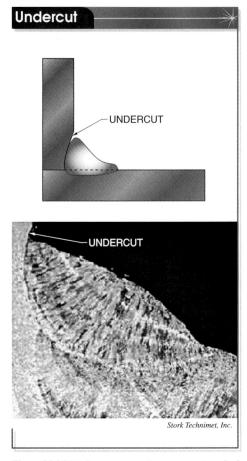

Stork Technimet, Inc.

Figure 34-26. *Undercut occurs when a groove is melted into the base metal adjacent to the weld toe and is left unfilled by weld metal.*

Undercut is sometimes repaired by grinding and blending or welding. Grinding should be performed with a pencil-type grinder, and the grinding marks should be transverse to the length of the weld with a 250 microinch finish or better.

Overlap

Overlap is the protrusion of weld metal built up beyond the weld toe or weld root. Overlap is an area of incomplete fusion that creates a stress concentration and can initiate premature failure under load. **See Figure 34-27.** Overlap is caused when current is set too low. It is detected by VT. Overlap is considered a defect that must be removed by grinding according to the applicable fabrication standard or code.

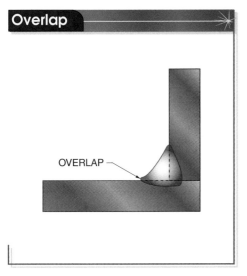

Figure 34-27. *Overlap is a protrusion of weld metal built up beyond the weld toe or the weld root.*

Overlap Prevention. Overlap is prevented by using a higher travel speed or welding current, reducing the electrode diameter, or changing the electrode angle so that the force of the arc will not push molten weld metal over unfused sections of base metal. **See Appendix.**

Excessive Weld Reinforcement

Excessive weld reinforcement is weld metal built up in excess of the quantity required to fill a joint. Excessive weld reinforcement can be of two types—excessive face reinforcement or excessive root reinforcement. **See Figure 34-28.**

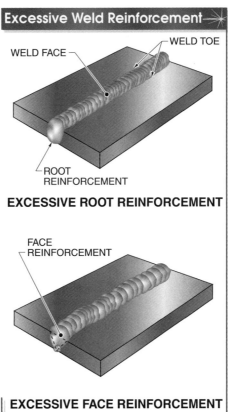

Figure 34-28. *Excessive weld reinforcement, while not a severe discontinuity, can stiffen a section of metal, causing stress concentrations along the toe of the weld. It is also more expensive due to the increased amount of filler metal needed, and can have an objectionable appearance.*

Filler metal added to make a weld must be as thick as the base metal. Slightly thicker filler metal is usually permitted to allow for discontinuities and to avoid the cost associated with grinding the weld metal flush with the base metal. Excessive weld reinforcement, though not as severe as overlap, is undesirable because it thickens and stiffens the section and establishes a stress concentration along the toes of the weld. The stress-concentrating effect is more severe for fillet welds than for butt welds. Excessive weld reinforcement is economically unsound and objectionable from the appearance point of view. It is typically caused by traveling too slow, or by setting amperage too low.

Fabrication standards and codes usually limit the allowable amount of excessive weld reinforcement. Various welding codes impose a maximum amount of reinforcement for the thickness of the material being welded. Thicknesses may vary from 1/16″ to 7/32″. Excessive weld reinforcement is detected by VT and measurement. If considered a defect, it must be removed by grinding.

Excessive Weld Reinforcement Prevention. Excessive weld reinforcement is prevented by use of the correct welding current, proper welding technique, and appropriate number of weld passes to fill the joint.

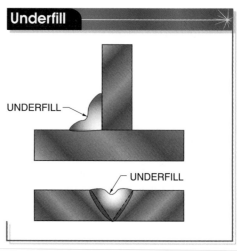

Figure 34-29. *Underfill is a discontinuity that extends below the adjacent surface of the base metal.*

Underfill

Underfill is a discontinuity in which the weld face or root surface extends below the adjacent surface of the base metal. Underfill occurs when a groove weld is not filled completely. Underfill reduces the cross-sectional area of the weld below the amount required in the design. **See Figure 34-29.** Underfill tends to occur primarily in the flat position in fillet welding and in the 5G and 6G pipe groove welding positions. Underfill creates a region susceptible to structural failure from insufficient cross section to support the load. In fillet welds, underfill is exhibited by a less than normal throat as measured by the length of the leg. Underfill is detected by VT.

Underfill Prevention. Underfill is prevented by reducing welding current and voltage, reducing arc length and arc travel speed, and adding sufficient filler metal.

Concave Root Surface

A *concave root surface* is a depression in the weld extending below the surface of the adjacent base metal caused by an underfill in the root pass of a weld. Concave root surface is detected by RT. If considered a defect, the surface may be suitably prepared or cleaned and additional weld metal added.

Concave Root Surface Prevention. Concave root surfaces are prevented in butt welds by reducing the root opening of the weld.

Melt-Through

Melt-through also called burn-through, is a discontinuity that occurs in butt welds when the arc melts through the bottom of the weld. Melt-through is different than melt-thru, which is visible root reinforcement produced in a joint that is welded from one side. **See Figure 34-30.** Melt-thru is often specified; melt-through is a discontinuity or defect. Melt-through is detected by RT as a region of excessive thickness (lower density) in the region of the weld root.

Melt-Through Prevention. Melt-through is prevented in butt welds by reducing the welding current and width of the root opening, and by increasing the arc travel speed.

MISCELLANEOUS DISCONTINUITIES

Miscellaneous discontinuities include weld discontinuities that do not fit into other categories of discontinuities. Miscellaneous discontinuities include arc strikes and spatter.

Melt-Through

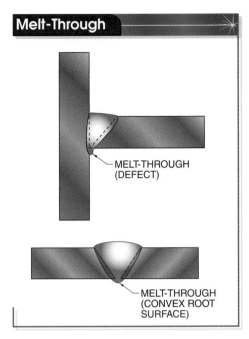

Figure 34-30. *Melt-through is a discontinuity produced in a joint when the arc melts through the bottom of the weld.*

Arc Strikes

An *arc strike* is a discontinuity that results from arcing of the electrode and consists of any localized remelted metal, heat-affected metal, or change in the surface profile of any base metal. An arc strike is a discontinuity on the base metal caused by striking the arc outside the weld joint. Arc strikes may be depressions or marks that occur on the surface of the weld by the welder accidentally striking the electrode on the base metal adjacent to the weld. Arc strikes may degrade base metal properties on hardenable steels like medium-carbon steels or low-alloy steels and may form a region of brittle martensite from the rapid quenching effect of the high temperature. **See Figure 34-31.** Arc strike detection is achieved by VT. Some fabrication standards and codes require arc strikes to be ground to a smooth contour and inspected to ensure soundness by an appropriate NDE test such as PT or MT.

Arc Strike Prevention. Arc strikes are prevented for certain types of work, such as pipe, by placing protective wrappings around the part to prevent accidental contact with the electrode.

Arc Strikes

Figure 34-31. *An arc strike results when the electrode strikes the base metal during welding, and it can degrade base metal properties.*

Spatter

Spatter is a discontinuity that occurs when metal particles are expelled during fusion welding and do not form part of the weld. Spatter appears as droplets of solidified weld metal on the base metal adjacent to the weld. **See Figure 34-32.** Spatter detection is achieved by VT.

Spatter

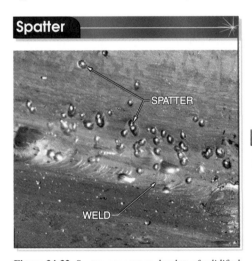

Figure 34-32. *Spatter appears as droplets of solidified weld metal on the base metal adjacent to the weld.*

Spatter Prevention. Spatter can be reduced or prevented by reducing the welding current, reducing the effect of arc blow, reducing the arc length, and ensuring the use of clean and undamaged electrodes. **See Appendix.** Anti-spatter spray is available for prevention of spatter for many welding applications.

> ✓ **Point**
>
> *Arc strikes may degrade base metal properties on hardenable steels like medium-carbon steels or low-alloy steels and may form a region of brittle martensite from the rapid quenching effect of the high temperature.*

Refer to Quick Quiz® on CD-ROM

Points to Remember

- Discontinuities are classified as defects when they exceed the minimum requirements permitted by the controlling code or standard.
- Weld stresses may be concentrated or enhanced by the presence of discontinuities, leading to failure under load.
- Cracks are fracture-type discontinuities and are not permitted in fabrication standards and codes.
- The higher the carbon content of steel, the higher the risk of cracking. This is why it is critical to pay proper attention to heat treatment.
- Cracks are classified according to their location in the weld.
- Cavities are rounded discontinuities within a weld or at the surface. The most common type of cavity is porosity.
- Inclusions consist of foreign matter in the weld metal, either from the base metal, filler metal, or nonconsumable electrode.
- Incomplete fusion and incomplete penetration are found in areas with incomplete melting between the base metal and the weld metal. Incomplete fusion is less desirable than incomplete penetration.
- Incorrect shapes, such as undercut, overlap, excess weld reinforcement, underfill, concave root surface, and melt-through, produce an unacceptable weld profile.
- Arc strikes may degrade base metal properties on hardenable steels like medium-carbon steels or low-alloy steels and may form a region of brittle martensite from the rapid quenching effect of the high temperature.

Questions for Study and Discussion

1. How can weld joint design be adjusted to prevent throat cracks?
2. How do crater cracks form?
3. How can crater cracks be prevented?
4. What causes toe cracks?
5. How can toe and root cracks be prevented?
6. What are the two main types of porosity?
7. What can be done to reduce porosity in a weld?
8. What are slag inclusions?
9. How can slag inclusions be prevented in multiple pass welds?
10. What causes tungsten inclusions?
11. Which process is more likely to produce incomplete fusion: SMAW or GMAW in short circuiting mode, and why?
12. What causes incomplete penetration?
13. What is overlap, and how can it be prevented?
14. What is melt-through, and how can it be prevented?
15. Why are arc strikes detrimental to medium-carbon or low-alloy steels?

 Refer to Chapter 34 in the *Welding Skills Workbook* for additional exercises.

Welding Procedure Qualification

Chapter

Welding procedure qualification encompasses not only the legal requirements of the applicable fabrication standard or code, but also the directions for making a consistent weld. Welding procedure qualification variables affect the weld and must be specifiedd

Welding procedure qualification determines, by preparation and testing of standard specimens, whether welding in accordance with a welding procedure specification (WPS) will produce sound welds and adequate joint properties. A WPS provides formal documentation for all qualified welding variables.

A welding procedure qualification record (WPQR) determines, by preparation and testing of standards specimens, whether welding in accordance with a WPS will produce sound welds and adequate joint properties. Much of the data required by the WPQR is the same information required in the WPS.

WELDING PROCEDURE QUALIFICATION

Welding procedures used by welders and welding operators require qualification to be in accordance with fabrication standards and codes. Welding procedure qualification encompasses not only the legal requirements of the applicable fabrication standard or code but also the directions for making a consistent, quality joint and weld. Differences, however subtle, between the requirements of various fabrication standards and codes make it essential that the applicable document be consulted for guidance.

Qualified welding procedures consist of welding procedure specifications (WPS) and welding procedure qualification records (WPQR). Both WPS and WPQR define applicable welding variables. **See Appendix.**

WELDING PROCEDURE QUALIFICATION VARIABLES

A *welding procedure qualification variable* is a fundamental condition (parameter) that affects the integrity of a weld joint. Welding procedure qualification variables must be indicated in the welding procedure qualification record. Essential variables are listed in the applicable fabrication standard or code. Welding procedure qualification variables for arc welding may consist of any or all of the following:

- welding process
- joint design
- base metal
- filler metal
- welding position
- preheat, interpass temperature control, and postheating
- shielding gas
- electrical characteristics
- welding technique

Oxyfuel welding, brazing, surfacing weld, and resistance welding require additional welding procedure qualification variables.

Welding Processes

Certain welding processes cannot be used with specific metals because the welding process used may affect the weldability

> **✓ Point**
>
> *Qualified welding procedures consist of the welding procedure specification (WPS) and the welding procedure qualification record (WPQR).*

of the metal. For example, titanium alloys are not typically welded by flux shielded welding processes such as SMAW. Titanium alloys are most often welded by gas shielded welding processes such as GMAW, GTAW, or RW.

> *Welding procedure specifications are typically developed by a welding engineer who has previous experience with the particular weld parameters, and who uses recommendations by suppliers of welding equipment such as the base metal, welding machine, and filler metals. The welding procedure specification must also meet applicable codes.*

Joint Design

Joint design is the shape, dimensions, and configuration of the joint. The joint is the junction of members or the edges of members that are to be joined. An effective joint design achieves welding at minimal cost. The joint design influences how much filler metal may be required to fill a joint, and the ease of adding filler metal. Welding procedure variables that affect joint design are weld type, edge preparation, and backgouging.

Weld Type. *Weld type* is the cross-sectional shape of the weld after filler metal is added to the joint. Basic weld types are groove weld and fillet weld. Each weld type can have several different configurations.

Edge Preparation. *Edge preparation* is the preparation of the workpiece edges by cutting, cleaning, or other methods. All fillet weld configurations can be made without additional edge preparation. Three groove weld configurations can be made without additional edge preparation. They are square groove, flare V-groove, and flare bevel groove. Edge preparation is done by shearing, thermal cutting, grinding, machining, or backgouging.

Shearing is the parting of material when one blade forces the material past an opposing blade. Shearing produces a square groove. Shearing is the most

economical method of edge preparation and is used for sheet metal.

Thermal (flame) cutting consists of a group of processes that remove metal by rapid oxidation. Thermal cutting is the most common method of edge preparation, and is used for most work with thickness greater than sheet metal. Thermal cutting is versatile and economical and may be manipulated to produce both square edges and added bevels. The heat produced by thermal cutting may alter the metallurgical structure of some metals. In such cases, the thermally cut surface must be dressed by grinding to remove a minimum of ⅛″ of affected base metal before any welding is performed.

Grinding is the mechanical removal of metal from the surface using hard, brittle grains of an abrasive material. Grinding is usually performed with a grinding wheel. Grinding is used for medium thicknesses of material and may be tooled up to provide reproducible geometries. **See Figure 35-1.**

Machining is precise shaping to a desired profile using special tools to remove material. Machining is used on thick-wall components to prepare J- and U-grooves and on circular components of all diameters and wall thicknesses. Machining is an accurate, final method of edge preparation.

Grinding

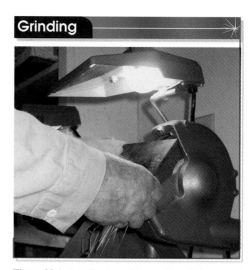

Figure 35-1. *Grinding is used for medium thicknesses of material to remove metal from the surface.*

Backgouging. *Backgouging* is the removal of weld metal and base metal from the weld root side of a welded joint to facilitate complete fusion and complete joint penetration when welding on that side is completed.

Backgouging is done when joints are welded from both sides and is used to produce final joints free from cracks and other unsound conditions. The backgouging method must be indicated on drawings when joints are to be welded from both sides. If backgouging requires an inspection method other than visual, the method should be indicated on the drawings. Methods of backgouging include chipping, grinding, air carbon arc gouging, or oxyfuel gouging. **See Figure 35-2.**

Base Metal

The base metal(s) must be properly identified. Two methods may be used: the base metal material specification and the base metal weldability classification. The base metal thickness range is also indicated.

Base Metal Material Specifications. A *base metal material specification* is the chemical composition or industry specification of the base metal. Any special condition of the base metal, such as heat treatment, cold working, or special cleaning must be indicated if it affects the metal's weldability, or if welding alters the base metal properties. For example, localized welding reduces the strength of a cold-worked metal in the heat-affected zone. The fact that the metal is cold-worked must be indicated on the drawings.

Base Metal Weldability Classifications. The *base metal weldability classification* is an alphanumeric system that groups base metals with similar welding characteristics. A welding procedure that provides excellent results with one base metal classification may prove completely inadequate with another classification. The base metal weldability classification system assigns a number to a base metal according to its chemical composition, weldability, and mechanical properties.

Backgouging

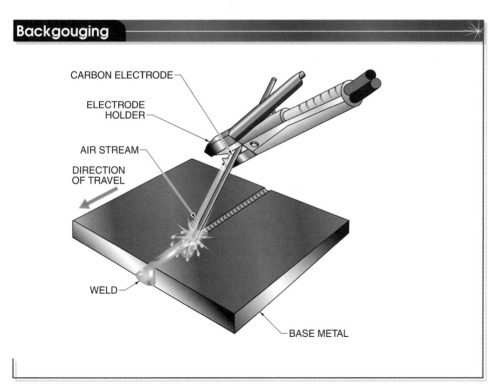

CARBON ELECTRODE

ELECTRODE HOLDER

AIR STREAM

DIRECTION OF TRAVEL

WELD

BASE METAL

Figure 35-2. *Backgouging is performed to improve the quality of the root pass.*

In the ASME Boiler and Pressure Vessel Code, base metal weldability classification consists of P-numbers assigned to base metals to indicate their characteristics. P-numbers are described in Section IX of the ASME Boiler and Pressure Vessel Code. Metals with the same P-number are covered under the same WPS. For example, P1 materials are low-carbon steels that generally do not require preheat. P4 materials are specific chrome-moly steels that require preheat to approximately 300°F. Welding procedures are qualified by grouping base metals according to their P-number, which reduces the number of welding procedure qualifications required. **See Figure 35-3.**

Base Metal Thickness Ranges. The *base metal thickness range* is a procedure qualification variable that indicates the range of base metal thicknesses covered in the welding procedure qualification record. For pipe, the pipe diameter range and pipe wall thickness must be indicated. In most cases, base metal thickness range is $\frac{1}{16}''$ or $\frac{3}{16}''$ to 2T, where T is the thickness of the test sample weld.

Filler Metal

Filler metal variables that must be considered are filler metal specification, filler metal usability classification, filler metal diameter, and filler metal quantity. A separate filler metal description is required for tack welding.

Filler Metal Specifications. *Filler metal specification* is identification of filler metal by AWS number or other specification designation. If required by the applicable fabrication standard or code, more details may be needed. Additional required information may include manufacturer; heat; lot or batch number of the filler metal or other welding consumable; and the results of supplementary identification such as X-ray fluorescence (XRF) analysis.

Filler Metal Usability Classifications. The *filler metal usability classification* is an alphanumeric method of grouping filler metals with similar characteristics. In AWS specifications and the ASME Boiler and Pressure Vessel Code, filler metals are given F-numbers to indicate their grouping. Filler metals

P-NUMBERS					
Spec. No.	Embedded Type & Grade	Welding P-No.	Brazing P-No.	Nominal Comp.	Product Form
SA-36	—	1	—	C-Mn-Si	Plate
SA-53*	Type E Gr. B	1	101	C-Mn	ERW Pipe
SA-53*	Type S Gr. B	1	101	C-Mn	Smls Pipe
SA-105	—	1	101	C-Si	Pipe Flange
SA-106	B	1	101	C-Si	Smls Pipe
A108	1018 CW	—	—	C	Bar
A134	A 285 B	—	—	C	Welded Pipe
SA-182†	F11, Cl. 2	4	102	1¼ Cr-Mo	Forging
SA-182†	F22, Cl. 1	5A	102	2¼ Cr-Mo	Forging
SA-182†	F304L	8	102	18Cr-8Ni	Forging < 5″
A211	A570 Gr. 30	—	—	C	Welded Pipe
SA-234	WPB	1	101	C-Si	Pipe Fitting
SA-234	WP5	5B	102	5Cr-Mo	Pipe
SA-240	Type 304L	8	102	18Cr-8Ni	Plate
SA-335	P22	5A	102	2¼ Cr-1Mo	Smls Pipe
SA-387	11,Cl. 1	4	102	1¼ Cr-½Mo	Plate
SA-516	Grade 60	1	101	C-Mn-Si	Plate
API5L	Grade B	—	—	C-Mn	Smls/welded

* SA-53 specifications have same UNS Number, but are different product forms.
† Materials have same specification number, but different nominal compositions.

Figure 35-3. *P-numbers reduce the number of welding procedures that must be developed by grouping metals that have similar weldability characteristics.*

with the same usability classification (F-numbers) generally may be substituted for one another, reducing the number of welding procedure specifications required. For ferrous weld metal, analysis numbers, or A-numbers, are additionally assigned to further segregate F-numbers. A-numbers, which range from 1-12, represent classifications of ferrous weld metal analysis for procedure qualification. A-numbers are essential variables for most welding processes. Filler metals with the same usability classification and the same A-numbers may be welded with the same welding procedure. **See Figure 35-4.** Filler metals with the same usability classification and different A-number require a new WPS to be qualified.

Filler Metal Diameter. The filler metal diameter influences welding current requirements and joint penetration ability. If the root opening is too tight, the groove angle too narrow, or the filler metal diameter too large, the welding electrode will not be able to deposit the weld metal at the root. Small-diameter filler metal is often required for the root pass to eliminate the chances of incomplete penetration, to prevent melt-through, and for heat control. Small-diameter filler metals also require less current than larger diameter filler metal. Filler metal diameter(s) required for welding different thicknesses of metal in different positions are also indicated.

> ✓ **Point**
>
> *Filler metals are grouped by usability classification to reduce the number of procedure qualification variables. Filler metals with the same usability classification may be substituted for one another with no effect.*

A- AND F-NUMBERS							
A-Numbers							
A-Number	**Type of Weld Deposit**	**Analysis**					
		C*	**Cr***	**Mo***	**Ni***	**Mn***	**Mn***
1	Mild Steel	.15	—	—	—	1.60	1
2	Carbon-Moly	.15	.50	.4 − .65	—	1.60	1
3	Cr-Mo (.4% to 2%)	.15	.40 − 2	.4 − .65	—	1.60	1
4	Cr-Mo (2% to 6%)	.15	2 − 6	.4 − .65	—	1.60	2

F-Numbers (Electrode and Welding Rod Groups for Qualification)		
F-Number	**ASME Specification Number**	**AWS Classification Number**
1	SFA-5.1 and 5.5	EXX 20, EXX 22, EXX 24, EXX 27, EXX 28
	SFA-5.4	EXX 25, EXX 26
2	SFA-5.1 and 5.5	EXX 12, EXX 13, EXX 14, EXX 19
3	SFA-5.1 and 5.5	EXX 10, EXX 11
4	SFA-5.1 and 5.5	EXX 15, EXX 16, EXX 18, EXX 48
	SFA-5.4 (other than austenitic and duplex)	EXX 15, EXX 16
5	SFA-5.4 (austenitic and duplex)	EXX 15, EXX 16
6	SFA-5.9	GTAW ERXX
	SFA-5.17	SAW FXX-EXX
	SFA-5.18	GMAW ERXXS-X
	SFA-5.20	FCAW EXXT-X
2X	Aluminum	GTAW ER 4043
3X	Copper	GTAW ER CuNi
4X	Nickel	SMAW ENiCrFe-3
5X	Titanium	GTAW ERTi-7
6X	Zirconium	GTAW ERZr3
7X	Weld Overlay	SMAW EXXX-X

* in percent (%)

Figure 35-4. *A- and F- numbers reduce the number of welding procedures that must be developed by grouping filler metals that have similar characteristics.*

Filler Metal Quantity. *Filler metal quantity* is the deposited weld metal thickness range for groove or fillet welds. Filler metal quantity is usually indicated by a sketch showing the location of each weld pass in the joint. The correct amount of deposited weld metal achieves the required joint strength. Overwelding (excess filler metal) not only increases cost, but may also create an undesirable stress concentration at the toe of the weld. **See Figure 35-5.** Methods of minimizing filler metal quantity include reducing the root opening; using a root face on groove welds; decreasing the groove angle; using single-U grooves; or using double-V or double-U grooves.

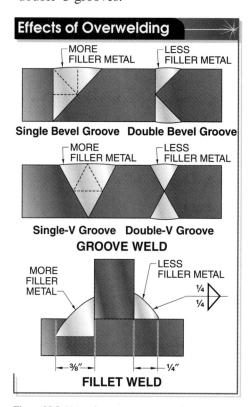

Figure 35-5. *Using the appropriate joint design ensures the use of the proper amount of filler metal.*

To calculate the weight of filler metal, multiply the cross-sectional area of the joint by the length of the weld, and multiply the result by the density of the filler metal.

Poor fit-up counteracts the optimizing benefits of the desirable filler metal quantity throughout a joint. Poor fit-up is a common problem with full- or partial-penetration fillet welds in T-joints

fabricated in the horizontal position. However, welding in a more difficult position may qualify a less difficult position.

Tack Welding. Tack welding is used to temporarily join parts in proper alignment until the final weld is made. Improperly made or improperly removed tack welds may affect the integrity of the final weld. Tack welding may require the use of designated procedures as indicated in the welding procedure specification. **See Figure 35-6.**

Figure 35-6. *Tack welding must comply with the welding procedure specification if it is incorporated into the final weld.*

Welding Position

Welding position is the relationship of the weld pool, joint, and base metals. Welding positions are flat, horizontal, overhead, and vertical. Welder accessibility must be considered when designing the joint and the assembly pattern to permit a comfortable working environment for the welder.

To achieve the best quality weld, a welder must be able to access the joints from both sides after all areas to be welded have been completely assembled and tack welded. The sequence of assembly may be adjusted to improve welder accessibility. Some welds cannot be accessed from both sides (box columns or small-diameter piping). Such joints are inaccessible and require one-sided welding. **See Figure 35-7.** When one-sided welding is done, backing material or consumable inserts can be used to ensure complete penetration on the backside of the weld.

When backing material or consumable inserts are not desired or feasible, open root welding must be done. Open root welding requires a higher welding skill than welding with backing and also requires good fit-up and joint preparation. Care must be taken to achieve the proper root weld without excessive penetration (excessive root reinforcement).

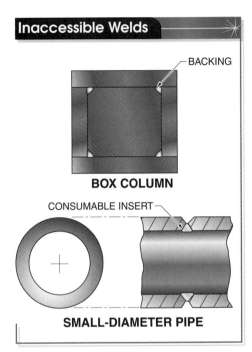

Inaccessible Welds

BACKING

BOX COLUMN

CONSUMABLE INSERT

SMALL-DIAMETER PIPE

Figure 35-7. *Welder accessibility is a key consideration in creating a sound joint. If a joint is inaccessible, backing material or consumable inserts can be used to ensure complete penetration.*

Preheat, Interpass Temperature Control, and Postheating

Preheat, interpass temperature control, and postheating are welding parameters that indicate the temperature to which the joint must be heated to improve the final properties of the joint. The temperature for each parameter varies depending on the metal to be welded.

Preheat and interpass temperature control are specified when applicable to ensure toughness of the heat-affected zone, particularly for heat-treatable steels. When preheat temperature controls are required, a minimum value must be specified. When interpass temperature control is required, a maximum value must be specified.

Postheating may be specified when welded structures require heat treatment after welding to develop required properties, maintain dimensional stability, reduce residual stress, or further improve toughness. The postheating procedure must be indicated either in the welding procedure specification or on a separate document, such as a shop heat-treating traveler. Postheating procedure requirements include rate of heating and cooling of the structure; time at temperature; and location of weld joint(s) to be postheated.

Shielding Gas

The shielding gas provides a gaseous protective atmosphere that prevents or reduces atmospheric contamination of the molten weld as it solidifies and cools. Shielding gas efficiency relates to the ability of a shielding gas to displace the atmosphere from the arc area. Shielding efficiency depends on shielding gas purity; the design of the nozzle; the distance from the nozzle to the work; the internal diameter or size of the nozzle; the gas flow rate; and side drafts.

Electrical Characteristics

Electrical characteristics should be documented when the welding involves the use of electric current. Electrical characteristics include current type, current level, polarity, and arc voltage. The proper current type and polarity must be defined in the welding procedure.

Welding Technique

The welding technique includes welding procedure details that are controlled by the welder or welding operator. Welding technique parameters include heat input, travel speed, interpass cleaning, and peening.

Heat Input. The heat input influences the weldability or as-welded properties of specific metals. Heat input details must be indicated whenever the heat input could influence the metal properties of the finished weld joint.

Alloys, such as nickel alloys, that are sensitive to hot cracking require heat input controls. When heat input controls are required, details such as using a straight bead or a weave bead must be specified.

A *straight bead* is a type of weld bead made without any appreciable weaving motion. A *weave bead* is a type of weld bead made with transverse oscillation. Using a straight bead or a weave bead can lead to either a reduction or an increase in heat input, respectively. Either bead type may be acceptable for certain types of metals. Many nickel alloys prefer a lower heat input, while chrome-moly steels prefer higher heat input.

Travel Speed. The travel speed used must be consistent throughout the joint to prevent altering the weld properties. Too low a travel speed may cause excessive heat input and impair the properties of a joint. Too fast a travel speed leads to a lack of complete fusion. Documentation of the acceptable travel speed range is always mandatory for automatic welding processes and often mandatory for semiautomatic welding processes.

Interpass Cleaning. Interpass cleaning is required to remove slag from the weld metal and to prepare it for the next pass. Ineffective interpass cleaning may leave slag inclusions in the completed weld and lead to rejection of the weld. Interpass cleaning methods include grinding, chipping, or wire brushing. Interpass cleaning methods are documented for welding processes that leave a slag residue, for example SMAW. **See Figure 35-8.**

Peening. *Peening* is the mechanical working of weld metal using impact blows. Peening reduces the effects of excessive residual stresses and distortion. Peening is used on highly restrained or thick welds to avoid warping or cracking of the weld or base metal. Peening must be performed immediately after completion of a bead length (approximately 9″), as soon as the weld has solidified. Peening is never applied to a root pass or cap pass. Details of peening must be specified to ensure correct application of the method.

Figure 35-8. *Interpass cleaning is required to remove slag from the weld and prevent slag inclusions.*

Oxyfuel Welding Qualification Variables

Oxyfuel welding qualification variables are similar to those for arc welding, where applicable. Unique qualification variables for oxyfuel welding are fuel gas requirements and welding tip size.

Fuel Gas Requirements. Fuel gas requirements that must be specified are fuel gas composition and gas pressure. The fuel gas composition is the combination of fuel gases that is to be used with oxygen to perform the welding. Acetylene is an example of a fuel gas. Oxygen is always used to support combustion in oxyfuel welding, so the pressure required at the regulators of both the fuel gas and the oxygen is indicated. The corresponding flame type (oxidizing, reducing, or neutral) must also be indicated.

Welding Tip Size. The welding tip size is the size of the orifice in the oxyfuel welding torch. The orifice is the point from which the oxyfuel welding gases issue. The size of the welding tip controls gas consumption during welding and must be documented on the WPQR.

> Welding procedure specifications are documents detailing all the applicable weld variables for a specific application, ensuring reliable repeatability of a welding task by properly trained welders.

Brazing Qualification Variables

Brazing qualification variables are the same as for arc welding, where applicable. Qualification variables unique to brazing are brazing temperature range; brazing flux; brazing joint design and clearance; brazing position; and brazing time. Brazing variables are indicated on a Brazing Procedure Specification. **See Figure 35-9.**

Brazing Temperature Ranges. *Brazing temperature range* is the temperature range within which the base metal is heated to enable filler metal to wet the base metal and form a brazed joint. The temperature range must melt the filler metal at a temperature below the melting point of the base metal(s). Filler metals for brazing are those that melt at temperatures above 840°F.

Brazing Procedure Specification

BRAZING PROCEDURE SPECIFICATION (BPS)

BPS No. _____ Date _____ B PQR NO. _____

Company _____

Brazing Process _____ Manual ☐   Mechanized ☐   Automatic ☐

Brazing Equipment _____

BRAZING CONDITIONS

BASE METAL:

    Identification _____ BM No. _____

    Thickness _____ Preparation _____

    Other _____

FILLER METAL:

    FM No. _____ AWS Classification _____

    Form _____ Method of Application _____

FLUX: AWS Type _____ Other _____

ATMOSPHERE: AWS Type _____ Other _____

TEMPERATURE: _____ TEST POSITION: _____

TIME: _____ CURRENT: _____

FUEL GAS: _____ TIP SIZE: _____

POSTBRAZE CLEANING: _____

POSTBRAZE HEAT TREATMENT: _____

OTHER: _____

JOINT:

    Type _____

    Clearance _____

    UTS _____

    Other _____

    _____

Approved for production by _____   JOINT SKETCH
                              Employer

Figure 35-9. *Brazing qualification variables are documented on a brazing procedure specification form.*

Brazing Flux. Brazing flux is intended to prevent or inhibit the formation of oxides during brazing. Brazing atmospheres include combusted fuel gas, hydrogen, or vacuum. Brazing flux constituents include borax, chloride, fluorides, or any combination of these with other chemicals. The chemicals within the flux are the agents that prevent or remove the oxides or other undesirable substances during brazing.

Brazing Joint Designs and Clearances. The two basic joints used for brazing are the lap joint and the butt joint. The lap joint is the most commonly used because it offers a large surface area for the greatest strength. Joint design is also based on joint clearance. Joint clearance has a major effect on the mechanical properties of a brazed joint. Adequate joint clearance should fall in the range between 0.001″ and 0.010″. Recommended joint clearances vary with the type of filler metal and the thickness of the base metal. Brazing joint design and clearance influence the strength of a brazed joint. Changes to joint design and clearance outside of tabulated values require requalification.

Brazing Positions. The brazing position, if altered, requires requalification, with certain exceptions. Basic brazing positions for plate are flat, vertical downflow, vertical upflow, and horizontal. Basic brazing positions for pipe are horizon-

tal, vertical downflow, and vertical upflow. **See Figure 35-10.** In the vertical downflow and vertical upflow positions, the joint faces are vertical and capillary flow of the filler metal is up and down, respectively. In the horizontal position for plate or pipe, the joint faces are also vertical, but the capillary flow of the filler metal is horizontal.

Brazing Time. The brazing time requirement must be requalified if it exceeds or fails to meet the brazing time indicated for the qualification test by a prescribed percentage.

Surfacing Weld Qualification Variables

Qualification variables for surfacing welds are the same as those for arc welding, where applicable. The qualification variable unique to surfacing welds is the chemical composition of the surfacing weld.

Chemical Composition of Surfacing Welds. The chemical composition of a surfacing weld influences wear resistance for hard facing and corrosion resistance for corrosion-resistant overlays. The chemical composition of a surfacing weld can be altered by dilution with the base metal. Different welding processes create different amounts of dilution. To overcome dilution, additional surfacing

QUALIFIED BRAZING POSITIONS								
Test Brazement Form	Test Brazing Position	Plate				Pipe		
		Flat Flow	Vertical Downflow	Vertical Upflow	Horizontal Flow	Horizontal Flow	Vertical Downflow	Vertical Upflow
Plate	Flat Flow	X	X	—	—	—	—	—
	Vertical Downflow	—	X	—	—	—	—	—
	Vertical Upflow	—	X	X	—	—	—	—
	Horizontal Flow	—	X	—	X	—	—	—
Pipe	Horizontal Flow	X	X	—	X	X	X	—
	Vertical Downflow	—	X	—	—	—	X	—
	Vertical Upflow	—	X	X	—	—	X	X

Figure 35-10. *The position of the test brazement may qualify one or more brazing positions.*

weld passes or modification of the welding procedure may be required. The chemical composition of the surfacing weld must be maintained on the surface layers, without excessive dilution. The measured chemical composition of the surfacing deposit must be within a prescribed percentage of the actual chemical composition of the base metal. It may be necessary to reduce heat input to reduce dilution.

Resistance Welding Qualification Variables

Resistance welding (RW) qualification variables are the same as those for arc welding, where applicable. Resistance welding qualification variables unique to RW are joint design; electrode type and size; weld size and strength; and surface appearance.

The joint design must account for contacting overlap, weld spacing, and the type and size of projection. Electrode variables include the alloy used, the contour, and the dimensions. If plates, dies, blocks, or other such devices are used whose properties would affect the quality of the welding, they should be specified. Weld size and strength must describe the extent of the joint and the anticipated strength value to be obtained by mechanical testing.

Surface appearance includes factors such as indentation, discoloration, or amount of upset. A general requirement for surface appearance may be sufficient. A statement such as "Surface shall be generally free of discoloration or indentations" is acceptable.

WELDING PROCEDURE SPECIFICATION (WPS)

A *welding procedure specification (WPS)* is a document providing the required welding variables for a specific application to ensure repeatability by properly trained welders and welding operators. The WPS provides formal documentation for all welding qualification variables.

The WPS is the "recipe" that must be followed when making the weld.

Information regarding test specifications and procedures are detailed in ANSI/AWS B2.1, *Welding Procedure and Performance Qualification.* As part of the procedure for qualification, forms are completed that specify all welding directives and requirements (Welding Procedure Specifications). **See Appendix.**

Fabrication standards and codes require an employer to prepare and qualify welding procedure specifications relevant to all fabrication work. The standards and codes define the details to be included in a WPS and refer only to the welding variables of the specific process that affect qualification. The user is allowed to determine what other variables and information should be included in the WPS.

Welding procedure specification items include WPS details, WPS variables, WPS conformance, WPS development, and standard WPSs.

WPS Details

WPS details describe all the welding qualification variables required by the applicable fabrication standard or code. The WPS details may be brief or long and detailed. Fabrication standards and codes usually contain suggested WPS forms on which to document qualification variables and other relevant information. For complex welded structures, the suggested WPS forms must be supplemented with additional notes or instructions, or new WPS forms are devised to suit specific requirements. **See Appendix.**

A WPS provides direction to the welder or welding operator and is an important control document. The WPS is given a specific reference number and must be signed by an authorized person, such as the fabricator's quality assurance manager, before release for production welding. Responsibility for the content, qualification status, and use of a WPS rests with the employer.

Welding Procedure Specification

WPS Variables

WPS variables are qualification variables that require documentation in a WPS. WPS variables are essential variables, supplementary essential variables, and nonessential variables.

Essential Variables. An *essential variable* is a welding qualification variable which, if altered, shall be considered to affect the mechanical properties of the weld. If an essential variable is altered in a welding procedure, the welding procedure is considered to be revised and the new procedure must be requalified. Essential variables are indicated in fabrication standards and codes. Essential variables differ depending on the welding process and the fabrication standards and codes. **See Figure 35-11.**

Supplementary Essential Variables. A *supplementary essential variable* is a qualification variable, for metals where impact testing is required, that requires a new welding procedure specification. The supplementary essential variable is a provision of some fabrication standards and codes, such as the ASME Boiler and Pressure Vessel Code.

Nonessential Variables. A *nonessential variable* is a qualification variable that may be changed in a WPS without requalification of the WPS. Nonessential variables differ for different welding processes and for different fabrication standards and codes.

All fabrication codes and standards indicate a specific level of conformance to welding performance or procedure qualification that must be met. Section IX of the ASME Boiler and Pressure Vessel Code requires the manufacturer or contractor to take responsibility for performing qualification testing of welding procedures for the weldments to be built under the code and for the performance of the welders who will carry out the welding. Section IX also requires the manufacturer or contractor to maintain an accurate, certified record of the results obtained during welding, as well as during procedure and performance qualification tests. Records must be available to authorized inspectors.

WPS Development

WPS development is generally the responsibility of the contractor in a given production shop. The end user or their representative specifies the properties desired in weldments in accordance with a code, specification, or special design requirements. The contractor then develops a welding procedure specification (WPS) that will produce the specified results, if an applicable procedure does not already exist. Certain fabrication codes and standards require welding procedure qualification.

WPS Conformance

Conformance with the WPS is required to meet the applicable fabrication standard or code. Many fabrication standards or codes reference the ASME Boiler and Pressure Vessel Code, Section IX, *Qualification Standard for Welding and Brazing Procedures — Welders, Brazers, and Welding and Brazing Operators.*

VARIABLE CHANGES THAT REQUIRE REQUALIFICATION. . .		
Procedure Variable*	Procedure Requalification Required	Welder/Welding Operator Requalification Required
Type, composition, or process condition of the base metal	• When the base metal is changed to one that does not conform to the type, specification, or process condition qualified • Some codes and specifications provide lists of materials that may be substituted without requalification	• Usually not required
Thickness of base metal	• When the thickness to be welded is outside the qualified range • Most codes provide for qualification on one thickness within a reasonable range • Some codes may require qualification on exact thickness or on min and max. thicknesses	• When the thickness to be welded is outside the qualified range • Most codes provide for an unlimited thickness test
Joint design	• When established limits of root openings, root face, and included angle of groove joints are increased or decreased • Some codes and specifications define upper and lower limits beyond which requalification is necessary. Others permit an increase in groove angle and root opening and a decrease in the root face without requalification • Requalification is often required when a backing or spacer strip is added or removed, or the basic type of material of backing or spacer strip is changed	• When changing from a double-welded joint or a joint using backing material to an open root, and vice versa • The addition or deletion of a consumable insert
Pipe diameter	• Usually not required • Some codes permit procedure qualification on plate to satisfy the requirements for welding on pipe	• When the diameter of piping or tubing is reduced below specified limits. • Smaller pipe diameters generally require more sophisticated techniques, equipment, and skills
Type of current or polarity (if DC)	• Usually not required for changes involving electrodes or welding materials adapted for the changed electrical characteristics • Sometimes required for change from AC to DC, or vice versa, or from one polarity to the other	• Usually not required for changes involving similar electrodes or welding materials adapted for the changed electrical characteristics
Electrode classification and size	• When electrode classification is changed • When the diameter is increased beyond allowable ranges specified in the relevant code	• When electrode classification grouping is changed • Sometimes when the electrode diameter is increased beyond specified limits
Welding current	• When the current is outside the range qualified	• Usually not required
Position or progression or both	• Usually not required, but desirable	• When the change exceeds the limits of the position(s) qualified or a change in progression
Deposition of filler metal	• When a marked change is made in the manner of filler metal deposition; e.g., from a small bead to large bead or weave arrangement or from an annealing pass to a no-annealing-pass arrangement, or vice versa	• Usually not required

* General requirements for requalification of welding procedures and welder performance. Not for use by inspector to determine necessity of requalification. Inspectors must reference the applicable code or standard for the work being inspected.

Figure 35-11...

... VARIABLE CHANGES THAT REQUIRE REQUALIFICATION		
Procedure Variable*	**Procedure Requalification Required**	**Welder/Welding Operator Requalification Required**
Preparation of root for second side welding	• When method or extent is changed	• Usually not required
Preheat and interpass temperatures	• When preheat or interpass temperature is outside the qualified range	• Usually not required
Postheating	• When adding or deleting postheating • When the postheating temperature or time cycle is outside the qualified range	• Usually not required

* General requirements for requalification of welding procedures and welder performance. Not for use by inspector to determine necessity of requalification. Inspectors must reference the applicable code or standard for the work being inspected.

...Figure 35-11. *Essential welding variables require requalification if they are changed.*

WELDING PROCEDURE QUALIFICATION RECORD (WPQR)

A *welding procedure qualification record (WPQR)* is documentation of the welding variables used to produce an acceptable test weld and the test results conducted on the weld to qualify a WPS. A welding procedure qualification record determines, by preparation and testing of standards specimens, whether welding in accordance with a WPS will produce sound welds and adequate joint properties. The test results are documented in the welding procedure qualification record (WPQR).

To support a WPS, it is necessary to test and certify the results in a WPQR. This is done by making the welds described in the WPS, machining them into test samples, and testing the samples in accordance with the applicable fabrication code and standard.

WPQR Details

Much of the data required by the WPQR is the same as the information referenced in the WPS. All essential variables and, when applicable, supplemental essential variables, must be included. Nonessential variables are optional, but when included must be accurate. The data on the front sheet of the WPQR and the WPS will often look very similar. A WPQR records exact data of what actually took place during the test. A WPS lists a range of al-

lowable variables. The back of the WPQR is a record of the mechanical test results. Mechanical tests that may be used include the tensile test, guided bend test, toughness test (when required), and fillet weld test (when required). **See Appendix.**

A change in any variable beyond the allowable limits of the applicable fabrication standard or code requires requalification of the WPS with a new WPQR. Any change within allowable limits requires only documentation in a revised WPS.

The applicable fabrication code or standard provides general guidance and specific acceptance-rejection criteria for evaluating test results. Minimum tensile strength, maximum number of inclusions, or the permissible level of other discontinuities may be specified. The acceptability of properties or conditions is based on engineering judgment and is especially important for service at high or low temperature, or in corrosive environments. WPQRs vary for the type of welding process. In some cases the type of fabrication may require mock-up tests or may allow the use of a prequalified WPS.

WPQR Steps

The steps involved in WPQR are welding a sample joint within the parameters of the WPS qualification variables; testing the sample joint using standardized protocols; and recording the test results in the WPQR.

Welding a Sample Joint. Welding a sample joint is usually done using pipe or plate samples, with a welding joint made to the qualification variables indicated in the WPS. The type, size, and thickness of the test sample are governed by the type, size, and thickness of the base metal to be welded in production, and by the nature of the pieces to be removed for test specimen preparation. Test specimen requirements are usually indicated in the applicable fabrication standard or code.

Testing a Sample Joint. Sample joint testing is performed on test specimens that have been removed from the sample weld joint. The type and number of test specimens depends on the requirements of the applicable fabrication standard or code. In most cases, the test specimens used are for tensile testing and guided bend testing. Exact testing requirements are indicated in the applicable fabrication standard or code.

Test specimens made from fillet welds are usually subjected to tensile-shear testing and macroetching. Testing determines the strength, ductility, soundness, and adequacy of fusion of the welds.

Nondestructive examination (NDE) of the sample joints is usually preferred before they are sectioned for test specimen preparation. Specific NDE procedures may be a requirement of the applicable fabrication standard or code.

If a fabricator has qualified a welding procedure, and at some later date wishes to make modifications in that procedure, it may be necessary to conduct requalification tests.

Requalification tests establish that the modified welding procedure will produce satisfactory results. Requalification tests are not usually required when only minor details of the original procedure are changed. They are required, however, if the changes might alter the properties of the resulting welds. The applicable fabrication standard or code provides guidance on whether requalification tests are required.

Recording Test Results. Recording test results in the WPQR is done when the qualifier is satisfied that the results are accurate. The WPQR is signed to certify the test results. If the test results meet the requirements of a job specification, the supported WPS may be issued for production welding.

A WPQR is a certified record of a qualification test and should not be revised. If information needs to be added later, it can be added in the form of a supplement or attachment. Additional qualification tests may be required if an employer later wishes to make changes to a WPS. A WPQR may support several WPSs, and a WPS may be supported by several WPQRs.

If changes become necessary in an established and qualified WPS, additional qualification tests may be needed to determine whether the modified WPS will yield satisfactory results. The applicable fabrication standard or code determines if requalification is needed.

Alternate WPQR Documentation

Alternate WPQR documentation encompasses various methods of qualifying welding procedures. Alternate WPQR documentation includes prequalified WPS, mock-up tests, brazing PQR, and resistance welding PQR.

Prequalified WPS. A *prequalified WPS* is a welding procedure specification that complies with the stipulated conditions of a particular fabrication standard or code and is acceptable for use under that code without requiring additional qualification testing. Prequalified welding procedures may be used as an alternate to testing by each employer.

In order to use a prequalified WPS, the employer prepares a written WPS conforming to the specific requirements of the applicable fabrication standard or code for the welding variables defined. The written WPS is a record of materials and welding

> **✓ Point**
> *Several WPQRs can support a single WPS, and several WPSs can be supported by a single WPQR.*

> **✓ Point**
> *A prequalified WPS is a WPS that complies with a specific fabrication code or standard and requires no qualification testing.*

procedure qualification variables that demonstrates that the joint welding procedure meets the requirements for prequalified status. For AWS D1.1, *Structural Welding Code — Steel,* this work is done under the requirements of AWS D1.1.

Welding procedure qualification tests need not be made if the requirements are followed in detail. The employer must accept responsibility for the use of prequalified WPSs. The use of prequalified welding procedures does not guarantee satisfactory production welds. The quality of all production welds should be verified by NDE during and after welding.

A standard WPS is a type of pre-qualified WPS. A standard WPS is one developed through analysis of thousands of qualified welding procedures that provide restricted ranges of welding variables to ensure a high probability of successful application by end users. Standard WPSs are approved for some fabrication codes, such as for sheet metal.

Mock-up Tests. Mock-up tests are used to simulate actual production welding conditions in certain types of fabrication jobs, usually under difficult or restricted welding conditions. Mock-up tests verify that proper tooling and inspection have been selected.

Certain variables such as joint geometry, welding position, and accessibility may not be considered as qualification variables. Often, the only way to gauge their effect is with mock-ups. Fabrication standards and codes do not usually require the fabrication of mock-ups for destructive examination unless they are to demonstrate that the welding procedures will produce the specified welds. For example, although mock-up tests are used to verify welding procedures for heat exchanger tube-tubesheet joints, the mock-up tests must be supported by a qualified WPS. Mock-up tests are a useful method of demonstrating expected quality levels under difficult or restricted welding conditions. **See Figure 35-12.**

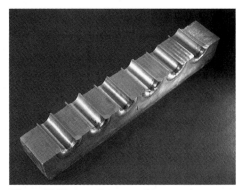

Figure 35-12. *A mock-up is used for a mock-up test to simulate actual production welding situations to ensure proper tooling and techniques are selected, such as for heat exchanger tube-tubesheet joints.*

Brazing PQR. Brazing procedure qualification testing consists of various destructive tests on test specimens obtained from braze samples made to the applicable brazing procedure specification. Test results are recorded on the Brazing Procedure Qualification Record and certified by the witnessing contractor representative.

Resistance Welding PQR. Resistance welding procedure qualification tests vary and depend largely on the type of work to be produced. When the welded part is small, the procedure may be qualified by making a number of finished pieces and testing them to destruction under service conditions, either simulated or real. In other cases, resistance welds can be made in test specimens that are tested in tension or shear, or inspected for other properties such as surface appearance and soundness.

A procedure qualification record determines, by preparation and testing of standards specimens, whether welding in accordance with a WPS will produce sound welds and adequate joint properties. Much of the data required by the PQR is the same as the information referenced in the WPS.

Points to Remember

- Qualified welding procedures consist of the welding procedure specification (WPS) and the welding procedure qualification record (WPQR).
- Welding procedure qualification variables are welding parameters that affect the integrity of a weld joint and must be indicated in the WPQR.
- Joint design is an example of a procedure qualification variable and may encompass weld type, edge preparation, and method of preparing the edge.
- Base metals are grouped by weldability classifications to reduce the number of procedure qualification variables. Base metals with the same weldability classification may be substituted for one another with no effect.
- Filler metals are grouped by usability classification to reduce the number of procedure qualification variables. Filler metals with the same usability classification may be substituted for one another with no effect.
- A WPS includes essential, supplementary essential, and nonessential variables.
- Essential variables are parameters which, if changed, could alter the mechanical properties of the weld. Requalification of the new variables is required.
- Supplementary essential variables are parameters which affect the impact properties (toughness) of the weld. Requalification of the new variables is required.
- Nonessential variables are parameters which, if changed, do not alter the mechanical properties of the weld and do not require requalification of the weld.
- A welding procedure qualification record (WPQR) is documentation of the welding variables used to produce an acceptable test weld and the test results conducted on the weld to qualify a WPS.
- WPQR development encompasses welding a sample joint within the applicable parameters of the WPS, testing the joint, and recording the results.
- Several WPQRs can support a single WPS, and several WPSs can be supported by a single WPQR.
- A prequalified WPS is a WPS that complies with a specific fabrication code or standard and requires no qualification testing.
- Mock-up tests are used to simulate actual welding jobs under difficult or restricted conditions, such as for heat exchanger tube-tubesheet joints.

1. What is one benefit of using an effective joint design?
2. What is the difference between base metal material specification and base metal weldability classification?
3. What is the range of base metal thicknesses covered in a welding procedure specification?
4. What is the difference between filler metal specification and filler metal usability classification?
5. Why are small-diameter electrodes preferable to large-diameter electrodes?
6. What is one benefit of postheating?
7. Why does a straight bead provide less heat input to a weld than a weave bead?
8. What are the effects of travel speeds that are too slow? Travel speeds that are too fast?
9. What is the effect of ineffective interpass cleaning?
10. What WPS variables require documentation in the WPS?
11. What are the three steps required in creating a WPQR?
12. How is a mock-up test useful when supported by a qualified WPS?

Refer to Chapter 35 in the *Welding Skills Workbook* for additional exercises.

Welder Performance Qualification

Chapter

The welder performance qualification (WPQ) test demonstrates a welder's ability to produce welds that meet a qualified welding procedure. Welder performance qualification tests are used to assess whether a welder has the required level of skill to produce a sound weld to the parameters of the applicable welding procedure specification (WPS).

The employer is responsible for ensuring that welder performance qualification tests meet the requirements of the applicable fabrication standard or code. Fabrication standards and codes contain similar methods of qualifying welders, welding operators, and tack welders, but differ in the requirement details. Welder performance qualification (WPQ) tests must be made in the most difficult position encountered in production. However, WPQ test results cannot predict how an individual will perform on a particular production weld. The quality of production welds should be determined by inspection both during and following completion of welding.

WELDER PERFORMANCE QUALIFICATION (WPQ)

A welder performance qualification (WPQ) contains three areas that must be certified for a welder to be approved for qualification: welder performance qualification, welder certification, and welder registration. The *welder performance qualification* is a test that demonstrates a welder's ability to produce welds that meet required standards. The welder performance qualification involves taking and passing a practical welding test.

Welder certification is a written statement that the welder has produced welds meeting a prescribed standard of welding performance. Welder certification implies that a testing organization, a manufacturer, a contractor, an owner, or a user has witnessed the preparation of the test welds, has conducted the prescribed testing of the welds, and has recorded the

successful results of a test in accordance with accepted standards. *Welder registration* is the act of approving a copy of the welder's certification document by an appropriate authority.

WPQ STANDARDS AND CODES

Many fabrication standards and codes exist, each having its own regulatory requirements. WPQ requirements vary between standards and codes, and the appropriate fabrication code or standard must be used when qualifying welders, welding operators, and tack welders.

> *AWS D1.1, Structural Welding Code—Steel, is an example of a standard that contains qualification requirements for welders, welding operators, and tack welders. Performance qualification requirements are found in Section 4, Qualification, Part C, Performance Qualification.*

> **✓ Point**
> *A welder performance qualification (WPQ) demonstrates a welder's ability to produce welds to meet the applicable welding procedure specification (WPS).*

The governing standard or code should be consulted for specific details. Requirements for the ASME Boiler and Pressure Vessel Code, AWS *Structural Welding Code—Steel*, AWS *Structural Welding Code—Sheet Steel*, and API *Cross Country Pipeline Welding* are typically specified. Qualification under one fabrication code or standard does not necessarily qualify a welder to weld under another code or standard, even though the qualification tests appear to be identical.

ASME Boiler and Pressure Vessel Code

Boiler and pressure vessel code requirements are contained in ASME Section IX, *Qualification Standard for Welding and Brazing Procedures—Welders, Brazers, and Welding and Brazing Operators*. Section IX requirements also apply to other structures such as elevated water storage tanks and oil storage tanks.

Per ASME requirements, the welder who prepares test samples for the WPQ must be personally qualified within ASME performance qualification variables. All other welders are qualified by specific welder qualification tests required by the welding procedure specification (WPS) that will cover the work. A welding procedure qualification record (WPQR) is used to document the ability of the welder or welding operator to meet the WPS.

A WPQR must include the essential welding variables, the type of test, the metal thickness ranges qualified, and the test results. When testing, RT may sometimes be substituted for mechanical tests, but not when GMAW with short circuiting transfer is used. RT cannot be used because incomplete fusion, a common discontinuity with GMAW in the short circuiting mode, may not be detected by RT. **See Appendix.**

Generally, welders who meet the requirements for groove welds are also qualified for fillet welds, but not vice versa. A welder qualified to weld in accordance with one qualified WPS is also qualified to weld in accordance with other qualified WPSs using the same welding process, within the limits of the indicated essential welding variables.

A qualified welder is given an identifying number, letter, or symbol that is used to identify his or her work. The qualification expires if the welder does not weld for a period of six months or more. Moreover, if there is reason to question the welder's ability to make welds meeting specifications, his or her qualification shall be considered expired.

AWS Structural Welding Code—Steel

Structural welding code WPQ requirements are contained in AWS D1.1, *Structural Welding Code—Steel*. AWS requirements are similar to those of the ASME Boiler and Pressure Vessel Code, but also contain provisions for prequalified welding procedures.

Under the AWS code, visual inspection, guided bend tests, fillet weld tests, and RT may be used to test sample welds. The Structural Welding Code—Steel also allows, at the engineer's discretion, acceptance of proper documented evidence of previous qualifications of welders. Unless the ability of a welder is questioned, their qualification to use a particular procedure lasts indefinitely under the AWS D1.1 code.

AWS Structural Welding Code—Sheet Steel

The structural welding code for sheet steel welder qualifications is contained in AWS D1.3, *Structural Welding Code—Sheet Steel*. The requirements are different from AWS D1.1 for structural steel in that qualification, when established for any one of the steels permitted by the code, allows the welder to be qualified to weld on any other steel permitted by the code, except for coated steels. Qualification on coated steels must be tested on coated steels.

Qualification is required in each position used. In the case of vertical position, uphill or downhill travel is qualified. Welders are qualified for all electrodes within a group designation (usability classification). Different combinations of electrode and shielding gas must be qualified separately. If any of the procedure qualification variables are changed, the procedure must be requalified under the new variables. **See Figure 36-1.** Check with the fabrication code and specification for actual essential variables.

ESSENTIAL WELDING VARIABLES*

Electrode/Filler Metal

- Electrode classification
- Electrode size
- Increase in filler metal strength
- Melting rate/current/wire feed speed
- Type of coating
- Coating thickness
- Use of flux (for SAW)

Position

- Change in position
- For vertical welding: uphill vs. downhill; downhill vs. uphill
- Welding from both sides to welding from one side only (for square butt joints)

Shielding Gas

- Type of shielding gas (for GMAW and GTAW)
- Flow rate (for GMAW and GTAW)

Current

- Current level/wire feed speed/melting rate
- Type of welding current, polarity

Base Metal

- Sheet steel thickness

Joint Design

- Root opening of square butt joints

Welding Process

- Mode of metal transfer (for GMAW)†

* require requalification if changed
† only essential when switching from short circuiting to spray transfer

Figure 36-1. *Essential welding variables are qualified as tested. Welders must be requalified when essential welding variables are changed.*

Cross-Country Pipeline Welding Code

Cross-country pipeline welder qualification requirements are contained in API standard 1104, *Standard for Welding Pipelines and Related Facilities.* The requirements are different from the previously described codes in that cross-country pipeline welder qualification and testing is usually done in the field. **See Figure 36-2.**

API allows for the use of tensile tests, bend tests, and nick-break tests. Welders can be qualified for a single qualification or multiple qualifications, depending on the results of each test attempted.

PRODUCT-SPECIFIC WPQS

Product-specific welder performance qualification tests are most commonly done for plate and structural member welding, pipe welding, sheet metal welding, and brazing.

Welder performance qualifications test the most difficult positions that will be encountered in production for welding and brazing. Qualification in a more difficult position usually also qualifies for welding or brazing in less difficult positions. A welder who qualifies in vertical, horizontal, or overhead positions is usually also considered qualified for welding or brazing in flat position. Qualification on a groove weld test will normally qualify that welder for the production of fillet welds in the same position. The applicable fabrication standard or code dictates the exact limits on production welding and brazing qualification test positions.

✓ **Point**

The WPQ must be developed for the most difficult position expected during welding or brazing.

AWS D1.1, *Structural Welding Code—Steel,* Table 4.10, *Welder and Welding Operator Qualification: Number and Type of Specimens and Range of Thickness and Diameter Qualified,* specifies the type of test welds, metal thickness, number of specimens, and the qualified dimensions for production welding.

PROCEDURE SPECIFICATION NO. _____

For _____ Welding of _____ Pipe and Fittings

Process _____

Material _____

Diameter and wall thickness _____

Joint design _____

Filler metal and No. of beads _____

Electrical or flame characteristics _____

Position _____

Direction of welding _____

No. of welders _____

Time lapse between passes _____

Type and removal of lineup clamp _____

Cleaning and/or grinding _____

Preheat/stress relief _____

Shielded gas and flow rate _____

Shielding flux _____

Speed of travel _____

Sketches and tabulations attached _____

Tested _____ Welder _____

Approved _____ Welding supervisor _____

Adopted _____ Chief engineer _____

ELECTRODE SIZE AND NUMBER OF BEADS

Bead Number	Electrode Size and Type	Voltage	Current and Polarity	Speed

Figure 36-2. *WPQ tests to meet API 1104 are usually performed in the field.*

A welder performance qualification test qualifies a welder based on the essential variables specified on the welding performance qualification record.

WPQ for Plate and Structural Members

Welder performance qualification for plate and structural members usually requires that the welder make one or more test welds on groove weld plate or fillet weld plate in accordance with the requirements of the qualified WPS. The welder would then be qualified to weld plates up to 2T (where T is the thickness of the qualification plate). Joint details should be in accordance with the qualified WPQR.

Groove weld qualifications usually qualify the welder to weld both fillet welds and groove welds in the positions qualified. Fillet weld qualifications limit the welder to fillet welding in only the position qualified and other specified positions of less difficulty. WPQ samples for groove and fillet welds are taken from key locations in the test joint.

WPQ for Pipe

Welder performance qualifications for pipe differ from plate and structural member welding principally in the test assemblies and the test positions used. When welding pipe, the root surface is inaccessible, requiring the use of backing rings or consumable inserts, or the production of a weld with an open root joint. An *open root joint* is an unwelded joint that does not use backing or consumable inserts.

Pipe welding requires more skill than welding plate or structural members with backing. To simulate the difficulties of production pipe welding, the WPQ for pipe requires that pipe samples be welded in the position, or positions, for which the welder is to be qualified. Space restrictions may also be placed on the welder during the test. Space restrictions measure the individual's ability to produce a satisfactory weld in locations where joint access is limited. Special joint designs are used for welder performance qualification to weld T-, K-, or Y-connections in pipe and for fillet or tack welds.

WPQ for Sheet Steel

Welder performance qualifications for sheet steel are based upon special requirements for joining thin members. Welding thin metals could result in holes burning through the sections. All fabrication standards and codes place limits on the minimum thickness that a welder can weld in production. The minimum thickness qualified is often the thickness used during qualification. **See Figure 36-3.**

WPQ for Brazing

Welder performance qualification for brazing is based upon the production of a joint arrangement in a similar position to that expected in production. Brazing operators are tested to verify their ability to operate mechanized or automatic brazing equipment according to a brazing procedure specification.

Acceptance of welder performance qualification tests on brazed joints may be based either on visual examination or on specimen testing. Welder performance qualification for brazing by specimen testing is done by making a standard test brazed joint consisting of a butt, scarf, lap, single- or double-spliced butt, or a rabbet joint in plate or pipe. **See Figure 36-4.**

The test pipe is sectioned and the exposed surfaces are polished and etched, and a macroetch test is done. A macroetch test is a way to examine a brazed joint at low magnification for discontinuities. Peel tests may be used in place of macroetch tests, or vice versa, using lap joints or spliced butt joints.

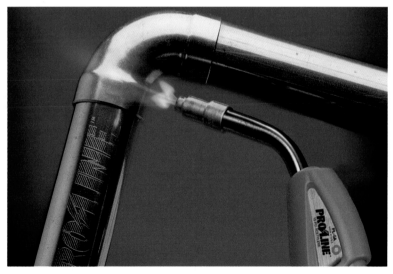

A WPQ for brazing requires that a joint be made in a position similar to that required in production. Brazing operators must be able to operate machinery or brazing equipment according to the brazing procedure specification.

WELDER PERFORMANCE QUALIFICATION TESTS FOR SHEET STEEL

Test Samples	Tested Welding Position	Qualified Welding Position	Qualified Weld Joint	Thickness	Type of Test*
Square groove butt joint, sheet to sheet	Flat Horizontal Vertical Overhead	Flat Flat, Horizontal Flat, Horizontal, Vertical Flat, Horizontal, Overhead	Square groove butt joint, sheet to sheet	Thickness tested	Bend
Arc spot weld, sheet to supporting member	Flat	Flat	Arc spot weld and arc seam weld, sheet to supporting member	Thickness tested	Twist
Arc seam weld, sheet to supporting member	Flat	Flat	Arc seam weld, sheet to supporting member	Thickness tested	Bend
Arc seam weld, sheet to sheet	Horizontal	Horizontal	Arc seam weld, sheet to sheet	Thickness tested	Bend
Fillet welded lap joint, sheet to sheet	Flat Horizontal Vertical Overhead	Flat Flat, Horizontal Flat, Horizontal, Vertical Flat, Horizontal, Overhead	Fillet welded lap joint, sheet to sheet, or sheet to supporting member	Thickness tested and thicker	Bend
Fillet welded lap joint, sheet to supporting member	Flat Horizontal Vertical Overhead	Flat Flat, Horizontal Flat, Horizontal, Vertical Flat, Horizontal, Overhead	Fillet welded lap joint, sheet to sheet, or sheet to supporting member	Thickness tested and thicker	Bend
Fillet welded T-joint, sheet to sheet	Flat Horizontal Vertical Overhead	Flat Flat, Horizontal Flat, Horizontal, Vertical Flat, Horizontal, Overhead	Fillet welded T- or lap joint, sheet to sheet, or sheet to supporting member	Thickness tested and thicker	Bend
Fillet welded T-joint, sheet to supporting member	Flat Horizontal Overhead	Flat Flat, Horizontal Flat, Horizontal, Overhead	Fillet welded T- or lap joint, sheet to supporting member	Thickness tested and thicker	Bend
Flare-bevel, sheet to sheet	Flat Horizontal Vertical Overhead	Flat Flat, Horizontal Flat, Horizontal, Vertical Flat, Horizontal, Overhead	Flare-bevel-groove weld, sheet to sheet, or sheet to supporting member; or Flare-V-groove weld, sheet to sheet	Thickness tested and thicker	Bend
Flare-bevel-groove, sheet to supporting member	Flat Horizontal Vertical	Flat Flat, Horizontal Flat, Horizontal, Vertical	Flare-bevel-groove weld, sheet to supporting member	Thickness tested and thicker	Bend
Flare -V-groove, sheet to sheet	Flat Horizontal Vertical Overhead	Flat Flat, Horizontal Flat, Horizontal, Vertical Flat, Horizontal, Overhead	Flare-V-groove weld, sheet to sheet; or Flare-bevel-groove weld, sheet to sheet, or sheet to supporting member	Thickness tested and thicker	Bend

* two tests required for certification

Figure 36-3. *For sheet steel welding, the position, weld joint, and thickness that are tested are typically the only variables for which the welder is qualified, per test.*

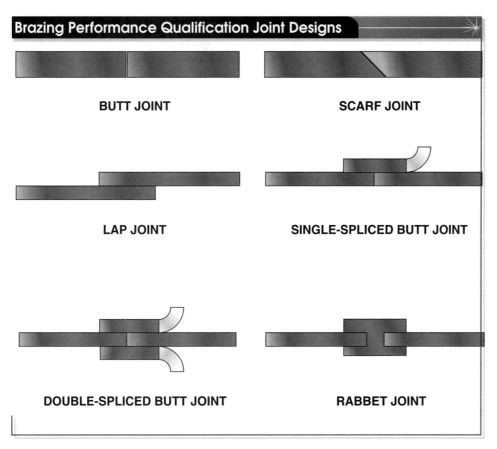

BUTT JOINT

SCARF JOINT

LAP JOINT

SINGLE-SPLICED BUTT JOINT

DOUBLE-SPLICED BUTT JOINT

RABBET JOINT

Figure 36-4. *WPQ for brazed joint is achieved by sectioning, polishing, and etching a test joint.*

 Refer to Quick Quiz® on CD-ROM

Points to Remember

- A welder performance qualification (WPQ) demonstrates a welder's ability to produce welds to meet the applicable welding procedure specification (WPS).
- Qualification under one fabrication code or standard does not necessarily qualify a welder to weld under another code or standard, even though the qualification tests appear to be identical.
- Structural welding code WPQ requirements are contained in AWS D1.1, *Structural Welding Code—Steel*.
- A WPQR must include the essential welding variables, the range qualified, the type of tests, and the test results.
- The WPQ must be developed for the most difficult position expected during welding or brazing.

1. Who is responsible for ensuring that qualification of welders meets applicable codes?
2. What is the difference between welder performance qualification and welder certification?
3. Are welders who meet the requirements for fillet welds automatically approved for groove welds?
4. When qualifying for pipe, in what position should the welder be qualified?

 Refer to Chapter 36 in the *Welding Skills Workbook* for additional exercises.

Welding Metallurgy

Chapter

Welding metallurgy is the study of the effect of welding on the metallurgical structure and properties of weld joints. Heat input during welding produces rapid heating, very high temperatures, and rapid cooling, leading to three distinct regions in a welded joint: weld metal, heat affected zone (HAZ), and base metal. The physical properties of the metals are influenced by their responses to the heat of welding. Mechanical properties of the metal, residual stresses, and corrosion resistance of metal are also affected by the heat of welding.

METALLURGICAL STRUCTURE

Metallurgy is the study of the influence of crystal and grain structure of metals on the mechanical, physical, and chemical properties of metals.

The crystal structure and grain structure of metals is known collectively as the metallurgical structure. *Metallurgical structure* is the arrangement of atoms in repeating patterns within a metal. The crystal structure is preserved in the grain structure of metals. The crystal structure may change as a metal is heated or cooled, or if the composition of the metal changes.

Crystal Structure

A *crystal structure* is a specific arrangement of atoms in an orderly and repeating three-dimensional pattern. All metals exhibit a crystal structure. Although 14 types of crystal structures are possible in nature, most metals exhibit one of three types: face-centered cubic, body-centered cubic, or close-packed hexagonal. **See Figure 37-1.**

The atomic arrangements in the different crystal structures lead to significant differences in the behavior of metals. Some metals, such as steel, may exhibit different crystal structures at different temperatures.

Grain Structure

Metals do not exist as a single crystal, but as a large number of grains. A *grain* is an assembly of crystals having different orientations of their crystal components. Grain structure develops as metals solidify from the molten state. The first atoms to solidify develop the characteristic crystal structure of the metal. Each solid crystal nucleus that forms develops its own orientation within the structure. The crystals grow by developing offshoots, but retain their orientation with respect to the other

> ✓ **Point**
>
> *Crystal structure is a specific arrangement of the building blocks of matter (atoms) in an orderly and repeating three-dimensional pattern.*

nuclei. The solidifying structures are called dendrites. As the dendrites grow, they fill the space between themselves with offshoots and branches until their extremities meet other dendrites. The dendrites continue to grow until the space between them is completely filled and solidification is complete. **See Figure 37-2.**

Heat Input

Heat is the most important element needed for welding. Heat (heat input) is required to melt the base metal and filler metal during welding. *Heat input is the amount of heat applied to the filler metal and the base metal surface at the required rate to form a weld pool,* plus the additional heat required to compensate for heat that is conducted away from the weld. Heat input during welding produces rapid heating, very high temperatures, and rapid cooling. **See Figure 37-3.**

The most common source of heat input in fusion welding is the electric arc. Other sources of heat input, such as burning oxygen and acetylene (oxyfuel welding), are also used. Controlling heat input is essential when welding because the heat input may affect the structure and properties of metals.

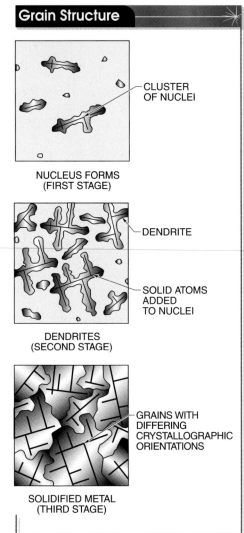

Grain Structure

NUCLEUS FORMS
(FIRST STAGE)

CLUSTER OF NUCLEI

DENDRITES
(SECOND STAGE)

DENDRITE

SOLID ATOMS ADDED TO NUCLEI

SOLIDIFIED METAL
(THIRD STAGE)

GRAINS WITH DIFFERING CRYSTALLOGRAPHIC ORIENTATIONS

Figure 37-2. *The grain structure of crystals develops from the solidification and growth of many nuclei that form and grow as molten metal cools.*

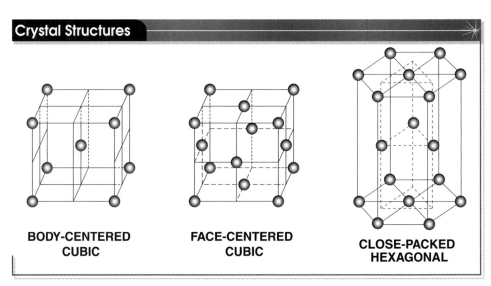

Crystal Structures

BODY-CENTERED CUBIC

FACE-CENTERED CUBIC

CLOSE-PACKED HEXAGONAL

Figure 37-1. *As atoms cool from a liquid to a solid state, they are arranged into one of three crystal structure patterns: body-centered cubic, face-centered cubic, or close-packed hexagonal.*

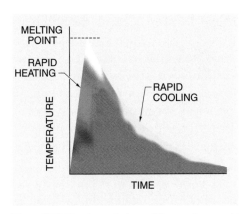

Figure 37-3. *Heat input during welding produces rapid heating, very high temperatures, and rapid cooling.*

Seventy percent to 85% of the heat generated in SMAW is used in making the weld. Most of the remaining heat is used to melt the base metal adjacent to the weld joint. The percentage of heat used to melt the filler metal varies with the welding process, welding procedure, base metal, and joint design. Additional heat is lost through heating the electrodes and flux, through weld spatter, and through convection to the surrounding atmosphere.

Calculating Heat Input. Heat input is measured in joules per linear inch of weld. The heat input produced by a moving electric arc is calculated using the following equation:

$$Q = \frac{WV \times WC}{WTS} \times 60$$

where

Q = heat input (in J/in.)

WV = welding voltage (in V)

WC = welding current (in A)

WTS = welding travel speed (in in./min)

What is the heat input when using SMAW at 29 V, 300 A, and a travel speed of 18 in./min?

$$Q = \frac{WV \times WC}{WTS} \times 60$$

$$Q = \frac{29 \times 300}{18} \times 60$$

$$Q = 483.33 \times 60$$

$$Q = \textbf{29,000 J/in.}$$

Heating Rate. *Heating rate* is the rate of temperature change of a weld joint over time from room temperature to the welding temperature. The heating rate is influenced by heat input, thermal conductivity, and the mass of the joint area.

Heat input exceeds heat loss during welding and the base metal becomes hotter. The temperature of the work near the arc rises and as soon as the arc moves on, the temperature begins to fall. If the weld pool becomes large and unmanageable, it can be cooled by reducing the current or breaking the arc, thus reducing heat input or cutting it off completely.

The maximum temperature achieved in a weld must be sufficient to cause melting of the base metal at the weld face. The amount by which the temperature must exceed the melting point of the filler metal depends on the welding process. The time the metal is at the maximum temperature can influence properties of both the filler metal and the base metal.

Cooling Rate. The *cooling rate* is the rate of temperature change of a weld joint over time from the welding temperature to room temperature. Weld joint cooling takes place at a much faster rate than any quenching process in heat treatment. The cooling rate rapidly decreases with distance from the weld, because the surrounding base metal acts as an effective heat sink.

The cooling rate is governed by factors such as heat loss, thermal conductivity of the base metal, and the amount of preheat and interpass temperature control required. *Preheat* is the heating of the joint area to a predetermined temperature in order to slow the cooling rate. *Interpass temperature control* is maintaining the temperature range within the weld between weld passes until welding is complete. Depending on the type of metal being welded, interpass temperature control may have an upper limit, a lower limit, or both.

When using medium-carbon and low-alloy steels, the rate of cooling must be controlled to maintain toughness of the heat-affected zone. There is a critical cooling rate for each type of steel, which, if exceeded, leads to loss of toughness. Using the proper preheat temperature, coupled with an upper limit on interpass temperature control, helps maintain the cooling rate below the critical cooling rate. The cooling rate of a weld also depends on the number of weld passes required. The root bead has the greatest preheating effect on the weld joint. The change in the cooling rate between subsequent passes is less significant.

Forced cooling may be used to accelerate cooling. *Forced cooling* is rapid cooling of a solidified weld joint between passes using water. Forced cooling is often used because it increases production. Forced cooling is most common with stainless steels, but is also used on other alloys. Abnormal stresses and other detrimental effects may be exerted on the joint integrity when forced cooling is used.

Slow Cooling of Steel. When steel is slow cooled from a high temperature, metallurgical structure changes occur under conditions of thermal equilibrium. *Thermal equilibrium* is a steady-state condition in which time is available for the diffusion of atoms. Austenite (which has a face-centered cubic crystal structure) transforms on cooling to a mixture of ferrite (which has a body-centered cubic structure) and iron carbide. Iron carbide is a compound formed from carbon that diffuses out of the austenite and combines with some of the ferrite. Slow cooling is used in heat treatment processes such as annealing that are designed to soften steel.

Rapid Cooling of Steel. Rapid cooling is used to strengthen steel. The steel is heated to a high temperature to produce austenite—a process called austenitizing—and then rapidly cooled, or quenched. When steel is rapidly cooled, an equilibrium-dependent structure change has no time to occur. The steel is then heated to an intermediate temperature, or tempered, to restore sufficient ductility while maintaining a stronger, harder product.

Welding produces metallurgical structure changes similar to the quenching stage of heat treatment. Consequently, as the carbon content of steel increases, the welding procedure must be manipulated to "cushion" the effect of quenching. This is achieved by preheating or blanket cooling. Steel may also be tempered after welding using postheat, which reduces hardness and residual stresses.

WELD REGIONS

The heat of welding creates three regions, with different metallurgical structures, within a weld joint. These regions are weld metal, heat-affected zone (HAZ), and base metal. **See Figure 37-4.** Additionally, surfacing and buttering procedures create regions with properties similar to the weld metal, HAZ, or base metal.

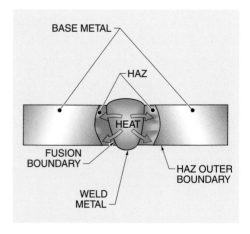

Figure 37-4. *The heat of welding creates three regions of metallurgical structures: weld metal, heat-affected zone (HAZ), and base metal.*

Weld Metal

Weld metal is the portion of a fusion weld that is completely melted during welding. Weld metal consists of solidified weld filler metal resulting from the addition of filler metal to the joint, plus a small amount of melted base metal at its boundaries, which creates the weld interface.

The *weld interface* is the boundary between the weld metal and the base metal in a fusion weld. In the liquid state, atoms of the metal move about energetically, which can be seen as a swirling motion. As the metal cools and solidifies to form grains, segregation will occur so that the chemical composition of the solidified weld metal is not homogeneous. Except for its faster cooling rate, the weld metal resembles a miniature casting. The melted base metal contributes to dilution if a filler metal of different composition is used.

Dilution modifies the chemical composition of filler metal because of mixing with the base metal or previously applied weld metal in the weld bead. Dilution may affect the selection of filler metal for dissimilar metal welding. *Dissimilar metal welding* is the joining of two metals of different composition using a compatible filler metal to ensure the weld meets required properties.

If filler metal is used, the solidified weld metal consists of the melted edges of the base metal and the filler metal. If the weld is made without filler metal it is said to be an autogenous weld and consists entirely of melted-out base metal. Examples of autogenous welds are edge and corner flange welds and welds made of very thin sheet metal or tubes. Autogenous welds depend on the quality of the base metal to ensure their soundness. Because the deoxidizers normally present in weld filler metals are missing, autogenous welds are sometimes prone to porosity and cracking.

Heat-Affected Zone (HAZ)

The *heat-affected zone (HAZ)* is a narrow band of base metal adjacent to the weld joint whose properties and/or metallurgical structure are altered by the heat of welding. With carbon steels, metallurgical structure changes can occur in any region of the base metal that exceeds 1350°F. With heat-treated aluminum alloys, any region heated above 600°F experiences metallurgical structure transformation. Welding a heat-treated aluminum alloy creates an HAZ that may be weaker and more susceptible to failure under service loads.

The strength and toughness of the HAZ depends on the type of base metal, the welding process, and the welding procedure. Base metals most affected by the heat of welding are those that are annealed or hardened by the temperatures experienced in the HAZ. For example, a cold worked material will lose some of its strength in the HAZ as it is softened by annealing.

The width of the HAZ is proportional to the amount of heat input during welding, and varies with the welding process used. It may extend from 0.06″ to 0.25″ into the base metal.

Base Metal

The base metal is the metal, after welding, that has not been structurally altered by exposure to heat. The boundary between the base metal and the HAZ depends on the temperature at which metallurgical structure transformation begins for any specific metal and is dependent on welding temperature.

Surfacing

A surfacing weld is applied to a surface, as opposed to a joint, to obtain the desired properties or dimensions. Surfacing can be applied using the SMAW, GTAW, and GMAW arc welding processes. Surfacing can also be applied using OFW or brazing. Arc welding processes generally produce the most dilution. It is usually necessary to apply two layers of surfacing weld to overcome dilution and ensure the second layer has the required chemical composition or other properties. **See Figure 37-5.**

> **✓ Point**
>
> *It is usually necessary to apply two layers of surfacing weld to overcome dilution and attain the required wear or corrosion resistance properties.*

> **✓ Point**
>
> *The HAZ is a narrow band of base metal adjacent to the weld joint. Most problems that occur during welding occur in the HAZ.*

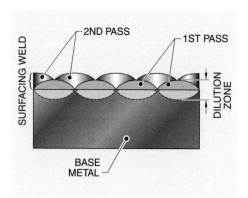

Figure 37-5. *Surfacing welds may require two or more passes to achieve the required chemical composition at the surface.*

EFFECT OF WELDING ON PHYSICAL PROPERTIES

Physical properties are the characteristic responses of metal to forms of energy such as heat, light, electricity, and magnetism. Some physical properties of metals significantly influence weldability of a metal, but are not altered by welding. Other properties may be altered by welding. Physical properties that influence the weldability of metals include melting point, thermal expansion, specific heat, thermal conductivity, electrical conductivity, magnetism, and oxidation.

Melting Point

Melting point is the temperature at which a metal passes from a solid state to a liquid (molten) state. Pure metals possess a specific melting point and pass from solid to liquid at a constant temperature. Alloys melt within a temperature range that depends on the alloy composition. The range of temperatures is bounded by the solidus and the liquidus. *Solidus* is the highest temperature at which an alloy is completely solid. *Liquidus* is the lowest temperature at which an alloy is completely molten. Melting begins at the solidus and is complete at the liquidus. Metals with low melting temperatures can be welded with low-temperature heat sources. **See Figure 37-6.**

Thermal Expansion

Thermal expansion is a measure of the dimensional change of a member due to heating or cooling. Most materials expand when heated and contract when cooled. The amount of change can be predicted through calculation using coefficients. A *coefficient of linear expansion* is the rate of change in a linear dimension

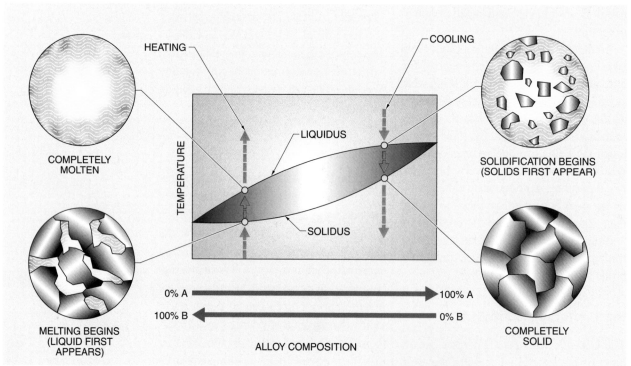

Figure 37-6. *The solidus and liquidus bracket the boundary temperatures between which an alloy is partially molten.*

(length, width, or height) of an object per degree of temperature. A *coefficient of volumetric expansion* is the rate of change in the volume of an object per degree of temperature. For solids, the coefficient of volumetric expansion is equal to three times the coefficient of linear expansion.

The amount of change per degree varies depending on the material, so each material has its own coefficients. **See Figure 37-7.** For example, aluminum increases its size at almost twice the rate as steel, so it has high coefficients.

Thermal expansion influences distortion control, fixture design, and workpiece fit-up because the heat of welding affects the size of a workpiece during welding and cooling. To calculate a change in linear dimension or volume due to thermal expansion, apply the following formulas respectively:

$$\Delta L = L_1 \times \alpha \times (T_2 - T_1)$$
$$\Delta V = V_1 \times \beta \times (T_2 - T_1)$$

where

ΔL = change in linear dimension (in in.)

L_1 = linear dimension at initial temperature (in in.)

α = coefficient of linear expansion (in cu in./cu in.-°F)

T_1 = initial temperature (in °F)

T_2 = final temperature (in °F)

ΔV = change in volume (in cu in.)

β = coefficient of volumetric expansion (in cu in./cu in.-°F)

For example, by how much will a 10″ × 5″ × 2″ workpiece made from 304 stainless steel expand in length as it is heated from 100°F to 300°F? How much will it expand in volume?

$$\Delta L = L_1 \times \alpha \times (T_2 - T_1)$$
$$\Delta L = 10 \times 0.00000916 \times (300 - 100)$$
$$\Delta L = 10 \times 0.00000916 \times 200$$
$$\Delta L = \textbf{0.01832 in.}$$

$$\Delta V = V_1 \times \beta \times (T_2 - T_1)$$
$$\Delta V = 100 \times 0.00002748 \times (300 - 100)$$
$$\Delta V = 100 \times 0.00002748 \times 200$$
$$\Delta V = \textbf{0.5496 cu in.}$$

COEFFICIENT OF THERMAL EXPANSION FOR VARIOUS METALS		
Alloy	**Linear***	**Volumetric†**
Aluminum 1100	13.1	39.3
Aluminum 3003	12.8	38.4
Aluminum, pure	13.1	39.3
Aluminum 6061	13.0	39.0
Aluminum 7075	12.9	38.7
Aluminum 356.0	11.8	35.4
Copper, pure	9.16	27.5
Copper, oxygen-free	9.83	29.5
Brass, 85%	10.4	31.2
Brass, 80%	10.6	31.8
Brass, 70%	11.1	33.3
Manganese Bronze	11.7	35.1
Phosphor Bronze, 8%	10.1	30.3
70-30 Copper-Nickel	9.00	27.0
90-10 Copper-Nickel	9.50	28.5
Aluminum Bronze	9.11	27.3
Iron, pure	6.56	19.7
Mild Steel (0.2%C)	6.50	19.5
Medium Carbon Steel (0.4%C)	6.39	19.2
304 Stainless Steel	9.16	27.5
Nickel, pure	7.39	22.2
Monel®	7.77	23.3
Inconel®	6.39	19.2
Hastelloy C	6.28	18.8
Hastelloy X	7.67	23.0
Titanium	4.67	14.0
Silver	10.9	31.8
Zirconium	3.25	9.75
Invar	1.11	3.33
Gold	7.89	23.7

* in μin./in.-°F
† in μin³/in³-°F

Figure 37-7. *The coefficient of linear expansion may be used to calculate the change in dimensions of a metal with heating.*

Specific Heat

Specific heat is the ratio of the quantity of heat required to increase the temperature of a unit mass of metal by 1°, compared with the amount of heat required to raise the same mass of water by the same temperature. Specific heat is a way of comparing the amount of heat required to melt various metals.

A metal with a low melting point and high specific heat requires as much heat input to melt as a metal with high melting point and low specific heat. Aluminum, with a low melting point and high specific heat, requires almost the same amount of heat to melt as steel, which exhibits a higher melting point but lower specific heat.

Thermal Conductivity

Thermal conductivity is the rate at which metal transmits heat. In welding, thermal conductivity provides a measure for the heat input required to compensate for the rate at which heat is conducted away from the weld. Copper has a high thermal conductivity and is difficult to weld with low-temperature heat sources.

Austenitic stainless steel, with one-eighth the thermal conductivity of copper, requires a significantly lower heat input. The high thermal conductivity of copper makes it an excellent backing for welding. The rapid conduction of heat through copper backing prevents it from sticking to weld metal.

Electrical Conductivity and Resistivity

Electrical conductivity is the rate at which electric current flows through a metal. The higher the electrical conductivity of the metal, the more easily current flows through it. Electrical conductivity decreases as temperature increases, but room temperature values of electrical conductivity may be used for comparison between metals.

Electrical resistivity (resistivity) is the electrical resistance of a unit volume of a material. Resistivity is the reciprocal of electrical conductivity. Resistivity is the common method of expressing electrical conductivity. Metals with low resistivity (high electrical conductivity) are more conducive to resistance welding.

Magnetism

Magnetism is the ability of a metal to be attracted by a magnet, or to develop residual magnetism when placed in a magnetic or electrical field. This property is also known as ferromagnetism. Most steels are magnetic and may contain residual magnetism that can occur during magnetic particle inspection or from lifting with a magnet. Parts may need to be demagnetized before welding to prevent problems such as arc blow during welding. Arc blow causes the welding arc to deflect from its normal path because of magnetic forces.

Oxidation

Oxidation is the combination of a metal with oxygen in the air to form metal oxide. Every metal forms a thin oxide layer at room temperature. As temperatures increase, the oxide layer thickens. At welding temperature, steps must be taken to remove the metal oxide layer to prevent it from interfering with weld quality. Using flux-coated or flux-cored filler metals and inert gas welding prevents oxides from entering the weld area.

EFFECT OF WELDING ON MECHANICAL PROPERTIES

The mechanical properties of metals are classified using standards established by the American Society of Testing and Materials (ASTM). A *mechanical property* is a property of metal that describes the behavior of metals under applied loads. Mechanical properties are influenced by the composition and treatment of the metal.

Welding may alter specific mechanical properties of metals, leading to premature failure under load. The joint designer must consider the mechanical properties of metals when specifying the welds required. Welders should be familiar with basic terms and concepts associated with the mechanical properties of metals, such as strength, ductility, malleability, toughness, embrittlement, hardness, fatigue, and creep. An understanding of these concepts is often directly related to the ability to produce sound welds.

Strength

Strength is the ability of a metal to resist deformation from mechanical forces exerted on it. Deposited filler metal is usually stronger than the base metals it joins. It is necessary to use only the minimum amount of filler metal specified. Excess filler metal may be detrimental and exaggerate residual stress problems. Properly executed weld test specimens do not fail in the weld metal or HAZ when mechanically tested, but fail in the base metal. **See Figure 37-8.**

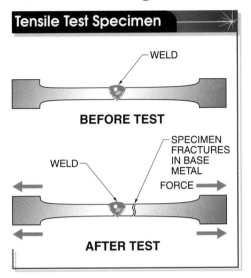

Tensile Test Specimen

WELD

BEFORE TEST

SPECIMEN FRACTURES IN BASE METAL

WELD

FORCE

AFTER TEST

Figure 37-8. *Weld mechanical test samples should fail in the base metal.*

In a structure, welds are classified as primary or secondary. A *primary weld* is a weld that is an integral part of a structure and that directly transfers a load. A primary weld must possess or exceed the strength of the structural members. A *secondary weld* is a weld used to hold joint members and subassemblies together. Secondary welds are subjected to less stress and less load than primary welds.

The strength properties of the metal being welded should be known, so that a strong, safe structure can be built. Likewise, when the strength of the weld is known as compared to the base metal, a weld joint strong enough to do the job can be produced.

Toughness

Toughness is the ability of a metal to absorb energy, such as impact loads, and deform rather than crack or fail catastrophically. Toughness is one of the most important metal mechanical properties. Weld procedures are designed to maintain toughness of the weld.

When heat-treatable steels are welded, the rapid cooling rate may cause an undesirable decrease in toughness of the HAZ. Proper methods of maintaining toughness

must be used, such as preheat, interpass temperature control, or postheating.

Toughness is difficult to measure, but with steels, toughness correlates inversely with hardness, which is relatively easy to measure. High hardness in the HAZ may indicate low toughness in steels. Crack-like discontinuities may provide a stress concentration effect that causes the crack to propagate rapidly when a load is applied. **See Figure 37-9.** For this reason, fabrication codes do not permit cracks or crack-like discontinuities.

Welding procedures for steels in specific applications may require impact testing requirements to ensure that there is no loss of toughness in the HAZ. *Impact testing* is special testing performed on small, notched specimens, to simulate a stress concentration effect.

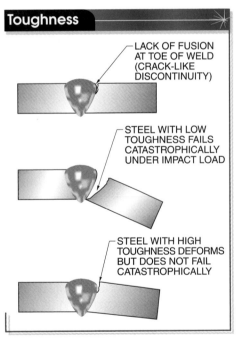

Toughness

LACK OF FUSION AT TOE OF WELD (CRACK-LIKE DISCONTINUITY)

STEEL WITH LOW TOUGHNESS FAILS CATASTROPHICALLY UNDER IMPACT LOAD

STEEL WITH HIGH TOUGHNESS DEFORMS BUT DOES NOT FAIL CATASTROPHICALLY

Figure 37-9. *Tough steel will absorb a sudden load, rather than crack catastrophically.*

Hardness

Hardness is the resistance of a material to deformation, indentation, or scratching. Hardness testing is one of the most widely used testing procedures because it is rapid, easy to use, and often nondestructive. Hardness is most often measured using indentation hardness tests, such as the Brinell test, the Rockwell test, or the Vickers test.

On steel, hardness can be used to estimate the toughness of a weld joint, especially where preheat, interpass temperature control, and/or post-heating are used to ensure integrity. **See Figure 37-10.** It can also be used to predict scratching or scuffing resistance of a material.

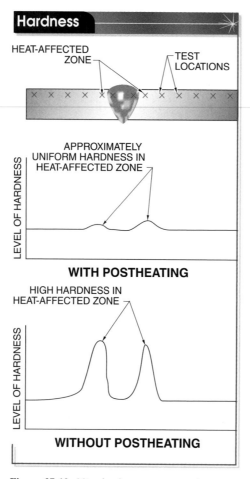

Hardness

HEAT-AFFECTED ZONE

TEST LOCATIONS

APPROXIMATELY UNIFORM HARDNESS IN HEAT-AFFECTED ZONE

LEVEL OF HARDNESS

WITH POSTHEATING

HIGH HARDNESS IN HEAT-AFFECTED ZONE

LEVEL OF HARDNESS

WITHOUT POSTHEATING

Figure 37-10. *Microhardness traverse made across a steel weld joint indicates whether there is a loss of toughness in the heat-affected zone.*

Ductility

Ductility is a measure of the ability of a metal to yield plastically under load, rather than fracture. High-ductility metals, such as copper, deform as the load on the metal is increased, eventually failing. Low-ductility metals, such as cast iron, deform only slightly and fail suddenly as the load is increased. Ductility is measured in tensile test samples by percentage elongation to failure, or percentage reduction of area to failure.

Embrittlement

Embrittlement is the complete loss of ductility and toughness of a metal, so that it fractures when a small load is applied. Embrittlement may be caused by applying the wrong brazing metal or when molten zinc contacts stainless steel. If galvanized steel is welded to stainless, the zinc adjacent to the weld region must be removed by sand blasting or careful grinding prior to welding. Embrittlement often occurs by penetration of the embrittling species into the grains of the metal (intergranular penetration).

Fatigue

Fatigue is failure of a material operating under alternating (cyclic) stresses at a value below the tensile strength of the material. Fatigue is a problem that affects the service life of any component that moves, rotates, vibrates, or is subject to thermal cycling. For example, a piston rod or an axle undergoes rapid and complete reversal of stresses from tension to compression.

Approximately 90% of all failures in engineering components are fatigue-related. Fatigue problems may be severe in welded structures since most welded joints have poor fatigue strength and finite fatigue life because of their shape, residual stresses, and discontinuities. All welding introduces stress concentrations into a weld, reducing fatigue strength; the effect is highest when the load is applied transversely to the weld.

Fatigue cracking initiates in the toe of the weld where stress concentrations are highest. Features that increase the strength of the weld, such as additional weld beads or inclusion of stiffeners, increase stress concentration and further reduce fatigue life. For this reason, attempting to fix a part that has failed in fatigue by adding a weld bead, or reinforcing with stiffeners welded to the structure, has the opposite effect and further reduces the life of the part. Although the weld itself is stronger under static load, weld discontinuities, coupled with the additional stress concentration, more than offset any strengthening effect. **See Figure 37-11.**

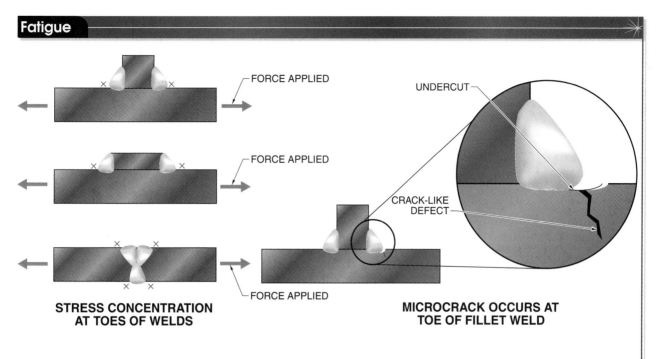

**STRESS CONCENTRATION
AT TOES OF WELDS**

**MICROCRACK OCCURS AT
TOE OF FILLET WELD**

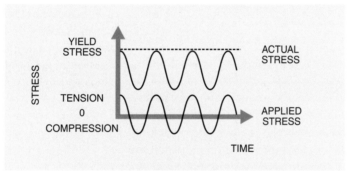

**EFFECT OF RESIDUAL STRESS
FROM WELDING ON APPLIED STRESS**

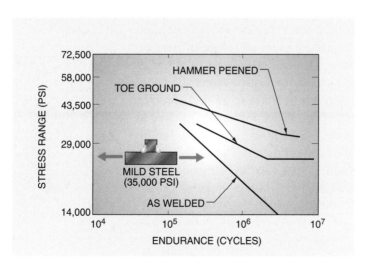

**EFFECT OF POST-WELD IMPROVEMENT
TECHNIQUES ON FATIGUE LIFE**

Figure 37-11. *Welding reduces the fatigue strength of structures.*

Fillet welds are particularly prone to failure by fatigue. During cooling of a fillet weld, the toe develops a microcrack about 0.005″ deep. The microcrack can grow into a full-scale fatigue crack and lead to premature failure. **See Figure 37-12.**

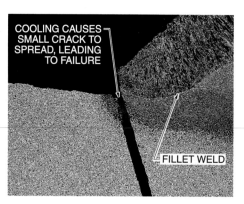

Figure 37-12. *Fillet welds are prone to fatigue failure. During cooling, a small crack can grow into a fatigue crack and lead to premature failure, as cooling can cause the crack to spread.*

If design improvement is not possible, it may be necessary to use post-weld improvement techniques such as grinding, peening, or GTAW plasma dressing of a fillet weld toe to remove microcracking. Post-weld improvements can increase fatigue life significantly, but must not introduce surface notch into the part.

Fatigue failures in welds are prevented by designing welds away from critical regions of high stress concentration. Welding in an area of high stress concentration is a leading cause of failure of rotating shafts. The area of high stress is where the shaft transitions to larger diameter. Welding or rebuilding by welding in a high stress concentration area, such as to rebuild a worn shaft, will lead to failure within a short period. The shaft must be rebuilt so that welding is carried out in locations away from the region of highest stress.

Creep

Creep is slow, plastic elongation that occurs during extended service under load above a specific temperature for that metal. Structural metals undergo creep at high temperatures. Creep-resistant alloys are used for high-temperature strength in petroleum refining, steam power generation, and other industries. Selecting the wrong filler metal or base metal may lead to premature failure from creep.

Malleability

Malleability is the ability of a metal to be deformed by compressive forces without developing defects such as those encountered in rolling, pressing, or forging.

Mechanical Force

Mechanical properties are characteristic responses of materials to mechanical forces. A *load* is an external mechanical force applied to a component. Standard terms used to describe the mechanical properties of solid metals include stress and strain. **See Figure 37-13.**

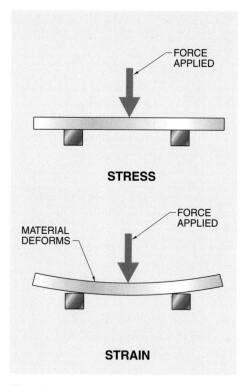

Figure 37-13. *Stress is the internal resistance of a material to an externally applied load. Stress is measured as the applied load over an area. Strain is the accompanying change in dimensions when a load induces stress in a material.*

Stress. *Stress* is the internal resistance of a material to an externally applied load. Stress is measured in terms of load divided by area. Every machine part or structural member is designed to safely withstand a certain amount of stress.

Strain. *Strain* is the accompanying change in dimensions when a load induces stress in a material. Strain is either elastic or plastic. Elastic strain occurs when a material is capable of returning to its original dimensions after removal of the load. For example, a spring with a normal load returns to its original length when the load is removed. Plastic strain occurs when a material is permanently deformed by the load. For example, an overloaded spring will develop a permanent set or an increase in length. As the load is steadily increased, a point is reached where the strain changes from elastic to plastic.

A *static load* is a load that remains constant. An example of a static load is a constant amount of water stored in a storage tank. An *impact load* is a load that is applied suddenly or intermittently. An example of an impact load is the action of a pile driver setting a pile.

A *cyclical (variable) load* is a load that varies with time and rate, but without the sudden change that occurs with an impact load. An example of a variable load is a revolving camshaft with a varying compressive and tensile load applied.

Mechanical Force Application. Mechanical force can be applied by five different methods: tension, compression, shear, torsion, and flexing. Combinations of methods may be applied under actual load conditions. **See Figure 37-14.**

Tension (tensile stress) is stress caused by two equal forces acting on the same axial line to pull an object apart. The magnitude of the stress depends on the amount of load placed on the object and the cross-sectional area of the object. The same load causes greater stress to an object with a small cross-sectional area than to an object with a large cross-sectional area.

Tensile strength is a measure of the maximum stress that a material can resist under tensile stress. Tensile stresses work to pull a material apart. The tensile strength of a metal is a primary factor to be considered in the evaluation of the metal. To find tensile stress, apply the formula:

$$\sigma_t = \frac{F}{A}$$

where
σ_t = tensile stress (in lb/sq in.)
F = force (in lb)
A = area (in sq in.)

For example, what is the tensile stress of an 8000 lb force applied to a square steel rod with a cross-sectional area of 0.50 sq in.?

$$\sigma_t = \frac{F}{A}$$

$$\sigma_t = \frac{8000}{0.50}$$

$$\sigma_t = \textbf{16,000 lb/sq in.}$$

Compression (compressive stress) is stress caused by two equal forces acting on the same axial line to crush an object. The deformation caused by compression consists of an increase in the cross-sectional area and a decrease in the original length of the object. *Compressive strength* is the ability of a material to resist being crushed. Nonmetallic materials, like brick, have high compressive strength compared to their tensile strength. To find compressive stress, apply the formula:

$$\sigma_c = \frac{F}{A}$$

where
σ_c = compressive stress (in lb/sq in.)
F = force (in lb)
A = area (in sq in.)

For example, what is the compressive stress of a 120,000 lb force applied to a rectangular cast iron bar with a cross-sectional area of 6 sq in.?

$$\sigma_c = \frac{F}{A}$$

$$\sigma_c = \frac{120,000}{6}$$

$$\sigma_c = \textbf{20,000 lb/sq in.}$$

✓ Point

Mechanical force may be applied by tension, compression, shear stress, torsion, or flexural stress.

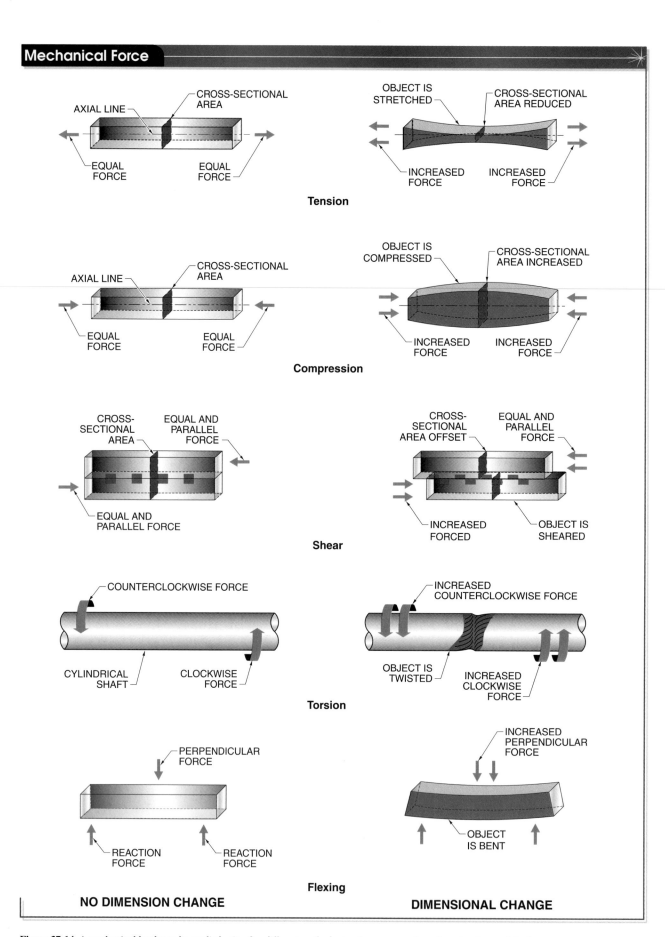

Figure 37-14. *A mechanical load may be applied using five different methods: tension, compression, shear, torsion, and flexing.*

The *modulus of elasticity* is a measure of the stiffness of an object under tension or compression. It is measured as the ratio of stress to strain for tensile or compressive forces that are within the elastic limit. Modulus of elasticity is an index of the ability of a solid material to deform when an external force is applied and then return to its original size and shape after the external force is removed. The less a material deforms under a given stress, the higher its modulus of elasticity.

The modulus of elasticity does not measure the amount of stretch a particular metal can take before breaking or deforming. It indicates how much stress is required to deform metal a given amount. **See Figure 37-15.** By checking the modulus of elasticity, the welder can ascertain the comparative stiffness of different materials. Rigidity (or stiffness) is an important consideration for many machine and structural applications. To find modulus of elasticity, apply the formula:

$$E = \frac{\sigma_s}{\varepsilon_n}$$

where

E = modulus of elasticity (in lb/sq in.)

σ_s = stress (in lb/sq in.)

ε_n = strain (in lb/sq in.)

For example, what is the modulus of elasticity of a 1″ square piece of metal subjected to 40,000 lb of tension (stress) and exhibiting 0.001 in./in. strain?

$$E = \frac{\sigma_s}{\varepsilon_n}$$

$$E = \frac{40,000}{0.001}$$

E = 40,000,000 lb/sq in.

Shear (shear stress) is stress caused by two equal and parallel forces acting upon an object from opposite directions. Shear stresses tend to cause one side of the object to slide in relation to the other side. Shear stress placed on the cross-sectional area of an object is parallel to the force. The strength of materials under a shearing stress

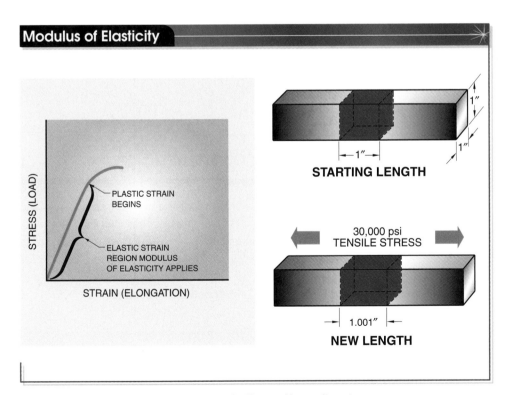

Modulus of Elasticity

STARTING LENGTH

30,000 psi
TENSILE STRESS

NEW LENGTH

STRESS (LOAD)

PLASTIC STRAIN
BEGINS

ELASTIC STRAIN
REGION MODULUS
OF ELASTICITY APPLIES

STRAIN (ELONGATION)

Figure 37-15. *Modulus of elasticity is a measure of stiffness and has no dimensions.*

is less than under a tensile stress or a compressive stress. To find shear stress, apply the formula:

$$\sigma_s = \frac{F}{A} \quad \text{or } F = \sigma_s \times A$$

where

σ_s = shearing stress (in lb/sq in.)
F = force (in lb)
A = area (in sq in.)

For example, a 0.750″ hole is to be punched in a steel plate 0.5″ thick. What is the required force of the press if the ultimate strength of the steel plate in shear is 42,000 lb/sq in.?

The shear cross-sectional area (A) is equal to the circumference of the hole times the thickness of the plate (3.14 × 0.750 × 0.5 = 1.1775).

$$F = \sigma_s \times A$$
$$F = 42,000 \times 1.1775$$
$$F = \textbf{49,455 lb}$$

Torsion (torsional stress) is stress caused by two forces acting in opposite twisting directions. Shafts used to transfer rotary motion are subject to torsional stress. The shafts are twisted by excessive torque, expressed in inch-pounds (in-lb). *Torque* is the product of the applied force (F) times the distance (L) from the center of the application. *Torsional strength* is the measure of a material's ability to withstand forces that cause it to twist. To find torque, apply the formula:

$$T = F \times L$$

where

T = torque (in in-lb)
F = force (in lb)
L = distance (in in.)

For example, what is the torque of a 160 lb force applied over a distance of 12″?

$$T = F \times L$$
$$T = 160 \times 12$$
$$T = \textbf{1920 in-lb}$$

Flexing (flexural or bending stress) is stress caused by equal forces acting perpendicular to the horizontal axis of an object. Bending stresses bend an object as the perpendicular force overcomes the reaction force. Bending stress is a combination of tensile stress and compressive stress. *Bending strength* is a

combination of tensile and compressive forces, and is a property that measures resistance to bending or deflection in the direction that the load is applied.

Bending stress is commonly associated with beams and columns. The deformation caused by bending stress changes the shape of the object and creates a deflection. To find bending stress, apply the formula:

$$\sigma_b = \frac{M \times C}{I}$$

where

σ_b = bending stress (in lb/sq in.)
M = maximum bending moment (in in.-lb)
c = distance from neutral axis to farthest point in cross section (in in.)
I = area moment of inertia (in in.[4])

For example, what is the bending stress of a 1″ solid shaft subjected to a bending moment of 1400 in.-lb? The distance from the neutral axis to the cross-sectional area is 0.5″, and area moment of inertia is 0.049 in.[4]

$$\sigma_b = \frac{M \times C}{I}$$

$$\sigma_b = \frac{1400 \times 0.5}{0.049}$$

$$\sigma_b = \textbf{14,286 lb/sq in.}$$

EFFECT OF WELDING STRESS ON WELDS

Welding creates significant stresses in joints, resulting in shrinkage stresses and residual stresses that may lead to cracking. Stress resulting from welding exerts a great influence on the behavior of welds in service. Stress types are shrinkage stress and residual stress.

Shrinkage Stress

Shrinkage stress is stress that occurs in weld filler metal as it cools, contracts, and solidifies. The solidifying filler metal is relatively weak and has difficulty accommodating the stresses that result from shrinkage. Additionally, the last part of weld filler metal to solidify contains the lowest melting point constituents, increasing the weakness of the weld. **See Figure 37-16.**

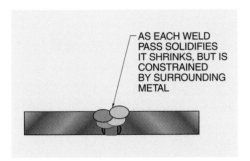

AS EACH WELD PASS SOLIDIFIES IT SHRINKS, BUT IS CONSTRAINED BY SURROUNDING METAL

Figure 37-16. *Shrinkage of the weld during solidification imposes severe stress on the weld when it is in a relatively weak condition.*

Shrinkage stress problems are made worse when contaminants react with the solidifying weld filler metal to form weak or brittle microconstituents, or when the joint restrains (stiffens) the base metal, hampering shrinkage of the solidifying weld metal. Shrinkage stresses can cause cracks (cracking).

Contamination of the weld metal or excessive heat input during welding increases the susceptibility of the part to hot cracks. Nickel alloys may hot crack from the presence of even trace amounts of sulfur on the surface. Copper alloys may hot crack from excessive heat input.

Hot cracks may also occur if insufficient weld metal is added to a joint. When welding heavy-wall pipe, the wall thickness dictates whether it is possible to radiograph or dye check the root bead of weld filler metal to monitor its quality and decide whether any repairs are required. Excessive shrinkage stresses in heavy-wall pipe may cause a root bead to crack as it cools to ambient temperature from restraint in the joint. Thus, in heavy-wall pipe, it is necessary to make several weld passes before cooling to ambient temperature in order to create sufficient volume of weld metal to accommodate shrinkage stresses without cracking.

Residual Stress

Residual stress is stress that occurs in a joint member or material after welding has been completed, resulting from thermal or mechanical conditions. Almost every fabrication process introduces residual stress into metals. Residual stress from welding is often significantly higher than other fabrication processes. Residual stress may also be introduced into parts by post-fabrication procedures such as installation and assembly, occasional service overloads, ground settlement, and repair or modification.

As solidified weld metal cools to room temperature, the stresses within it increase and eventually exceed the yield strength of the base metal and the HAZ. *Yield strength* is the level of stress within a metal that is sufficient to cause plastic flow. Residual stress may cause cold cracking or distortion if the welded structure deforms to accommodate it. Cold cracking may be delayed hours or even days after the weld is finished. *Distortion* is the undesirable dimensional change of a fabrication. Distortion leads to out-of-specification dimensions or shape. **See Figure 37-17.**

The Lincoln Electric Company

Residual stresses must be controlled during welding and during postproduction procedures, such as installation and assembly, to prevent defects such as cold cracking and distortion.

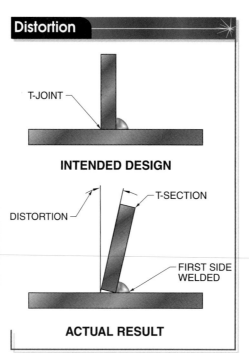

Distortion

T-JOINT

INTENDED DESIGN

DISTORTION — T-SECTION

DISTORTION —

FIRST SIDE WELDED

ACTUAL RESULT

Figure 37-17. *Residual stress leads to many problems, such as distortion or loss of fatigue strength.*

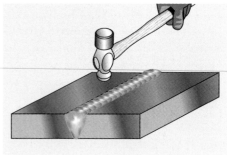

Figure 37-18. *Peening relieves internal stresses in a weld and helps the welded joint stretch as it cools.*

stress in the surface layers of a weld. Peening is performed for each weld pass immediately after solidification with impact blows. Peening induces compressive stresses and improves resistance to fatigue failure. Peening is not a substitute for the postheat required to restore toughness to a weld joint.

Residual Stress Reduction. To accommodate residual stresses and prevent distortion, welding procedures are designed to balance residual stresses across different parts of the weld. Methods of reducing residual stress include postheating, peening, and vibratory stress relief.

> 🛡 *Residual stresses in welds must be controlled to prevent the occurrence of distortion in the weldment, premature failure of the weldment, or both.*

Postheat is the reheating of the weld area to a high temperature, holding for a predetermined time at temperature, and cooling at a specified rate. Postheat is used to prevent cold cracking from residual stresses. Postheat also stress-relieves the joint, reducing the possibility of distortion or cracking in service. With steels, postheat additionally tempers (softens and toughens) the weld. Postheat is often specified in conjunction with preheat and interpass temperature control.

Peening using a ball peen hammer relieves stresses in the metal by helping the metal stretch (yield) as it cools. **See Figure 37-18.** Peening reduces residual

Vibratory stress relief is the application of subresonant vibration during welding to control distortion, or after cooling to provide stress relief. *Subresonant vibration* is vibration frequency less than the resonant frequency of the weld. Vibratory stress relief may control distortion during welding, but does not offer any significant stress relief. It should not be substituted for any specified preheat, interpass temperature control, or postheating procedure.

EFFECT OF WELDING ON CORROSION RESISTANCE

The heat of welding can reduce the corrosion resistance of most metals. The loss of corrosion resistance may be caused by chemical inhomogeneity, residual stress, excessive hardness, or an undesirable microstructure.

Chemical Inhomogeneity

Welding creates chemical inhomogeneity, or segregation, in the weld joint. *Chemical inhomogeneity* is any disturbance in the chemical composition gradient of a metal. Chemical inhomogeneity leads to a loss of chemical resistance in corrosion-resistant alloys. Corrosion-resistant alloys must be welded with filler metals that do not reduce their corrosion resistance.

When similar base metals are welded, filler metal with a chemical composition similar to or slightly more corrosion-resistant than the base metal should be used. When dissimilar metals are welded, the filler metal must exceed the corrosion resistance of both metals. Dilution or segregation must not result in reduced corrosion resistance of the joint.

Segregation. *Segregation* is any concentration of alloying chemical elements in a specific region of a metal and is similar to inhomogeneity. Segregation can be an increased concentration or a depletion of chemical elements in the region. For example, molybdenum is added to stainless steels to improve their resistance to chloride-containing environments. When stainless steel base metal that contains 4.5% molybdenum or greater is joined, matching filler metal with 4.5% molybdenum is not sufficient. Molydenum segregation occurs in the weld bead, leading to small molybdenum-depleted regions with inferior corrosion resistance. In this instance, filler metal with a molybdenum content higher than 4.5% must be used to compensate for segregation.

Residual Stress

Weld joints with high residual stress may be susceptible to corrosion in specific environments. Such welds are stress-relieved when necessary to prevent premature failure. Weld repair or burning is not permitted on stress-relieved equipment unless a welding procedure that incorporates stress relief is used. **See Figure 37-19.**

Excessive Hardness

An excessively hard HAZ, produced by rapid cooling from welding, may crack in certain chemical environments. Hard HAZs are also susceptible to hydrogen-assisted cracking from corrosion in service. *Hydrogen-assisted cracking* is loss of toughness in steels resulting from hydrogen atoms created at the surface of the metal by corrosion that diffuse into the HAZ and the base metal. Hydrogen diffusion interferes with the metal's ability to yield under stress, reducing its ductility and toughness.

When a corrosion reaction produces hydrogen atoms on the metal surface, the hydrogen atoms may or may not combine with one another. If they combine, hydrogen molecules are produced, which harmlessly dissipate from the metal surface. If they do not combine, the hydrogen atoms are extremely active and diffuse into the metal to cause hydrogen-assisted cracking.

Some species contained in corrosive environments, called poisons, are very harmful because they prevent, or "poison," the recombination of hydrogen atoms to hydrogen molecules. Poisons include sulfides such as hydrogen sulfide. Sulfide stress cracking is a form of hydrogen-assisted cracking that is a problem in the oil and gas production industry. Sulfide stress cracking is caused by the presence of hydrogen sulfide. Susceptibility of steels to hydrogen-assisted cracking increases with hardness of the steel.

In environments containing poisons, carbon and low-alloy steels require proper preheat and postheating to reduce hardness in the HAZ to a value below Rockwell C 22 (22 HRC). Weld repairs that do not use adequate preheat and postheating may create an HAZ with excessive hardness.

Welding operations can significantly affect the corrosion resistance of metal. How the welding process will affect the corrosion resistance of the metal must be considered before a particular metal is selected. Stress-relief must be performed on metals whose corrosion resistance may be affected during welding.

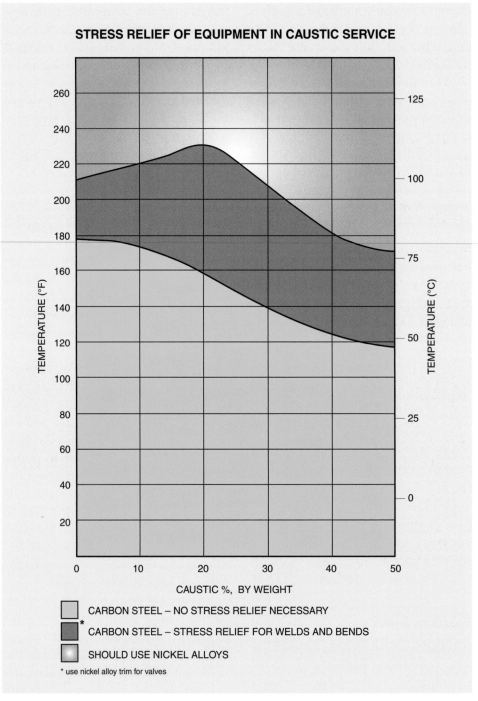

STRESS RELIEF OF EQUIPMENT IN CAUSTIC SERVICE

CARBON STEEL – NO STRESS RELIEF NECESSARY

*CARBON STEEL – STRESS RELIEF FOR WELDS AND BENDS

SHOULD USE NICKEL ALLOYS

* use nickel alloy trim for valves

Figure 37-19. *Stress-relieved equipment should not be welded without a procedure that includes postheating.*

Undesirable Microstructure

A *microstructure* is the appearance of the metallurgical structure of metals when they are specially prepared to reveal their features. **See Figure 37-20.** Microstructure is examined on polished and etched samples of metals, with a metallurgical microscope producing magnification from 100x to 1000x. The metallurgical structure of weld joints is revealed by examining their microstructure.

Undesirable microstructure is the creation, through the heat of welding, of microstructures that are preferentially attacked in a corrosive environment. For example, 304 or 316 stainless steels may develop an undesirable microstructure in the HAZ, known as sensitization, during welding. If conditions are favorable for sensitization, chromium and carbon within the stainless steel combine rapidly in the temperature range of 800°F to 1500°F, and most rapidly at 1200°F.

Chromium carbide within stainless steel reduces the corrosion resistance of the stainless steel. The reduced corrosion resistance of the stainless steel results in a line of deep corrosion in the HAZ when it is exposed to certain corrosive environments. An extra-low-carbon grade of 304 or 316, such as 304L or 316L, or specially formulated grades that are immune to sensitization during welding should be used. In the extra-low-carbon grades, the carbon content is reduced to a level that is insufficient to combine with the chromium in the metal during the time spent in the sensitization temperature range.

Metallurgical Microstructure

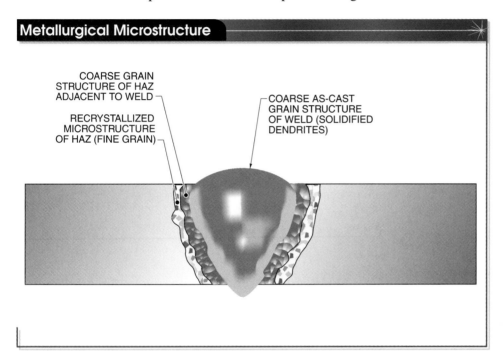

Figure 37-20. *A metallurgical microstructure is the appearance of the metallurgical structure of metal when specially prepared to reveal its features.*

 Refer to Quick Quiz® on CD-ROM

Points to Remember

- Crystal structure is a specific arrangement of the building blocks of matter (atoms) in an orderly and repeating three-dimensional pattern.
- Heat input is the most important element for welding. Heat (heat input) is required to melt the base metal and filler metal during welding.
- The three key regions of a weld are the weld metal, the base metal, and the heat-affected zone.
- It is usually necessary to apply two layers of surfacing weld to overcome dilution and attain the required wear or corrosion resistance properties.
- The HAZ is a narrow band of base metal adjacent to the weld joint. Most problems that occur during welding occur in the HAZ.
- Physical properties of metal include melting point, thermal expansion, specific heat, thermal conductivity, electrical conductivity, magnetism, and oxidation.
- Mechanical properties describe the behavior of metals under mechanical loads and include strength, toughness, hardness, ductility, fatigue, creep, and malleability.
- Mechanical force may be applied by tension, compression, shear, torsion, or flexural stress.
- Welding creates significant stresses in joints, resulting in shrinkage stresses and residual stresses that may lead to cracking.
- Residual stresses may be reduced using intermittent welding, low heat input welding, postheating, or peening.
- Welding creates chemical inhomogeneity, or segregation, in the weld joint, which leads to a loss of chemical resistance.
- An excessively hard HAZ, produced by rapid cooling from welding, may crack in specific chemical environments.
- Chromium carbide within stainless steel reduces the corrosion resistance of the stainless steel. The reduced corrosion resistance of the stainless steel results in a line of deep corrosion in the HAZ when it is exposed to certain corrosive environments.

Questions for Study and Discussion

1. When does the grain structure of a metal begin to develop?
2. What is the value of heat input with a welding current of 400 A at 45 V and a travel speed of 12 in./min?
3. What is the effect of preheat on the cooling rate of the weld?
4. What is the effect of the heat of welding in the HAZ of an alloy that has been heat-treated?
5. Why must two layers of surfacing weld be used when applied using arc welding processes?
6. Why is copper a good material for use as a backing material?
7. What is the difference between strength and toughness?
8. How does welding reduce the fatigue strength of metals?
9. Name three ways by which mechanical force may be applied.
10. Why must extra-low carbon grades of stainless steel be used when welding stainless steel for corrosive surfaces?

Refer to Chapter 37 in the *Welding Skills Workbook* for additional exercises.

Metal Identification

Chapter

Metal identification verifies that as-received base metals and filler metals meet specifications. Metal identification is also required when the materials test report has been lost or physical identification markings have disappeared because of environmental wear. For critical welding work, supplementary metal identification may be required to verify conformance with purchase specifications.

Metals used in fabrication are typically specified on the weld prints. If a metal is not specified, qualified personnel must determine the metal to be used. Welders may be required to identify appropriate metals without assistance from qualified personnel during maintenance and repair tasks.

Many metal products such as pipe or plate are often purchased and stored for future use. Metals and filler metals can be identified before welding using visual identification, qualitative identification, semi-quantitative identification, and quantitative identification.

MANUFACTURER PAPERWORK

Manufacturer and supplier paperwork provides the initial means of checking specification compliance. *Paperwork* is physical certification or documentation provided by a product manufacturer or supplier. Paperwork may be hard copy or soft copy (computerized). The paperwork supplied by the manufacturer includes a materials test report (MTR), product analysis, and certificate of compliance (COC).

Materials Test Reports

A *materials test report (MTR)* is a certified statement issued by the primary manufacturer indicating the chemical analysis and mechanical properties of the metal. An MTR is also called a certificate of analysis (COA). Although an MTR is not formally required for all types of ASME code-approved metals used for code work, many companies require that an MTR accompany the metal.

An MTR allows the end user to ensure that the metal meets specified chemical composition and mechanical property requirements.

Product Analysis

A *product analysis* is a chemical report that a particular metal, such as tubing or piping, is made from a particular heat of metal. Product analyses ensure that substitutions have not been made during processing of the metal. Product analyses are called out as supplemental requirements in ASTM specifications.

> ✓ **Point**
>
> *Manufacturers supply three types of paperwork to identify their products: materials test report, product analysis, and certificate of compliance.*

Certificate of Compliance

A *certificate of compliance (COC)* is a statement by a manufacturer, without supporting documentation, that the supplied metal meets specifications. A COC contains no test reports; it only states that, from the records, the manufacturer is confident no substitutions have been made. A COC can be issued for any metal.

Media Clip

Visual Identification

MATERIALS NONCONFORMANCE REPORT

A *materials nonconformance report* is a form created by the receiver of the metal to audit manufacturer paperwork regarding supplied metals. Analysis of materials nonconformance reports allows problem areas in metals acquisition to be identified, corrected, and prevented in the future. Materials nonconformance reports are only valuable if followed up by corrective programs. **See Figure 38-1.**

Consequences of Improper Materials Substitution

If improper metal substitutions are made, significant damage to equipment or injury to workers may result. For example, chrome-moly steels have a key use in critical applications, such as piping for handling high-temperature steam or hydrogen. Chrome-moly steels can easily be mistaken for carbon steels. They are similar in appearance to carbon steels, are magnetic like carbon steel, and rust like carbon steel if stored outdoors unprotected. However, substituting carbon steels for chrome-moly steels may result in catastrophic failure because, in a critical application, carbon steel is likely to fail before chrome-moly fails. Also, substituting the wrong type of metal, such as medium-carbon steel for low-carbon steel, nullifies the welding procedure and increases the chance of cracking.

VISUAL IDENTIFICATION

Visual identification is metal identification that consists of checking the appearance of the base metal or filler metal for key features that identify the metal type. Visual identification is performed by checking the appearance, color, nameplate, and markings of the metal.

Appearance

The appearance and shape of a metal may indicate the type of metal. Appearance includes the form and dimensions of metal components and parts. A hot-rolled structural shape in a steel-frame building would be low-carbon steel.

A rail would be identified by its shape as high-carbon steel. Many machine parts for light- and medium-duty industrial equipment and agricultural equipment are made of cast iron. Castings for heavy-duty work such as brake presses are commonly made of medium-carbon steel.

Harrington Hoists, Inc.

Materials can be identified by color and appearance, by a nameplate, or by markings stenciled on the end of the metal.

Materials Nonconformance Report

To be completed by field inspector or whoever discovers problem.
Keep one copy in the component file and submit one copy to
_____ (appropriate area resource)

Equipment name_____ Number_____

Component name _____ Number_____

Reason(s) for nonconformance _____

Supplier or replicator _____

Rebuilder (if applicable) _____

If rebuilt, by whom?_____

Specification(s) _____

Shipping procedures_____

Receiving/stores procedures _____

Inspection procedures_____

Material type _____

Dimensional requirement(s)_____

Tolerances _____

Improper repairs or modifications _____

Installation Procedures _____

Other _____

_____

Reported by _____

Figure 38-1. *A materials nonconformance report is a form created by the receiver of the metal to audit manufacturer paperwork regarding supplied metals.*

Color

The color of a metal is any specific hue that the metal typically exhibits. Some metals are relatively easy to distinguish by their color. Copper is reddish in color and easily identifiable. **See Figure 38-2.** Heat tint from heat-treating operations and surface scales and tarnishes from exposure to the environment may hide a metal's true color. To be sure of the true color, a small area of the surface must be cleaned by filing or rubbing with coarse abrasive paper. Color identification must not be used on metals that have suffered corrosion or oxidation that has resulted in surface color changes. **See Figure 38-3.**

> ✓ **Point**
>
> *Visual identification includes appearance, color, nameplate, and markings to determine key features that identify the metal type.*

Figure 38-2. *Color is one key feature that can be used to visually identify metals such as copper.*

forgings, the stamped impression is produced with a metal die. The impression is usually located on the outside surface of the forging and consists of the ASTM or other materials standard, the pressure and temperature rating, and the forge shop logo.

Fasteners are identified by an embossed or stamped marking on either end of the fastener. Space is limited on fasteners, so a code is used to identify the standards organization and manufacturer. **See Figure 38-4.** The Industrial Fasteners Institute publishes a list of fastener manufacturers' logos. Metal markings consist of foundry marks, color-coding, and stencil marking.

Nameplate

Fabricated equipment, such as heat exchangers or pressure vessels, must have a plaque or nameplate fixed to the exterior. The nameplate identifies the design, pressure and temperature rating, test pressure, and materials of construction. The nameplate must not be covered, damaged, or removed during the life of the equipment.

Markings

Markings may be embossed, stamped, stenciled, or attached to a part. Stamping and embossing are surface identification markings created by mechanical deformation on wrought products. On

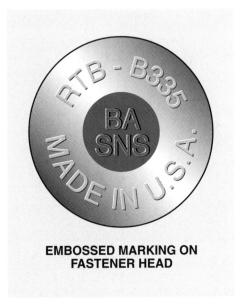

EMBOSSED MARKING ON FASTENER HEAD

Figure 38-4. *An embossed marking on the head or other end of a fastener is one method of identifying fasteners.*

CHARACTERISTIC COLOR GROUPINGS

Color	Metal
Red or Reddish	Copper, >85 Copper Alloys
Light Brown or Tan	90% Cu/10% Ni (Copper-nickel)
Dark Yellow	Bronzes and Gold
Light Yellow	Brasses
Bluish or Dark Gray	Lead, Zinc, and Zinc Alloys
Silvery White with soft luster	Aluminum
Silvery White with bright luster	Stainless Steels
Gray	Carbon and Low-alloy Steels, 70% Cu/30% Ni (Copper-Nickel)
White or Gray	Nearly all others

Figure 38-3. *Metals can be identified and grouped by the characteristic colors that the metal exhibits.*

Foundry Marks. A *foundry mark* is an identification marking embossed on the exterior of castings. Foundry marks are incorporated into the casting mold. Identification information includes the ASTM grade number, foundry name or logo, heat number, and foundry shorthand description for the alloy. When identifying castings by their foundry marks, the manufacturer's alloy codes must be known. **See Figure 38-5.**

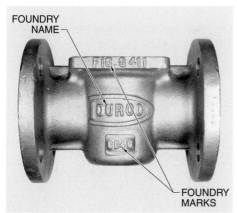

The Duriron Company, Inc.

Figure 38-5. *Foundry marks are identification markings that are embossed on the exterior of castings.*

Color-Coding. *Color-coding* is an identification marking that consists of colored stripes painted on one end of metal to allow for permanent storage or temporary storage and subsequent retrieval from a metal service center or a user's storeroom.

Color-coding systems must clearly identify each metal. Since color-coding is set up to identify specific metals stored at a particular location, there is no universal color-coding system. To retain the color-coding system, metal must be cut from the end opposite the colored end. **See Figure 38-6.**

Stencil Marking. *Stencil marking* is an identification marking that consists of continuous or repeated ink markings on the metal. Stencil markings indicate alloy type, conformance to standards, and the dimensions of the metal. Stencil markings are repeated at regular intervals along the metal so the identification is not lost when the metal is cut or sectioned. Stencil markings are not permanent and may degrade during service or if stored outdoors.

Some chemical elements found in materials used for color-coding or stencil marking are potentially harmful. Chlorine (Cl), sulfur (S), and zinc (Zn) are some potentially harmful chemical elements that may be present. These chemical elements may cause catastrophic cracking in susceptible alloys such as stainless steel or high nickel alloy. Cracking is likely to occur when the paint or marking material on the metal is exposed to the heat of welding, to high-temperature service, or to corrosive environments in service. Marking materials that are used on susceptible alloys must contain low quantities (measured in parts per million, or ppm) of harmful chemical elements. No more than 250 ppm is allowable.

COLOR-CODING FOR SELECTED STEELS			
AISI-SAE Designation	**Color**	**AISI-SAE Designation**	**Color**
1010 Carbon Steel	White	4640 Molybdenum Steel	Green and Pink
1025 Carbon Steel	Red	3125 Nickel-Chromium Steel	Pink
1112 Free-cutting Steel	Yellow	3325 Nickel-Chromium Steel	Orange and Black
1120 Free-cutting Steel	Yellow and Brown	5120 Chromium Steel	Black
2015 Nickel Steel	Red and Brown	6115 Chromium-Vanadium Steel	White and Brown
2330 Nickel Steel	Red and White	7260 Tungsten Steel	Brown and Aluminum
4130 Molybdenum Steel	Green and White	9255 Silicon-Manganese	Bronze

Figure 38-6. *Color-coding allows easy and rapid identification of metals.*

Material markings are based on the materials designations assigned by classification societies such as the American Iron and Steel Institute (AISI), the Society of Automotive Engineers (SAE), or the American Society for Testing and Materials (ASTM).

Marking materials should be removed from the area of the metal to be welded, brazed, or soldered using an approved solvent. A marker with a fiber tip may be used to mark a metal. Because markers leave no solid residue that may lead to cracking, a solvent is not needed for removal.

QUALITATIVE IDENTIFICATION

Qualitative identification is metal identification by a qualified person to confirm the identity of an unknown metal. Qualitative identification has a relatively high degree of certainty for many applications. Qualitative identification techniques include magnetic response testing, chisel testing, torch testing, and file testing.

Magnetic Response Testing

Point

Magnetic force is categorized as strong attraction, weak attraction, or no attraction. The category of magnetic response allows the unknown metal to be placed into a specific identification grouping.

Magnetic response testing is a qualitative identification method in which a magnet is laid on the surface of an unknown metal to test for a magnetic force. Magnetic force is categorized as strong attraction, weak attraction, or no attraction. **See Figure 38-7.** The category of magnetic response allows the unknown metal to be placed into a specific identification grouping.

The magnetism of a metal can change with temperature. As the temperature of some metals increases, the magnetism decreases. The point at which this change occurs is known as the Curie temperature. *Curie temperature* is the temperature of magnetic transformation, above which a metal is nonmagnetic, and below which it is magnetic. For some alloys, the magnetic transformation temperature may occur over a range of Curie temperatures. The effect of temperature on magnetic properties is illustrated in alloy 400 (Monel® 400), which is slightly magnetic at ambient temperature. Monel® 400 is nonmagnetic if its temperature is raised above the boiling point of water.

Magnetic behavior of some metals may change with mechanical processing. For example, 302 and 304 stainless steels, nonmagnetic in the annealed (soft) condition, become increasingly magnetic as they are cold-worked.

MAGNETIC FORCE

Strong Attraction:

- Carbon Steels
- Cast Irons
  - Gray
  - Ductile
  - Malleable
- Cobalt
- Iron-Silicon Alloys (.05% Si to 4.5% Si)
- Iron-Cobalt Alloys
- Iron-Molybdenum Alloys
- Low-Alloy Steels
- Nickel
- Stainless Steels
  - Ferritic
  - Martensitic (400 series)
  - Martensitic precipitation hardening
- Tool Steels

Weak Attraction:

- Stainless Steels
  - Cast 300 series
  - Cold-worked 302
  - Cold-worked 304
  - 308 weld metal
  - 309 weld metal
  - 329
- Monel® 400 (becomes nonmagnetic in boiling water)

No Attraction:

- Alloy 20 types
- Commercially pure nonferrous metals (except nickel and cobalt)
- Copper-Nickels
- Hastelloys®
- Incoloys®
- Inconels®
- Stainless Steels
  - Austenitic (other 300 series)
- Stellite®

Figure 38-7. *Metals can be identified and grouped by the magnetic force they produce.*

Minor microstructural differences between cast and wrought stainless steels can alter magnetic behavior. For example, E308 or ER308 filler metal for welding nonmagnetic 304 stainless steel may be slightly magnetic due to compositional adjustment. Despite these minor complications, magnetic response testing is a convenient and rapid method of qualitative identification.

Chisel Testing

Chisel testing is a qualitative identification method that identifies metal by the shape of the chips it produces. Chisel testing consists of producing a chip by striking the edge or corner of the unknown metal with a chisel and hammer. Metal can be identified by the type of chips that result during chiseling. **See Figure 38-8.** The ease with which the chip breaks from the metal is an indication of the metal's hardness. The continuity of the chip indicates the metal's toughness. Long and curled chips result from mild steel and soft metals such as aluminum. Short, broken chips result from cast steel. High-carbon steels do not break easily and sample chips are difficult to obtain.

Torch Testing

Torch testing is a qualitative identification method that identifies a metal by the melting rate, the appearance of the metal when heat is applied, and the action of the molten metal. **See Figure 38-9.** These factors provide clues to the identity of the metal. Torch testing requires heating a small area of the surface of the unknown metal with a high-temperature oxyacetylene flame to cause local melting. To distinguish aluminum from magnesium, apply a torch to the filings. Magnesium burns with a sparking white flame.

CHISEL TEST IDENTIFICATION		
Type of Chip	**Type of Material**	**Possible Metal Type**
Continuous, easily removed	Ductile	Aluminum, Low-Carbon Steel, Malleable Iron
Fragmented small pieces, easily removed	Brittle	Gray Cast Iron
Fragmented or continuous, hard to remove	Brittle	High-Carbon Steel

Figure 38-8. *Metal can be identified by the type of chips that result during chiseling.*

TORCH TEST IDENTIFICATION		
Melting Rate	**Appearance of Metal After Heating**	**Possible Metal Type**
Slow, melts only after sufficient heat input	White metal	Aluminum
Fast, melts with little heat input	White metal	Zinc
Slow, melts only after sufficient heat input	Reddish metal	ETP Copper
Faster, melts with relatively little heat input	Reddish metal	Deoxidized Copper
Boils while melting	Reddish metal	Leaded Copper

Figure 38-9. *Torch testing identifies a metal by the melting rate and the appearance of the metal after heating.*

Point

Semi-quantitative identification methods use a physical stimulus to provide a signal that may be compared with a set of standards. Semi-quantitative identification methods include density testing, spark testing, chemical spot testing, thermoelectric potential sorting, and optical emission spectroscopy.

When using the torch test, care must be taken to prevent damaging the sample. Heat input required to heat the sample varies depending on the type of metal being tested. If aluminum and zinc are being separated, the aluminum will not melt until sufficient heat has been applied because of its high thermal conductivity, whereas with zinc a sharp corner will melt quickly because zinc is not a good thermal conductor. In the case of leaded copper alloys, the surface will boil as the lead comes off.

File Testing

File testing is a qualitative identification method in which a file is used to indicate the hardness of steel compared with that of the file. File testing consists of assessing the degree of bite when a sharp mill file is drawn across the surface or edge of the unknown metal. **See Figure 38-10.** The file test provides a rapid and approximate method of estimating the hardness of steel. The easier the degree of bite, the softer the steel. The hardness of steel is a useful indicator of its weldability. The file test must be used with caution and only by experienced personnel.

SEMI-QUANTITATIVE IDENTIFICATION

Semi-quantitative identification is metal identification by applying a physical stimulus to an unknown metal to produce a signal that is interpreted against a set of standards. Semi-quantitative identification methods supported by documentation may be used in a formal quality control program. Semi-quantitative identification methods include density testing, spark testing, chemical spot testing, thermoelectric potential sorting, electrical resistivity testing, and optical emission spectroscopy.

Density Testing

Density testing is a semi-quantitative identification method that measures the density of an unknown metal. Density is measured by obtaining a small specimen of metal (½″ cube), a length of fine wire, an analytical balance, a small bench to straddle the analytical balance pan, and a 250 ml beaker that is filled approximately two-thirds full of distilled water.

FILE TEST IDENTIFICATION

File Effect on Metal	Brinell Hardness Number (BHN)	Possible Steel Type
File easily bites into metal	100	Mild (Low-Carbon) Steel
File bites into metal with pressure	200	Medium-Carbon Steel
File only bites into metal with extreme pressure	300	High Alloy Steel-High Carbon Steel
Metal filed with difficulty	400	Unhardened Tool Steel
File leaves marks on metal but metal is nearly as hard as file	500	Hardened Tool Steel
Metal is harder than file	600+	Carbide Tool

Figure 38-10. *File testing consists of assessing the degree of bite when a sharp mill file is drawn across the surface or edge of the unknown metal.*

Dirt and foreign matter are thoroughly removed from the surface of the specimen. The specimen is washed with acetone and allowed to dry for 2 min to 3 min. The specimen is then weighed on an analytical balance to ± 0.001 g. The fine wire is also weighed to 0.001 g. The beaker containing the distilled water is placed on the small bench that straddles the balance pan. One end of the fine wire is tied firmly around the metal specimen.

The other end is attached to the balance hook so that the metal specimen is suspended and totally immersed in the distilled water. The metal specimen is reweighed when it is completely immersed in the distilled water. **See Figure 38-11.** The density of a metal is found by applying the following formula:

$$D = \frac{W_a}{W_a - (W_d - W_w)}$$

where

D = density (in g/cm³)

W_a = weight of specimen (in g)

W_d = weight of specimen in distilled water (in g)

W_w = weight of fine wire (in g)

For example: What is the density of a specimen of 304 stainless steel that weighs 18.102 g in air, and weighs 15.960 g in the distilled water of an analytical balance, and that has a fine wire that weighs .151 g?

$$D = \frac{W_a}{W_a - (W_d - W_w)} \text{ g/cm}^3$$

$$D = \frac{18.102}{18.102 - (15.960 - .151)} \text{ g/cm}^3$$

$$D = \frac{18.102}{18.102 - 15.809} \text{ g/cm}^3$$

$$D = \frac{18.102}{2.293} \text{ g/cm}^3$$

$$D = \textbf{7.89 g/cm}^3$$

The four density groupings for metals are very high density, high density, average density, and low density. From the figured value of density, metals are placed in one of the groupings. Depending on the separation of their density values, metals within the same group are distinguished from each other by checking the figured densities against a table of known density values. **See Figure 38-12.**

Spark Testing

Spark testing is a semi-quantitative identification method that identifies metals by the shape, length, and color of the spark produced when the metal is held against a grinding wheel rotating at high speed. The chemical composition of the unknown steel influences the form of the spark stream produced. Spark stream characteristics are compared to standard spark stream charts to identify the unknown metal. **See Figure 38-13.**

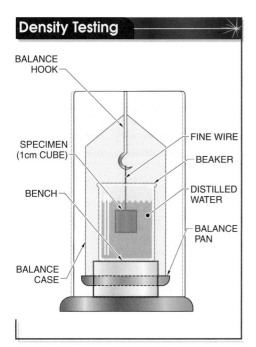

Density Testing

BALANCE HOOK

SPECIMEN (1cm CUBE)

BENCH

BALANCE CASE

FINE WIRE

BEAKER

DISTILLED WATER

BALANCE PAN

Figure 38-11. *Metals can be identified by measuring the density of the metal with an analytical balance.*

DENSITY IDENTIFICATION GROUPINGS		
Grouping	**Density Range***	**Metals**
Very High Density	12 to 22	Gold Rhodium Iridium Ruthenium Osmium Tantalum Palladium Tungsten Platinum Uranium (depleted)
High Density	9.8 to 11.9	Bismuth Molybdenum Lead Silver
Average Density	6 to 9.7	Antimony Nickel alloys Cadmium Stainless steels Cast Iron Steels Copper alloys Tin Nickel Zinc
Low Density	1 to 5.9	Aluminum Magnesium Aluminum alloys Magnesium Beryllium alloys Beryllium alloys Titanium Titanium alloys

* (g/cm³)

Figure 38-12. *Figured density values can be used to place a metal in one of four groupings.*

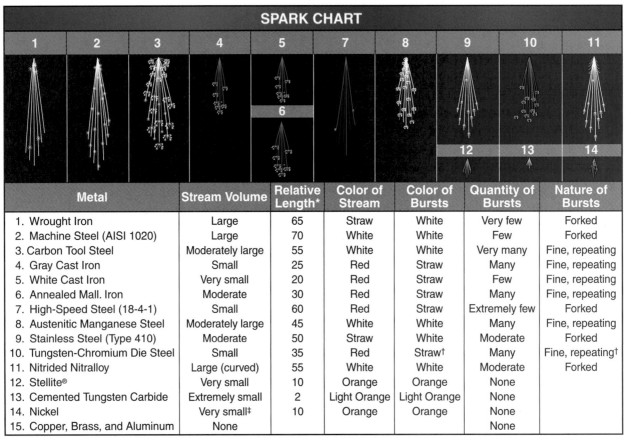

Metal	Stream Volume	Relative Length*	Color of Stream	Color of Bursts	Quantity of Bursts	Nature of Bursts
1. Wrought Iron	Large	65	Straw	White	Very few	Forked
2. Machine Steel (AISI 1020)	Large	70	White	White	Few	Forked
3. Carbon Tool Steel	Moderately large	55	White	White	Very many	Fine, repeating
4. Gray Cast Iron	Small	25	Red	Straw	Many	Fine, repeating
5. White Cast Iron	Very small	20	Red	Straw	Few	Fine, repeating
6. Annealed Mall. Iron	Moderate	30	Red	Straw	Many	Fine, repeating
7. High-Speed Steel (18-4-1)	Small	60	Red	Straw	Extremely few	Forked
8. Austenitic Manganese Steel	Moderately large	45	White	White	Many	Fine, repeating
9. Stainless Steel (Type 410)	Moderate	50	Straw	White	Moderate	Forked
10. Tungsten-Chromium Die Steel	Small	35	Red	Straw†	Many	Fine, repeating†
11. Nitrided Nitralloy	Large (curved)	55	White	White	Moderate	Forked
12. Stellite®	Very small	10	Orange	Orange	None	
13. Cemented Tungsten Carbide	Extremely small	2	Light Orange	Light Orange	None	
14. Nickel	Very small‡	10	Orange	Orange	None	
15. Copper, Brass, and Aluminum	None				None	

* actual length varies with grinding wheel, pressure, etc.
† blue-white spurts
‡ some wavy streaks

Figure 38-13. *Spark charts are compared with spark stream characteristics to identify unknown metals.*

Spark testing heat treats the surface layer of the metal, leading to localized hardening and possible cracking. Stock is discarded any closer than ¼″ from the area of contact with the grinding wheel because of possible failure.

Spark Test Preparation. The area of metal selected for spark testing must be free of scale and representative of the chemical composition of the metal. Before conducting a spark test, the grinding wheel is cleaned with a diamond wheel dresser to remove particles of metal from previous tests. If these particles are not removed, the spark stream of the specimen being examined would be contaminated by sparks from previous tests.

Small, portable grinders are most often used for spark testing, because they can be transported to the field. Stationary grinders may be used if convenient. **See Figure 38-14.**

The pressure between the grinding wheel and the specimen must be sufficient to maintain steady contact. The spark stream should be given off approximately 1 ft horizontally and at right angles to the line of vision. The tester must have a clear, unobstructed view of the spark stream. Conditions for spark testing must be standardized and testing should be conducted in diffuse daylight, not bright sunlight or darkness. The spark should be tested away from air drafts that may cause the tail of the spark stream to hook, which leads to an erroneous interpretation.

Grinding Wheel Rotation. The speed of the wheel in feet per minute (fpm) equals the circumference in inches multiplied by the revolutions per minute at which the wheel turns, divided by 12. To provide a satisfactory spark stream, the grinding wheel must rotate at high speeds (15,000 fpm or greater) and must be hard (for example, 40 grain alumina wheel).

Figuring wheel rotation. Is a rotation speed of 16,000 rpm suitable for a 2″ diameter portable grinder?

$C = \pi d$

where

C = Circumference of wheel

π = pi (3.142)

d = diameter

$C = 3.142 \times 2$

$C = \mathbf{6.248''}$

$S = \dfrac{C \times R}{12}$

where

S = Speed of wheel (in fpm)

C = Circumference of wheel (in in.)

R = Rotation speed (rpm)

12 = constant

$S = \dfrac{6.284 \times 16,000}{12}$

$S = \dfrac{100,544}{12}$

$S = \mathbf{8378.6 \ fpm}$

The wheel rotation speed is not suitable for spark testing.

> **⚠ WARNING**
>
> Protective goggles and protective clothing must be worn when spark testing.

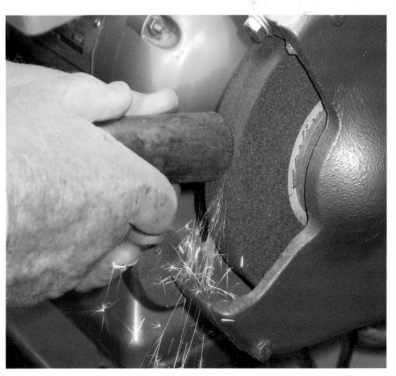

A grinding wheel is used for spark testing and should be kept clean to prevent contaminants from interfering with the spark stream.

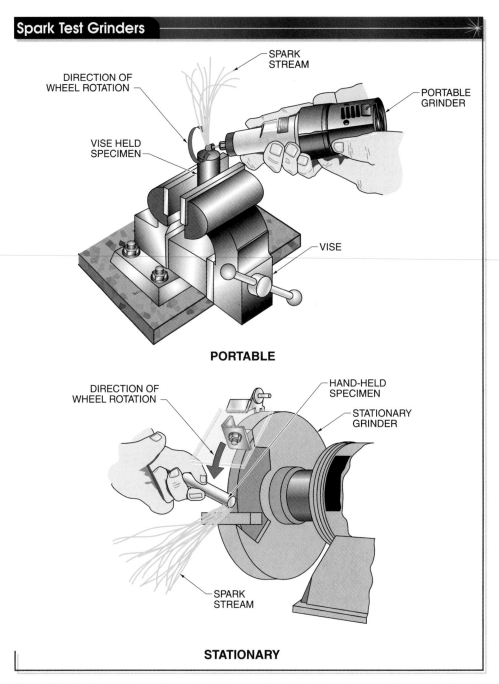

SPARK
STREAM

DIRECTION OF
WHEEL ROTATION

PORTABLE
GRINDER

VISE HELD
SPECIMEN

VISE

PORTABLE

DIRECTION OF
WHEEL ROTATION

HAND-HELD
SPECIMEN

STATIONARY
GRINDER

SPARK
STREAM

STATIONARY

Figure 38-14. *Spark testing is most often performed using portable grinders, but stationary grinders may also be used.*

Spark Stream Identification. The spark stream must be closely examined for its characteristic features. Characteristic features include carrier lines, forks, bursts, and arrowheads. A carrier line is an incandescent (glowing) streak that traces the trajectory (path) of each particle (spark). A fork is a simple branching of the carrier line. A burst is a complex branching of the carrier line. An arrowhead is a termination of the carrier line in the shape of an arrowhead. **See Figure 38-15.**

By learning to identify the different portions of the spark stream, and by making tests on known samples, it is possible to acquire sufficient experience to make relatively accurate determination of the metal being investigated.

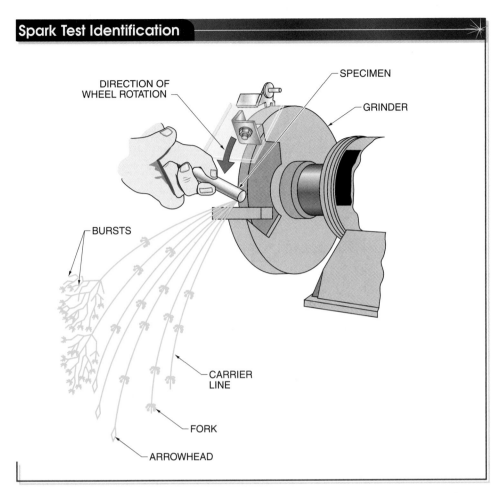

DIRECTION OF
WHEEL ROTATION

SPECIMEN

GRINDER

BURSTS

CARRIER
LINE

FORK

ARROWHEAD

Figure 38-15. *Characteristic features of spark streams include carrier lines, forks, bursts, and arrowheads.*

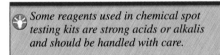
Some reagents used in chemical spot testing kits are strong acids or alkalis and should be handled with care.

Chemical Spot Testing

Chemical spot testing is a semi-quantitative identification method that uses chemicals that react when placed on certain types of metals. Chemical spot testing is used to identify metals by the color changes that occur to the metal when a metal is contacted with specific chemical reagents.

The solution is often produced using electric current (electrographic technique) to dissolve small amounts of the metal in a chemical reagent. When the solution reacts with the chemical reagent, a color change occurs, which is used to identify the unknown metal.

The most common chemical spot test is the electrographic chemical spot test. In electrographic chemical spot testing, a metal surface is first prepared by dressing it with a file or emery paper to remove scale or unnecessary roughness, after which the metal surface is degreased. A filter paper wetted with measured drops of chemical reagent is placed on the metal surface. The unknown metal (anode) is electrically connected to the positive terminal of a 6 VDC battery. An aluminum cathode, connected to the negative terminal of the battery, is pressed against the wet filter paper. This connection completes the electrical circuit and allows current to flow until the cathode is removed and the circuit is disconnected. **See Figure 38-16.**

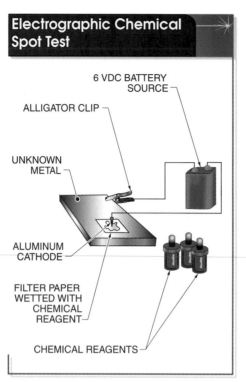

6 VDC BATTERY SOURCE

ALLIGATOR CLIP

UNKNOWN METAL

ALUMINUM CATHODE

FILTER PAPER WETTED WITH CHEMICAL REAGENT

CHEMICAL REAGENTS

Figure 38-16. *The electrographic chemical spot test is the most commonly used chemical spot test.*

The filter paper, which is soaked in a small amount of the metal solution, is lifted from the surface. Measured drops of a second chemical reagent are applied to the wet filter paper, causing a color change. The color of the filter paper identifies the metal. Supplementary reagents may be applied to the filter paper to cause additional color changes, which further identify the metal. When the test is complete, the metal surface is thoroughly cleaned to remove excess chemical reagent.

⚠ **WARNING**

Avoid accidental contact with hot metal surfaces during thermoelectric potential sorting.

Thermoelectric Potential Sorting

Thermoelectric potential sorting is a semi-quantitative identification method that uses measurement of the electric potential generated when two metals are heated. The common and reproducible method of thermoelectric potential sorting is to standardize on the voltage generated by the heated junction of two dissimilar metals. Standardizing on the voltage generated by the junction of the two metals allows a significantly greater amount of heat to be generated, which increases sensitivity. To carry out identification, the unknown metal is put in contact with a heated probe. The thermoelectric potential generated is indicated on a digital or analog readout. This value of thermoelectric potential is compared with values obtained under identical conditions using known metal samples. Thermoelectric potential sorting is described in ASTM E977, *Standard Practice for Thermoelectric Metal Sorting.*

Null Point Method. The *null point method* is an alternative method of thermoelectric potential sorting. The null point method is used for identifying an unknown metal or distinguishing it from other metals. In the null point method, a known standard specimen and a probe are electrically connected and the deflection of the meter caused by the resulting potential is recorded. The resulting potential is calibrated to read zero on a meter.

An unknown metal that is the same as the known specimen will produce no deflection of the meter. If the unknown metal is different from the known specimen, the meter will deflect to either side of zero.

Electrical Resistivity Testing

Electrical resistivity testing is a semi-quantitative identification method that uses differences in electrical resistivity to identify metals. With electrical resistivity testing, a small probe containing four electrodes is placed on the metal surface and an electric current is passed through the metal. The current passing through the metal causes a number to register on the panel of the instrument. This number (ohms) is a measure of the resistivity of the unidentified metal. The surface must be prepared with a file if it is excessively rough or corroded.

For materials over 0.1″ thick, the instrument is self-compensating. For materials less than 0.1″ thick, the tester must apply a correction factor based on the metal thickness. The instrument is also sensitive to the area of metal beneath the probe. Two differing measurements may be displayed on different parts of a component exhibiting different thicknesses, such as a casting. The tester must calibrate the instrument readings against known metal samples to prevent misinterpretation of the data.

The electrical resistivity method provides rapid metal sorting or identification. The relatively small probe head of the electrical resistivity instrument allows it to be used for examining hard to reach areas such as the internal components of valves.

Optical Emission Spectroscopy

Optical emission spectroscopy is a semi-quantitative identification method that separates and analyzes the light emitted from an unknown metal surface when it is arced by an electric current. An *optical emission spectrometer* is an instrument used for optical emission spectroscopy that is placed on the surface of an unknown metal. A small area of the surface is intermittently sparked by striking an arc between the surface and a tungsten electrode using a power source of 25 V to 40 V. When the electric arc is struck on a metal surface, the light emitted is composed of various wavelengths. The chemical elements in the metal determine the component wavelengths produced. The intensity of each component wavelength is proportional to the concentration of its corresponding chemical element. **See Figure 38-17.**

All light emitted from the arc is passed through a glass prism, which diffracts it into its component wavelengths. *Diffraction* is a modification of light in which the rays appear to be deflected to produce fringes of parallel light and dark colored bands. The separated wavelengths are viewed as a series of lines of varying intensity and color.

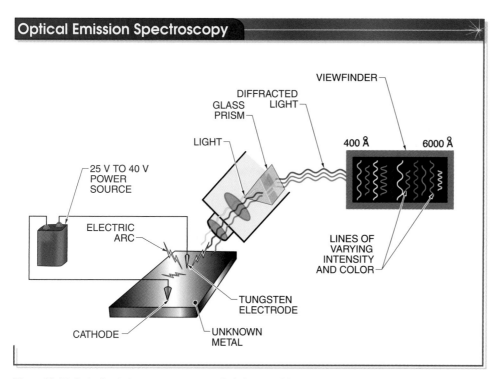

Figure 38-17. *Optical emission spectroscopy uses the light emitted from unknown metal surfaces to identify a metal.*

The wavelength lines are compared with those obtained from standard elements. A camera that is connected to the eyepiece of the optical emission spectrometer permanently records the lines. The camera improves the sensitivity of the instrument because it records lines that are too faint for detection by the human eye. The camera also records the lines from the ultraviolet spectrum.

The chemical elements detectable by optical emission spectroscopy are limited to those elements that have observable light wavelengths after diffraction and are not vaporized by the heat of the arc. Low percentages of chemical elements may be undetected if the line obtained by diffraction is faint. Optical emission spectroscopy can detect nickel, chromium, molybdenum, titanium, manganese, vanadium, copper, zinc, tungsten, magnesium, cobalt, lead, and niobium.

QUANTITATIVE IDENTIFICATION

✓ Point

Quantitative identification methods sep-arate and identify metals by measuring the amounts of chemical elements present in a metal.

Quantitative identification techniques are essentially nondestructive tests that identify metal composition by percentage. Although these tests do not analyze for every chemical element that may be present, they are often comprehensive enough to identify many unknown metals with a high degree of accuracy. However, the hardware for quantitative identification is significantly more expensive than the hardware for qualitative identification.

Quantitative identification techniques include X-ray fluorescence (XRF) analysis and chemical analysis. XRF analysis uses two types of instruments: energy dispersive and wavelength dispersive. Certain types of XRF instruments are portable and easily taken to the job site. They can analyze for 22 chemical elements, but not for carbon.

X-Ray Fluorescence Spectrography (XRF)

X-ray fluorescence spectrography (XRF) is a nondestructive quantitative identification method that uses a gamma ray to identify an unknown metal. X-ray fluorescence analysis instruments consist of three major components: a source, such as a radioisotope or X-ray tube, a detector, and a data processing unit. The source causes the atoms of the chemical elements in the metal to exhibit fluorescent X-rays. The energy levels and wavelengths of the X-rays denote specific elements. A detector converts the emitted fluorescent X-rays into measurable electronic signals. A data processing unit records the emissions and calculates the concentration of chemical elements in the sample. **See Figure 38-18.**

Fluorescent X-rays are created by reactions to either a gamma ray from a radioisotope or an X-ray beam from a low-power X-ray tube. The decay of a radioisotope over time necessitates its periodic replacement, whether the instrument is used or not. Radioisotope instruments may also be subject to regulations that restrict their transportation. X-ray tubes emit energy only when turned on, so they have much longer useful lives.

Energy-Dispersive X-Ray Fluorescence Analysis. The probe of an energy dispersive X-ray fluorescence analysis (EDXA) instrument contains a shutter that opens for a specific length of time to allow the gamma rays or the X-rays to be beamed to the unknown metal. The energy levels of the emitted fluorescent X-rays are measured by the EDXA detector, which identifies various chemical elements and their concentrations. The microprocessor calculates and displays the percentages of the elements present in the unknown metal and also specific alloys that correspond to the measured composition.

When exposed to X-rays or gamma rays, some of a material's atoms are ionized, which is the ejection of one or more electrons. If electrons are expelled from an inner orbital, the electronic structure of the atom becomes unstable, and electrons in higher orbitals fall into the hole left behind. While the electrons are falling, energy is released in the form of a photon, which has characteristics identifying the atom.

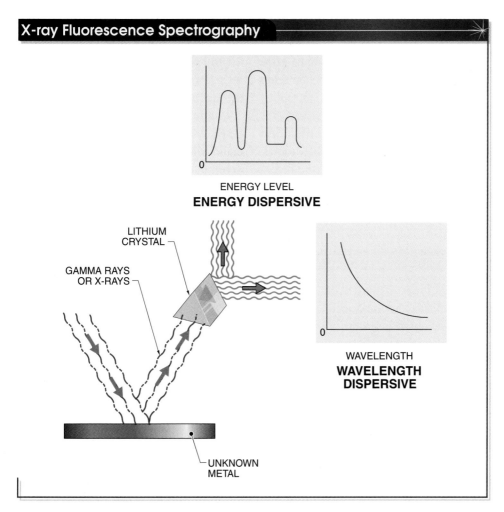

Figure 38-18. *X-ray fluorescence spectrography uses a detector that separates and identifies energy levels or energy wavelengths.*

Wavelength-Dispersive X-Ray Fluorescence Analysis. In wavelength dispersive X-ray fluorescence analysis (WDXA) instruments, only an X-ray beam is used to create fluorescent X-rays. The detector system uses a lithium crystal to disperse the fluorescent X-rays by wavelength. The wavelengths identify the chemical elements present in the unknown metal, and the relative intensities of each wavelength are measured to determine the proportions of the elements.

The X-ray generator used for WDXA is relatively large and must be water cooled, which makes the instrument less portable than EDXA instruments. Because WDXA instruments have less mobility, the unknown metal specimen is usually brought in from the field to be identified.

Chemical Analysis

Chemical analysis is a destructive quantitative identification method that requires removal of a small sample (1 g to 2 g) of metal for chemical analysis of its constituent elements. Chemical analysis is destructive and time-consuming and is used when auditing of a product is required or the analysis must be checked against a materials test report (MTR). Wet chemical analysis is the only method of obtaining the amount of carbon present in an alloy.

Filler Metal Identification

Welding filler metals (wire or rod) and electrodes use a mixture of marking systems. Filler metal and electrodes are

⚠ WARNING

Adequate precautions against exposure to radiation must be taken when using X-ray fluorescence equipment.

identified by markings that are attached, stamped, or stenciled on filler metals. Identification markings on bare wire usually consist of the AWS designation printed on a paper tag glued to one or both ends of the wire. On large-diameter nonferrous filler metal, identification markings are stamped on the filler metal. Identification markings on covered electrodes consist of the AWS designation stenciled on the flux coating at one end of the filler metal. In all cases, additional identification is provided on a label attached to the container or spool holding the electrode or filler metal.

Some marking products may be potentially harmful to metals. Care must be taken to identify harmful marking products and thoroughly remove them before welding, brazing, or soldering.

Refer to Quick Quiz® on CD-ROM

Points to Remember

- Manufacturers supply three types of paperwork to identify their products: materials test report, product analysis, and certificate of compliance.
- A materials nonconformance report helps the end user document problems in received materials so that problem areas can be identified, corrected, and prevented in the future.
- Visual identification includes appearance, color, nameplate, and markings to determine key features that identify the metal type.
- Magnetic force is categorized as strong attraction, weak attraction, or no attraction. The category of magnetic response allows the unknown metal to be placed into a specific identification grouping.
- Semi-quantitative identification methods use a physical stimulus to provide a signal that may be compared with a set of standards. Semi-quantitative identification methods include density testing, spark testing, chemical spot testing, thermoelectric potential sorting, and optical emission spectroscopy.
- Metals are categorized as one of four density groupings, very high density, high density, average density, and low density, based on their figured density value.
- Quantitative identification methods separate and identify metals by measuring the amounts of chemical elements present in a metal.
- Filler metals are identified by paper labels or identification markings stamped on one end of the filler metal.

Questions for Study and Discussion

1. What type of information is contained in an MTR?
2. What is a certificate of compliance (COC)?
3. When should visual identification by color not be used?
4. What information is included in a foundry marking?
5. What change in the welding process can affect the magnetic response of metals?
6. Metal can be placed into which four density groupings?
7. What are the four primary characteristics of a spark stream?
8. What is the most common type of chemical spot test?
9. What is an important element that X-ray fluorescence spectrography (XRF) fails to detect?
10. How are welding filler metals identified?

Refer to Chapter 38 in the *Welding Skills Workbook* for additional exercises.

Weldability of Carbon and Alloy Steels

39

Chapter

Weldability is the capacity of a metal to be welded under imposed fabrication conditions. When welding carbon steels, carbon is the principal alloying element affecting weldability. Carbon steels are grouped according to their carbon content. Other alloy steels are grouped according to their alloying elements. Factors that affect the weldability of carbon and alloy steels must be understood to ensure the desired quality during fabrication.

WELDABILITY OF CARBON AND ALLOY STEELS

Steels are broadly classified as carbon steels or alloy steels based on their alloying elements. Carbon steels are alloys of iron, carbon, and manganese. Carbon is the principal alloying element that affects the mechanical properties and metallurgical structure of carbon steels. Carbon steels are grouped according to their carbon content and include low-carbon steels, medium-carbon steels, and high-carbon steels. Free-machining steels are an additional group. **See Figure 39-1.** The weldability of carbon steels decreases as the carbon content increases.

Low-carbon steels contain up to 0.3% carbon and up to 1.2% manganese. They are not strengthened by heat treatment but may be surface hardened by carburizing. Low-carbon steels are used for structural applications such as building framework, pressure vessels, and automobile bodies.

Nickel steels are low-carbon steels that contain 2% to 9% nickel for service

at low temperatures for applications from 32°F to −320°F. Nickel steels are used for storage tanks for liquefied hydrocarbon gases and machinery designed for use in cold climates.

Medium-carbon steels contain 0.3% to 0.6% carbon and 0.6% to 1.65% manganese. Medium-carbon steels are stronger than low-carbon steels. They form high hardness martensite in the HAZ when rapidly cooled and are susceptible to hydrogen cracking. They may require heat treatment after welding to achieve the specified strength and hardness. Wear resistance may be improved by surface treatments such as chrome plating or nitriding. The surface coating must be removed by grinding if any weld repair is to be performed. Medium-carbon steels are used in machinery parts such as tractors, derricks, and pumps.

High-carbon steels contain more than 0.6% carbon. High-carbon steels are usually not welded. They are used for their hardness and strength, especially where a cutting edge is required, such as on drill bits and files.

> **✓ Point**
>
> *Carbon steels include low-carbon steels, medium-carbon steels, and high-carbon steels. The weldability of carbon steels decreases as the carbon content increases.*

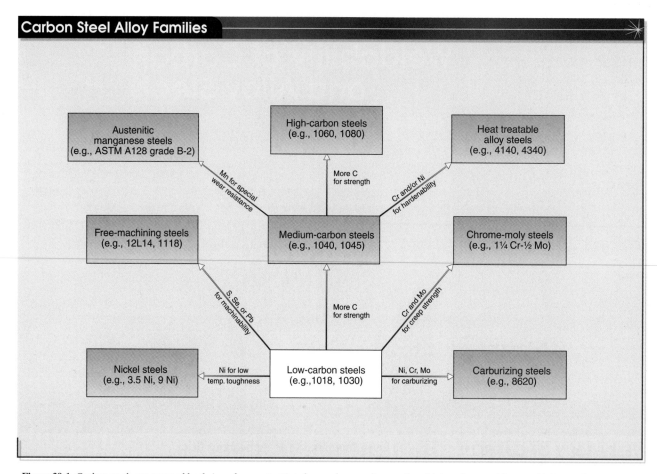

Figure 39-1. *Carbon steels are grouped by their carbon content into low-carbon, medium-carbon, high-carbon, and free-machining steels.*

Free-machining steels are low-carbon steels that contain small amounts of sulfur, phosphorus, or lead, which are added to improve their machinability. Free-machining steels are used where high production machining is required. Free-machining steels are not typically welded.

Alloy steels contain specified quantities of alloying elements other than carbon and manganese. Alloy steels are grouped according to their alloying elements. The presence of one or more alloying elements in alloy steels leads to better mechanical properties than carbon steels. The principal benefits of alloy steels over carbon steels are higher strength and greater capacity for strengthening in thick sections (hardenability). Alloy steels consist of low-alloy steels, chrome-moly steels, and austenitic manganese steels.

Low-alloy steels are medium-carbon steels that contain small percentages of nickel, chromium, and molybdenum to achieve optimum mechanical properties in the quenched and tempered condition. The wear resistance of low-alloy steels may be improved by surface treatments such as chrome plating or nitriding. The hardened surface coating must be removed by grinding if any weld repair is to be performed. Low-alloy steels are used for mechanical components such as shafts and machinery where high strength and toughness are required, particularly where the section thickness exceeds 2″.

Chrome-moly steels contain approximately 0.2% carbon, 0.5% to 9% chromium (Cr), and 0.5% to 1% molybdenum (Mo). Scaling resistance increases as the chromium content is increased. Molybdenum increases strength at elevated temperatures and provides resistance to graphitization. *Graphitization* is the formation of iron carbide that results in loss of ductility. The carbon content of chrome-moly steels is kept below 0.2% to maintain weldability. Chrome-moly steels are widely used for piping and vessels operating at temperatures up to 1000°F in the petroleum refining industry and in steam power generation. Chrome-moly steels are identified by their nominal percentages of chromium and molybdenum, for example 1¼Cr-½Mo, or 2¼Cr-1Mo. **See Appendix.**

Austenitic manganese steels contain 11% to 14% manganese and .7% to 1.4% carbon. Austenitic manganese steels are nonmagnetic alloy steels noted for high strength, excellent ductility, toughness, and outstanding wear resistance. Austenitic manganese steels are used for crushing, earth-moving, and material handling equipment; railroad track parts; and electrical equipment where nonmagnetic properties are important.

Nickel steels contain various quantities of nickel to improve toughness at low temperature. Nickel steels are frequently used for transporting cryogenic gases as liquids. The greater the percentage of nickel, the lower the service temperature.

Steel Deoxidation

Steel deoxidation is the process of removing a controlled amount of oxygen from steel during steelmaking. The deoxidation practice determines the amount of deoxidation performed and the basic steel type that is produced. Steel deoxidation results in four types of steel: killed, semikilled, rimmed, and capped. **See Figure 39-2.**

Killed steel is steel that is completely deoxidized during steel production by adding silicon or aluminum in the furnace ladle or to the mold. Aluminum and silicon cause the steel to solidify quietly and suppress (kill) the gas evolution that would result from combining carbon and oxygen and forming carbon monoxide. Killed steel is homogeneous, has a smooth surface, and contains no blowholes. Killed steels are commonly used where improved strength and toughness are important.

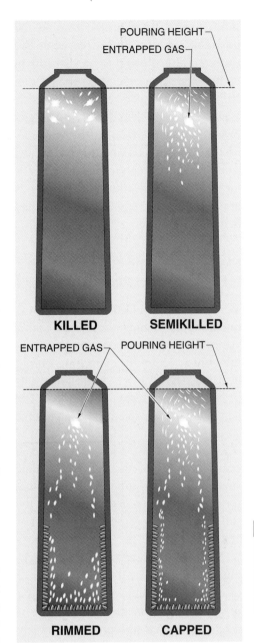

Figure 39-2. *The steel deoxidation practice influences the type of steel produced.*

Semikilled steel is steel in which deoxidizers only partially kill the oxygen-carbon reaction. Semikilled steel is more uniform in composition throughout the cross section and is suitable for applications involving carburizing and heat treating.

Rimmed steel is steel with little or no deoxidizer addition. The molten metal briskly bubbles as oxygen evolves from it when it is poured into a mold. The evolving oxygen reacts with the carbon at the boundary between the solidified metal adjacent to the mold and the remaining molten metal, forming carbon monoxide gas. This reaction causes the outer rim of the solidified metal to be very low in carbon and, consequently, very ductile. Rimmed steel may be rolled to produce a very sound surface and is used for sheet products such as automobile bodies.

Capped steel is a variation of rimmed steel, providing a surface condition similar to rimmed steel, but other properties are intermediate between rimmed steel and semikilled steel.

General Welding Considerations for Carbon and Alloy Steels

When steels are welded, the rapid cooling rate from the welding temperature causes a metallurgical transformation similar to that which occurs in quenching during heat treatment. This transformation results in martensite formation in the HAZ. The hardness of martensite increases with the carbon content of steel, reducing toughness and increasing susceptibility to cold cracking (hydrogen cracking) from residual stress in the weld joint. Martensite that forms in low-carbon steels is generally too soft and ductile to cause embrittlement or cracking.

Steels that are susceptible to cracking must be preheated to reduce the rate of cooling and decrease the possibility of martensite formation. Postheating is used to improve the toughness of any martensite that does form by tempering it to reduce (relieve) residual stress and eliminate hydrogen.

Alloy steels form either martensite or bainite as they cool, depending on the cooling rate. Bainite is not as brittle as martensite and forms at a slower cooling rate. A high preheat temperature is used to slow the cooling rate of alloy steels to form bainite rather than martensite. Bainite is less likely to crack than martensite, allowing time for postheating to be performed.

Hydrogen Cracking. Hydrogen cracking is caused by atomic hydrogen that may be present on carbon and alloy steels.

Sources of atomic hydrogen are organic material such as grease; chemically bonded or absorbed water in the electrode coating; and moisture on the steel surface at the weld location.

Atomic hydrogen is created at welding temperature and diffuses rapidly into molten weld metal. As the weld metal solidifies, the hydrogen tries to escape because solidified metal accommodates significantly less hydrogen than liquid metal. Some hydrogen escapes into the atmosphere; however, some hydrogen escapes into the HAZ.

Martensite formed in the HAZ by rapid cooling of the weld is extremely susceptible to embrittlement from the hydrogen that escapes into it. Hydrogen cracking occurs when the brittle martensite fails to yield (stretch) to accommodate the residual stresses that develop as the weld cools. Hydrogen cracking may occur several days after the weld has cooled. Hydrogen cracking is often located below the surface and may not be detected by common nondestructive examination techniques. **See Figure 39-3.** Methods of preventing hydrogen cracking are:

- using low-hydrogen electrodes and storing electrodes in a low-temperature oven
- heating surface before welding to remove moisture
- postheating immediately after welding to drive out hydrogen

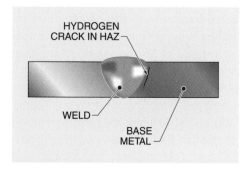

Figure 39-3. *Hydrogen cracking may not be detected by nondestructive examination because it commonly occurs below the surface.*

Carbon Equivalent. *Carbon equivalent* is a formula based on the chemical composition of a steel, which provides a numerical value to indicate whether preheat and postheating are required. The greater the numerical value of carbon equivalent, the greater the tendency for cold cracking and the greater the need for preheat and postheating.

Carbon is the most significant alloying element that is added to iron in steels, contributing to overall strength and hardness. Other alloying elements contribute to hardness, but to a lesser extent than carbon. The carbon equivalent is the sum of the carbon percentage, plus the weighted percentages of each alloying element on martensite formation. With carbon steels, manganese is the only other element whose influence is weighted. For alloy steels, the weightings of individual alloying elements are added. The higher the carbon equivalent, the greater the need for preheat and postheating to prevent embrittlement by martensite.

To find the carbon equivalent for caron steels, apply the formula:

$$CE = C + \frac{Mn}{6}$$

where

CE = carbon equivalent
C = percent carbon
Mn = percent manganese
6 = constant

For example, what is the carbon equivalent of a steel thtcontains 0.28% C and 0.7% Mn?

$$CE = \%C + \frac{\%Mn}{6}$$

$$CE = 0.28 + \frac{0.7}{6}$$

$$CE = 0.28 + 0.12$$

$$CE = \mathbf{0.4\%}$$

Carbon steel with a carbon equivalent less than 0.4% is weldable without preheat or postheating, depending on joint member thickness. For alloy steels, the carbon equivalent is ound by applying the formula:

$$CE = \%C + \frac{\%Mn}{6} + \frac{\%Ni}{20} + \frac{\%Cr}{10} + \frac{\%Cu}{40} - \frac{\%Mo}{50} - \frac{\%V}{10}$$

Joint Member Thickness. Joint member thickness also influences preheat. With increasing joint member thickness, the preheat temperature must be increased to reduce the cooling rate and the tendency to form martensite. Since the ductility of martensite depends on its hardness, which is a function of the carbon content of the steel, the formula for calculating preheat is based upon the thickness of the steel and its carbon content:

$$P = 1000\,(C - 0.11) + 18t$$

where

P = preheat temperature (in degrees Fahrenheit)
C = carbon content of steel
t = thickness of joint (in inches)

For example, what is the preheat temperature for a joint 2″ thick made of steel containing 0.35% carbon?

$$P = 1000\,(0.35 - 0.11) + 18(2)$$

$$P = 1000 \times 0.24 + 36$$

$$P = 240 + 36$$

$$P = \mathbf{276°F}$$

As a rule, preheat is usually unnecessary for steels with a carbon content less than 0.2% if the joint thickness is less than:

- 1½″, for wrought pressure vessel plate
- ¾″, for wrought pipe
- ½″, for castings

However, the weld area should always be heated to hand warmth before welding.

Heat Requirements. Heat requirements include preheat, interpass temperature control, and postheating. Preheat heats the base metal to a relatively low temperature before welding starts. The main purpose of preheat is to lower the cooling rate of the weld, thus allowing slower withdrawal of heat from the weld area, which lessens the tendency for martensite to form. Consequently, there is less likelihood for a hard zone to develop in the surrounding weld area than if a weld joint is made without preheat.

Preheat prevents cold cracks, reduces hardness in the HAZ, reduces residual stresses, and reduces distortion. Preheat also burns grease, oil, and scale out of the joint, ensuring a clean welding surface and allowing a more rapid welding speed. Preheat can be accomplished by moving an oxyacetylene flame over the surface or by placing the part in a heating furnace. **See Figure 39-4.** Preheat temperatures for carbon steel range from 200°F to 700°F, depending on the carbon content. The greater the carbon content, the higher the preheat temperature.

Point

Temperature must be maintained between the minimum and maximum interpass temperatures for each succeeding welding pass.

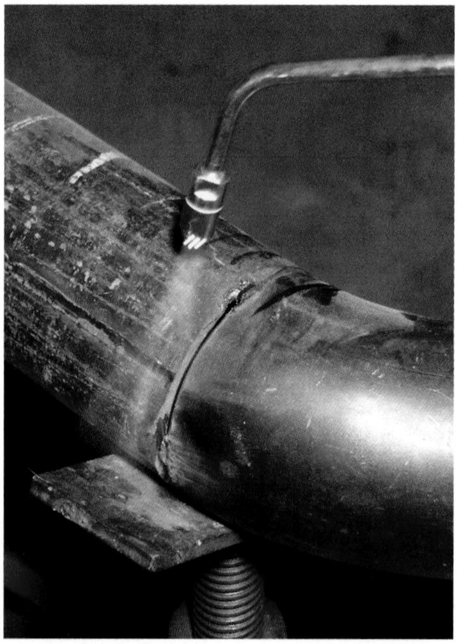

Figure 39-4. *A common preheat method is to move an oxyacetylene flame over the surface of the metal.*

The interpass temperature is the temperature of the weld area between passes of a multiple-pass weld. For most steels, the large volume of heat produced by the welding process is often a more economical method of maintaining interpass temperature. Using a high current and slow travel rate causes considerable heat to build up in the metal, slowing the rate of cooling after welding and preventing martensite from forming near the weld area.

In multiple-pass welding, the first pass preheats the base metal. Heat from the second pass tempers the base metal adjacent to the first pass. Each successive pass produces enough heat to prevent hardening caused by rapid cooling. The interpass temperature must be carefully regulated. Minimum and maximum interpass temperatures are generally specified for multiple-pass welds. Temperature must be maintained between the minimum and maximum interpass temperatures for each succeeding welding pass. On small parts, the temperature during welding can increase to undesirable levels. A welder must allow time for the workpiece to cool between weld passes.

Postheating is a stress-relief treatment for welding medium- and high-carbon steels. Postheating is as important as preheat. Although preheat controls the cooling rate, the possibility of stresses being locked into the weld area is always a factor. Postheating is especially necessary for thick metal or when the part is restrained in a jig or fixture during welding. Unless stresses are removed, cracks may develop, or the part may become distorted when it cools completely, especially during machining operations.

Postheating temperatures for stress relief should be in the range of 900°F to 1250°F. The post-heating period normally runs about 1 hr per inch of metal thickness.

Specific Welding Considerations for Carbon and Alloy Steels

Factors influencing the welding of carbon steel depend upon the group to which the specific steel belongs. Weldability decreases with increased carbon content, and the need for preheat and postheating increases with increased carbon content. **See Figure 39-5.**

Low-Carbon Steels. Low-carbon steels that contain less than 0.2% carbon and less than 1% manganese (carbon equivalent 0.36) are weldable without preheat or postheating when joint thickness is less than 1″ and joint restraint is not severe.

Low-carbon steels can be welded by arc welding and OFW processes. Low-carbon steels are the easiest to weld since no special welding preparations are necessary. For SMAW, E60XX filler metal is suitable provided there is sufficient weld metal in the joint to provide adequate strength. The choice of filler metal is determined by depth of penetration, type of current, position of the weld, joint design, and deposition rate. When slightly higher strength filler metal is desirable, or low-hydrogen filler metals are necessary, E70XX regular or E70XX low-hydrogen filler metals must be used. **See Appendix.**

When using GMAW or GTAW, filler metal selection depends on the deoxidation practice of the steel. Rimmed or capped steels create porous welds unless the filler metal contains deoxidizers. A suitable filler metal for these applications is ER70S-2. For killed or semikilled steels, in addition to ER70S-2, E70S-6 or E70S-7 filler metal may be used.

OFW requires steel filler metal that matches the strength of the base metal. Type R45 deposits low-carbon steel weld metal. Higher strength R60 is used to weld low-carbon steels with tensile strengths from 50 ksi to 65 ksi.

Medium-Carbon Steels. As the carbon content increases beyond 0.3% and the manganese content increases to 1.4% (carbon equivalent 0.53), susceptibility to hydrogen cracking increases so that welding with low-hydrogen filler metal is necessary. Nevertheless, steels containing about 0.3% carbon and a relatively low manganese content have good weld-ability. However, a pronounced change in weldability occurs when the carbon content is in the 0.3% to 0.5% range. As the carbon content of the steel is increased, the welding procedure must be altered to prevent the formation of hard martensite in the HAZ. The required preheat temperature increases as the carbon equivalent increases. With a carbon equivalent between 0.45% and 0.60%, a preheat temperature between 200°F and 400°F is required, depending on joint thickness. The interpass temperature should equal the preheat temperature. Postheating between 1100°F and 1250°F is recommended immediately after welding. If postheating is not possible, the temperature of the joint should be maintained after welding at slightly above the specified preheat temperature for 2 hr to 3 hr per inch of thickness to promote the diffusion of hydrogen into the base metal from the weld bead.

CARBON STEELS			
Steel	**Carbon Content**	**Weldability**	**Uses**
Low-Carbon	Up to .3%	Excellent	• Piping • Industrial Fabrication
Medium-Carbon	.3% to .6%	Fair	• Machine Parts
High-Carbon	.6% or Higher	Poor	• Railroad Track Lengths • Machine Dies

Figure 39-5. *Weldability decreases with increased carbon content, and the need for preheat and postheating increases with increased carbon content.*

Most medium-carbon steels are relatively easy to weld by arc and gas welding processes. For SMAW, E7018 or E7024 filler metal is frequently used because they have a high tensile strength and less tendency to produce underbead cracking, particularly when no preheat can be applied. However, medium-carbon steels must typically be preheated and/or postheated. E6012 or E6020 filler metal can also be used if precautions are taken and the cooling rate is sufficiently slowed to prevent excessive hardening of the weld.

For GTAW and GMAW, any of the ER70S-X series filler metals may be used if precautions are taken to prevent hydrogen entry into the weld from rusty or contaminated surfaces or from contaminated shielding gases. For OFW, a high-strength filler metal that matches the strength of the base metal, such as R60 or R65, should be used.

High-Carbon Steels. High-carbon steels are significantly more difficult to weld than other carbon steels and are not usually welded. They form hard martensite when quenched and are extremely sensitive to cracking. When high-carbon steels must be welded, high-strength filler metals in the E80XX, E90XX, or E100XX groups are preferred because they minimize underbead cracking. Preheat must also be used to prevent cracking. The postweld cooling rate must be kept as slow as possible.

Stainless steel filler metals such as the E310-15 type are frequently recommended for welding high-carbon steels because of their high ductility, provided weld strength is not an issue. Low-hydrogen filler metals with iron-powder coatings produce a ductile weld with minimum penetration.

Free-Machining Steels. Free-machining steels have poor weldability because they are susceptible to hot cracking from the formation of low-melting-point sulfur- and phosphorus-containing compounds. Lead in free-machining steels can melt during welding, emitting weld fumes and creating a health hazard. Lead may also cause porosity and embrittlement under certain welding conditions. Free-machining steels are not usually welded unless absolutely necessary.

Certain precautions must be taken if free-machining steels must be welded. For SMAW, low-hydrogen filler metals of the EXXX-18 group are used. For FCAW or GMAW, the same type of electrode as for the corresponding regular grade (non-free-machining) steel is used. GTAW is not normally used to weld free-machining steels. A low welding current is used to minimize dilution, porosity, and cracking; however, the low welding current leads to reduced welding speed. The work area must be adequately ventilated when welding free-machining steels that contain lead.

Low-Alloy Steels. Low-alloy steels are welded by arc welding and gas welding processes if they have been annealed or normalized. They are then quenched and tempered to achieve the desired properties. If quenching or tempering is not possible—for example, with complex parts where distortion might occur—preheat at 600°F or higher is used. High preheat temperatures slow the cooling rate, allowing the formation of soft bainite rather than hard martensite, and permitting handling of the part between welding and postheating.

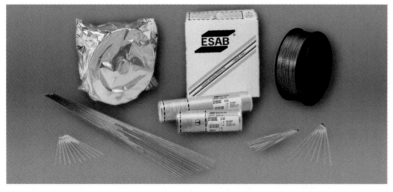

ESAB Welding and Cutting Products

Filler metals used for welding carbon and alloy steels are selected based on the metal composition and the desired properties of the metal after welding.

The recommended preheat temperature is about 50°F (28°C) above the temperature at which martensite begins to form on cooling. The preheat temperature may also be influenced by the thickness of the joint, alloy composition, and joint restraint.

Both preheat and postheating prevent weld cracks caused by shrinkage stresses. By reducing the rate of cooling, the stresses are distributed more evenly throughout the weld and released while the metal is still hot.

When the proper preheat temperature cannot be determined, the clip test can be used as a quick check. The clip test is not applicable to thin steels but produces good results on sections up to ⅜″ thick.

The clip test involves welding a piece of low-carbon steel to the steel workpiece that is being checked for preheat temperature. A convex fillet weld is made using an electrode and welding current similar to those required for the welding job. The weld is allowed to cool for 5 min and then the welder, wearing safety glasses, hammers the lug until it breaks off. If the lug breaks through the weld after a number of blows, the test indicates that no serious underbead cracking will result when welding is carried out in the same manner at normal room temperature. If the lug breaks and pulls out some of the base metal, the test indicates that the particular steel must be preheated. **See Figure 39-6.**

Low-alloy, high-strength filler metals E70XX, E80XX, E90XX, and E100XX are used for welding low-alloy, chrome-moly, and nickel steels when full strength is required. In addition to the standard symbols, low-alloy, high-strength steel filler metals carry a suffix in the form of a letter and a final digit. The letter indicates the chemical composition of the deposited metal. The final digit designates the exact composition of the broad chemical classifications. Low-alloy, high-strength steel arc welding filler metals are designated as E7010-A1, E8016-B2, etc. When welding any alloy steel, contact the filler metal manufacturer for proper filler metal selection.

The reaction of filler metals to heat treatment for alloy steels must match the reaction of the base metal. The carbon, phosphorus, and sulfur contents of the filler metal are generally maintained at low values to reduce hot cracking susceptibility and improve weld metal ductility. Filler metals with comparable composition but lower carbon content may be satisfactory where lower joint strength is acceptable.

Figure 39-6. *The need for preheating may be indicated by the clip test.*

Chrome-Moly Steels. Chrome-moly steels are air-hardening and form martensite on cooling. The martensite is relatively soft because of the low carbon content, but all chrome-moly steels require preheat, interpass temperature control, and postheating to produce a tough weld joint. **See Figure 39-7.** Preheat and postheating temperatures vary depending on the alloy content of the steel. Postheating for chrome-moly steels is usually completed immediately after welding.

Point

The recommended preheat temperature for low-alloy steels is about 50°F (28°C) above the temperature at which martensite begins to form on cooling.

Point

Before welding any alloy steel, check with the manufacturer for the proper filler metal.

Postheating temperatures for chrome-moly steels are higher than for carbon and low-alloy steels, because chrome-moly steels are more creep-resistant and require higher temperatures to cause them to yield. A postheating temperature of 1300°F to 1350°F is commonly used.

Chrome-moly steels may be joined by arc welding processes. A low welding current and rapid welding speed should be used, without extensive preheat. Care must be taken to prevent an excessive amount of base metal from mixing into the weld. Some preheat is advisable to reduce underbead cracking. Postheating is recommended for stress relief.

When an interrupted welding procedure is required — for example, to radiograph the partially completed joint in a heavy-wall pipe — the welding should not be interrupted until a distance equal to one-third the wall thickness has been welded, or not less than two weld passes for pipe less than 1″ thick. These precautions prevent cracking of the partially completed joint from residual stresses as it cools in order to be radiographed.

Filler metals must match the base metal composition, except that carbon content slightly lower than that of the base metal is needed to reduce cracking susceptibility. To limit the number of filler metals required when several chrome-moly steels are used on one job, filler metals of the same or slightly higher alloy content can be used. For example, 1¼Cr-½Mo filler metals can be used for welding ½Cr-½Mo and 1¼Cr-½Mo. For SMAW, low-hydrogen filler metals are used. **See Figure 39-8.**

Stainless steel filler metals E309 and E310 may be used for minor repair welding of chrome-moly steels. They are preferred for applications where the weld joint cannot be postheated. Stainless steel filler metals are weaker than chrome-moly electrodes and possess excellent as-welded ductility, yielding easily and relieving the majority of residual stresses. However, the selection of a stainless steel filler metal must be made carefully, especially if the weld joint is operating in cyclic temperature service where premature failure might occur due to the higher expansion of the stainless steel.

	RECOMMENDED PREHEAT TEMPERATURES FOR CHROME-MOLY STEELS*					
Steel†	**Thickness**					
	Up to .5″ (13 mm)		**.5″ to 1″ (13 mm to 25 mm)**		**Over 1″ (25 mm)**	
	°F	°C	°F	°C	°F	°C
½Cr-½Mo	100	38	200	93	300	149
1Cr-½Mo	250	121	300	149	300	149
1¼Cr-½Mo						
2Cr-½Mo	300	149	350	177	350	177
2¼Cr-1Mo						
3Cr-1Mo						
5Cr-½Mo	350	177	400	204	400	204
7Cr-½Mo						
9Cr-1Mo						
9Cr-1Mo V+Nb+N						

* Welding with low-hydrogen covered electrodes
† Maximum carbon content of .15%. For higher carbon steels, preheat temperature should be increased 100°F to 200°F (38°C to 93°C). Lower preheat temperatures may be used with gas tungsten arc welding.

Figure 39-7. *The required preheating temperature for chrome-moly steels varies according to the alloy content.*

FILLER METALS FOR CHROME-MOLY STEELS*				
Chrome-Moly Content	GTAW† and GMAW	SMAW‡	FCAWS§	SAW‖
½Cr-½Mo	#	E801X-B1	E7XT5-A1 or E8XT1-A1	F8XX-EXXX-B1
1Cr-½Mo, 1¼Cr-½Mo	ER80X-B2 or ER70X-B2L	E801X-B2 or E701X-B2L	E8XTX-B2 or E8XTX-B2L or E8XTX-B2H	F8XX-EXXX-B2 or F8XX-EXXX-B2H
2¼Cr-1Mo	ER90X-B3 or ER80X-B3L	E901X-B3 or E801X-B3L	E9XTX-B3 or E9XTX-B3L or E9XTX-B3H	F9XX-EXXX-B3
3Cr-1Mo	**	**	**	**
5Cr-½Mo	ER502†† or ER80X-B6	E502-1X‡‡ or E801X-B6 or E801X-B6L	E502T-1 or 2 or E6XT5-B6	F9XX-EXXX-B6 or F9XX-EXXX-B6H
7Cr-½Mo	§§	E7Cr-1X‡‡ or E801X-B7 or E801X-B7L	§§	§§
9Cr-1Mo	ER505†† or ER80X-B8	E505-1X‡‡ or E801X-B8 or E801X-B8L	E505T-1 or 2 or EX15-B8 or E6XT5-B8L	F9XX-EXXX-B8
9Cr-1Mo and V+Nb+N	ER90X-B9	E901X-B9	———	F9XX-EXXX-B9

* by welding process
† per ANSI/AWS A5.28, *Specification for Low-Alloy Steel Filler Metals for Gas Shielded Arc Welding* (unless indicated)
‡ per ANSI/AWS A5.5, *Specification for Low-Alloy Steel Covered Arc Welding Electrodes* (unless indicated)
§ per ANSI/AWS A5.29, *Specification for Low-Alloy Steel Electrodes for Flux Cored Arc Welding* (use with CO_2 or Ar-CO_2 mixture)
‖ per ANSI/AWS A5.23, *Specification for Low-Alloy Steel Electrodes and Fluxes for Submerged Arc Welding*
no match, consider higher alloy than base metal
** no match, use between 2¼Cr-1Mo and 5Cr-½Mo
†† per ANSI/AWS A5.9, *Specification for Bare Stainless Steel Welding Electrodes and Rods*
‡‡ per ANSI/AWS A5.4, *Specification for Covered Corrosion-Resistant Chromium and Chromium-Nickel Steel Welding Electrodes*
§§ no match, use between 5Cr-1Mo and 9Cr-½Mo

Figure 39-8. *Filler metals for welding chrome-moly steels may be slightly more alloyed in order to minimize the number of filler metal types required.*

Nickel Steels. To weld nickel steels where the tensile strength of the weld must be equal to that of the base metal, low-alloy nickel filler metals in the E80XX series are generally used. Examples are E8016-C1, E8018-C2, and E8018-C3. On thick metal, preheat to a dull red is generally advisable.

Austenitic Manganese Steels. Austenitic manganese steels require care when welding as they experience loss of ductility when reheated. A low welding current and rapid welding speed must be used, without extensive preheat. Care must be taken to prevent an excessive amount of base metal from mixing into the weld. A slight preheat is advisable to reduce underbead cracking. Postheating is recommended for stress relief. Use the following guidelines to ensure quality welds when welding austenitic manganese steel:

- V the joint and clean the surfaces carefully and thoroughly.
- Use the lowest possible current to prevent the formation of a brittle zone next to the weld.
- Use a stainless steel 18-8 type electrode.

Other types of filler metals used for welding austenitic manganese steel are molybdenum-copper-manganese and nickel-manganese. However, more skill is needed to produce good welds with these filler metals. Do not weld in a localized area for an extended time unless the temperature of the metal is below 750°F. Use temperature-indicating crayons to determine temperature by marking the base metal ⅜″ to ½″ from the weld. The welder should be able to place a hand within 6″ to 8″ of the weld at any time. If necessary, place wet rags on

> ✓ **Point**
>
> *Use the lowest possible current when welding austenitic manganese steel.*

areas adjacent to the weld to control heat. The high thermal expansion of austenitic manganese steel may cause residual stresses to develop as the weld cools, and cracks may develop during contraction. To reduce cracking, peen each weld pass when it is completed.

 Refer to Quick Quiz® on CD-ROM

Points to Remember

- Carbon steels include low-carbon steels, medium-carbon steels, and high-carbon steels. The weldability of carbon steels decreases as the carbon content increases.
- The principal benefits of alloy steels over carbon steels are higher strength and greater capacity for strengthening in thick sections (hardenability).
- The deoxidation practice determines the amount of deoxidation performed and the basic steel type produced.
- Hydrogen cracking is often located below the surface and may not be detected by common nondestructive examination techniques.
- Steels that are susceptible to cracking must be preheated to reduce the rate of cooling and decrease the possibility of martensite formation.
- Temperature must be maintained between the minimum and maximum interpass temperatures for each succeding welding pass.
- Low-hydrogen filler metals should be used when welding medium-carbon steels.
- Low-hydrogen filler metals with iron powder coatings usually minimize cracking in welding high-carbon steel.
- Free-machining steels are not usually welded unless special precautions are taken.
- The recommended preheat temperature for low-alloy steels is about 50°F above the temperature at which martensite begins to form on cooling.
- Before welding any alloy steel, check with the manufacturer for the proper filler metal.
- Use the lowest possible current when welding austenitic manganese steel.

Questions for Study and Discussion

1. When is steel classified as medium-carbon steel?
2. What is the difference between killed and semikilled steel?
3. Why is some form of preheat recommended when arc welding alloy steels?
4. What are some of the basic characteristics of austenitic manganese steel?
5. What is the function of postheating? At what temperature should postheating be done?
6. What type of filler metal is required for welding medium-carbon steel?
7. Why are high-carbon steels more difficult to weld?
8. What is the purpose of a clip test and how is it conducted?
9. Why must the lowest possible current be used when welding austenitic manganese steel?

 Refer to Chapter 39 in the *Welding Skills Workbook* for additional exercises.

Weldability of Tool Steels and Cast Irons

Chapter

Weldability is the ability of a metal to be welded under imposed conditions. Tool steels and cast irons are metals that require special consideration when welding. Tool steels are the most highly alloyed steels and in general are the hardest and strongest steels available. In most cases, the welding of tool steels encompasses the repair of tools or dies that have been hardened and machined to final shape and have failed from wear, chipping, or cracking.

Cast irons are alloys of iron with significant amounts of carbon, silicon, and occasionally other elements. The primary consideration for joining cast irons is to accommodate their poor weldability, the principal cause of which is their high carbon content.

WELDABILITY OF TOOL STEELS

Tool steels are the most highly alloyed steels and in general are the hardest and strongest steels available. Tool steel groups are named for their response to heat treatment or their major end use. The chemical composition and metallurgical structure of tool steels are designed for specific end uses. Tool steel groups consist of water hardening, cold work, shock resisting, hot work, high-speed, mold, and special purpose. **See Figure 40-1.** Tool steels are very difficult to weld and are not usually welded unless for repair or rebuilding. **See Figure 40-2.**

> *Tool steels should be welded in the annealed condition (as they are received from the manufacturer) as this minimizes the tendency to crack on welding.*

Water hardening tool steels, Group W, are high-carbon steels that contain between 0.6% and 1.4% carbon, plus small amounts of chromium and vanadium to increase hardenability and maintain fine

grain size to improve toughness. Water hardening tool steels are the least costly and have many applications.

Cold work tool steels (Groups O, A, and D) generally contain between 1% and 2% carbon, and can range from 0.5% to 2.35% for some alloys. Cold work tool steels have alloy compositions designed to provide moderate-to-high hardenability and good dimensional stability during heat treatment. They have high wear resistance, and poor-to-fair toughness. Cold work tool steels begin to soften at temperatures above 400°F and are generally limited to working temperatures below 900°F. The majority of tool applications can be served by one or more cold work tool steels.

Shock resisting tool steels, Group S, have a relatively low carbon content—between 0.4% carbon and 0.65% carbon—and contain manganese, silicon, tungsten, and molybdenum. Shock resisting tool steels are used in applications involving impact loading because of their high strength and toughness under repeated shock and low-to-medium wear resistance.

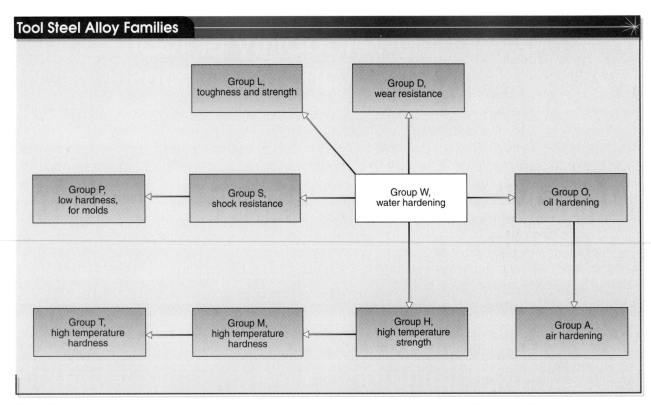

Figure 40-1. *Tool steel groups consist of water hardening, cold work, shock resisting, hot work, high-speed, mold, and special purpose.*

TOOL STEELS		
Tool Steel Group	**Properties and Characteristics**	**Common Use**
Group W	Tough core and hard and wear resistant surface	Cutlery, forging dies, and hammers
Group O	Wear resistant to moderate temperatures	Dies and punches
Group A	Minimum distortion and cracking on quenching	Dies, punches, and forming rolls
Group D	High hardness and excellent wear resistance	Long run dies and brick molds
Group S	Excellent toughness and high strength	Chisels, rivet sets, and structural applications
Group H	Good resistance to softening at elevated temperatures and good toughness	High stressed components and high-temperature extrusion dies
Group T	High hardenability and high hardness	Cutting tools and high-temperature structural components
Group M	High hardenability and high hardness	Cutting tools
Group L	High toughness and good strength	Arbors, cams, and chucks
Group P	Low hardness and low resistance to work hardening	Dies and molds
Group F	Tough core, hard surface, and galling resistance	Burnishing tools and tube-drawing

Figure 40-2. *Diversification of properties and characteristics influence the number and type of uses of tool steels.*

Hot work tool steels, Group H, have a medium carbon content—0.35% carbon to 0.45% carbon—with chromium, tungsten, molybdenum, and vanadium added for total alloying between 6% and 25%. The alloying elements contribute to good hardenability, toughness, and resistance to softening (red hardness) on continuous exposure up to 1000°F. *Red hardness* is the capacity to resist softening in the red heat temperature range. Hot work tool steels are used for hot die work.

High-speed tool steels, Groups T and M, have high carbon content and relatively large amounts of expensive alloying elements, particularly tungsten (Group T) and molybdenum (Group M). They are resistant to softening up to 1000°F but have relatively low toughness. High-speed tool steels are used for high-speed cutting operations because the alloy carbides in their metallurgical structure allow these steels to maintain their cutting edge at high temperatures.

Mold steels and special purpose tool steels, Groups L and P, and Group F, are minor tool steel groups whose properties are tailored to specific applications. Mold steels have a low to medium carbon content and contain chromium and nickel as the principal alloying elements for a total alloy content of 1.5% to 5%. Mold steels exhibit low hardness and low resistance to work hardening in the annealed (softened) condition, which facilitates the formation of mold impressions for cold hobbing operations. Special purpose tool steels contain small amounts of chromium, vanadium, and nickel and are used in applications requiring good strength, toughness, scratch resistance, and galling resistance.

General Welding Considerations for Tool Steels

In most cases, the welding of tool steels encompasses the repair of tools or dies that have been hardened and machined to final shape and have failed by wear,

chipping, or cracking. Tools or dies may also be welded to alter the tool or die to accommodate design changes. Due to their high carbon and alloy content, they are extremely prone to hydrogen cracking in the HAZ if rapidly cooled. If high heat input and slow cooling are used to counteract cracking, the weld may be too soft. Welding procedures must be carefully controlled.

Preheat and Postheating Requirements. Tool steels are always preheated for welding. The required preheat temperature depends on the specific alloy, heat-treated condition, and section thickness. When preheating a hardened tool steel, the preheat temperature should not exceed the tempering temperature of the tool steel, or it will soften. The preheat temperature should be maintained between weld passes. After welding, the workpiece should be cooled to about 150°F and immediately postheated at the recommended temperature. **See Figure 40-3.**

Welding Processes. The welding process for tool steels must be carefully selected to produce a quality weld. Welding processes that can be used for tool steels include SMAW, FCAW, and GMAW. SMAW is the most versatile for repair welding small areas. Large areas may be more economically welded with FCAW or GMAW. OFW should not be used for tool steels because it is too slow and introduces excessive heat into the base metal, leading to distortion, softening of hardened metal, embrittlement of annealed metal, or cracking.

Filler Metals. Filler metals used for tool steels must be carefully selected to ensure a quality weld. Filler metals for welding tool steels fall into three categories: matching, low-alloy steel, and soft. Filler metals that produce deposits matching the basic tool steel type should be used because they produce a surface that matches the wear resistance of the tool steel. However, filler metals are not available to match all tool steel compositions. Although an exact match may not always be available, using manufacturer trade name products and their recommended procedures usually produces a quality weld.

When matching filler metal is not available, filler metals that produce deposit compositions similar to low-alloy steel may be used as they exhibit moderate hardness. Toughness may be improved by peening. Soft filler metals such as stainless steels, nickel, nickel-copper alloys, and copper-nickel alloys may be used to build up worn parts, followed by a hard wear-resistant deposit that matches the base tool steel composition. Using a soft buildup material minimizes cracking.

PREHEAT AND POSTHEATING TEMPERATURES FOR TOOL STEELS						
		Annealed Base Metal		Hardened Base Metal		
Type	Group	Preheat and Postheating*	Deposit HRC†	Preheat and Postheating*	Tempering Temperature*	Deposit HRC‡
W1, W2	Water-hardening	250–450 (121–232)	50–64	250–450 (121–232)	350–650 (177–343)	56–62
S1	Shock-resisting	300–500 (149–260)	40–58	300–500 (149–260)	400–1200 (204–649)	52–56
S5	Shock-resisting	300–500 (149–260)	50–60	300–500 (149–260)	350–800 (177–426)	52–56
S7	Shock-resisting	300–500 (149–260)	47–58	300–500 (149–260)	400–1100§ (204–621)	52–56
O1	Oil-hardening	300–400 (149–204)	57–62	300–400 (149–204)	350–500 (177–260)	56–61
O6	Oil-hardening	300–400 (149–204)	58–63	300–400 (149–204)	350–600 (177–316)	56–61
A2	Air-hardening	300–400 (149–204)	57–62	300–400 (149–204)	350–1000§ (177–538)	56–61
A4	Air-hardening	300–400 (149–204)	54–62	300–400 (149–204)	350–800§ (177–426)	60–62
D2	Air-hardening	700–900 (371–482)	54–61	700–900 (371–482)	400–1000§ (204–538)	58–60
H12, H13, H19	Hot work	700–1000 (371–538)	38–56	700–1000 (371–538)	1000–1200§ (538–649)	46–54
M1	High-speed	950–1050 (510–566)	60–65	950–1050 (510–566)	1000–1100§ (538–593)	60–63
M2	High-speed	950–1050 (510–566)	60–65	950–1050 (510–566)	1000–1100§ (538–593)	60–63
M10	High-speed	950–1050 (510–566)	60–65	950–1050 (510–566)	1000–1100§ (538–593)	60–63
T1, T2, T4	High-speed	950–1050 (510–566)	60–66	950–1050 (510–566)	1000–1100§ (538–593)	61–64
P20	Mold steel	800–1000 (426–538)	28–42	800–1000 (426–538)	900–1100 (480–595)	28–37

* °F (°C)
† hardness varies with heat input and cooling rate
‡ after postheating and tempering, varies with heat input and cooling rate
§ double temper

Figure 40-3. *The required preheat temperature for tool steels depends on the specific alloy, heat-treated condition, and metal thickness.*

Repair Welding. Repair welding requires adequate preparation. Preparation for repairs requires first grinding the damaged area to a uniform depth to allow for buildup of a deposit with the required hardness and wear resistance. A groove depth of ⅛″ is common. Small weld passes are used to fill the groove. The bead size of the final pass should be adjusted so that the repair is as close as possible to final size to minimize the final grinding operation. Welding is done in flat position with minimum heat input. Intermittent welding is used on symmetrical repairs to ensure uniform heat distribution. The weld should be cleaned frequently by chipping and brushing. Warpage and distortion are counteracted by preheat and peening. When repairing the cutting edge of a tool or die, the edge to be welded should be grooved approximately 45° for sufficient depth. **See Figure 40-4.**

WELDABILITY OF CAST IRONS

Cast irons are alloys of iron that contain significant amounts of carbon and silicon, and occasionally other elements. Cast irons can be easily poured into complex shapes; however, they are difficult to weld. When casting defects occur, or components break in service, repairs can be made (except on white iron) by welding or braze welding.

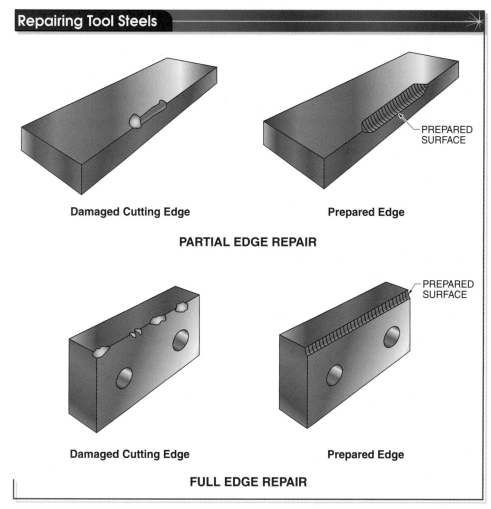

Repairing Tool Steels

Damaged Cutting Edge

Prepared Edge — PREPARED SURFACE

PARTIAL EDGE REPAIR

Damaged Cutting Edge

Prepared Edge — PREPARED SURFACE

FULL EDGE REPAIR

Figure 40-4. *The buildup area is grooved when repair welding tool steels to allow sufficient metal to be deposited.*

Cast irons are used for wear and corrosion resistance and for general applications where good castability is needed. Cast irons are grouped according to their metallurgical structure into gray irons, white irons, malleable irons, ductile irons, compacted graphite irons, and alloy irons. **See Figure 40-5.**

A gray iron microstructure consists of pearlite (iron carbide and ferrite), ferrite, or martensite. All three types of gray iron microstructures contain an even distribution of graphite flakes. The graphite flakes make gray irons extremely brittle but also provide for the highest damping capacity of any engineering material, high temperature scaling resistance, and thermal shock resistance.

Gray iron is the most widely used cast iron. Gray iron can be identified by its dark gray, porous structure on the fracture face. When using the spark test to identify gray iron, short, brick-red streamers are given off that follow a straight line and have numerous fine, repeating yellow sparklers. Gray iron is relatively easy to arc weld.

Gray irons are used for bases and supports for moving components to dampen vibrations; in pressure applications such as cylinder blocks; for wear-resistant and scuff-resistant materials in cylinder sleeves; and for general municipal applications such as manhole covers and hydrants.

White irons are formed when carbon does not precipitate as graphite during solidification but combines with iron or alloying elements such as molybdenum, chromium, or vanadium to form iron carbide or alloy carbide. This combining occurs because of fast cooling of the molten metal in the mold. Thus, white iron can be formed if the mold contains coolers that accelerate the cooling rate. The carbides make white iron extremely hard, wear-resistant, and brittle.

Cast Iron Alloy Families

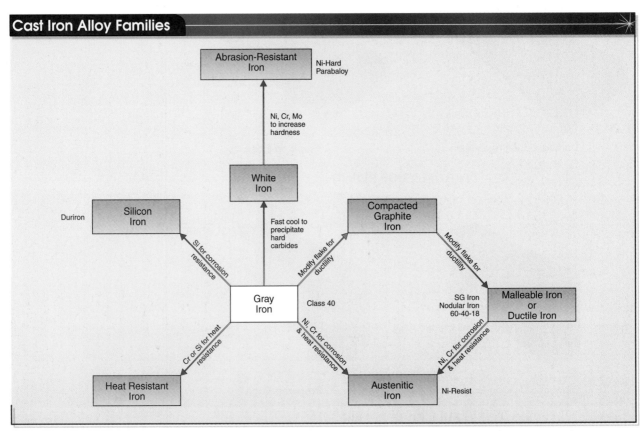

Figure 40-5. *Cast irons are grouped according to their metallurgical structure into gray irons, white irons, malleable irons, ductile irons, compacted graphite irons, and alloy irons.*

White irons have a fine, silvery white, silky, crystalline fracture face. When spark tested, white iron shows short, red streamers. There are fewer sparklers than in gray cast iron and these are small and repeating. Welding is not recommended for white irons. White iron is used for wear plates.

Malleable irons are produced by heat treating white iron. Malleable iron can be welded. However, the metal must not be heated above its critical temperature (approximately 1382°F). If it is heated above the critical temperature, the metal reverts to the original characteristics of white iron.

Heat treatment transforms graphite flakes into nodules, leading to increased ductility. Improved ductility creates many uses for malleable iron. These include axle and differential housings, camshafts, and crankshafts in automobiles; and gears, chain links, sprockets, and elevator brackets in conveying equipment.

Malleable irons exhibit a white crystalline fracture face with a dark center. A spark test shows a moderate number of short, straw-yellow streamers with numerous sparklers that are small and repeating.

Ductile irons are similar to malleable irons in that the shape of the graphite flakes is nodular, or spheroidal. Ductile irons are sometimes referred to as SG irons for spheroidal graphite.

Spheroidization of the graphite is achieved by adding small amounts of magnesium or cerium to molten iron before it cools and solidifies. The cost of the elements added to the iron makes ductile iron more expensive, but prolonged heat treatment is not required, so that its cost is comparable to malleable iron. Grade for grade, ductile iron has strength equivalent to gray iron, but ductile iron has significantly greater elongation.

Ductile irons can be arc welded, provided adequate preheat and postheating are used; otherwise, some of the original properties are lost. Ductile iron is used for many structural applications, particularly those requiring strength and toughness, and combines good machinability at low cost. Ductile iron is used for items such as crankshafts, front wheel spindle supports, steering knuckles, and pumps. Piping such as culvert, sewer, and pressure pipe is another application for ductile irons.

Compacted graphite iron exhibits a graphite shape between that of the flakes in gray iron and the spheroids in ductile iron. Compacted graphite irons are produced by adding specific elements to the molten metal in a way similar to ductile irons. The resulting graphite is in the form of interconnected flakes with blunted edges and a relatively short span. The intermediate shape of the graphite results in a combination of properties between gray and ductile iron. Compacted graphite irons are used for specific applications, such as disc brake rotors and diesel engine heads.

Alloy irons are cast irons that contain one or more added alloying elements, such as chromium, nickel, copper, molybdenum, vanadium, and silicon, to a total of up to 30% of the final composition. The three subgroups of alloy irons are abrasion-resistant irons, corrosion-resistant irons, and heat-resistant irons.

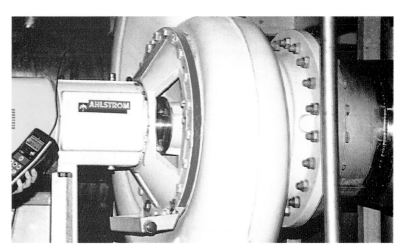

SPM Instrument, Inc.

Ductile irons are commonly used for pumps in industrial applications.

Abrasion-resistant irons are alloys of white iron and include the Ni-Hard® (nickel-containing) irons and the chromium irons. Abrasion-resistant irons are used for abrasive materials handling, as in slurry pumps, grinding equipment, and mud pump liners in well drilling.

Corrosion-resistant irons may be nickel-containing types such as the Ni-Resist® series or silicon-containing types such as the Duriron® series. Nickel-containing types are used in many corrosion-resistant applications, such as pump impellers and casings for seawater, acids, and sour gas. Silicon-containing types are brittle but possess exceptional corrosion resistance and are used for pumps, agitators, mixing nozzles, and valves.

Heat-resistant irons are gray or ductile irons containing alloying elements to improve high-temperature strength and oxidation resistance. They are used for turbine diaphragms, valves, and nozzle rings; manifolds and valve guides for heavy-duty engines; burner nozzles; glass molds; and valve seats for engines.

Most alloy irons can be arc welded, but precautions must be taken during preheat and postheating to prevent compromising desired metallurgical properties.

General Welding Considerations for Cast Irons

Cast irons are difficult to weld and heat input and joint preparation must be carefully controlled. Welding or braze welding can be used to repair broken castings, correct machining errors, fill defects, or weld cast iron to steel.

The primary consideration for joining cast irons is to accommodate their poor weldability, the principal cause of which is their high carbon content. A high carbon content can lead to the formation of very hard martensite in the HAZ, which, coupled with low ductility and the presence of residual stress, increases the susceptibility of cast iron to cracking.

The feasibility of repairing cast irons that have been in service depends on the service conditions. For example, it is generally not recommended to repair weld gray iron castings that are subject to repeated heating and cooling in normal service, especially if the temperature range exceeds 400°F. Unless cast iron is used as filler metal, the different coefficient of expansion between the weld metal and filler metal causes stresses that lead to cracking.

Mechanical joining techniques or braze welding are often effective alternatives to welding on cast irons. Mechanical joining methods may be used for joining cast iron if pressure retention is not a concern. The principal mechanical joining method for cast irons is cold mechanical repair. Parameters that must be considered before welding cast irons include joint preparation, heat requirements, welding processes, filler metals, repairing cracked castings, and studding broken castings.

Joint Preparation. Cast iron must be completely cleaned of contaminants around the weld area before welding. Joint preparation must ensure that the filler metal can thoroughly bond to the base metal. All casting skin and foreign matter must be removed from the joint surface and adjacent areas. Where possible, the casting is heated uniformly using an oxyacetylene torch at 700°F for 30 min, or less than 30 min at 1000°F. Graphite on the surface of gray iron can be oxidized by searing the surface with an oxidizing flame or by heating the casting with a strongly decarburizing flame, followed by wire brushing to remove debris.

To prepare cast iron for welding, grind a narrow strip along each edge of the joint to remove the surface film or casting skin. V the edges of the weld area. On metal less than ³⁄₁₆″ thick, no V is necessary. On metal ³⁄₁₆″ to ³⁄₈″ thick, a single-V joint is required with a groove angle of approximately 60°. Metal ³⁄₈″ thick or more requires a double-V joint with a ¹⁄₁₆″ to ³⁄₃₂″ root face. The groove angle should be 60°. **See Figure 40-6.**

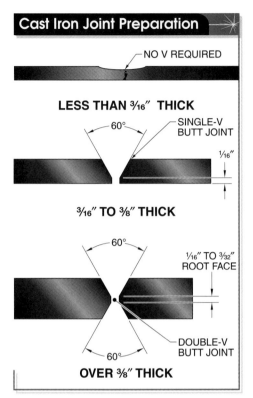

Cast Iron Joint Preparation

NO V REQUIRED

LESS THAN ³⁄₁₆″ THICK

60° — SINGLE-V BUTT JOINT

¹⁄₁₆″

³⁄₁₆″ TO ³⁄₈″ THICK

60°

¹⁄₁₆″ TO ³⁄₃₂″ ROOT FACE

DOUBLE-V BUTT JOINT

60°

OVER ³⁄₈″ THICK

Figure 40-6. *Careful joint preparation is required when welding cast iron.*

Heat Requirements. Preheat and postheating help minimize cracking and relieve residual stress. The preheat temperature depends on the type of cast iron, its mass, the welding process, and the welding filler metal. The preheat temperature should be monitored with a contact pyrometer, temperature-indicating crayon, or thermocouple to ensure accuracy. With large or complex castings, the preheat rate must be slow and uniform to prevent cracking from unequal expansion. The casting should be maintained at a constant temperature until the weld is completed. If possible, preheat the entire welded section with an oxyacetylene torch.

When a high preheat temperature such as 1200°F is used, post-heating may not be necessary if the casting is allowed to cool slowly from the welding temperature. Slow cooling may be achieved by covering the casting with an insulating blanket, vermiculite, or sand. If slow cooling is not possible, postheating is required.

When it is impossible to preheat the workpiece, the weld temperature can be controlled by depositing short weld beads 2″ to 3″ long. After a bead is deposited, allow it to cool until it can be touched with the hand. Consecutive beads should not be started until the previous bead has cooled sufficiently. As the weld bead cools, peen it by striking it lightly with a hammer. Peening helps to put the weld in compression and relieve stresses in the weld. Peening can be done only on the machinable weld deposit and heat affected zone, not on the entire casting. **See Figure 40-7.**

A CAUTION
Never postheat cast iron above a dull red color or above a temperature of 1200°F (649°C).

Figure 40-7. *Peening helps to relieve stresses when welding cast iron.*

Postheating is mandatory to stress-relieve fully restrained welds or welds intended for severe service. Postheating is performed immediately after welding by increasing the temperature to 1000°F to 1150°F, followed by holding the casting at temperature for about 1 hr per inch of thickness. The cooling rate must be kept to 50°F per hour until the casting reaches 700°F. Postheating slows the cooling rate, reduces hardness in the weld, and improves the machinability of the HAZ.

Welding Processes. Welding processes that can be used for cast irons include SMAW, GMAW, FCAW, and OFW. Brazing is used for some applications; soldering is not commonly used. SMAW is the most versatile process and can use all types of filler metals. GMAW with short circuiting transfer is suitable for joining ductile iron. Because of the relatively low heat input with GMAW, the hard portion of the HAZ is confined to a thin film next to the weld metal, so that the strength and ductility of the weld joint are about the same as the base metal.

OFW requires extensive heat input during preheat and welding. The high heat input is a limitation of OFW when welding finished or semi-finished castings because it may distort the metal. However, the slower cooling rate lessens the tendency toward brittleness. A high preheat temperature of 1100°F to 1200°F is required for OFW to compensate for the low welding heat obtained during OFW.

Buttering is used to provide good weld joint ductility and eliminates the need for postheating for the entire completed joint. Buttering places the HAZ of the welded joint in the buttered layer rather than in the cast iron. A layer of weld metal about 0.03″ thick is deposited (buttered) on the joint faces and the part is immediately postheated.

Filler Metals. Filler metals used for welding cast irons can be nickel alloy, carbon steel, or cast iron. Copper alloy filler metals are used for braze welding cast iron. Nickel alloy filler metals may also be used for buttering. Filler metals that match the base metal may be used, although a filler metal that minimizes cracking should always be used. The composition of the filler metal used to weld cast iron varies depending on the requirements of the weld. **See Figure 40-8.** Factors that influence filler metal selection for cast iron include the following:

- type of cast iron, mechanical properties desired in the joint
- the need for the filler metal to deform plastically and relieve welding stresses
- machinability of the joint
- color matching between the base metal and the filler metal
- allowable dilution
- cost

Nickel alloy filler metals are specially designed for welding cast irons. They dilute with the base metal and expand on cooling to minimize solidification shrinkage and reduce residual stress. Nickel alloy filler metals are machinable. The use of nickel alloy filler metals reduces preheat to minimal values, except in highly restrained sections. Peening of the hot weld bead helps to reduce residual stresses and maintain dimensions. Two common nickel alloy filler metals are:

- ENi-Cl (nickel) filler metal. DCEP or AC, general-purpose welding, used for thin and medium cast iron sections, castings with low phosphorus content, and where little or no preheat is used.
- ENi-FeCl (nickel-iron) filler metal. DCEP or AC, for welding heavy cast iron sections, high-phosphorus castings, high-nickel alloy castings where high-strength welds are required, and for welding nodular iron.

Carbon steel filler metals are nonmachinable and are used primarily to repair small, cosmetic casting defects with SMAW, where a fair color match is acceptable and machining is not of major concern. Carbon steel filler metals are prone to embrittlement from carbon pickup by dilution with the base metal and should not be used where the joint is loaded in tension or bending. The welding procedure is designed to minimize heat input to keep dilution to an acceptable level.

Carbon steel filler metals consist of a low-carbon steel core and a heavy coating that melts at low temperatures, allowing a low welding current. Carbon steel filler metals leave a very hard deposit and are used only when the welded section is not to be machined afterward. Nonmachinable carbon steel filler metals produce a tight and nonporous weld, making them ideal for repairing motor blocks, transmission cases, compressor blocks, pulley wheels, pump parts, mower wheels, and other similar structures.

Cast iron filler metals are also nonmachinable and have limited application for repair welding of gray iron and ductile iron castings. Cast iron filler metals are most often used for SMAW or OFW. For SMAW, cast iron filler metals have the composition of gray iron and are used to repair gray iron castings. For OFW, filler metal compositions matching gray iron or ductile iron are used, with other elements added to improve specific properties. A flux is also required with OFW to increase fluidity and remove slag that forms in the weld pool. Extensive heat input is required before, during, and after welding to prevent cracking.

✓ **Point**

Extensive heat input is required before, during, and after welding cast iron to prevent cracking.

FILLER METALS FOR ARC WELDING OR BRAZE WELDING OF CAST IRON			
Description	Filler Metal Form	Classification	Welding (or Braze Welding) Process*
NICKEL ALLOY FILLER METALS			
93 Ni	Bare	ERNi-Cl	GMAW
95 Ni	Covered	ENi-Cl, ENi-Cl-A	SMAW
53 Ni-45 Fe	Covered	ENiFe-Cl, ENiFe-Cl-A	SMAW
53 Ni-45 Fe-4.5 Mn	Flux Cored	ENiFeT3-Cl	FCAW
55 Ni- 40 Cu-4 Fe	Covered	ENiCu-A	SMAW
65 Ni-30 Cu-4 Fe	Covered	ENiCu-B	SMAW
44 Ni-44 Fe-12 Mn	Bare	ERNiFeMn-Cl	GMAW
53 Ni-45 Fe-4.5 Mn	Flux Cored w/Flux 5†	ENiFeT3-Cl	SAW
44 Ni-44 Fe-12 Mn	Bare w/Flux 6†	ERNiFeMn-Cl	SAW
44 Ni-44 Fe-12 Mn	Bare	ERNiFeMn-Cl	GTAW
44 Ni-44 Fe-12 Mn	Covered	ENiFeMn-Cl	SMAW
CARBON STEEL FILLER METALS			
Carbon Steel	Covered	ESt	SMAW
Carbon Steel	Covered	E7018	SMAW
Carbon Steel	Bare	E70S-2	GMAW
CAST IRON FILLER METALS			
Gray Iron	Welding Rod	RCI	OAW
Alloy Gray Iron	Welding Rod	RCI-A	OAW
Ductile Iron	Welding Rod	RCI-B	OAW
COPPER ALLOY FILLER METALS			
Low-fuming Brass	Welding Rod	RCuZn-B	OAW
Low-fuming Brass	Welding Rod	RCuZn-C	OAW
Nickel-Brass	Welding Rod	RBCuZn-D	OAW
Copper-Tin	Covered	ECuSn-A	SMAW
Copper-Tin	Bare	ERCuSn-A	GMAW
Copper-Aluminum	Covered	ECuAl-A2	SMAW
Copper-Aluminum	Bare	ERCuAl-A2	GMAW

* OAW, oxyacetylene welding; SMAW, shielded metal arc welding; GMAW, gas metal arc welding; SAW, submerged arc welding; GTAW, gas tungsten arc welding
† Incoflux 5 and Incoflux 6 are proprietary fluxes from Special Metals, Inc.

Figure 40-8. *Filler metals for welding or braze welding cast iron are selected based on the requirements of the weld and the welding process.*

The correct current setting for cast iron welding should always be used. The correct current is commonly suggested by the filler metal manufacturer. Generally, the current setting for welding cast iron is lower than for welding carbon steel. The heat applied to cast iron during welding must be kept to a minimum. To ensure that only the minimum heat necessary is used, always use small-diameter filler metals. Welders seldom use filler metals greater than 1/8″ in diameter.

For SMAW, welding current should be as low as possible but within the manufacturer's recommended range for consistent, smooth operation, desired bead contour, and good fusion. When welding in vertical position, current should be reduced by about 25%. When welding in overhead position, current should be reduced by about 15%. For FCAW and GMAW of cast iron, as-deposited nickel alloy filler metal compositions are similar to those used with SMAW.

Copper alloy filler metals are used for braze welding cast iron. The joint is soft and ductile when hot and yields during cooling, so that residual stress is reduced.

To perform braze welding, a brazing filler metal is deposited into the weld joint. There is no melting of the base metal adjacent to the joint. After the surface is heated, a thin layer of filler metal is added to the surface to help ensure a satisfactory bond. A welding tip with high heat output at low gas pressure should be used to provide a soft flame that will not blow the flux away from the joint. After braze welding, the metal should be cooled slowly to prevent white cast iron from forming. The deposited metal is machinable but does not provide a color match. The quality of braze welding with copper alloy filler metals depends upon the following:

- using wide grooves
- thoroughly cleaning joints of moisture, grease, oil, and dirt before welding

- using a preheat temperature between 250°F and 400°F
- using the lowest possible current for good bonding
- welding at a fast speed to minimize dilution
- preventing puddling
- cooling the part slowly after braze welding

Three common filler metals for brazing cast iron are ECuSn-A and ECuSn-C (copper-tin classification), and ECuAl-A2 (copper-aluminum classification). The main difference between ECuSn-A and ECuSn-C is the amount of tin they contain. The ECuSn-C filler metal has a higher percentage of tin (8%) than ECuSn-A filler metal (5%), thereby producing welds with greater hardness, tensile strength, and yield strength. Both are used with DCEP and normally require that the area to be brazed be preheated to 400°F.

The ECuAl-A2 filler metal has a relatively low melting point and high deposition rate at lower current, which permits fast welding. A faster welding speed minimizes distortion and the formation of white cast iron in the weld zone. The tensile strength and yield strength of the deposits is nearly double that of copper-tin deposits.

Repairing Cracked Castings. If a crack in a casting is to be welded, V the crack approximately 1/8″ to 3/16″ deep with a diamond-point chisel or with a grinder. On sections that are less than 3/16″ thick, V only one-half the thickness of the cast iron. Fine, hairline cracks in a casting can be made more visible by rubbing a piece of white chalk over the surface. The chalk leaves a visible line where the crack is located. Cracks have a tendency to extend during welding because of heat expansion. To prevent crack expansion, drill a 1/8″ hole a short distance beyond each end of the crack.

Start the weld about 1/8″ before the end of the crack and weld back to the hole, filling the hole; then move slightly beyond the hole. Next, move to the other end of the crack and repeat. Continue to alternate the weld on each end, limiting the length of each weld to 1″ to 1½″ on thin cast iron and 2″ to 3″ on thick cast iron. Allow each section of weld to cool before starting the next section and peen each short bead.

For a crack near the edge of the part, grind open the crack to allow for adequate filler metal access. Weld back from the drilled hole to the start of the crack (the edge of the part). If the crack is longer than 1″ or 2″, use skip welding, otherwise use continuous welding. **See Figure 40-9.**

Studding Broken Castings. When a casting is 1½″ thick or more and is subjected to heavy stresses, steel studs are used to strengthen the joint. Studding is not advisable on castings less than 1½″ thick because it tends to weaken rather than strengthen the joint.

To apply studs, V the crack and drill and tap ¼″ or ⅜″ holes in the casting at right angles to the sides of the V. Space the holes so the center-to-center distance is equal to three to six times the diameter of the stud. Screw the studs into the tapped holes. The threaded end of the studs should be about ⅛″ to ⅜″ in length and should project approximately ¼″ to ⅜″ above the casting. **See Figure 40-10.** Deposit beads around

the base of the studs, welding them thoroughly to the casting. Remove the slag and deposit additional layers of beads to fill the V.

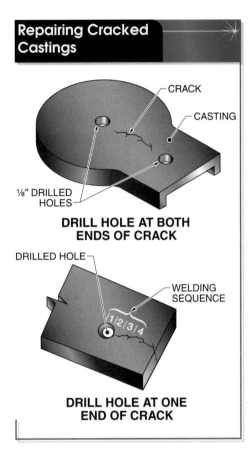

Figure 40-9. *When welding cracks in cast iron, holes are drilled to prevent cracks from extending during welding.*

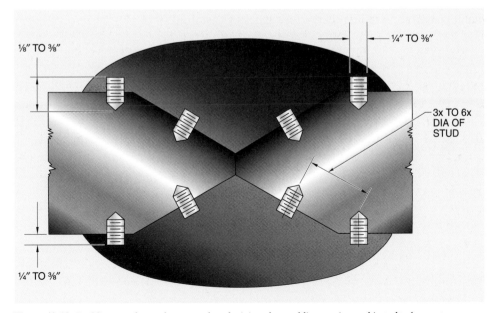

Figure 40-10. *Studding may be used to strengthen the joint when welding castings subjected to heavy stresses.*

Refer to Quick Quiz® on CD-ROM

 Points to Remember

- Tool steels are the most highly alloyed steels and in general are the hardest and strongest steels available.
- OFW should not be used for tool steels because it is too slow and introduces excessive heat into the base metal, leading to distortion, softening of hardened metal, embrittlement of annealed metal, or cracking.
- Cast irons are used for wear and corrosion resistance and for general applications where good castability is needed.
- All casting skin and foreign matter must be removed from the joint surface and adjacent areas of cast irons before welding.
- Filler metals used for welding cast irons can be composed of nickel alloy, carbon steel, or cast iron.
- Extensive heat input is required before, during, and after welding cast iron to prevent cracking.

 Questions for Study and Discussion

1. What are some basic characteristics of water hardening tool steels?
2. What alloying elements are added to carbon in hot work tool steels? To what total alloying amount?
3. How should the joints be prepared for welding cast iron?
4. What can be done to make fine cracks in castings more visible?
5. How can cracks in castings be prevented from spreading?
6. What type of filler metal is used for welding the various types of cast iron?

 Refer to Chapter 40 in the *Welding Skills Workbook* for additional exercises.

Weldability of Stainless Steels

Weldability is the ability of a metal to be welded under imposed conditions. The weldability of a stainless steel depends on its metallurgical structure. The five groups of stainless steels as defined by metallurgical structure are austenitic, martensitic, ferritic, duplex, and precipitation hardening.

Stainless steels owe their corrosion resistance to the presence of chromium. Stainless steels contain from 12% to 30% chromium plus other alloying elements, such as up to 25% nickel and up to 6.5% molybdenum. Welding conditions for stainless steels are dictated by the specific alloy family, which will influence factors such as heat input during welding, preheat, and postheating.

WELDABILITY OF STAINLESS STEELS

Stainless steels contain from 12% to 30% chromium (Cr) plus other alloying elements, such as up to 25% nickel (Ni) and up to 6.5% molybdenum (Mo). These create a variety of metallurgical structures, making stainless steels the most versatile family of metals. They are used for their heat resistance, corrosion resistance, and low-temperature toughness. The weldability of stainless steels depends on their metallurgical structure.

Wrought stainless steels consist of five groups named for their metallurgical structure: austenitic, martensitic, ferritic, duplex, and precipitation hardening. Wrought stainless steels are usually identified by a three-digit AISI designation, such as 410 or 316.

Austenitic stainless steels are the largest group and have the widest usage of all stainless steels. They exhibit excellent corrosion resistance, weld-ability, high-temperature strength, and low-temperature toughness. They cannot be hardened by quenching and are only strengthened and hardened by cold working.

Austenitic stainless steels have varying amounts of Cr and Ni. The basic austenitic stainless steel composition is 18% Cr and 8% Ni although amounts can range from 16% to 26% Cr and 3.5% to 37% Ni. The austenitic structure is achieved by the addition of nickel. Other elements that contribute to the austenitic structure are manganese and nitrogen. Carbon contributes to the austenitic structure, but it is not used in large amounts because it reduces corrosion resistance.

Compared with martensitic stainless steels, austenitic stainless steels are relatively weak. Solution annealing heat treatment is used primarily to improve corrosion resistance. Stress-relief heat treatment, conducted at lower temperatures, usually causes distortion and loss of corrosion resistance. Molybdenum is added to austenitic stainless steels to improve corrosion resistance. Standard grades of austenitic stainless steels are the 200 and

> ✓ **Point**
>
> *Stainless steels are used for their heat resistance, corrosion resistance, and low-temperature toughness.*

> ✓ **Point**
>
> *Wrought stainless steels are usually identified by a three-digit AISI designation, such as 410 or 316.*

300 series. The basic austenitic stainless steel is type 302. **See Figure 41-1.**

Martensitic stainless steels contain up to 18% chromium and up to 1.5% carbon. They are air hardened and tempered to develop high strength and wear resistance. Martensitic stainless steels contain no alloying elements other than chromium and have the lowest corrosion resistance of the stainless steels. The basic martensitic stainless steel is type 410. **See Figure 41-2.**

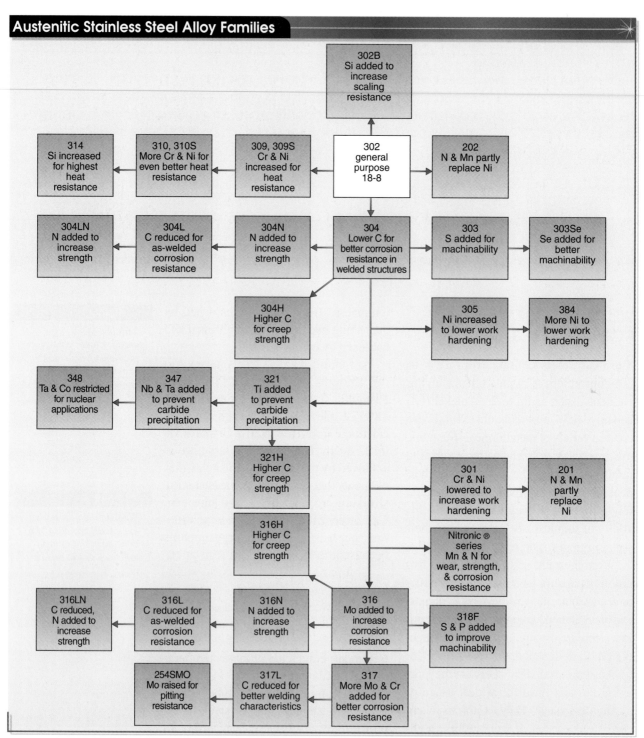

Austenitic Stainless Steel Alloy Families

302B Si added to increase scaling resistance	

314 Si increased for highest heat resistance ← **310, 310S** More Cr & Ni for even better heat resistance ← **309, 309S** Cr & Ni increased for heat resistance ← **302** general purpose 18-8 → **202** N & Mn partly replace Ni

304LN N added to increase strength ← **304L** C reduced for as-welded corrosion resistance ← **304N** N added to increase strength ← **304** Lower C for better corrosion resistance in welded structures → **303** S added for machinability → **303Se** Se added for better machinability

304H Higher C for creep strength

305 Ni increased to lower work hardening → **384** More Ni to lower work hardening

348 Ta & Co restricted for nuclear applications ← **347** Nb & Ta added to prevent carbide precipitation ← **321** Ti added to prevent carbide precipitation

321H Higher C for creep strength

301 Cr & Ni lowered to increase work hardening → **201** N & Mn partly replace Ni

316H Higher C for creep strength

Nitronic ® series Mn & N for wear, strength, & corrosion resistance

316LN C reduced, N added to increase strength ← **316L** C reduced for as-welded corrosion resistance ← **316N** N added to increase strength ← **316** Mo added to increase corrosion resistance → **318F** S & P added to improve machinability

254SMO Mo raised for pitting resistance ← **317L** C reduced for better welding characteristics ← **317** More Mo & Cr added for better corrosion resistance

Figure 41-1. *Austenitic stainless steels are nonmagnetic and are strengthened and hardened by cold work.*

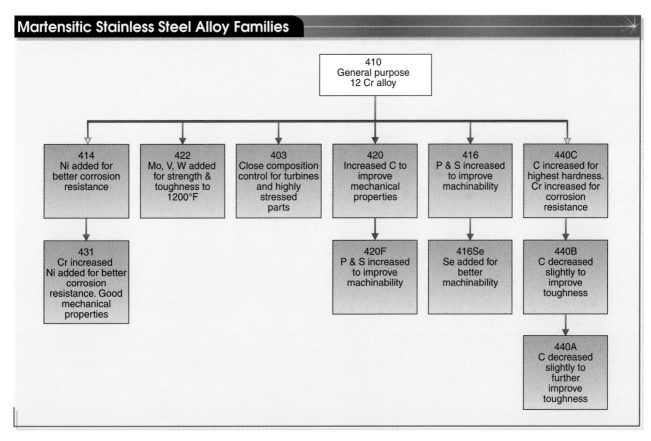

Figure 41-2. *Martensitic stainless steels may be quenched and tempered to improve their strength, and they have the lowest corrosion resistance of all stainless steels.*

Ferritic stainless steels contain more chromium than martensitic stainless steels, which improves their corrosion resistance. Ferritic stainless steels cannot be hardened by quenching and tempering. They are used chiefly for their corrosion and scaling resistance. Ferritic stainless steels are divided into regular ferritics and low-interstitial ferritics. The basic ferritic stainless steel is type 430. **See Figure 41-3.**

Duplex stainless steels are composite materials whose metallurgical structure consists of approximately equal quantities of austenite and ferrite. The properties of duplex stainless steels are achieved by maintaining a balance of the austenite and ferrite. Properly balanced duplex stainless steels possess certain desirable qualities that austenitic and ferritic stainless steels do not. For example, duplex stainless steels have higher strength and chloride stress-cracking resistance than austenitic stainless steels. They also have better fabricability and toughness than ferritic stainless steels. Heat treatment and fabrication practices for duplex stainless steels must be carefully controlled or significant loss of toughness and/or corrosion resistance may occur. The addition of between 0.15% N and 0.25% N helps ensure a balance between austenite and ferrite in duplex stainless steels, especially during welding.

The basic duplex stainless steel is type 329. However, alloying additions, particularly nitrogen, have been carefully controlled to yield a second generation of duplex stainless steels with better control of austenite to ferrite balance during welding operations. These second-generation alloys include types 2205 and 2307. **See Figure 41-4.**

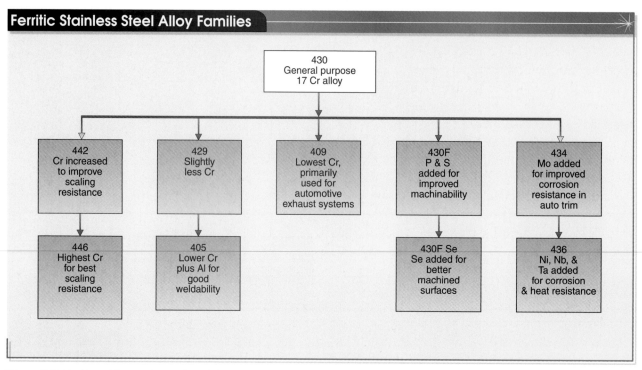

Figure 41-3. *Ferritic stainless steels are relatively low strength alloys that cannot be hardened or strengthened by heat treatment.*

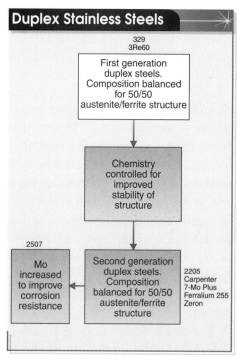

Figure 41-4. *Duplex stainless steels are stronger than ferritic or austenitic stainless steels and consist of a composite structure of austenite and ferrite.*

Precipitation hardening stainless steels are specially alloyed stainless steels that are heat-treated by precipitation hardening to much higher strengths than other stainless steels. They are relatively weak and soft when quenched from the solution annealing temperature. After machining, working, or stamping to the desired shape, they are precipitation hardened to achieve the desired strength and hardness with very little distortion or scaling. The lower the precipitation hardening temperature, the higher the strength, but strength is achieved with a loss of toughness. Higher precipitation hardening temperatures reduce strength but increase toughness. Precipitation hardening stainless steels fall into three subgroups: martensitic, semi-austenitic, and austenitic. **See Figure 41-5.**

Cast stainless steels exhibit the various metallurgical structures of their wrought stainless steel equivalents. Cast stainless steels are identified by alphanumeric designations such as CF-3M or HK-40.

Precipitation hardening stainless steels exhibit the highest strengths of all stainless steels.

Cast stainless steels may have a martensitic, ferritic, or duplex metallurgical structure. They are divided into corrosion-resistant (C series) and heat-resistant (H series). **See Figure 41-6.** Corrosion-resistant castings are designated by the uppercase letter C, followed by a letter that indicates the approximate alloy content. The higher the letter (with A considered the lowest and Z considered the highest), the greater the alloy content. Numbers and letters following a dash indicate carbon content and the presence of alloying elements.

Heat-resistant castings include a range of alloy compositions that include stainless steels and nickel alloys. Heat-resistant castings are designated by the letter H followed by a letter indicating the approximate alloy content. The higher the letter, the greater the percentage of alloying elements.

General Welding Considerations for Stainless Steels

Welding of stainless steels is influenced by physical properties, cleaning and joint preparation, and removal of heat tint. Wrought and cast stainless steels can be welded using arc welding or OFW processes. Filler metals of the same or higher alloy content as the base metal must be used to maintain corrosion resistance or mechanical properties.

Effect of Physical Properties. The coefficient of thermal expansion of martensitic and ferritic stainless steels is approximately equal to that of carbon steel. Consequently, the allowances for thermal expansion are practically the same as those for carbon steel. Austenitic stainless steels have about a 50% to 60% greater coefficient of thermal expansion than carbon steel, and are therefore more prone to distortion during welding or heat treatment.

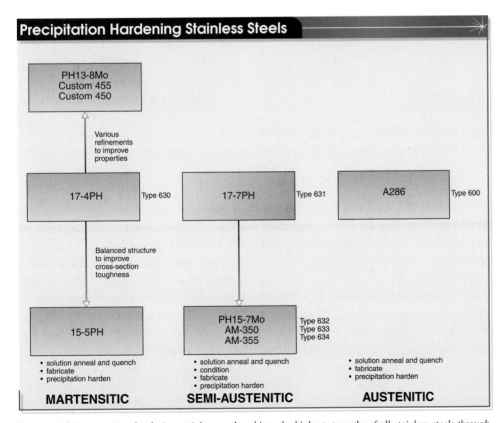

Figure 41-5. *Precipitation hardening stainless steels achieve the highest strengths of all stainless steels through heat treatment.*

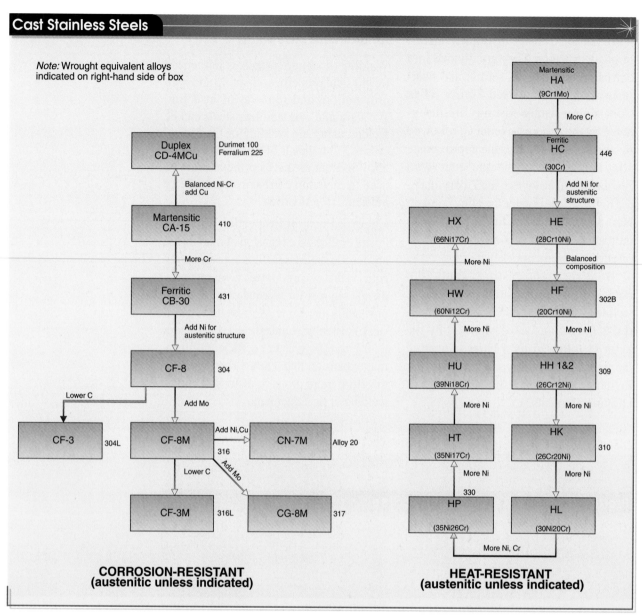

Figure 41-6. *Cast stainless steels exhibit the metallurgical structure of their wrought counterparts and are divided into corrosion-resistant and heat-resistant types.*

The thermal conductivity of ferritic and martensitic stainless steels is approximately 50% that of carbon steel, and that of austenitic stainless steels is about 33%. Consequently, heat is conducted away more slowly. As a result, stainless steels take longer to cool. This can be a particular problem when welding thin-gauge steels since there is greater danger of burning through the metal.

Unfavorable effects of heat can be reduced substantially by using chill plates. A *chill plate* is a metal plate used to prevent overheating during welding.

The use of chill plates, such as copper plates, helps conduct heat away from the weld area.

Jigs and fixtures should be used whenever possible, especially for austenitic stainless steels. When stainless steels are held in a jig or fixture during cooling, warping and distortion are practically eliminated. If a jig or fixture cannot be used, special welding procedures are necessary to counteract expansion forces. A common practice is to use intermittent welding or back-step welding.

✓ **Point**

When welding stainless steels, using chill plates made of copper helps conduct heat away from the weld area, reducing the unfavorable effects of heat on the alloy.

Cleaning and Joint Preparation. Cleaning and joint preparation are critical to ensuring a quality weld. Surface contaminants affect stainless steel welds to a greater degree than carbon and alloy steel welds. The surface of the weld area must also be completely cleaned of all hydrocarbon-containing contaminants such as oils, or chloride-containing cleaning fluids. Contamination from grease and oil must be prevented so that corrosion resistance is not reduced through carbon pickup during welding. In addition to brushing with a clean stainless steel wire brush, other acceptable methods of surface preparation include blasting with clean sand or grit, and machining or grinding with chloride-free cutting fluid.

The area to be cleaned must include the weld groove and adjacent faces for at least ½″ on each side of the groove. Cleaning a wider area is recommended for plate thicker than ⅜″. The surfaces of parts to be resistance welded, spot welded, or seam welded must also be cleaned. The degree of cleaning depends on the welding process. For example, special care is required for cleaning surfaces for gas shielded welding because of the absence of flux, which acts as a cleaning agent. Carbon contamination can adversely affect the metallurgical structure, corrosion resistance, or both. Clean stainless steel wire brushes must be used to prevent carbon and iron pickup. Thorough post-weld wire brushing is used to remove welding slag after welding.

Removal of Heat Tint (Heat Discoloration). Stainless steels obtain their corrosion resistance from a surface film composed largely of chromium and oxygen (chromium oxide). The film forms spontaneously in air or water on alloys that contain more than 10% chromium. The quality of the film must be preserved during fabrication. The physical appearance of the chromium oxide film does not necessarily indicate the overall corrosion resistance of the alloy.

When stainless steel equipment is welded, the chromium oxide film adjacent to the weld thickens from the localized heating effect and changes color due to diffraction of light. The color change is known as heat tint. The presence of heat tint often prompts questions about quality from receivers of stainless steel equipment.

Although heat tint may cause a slight overall chromium depletion in the surface film, it does not usually compromise the ability of the surface film to provide corrosion resistance, unless the stainless steel provides borderline corrosion resistance in the expected service environment. In highly corrosive service environments, it might be necessary to use a more corrosion-resistant stainless steel or nickel alloy. **See Figure 41-7.**

HEAT TINT

Figure 41-7. *Heat tint is formed on stainless steel during welding but does not usually compromise the corrosion resistance of the stainless steel.*

It may be difficult to remove heat tint, especially from inside corners. Grinding may be used, but is often impractical or expensive. Commercial stainless steel chemical cleaners are available that typically consist of a paste that is painted on the weld seam and allowed to soak for 10 min to 15 min, after which it is removed with a stainless steel wire brush.

Arc Welding Processes. Arc welding processes that can be used for stainless steels include SMAW, GTAW, GMAW, and SAW. OFW is sometimes used for welding 19-gauge and lighter stainless steel sheets. For SAW, a flux suitable for stainless steel welding must be selected.

GTAW or GMAW is used to weld stainless steel because of the ease with which welds can be made. GTAW and GMAW do not significantly reduce the corrosion-resistant properties of stainless steel. **See Figure 41-8.** GTAW is mostly used for thin sections of stainless steel. A 2% thoriated tungsten electrode is used and is ground to a taper. Argon is normally used as the shielding gas.

Miller Electric Manufacturing Company

Figure 41-8. *GTAW and GMAW welding processes are preferred for welding stainless steel because they minimize heat input and help retain the metal's corrosion-resistant properties.*

GMAW is used for thick stainless steels. Spray transfer is used for flat position welding and requires argon shielding with 2% to 5% oxygen or special mixtures. When welding thin stainless steel with GMAW, short circuiting transfer can be used, in which case 90% He/7.5% Ar/2.5% CO_2 (tri-mix) is the best choice and provides the best weld appearance. When welding extra-low-carbon grades of stainless steel, or if the stainless steel is used in a highly corrosive environment, Ar-CO_2 mixtures should not be used and the CO_2 level must be kept low enough so that the corrosion resistance of the material is not affected.

Joint Design. For thin metal, the flange-type joint is probably the most suitable design. Slightly thicker sheets up to ⅛″ may use a butt joint. For plates thicker than ⅛″, the edges should be beveled to form a V so fusion can be obtained through to the bottom of the weld.

Filler Metals. Filler metal selection for stainless steel welding depends on the base metal. Not every stainless steel has a matching weld filler metal, but there are usually several choices of filler metal for welding any particular stainless steel.

The alloy content of the filler metal should be higher than or the same as that of the base metal to compensate for expected alloy loss. A columbium-bearing filler metal must be used for both the columbium (Type 347) and the titanium (Type 321) stabilized grades of stainless steel. Chromium-nickel filler metals are often used to weld chromium-grade stainless steels because they provide a ductile weld metal. Covered filler metals for SMAW must be stored in heated ovens at 300°F because they are low-hydrogen filler metals and are susceptible to moisture pickup.

Stainless steel filler metals are identified differently from carbon steel filler metals. For example, a standard 18-8 filler metal for AC/DC current is designated as E308-16. The prefix E indicates the filler metal is an arc welding electrode. The next three digits are the AISI symbols for a particular type of metal. Thus, 308 represents a metal containing 18% chromium and 8% nickel. The last two digits following the dash may be either 15 or 16; the 1 indicates all-position welding, and the 5 or 6 specifies the type of covering and applicable welding current. The 5 designates a lime-coated filler metal. The lime type can be used only with DCEP. The 6 indicates a titanium-type covering. The titanium-type can be used with AC and DCEP. **See Figure 41-9.**

FILLER METALS FOR WELDING STAINLESS STEEL	
Stainless Steel Grade	Filler Metal
AUSTENITIC	
201, 202	308-15, 308-16
301, 302, 304, 305, 308	308-15, 308-16
309	309-15, 309-16
310, 314	310-15, 310-16
316	316-15, 316-16
321, 347, 348	347-15, 347-16
FERRITIC	
405, 409	410-15, 410-16
430, 434, 436, 442	430-15, 430-16
446	446-15, 446-16
MARTENSITIC	
408, 410, 414, 416, 420	410-15, 410-16
431	430-15, 430-16

Figure 41-9. *Electrodes for welding stainless steels are typically identified by their AISI classification.*

Selecting the proper filler metal for stainless steel is, in most cases, a more critical choice than with carbon steel because of the number of types and grades of stainless steel and the varying degrees of severity of heat, corrosion media, etc., to which the weld joint will be subjected. Selecting the right filler metal for satisfactory results requires analyzing all of the conditions that apply to a particular job. To determine the right type and size of filler metal for a given set of conditions, the following factors must be considered:

- chemical composition of the base metal to be welded
- dimensions of the section to be welded
- type of welding current required
- welding position(s) to be used
- fit-up of the section to be welded
- specific properties of the weld deposit
- specific fabrication code requirements

Filler metals must be selected carefully because of the high cost of the material to be welded. The stainless steel weld must have tensile strength, ductility, and corrosion resistance equivalent to the base metal.

Welding Current. Both AC and DC current can be used for arc welding stainless steel. DCEP produces deeper weld penetration and a more consistent fusion when used on stainless steel sheets and light plates.

Since stainless steels have a lower melting point than carbon steel, at least 20% less current is recommended than would ordinarily be used for carbon steel. The low thermal conductivity of stainless steel localizes the heat from the arc along the weld, further reducing current requirements.

Welding Technique. To produce quality welds, square butt joints should be used for stainless steel sheets 18-gauge and less and are fitted up with no gap. Heavier gauge sheets and plates are fitted up with a beveled joint edge preparation and a gap to allow penetration. Metals must be free from scale, grease, and dirt to prevent weld contamination.

To begin arc welding, the filler metal is touched to the work and quickly withdrawn a short distance (enough to maintain the proper arc). To maintain the arc, the filler metal should be fed continuously into the molten weld pool to compensate for metal deposited, and moved rapidly and continuously in the direction of welding. To finish the weld or break the arc, the filler metal should be held close to the work to shorten the arc, then moved quickly back over the finished bead. To reduce weld oxidation and porosity, the arc should be kept as short as possible during welding. Too long an arc is inefficient and increases spattering.

After welding, all slag and scale should be completely removed from the weld bead and the adjacent base metal. Scale or oxide can be removed by grinding, pickling, or sandblasting. Discoloration (heat tint) should be removed if required in the specification. Light weld discoloration may be removed electrolytically. When grinding, refinish with progressively finer grits. The smoother and cleaner the surface of any stainless part, the better the corrosion resistance.

Ferritic stainless steels can be arc welded by GTAW, GMAW, and FCAW. Welding processes that tend to increase carbon pickup are not recommended. This would include OFW, carbon arc, and GMAW with CO_2 shielding gas.

Matching ferritic stainless steel, austenitic stainless steel, and nickel alloy filler metal are used. Matching ferritic stainless steel filler metals are commonly available as type 409 and type 430, and are available as solid and flux cored electrode.

Ferritic stainless steels can be joined to themselves or to other metals using austenitic stainless steel filler metals. Type 309 and type 310 stainless steel are most often used. Nickel alloy filler metals such as ENiCrFe-3 can also be used (with SMAW). The advantage of austenitic stainless steel filler metals and nickel alloy filler metals is better as-welded toughness than matching ferritic stainless steel filler metals.

Austenitic Stainless Steels. Austenitic stainless steels are weldable except for type 303 and type 303Se, which contain sulfur and selenium, respectively, to make them free-machining. Austenitic stainless steels do not harden on quenching and do not require preheat or postheating to improve their mechanical properties. However, cold-worked austenitic stainless steel will lose strength in the HAZ when welded.

Austenitic stainless steels are susceptible to hot cracking. A completely austenitic metallurgical structure possesses insufficient ductility when solidifying from the molten state and may hot crack from an inability to accommodate shrinkage stress. Thus, filler metals used to weld austenitic stainless steels must have modified chemical compositions to produce a small amount of ferrite in the metallurgical structure, which counteracts hot cracking. The chemical composition adjustment is indicated by the ferrite number of the filler metal. Filler metal suppliers indicate a ferrite number (FN) on electrodes and wire used for welding austenitic stainless steels. A ferrite number between 2FN and 12FN is required when welding austenitic stainless steels. The lower the FN, the less the amount of ferrite. Ferrite makes a weld slightly magnetic.

Austenitic stainless steels can be welded using most arc welding processes, including GMAW, GTAW, SMAW, FCAW, PAW, and SAW. OFW is infrequently used, but may be used when arc welding equipment is not available. In general, the deposited weld metal composition should nearly match the base metal composition when welding austenitic stainless steels to themselves. Other austenitic stainless steel filler metals may be used provided the selected filler metal has suitable corrosion resistance, mechanical properties, or both. Alternate filler metals are usually more highly alloyed than the base metal to provide superior corrosion resistance.

Consumable inserts are available for welding austenitic stainless steels. They are used as preplaced filler metal in the root opening for the first weld pass, and are completely fused into the root of the joint. Consumable inserts should not be used where the presence of a crevice between the insert and the base metal creates a condition for corrosion.

Austenitic stainless steels require less heat input and less current than carbon steel because of their lower melting point and higher electrical resistivity. Their high coefficient of thermal expansion coupled with low thermal conductivity increases the possibility for distortion and warpage. When welding austenitic stainless steels less than ¼″ thick, distortion or warpage may be a serious problem. Rigid fixturing of the workpieces can help control distortion of thin sheets during welding. Metals more than ¼″ thick may require special welding techniques to counteract distortion. Back-step welding and intermittent welding help overcome the problems of distortion and warpage.

Stress relief heat treatment is not recommended for austenitic stainless steels. Stress relief heat treatment between 1200°F and 1600°F can result in significant distortion and loss of corrosion resistance from sensitization. However, a low-temperature stress relief heat treatment between 400°F and 800°F helps improve dimensional stability and helps reduce peak stresses, but does not reduce corrosion resistance. This heat treatment is sometimes performed on items that must be straight, such as shafts, by suspending them vertically at temperature for many hours.

With austenitic stainless steels, sensitization occurs between 800°F and 1500°F, and most rapidly above 1200°F. When heated to the sensitization temperature range, carbon and chromium in the alloy combine to form chromium carbide, reducing corrosion resistance.

To prevent sensitization when welding austenitic stainless steels, low-carbon stainless steel grades or dual-marked stainless steel grades are used. Low-carbon grades have the suffix letter L in their designation, for example 304L or 316L. Their carbon content is reduced to prevent chromium carbide precipitation during the temperate-time cycle of welding operations. Dual-marked stainless steels such as 304/304L or 316/316L are also low in carbon to prevent sensitization, but contain nitrogen to counteract the strength loss from the lower carbon content. Dual-marked stainless steels exhibit the superior mechanical properties of the higher carbon grade and the superior corrosion resistance of the lower carbon grade. Another method of preventing sensitization is to use austenitic stainless steels, such as 321 and 347, which contain alloying elements that counteract the formation of chromium carbide and are resistant to sensitization.

Duplex Stainless Steels. Welding of duplex stainless steels can upset the balance of austenite and ferrite, leading to loss of corrosion resistance and loss of toughness in the HAZ. The welding procedure must not allow an imbalance by increasing the ferrite content of the HAZ. To prevent problems, weld with low heat input and control the cooling rate. Low heat input minimizes dwell time in the "red heat" temperature zone. Controlling the cooling rate prevents the formation of excessive ferrite (with rapid cooling) or excessive austenite (with slow cooling). The maximum interpass temperature should be 240°F for thin metal and 300°F for thick metal, to promote the proper cooling rate. Preheat and postheating are not usually performed on duplex stainless steels.

Duplex stainless steels can be welded using any arc welding process. Duplex stainless steels should always be welded with filler metal added. Without filler metal, the weld and the HAZ contain too much ferrite and the joint properties are inadequate. Autogenous welds should not be used. Matching filler metal is usually recommended. In some cases filler metals with more chromium and molybdenum than the base metal may be used to enhance corrosion resistance. For each welding job and type of duplex stainless steel, it is necessary to develop the appropriate welding procedure and technique.

Precipitation Hardening Stainless Steels. Martensitic and semi-austenitic precipitation hardening stainless steels are weldable without preheat. Austenitic precipitation hardening stainless steels are susceptible to hot cracking and have poor weldability. If maximum strength is required in martensitic and semi-austenitic precipitation hardening stainless steels, matching filler metal is required and the complete heat treatment cycle must be repeated. If the complete cycle cannot be repeated, the parts should be solution annealed before welding and precipitation hardened after. For martensitic precipitation hardening stainless steels, a repeat of the precipitation hardening after welding may be adequate to restore mechanical properties.

If full strength is not required, ductile 309 or 310 austenitic stainless steel or nickel alloy filler metal may be used on martensitic precipitation hardening stainless steels. Follow manufacturer instructions when welding precipitation hardening stainless steels. Nickel alloy or austenitic stainless steel filler metals are normally used for welding austenitic precipitation hardening stainless steels.

Cast Stainless Steels. Weldability of cast stainless steels varies according to their metallurgical structure; guidelines for the corresponding wrought alloys should be followed. Cast stainless steels are usu-

ally welded to repair casting defects or damage from service. Heat-resistant cast stainless steels that have been in service at elevated temperatures tend to lose ductility and may crack during welding. High-temperature solution annealing heat treatment at 2000°F should be performed prior to welding to restore as-cast ductility. Solution annealing dissolves alloy carbides that precipitate during service at elevated temperature and cause the reduction in ductility.

Regular carbon grades of corrosion-resistant cast stainless steels that require remedial welding to repair casting defects must be solution annealed after welding to fully restore corrosion resistance. Regular carbon grades that include CF-8 or CF-8M sensitize on welding. Solution annealing of finish-machined castings may lead to distortion so that remedial welding must be carried out before final machining.

Cast stainless steels are more prone to hot cracking than cast steel so the weld bevel angle should be wider than that used for cast steels. It is common to have a bevel angle up to 90° for cast stainless steels versus the 45° that is common for cast steels. Low heat input also helps reduce hot cracking. Because of the tendency of cast stainless steels toward hot cracking, it may be necessary to butter the weld bevel for certain types of repairs.

The Lincoln Electric Company

Stainless steel weldability varies depending on the metallurgical structure. Service requirements of the stainless steel must be considered when developing welding procedures.

Refer to Quick Quiz® on CD-ROM

Points to Remember

- Stainless steels are used for their heat resistance, corrosion resistance, and low-temperature toughness.
- Wrought stainless steels are usually identified by a three-digit AISI designation, such as 410 or 316.
- When welding stainless steels, using chill plates made of copper helps conduct heat away from the weld area, reducing the unfavorable effects of heat on the alloy.
- Heat tint may cause a slight overall chromium depletion in the surface film, but it does not usually compromise the ability of the surface film to provide corrosion resistance.
- When welding stainless steel, use a short arc with only a slight weaving motion.
- In uphill welding of stainless steel, avoid any whipping action of the filler metal. Instead, use a motion in the form of a V.

Questions for Study and Discussion

1. How are stainless steels classified?
2. What are the qualities of stainless steel that make this metal so valuable?
3. Why are chill plates frequently used when welding stainless steel?
4. Why is less current required in welding stainless steel?

 Refer to Chapter 41 in the *Welding Skills Workbook* for additional exercises.

Weldability of Nonferrous Metals

Chapter

Weldability is the ability of a metal to be welded under imposed conditions. Nonferrous metals include nickel alloys, copper alloys, aluminum alloys, magnesium alloys, and titanium alloys.

Nickel is incorporated as a major or minor constituent in approximately 3000 alloys. Nickel alloys are generally easy to weld provided the joint is clean. Copper alloys are wrought or cast and consist of several families of alloys. Copper alloys are difficult to weld because of their high thermal conductivity, high coefficient of thermal expansion, hot cracking susceptibility, and high fluidity.

Aluminum alloys are generally easy to weld. Magnesium alloys result in metals with high strength-to-weight ratios that are easy to weld. Titanium alloys are difficult to weld because of the need for high purity.

WELDABILITY OF NICKEL ALLOYS

Nickel is incorporated as a major or minor constituent in approximately 3000 alloys. The principal alloying elements added to nickel are copper, iron, molybdenum, chromium, and cobalt. **See Figure 42-1.** The major nickel alloy systems are based on nickel and nickel-chromium. Nickel alloys have an austenitic structure and behave similarly to austenitic stainless steels in many ways. Nickel alloys are strengthened by cold working or precipitation hardening and may be heat-treated to improve corrosion resistance.

General Welding Considerations for Nickel Alloys

Weldability factors for nickel alloys include joint cleanliness, distortion, heat requirements, welding processes, and filler metals.

Joint Cleanliness. Joint cleanliness is the single most important requirement for welding nickel alloys. Nickel alloys are extremely sensitive to cracking from contamination. Sulfur, present in grease and oil, is particularly harmful to nickel. Oxides can inhibit wetting, prevent fusion of the base metal and filler metal, and can cause subsurface inclusions and poor bead contour. A region approximately 1″ on both sides of the joint should be thoroughly degreased to prevent contamination by sulfur, and mechanically cleaned to remove oxides before welding. Mechanical cleaning is accomplished by grinding, abrasive blasting, or machining and pickling.

Distortion. Distortion is possible in nickel alloys because of their relatively low thermal conductivity. Low thermal conductivity causes heat to be retained in the weld rather than be dissipated into the base metal. The exception is

> **Point**
>
> *Nickel alloys are extremely sensitive to cracking from contamination, and the joint must be thoroughly cleaned before welding.*

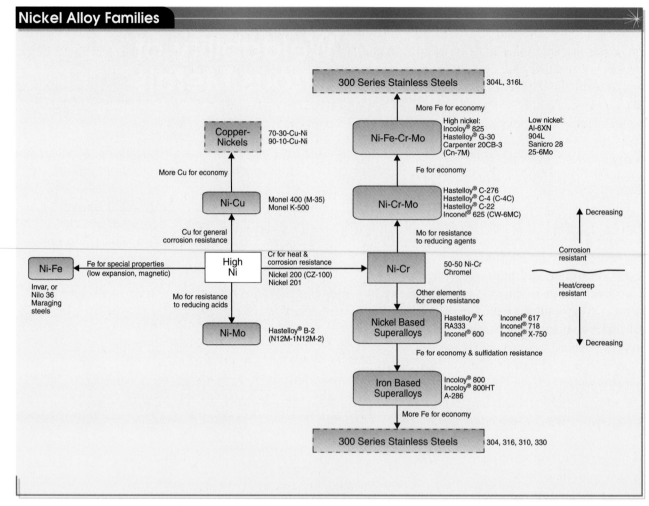

Figure 42-1. *Nickel is incorporated as a major or minor constituent in approximately 3000 alloys.*

commercially pure nickel, which has relatively high thermal conductivity. However, since the coefficients of thermal expansion of nickel alloys are similar to those of carbon steels and low-alloy steels, the welding of nickel alloys does not present significant distortion problems. Nickel-iron alloys are an exception because they have very low coefficients of thermal expansion.

Heat Requirements. Since nickel alloys are sensitive to high heat input, low heat input should be used when welding nickel alloys. High heat input can lead to hot cracking, loss of corrosion resistance, or both. If hot cracking is anticipated, such as might occur in a highly restrained joint, the welding technique is modified to decrease heat input, or a welding process with lower heat input is substituted. Preheat is not required for nickel alloys; however, the joint area is heated to about 60°F to eliminate moisture condensation that could lead to porosity. The interpass temperature should be low to minimize total heat input. Cooling methods that reduce the interpass temperature should not introduce contaminants that cause weld metal cracking.

Postheating is not required to restore mechanical properties, except for precipitation hardening alloys. Alloys that sensitize when welded may require postheating consisting of solution annealing and quenching to restore corrosion resistance.

Welding Processes. Almost all arc welding processes can be used for welding nickel alloys; however, SMAW, GTAW, and GMAW are the most common processes used. Not all arc welding processes are applicable to every alloy because of metallurgical characteristics and/or availability of suitable filler metals. OFW should only be used when arc welding equipment is not available. The welding of nickel alloys is similar to the welding of austenitic stainless steels except that cleanliness requirements are more stringent and groove openings are increased to allow for the lower penetration of nickel alloys.

Filler Metals. Filler metals for welding nickel alloys should have a chemical composition that is similar to the base metal. Covered filler metals for SMAW normally contain additions of deoxidizing elements such as titanium, manganese, and columbium to prevent weld metal cracking.

Precipitation hardenable filler metal will respond to the precipitation hardening treatment used for the base metal. However, the response is usually less pronounced and the weld joint is generally lower in strength than the base metal after the precipitation hardening treatment.

Fluxes are available for SAW for many nickel alloys. The flux composition must be suited to both the filler metal and the base metal. An improper flux can cause slag adherence, inclusions, poor weld bead contour, and undesirable changes in weld metal composition.

Welding Monel® and Inconel®

Monel® and Inconel® are trademarks for two groups of nickel alloys. When used without qualification, they refer to alloy 400 (Monel® alloy 400) or alloy 600 (Inconel® alloy 600). Monel® and Inconel® can be satisfactorily welded using SMAW. Welding of Monel® and Inconel® is performed almost as easily as welding low-carbon steel. Although Monel® and Inconel® can be welded in any position, better results are obtained

if welded in flat position. In general, SMAW should not be used on sheet less than .050″ (18-gauge) thick; GTAW is best used on thin gauges. No preheat is necessary to weld Monel® or Inconel®. The procedure for welding Monel® or Inconel® is as follows:

1. Remove the thin, dark-colored oxide film from around the area to be welded. The oxide can be removed by grinding, sandblasting, rubbing with emery cloth, or pickling.
2. Use a heavily coated filler metal specially designed for welding Monel® and Inconel®. Use DCEP current.
3. Hold the filler metal at a travel angle of about 20° from the vertical and ahead of the weld pool when welding in flat position, as it is easier to control the molten flux and to estimate slag trappings. To make welds in other positions, hold the filler metal at approximately a right angle to the workpiece.
4. Withdraw the filler metal slowly from the crater to permit a blanket of flame to cover the crater, protecting it from oxidation while the metal solidifies.
5. Use a minimum of weaving to prevent depositing wide weld beads.

WELDABILITY OF COPPER ALLOYS

Copper can be combined with many elements to produce various alloys. Copper alloys can be strengthened by cold working or precipitation hardening and generally possess good thermal and electrical conductivity, which affect their weldability.

Copper alloys are wrought or cast and consist of commercially pure coppers, modified coppers, beryllium coppers, brasses, tin bronzes, aluminum bronzes, copper-nickels, and nickel-silvers. **See Figure 42-2.** Many copper alloys have leaded equivalents, which contain a small amount of lead to improve their machinability.

✓ **Point**

Remove the oxide film from the surface to be welded and use heavily coated filler metal specially designed for Monel® and Inconel®.

✓ **Point**

Commercially pure coppers are wrought or cast and are used primarily for their high electrical conductivity.

Commercially pure coppers are wrought or cast. Wrought commercially pure coppers contain at least 99.9% copper. They are used primarily for their high electrical conductivity. Cast commercially pure coppers have lower electrical and thermal conductivity than equivalent wrought alloys because the elements that must be added to ensure a sound casting, such as silicon, decrease conductivity. Commercially pure coppers are soft, weak, and very ductile. They include oxygen-free coppers, deoxidized coppers, and tough pitch coppers.

Beryllium coppers are wrought and cast copper alloys that contain small amounts of beryllium. Beryllium coppers are precipitation hardened to extremely high levels of tensile and fatigue strength, comparable to low-alloy steels. Small amounts of cobalt or nickel may be added to refine the grain size.

Brasses are wrought alloys of copper and zinc, with 5% to 50% zinc content. Some brasses also contain other elements. Brasses are the most popular and least expensive of the copper alloys. They display a wide range of mechanical properties, are easy to work, have a pleasing color, and exhibit good corrosion resistance. Brasses consist of alpha and beta brasses, tin brasses, and leaded brasses.

Casting brasses contain specific alloying elements to improve their castability and strength beyond that of regular wrought brasses. They consist of combinations of tin, lead, iron, manganese, aluminum, and nickel. Casting brasses can be poured into complex shapes with low porosity and good mechanical properties.

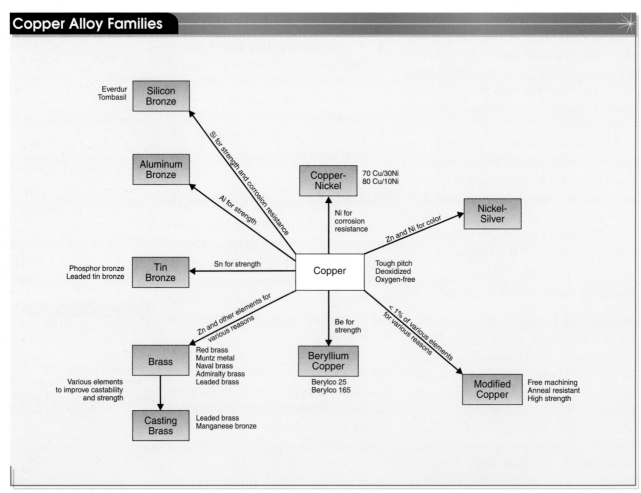

Copper Alloy Families

Figure 42-2. *Many copper alloys have leaded equivalents that contain a small amount of lead to improve their machinability.*

Tin bronzes (phosphor bronzes) are wrought and cast alloys of copper and tin. Tin bronzes contain from 1.25% to 10% tin, plus lead, zinc, nickel, and phosphorus. Phosphorus can be added as a deoxidizer in castings to improve soundness and cleanliness. Tin bronzes have high strength, good toughness, high corrosion resistance, and a low coefficient of friction, making them suitable for bearings operating under high loads.

Aluminum bronzes are wrought and cast alloys of copper that contain between 7% and 13.5% aluminum, plus small amounts of manganese, nickel, and iron. Aluminum bronzes have good strength and excellent corrosion and wear resistance. Nickel is added to aluminum bronzes to further improve corrosion resistance. Aluminum bronzes are used for bushings and corrosion-resistant parts.

Silicon bronzes are wrought and cast alloys of copper that contain between 1% and 5% silicon and additions of manganese, iron, and zinc. Some wrought and cast silicon bronzes have leaded equivalents. Silicon bronzes have high strength similar to carbon steel, good toughness, and excellent corrosion resistance. They are used for bearings, and pump and valve components.

Copper-nickels are wrought and cast alloys of copper containing up to 30% nickel, plus minor additions of iron, chromium, tin, or beryllium. Iron is added for increased resistance to erosion-corrosion in water. *Erosion-corrosion* is the detrimental effect of velocity or turbulence in a corrosive environment. Copper-nickels have moderate strength and better corrosion resistance than other copper alloys. They are used for seawater components.

> *Copper is resistant to oxidation, moisture, and some organic chemicals, making it useful for electrical conductors, water tubing, heat exchangers, and chemical equipment.*

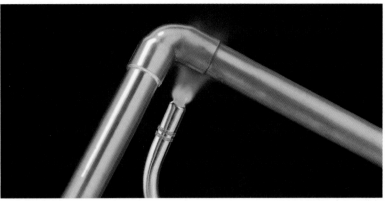

Thermadyne Industries, Inc.

Copper tubing has high thermal conductivity and is commonly joined by sweat soldering.

Nickel-silvers are wrought and cast alloys of copper that contain between 5% and 45% zinc, and from 5% to 30% nickel. Nickel has a strong decolorizing effect on copper-zinc alloys (brasses). With greater than 20% nickel, the color turns to silver-white and the alloy takes on a brilliant polish. Nickel-silvers are used for valve trim, zippers, and camera parts.

General Welding Considerations for Copper Alloys

Copper and copper alloys are difficult to weld because of their high thermal conductivity, high coefficient of thermal expansion, hot cracking susceptibility, and high fluidity. *Fluidity* is a measure of the viscosity or flowability of a liquid or molten solid. Leaded copper and copper alloys should not be welded because the lead creates porosity and promotes cracking within a weld.

Welding Processes. Most arc welding processes as well as OFW can be used to weld copper alloys. The low heat input of the oxyacetylene flame makes OFW a relatively slow process compared with arc welding. Coppers and certain high-copper alloys are very difficult to resistance spot and seam weld because of their high electrical and thermal conductivities. Copper alloys can be readily joined by brazing and soldering.

Filler Metals. Copper alloys are generally welded with matching filler metals. Filler metals that can be used to weld copper alloys include covered and bare filler metal. These may be used to weld copper alloys to themselves or to other metals. Many of these filler metals meet AWS classifications. Silver alloys and copper-phosphorus filler metals are most commonly used for brazing copper alloys. **See Figure 42-3.**

Specific Welding Considerations for Copper Alloys

Silicon bronzes have relatively low thermal conductivity and only require preheat when joint thickness is more than 2″. Aluminum bronzes are susceptible to hot cracking, especially with less than 7% aluminum. Alloys with higher aluminum content are weldable with adequate preheat. Copper-nickels have thermal and electrical conductivities similar to carbon steel and are relatively easy to weld. Cleanliness is essential and preheat is not required. Nickel-silvers are similar to brasses because of their high zinc content and should be brazed rather than welded.

Commercially Pure Coppers. Commercially pure coppers require preheat from 250°F to 1000°F, depending on joint thickness. High-strength anneal-resistant coppers are welded with less preheat than is required for other commercially pure coppers to preserve their strength.

FILLER METALS FOR WELDING COPPER ALLOYS			
Base Metal	**Name**	**Covered***	**Bare†**
Copper	Copper	ECu	ERCu
Silicon Bronzes, Brasses	Silicon Bronze	ECuSi	ERCuSi-A
Phosphor Bronzes, Brasses	Phosphor Bronze	ECuSn-A	ERCuSn-A
Phosphor Bronzes, Brasses	Phosphor Bronze	ECuSn-C	ERCuSn-A
Copper-Nickels	Copper-Nickel	ECuNi	ERCuNi
Aluminum Bronzes, Brasses, Silicon Bronzes, Manganese Bronzes	Aluminum Bronze	ECuAl-A2	ERCuAl-A1 ERCuAl-A2
Aluminum Bronzes	Aluminum Bronze	ECuAl-B	ERCuAl-A3
Nickel-Aluminum Bronzes	—	ECuNiAl	ERCuNiAl
Manganese-Nickel-Aluminum Bronzes	—	ECuMnNiAl	ERCuMnNiAl
Brasses, Copper	Naval Brass	—	RBCuZn-A
Brasses, Manganese Bronzes	Low-fuming Brass	—	RBCuZn-B
Brasses, Manganese Bronzes	Low-fuming Brass	—	RBCuZn-C
Nickel-Silvers	—	—	RBCuZN-D

* ANSI/AWS A5.6, *Specification for Covered Copper and Copper Alloy Arc Welding Electrodes*
† ANSI/AWS A5.7, *Specification for Copper and Copper Alloy Bare Welding Rods and Electrodes.* ANSI/AWS A5.8, *Specification for Filler Metals for Brazing and Braze Welding*

Figure 42-3. *Filler metals that can be used to weld copper alloys include covered and bare filler metal and bare rods. Many of these filler metals meet AWS classifications.*

Oxygen-free coppers are welded as rapidly as possible to minimize oxygen pickup. Deoxidized coppers are the most commonly used type of copper for fabrication by welding. Deoxidized copper is susceptible to oxygen pickup and requires silicon-containing filler metal to minimize the effects of oxygen pickup. Since copper has a very high coefficient of expansion, precautions must be taken to prevent contraction of the joint. Jigs and fixtures must be used to prevent movement during cooling. However, even when jigs are used, contraction forces can cause cracking during cooling.

Special coated metal arc filler metals have been developed to weld sheet copper. The most common are phosphor bronze (ECuSn-A) and aluminum bronze (ECuAl-A). The joint design used for deoxidized coppers must include a relatively large root opening and groove angle. Tight joints should be avoided to prevent buckling, poor penetration, slag inclusions, undercutting, and porosity. Copper backing strips are often advisable.

Tough pitch coppers contain a uniform distribution of copper oxide, which is insufficient to affect ductility, but can cause problems when welding. When heated above 1680°F for prolonged periods, the copper oxide tends to migrate to the grain boundaries, leading to a reduction in strength and ductility. Additionally, the copper absorbs carbon monoxide and hydrogen, which react with the copper oxide and release carbon dioxide and water vapor. Carbon dioxide and water vapor are not soluble in copper and exert pressure between the grains, producing internal cracking and embrittlement.

Tough pitch coppers are not recommended for gas welding because gas welding causes embrittlement; brazing or soldering should be used. However, some welds can be made with SMAW in situations where tensile strength requirements are extremely low (19,000 psi or less), provided a high welding current and high travel speed are used. The high current and travel speed do not allow embrittlement to develop.

Beryllium Coppers. Beryllium coppers form an oxide film that inhibits wetting and fusion during welding. An absolutely clean joint surface is required and may be achieved by abrading the surface. Beryllium coppers are welded in the soft annealed condition and then precipitation hardened to achieve the required strength.

Brasses. Since the application of heat tends to vaporize zinc, arc welding on brass is difficult. When zinc volatilizes, the zinc fumes and oxides often obscure vision and make welding hard to perform. Furthermore, the formation of oxides produces a dirty surface that ruins the wetting properties of the molten metal. To arc weld brasses, use heavily coated phosphor-bronze filler metals and make small deposits of metal. Preheat should be eliminated and a lower welding current used.

Zinc vapors can be minimized by decreasing or eliminating preheat, or by using lower welding currents. High-zinc brasses have lower thermal conductivity and require less preheat than low-zinc brasses.

Tin Bronzes. Since the thermal conductivity of tin bronze is similar to that of steel, it can be easily welded. When using SMAW, a heavily coated phosphor-bronze filler metal should be used, with DCEP current. The metal must be absolutely clean to ensure a sound weld.

Tin bronzes are very susceptible to hot cracking. To prevent hot cracking, tin bronzes should be preheated to between 300°F and 400°F. High welding currents and high travel speeds are used and each weld pass is peened.

> Some copper alloys include volatile and toxic elements that are released to the atmosphere during welding, so it is imperative to use an effective ventilation system to collect and dispose of noxious fumes, powders, and dust.

⚠ WARNING

When welding brass, ensure proper ventilation of the work area to remove harmful zinc oxide fumes.

⚠ CAUTION

To keep airborne concentrations of beryllium within allowable limits, proper safety precautions must be taken when melting, welding, flame cutting, polishing, buffing, grinding, and machining beryllium coppers.

WELDABILITY OF ALUMINUM ALLOYS

Aluminum alloys can be wrought or cast. Various elements are added to aluminum to produce alloys with specific properties. Aluminum alloys can be strengthened by work hardening or precipitation hardening. The weldability of aluminum alloys is influenced by cleanliness requirements, heat requirements, and desired appearance.

Aluminum alloys have low density, good corrosion resistance, and good weldability. Cold-worked alloys suffer a loss of strength in the HAZ during welding. Precipitation hardened alloys must be heat-treated after welding to restore their strength.

Aluminum alloys consist of various families (series) of wrought or cast alloys. Each series is identified by a sequence of numbers. For example, wrought aluminum manganese alloys are identified by

the 3XXX series, such as alloy 3003. Cast aluminum-silicon alloys are identified by the 3XX.X series, for example alloy 356.0. **See Figure 42-4.**

Temper designations are alphanumeric notations that indicate the final condition of cold-worked (H) or heat-treated (T) metal. A number following the letter indicates the condition. Temper designation is separated from the alloy identification number by a hyphen. For example, 3003-H2 designates quarter hard aluminum manganese alloy.

General Welding Considerations for Aluminum Alloys

General welding considerations for all aluminum alloys include appearance, cleaning requirements, heat requirements, welding processes, and filler metals.

Aluminum Alloy Families

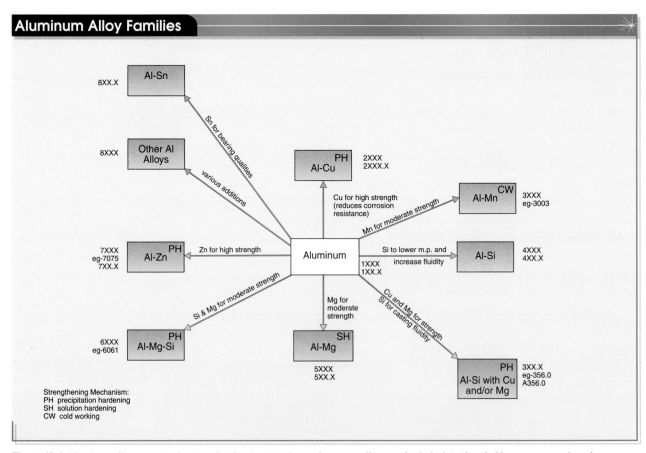

Figure 42-4. *Aluminum alloys consist of various families (series) of wrought or cast alloys, each of which is identified by a sequence of numbers.*

Appearance. The appearance of aluminum after welding is often of great importance. GMAW and GTAW can provide the best as-welded bead appearance. Welded parts may be given a chemical or electrochemical (anodic) surface treatment to provide corrosion resistance, coloring, or both. All flux must be removed from brazed, soldered, or welded joints prior to surface treatment. Filler metals that contain a large amount of silicon darken during anodic treatment.

Cleaning Requirements. Cleaning requirements for aluminum alloys are stringent because, during welding, the naturally formed aluminum oxide surface film thickens and becomes a hindrance. The surface film must be removed before fusion or resistance welding, and must be prevented from re-forming by means of an inert gas shield or by pressure between the joint surfaces. If the surface film is not removed, small particles of unmelted oxide will be trapped in the weld, causing a reduction in ductility, lack of fusion, and weld cracking.

The aluminum oxide surface film may be removed electrically, mechanically, or chemically. Electrical cleaning occurs during welding. The surface film is blasted away by cathodic bombardment during the positive half-cycle of the sine wave, making electrical cleaning a good method of in situ cleaning.

Mechanical cleaning is usually done immediately before welding by scraping the surface using a clean stainless steel wire brush with light pressure to prevent burnishing or contaminating the surface. Chemical cleaning requires a chemical solution to dissolve the surface film. Chemical attack of the metal must be prevented during cleaning by minimizing exposure time in the solution and, after welding, by immediate removal of residual flux.

Heat Requirements. The high thermal conductivity and high thermal expansion coefficient of aluminum influence its weldability. Aluminum alloys conduct heat three to five times faster than steel so that more heat input is required than for steel, even though the melting point of aluminum is significantly lower than that of steel. Preheat is often required for thick joints, but must not exceed 400°F to prevent detrimental effects to the weld joint. High-speed welding processes with high heat input, such as GMAW, are favorable for aluminum welding. The high thermal conductivity of aluminum is beneficial in all-position welding because the rapid cooling of the weld, coupled with its surface tension when molten, results in rapid solidification.

Distortion in aluminum alloys during welding is about twice as great as when welding steel. The amount of distortion is inversely proportional to the speed of welding. Additionally, a volume shrinkage of about 6%, which occurs during solidification, increases the chance of hot cracking in fully restrained joints. Fixtures for welding aluminum alloys must be designed to accommodate both expansion and contraction, and yet maintain the proper geometric position for welding.

Welding Processes. The welding processes commonly used for aluminum are GTAW, GMAW, and resistance welding. GTAW is used for thin joints. AC current is generally used because it provides a cleaning action on the positive half-cycle of the sine wave. Argon is commonly used when welding aluminum and is used at a low flow rate. Helium increases penetration, but a higher flow rate is required. Filler metal must be clean and free of oxide; otherwise, the weld will be porous.

GMAW is applied to thick joints and is much faster than GTAW. Pure argon is normally used for shielding. The filler metal must be kept clean to prevent porosity. All aluminum alloys may be resistance welded.

Filler Metals. Filler metals for welding aluminum alloys are classified by the same four-digit system used to designate wrought and cast aluminum alloys. Filler metals for joining aluminum alloys fall into the 1XXX, 2XXX, 3XXX, 4XXX, or 5XXX groups. The 1XXX and 4XXX groups are the only two recommended for oxyacetylene welding. Filler metal selection for welding aluminum depends on a number of factors, including base metal composition, strength requirements, ductility requirements, color match after anodizing, corrosion resistance, and cracking tendency. Generally, one type of filler metal usually satisfies several requirements for a specific alloy. **See Figure 42-5.**

FILLER METALS FOR WELDING ALUMINUM ALLOYS...									
Base Metal	Filler Metal Types[a,b,c]								
	201.0, 206.0, 224.0	319.0, 333.0, 354.0, 355.0, C355.0	356.0, A356.0, 357.0, A357.0, 413.0, 443.0, A440.0	511.0, 512.0, 513.0, 514.0, 535.0	7004, 7005, 7039, 701.0, 712.0	6009, 6010, 6070	6005, 6061, 6063, 6101, 6151, 6201, 6351, 6951	5456	5454
1060, 1070, 1080, 1350	ER4145	ER4145	ER4043[d,e]	ER5356[e,f,g]	ER5356[e,f,g]	ER4045[d,e]	ER4043[e]	ER5356[g]	ER4043[e,g]
1100, 3003, Alc. 3003	ER4145	ER4145	ER4043[d,e]	ER5356[e,f,g]	ER5356[e,f,g]	ER4043[d,e]	ER4043[e]	ER5356[g]	ER4043[e,g]
2014, 2036	ER4145[h]	ER4145[h]	ER4145	—	—	ER4145	ER4145	——	——
2219	ER2319[d]	ER4145[h]	ER4145[e,f]	ER4043[e]	ER4043[e]	ER4043[d,e]	ER4043[d,e]	——	ER4043[e]
3004, Alc. 3004	—	ER4043[e]	ER4043[e]	ER5356[i]	ER5356[i]	ER4043[e]	ER4043[e,i]	ER5356[g]	ER5356[i]
5005, 5050	—	ER4043[e]	ER4043[e]	ER5356[i]	ER5356[i]	ER4043[e]	ER4043[e,i]	ER5356[g]	ER5356[i]
5052, 5652	—	ER4043[e]	ER4043[e,i]	ER5356[i]	ER5356[i]	ER4043[e]	ER5356[f,i]	ER5356[i]	ER5356[i]
5083	—	—	ER5356[e,f,g]	ER5356[g]	ER5183[g]	—	ER5356[g]	ER5183[g]	ER5356[g]
5086	—	—	ER5356[e,f,g]	ER5356[g]	ER5356[g]	—	ER5356[g]	ER5356[g]	ER5356[g]
5154, 5254	—	—	ER4043[e,i]	ER5356[i]	ER5356[i]	—	ER5356[i]	ER5356[i]	ER5356[i]
5454	—	ER4043[e]	ER4043[e,i]	ER5356[i]	ER5356[i]	ER4043[e]	ER5356[f,i]	ER5356[i]	ER5554[h,i]
5456	—	—	ER5356[e,f,g]	ER5356[g]	ER5556[g]	—	ER5356[g]	ER5556[g]	—
6005, 6061, 6101, 6151, 6201, 6351, 6951	ER4145	ER4145[e,f]	ER4043[e,i,j]	ER5356[i]	ER5356[e,f,i]	ER4043[d,e,j]	ER4043[e,i,j]	—	—
6009, 6010, 6070	ER4145	ER4145[e,f]	ER4043[d,e,j]	ER4043[e]	ER4043[e]	ER4043[e,i,j]	—	—	—
7004, 7005, 7039, 710.0 712.0	—	ER4043[e]	ER4043[e,i]	ER5356[i]	ER5356[g]	—	—	—	—
511.0, 512.0, 513.0, 514.0, 535.0	—	—	ER4043[e,i]	ER5356[i]	—	—	—	—	—
356.0, A356.0, 357.0, A357.0, 413.0, 443.0 A444.0	ER4145	ER4145[e,f]	ER4043[e,k]	—	—	—	—	—	—
319.0, 333.0, 354.0, 355.0, C355.0	ER4145[h]	ER4145[e,f,k]	—	—	—	—	—	—	—
201.0, 206.0, 224.0	ER2319[d,k]	—	—	—	—	—	—	—	—

a. Service conditions may limit the choice of filler metals. Filler metals ER5183, ER5356, ER5556, and ER5654 are not recommended for sustained elevated-temperature service.
b. For gas shielded arc welding processes only. For OAW ER1188, ER1100, ER4043, ER4047, and ER4145 filler metals are used.
c. Where no filler metal is listed, the base metal combination is not recommended for welding.
d. ER4145 may be used for some applications.
e. ER5183, ER5356, ER5554, ER5556, and ER5654 may be used. They may provide improved color match after anodizing treatment, highest weld ductility, and higher weld strength. ER5554 is suitable for sustained elevated-temperature service.
f. ER4043 may be used for some applications.
g. ER5183, ER5356, or ER5556 may be used.
h. ER2319 may be used for some applications to supply high strength when the weldment is postweld solution heat-treated and aged.
i. ER4047 may be used for some applications.
j. ER4643 provides high strength in ½″ and thicker groove welds in 6XXX alloys when postweld solution heat-treated and aged.
k. Filler metal of the same composition as the base metal may be used.

Figure 42-5...

...FILLER METALS FOR WELDING ALUMINUM ALLOYS										
	Filler Metal Types[a,b,c]									
Base Metal	5154, 5254	5086	5083	5052, 5652	5005, 5050	3004, Alc. 3004	2219	2014, 2036	1100, 3003, Alc. 3003	1060, 1070, 1080, 1350
1060, 1070, 1080, 1350	ER5356[e,f,g]	ER5356[g]	ER5356[g]	ER4043[e,g]	ER1100[e,f]	ER4043[e,g]	ER4145[e,f]	ER4145	ER1100[e,f]	ER1188[e,f,k]
1100, 3003, Alc. 3003	ER5356[e,f,g]	ER5356[g]	ER5356[g]	ER4043[e,g]	ER1100[e,f]	ER4043[e,g]	ER4145[e,f]	ER4145	ER1100[e,f]	—
2014, 2036	—	—	—	—	ER4145	ER4145	ER4145[h]	ER4145[h]	—	—
2219	ER4043[e]	—	—	ER4043[e,g]	ER4043[d,e]	ER4043[d,e]	ER2319[d]	—	—	—
3004, Alc. 3004	ER5356[i]	ER5356[g]	ER5356[g]	ER5356[e,f,i]	ER5356[f,i]	ER5356[f,i]	—	—	—	—
5005, 5050	ER5356[i]	ER5356[g]	ER5356[g]	ER5356[e,f,g]	ER5356[f,i]	—	—	—	—	—
5052, 5652	ER5356[i]	ER5356[g]	ER5356[g]	ER5654[f,i]	—	—	—	—	—	—
5083	ER5356[g]	ER5356[g]	ER5183[g]	—	—	—	—	—	—	—
5086	ER5356[g]	ER5356[g]	—	—	—	—	—	—	—	—
5154, 5254	ER5654[i]	—	—	—	—	—	—	—	—	—

. . . **Figure 42-5.** *Filler metals for welding aluminum alloys are selected for the type of base metal to be welded.*

WELDABILITY OF MAGNESIUM ALLOYS

Magnesium is one of the lightest commercial metals. Magnesium is alloyed with many chemical elements to create products with a high strength-to-weight ratio. Some magnesium alloys have strength-to-weight ratios comparable to some aluminum alloys and high-strength steels, making them suitable for high-strength applications where low weight is advantageous. Some wrought magnesium alloys are strengthened by cold working or precipitation hardening, and some cast magnesium alloys are strengthened by precipitation hardening.

Magnesium alloys are grouped broadly according to their cost. The lower cost group of magnesium alloys contain from 2% to 10% aluminum, plus minor amounts of manganese, silicon, and zinc. The second group contains manganese, zinc, rare earth elements, and thorium, plus small amounts of zirconium to refine the grain size. The second group has better properties at higher temperatures, is more difficult to produce, and is much more expensive. Magnesium

alloys are identified by a four-part numbering system indicating chemical composition and temper designation. **See Figure 42-6.**

The temper designation of magnesium alloys is included in the alloy designation and is similar to the codes used to describe aluminum alloys. For example, T6 describes a temper which is solution treated and artificially aged (precipitation hardened).

Care must be taken when preparing or repairing magnesium. Magnesium can heat to a combustion point and will ignite.

General Welding Considerations for Magnesium Alloys

The best weldability is achieved with magnesium alloys that contain aluminum and zinc, rare earth elements, or thorium. These alloys are represented by the AM, AZ, ZE, EZ, HK, HM, and HZ series. Alloys with zinc as the major alloying element are more difficult to weld. These alloys are represented by the ZE, ZH, and ZK series.

✓ **Point**

The best weldability is achieved with magnesium alloys that contain aluminum and zinc, rare earth elements, or thorium.

ASTM MAGNESIUM ALLOY AND TEMPER DESIGNATION

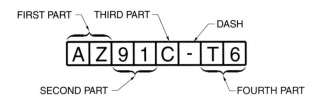

First Part	Second Part	Third Part	Fourth Part
Indicates the two principal alloying elements	Indicates the amounts of the two principal alloying elements	Distinguishes between different alloys with the same percentages of the two principal alloying elements	Indicates temper condition
A-Aluminum E-Rare Earth Elements H-Thorium K-Zirconium M-Manganese Q-Silver S-Silicon T-Tin Z-Zinc	Whole numbers	Letters of alphabet except I and O	F-As-fabricated O-Annealed H10 and H11-Slightly strain hardened H23, H24, and H26-Strain hardened and partially annealed T4-Solution heat-treated T5-Precipitation hardened only T6-Solution heat-treated and precipitation hardened T8-Solution heat-treated, cold worked, and precipitation hardened
Two main alloying elements in order of decreasing percentage or alphabetically if percentages are equal	Percentages of the two main alloying elements and arranged in same order as alloy designations in first part	Letter of the alphabet assigned in order as compositions become standard	Letter followed by a number

Figure 42-6. *The designations for magnesium alloys consist of a four-part numbering system.*

Weldability factors that must be considered before welding magnesium include surface preparation, heat requirements, welding processes, and filler metals.

Surface Preparation. Surface preparation is required to remove the oxide film before welding magnesium. The surface film thickens as the temperature increases, becoming a hindrance to welding. The surface must be thoroughly degreased to remove surface preservatives and then chemically or mechanically cleaned to remove the oxide film. Various chemical cleaning solutions can be used. For critical work, chemical cleaning is followed by mechanical cleaning with a clean stainless steel wire brush, using light pressure to prevent gouging.

Heat Requirements. Heat requirements for welding magnesium alloys are dictated by their high thermal conductivity and high coefficient of thermal expansion. Because of these factors, thick workpieces and highly restrained joints generally require preheat to prevent weld cracking.

Welding Processes. GTAW and GMAW are commonly used for welding magnesium. GTAW is generally used for thin sections and GMAW for medium to thick sections. Argon is the most common shielding gas, but argon-helium mixtures are also used. Most wrought alloys can be readily resistance spot welded. In resistance spot welding, the metal is molten for a very short time and the cooling rate is very high, so there is little time for harmful metallurgical changes to occur.

Filler Metals. Filler metals with a lower melting point and a larger freezing range than the base metal provide good weldability and minimize weld cracking in magnesium alloys. **See Figure 42-7.**

WELDABILITY OF TITANIUM ALLOYS

Titanium alloys vary from low-strength to high-strength, depending on their metallurgical structure. High-strength titanium alloys have a high strength-to-weight ratio. Strict attention to cleanliness and gas atmosphere is required when welding titanium.

Titanium alloys are grouped into alpha, alpha-beta, and beta alloys according to their metallurgical structure. Alpha alloys are generally the lowest strength. Alpha-beta alloys have higher strength than alpha alloys and are annealed or precipitation hardened. Beta alloys develop extremely high strengths through cold working or precipitation hardening.

Alpha titanium alloys consist of three groups: commercially pure titanium, alpha alloys, and near-alpha alloys. Commercially pure titanium contains very small amounts of interstitial elements.

Magnesium Alloy	Filler Metal Types				
	Base Metal	ERAZ61A	ERAZ92A	EREZ33A	ERAZ101A
FILLER METALS FOR WELDING MAGNESIUM ALLOYS*					
WROUGHT MAGNESIUM ALLOYS					
AZ10A		X	X		
AZ31B		X	X		
AZ61A		X	X		
AZ80A		X	X		
ZK21A		X	X		
HK31A				X	
HM21A				X	
HM31A				X	
M1A	X				
CAST MAGNESIUM ALLOYS					
AM100A	X		X		X
AZ63A	X		X		X
AZ81A	X		X		X
AZ91A	X		X		X
AZ92A	X		X		X
EK41A	X			X	
EZ33A	X			X	
HK31A	X			X	
HZ32A	X			X	
K1A	X			X	
QH21A	X			X	
ZE41A	X			X	
ZH62A	X			X	
ZK51A	X			X	
ZK61A	X			X	

* ANSI/AWSA 5.19, *Specification for Magnesium Alloy Welding Electrodes and Rods*

Figure 42-7. *Filler metals with a lower melting point and larger freezing range than the base metal provide good weldability for magnesium alloys.*

An *interstitial element* is a chemical element added in small amounts, whose atomic size is significantly less than the major elements present in the metal. The primary difference between the various grades of commercially pure titanium is the interstitial element content. Alloys with higher purity (grades 1 and 2) have lower strength and lower hardness than alloys of lower purity (grades 3 and 4). Alpha and near-alpha alloys have improved strength over commercially pure titanium and have high-temperature strength. **See Figure 42-8.**

Alpha-beta titanium alloys can be strengthened by solution treatment and precipitation hardening to achieve high strengths. Beta titanium alloys are heat-treated to high strength levels by solution treatment and precipitation hardening. Beta alloys also have exceptional work hardening characteristics.

General Welding Considerations for Titanium Alloys

General weldability considerations for titanium alloys are cleaning and shielding, welding processes, and brazing. Cleanliness is the single most important requirement for welding titanium alloys, including effective inert gas shielding to ensure that no atmospheric contaminants enter the material during welding and during cooling from the welding temperature. When using a coated electrode to weld titanium, welding can be performed in a normal atmosphere.

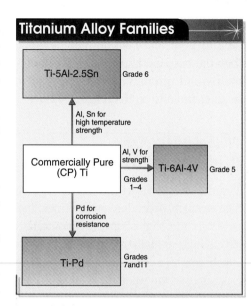

Titanium Alloy Families

Figure 42-8. *Titanium alloys provide a combination of light weight and relatively high strength.*

Titanium is the fourth most abundant metal, but the difficulty in extracting it results in increased cost.

Cleaning and Shielding Requirements. Cleaning and shielding requirements before and during welding are of paramount importance when welding titanium alloys. Contamination by impurities such as oxygen or nitrogen must be carefully controlled to prevent brittle welds. Oil, fingerprints, grease, paint, and other foreign matter should be removed using a suitable solvent cleaning method. Chloride-containing solvents leave residues that can cause cracking. Hydrocarbon residues can result in oil contamination and embrittlement. Only stainless steel wire brushes should be used to remove residues.

Welding Processes. Welding processes used to weld titanium alloys are GTAW, GMAW, electron beam welding, laser beam welding, or resistance welding. Preheat is not required for titanium alloys.

Welding must be performed with an inert shielding gas such as argon to prevent oxygen and nitrogen pickup. The argon shield must be maintained on all metal surfaces above a temperature of 1000°F. The shielding gas used must be free of harmful contaminants and

The Lincoln Electric Company

GMAW is commonly used for welding aluminum because welding can be performed rapidly, keeping heat input low.

must completely envelop both sides of the metal, both during welding and as the weld cools. The metal as it cools from welding temperature must also be protected by a trailing shield. **See Figure 42-9.**

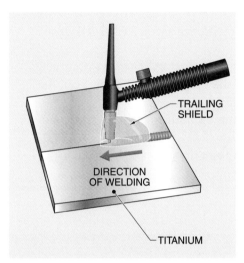

Figure 42-9. *The metal cooling from welding temperature must also be protected by a trailing shield.*

Brazing. Brazing may be performed on titanium alloys. Brazing has very little effect on the properties of alpha alloys. The mechanical properties of alpha-beta alloys can be severely reduced by brazing. The brazing temperature must be below 1650°F to prevent reduction in mechanical properties. Beta alloys are unaffected when used in the annealed condition.

Specific Welding Considerations for Titanium Alloys

Most titanium alloys do not require heat treatment after welding to restore mechanical properties. Specific welding considerations depend on the alloy group.

> *When burning titanium, wear a dark lens helmet to protect from the brightness of the material.*

Alpha Titanium Alloys. Alpha titanium alloys have good weldability because they are ductile. Welding or brazing operations have little effect on the mechanical properties of annealed material. Commercially pure titanium is usually welded with a filler metal one grade below that of the base metal because welding operations lead to slight pickup of oxygen and nitrogen. For example, grade 2 titanium (0.25% O) is welded with grade 1 titanium filler metal (0.18% O).

Alpha-Beta Titanium Alloys. Alpha-beta titanium alloys may undergo harmful strength, ductility, and toughness changes when welded. Ti-6Al-4V has the best weldability of the alpha-beta alloys and can be welded in either the annealed condition or the partially precipitation hardened condition. Precipitation hardening may be completed during postweld stress-relief heat treatment. Alpha-beta titanium alloys may suffer significant mechanical property loss during welding. Alpha-beta titanium alloys may be welded with commercially pure titanium filler metals to increase joint ductility.

Beta Titanium Alloys. Beta titanium alloys are weldable in either the annealed or the heat-treated condition. Weld joints have good ductility but relatively low strengths as welded. Beta titanium alloys are welded with matching filler metals. They are not usually heat-treated after welding, because even though filler metals match the base metals in chemical composition, their response to heat treatment is different.

Refer to Quick Quiz® on CD-ROM

Points to Remember

- Nickel alloys are extremely sensitive to cracking from contamination, and the joint must be thoroughly cleaned before welding.
- Remove the oxide film from the surface to be welded and use heavily coated filler metals specially designed for Monel® and Inconel®.
- Commercially pure coppers are wrought or cast and are used primarily for their high electrical conductivity.
- Use heavily coated phosphor-bronze filler metals when welding tin bronze and make small deposits of beads at a time.
- The surface film must be removed from aluminum alloys before fusion or resistance welding, and must be prevented from re-forming by means of an inert gas shield or by pressure between the joint surfaces.
- The best weldability is achieved with magnesium alloys that contain aluminum and zinc, rare earth elements, or thorium.
- Stainless steel wire brushes should be used to remove residues from titanium alloys.

Questions for Study and Discussion

1. What is the copper content of commercially pure coppers?
2. Brass is an alloy consisting of what elements?
3. What are the principal alloying ingredients in bronze?
4. What is fluidity?
5. What are some of the outstanding properties of aluminum?
6. Why should the surface film be removed before welding on aluminum?
7. Which alloying elements provide the best weldability when added to magnesium?
8. Why must titanium alloys be absolutely clean before and during welding?
9. Why does distortion occur in nickel alloys?
10. What must be done to the surface of Monel® and Inconel® before welding?
11. When welding Monel® or Inconel®, what type of current should be used?

 Refer to Chapter 42 in the *Welding Skills Workbook* for additional exercises.

Dissimilar Metal Welding

Chapter

W hen welding dissimilar metals, the properties of the two base metals being joined and the weld filler metal joining them must be considered. The properties may be mechanical, physical, or chemical. A successful dissimilar metal joint must meet or exceed the minimum property requirements of the base metals being joined and not fail prematurely in service. A dissimilar metal joint is more complex than a joint between the same metals and requires metallurgical study and the development of applicable procedures.

DISSIMILAR METAL WELDING

Dissimilar metal welding is the application of a welding process to join two different base metals. **See Figure 43-1.** The differences between the base metals may be significant if the metals belong to two different alloy families, such as brass and carbon steel. Or, the differences may be minor if the base metals belong to the same alloy family, but they may differ in chemical composition, such as 304 and 310 stainless steel.

Most base metal combinations may be joined by adhesive bonding, brazing, soldering, or welding. Adhesive bonding is compatible with many materials and avoids the undesirable effects of heat or dilution, but may not provide sufficient strength. Brazing and soldering are effective methods of joining many combinations of metals, unless the braze filler metal or solder causes embrittlement of one or both of the base metals. Brazing and soldering are done below the melting temperatures of the base metals being joined, avoiding the undesirable effects of dilution.

Welding is limited to certain combination of metals. The problem with welding dissimilar metals is related to the transition zone between the weld and the dissimilar metals, where undesirable brittle compounds can form that reduce ductility. Therefore, limited combinations of metals can be joined by welding. Proper filler metal selection is also vital. Arc welding, such as SMAW and GTAW are most typically used for dissimilar metal welding. GMAW is sometimes used and OFW is typically not recommended.

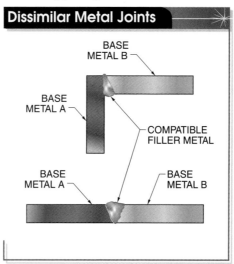

Figure 43-1. *Dissimilar metal welding is the joining of two different base metals, which requires the careful selection of a compatible filler metal.*

Dilution

Dilution is the change of the chemical composition of filler metal at the weld interface when mixed with the base metal or previously deposited layers of weld metal. For a proper weld, the filler metal must alloy readily with the base metals and be capable of being diluted by the base metals without the formation of brittle intermetallic phases. Since it changes the final alloy composition of the weld metal, dilution affects its properties and characteristics and is a particularly important consideration for dissimilar metal welding.

Greater heat input increases the degree of dilution. As different welding processes employ different amounts of heat, dilution is a significant characteristic of a welding process. Each process has a dilution rate, which estimates the percentage of base metal included in the final weld composition. For dissimilar metal welds, the two base metals are considered to contribute equally to fulfilling this percentage. These percentages are then used to calculate the alloy composition of the weld metal.

For example, the dilution rate for SMAW in the horizontal position is 30%. That is, 70% of the completed weld bead is supplied by the filler metal and 15% is supplied by each of the two base metals. **See Figure 43-2.** If alloy 400 (67% nickel, 32% copper) and 304 stainless steel (8% nickel, 18% chromium, 74% iron) are welded with ENiCrFe-2 (70% nickel, 15% chromium, 8% iron) filler metal, the alloy composition of the diluted weld metal can be calculated. The filler metal composition is multiplied by 70%, each base metal composition is multiplied by 15%, and the corresponding element amounts are added.

> *Percentage dilution of the weld is affected by the welding process. In SMAW, it is about 30%, meaning that 70% of the composition of the weld is provided by the filler metal and 15% by each base metal. Heat input also affects dilution. Since weld metal composition is affected by dilution, it is important to select the correct filler metal to avoid brittle compositions.*

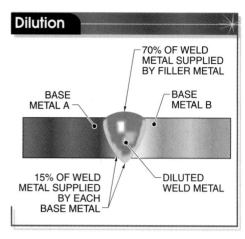

Figure 43-2. *Dilution is the change in the chemical composition of a weld metal from the pure filler metal due to contributions of the base metals.*

Contribution to weld metal by ENiCrFe-2 filler metal:

70% × 70% nickel = 49% nickel

70% × 15% chromium = 10.5% chromium

70% × 8% iron = 5.6% iron

Contribution to weld metal by alloy 400 base metal:

15% × 67% nickel = 10% nickel

15% × 32% copper = 4.8% copper

Contribution to weld metal by 304 stainless steel base metal:

15% × 8% nickel = 1.2% nickel

15% × 18% chromium = 2.7% chromium

15% × 74% iron = 11.1% iron

The calculated composition of the weld metal is obtained by adding the filler metal contribution to the base metal contributions.

49% + 10% + 1.2% = 60.2% nickel

10.5% + 2.7% = 13.2% chromium

5.6% +11.1% = 16.7% iron

4.8% copper

The composition does not equal exactly 100% because minor percentages of chemical elements in the base metal and filler metal are not included in the calculation. In a multiple-pass weld, the root bead is diluted equally by the base metals being welded. Subsequent passes are diluted partially by the base metal and partially by the previous weld bead.

Buttering

Because of its unique challenges, dissimilar metal welding may require special techniques to successfully fuse the two base metals. Buttering is a surfacing weld technique that applies weld metal on one or more joint surfaces to provide compatible base metal for subsequent completion of the weld. Buttering overlays one base metal surface with a third material, which has properties between those of the two base metals. **See Figure 43-3.** The intermediate properties reduce the gaps between the properties of the metals to be fused together.

Buttering is particularly important for welding base metals with very different coefficients of thermal expansion or melting temperatures. For example, a particular surfacing metal for buttering may have a melting temperature in between the melting temperatures of the base metals. The buttered surface is then ground flat to prepare it for welding to the other base metal. This effectively provides a new base metal with properties closer to the other base metal, making the welding easier and more likely to be successful.

Buttering is also used to reduce the need for preheating and postheating when welding two components, such as an alloy steel fixture and a low-carbon steel part. The alloy steel fixture is buttered with austenitic stainless steel filler metal and postheated to restore toughness. The buttered alloy steel may then be welded to the low-carbon steel without further preheat or postheating. The desired chemical composition must be obtained in the buttered deposit so there that are no problems in completing the weld to the second base metal.

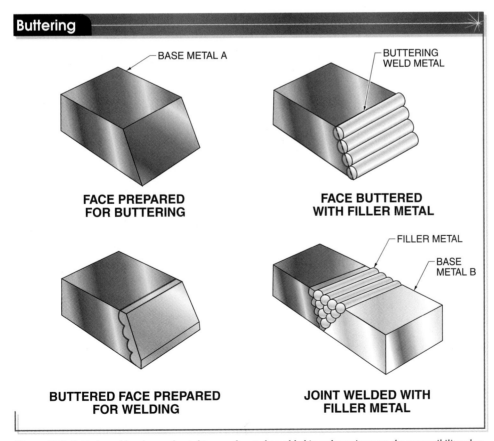

Buttering

BASE METAL A

FACE PREPARED FOR BUTTERING

BUTTERING WELD METAL

FACE BUTTERED WITH FILLER METAL

BUTTERED FACE PREPARED FOR WELDING

FILLER METAL

BASE METAL B

JOINT WELDED WITH FILLER METAL

Figure 43-3. *Buttering adds a layer of metal to a surface to be welded in order to improve the compatibility when welding two dissimilar metals.*

✓ Point

Strength, ductility, and toughness of the welded joint influences filler metal selection in dissimilar metal welding.

WELDABILITY FACTORS

The weldability of dissimilar base metal combinations depends on the characteristics of the individual base metals. These characteristics also affect the choice of filler metal. In some cases, multiple filler metals may be needed to weld the joint in stages. Factors affecting the weldability of base metals and the selection of a filler metal include solubility, mechanical properties, thermal expansion, thermal conductivity, melting temperature, corrosion resistance, and heat treatment.

Mechanical Properties

Dissimilar metals to be welded may have very different mechanical properties, such as strength, ductility, and toughness. The choice of filler metal is critical to preserving the desired mechanical properties of the base metals in the welded assembly, especially if used at high temperatures or in temperature cycling (thermal fatigue) services. The filler metal must alloy with both base metals to produce a weld that is ductile and at least as strong as the weaker of the base metals so that the joint will not fail in service.

✓ Point

Thermal expansion and thermal conductivity of the base metal influences welding and service behavior, particularly at high temperature or under conditions of thermal cycling.

Solubility

Solubility is the mutual intermixing of atoms of one metal or alloy with another without the formation of separate phases. Solubility between base metals creates a ductile joint and a proper weld. For example, copper and nickel are mutually soluble base metals.

Insolubility between base metals creates intermetallic phases between the metals and results in a brittle joint. An *intermetallic phase* is a chemical compound formed between metallic chemical elements that is nonmetallic (brittle). To avoid intermetallic phases, filler metal must be selected that permits mutual solubility. If a suitable filler metal is not available, the base metals cannot be joined by welding. For example, zinc (such as on the surface of galvanized steel) and stainless steels are mutually insoluble base metals. **See Figure 43-4.**

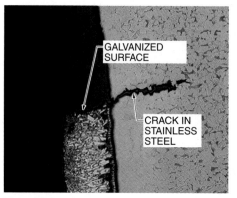

GALVANIZED SURFACE

CRACK IN STAINLESS STEEL

Figure 43-4. *When there is mutual insolubility between base metals, the joint is brittle and may fail.*

Thermal Expansion

Thermal expansion characteristics affect whether the base metals can withstand the stresses of the welding process and any subsequent thermal changes. Many dissimilar metal welding problems are due to high temperatures or excessive temperature cycling. If there is a difference between the coefficients of thermal expansion of the base metals, the metal with the lower coefficient is subject to tensile stress when heated, while the metal with the higher coefficient is subject to compressive stress. A greater difference in coefficients results in greater stresses, which may cause a hot crack during welding if there is excessive joint restraint. The joint may also cold crack after welding or later in service if the joint is not stress relieved.

When metals with different thermal expansion rates are welded together and used in high-temperature service, a filler metal with a coefficient of thermal expansion between the two base metals is selected. Alternatively, one of the base metals can first be buttered with a filler metal with an intermediate coefficient of thermal expansion.

For example, chrome-moly (2¼Cr-½Mo) steel and 304H stainless steel have very different coefficients of thermal expansion. These materials may be joined by welding for certain high-temperature applications, such as piping for steam service. At ambient temperature, with little or no thermal stress present, a stainless steel filler metal having a high coefficient of thermal expansion is acceptable for the combination. However, at high-operating temperatures, the different rates of thermal expansion cause a line of high stress on the chrome-moly steel side of the weld. Since chrome-moly steel is weaker than stainless steel at elevated temperatures, a failure is likely to occur on the chrome-moly steel side of the joint. One possible remedy is to use a nickel alloy filler metal with a coefficient of expansion between carbon steel and stainless steel, which will reduce stress on the chrome-moly steel side of the joint.

Thermal Conductivity

Thermal conductivity affects the conduction of heat from the weld pool by the base metals. High conductivity increases the energy input required to melt the base metal adjacent to the weld. To weld two base metals with a significant difference in thermal conductivity (such as carbon steel and copper), the welding heat source must be directed toward the metal having the higher thermal conductivity. This helps balance the heat between the two base metals. Preheating the metal with the higher thermal conductivity may also help balance the heat. For example, for welding copper to carbon steel, the copper should be preheated to between 400°F and 1000°F.

Melting Temperature

A significant difference between the melting temperatures of the base metals complicates welding. As the heat is applied, the metal with the lower melting temperature becomes liquid while the metal with the higher melting temperature is still solid. Increasing the heat in order to melt the metal with the higher melting temperature may cause the other metal to become so fluid that it flows out of the weld area.

If the welder still manages to weld the metals, the metal with the lower melting temperature may rupture while cooling. During cooling, the metal with the higher melting temperature solidifies first, and while the metal with the lower melting temperature is still molten, it is mechanically weak and susceptible to hot cracking from the thermal stress.

Buttering may be required to compensate for the melting temperature difference. The base metal with the higher melting temperature is surfaced with a metal with an intermediate melting temperature. Also, the melting temperature of the filler metal must be between the melting temperatures of the base metals.

For example, when welding 304 stainless steel to 90-10 copper-nickel (Cu-Ni), the large difference in the melting temperature may lead to hot cracking on the copper-nickel side of the joint as the stainless steel cools. **See Figure 43-5.** This may be prevented by buttering the copper-nickel with 70-30 Cu-Ni, which has a melting temperature between those of the base metals.

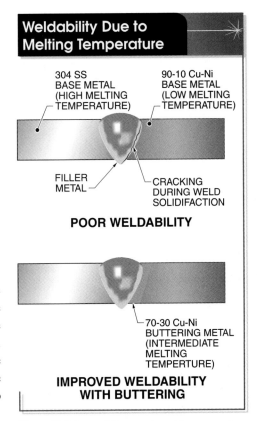

Figure 43-5. *Large differences in melting temperature between base metals can be accommodated by buttering one base metal with a metal that has an intermediate melting temperature.*

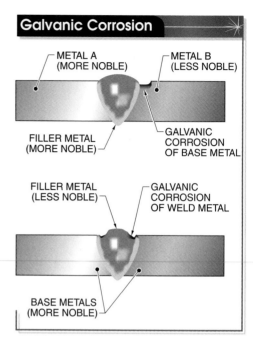

Corrosion Resistance

The filler metal must provide the corrosion resistance required when diluted with the base metals. Factors that influence corrosion of dissimilar metal joints are weld segregation and galvanic corrosion.

Weld Segregation. Welds exhibit a coarse as-cast structure. The composition of the solidified filler metal is usually segregated, or nonuniform. Specific chemical elements necessary for corrosion resistance may be locally higher or lower than the nominal composition, leading to microscopic regions with diminished corrosion resistance.

For example, molybdenum is an alloying element in corrosion-resistant stainless steels, and stainless steels like 904L and AL6X-N® contain more molybdenum than conventional stainless steels for enhanced corrosion resistance. However, molybdenum in stainless steel filler metal may segregate in the weld. To maintain corrosion resistance in the weld region, alloys 904L and AL6X-N may be welded with a higher molybdenum nickel alloy, like ERNiCrMo-10, to compensate for molybdenum segregation. Another method of improving the corrosion resistance of the filler metal is to cap it with a more corrosion-resistant surface layer.

Galvanic Corrosion. When dissimilar metals are welded and exposed to a corrosive environment, the potential for galvanic corrosion must be considered. *Galvanic corrosion* is the acceleration of corrosion that occurs due to an electrochemical reaction between two dissimilar metals. **See Figure 43-6.** A small electrical current forms between the metals when in the presence of an electrolyte, such as seawater. The subsequent movement of ions causes corrosion.

Figure 43-6. *Galvanic corrosion attacks the less noble of joined metals, especially in corrosive environments.*

Different metal combinations have different susceptibilities to galvanic corrosion. Each metal has a nobility property that indicates its electrochemical potential. The least noble of a pair of dissimilar metals is the metal that is likely to corrode due to galvanic corrosion. Also, a greater potential difference between the two increases the tendency towards galvanic corrosion. A galvanic series lists many common metals in order of nobility for a particular electrolyte, such as seawater. **See Figure 43-7.** For example, if the pair consisted of aluminum and steel, the aluminum would likely corrode, but if the pair were steel and copper, the steel would likely corrode, and more severely.

The galvanic corrosion effect may be localized (confined to the immediate area of the joint) in less conductive corrosive environments. This may even be more threatening to structural integrity than widespread corrosion.

Other factors that influence the likelihood for galvanic corrosion include temperature, relative surface areas, polarization behavior, amount of contact and exposure to electrolyte, and the electrolyte's electrical conductivity.

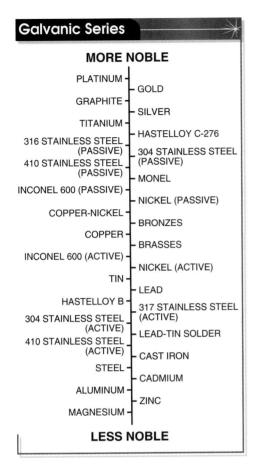

MORE NOBLE

PLATINUM
GRAPHITE
TITANIUM
316 STAINLESS STEEL (PASSIVE)
410 STAINLESS STEEL (PASSIVE)
INCONEL 600 (PASSIVE)
COPPER-NICKEL
COPPER
INCONEL 600 (ACTIVE)
TIN
HASTELLOY B
304 STAINLESS STEEL (ACTIVE)
410 STAINLESS STEEL (ACTIVE)
STEEL
ALUMINUM
MAGNESIUM

GOLD
SILVER
HASTELLOY C-276
304 STAINLESS STEEL (PASSIVE)
MONEL
NICKEL (PASSIVE)
BRONZES
BRASSES
NICKEL (ACTIVE)
LEAD
317 STAINLESS STEEL (ACTIVE)
LEAD-TIN SOLDER
CAST IRON
CADMIUM
ZINC

LESS NOBLE

Figure 43-7. *A galvanic series lists metals in order of nobility. The less noble of the pair of metals is susceptible to galvanic corrosion.*

The relative surface area of the metals affects galvanic corrosion. For example, a smaller area for the less noble metal increases the potential for corrosion concentrating in that area. Since welds joining dissimilar metal have small areas compared with the metals they join, the relative-area effect may be very important.

Polarization behavior describes the potential change that occurs when two dissimilar metals are electrically connected in a specific solution. Polarization is an important indicator of whether a more noble metal in the galvanic series will actually cause corrosion of a less noble metal. For example, stainless steels are readily polarized and therefore do not exert a strong galvanic corrosion effect on less noble metals.

Heat Treatment

Preheat and postheat processes are often part of a welding procedure, such as for stress relieving the weld and base metal. However, when welding dissimilar metals, the preheat and postheat requirements of the metals may conflict. For example, when alloy steel is joined to stainless steel, postheat may be required to stress relieve the alloy steel component of the joint, but this heat treatment may sensitize the stainless steel component, reducing its corrosion resistance. In this case, it may be necessary to butter the alloy steel, stress relieve the combination, and then weld the buttered alloy steel component to the stainless steel component.

WELDABILITY OF SPECIFIC DISSIMILAR METAL COMBINATIONS

Many dissimilar metals combinations can be successfully welded provided their metallurgical, mechanical, and physical property differences are recognized and accounted for. These properties will affect the selection of compatible filler metal, the welding technique used, and the preheat/postheat procedures. It may be necessary to butter one of the metals first.

Common dissimilar metal combinations that can be successfully welded include stainless steels to carbon steel; nickel alloys to carbon steel, austenitic stainless steel, and copper alloys; copper alloys to carbon steel; and aluminum alloys to carbon steel.

Stainless Steels

Martensitic, ferritic, and austenitic stainless steels are welded to carbon steel using nickel alloy or stainless steel filler metals. **See Figure 43-8.** Filler metal selection is primarily influenced by coefficient of thermal expansion differences, hot cracking tendency, and preheat or postheat requirements.

Martensitic Stainless Steel to Carbon Steel. Martensitic stainless steels are hardened by heat treatment, so the weld may require preheat to prevent cracking and postheat to temper the stainless steel and stress relieve the joint. Alternatively, the stainless steel can be buttered with a nickel alloy filler metal, like ERNiCr-3 or ENiCrFe-2, heat treated, and then joined to the carbon steel component without the need for preheat or postheat.

Ferritic Stainless Steel to Carbon Steel. Ferritic stainless steels cannot be hardened by heat treatment. They can be joined by austenitic stainless steel, ferritic stainless steel, or nickel alloy filler metals. However, they are subject to loss of toughness from grain growth when heated to the welding temperature. Preheating to between 400°F and 600°F should be used for highly restrained joints to prevent cracking. Welding is followed by postheating at 1400°F to 1450°F for four hours, followed by furnace cooling to 1100°F, and then air-cooling.

Austenitic Stainless Steel to Carbon Steel. Austenitic stainless steel has twice the coefficient of thermal expansion of carbon steel, so a weld joint may fail when exposed to repetitive thermal cycling. Carbon steel cannot be used as a filler metal because dilution creates a hard and brittle joint. These welds must use austenitic stainless steel filler metals with the appropriate coefficient of thermal expansion and ferrite content.

The deposited filler metal must contain a small amount of ferrite within the austenite matrix to prevent hot cracking as the weld metal solidifies and is subject to thermal stresses. The required amount of ferrite is determined from the composition of the stainless steel base metal and the amount of dilution.

Various diagrams have been developed to determine the amount of ferrite required from the composition of the base metals. These diagrams are applicable to different composition ranges of stainless steel. Some, like the DeLong diagram, specify the amount of required ferrite as a percentage. Alternatively, the Welding Research Council (WRC) uses ferrite numbers. A *ferrite number (FN)* is a standard value assigned to indicate a specific ferrite content. The ferrite number of deposited weld metal is measured with a Severn gage.

These diagrams plot nickel equivalents on the y-axis against chromium equivalents on the x-axis. **See Figure 43-9.** The nickel equivalent is the weighted sum of elements in a stainless steel that, like nickel, promote austenite formation. These elements include nickel, carbon, nitrogen, and manganese. The nickel equivalent is calculated as $\%Ni + (30 \times \%C) + (30 \times \%N) + (0.5 \times \%Mn)$. The chromium equivalent is the weighted sum of the elements in a stainless steel that promote ferrite or martensite formation. These elements include chromium, molybdenum, silicon, and niobium. The chromium equivalent is calculated as $\%Cr + \%Mo + (1.5 \times \%Si) + (0.5 \times \%Nb)$.

FILLER METALS FOR WELDING STAINLESS STEELS TO CARBON STEELS		
Stainless Steels	**Low Carbon Steel**	**Low Alloy Steel**
410 SS	E309, ENiCrFe-3, ER309, ENiCr-3	E309, ENiCrFe-3, ER309, ENiCr-3
430 SS	E309, ENiCrFe-3, E430, ER309, ERNiCr-3 ER430	E309, ENiCrFe-3, E430, ER309, ERNiCr-3 ER430
304 or 316 SS	E309, ENiCrFe-3, ER309, ERNiCrFe-5	E309, ENiCrFe-3, ER309, ENiCrFe-5
304L or 316L SS	E309L, ENiCrFe-3, ER309L, ERNiCrFe-5	E309L, ENiCrFe-3, ER309L, ENiCrFe-5

Figure 43-8. *Numerous filler metals are available that can successfully weld stainless steels to carbon steels.*

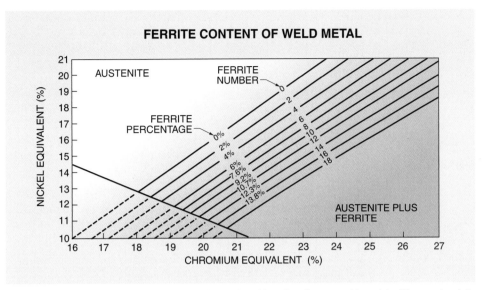

FERRITE CONTENT OF WELD METAL

Figure 43-9. The amount of ferrite in weld metal is predicted based on the composition of the filler metal and the stainless steel base metal.

Filler metal compositions with ferrite numbers 4 to 8 should be suitable for most conditions. However, presence of ferrite leads to reduced ductility for low-temperature service and for service temperatures in the range 900°F to 1700°F. In these cases, a nickel alloy filler metal is preferred.

Nickel Alloys

Nickel alloys can be welded to carbon steel, stainless steel, and copper alloys using nickel alloy filler metals and any arc welding process. **See Figure 43-10.** Selection of the filler metal depends on the particular dissimilar metal combination.

Nickel Alloys to Carbon Steel. Nickel alloys tolerate a high level of dilution with iron. However, nickel alloys may crack in the presence of sulfur and phosphorus, so that steels with high levels of these elements (such as free-machining steels) can cause problems when welded.

Steel filler metals should not be used when welding to nickel alloys because brittle intermetallic compounds can form. Austenitic stainless steel filler metals may crack because dilution with nickel removes the ferrite that helps avoid hot cracking. Nickel alloy filler metals are recommended.

Monel® to Carbon Steel. Monel® contains approximately 66% nickel and 44% copper. It may require buttering with nickel or nickel-copper filler metal for GTAW and GMAW, which are less tolerant of dilution than flux-shielded processes like SMAW. Preheating and postheating are typically not required, but postheating may improve the ductility of joints made with GTAW and GMAW.

Nickel Alloys to Austenitic Stainless Steel. Nickel alloy filler metals are preferred when joining nickel alloys to austenitic stainless steel. These filler metals tolerate more dilution, provide greater joint strength, and avoid potential service problems, especially at elevated temperatures. Base metal dilution is easily controlled with SMAW and GMAW.

Nickel Alloys to Copper Alloys. Nickel alloys can be welded to copper alloys because they have mutual solubility and pose few of the problems typical of welding other dissimilar metals. Either copper-nickel or nickel-copper filler metal can be used. Nickel filler metal is preferred when welding copper or copper-nickel to nickel alloys containing chromium to control dilution. The copper face may also be buttered with nickel for the same purpose. However, autogenous welding (welding without filler metal) should not be used because it promotes porosity.

FILLER METALS FOR WELDING NICKEL ALLOYS TO STEEL OR COPPER ALLOYS				
Nickel Alloys	Carbon Steels	304 or 316 SS	304L or 316L SS	Copper Alloys
Alloy 400	ENiCu-2, Eni-1, ENiCrFe-3, ERNiCu-7, ERNi-1, ERNiCrFe-5	ENiCrFe-2, ENiCrFe-3, ENi-1, ENiCu-7, ERNiCr-3, ERNi-1, ERNiCu-7	ENiCrFe-2, ENiCrFe-3, ENi-1, ENiCu-7, ERNiCr-3, ERNi-1, ERNiCu-7	ERCuSA, ERCuAl-A2, ERCuNi, ERCuNi-7, ERNi
Nickel	ENi-1, ENiCrFe-3, ENiCrFe-2, ERNi-1, ERNiCr-3	ENiCrFe-3, ENi-1 ENiCrFe-2, ERNiCr-3, ERNi-1, ERNiCrFe-5	ENiCrFe-3, ENi-1 ENiCrFe-2, ERNiCr-3, ERNi-1, ERNiCrFe-5	ERCuSn-A, ERCuAl-A2, ERCuNi, ERCuNi-7, ERNi
Alloy 600	ENiCrFe-3, ERNiCr-3	ENiCrFe-2, ENiCrFe-3, ENi-1, ERNiCr-3, ERNi-1,	ENiCrFe-2, ENiCrFe-3, ENi-1, ERNiCr-3, ERNi-1,	Consult Manufacturer
Alloy C-276 or 625	E309, ENiCrMo-4, E316, ENiCrFe-3, ENiCrMo-3, ENi-1, ER309, ERNiCrMo-4, ER316, ERNiCrMo-3, ERNi-1	E309, ENiCrMo-4, E316, ENiCrFe-3, ENiCrMo-3, ENi-1, ER309 ERNiCrMo-4, ER316, ERNiCrMo-3, ERNi-1	E309L, ENiCrMo-4, E316, ENiCrFe-3, ENiCrMo-3, ENi-1, ER309L, ERNiCrMo-4, ER316, ERNiCrMo-3, ERNi-1	Consult Manufacturer

Figure 43-10. *Nickel alloy filler metals are used to weld nickel alloys to steel or copper alloys.*

Copper Alloys

In addition to nickel alloys, copper alloys can be welded to carbon steel and stainless steel. **See Figure 43-11.**

Copper to Carbon Steel. The welding of copper to carbon steel is dictated by the formation of intermetallic compounds and the relatively high thermal conductivity of copper compared to steel. Intermetallic compound formation can be prevented by buttering the copper with a nickel alloy because copper and nickel are mutually soluble. Buttering may be also be used in heavier sections to avoid excessive penetration into the steel part of the joint, which can result in iron pickup and embrittlement. Nickel and carbon steel are mutually soluble and create a good weld joint.

The problems associated with the significant difference in thermal conductivity can be mitigated by preheating the copper to between 400°F and 1000°F.

The arc should be directed to the copper component to minimize iron pickup.

Copper-Nickel to Carbon Steel. When welding to copper-nickel, carbon steel is typically first buttered with nickel. The copper-nickel is welded to the buttered carbon steel using 70-30 copper-nickel or Monel filler metal. It is also possible to use Monel filler metals without buttering. Copper-nickel alloys may suffer from hot shortness. Heat input should be kept low with the preheating and interpass temperatures controlled to a maximum of 150°F.

Aluminum Bronze to Carbon Steel. Aluminum bronze can be welded to steel using nickel or aluminum bronze filler metal. The preheat temperature should be between 300°F and 500°F. The joint should be cleaned between each pass with a clean stainless steel brush to remove oxides that build up on the surface due to heat. In GTAW, alternating current provides the cleaning of the joint.

FILLER METALS FOR WELDING COPPER ALLOYS TO STEELS OR STAINLESS STEELS			
Copper Alloys	Carbon Steels	304 or 316 SS	304L or 316L SS
Copper-Nickel	ECu-7, ERNi-1	Not Recommended	ECuAl-A2
Copper	ECu, RCu	ECuAl-A2 ERuAl-A2	ECuAl-A2 ERuAl-A2
Brass	ECuAl-A2, ERCuAl-A2,	ERCuAl-A2, ERCuSn-A	ERCuAl-A2, ERCuSn-A
Aluminum Bronze	ECuAl-A2 ERuAl-A2	ERCuAl-A2	ERCuAl-A2

Figure 43-11. *Copper alloys can be welded to carbon steel or stainless steel using copper alloy filler metals.*

Brass to Carbon Steel. Brass with less than 20% zinc can be welded to carbon steel using GTAW. The steel must be buttered with a copper-tin filler metal using DCEN polarity. Completion of welding (weldout) is accomplished with the same filler metal, but with AC to promote cleaning. To avoid zinc fuming, porosity, and excessive iron dilution, the arc should not impinge on the filler metal or directly on the brass or buttered layer.

Copper Alloys to Austenitic Stainless Steel. Factors that influence the weldability of copper alloys to austenitic stainless steels are differences in melting temperatures and thermal conductivity and a tendency for hot cracking. Copper alloy filler metals should be used. Preheating and interpass temperature control help minimize problems associated with differences in melting temperatures and thermal conductivity. Buttering can be used to minimize dilution.

Copper-Nickel to Austenitic Stainless Steel. When welding to copper-nickel, a stainless steel component should be buttered with a nickel or a Monel alloy to limit dilution of the weld metal by the chromium in the stainless steel. The joint can then be completed with copper-nickel or nickel-copper filler metal. Copper-nickel alloys may suffer from hot shortness. Heat input should be kept low with the preheating and interpass temperatures controlled to a maximum of 150°F.

Aluminum Alloys

Aluminum alloys cannot be directly welded to other alloys because of the formation of intermetallic compounds. However, it is possible to fusion weld aluminum to specific metals if a compatible coating is first applied to one of the metals or a bimetallic transition insert is used.

Aluminum Alloys to Carbon Steel. There is a wide difference between the melting temperatures of aluminum and carbon steel, which are 1200°F and 2800°F respectively. When welded, the aluminum melts and flows before the steel melts, and brittle intermetallic compounds are formed. Additionally the significant difference in the coefficients of thermal expansion introduces significant thermal stresses on cooling.

Aluminum alloys can be welded to steel if the steel is first hot-dip-coated with aluminum. A high silicon-aluminum alloy filler metal may be applied using GTAW. The arc should be directed towards the aluminum, and pulsing is beneficial.

An alternative method is to use a bimetallic transition insert. Bimetallic transition inserts are combinations of metal made by processes such as rolling or explosion welding. One side of the insert is steel and the other side aluminum. Transition inserts can be fusion welded to their respective sides of the joint. They are used for applications such as attaching aluminum structures to steel-hulled ships and fabricating aluminum-tube heat exchangers with steel or stainless steel tubesheets.

! Points to Remember

- Dilution of the base metal is the greatest single factor influencing dissimilar metal welding.
- Buttering is a method of joining two incompatible base metals by overlaying (surfacing) one of them before completing the weld.
- Strength, ductility, and toughness of the welded joint influences filler metal selection in dissimilar metal welding.
- Thermal expansion and thermal conductivity of the base metal influences welding and service behavior, particularly at high temperature or under conditions of thermal cycling.
- Welds usually have poorer corrosion resistance than the base metals they join unless they are over-alloyed.

? Questions for Study and Discussion

1. What is dilution?
2. What is the benefit of buttering heat-treatable alloys?
3. What is the effect on the mechanical properties of the joint if intermetallic phases form in the filler metal?
4. Why is the thermal expansion of austenitic stainless steel a detrimental factor in welding base metals combinations that contain it?
5. How may buttering be used to compensate for melting temperature differences between two base metals?
6. Why are welds chemically inhomogeneous?
7. What is the galvanic series and how does it predict performance of welded joints?
8. How much ferrite is required on the filler metal to create an acceptable joint between carbon steel and austenitic stainless steel?
9. Why are nickel alloys easy to join to copper and vice versa?
10. Can aluminum-coated steel be welded to aluminum?

Refer to Chapter 43 in the *Welding Skills Workbook* for additional exercises.

Distortion Control

Chapter

Distortion is the undesirable dimensional change of a part during fabrication. Welding distortion occurs because of the nonuniform expansion and contraction of weld metal and adjacent base metal from the welding process. Distortion makes it difficult to maintain proper fit-up as welding progresses. Expensive remedial work may be required to correct a job after completion. Residual stresses in welding also contribute to distortion. Distortion is controlled by welding procedure, welding sequence, restraining methods, and heat shaping methods. Fabrication codes and standards have requirements for maximum allowable distortion.

DISTORTION

Distortion is the undesirable dimensional change of a part during fabrication. Distortion in welding arises from shrinkage of the weld metal and base metal during cooling, leading to high residual stresses in the metal. *Residual stress* is locked-in stress in materials that occurs as a result of manufacturing processes such as casting, welding, forming, or heat treatment.

Residual stresses can be detrimental to metals, both alone and under normal service stresses, and can contribute to fatigue and other mechanical failure. Residual stresses can also lead to stress corrosion cracking in some materials in specific corrosive environments. For example, welded carbon steel equipment and piping that is operating in hot caustic service must be given stress-relief heat treatment to prevent caustic stress cracking at the weld. The presence of residual stresses generally goes unrecognized, so welders must be cautious to avoid them. The part may be forced out of alignment as residual stresses in the weld joint ease and cause the part to move. Distortion is related to the direction of the weld and varies with the weld joint configuration.

Weld Metal Shrinkage

Weld metal shrinkage occurs as metal cools, producing distortion in the weld. As a weld begins to solidify, it expands to its maximum volume. Then, the metal cools and attempts to contract in volume, but adjacent base metal restrains it.

The restraint causes increasing stresses within the weld. Once the yield strength of the weld metal is exceeded, it begins to stretch, thinning out and adjusting to the new volume requirements. Only the stresses that exceed the yield strength are relieved by this accommodation. By the time the weld reaches ambient temperature—assuming it is completely restrained by the base metal and cannot move—the weld contains locked-in tensile stresses approximately equal to its yield strength.

Shrinkage in weld metal may be transverse or longitudinal. **See Figure 44-1.** *Transverse shrinkage* is shrinkage that occurs perpendicular to the weld axis. Transverse shrinkage depends on the volume of weld metal and the amount of the root opening, though some shrinkage occurs regardless of whether the joint is made with a root opening. *Longitudinal shrinkage* is shrinkage that occurs parallel to the weld axis.

✓ **Point**

Distortion in welding is caused by shrinkage in the weld metal and the base metal that occurs during cooling and by creating restraint that exceeds the yield strength of the material.

✓ **Point**

Residual stress contributes to distortion.

✓ **Point**

Distortion of welded structures is either transverse (at 90° to the weld axis) or longitudinal (along the length of the weld axis).

✓ **Point**

Welders must protect against residual stresses. The presence of these stresses generally goes unrecognized until failure occurs.

Weld Metal Shrinkage

(T) = TRANSVERSE SHRINKAGE
(L) = LONGITUDINAL SHRINKAGE

Figure 44-1. *Distortion is caused by weld shrinkage, which can be longitudinal (L) or transverse (T).*

Distortion in Groove Welds

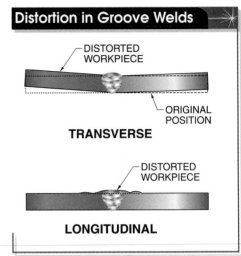

DISTORTED WORKPIECE

ORIGINAL POSITION

TRANSVERSE

DISTORTED WORKPIECE

LONGITUDINAL

Figure 44-2. *Residual stresses in groove welds cause transverse and longitudinal distortion.*

Groove Welds. Several weld passes are often necessary to complete a groove weld. The root pass creates little or no distortion, but restrains the two components being joined. As the second pass solidifies, it shrinks, but the solidified root pass offers restraint (no movement), so that the shrinkage must occur within the second pass and toward its upper surface. Successive passes are larger and wider and there is a greater mass of weld metal shrinking.

As each successive pass goes through a solidifying shrinkage cycle as it cools, the previous pass acts as a restraint. The weld distortion works like a hinge: the weld root is like a hinge pin and the faces of the joint are drawn to one another with the shrinkage of each pass. The result is transverse shrinkage. **See Figure 44-2.** For groove welds in carbon steel with a 60° groove angle, the transverse shrinkage rate is typically 1/16″ to 1/8″ per weld pass.

In a groove weld, the joint is also strained in tension in the longitudinal direction. The resulting distortion is observed as longitudinal contraction of the weld. Longitudinal shrinkage is less of a problem in groove welds than transverse shrinkage.

Fillet Welds. Distortion in fillet welds is more complex than in groove welds. The distortion effect depends on the type of fillet weld, the number of passes, and the location of the center of gravity of the workpieces in relation to the fillet weld.

A single fillet weld creates transverse shrinkage. For fillet welds in carbon steel, the transverse shrinkage rate is 1/32″ per weld pass (where the leg length of the weld does not exceed three-quarters of the base metal thickness). When the root pass is laid, the workpieces being joined become integral to each other. As more passes are laid, there is more shrinkage at the face of the fillet than at the root because the face receives a greater amount of filler metal. To accommodate the weld metal shrinkage, the workpieces move toward one another, creating transverse shrinkage. **See Figure 44-3.** As more weld passes are added, the distortion increases.

A double-fillet weld, if properly made, does not exhibit transverse shrinkage, as each weld compensates for the transverse stress caused by the weld on the other side. However, the weld is still susceptible to longitudinal shrinkage. **See Figure 44-4.** If the fillet weld of the T-joint is above the center of gravity of a welded structure, the metal distorts longitudinally upward at its ends. If the fillet weld is below the center of gravity, it distorts longitudinally downward at its ends.

Residual stresses are stresses that remain after the original cause of the stresses has been removed, such as external forces or temperature change.

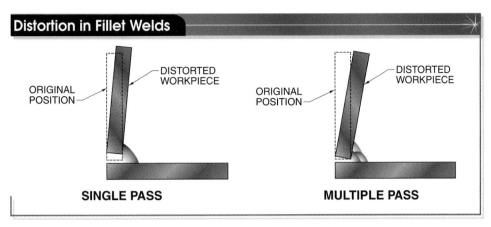

Figure 44-3. *When weld beads are deposited in a single-fillet weld, shrinkage occurs at the face of the fillet because of the amount of filler metal deposited.*

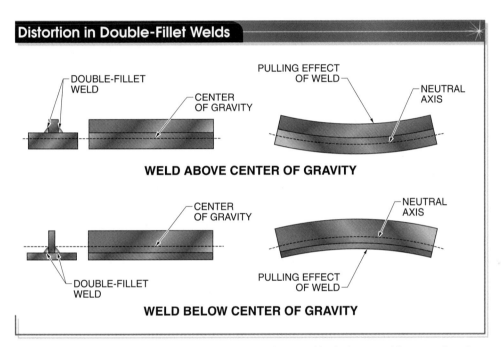

Figure 44-4. *In a double-fillet weld, longitudinal distortion is determined by the location of the center of gravity.*

Base Metal Shrinkage

During welding, the HAZ is heated close to its melting point and the metal expands in volume. Less than an inch away, however, the base metal is less affected and the temperature is substantially lower. The intense temperature difference causes movement of the base metal. As the welding arc moves along the joint, the formerly heated part of the base metal (the HAZ) begins to cool and shrink. If the less-affected base metal restrains it from contracting, residual stresses build up in the HAZ. These stresses, combined with stresses that develop in the weld metal as it cools, increase distortion in the base metal.

DISTORTION CONTROL

Maximum allowable distortion is often specified by fabrication codes and standards. Residual stress is a significant contributor to distortion. Distortion control includes methods to minimize or eliminate residual stress and the resulting distortion. Distortion control is necessary to overcome poor fit-up and undesirable stresses and to meet specific dimensional requirements. Distortion control methods include modifying the welding procedure, choosing the appropriate welding sequence, restraining, and heat shaping.

Welding Procedures

Welding procedures may be modified to minimize or control distortion. Aspects of welding procedures that may be modified to prevent distortion include fit-up and edge preparation, presetting, preheat, backing bars, heat input, welding process, weld passes, weld metal deposition, and postheat.

Point

Proper fit-up is essential on thin metals. Closely spaced tack welds must be used to control distortion.

Fit-Up and Edge Preparation. Proper fit-up and edge preparation help ensure that the correct amount of weld metal is used in a joint. If gaps occur in a joint, the welder must slow down to fill them, using more filler metal than specified and increasing contraction across the joint. Proper fit-up and edge preparation avoids the need for excessive filler metal and increased joint contraction. Undercut spots in butt joints can be filled by weld buildup on the edge of the base metal before welding to improve poor fit-up. Joint preparation cannot be manipulated for fillet welds as it can for groove welds.

For a butt joint, a minimum root opening of ⅛″ is desired. A 60° groove angle allows for complete penetration at the root yet requires minimal weld metal. **See Figure 44-5.** For thick metal, increasing the root opening to 3⁄16″ allows the groove angle to be decreased. Alternatively, a J-groove may be used to reduce the amount of weld metal required. A double-V-groove may be used, which reduces by half the amount of weld metal necessary as compared with a single-V-groove.

If the root opening is increased and the groove angle is reduced, the amount of metal deposited at the root and at the face of the weld is more equal, reducing transverse shrinkage. A square groove on thin metals reduces distortion but does not completely eliminate it.

On thin stock such as sheet metal, small, closely spaced tack welds are the only means of controlling distortion. After tack welding, the entire joint should be lightly hammered before welding. On very thin material (26-gauge and thinner), almost continuous tack welds may be required.

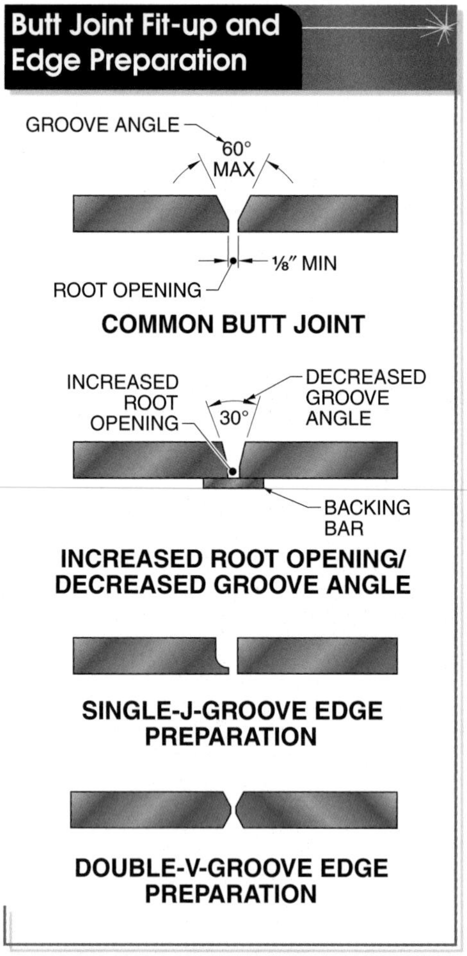

Figure 44-5. *Butt joint fit-up and edge preparation can be altered to help control distortion.*

Presetting. *Presetting* is a technique that sets the parts to be welded in an out-of-alignment position and relies on the forces that cause distortion to pull the joint into the correct alignment. **See Figure 44-6.** The main advantage of presetting is that no expensive equipment is needed and there is lower residual stress in the structure compared with the use of mechanical restraints.

However, it is difficult to predict the amount of presetting needed to accommodate weld shrinkage, and a number of trial welds are often needed using a mock-up that simulates the level of distortion that will likely occur. For this reason, presetting is suitable for simple components or assemblies.

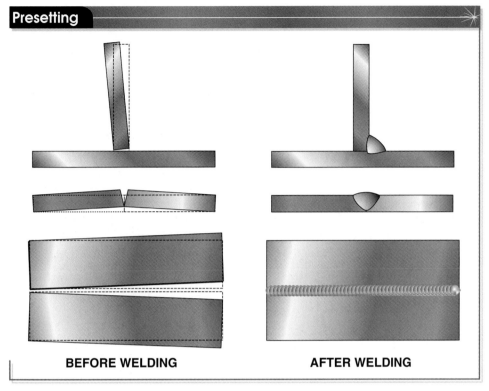

BEFORE WELDING   **AFTER WELDING**

Figure 44-6. *Presetting results in correct alignment after welding by misaligning the workpieces before welding.*

Preheat. Properly applied preheat can help reduce weld distortion and residual stresses because it lowers the temperature gradients in the metal around the weld. Preheating causes the workpiece volumes to expand uniformly, avoiding significantly unequal expansion forces when the joint is heated during welding. It also allows the work to contract evenly during cooling. On steel, preheat also reduces the tendency for cracking in the HAZ or the weld metal.

Preheat is more effective for distortion control in thicker sections because it often offsets heat losses by conduction. However, it has little use in distortion control when welding thin sheet metal. Preheated workpieces also require less heat input from the welding process, allowing smaller nozzle sizes when gas welding and lower current settings when arc welding. Welding speeds may be increased, making it easier to obtain good penetration and fusion.

Backing Bars. Backing bars, also known as chills, are thick blocks placed behind the weld to conduct heat away from the joint. Backing bars keep heat from spreading to the base metal. **See Figure 44-7.** A backing bar should not contact the weld zone, so it is positioned or notched in a way that allows space just below the weld. Backing bars are often made of copper, which is a good conductor of heat. Aluminum or steel can also be used and conduction of heat away from the weld is assisted by the passage of water through holes drilled in the backing bars. Some backing bars include channels for connection to a water cooling system, which increases the rate of heat transfer away from the workpieces.

Heat Input. Minimum heat input should be applied to provide a stable arc with good fusion and penetration. To solve the problem of overheating, welding should be done quickly. The faster the weld is made, the lower the amount of heat absorbed by the base metal. Use of stringer beads versus weave beads also reduces heat input.

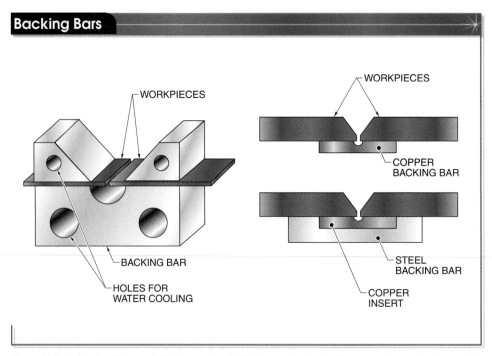

Figure 44-7. *Backing bars help conduct heat away from the joint, preventing the heat from spreading into the base metal.*

Welding Processes. With manual welding processes, GMAW produces less distortion than GTAW, and PAW produces less distortion than GMAW. In addition, PAW can be used to weld thicker metals than GTAW or GMAW.

Oxyfuel welding processes generally produce more distortion than arc welding processes because heating of the base metal is slow and more heat is required to offset the heat loss from diffusion.

Automatic and semiautomatic welding processes use higher travel speeds and greater deposition rates per pass than manual welding processes, resulting in less distortion. Additionally, with automatic welding processes, progressive shrinkage of the weld as it cools (which occurs in manual welding during the interval between each weld pass) is eliminated.

Weld Passes. A large number of weld passes increases shrinkage, and therefore distortion. Where transverse distortion might be a problem, a few passes made with a large-diameter filler metal are preferable to many passes made with a small-diameter filler metal. **See Figure 44-8.** Shrinkage caused by each pass is cumulative. Making fewer passes also reduces welding time, which reduces the amount of heat at the weld, causing less expansion of the metal surrounding the weld.

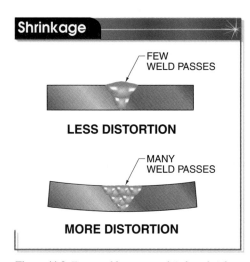

Figure 44-8. *Fewer weld passes result in less shrinkage in the weld, which minimizes distortion.*

Weld Metal Deposition. The amount of weld metal deposited in the joint affects the potential shrinkage. More weld metal increases shrinkage. To minimize shrinkage, only the required amount of weld metal should be used. **See Figure 44-9.**

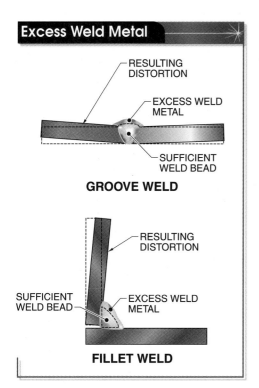

Excess Weld Metal

RESULTING DISTORTION

EXCESS WELD METAL

SUFFICIENT WELD BEAD

GROOVE WELD

RESULTING DISTORTION

SUFFICIENT WELD BEAD

EXCESS WELD METAL

FILLET WELD

Figure 44-9. *Excess weld metal increases distortion in a weld.*

The effective throat in a fillet weld determines the weld joint strength. A fillet weld that yields an effective throat size that is sufficient for the strength required by the weld design is preferred. In a butt joint, excess weld metal in a highly convex bead does not increase the allowable strength of the weld in the design code, but does increase shrinkage and distortion.

Postheat. Postheat involves reheating of the weld area to a high temperature, holding it for a predetermined time at that temperature, and cooling it at a specified rate. Postheating is used to prevent cold cracking from residual stresses. Postheating also stress-relieves the joint, reducing the possibility of distortion or cracking in service. With steels, postheating additionally tempers (softens and toughens) the weld. Postheating is often specified in conjunction with preheat and interpass temperature control.

Postheat is most effective when used with joints that have been mechanically restrained to prevent distortion using

welding. It removes any stresses that have been absorbed by the elasticity of the joint and prevents the joint for springing out of shape. Postheat should be conducted with the mechanical restraint in place to prevent cracking.

Welding Sequence

Various welding sequences may be used to balance shrinkage stresses and control distortion. Typical welding sequences used include back-step welding, intermittent welding, and balanced welding.

Back-Step Welding. *Back-step welding* is a welding sequence in which short weld passes are made in the direction opposite of the overall progress of welding. This method of distortion control creates an expansion force from the deposited metal that counteracts the contraction force of a weld that has just been deposited and is cooling. Each weld pass locks the parts being joined. The greatest amount of metal expansion occurs when the first weld bead is deposited. Metal expands less with each successive weld bead because of the locking effect of previous back-step welds. **See Figure 44-10.** Back-step welding cannot be performed with automatic welding processes.

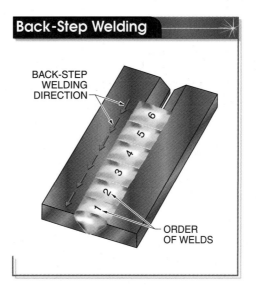

Back-Step Welding

BACK-STEP WELDING DIRECTION

1 2 3 4 5 6

ORDER OF WELDS

Figure 44-10. *Back-step welding is a welding process in which weld passes are made in the direction opposite of the progression of welding.*

Intermittent Welding. Intermittent welding, also known as skip welding, is an alternative to back-step welding. *Intermittent welding* is a stress-reduction technique in which the continuity of the weld is broken by recurring spaces between welds. Intermittent welding is performed by depositing weld metal in evenly spaced increments. **See Figure 44-11.** The intermittent weld sequence brings the entire joint to an approximately even temperature by depositing welds in a planned sequence.

On a T-joint, welds are alternated on either side of the joint. Three short (usually 1″) weld beads are made, and then two longer beads (approximately three times the length of the short beads) are made on the other side of the T-joint. The direction of welding should remain the same throughout the process, but it is not necessary for the direction to be opposite of the general progression, as with back-step welding.

Intermittent welding may be used to reduce the amount of weld metal required and must be allowed by the design code. Intermittent welding is not used where corrosion is an issue because welding intermittent fillet welds creates crevices that may allow corrosives to enter the weld area.

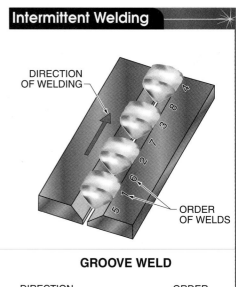

Intermittent Welding

GROOVE WELD

DOUBLE-FILLET WELD

Figure 44-11. *In intermittent welding, weld metal is deposited in evenly spaced increments.*

Intermittent welding is particularly useful for welding flush patches. A *flush patch* is a replacement of a thin metal component that is applied to provide a smooth transition with the component. A flush patch is set much like a window so that there is no ledge or raised surface between the patch and the component. A flush patch is usually applied to repair or replace metal that is corroded and excessively thinned and must be removed.

When a flush patch is welded, the surface may become distorted from shrinkage stresses. Distortion can be minimized using an intermittent welding technique and a slightly dished flush patch. **See Figure 44-12.** Dishing of the patch allows it to draw in and settle relatively free from stress. The amount of dishing should be about equal to the thickness of the metal being welded.

Balanced Welding. *Balanced welding* is the specific ordering of welds to be completed on a component in order to compensate for distortion-inducing stresses.

Intermittent welding may be used for a groove weld to reduce stress in the weld.

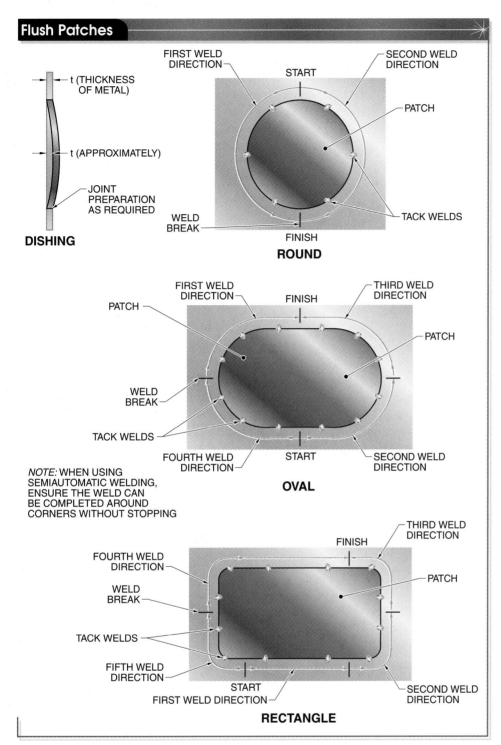

Figure 44-12. *Flush patches on thin surfaces may be welded without resulting in distortion by using an intermittent welding technique and a slightly dished flush patch.*

If the order is not already specified in the weld requirements, a specific sequence of welds can be chosen to address the anticipated distortion.

For example, distortion may occur when welding pipe branch connections, which can cause piping to bow due to shrinkage on one side. **See Figure 44-13.**

Branch welds must be welded in sequence to minimize distortion of the pipe. The branches furthest from the center of the pipe assembly are welded first because they cause less distortion. If the pipe is bent because of welds at the first two branches, the third branch welding may straighten the pipe.

Branch Welds

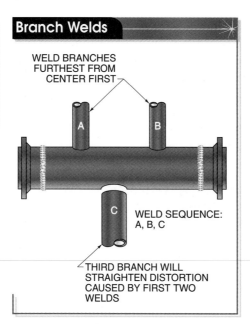

WELD BRANCHES FURTHEST FROM CENTER FIRST

WELD SEQUENCE: A, B, C

THIRD BRANCH WILL STRAIGHTEN DISTORTION CAUSED BY FIRST TWO WELDS

Figure 44-13. *Balanced welding of branch welds reduces the tendency of pipe to distort.*

Distortion can also be caused by transverse shrinkage of a groove weld in adjoining pipe sections, which causes a reduction in the overall length of the pipe. In most cases, carbon steel shrinkage is approximately $\frac{1}{16}''$ (plus or minus $\frac{1}{32}''$) per butt joint. This shrinkage in the overall length of the piping assembly should be accounted for.

Restraining

Restraining is the use of a mechanical fixture on a part to restrict movement and counteract shrinkage stresses that occur during welding. The restraint holds the part in the desired position until welding is completed. The restraint causes a buildup of internal stresses in the weld until the yield stress of the weld is exceeded. The part is restrained until it cools and the internal stresses are eased. Once the part has cooled, the restraint is removed, with little resulting distortion or movement. Movement does not occur when the restraint is removed because cooled, solid metal is under less strain than hot, restrained metal. Typical mechanical restraint methods are strongbacks, internal restraints, back-to-back positioning, and prebending.

Strongbacks. A *strongback* is a mechanical restraint that is attached by welding or clamping to both workpieces of the weld joint to maintain alignment during welding and prevent distortion. The parts are placed in position and held under restraint of the strongback to minimize any movement during welding. **See Figure 44-14.**

Strongbacks are only removed when the joint has cooled to room temperature. Welded-on strongbacks are removed by grinding. The area is liquid-penetrant or magnetic-particle inspected to ensure there are no cracks where the strongbacks were welded on. As significant stresses can be generated across the weld, increasing the tendency for cracking, care should be taken in the use of strongbacks. Also, relatively small amount of movement may still occur due to locked-in stresses, which can be cured by applying stress relief heat treatment to the assembly before removal of the strongbacks.

Strongbacks

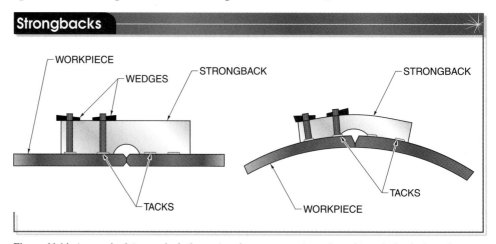

WORKPIECE

WEDGES

STRONGBACK

STRONGBACK

TACKS

WORKPIECE

TACKS

Figure 44-14. *A strongback is a method of restraint where a strong piece of metal is tacked to both workpieces to hold them securely during welding. The strongback is removed after the weld cools.*

Internal Restraints. Depending on the size and configuration of the workpieces, fastened or welded-on restraints may not be needed. Restraints may be fitted inside certain parts in order to maintain their correct size and shape during and after welding. Like strongbacks, these restraints are removed after cooling.

For example, distortion may occur during the placement of nozzles on equipment, such as on small-diameter heat exchangers, because the shell thickness is generally less than ¾″. **See Figure 44-15.** Distortion increases as the metal cross section (shell thickness) becomes thinner. Distortion on an equipment nozzle appears as a flat spot on the shell where the nozzle is welded. Distortion also causes the nozzle to sink into the shell. When welding to thin parts, an internal restraint, such as a jack, should be used to prevent the shell from collapsing.

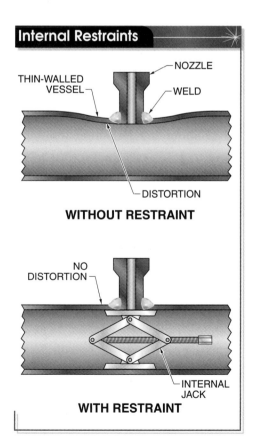

Figure 44-15. *Internal restraints can be used to prevent distortion of thin-walled vessels, such as pipes and tanks.*

Back-to-Back Positioning. *Back-to-back positioning* is a mechanical restraint method that places identical weldments back-to-back and clamps them together. **See Figure 44-16.** The welds are completed and both weldments are allowed to cool before the clamps are released. The welding stresses on the weldments counteract each other and cancel out distortion.

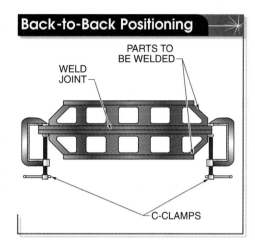

Figure 44-16. *Back-to-back positioning counteracts shrinkage in two identical components.*

Prebending. *Prebending* is a mechanical restraint method that holds workpieces out of position before welding in such a way that compensates for the stresses of angular distortion. **See Figure 44-17.** Releasing the wedges after welding allows the parts to move back into alignment. Residual stresses in the part cause the part to straighten. This technique is also known as prespringing. Prebending uses strongbacks and wedges, fastened or welded on, to hold the workpieces in position.

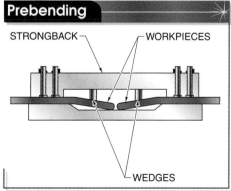

Figure 44-17. *With prebending, workpieces are forced out of position before welding so that shrinkage stresses later pull them into alignment.*

Heat Shaping

Heat shaping is the application of localized heating to cause movement of a distorted part and restore its dimensions. Heat shaping is applied using an oxyacetylene flame. It requires temperature monitoring and measurement of the movement achieved. In some cases, movement may be assisted with mechanical devices. For complete correction of distortion, mechanical restraints may be used with heat shaping. Four basic heating patterns are used when heat shaping: line, spot, V, and block heating. **See Figure 44-18.**

Heat-Shaping Patterns

LINE HEATING

SPOT HEATING

V HEATING

BLOCK HEATING

Figure 44-18. *The four basic heat-shaping patterns are line, spot, V, and block heating.*

The line-heating pattern can be used on a metal plate. The metal is heated on the convex (high) side that is to be bent down. A slightly oscillating torch follows the line, with the oscillations about as wide as the metal thickness. **See Figure 44-19.** The torch progresses across the metal at a constant speed to bring the plate to temperature. Movement in a line-heating pattern progresses in a linear fashion with relatively little width compared to its length.

The spot-heating pattern concentrates heat in one area in a circular motion and is applied with little, if any, lateral motion. The V-heating pattern starts at a point and moves in a linear fashion along a marked axis, becoming progressively wider as it weaves back and forth. The block-heating pattern moves in a linear fashion, weaving back and forth to create a rectangular area. The V- and block-heating patterns can be used on structural steel shapes such as channels, I-beams, and angles. **See Figure 44-20.** The patterns are applied alternately to achieve straightening. Two torches may be applied opposite one another in specific cases.

To perform heat shaping, the area to be shaped is marked with soapstone, a paint stick, or another marking material that is insensitive to heat. When heat shaping stainless steels and nickel alloys, the marking material can contain only minimal amounts (less than 50 parts per million) of chlorides, sulfur, or other harmful elements such as zinc. Otherwise, cracking may occur during heat application.

The oxyacetylene torch is ignited and the flame adjusted. A small area is quickly heated, with the point of the flame far enough above the surface to prevent the surface from melting. The torch is weaved slightly, but not advanced in a heating pattern until the starting point reaches the specified temperature. Heat is progressively applied to the marked area, maintaining the desired temperature at the point of the flame. The flame is not backtracked over any area already heated. A temperature-indicating crayon or a contact pyrometer may be used to monitor the temperature.

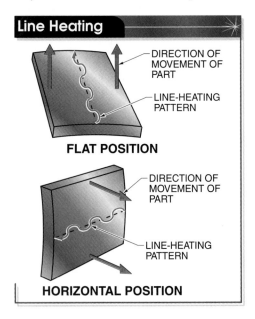

Line Heating

DIRECTION OF MOVEMENT OF PART

LINE-HEATING PATTERN

FLAT POSITION

DIRECTION OF MOVEMENT OF PART

LINE-HEATING PATTERN

HORIZONTAL POSITION

Figure 44-19. *Line heating is used to bring a distorted piece of metal back into alignment.*

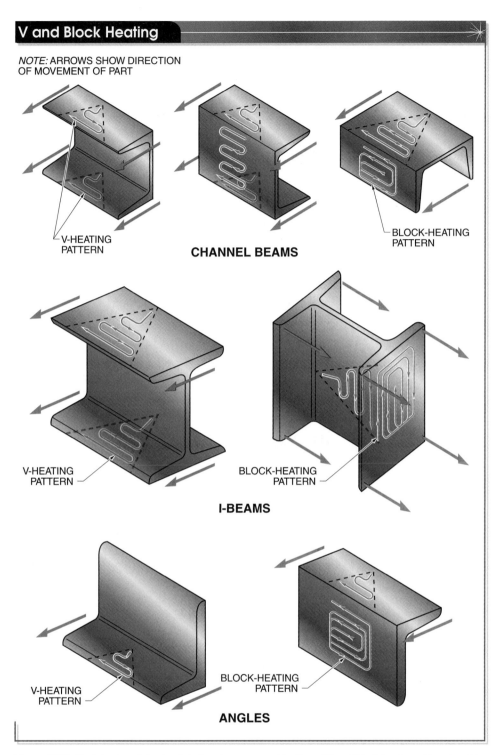

NOTE: ARROWS SHOW DIRECTION
OF MOVEMENT OF PART

V-HEATING
PATTERN

BLOCK-HEATING
PATTERN

CHANNEL BEAMS

V-HEATING
PATTERN

BLOCK-HEATING
PATTERN

I-BEAMS

V-HEATING
PATTERN

BLOCK-HEATING
PATTERN

ANGLES

Figure 44-20. *Heat shaping of structural steel sections uses combinations of V- and block-heating patterns to achieve straightening.*

Refer to Quick Quiz® on CD-ROM

- Distortion in welding is caused by shrinkage in the weld metal and the base metal that occurs during cooling and by creating restraint that exceeds the yield strength of the material.
- Residual stress contributes to distortion.
- Distortion of welded structures is either transverse (at 90° to the weld axis) or longitudinal (along the length of the weld axis).
- Welders must protect against residual stresses. The presence of these stresses generally goes unrecognized until failure occurs.
- Modifying the welding procedure, using special welding techniques, using mechanical restraints, or heat shaping can help control distortion.
- Proper fit-up is essential on thin metals. Closely spaced tack welds must be used to control distortion.
- The greatest amount of metal expansion occurs when the first weld bead is laid. Metal expands less with each successive weld bead because of the locking effect of previous back-step welds.
- Mechanical restraints cause a buildup of internal stresses in the weld to the point that the yield stress of the weld is exceeded.
- Heat shaping is the application of localized heat to a structure to cause beneficial movement of a part in order to counteract distortion.
- To completely correct distortion, mechanical restraints may be used with heat shaping.
- The four basic heating patterns used when heat shaping metals are line, spot, V, and block heating.

1. How does the heat of welding cause distortion?
2. Why is residual stress harmful?
3. What happens to molten weld metal as it cools that contributes to distortion?
4. In which two directions does weld metal shrinkage occur?
5. How does preheat help to reduce distortion?
6. Why is it important to use the minimum thickness of weld filler metal prescribed by the applicable fabrication code?
7. Is distortion more likely to occur in multiple-pass welds or single-pass welds?
8. What is the difference between back-step welding and intermittent welding as distortion prevention methods?
9. Name two methods of mechanical restraint used to prevent distortion.
10. What is heat shaping?

Refer to Chapter 44 in the *Welding Skills Workbook* for additional exercises.

When fabricating metal products, a welder may use a print that details product specifications. The print specifies where welds are to be located, the type of joints, and correct weld sizes. Information is indicated by a set of symbols that have been standardized by the American Welding Society (AWS).

WELDING SYMBOLS

A *welding symbol* is a graphical representation of the specifications for producing a welded joint. A welding symbol has instructions attached that indicate the type of weld required (weld symbol), the location of the weld, whether it is to be performed in the field, and other reference data that are necessary to prepare the joint and deposit the weld correctly. The welding symbol is designed so that specific information (weld symbol, weld size, groove angle, and depth of bevel) has a designated location on the symbol. **See Appendix.**

Reference Line

The foundation of the welding symbol is a reference line. An arrow is attached to the reference line and points to the location of the weld. Instructions regarding the type of weld to be deposited are attached above, below, or on both sides of the reference line. A tail may also be added to the reference line. The tail of the welding symbol contains information such as the type of welding process to be used or a reference to notes located elsewhere on the drawing. **See Figure 45-1.**

The type of weld may be accompanied by other reference data such as weld size, groove angle, root opening, depth of bevel, weld length, surface contour, and how the weld is to be finished. All welding symbol data are indicated with geometric figures, numerical values, and abbreviations.

> ✓ **Point**
>
> *A welding symbol is a graphical representation of the specifications for producing a welded joint.*

> ✓ **Point**
>
> *Instructions regarding the type of weld are indicated either above or below the reference line.*

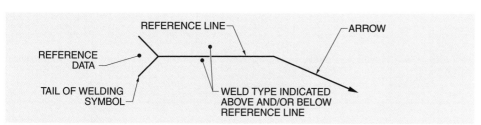

Figure 45-1. *The foundation of the welding symbol is a reference line with an arrow at one end.*

Designating Types of Welds

The most important element in a welding symbol is the type of weld. Types of welds are fillet, groove, plug or slot, spot or projection, and seam. Weld types are indicated by a weld symbol. A *weld symbol* is a graphic symbol connected to the reference line of a welding symbol specifying the weld type. Groove welds can be further classified according to the particular shape of the grooved joint. **See Figure 45-2.**

Each weld has its own specific symbol. For example, a fillet weld is designated by a right-angle triangle, and a plug weld is designated by a rectangle. The type of weld specified is directly related to the type of joint. For example, groove welds are used in butt joints, fillet welds are used in lap and T-joints, and edge welds are used in edge joints.

Symbol Location

The location of the weld symbol on the reference line specifies which side of the joint the weld is made on. The arrow attached to the reference line points to the joint where the weld is deposited. The arrow can be directed to either side of a joint and can extend up or down. A weld is said to be on either the "arrow side" or the "other side" of a joint. The *arrow side* is the side of a joint to which the arrow points. The *other side* is the side of a joint opposite the arrow side.

If the weld is to be made on the arrow side of the joint, the appropriate weld symbol is placed below the reference line. If the weld is to be located on the other side of the joint, the weld symbol is placed above the reference line. When both sides of the joint are to be welded, the same weld symbol appears above and below the reference line. **See Figure 45-3.**

A complete treatment of symbols as they apply to all forms of manual and mechanized welding can be found in AWS A2.4, *Standard Symbols for Welding, Brazing, and Nondestructive Examination,* published by the American Welding Society. **See Appendix.**

The only exception to the indication of weld location on the reference line is in spot and seam welding. With spot or seam welds, the arrow points to the centerline of the weld seam and the appropriate weld symbol is centered above or below the reference line. **See Figure 45-4.** If there is no side significance, the symbol is placed astride the reference line to indicate this condition.

Welding and nondestructive examination systems originated from AWS committee work begun in the 1940s. AWS continues to update and refine these symbols in the standard ANSI/AWS A2.4 Standard Symbols for Welding, Brazing, and Nondestructive Examination.

Weld Types

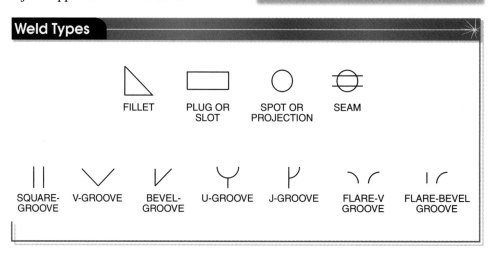

Figure 45-2. *Types of welds are fillet, plug, spot or projection, seam, and groove. Groove welds can be subdivided by the particular shape of the grooved joint.*

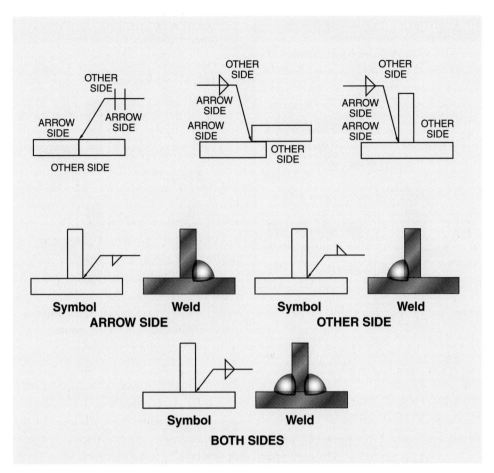

Figure 45-3. *The location of the weld symbol determines where the weld is made.*

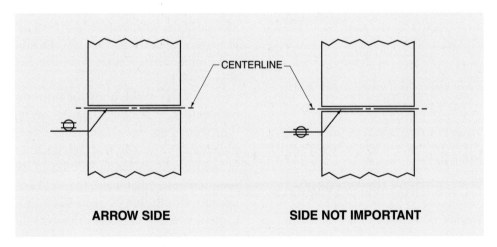

Figure 45-4. *For a seam weld symbol, the arrow points to the centerline of the weld seam, with the appropriate symbol above or below the reference line. If there is no side significance, the symbol is placed astride the reference line.*

On beveled joints, it is often necessary to show which weld part is to be beveled. In such cases, the arrow points with a definite break toward the part to be beveled.

See Figure 45-5. Information on welding symbols is placed to read from left to right along the reference line in accordance with the conventions of drafting.

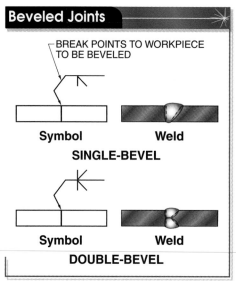

Beveled Joints

BREAK POINTS TO WORKPIECE TO BE BEVELED

Symbol   Weld

SINGLE-BEVEL

Symbol   Weld

DOUBLE-BEVEL

Figure 45-5. *The arrow break points toward the joint that must be beveled.*

Fillet, single-bevel-groove, J-groove, and flare-bevel-groove weld symbols are always shown with the perpendicular leg on the left. **See Figure 45-6.**

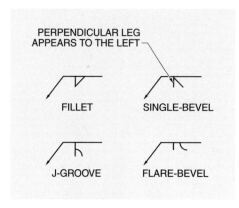

PERPENDICULAR LEG APPEARS TO THE LEFT

FILLET   SINGLE-BEVEL

J-GROOVE   FLARE-BEVEL

Figure 45-6. *Fillet, single-bevel-groove, J-groove, and flare-bevel-groove weld symbols appear to the left of the weld symbol.*

COMBINING WELD SYMBOLS

During fabrication of a product, it may be necessary to perform more than one operation on a joint. For example, if a fillet weld is required in a T-joint with thick members, it may be necessary to bevel one or both sides of the perpendicular member before welding to achieve the desired penetration. In such cases, both the type of joint preparation, and the type of weld can be combined in the weld symbol. **See Figure 45-7.**

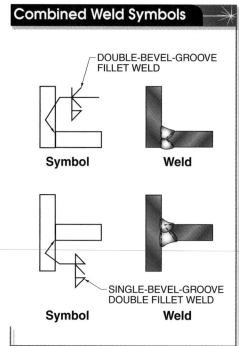

Combined Weld Symbols

DOUBLE-BEVEL-GROOVE FILLET WELD

Symbol   Weld

SINGLE-BEVEL-GROOVE DOUBLE FILLET WELD

Symbol   Weld

Figure 45-7. *A joint that requires more than one type of weld is represented by a combined weld symbol.*

Fillet Welds

A fillet weld is represented by a right-angle triangle. A fillet weld has two legs. A *fillet weld leg* is the distance from the joint root to the toe of the weld. The weld toe is formed by the junction between the surface of the weld and the base metal. A dimension appearing to the left of a fillet weld symbol specifies the weld size. **See Figure 45-8.**

In a fillet weld, joint penetration and weld size are the same. If the leg dimensions are equal, the weld size is indicated by a single fraction, decimal, or metric unit. When both sides of the joint require a fillet weld, the weld size appears to the left of each fillet weld symbol whether the dimensions are the same or different.

When a fillet weld with unequal legs is required, the dimension of each leg size is placed to the left of the weld symbol. The location of the longest leg is not specified by the weld symbol. However, a note can be added in the tail of the welding symbol to specify the orientation of the weld. The length of the weld is shown to the right of the weld symbol.

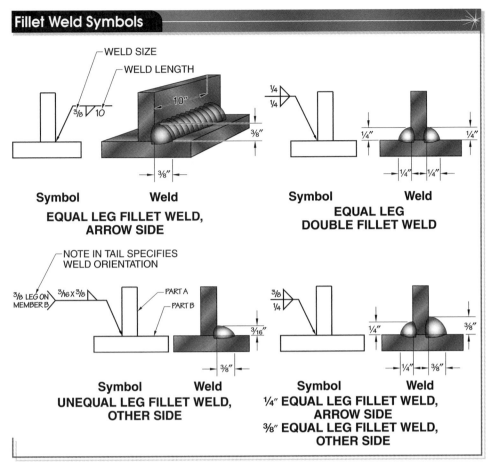

WELD SIZE
WELD LENGTH

Symbol | Weld
EQUAL LEG FILLET WELD, ARROW SIDE

Symbol | Weld
EQUAL LEG DOUBLE FILLET WELD

NOTE IN TAIL SPECIFIES WELD ORIENTATION

PART A
PART B

Symbol | Weld
UNEQUAL LEG FILLET WELD, OTHER SIDE

Symbol | Weld
¼″ EQUAL LEG FILLET WELD, ARROW SIDE
⅜″ EQUAL LEG FILLET WELD, OTHER SIDE

Figure 45-8. *The leg size (width) of the fillet weld is expressed as a fraction, decimal, or metric unit to the left of the weld symbol. The length is indicated by the actual numerical value to the right of the weld symbol.*

Intermittent Fillet Welds. An *intermittent weld* is a weld in which continuity is interrupted by recurring unwelded spaces. The length and pitch of intermittent fillet weld segments are shown to the right of the weld symbol separated by a dash. The first number indicates the length of each weld segment and the second number represents the pitch (center-to-center spacing) of adjacent weld segments on one side of the joint. **See Figure 45-9.**

Groove Welds

Groove welds join members edge to edge. Edge preparation may be required on thick members to allow the desired amount of penetration. The type of edge preparation determines the type of groove joint and the groove weld. Common types of groove welds are single-square-groove welds, double-square-groove welds, single-V-groove welds, double-V-groove welds, single-bevel-groove welds, and double-bevel-groove welds. **See Figure 45-10.**

The Lincoln Electric Company

A fillet weld may be specified with the proper weld symbol and dimension.

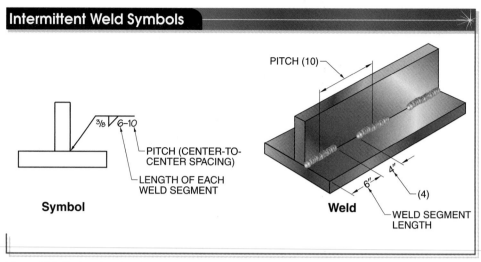

PITCH (CENTER-TO-CENTER SPACING)

LENGTH OF EACH WELD SEGMENT

Symbol

PITCH (10)

Weld

WELD SEGMENT LENGTH

Figure 45-9. *The length and pitch of intermittent welds are shown to the right of the weld symbol.*

COMMON GROOVE WELD TYPES					
Type	**Symbol**	**Weld**	**Type**	**Symbol**	**Weld**
SINGLE-SQUARE-GROOVE WELD			DOUBLE-SQUARE-GROOVE WELD		
SINGLE-V-GROOVE WELD			DOUBLE-V-GROOVE WELD		
SINGLE-BEVEL-GROOVE WELD			DOUBLE-BEVEL-GROOVE WELD		

Figure 45-10. *Groove welds are identified by the edge shape of the joint members as they appear in cross section.*

The finished surface of the weld produced by the cover pass is called the weld face. In groove welds, the height of the weld above the base metal is called the face reinforcement. Face reinforcement of $\frac{1}{16}''$ to $\frac{1}{8}''$ is considered acceptable. Excessive reinforcement is generally unacceptable, because it adds weight to the weldment and also wastes filler metal. **See Figure 45-11.**

The weld reinforcement opposite the side from which the weld is performed is called the root reinforcement. Root reinforcement should not exceed $\frac{1}{16}''$. The sloped surfaces formed by the bevels of a joint where weld metal is deposited are called groove faces.

A dimension in parentheses to the left of a groove weld symbol indicates weld size. Typically, weld size is provided for partial penetration welds. If weld size is not indicated, it is assumed that the groove weld is a complete penetration weld. A

dimension to the left of a weld symbol that is not in parentheses indicates the depth of bevel. When both weld size and depth of bevel are specified, the depth of bevel is located to the left of the weld size. **See Figure 45-12.**

Groove Weld Elements

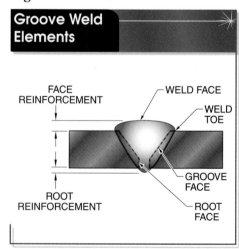

FACE REINFORCEMENT

ROOT REINFORCEMENT

WELD FACE

WELD TOE

GROOVE FACE

ROOT FACE

Figure 45-11. *The sloped surfaces formed by the bevels of the joint on which the weld metal is deposited are called groove faces.*

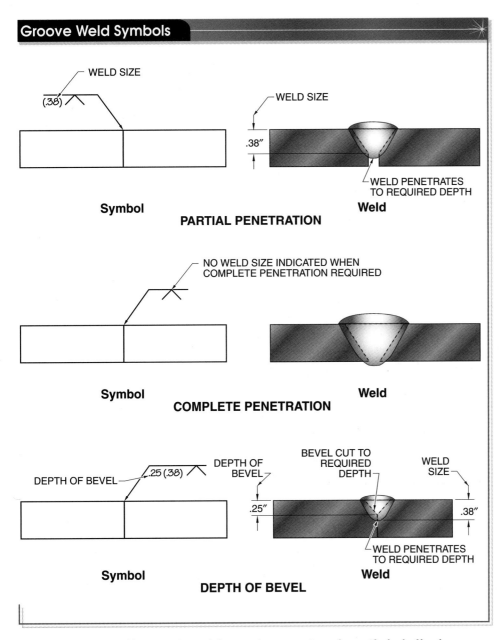

Figure 45-12. *Groove welds may require partial or complete penetration and a specific depth of bevel.*

The dimension for the root opening of a square groove butt joint is located inside the square groove weld symbol. If a single-bevel-groove weld or single-V-groove weld is to be deposited on the arrow side of the joint, the dimension for the root opening is indicated below the weld symbol. The groove angle for a groove weld is located under the dimension for the root opening. When a groove weld is specified on both sides of a joint, each groove angle is dimensioned, but the root opening is specified only once. **See Figure 45-13.** The size of a flare-groove weld is considered to extend only to the tangent points as indicated by dimensional lines. **See Figure 45-14.**

Plug or Slot Welds

The size of a plug or slot weld is shown to the left of the weld symbol. The depth, when less than complete penetration, is shown on the inside of the weld symbol. The center-to-center spacing (pitch) is shown to the right of the weld symbol, and the groove angle of countersink is shown below the weld symbol. Plug and slot welds are often used on lap joints. **See Figure 45-15.**

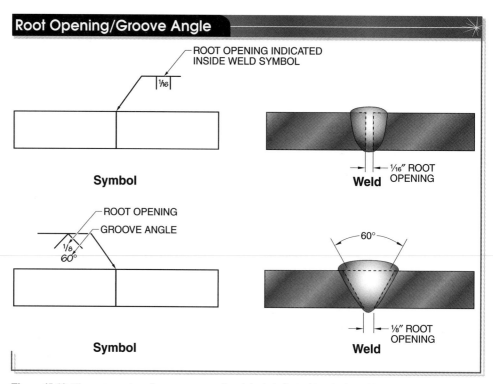

Figure 45-13. *The root opening of a square-groove butt joint is indicated inside the weld symbol. The root opening for beveled groove joints is indicated below the weld symbol. The groove angle is located under the dimension for the root opening.*

Flare-Groove Weld Symbols

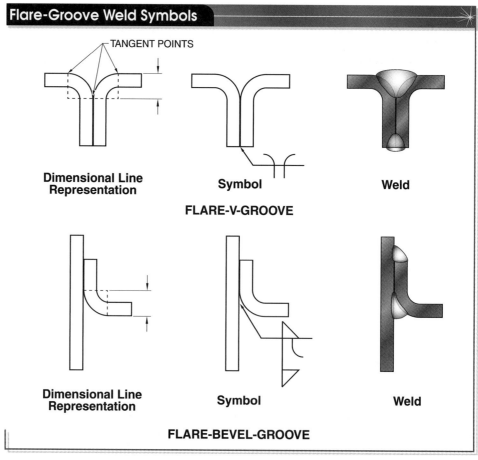

Figure 45-14. *The sizes of flare-V-groove and flare-bevel-groove welds are indicated by dimensional lines.*

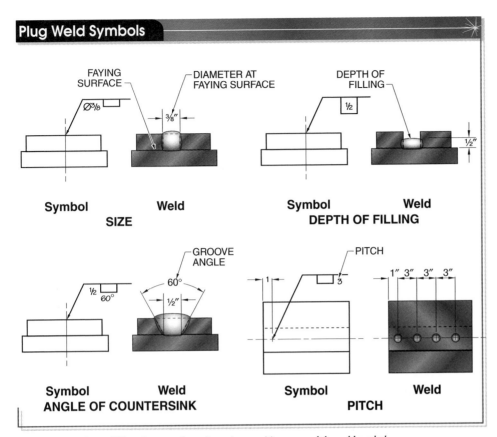

Figure 45-15. *Plug weld locations are shown in various positions around the weld symbol.*

Spot or Projection Welds

Spot welds are dimensioned by either size or by strength. Size is designated as the diameter of the weld and is expressed in fractions, decimals, or millimeters and placed to the left of the weld symbol. The strength requirement, when used, is placed to the left of the weld symbol and is expressed as the required minimum shear strength in pounds per spot weld. **See Figure 45-16.**

The spacing of spot welds is shown to the right of the weld symbol. When a definite number of spot welds are needed in a joint, this number is indicated in parentheses either above or below the reference line.

Seam Welds

Seam welds are dimensioned either by size or strength. Location and designation of sizes are similar to those used for fillet welds. Size is designated as the width of the weld in fractions, decimals, or millimeters and is shown to the left of the weld symbol. **See Figure 45-17.**

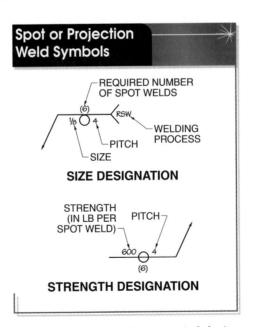

Figure 45-16. *Spot weld designations include size, strength, spacing, and number of spot welds.*

Spot, projection, and seam welds are welds produced by fusion between or on joint members using heat and pressure. A weld nugget is formed where the fusion occurs.

Seam Weld Symbols

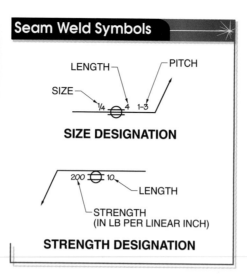

SIZE DESIGNATION

STRENGTH DESIGNATION

Figure 45-17. *Seam weld designations include size, strength, length of weld seam, and pitch of weld.*

The length of the weld seam is placed to the right of the weld symbol, and the pitch is shown to the right of the length dimension. The strength of the weld, when used, is located to the left of the weld symbol and is expressed as the minimum acceptable shear strength in pounds per linear inch.

Weld-All-Around Symbol

When a weld is to extend completely around a joint, a small circle is placed around the junction of the reference line and arrow, indicating that the weld should be performed all-around. **See Figure 45-18.**

Field Weld Symbol

The field welding symbol indicates welds to be made in the field (not in a shop or at the place of initial construction). Field welding symbols are shown by a darkened triangular flag at the juncture of the reference line and arrow. **See Figure 45-19.**

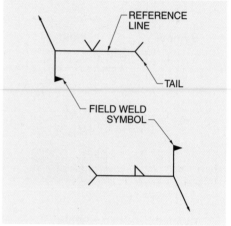

Figure 45-19. *The field weld symbol is placed at a right angle to the reference line at the junction with the arrow.*

Welding Symbol Tail

The welding symbol tail contains additional information about a weld that does not belong on the reference line. **See Appendix.** Information on the tail of a welding symbol may include the type of welding or cutting process to be used, a note pertaining to the orientation of an unequal leg fillet weld or a reference to a

Weld-All-Around Symbol

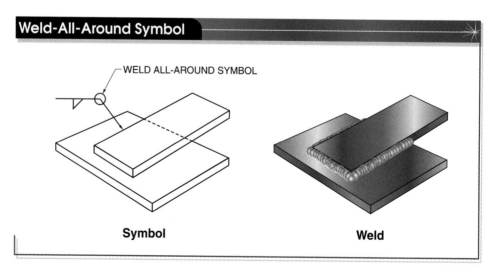

Symbol

Weld

Figure 45-18. *A small circle appears where the arrow connects the reference line to denote "weld-all-around."*

note located on the drawing, or a reference to a particular welding code, specification, or procedure. **See Figure 45-20.**

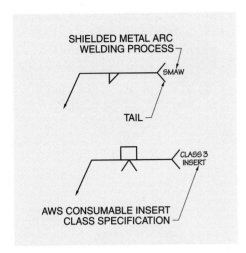

Figure 45-20. *The tail of the welding symbol is used when specific details or weld processes are required.*

Surface Contour of Welds

When bead contour is important, a special flat, concave, or convex contour symbol is added to the welding symbol. Welds that are to be mechanically finished also carry a finish symbol along with the contour symbols. **See Figure 45-21.**

Back Weld and Backing Weld

A back weld symbol and a backing weld symbol are both indicated by an unfilled semicircle opposite a groove weld symbol. The sequence of welding determines whether the weld is a back weld or a backing weld. A *back weld* is a weld deposited on the other side of a butt joint after a groove weld is made on the arrow side. Back welds are occasionally specified to ensure adequate penetration and provide additional strength to a joint. A *backing weld* is a weld applied to the other side of a butt joint before a groove weld is made on the arrow side. Backing welds prevent excessive penetration of the weld metal on the root side of the joint. **See Figure 45-22.**

When a single reference line is used, "back weld" or "backing weld" is specified in the tail of the welding symbol to indicate the sequence of welding. When multiple reference lines are used, a back weld is shown following a groove weld, while a back weld is shown preceding a groove weld. No dimensions other than height of reinforcement are shown on the welding symbol for back or backing welds.

SURFACING CONTOURS				
LETTER	MECHANICAL METHOD	SYMBOL		
		Flat	Convex	Concave
C	Chipping	$\sqrt{}_C$	$\sqrt{}_C$	$\sqrt{}_C$
H	Hammering	$\sqrt{}_H$	$\sqrt{}_H$	$\sqrt{}_H$
G	Grinding	$\sqrt{}_G$	$\sqrt{}_G$	$\sqrt{}_G$
M	Machining	$\sqrt{}_M$	$\sqrt{}_M$	$\sqrt{}_M$
R	Rolling	$\sqrt{}_R$	$\sqrt{}_R$	$\sqrt{}_R$
U	Unspecified	$\sqrt{}_U$	$\sqrt{}_U$	$\sqrt{}_U$
Symbol ($\sqrt{}_G$)			Weld — FLAT CONTOUR OBTAINED BY GRINDING WELD	

Figure 45-21. *A flat, concave, or convex symbol added to the welding symbol indicates how the surface should be contoured.*

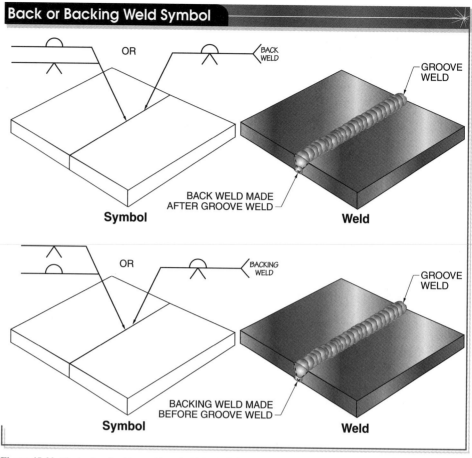

Back or Backing Weld Symbol

Symbol

OR

BACK WELD

BACK WELD MADE AFTER GROOVE WELD

Weld

GROOVE WELD

Symbol

OR

BACKING WELD

BACKING WELD MADE BEFORE GROOVE WELD

Weld

GROOVE WELD

Figure 45-22. *The back or backing weld symbol is included opposite the weld symbol with a note included in the tail. When multiple reference lines are used, back and backing welds are shown in the appropriate sequence.*

Melt-Through Symbol Welds

A solid semicircle on the reference line opposite the weld symbol indicates complete penetration or melt-through. The melt-through symbol is used when welds are made from one side only. No dimension of melt-through, except height of reinforcement, is shown on the welding symbol. **See Figure 45-23.**

Surfacing Welds

Welds whose surfaces must be built up by single- or multiple-pass welding are denoted by a surfacing weld symbol. The height of the built-up surface is indicated by a dimension placed to the left of the surfacing symbol. **See Figure 45-24.** The extent, location, and orientation of the area to be built up are normally indicated on the drawing.

> ✦ *Surfacing is often used to apply a hard wear surface to a part.*

NONDESTRUCTIVE EXAMINATION SYMBOLS

Nondestructive examination (NDE) symbols are symbols that specify examination methods and requirements to verify weld quality. The method of examination required can be specified on a separate reference line of the welding symbol or as a separate NDE symbol.

Whether the NDE method is specified on the same reference line as the weld symbol or on a separate reference line, the order of operations is the same as for multiple welding operations. The operation on the reference line nearest the arrowhead is performed first. The reference line furthest from the arrowhead indicates the last operation to be performed. When used separately, NDE symbols include an arrow, reference line, examination letter designation, dimensions, areas, number of examinations, supplementary symbols, tail, and specifications and other references. **See Appendix.**

Melt-Through Weld Symbol

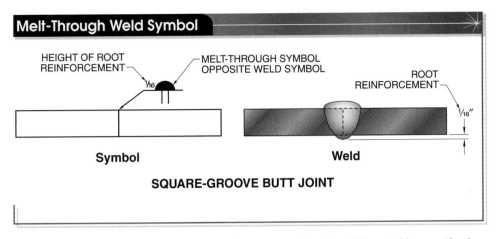

SQUARE-GROOVE BUTT JOINT

Figure 45-23. *A melt-through symbol indicates that complete joint penetration of the weld is required from one side only.*

Surfacing Welds

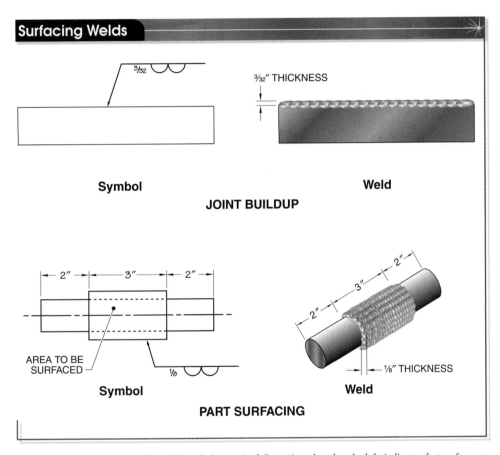

Figure 45-24. *A surfacing weld symbol, with the required dimension placed to the left, indicates that surfaces are to be built up by welding.*

 **Refer to Quick Quiz® on CD-ROM**

- A welding symbol is a graphical representation of the specifications for producing a welded joint.
- Instructions regarding the type of weld are indicated either above or below the reference line.
- The arrow side of the joint is the side of the joint to which the arrow points. The other side is the side of the joint opposite the arrow side.
- When more than one operation is required on a joint, a symbol is shown for each operation.
- Welds to be made in the field are shown by a darkened triangular flag at the juncture of the reference line and arrow.
- Nondestructive examination (NDE) symbols are symbols that specify examination methods and requirements to verify weld quality.

? **Questions for Study and Discussion**

1. What is meant by the arrow side of the welding symbol?
2. What is meant by the other side of the welding symbol?
3. Indicate the meaning of the following welding symbols.

4. What type of weld do these symbols indicate?

5. These symbols represent which weld specifications?

6. These symbols represent which weld specifications?

7. Draw completed welding symbols, including necessary information, to describe the following welds.

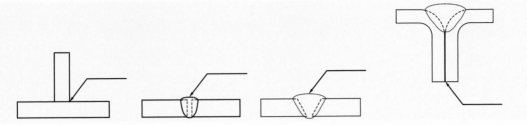

8. What do these welding symbols mean?

9. What do these welding symbols represent?

10. Using the appropriate table in the appendix, identify the parts of the master welding symbol shown.

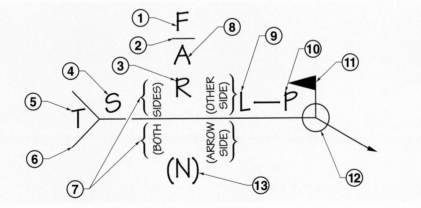

 Refer to Chapter 45 in the *Welding Skills Workbook* for additional exercises.

Materials and Fabrication Standards and Codes

46

Chapter

Materials and fabrication standards and codes provide a common language for ensuring consistency among products of various manufacturers. Purchase orders for materials must refer to materials standards. Certification accompanying products must be checked to ensure that the materials conform to indicated standards. Fabrication standards and codes ensure that materials and welded products meet specified mechanical property and quality requirements. Welders who work within specific industries must be aware of the codes and standards that affect them.

Quality requirements specified in materials and fabrication standards and codes are accepted by manufacturers, suppliers, and users as the basis for ordering and fabricating materials. The steps involved in specifying, procuring, and fabricating materials are addressed by materials and fabrication standards and codes.

Quality requirements for welding are based on the possible risks and consequences of failure of the equipment or component. Quality requirements for welding are established by industry groups and ensure the necessary quality at a reasonable cost.

MATERIALS STANDARDS

Materials standards are classified according to the kind of information they contain. Various organizations are responsible for the development of materials standards. Materials standards are developed and reviewed by qualified people organized into committees of producers, end users, and general interest groups.

Classification of Materials Standards

A *standard* is a document that, by agreement, serves as a model for the measurement of a property or the establishment of a procedure. "By agreement" means that all parties involved in the product, including manufacturers, suppliers, and end users, must agree to the use of the standard as being fair and practical. Materials standards are classified as specifications, recommended practices, and codes. **See Figure 46-1.**

A *specification* is a type of standard that indicates the technical and commercial requirements for a product. Material requirements are most often described by means of specifications. For example, ASTM A36 is a specification for structural steel members used in riveted, bolted, or welded construction of bridges and buildings, and for general structural purposes. ASTM A36 indicates acceptable methods of manufacture and minimum acceptable properties of structural steel members.

A *recommended practice* is a type of standard that provides instructions for performing one or more repetitive technical functions. For example, ASTM E165 is a recommended practice for conducting liquid penetrant testing. ASTM E165 indicates standard test parameters that should be followed to allow comparison between liquid penetrant tests performed on different welds or other items.

> ✓ **Point**
>
> *Materials standards and codes are developed by consensus (agreement) between parties representing producers, end users, and general interest groups.*

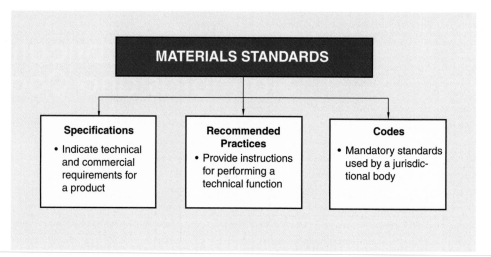

Figure 46-1. *Materials standards are classified as specifications, recommended practices, and codes.*

A *code* is a type of standard that is mandatory and is used by a jurisdictional body. A code indicates what "shall" be done rather than what "may" be done. For example, ASME (American Society of Mechanical Engineers) International administers the code for pressure piping. The code for pressure piping covers specific types of piping, such as for steam or petroleum products, and contains regulations for the design and fabrication of piping for the specific service category to achieve safe and reliable operation.

Standards Development

Standards are developed by standards committees. Standards committees consist of a balanced representation of producers, end users, and certain general interest groups to represent all interested parties. **See Figure 46-2.** Balanced representation ensures that standards are created that are acceptable to all representatives. Standards committees meet regularly, generally every six months, to consider actions on standards for which they are responsible. Actions on standards include new standards development or revision of existing standards.

New Standards Development. New standards development is initiated by task groups within standards committees. New standards development is a relatively slow and deliberate process.

The objective is to create documents that are acceptable to the majority of producers and end users whose businesses are affected by them.

Task groups develop draft documents. Draft documents are the starting point for new standards. The applicable standards committee reviews the draft document and suggestions for improvement are balloted by the committee. *Balloting* is a formal method of documenting and voting upon the reviewers' suggestions. Once the draft document is revised according to the ballot, the task group is disbanded. The revised draft standard becomes the responsibility of the standards committee. **See Figure 46-3.**

It is not necessary to ballot all reviewers' suggestions. For example, editorial content items and nonrelevant technical suggestions are not necessarily balloted. Editorial content items are proposed segments of a standard that do not affect technical content. Nonrelevant technical suggestions are proposed segments of a standard which, although technical, are not within the scope of the standard.

Several ballots are usually required before a draft standard is ready for review outside the standards committee. Outside review is also done through balloting. Supplementary review(s) may result in the standard being returned to the committee for further work, and so on.

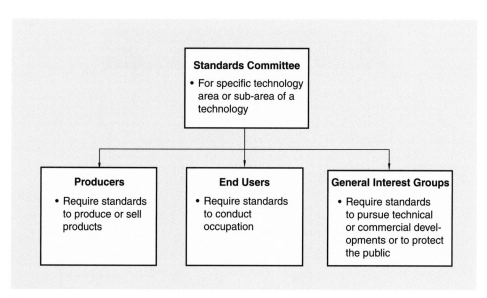

Figure 46-2. *Standards committees consist of a balanced representation of producers, end users, and certain general interest groups to represent all interested parties.*

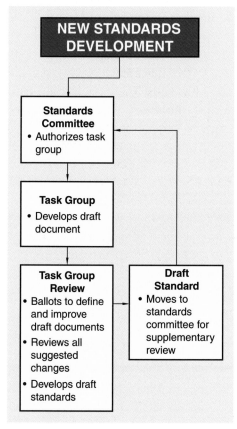

Figure 46-3. *The objective of new standards development is the creation of documents that are acceptable to the majority of producers and users whose business is affected by them.*

Existing Standards Revision. Existing standards revision is the job of the responsible committee. If necessary, responsibility may be transferred to another committee more closely aligned with the contents of the standard.

Standards must be reviewed regularly to maintain their relevance to current technical and commercial practices. The formal time interval for standards review varies from two years to five years, although review may be carried out whenever there is anything significant to address. The process for existing standards revision is similar to that for new standards development. However, existing standards revision is usually confined to specific segments of the existing standard that may have become irrelevant or obsolete through changes in technical or commercial practices. The specific segments are revised and balloted. **See Figure 46-4.**

The balloting process for existing standards revision is more rapid than new standards development because fewer parts of the existing standard are reviewed. If the entire standard is acceptable without changes, it is reissued as a reaffirmed standard. If the standard is modified, it is issued as a revised standard. Reaffirmed or revised standards carry the most current date or revision number. The latest issue of any standard supercedes previous issues. This rule applies to most, but not necessarily all, standards. Work with the latest issue of any standard, unless otherwise directed.

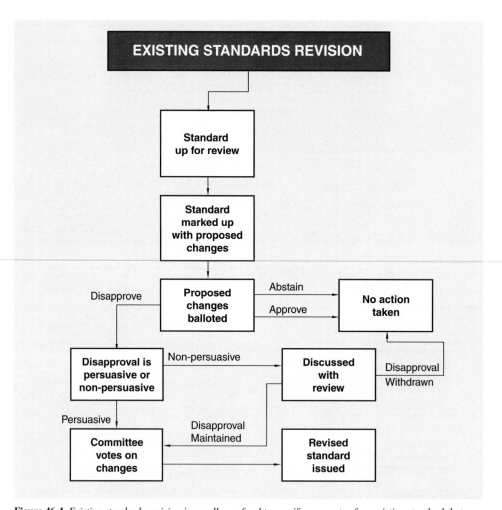

EXISTING STANDARDS REVISION

Standard up for review

Standard marked up with proposed changes

Proposed changes balloted — Abstain / Approve → No action taken

Disapprove

Disapproval is persuasive or non-persuasive — Non-persuasive → Discussed with review — Disapproval Withdrawn

Persuasive

Committee votes on changes — Disapproval Maintained

Revised standard issued

Figure 46-4. *Existing standards revision is usually confined to specific segments of an existing standard that may have become irrelevant or obsolete through changes in technical or commercial practices.*

User Enquiry

A *user enquiry* is a formal procedure developed by standards committees and code-creating organizations to help users interpret issues and offer suggestions. The intent of all user enquiry procedures is to maintain a channel of official communication between a standard- or code-writing committee and end users on questions or problems arising from the use of a standard or code. User enquiry procedures include scope, purpose, content, and proposed reply.

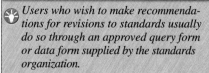

Users who wish to make recommendations for revisions to standards usually do so through an approved query form or data form supplied by the standards organization.

- Scope identifies the segment of the standard or code relevant to the enquiry. One item is addressed per enquiry.
- Purpose states the intent of the user enquiry—for example, to obtain an interpretation of a code requirement or to request revision of a particular segment of a standard.

- Content lists relevant paragraphs, figures, sketches, and tables in the code or standard that bear upon the user enquiry, with complete documentation to permit the standards committee to quickly and fully understand the enquiry. Technical justification must be provided if the user wants to obtain revision of the standard or code.
- Proposed reply to the user enquiry should be indicated when necessary. For example, when a revision of a particular segment of the standard is requested, the wording of a proposed revision must be supplied by the end user proposing the change.

The result of a user enquiry may be a temporary addendum to the standard or code to permit usage of the suggested modification. Temporary addenda contain a time limit for the proposed modification before it is formally balloted as a revision to the current version of the standard or code.

MATERIALS STANDARDS ORGANIZATIONS

Materials standards organizations that produce standards for base metals and welding consumables include ASTM International (ASTM), the Society of Automotive Engineers (SAE), Aerospace Material Specifications (AMS), the American Welding Society (AWS), ASME International, the American Petroleum Institute (API), and the American National Standards Institute (ANSI). Additionally, the Canadian Standards Association (CSA), the European Standards Council (CEN), and the International Organization for Standardization (ISO) develop standards globally, or for other countries.

ASTM International (ASTM)

ASTM International is the largest source of materials standards. From the work of over 130 standards-writing committees, ASTM International publishes standard test methods, specifications, practices, guides, classifications, and terminology. ASTM International standards cover metals, paints, plastics, textiles, petroleum, construction, energy, the environment, consumer products, medical services and devices, computerized systems, and electronics. ASTM International has no technical, research, or testing facilities. Such work is done voluntarily by 35,000 technically qualified ASTM International members worldwide.

More than 12,000 standards are published each year in 80 volumes of the *Annual Book of ASTM Standards.* These standards and related information are sold worldwide. ASTM International

standards used for base metals in welding contain information on the manufacturing practices and performance characteristics of materials in various product forms such as plate, bar, pipe, and rod.

Standards organizations may develop standards that are applicable in their home country only, or, as in the case of ASTM International, that have been adopted worldwide.

ASTM International Standards Designation. The ASTM International standards designation is based on a letter-number combination, such as A36 or B315. If the standard is tentative (issued on a trial basis), the year is followed by the letter T. If the standard is revised a second time in the same year, the date is followed by the letter a. If it is revised a third time in the same year, the date is followed by the letter b, etc. If the standard is a metric equivalent of another standard, the serial number is followed by an M. A *metric equivalent standard* is a version of a standard in which all the units are indicated in metric (SI) values. **See Figure 46-5.**

Embedded designations are unique materials identifications that are part of the standard. In most cases, an embedded designation must be coupled with ASTM International or other specification number to uniquely define a material. ASTM International and other materials standards usually refer to several different materials that are described by the prefix, grade, type, or class followed by a unique designation. For example, ASTM A193 is a specification for alloy and stainless steel bolting materials. However, to specify ASTM A193 alone is not enough. ASTM A193 includes embedded materials such as:

- Grade B7 (high-strength, low-alloy steel)
- Grade B8 class 1 (304 annealed stainless steel)
- Grade B8 class 2 (304 cold worked stainless steel)

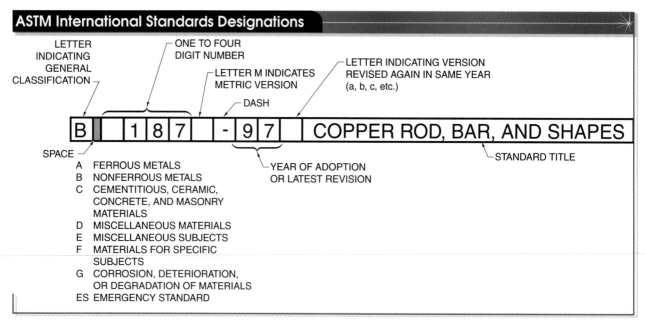

LETTER
INDICATING
GENERAL
CLASSIFICATION

ONE TO FOUR
DIGIT NUMBER

LETTER M INDICATES
METRIC VERSION

LETTER INDICATING VERSION
REVISED AGAIN IN SAME YEAR
(a, b, c, etc.)

DASH

B 1 8 7 - 9 7 COPPER ROD, BAR, AND SHAPES

SPACE

STANDARD TITLE

YEAR OF ADOPTION
OR LATEST REVISION

A   FERROUS METALS
B   NONFERROUS METALS
C   CEMENTITIOUS, CERAMIC,
    CONCRETE, AND MASONRY
    MATERIALS
D   MISCELLANEOUS MATERIALS
E   MISCELLANEOUS SUBJECTS
F   MATERIALS FOR SPECIFIC
    SUBJECTS
G   CORROSION, DETERIORATION,
    OR DEGRADATION OF MATERIALS
ES EMERGENCY STANDARD

Figure 46-5. *The ASTM International standards designation is based on a letter-number combination.*

Unified Numbering System. The *unified numbering system (UNS)* is a common embedded designation system that unifies all families of metals and alloys. The UNS uniquely identifies the chemical composition of alloys that have been fixed by other specification bodies. If the alloy is proprietary (produced by a limited number of suppliers), the chemical composition is established by the producer. The UNS consists of a capital letter followed by five numbers. The capital letter identifies the alloy family and, where possible, the five numbers are related to the pre-UNS designation of the alloy. **See Figure 46-6.**

Society of Automotive Engineers (SAE) and Aerospace Material Specifications (AMS)

The Society of Automotive Engineers (SAE) and Aerospace Material Specifications (AMS) follow standards for engineering materials used in on- and off-road vehicles, aircraft, and spacecraft. The SAE is a major source of technical information and expertise used in designing, building, maintaining, and operating self-propelled vehicles, whether land-, sea-, air-, or space-based. SAE collects, organizes, stores, and

disseminates information on cars, trucks, aircraft, space vehicles, marine equipment, and engines of all sizes.

SAE and AMS standards are administered by SAE and describe quality levels required for end use. AMS standards generally contain the most stringent quality requirements of any standards because they define the requirements for use in extremely critical services. AMS standards may be used in applications outside of the aerospace industry where stringent quality requirements justify the additional cost. For example, critical forgings for extreme cyclic (fatigue) applications may require materials manufactured to AMS specifications because the high degree of internal cleanliness required of materials that meet AMS specifications ensures high fatigue resistance.

American Welding Society (AWS)

American Welding Society (AWS) standards cover automatic, semiautomatic, and manual welding, as well as brazing, soldering, ceramics, lamination, robotics, and safety and health issues. AWS is organized into more than 180 committees, 125 of which are technical committees, involving 1400 members in the production of standards.

UNIFIED NUMBERING SYSTEM	
UNS Number	Type of Metal
Axxxxx	Aluminum and Aluminum alloys
Cxxxxx	Copper and Copper alloys
Exxxxx	Rare Earth and similar metals and alloys
Fxxxxx	Cast Irons
Gxxxxx	AISI and SAE Carbons and alloy Steels
Hxxxxx	AISI and SAE H-Steels
Jxxxxx	Cast Steels (except tool steels)
Kxxxxx	Miscellaneous Steels and ferrous alloys
Lxxxxx	Low melting metals and alloys
Mxxxxx	Miscellaneous nonferrous metals and alloys
Nxxxxx	Nickel and Nickel alloys
Pxxxxx	Precious metals and alloys
Rxxxxx	Reactive and refractory metals and alloys,
Sxxxxx	Heat and corrosion resistant steels (including stainless) Valve Steels, and Iron-base "superalloys"
Txxxxx	Tool Steels, wrought and cast
Wxxxxx	Welding filler metals
Zxxxxx	Zinc and Zinc alloys

Figure 46-6. *The unified numbering system consists of a capital letter followed by five numbers. The capital letter identifies the alloy family and the five numbers indicate the pre-UNS designation of the alloy.*

AWS standards also cover welding consumables. Filler metals are one category of welding consumables. Most commercial filler metals are identified by an AWS designation. Whenever possible, welding consumables should be referred to by AWS designations rather than commercial names.

Welding consumable requirements are standardized by AWS in a series of specifications based on the material family. For example, AWS A5.1 describes standard carbon steel covered arc welding electrodes. Embedded welding consumables are identified by letter-number designations within each specification.

AWS specifications indicate chemical compositions of materials and mechanical properties of the deposited weld metal using standardized welding procedures in a specified joint detail to produce weld specimens for testing. When required, specifications may also indicate other properties such as toughness or an acceptable amount of porosity. Most specifications include usability parameters such as the weld position for which the filler metal is designed, welding current that should be used, and in the case of covered electrodes, the type of coating. Size and packaging information is also provided. AWS publication FMC: *Filler Metal Comparison Charts,* lists commercial names for AWS filler metal designations. **See Figure 46-7.**

The AWS identification of welding filler metals consists of letters and numbers. The letters include R for rod, E for electrode, RB for rod or wire, and ER for electrode rod or wire. Rod is welding wire that is cut and straightened. Rod may be flux-coated or bare. Electrodes may be flux cored (tubular), consisting of a metal sheath packed with fluxes and alloying elements. Fluxes, when used separately from filler metals, are also classified. Since the welding consumable identifications embedded within AWS specifications are unique, they are often referred to without their specification number, such as E7018 or ER308.

ASME International (ASME)

ASME publishes codes and standards for the design, manufacture, and installation of mechanical devices.

A5.1, CARBON STEEL Covered Arc Welding Electrodes

See ANSI/AWS A5.1, Specification for Carbon Steel Electrodes for Shielded Metal Arc Welding

E6010

SOURCE	PRODUCT
AIR LIQUIDE CANADA INC.	LA 6010
Airco Filler Metals	Pipe-Craft
American Filler Metals Company	AFM 6010
American Welding Alloys	AWA 6010
Arcweld Products, Ltd.	EASYARC 10, EASYARC 10+
Askaynak Kaynak Teknigi Sanayi Ticaret A.S.	AS S-6010
Bohler Thyssen Welding USA, Inc.	Thyssen Cel 70, Bohler Fox Cel
CARBO-WELD Schweissmateriallen GmbH	CARBO RC 3
Champion Welding Products	CHAMPION E6010
D&H Secheron Electrodes Limited	CELLUTHERME
Electromanufacturas S.A.	West Arco XL-610, ZIP 10-T, West Rode 600/10
ESAB AB	Pipeweld 6010
ESAB WELDING & CUTTING PRODUCTS	SUREWELD 10-P, SUREWELD AP-100
EXSA S.A. - División Soldaduras OERLIKON	Cellocord P, PT
EXSA S.A. - División Fontargen	FON E 51 A, FON E 51 AT
HILARIUS HAARLEM HOLLAND BV	HILCO Pipeweld 6010
Hobart Welding Products	PIPEMASTER 60
HYUNDAI WELDING PRODUCTS, INC.	S-6010 D
Indura S.A. Industria y Comercio	INDURA 6010
Industrial Welding Corporation	NIHONWELD N-6010

American Welding Society

Figure 46-7. *AWS publication FMC: Filler Metal Comparison Charts, lists commercial names for AWS filler metal designations.*

ASME International Boiler and Pressure Vessel Code materials utilize selected ASTM and AWS specifications for base metals and welding consumables, but with minor changes to those specifications where they are too broad for boiler and pressure vessel applications. ASME International Boiler and Pressure Vessel Code-approved materials and welding consumables are assigned the prefix letter S to indicate approval. Only ASME Code-approved materials and welding consumables may be used for fabrication or repair of equipment built to the ASME International Boiler and Pressure Vessel Code.

ASME Pressure Piping Code materials carry ASTM and AWS specifications for base metals and welding consumables, respectively. Specific ASME pressure piping codes indicate which ASTM and AWS specifications are approved.

The ASME International Boiler and Pressure Vessel Code consists of 11 Sections. Each Section covers aspects of design, fabrication and inspection, care and operation, materials specifications, nondestructive testing, and welding and brazing qualifications. Some Sections consist of sub-parts known as Divisions.

The ASME International Boiler and Pressure Vessel Code is unique in that it requires third-party inspection independent of the fabricator and the user. Inspectors are commissioned by examination by the National Board of Boiler and Pressure Vessel Inspectors (NB). These authorized inspectors (AI) are employed by inspection agencies such as insurance companies or jurisdictional authorities. Users who are qualified to carry out pressure vessel fabrication and repair submit applications to have their own third-party inspectors, or owner-user inspectors.

A company must exhibit a quality control system and quality manual before fabricating a boiler or pressure vessel. The quality control system is audited by the authorized inspection agency and either the jurisdictional authority or the National Board. Based on successful audit of the fabricator's quality system, ASME may issue the fabricator a Certificate of Authorization and a code symbol stamp. The authorized inspection agency is involved in monitoring fabrication and field erection of boilers and pressure vessels. The AI must be satisfied that all applicable provisions of the ASME International Boiler and Pressure Vessel Code have been followed before allowing the fabricator to apply its code symbol stamp to the vessel nameplate.

Manufacturers and contractors who regularly build or install pressure vessels or pressure piping are required to have an ASME symbol stamp, indicating they have been approved by ASME as an authorized manufacturer of the type of equipment specified. Symbol stamps consist of letters designating the type of construction permitted.

American Petroleum Institute (API)

The American Petroleum Institute (API) develops materials standards applicable to petroleum storage and natural gas and petroleum transmission by pipeline. Pipe steels are low-carbon steels used in the oil and gas industries and include drill pipe, casing, tubing, and line pipe. API 5D, *Specification for Drill Pipe,* covers drill pipe. Casing is used to structurally restrain the walls of oil wells or gas wells, to exclude undesirable fluids, and to confine oil or gas to subsurface layers. Tubing is used within the casing of oil wells to conduct oil and gas to ground level. API 5CT, *Specification for Casing and Tubing,* covers casing and tubing. Line pipe (transmission pipe) is welded or seamless pipe used principally for conveying gas and oil. API 5L, *Specification for Line Pipe,* covers line pipe.

American National Standards Institute (ANSI)

The American National Standards Institute (ANSI) is a standards organization that adopts standards written and approved by member organizations. ANSI connects its member organizations by unifying their adopted standards. ANSI standards have been formally adopted at the national level. ANSI functions as coordinator of American national standards. ANSI also manages United States participation in international standards activities. An ANSI-approved standard retains its sponsor organization designation but additionally carries on the title page a descriptor indicating it is an American National Standard. For example, ASTM A36/A36M, *Standard Specification for Carbon Structural Steel,* is also an ANSI standard.

Canadian Standards Association (CSA)

The Canadian Standards Association (CSA) develops standards and certification requirements used throughout Canada. **See Figure 46-8.** CSA standards for filler metals are in general agreement with AWS specifications. Since Canada uses the metric (SI) system of units, the familiar AWS embedded designations such as E60XX or E70XX that are related to the tensile strength of the filler metal in ksi are changed to three-digit numbers corresponding to tensile strength in megapascals (MPa). CSA standards also indicate the diameter of the core wire in millimeters (mm).

CANADIAN STANDARDS	
Number	**Title**
CSA W47.1	"Certification of Companies for Fusion Welding of Steel Structures"
CSA W47.2	"Aluminum Welding Qualification Code"
CSA W117.1	"Code for Safety in Welding and Cutting (Requirements for Welding Operators)"
CSA W178	"Qualification Code for Welding Inspection Organizations"
CSA W55.3	"Resistance Welding Qualification Code for Fabricators of Structural Members Used in Buildings"
CSA S244	"Welded Aluminum Design and Workmanship (Inert Gas Shielded Arc Process)"
CSA W59	"Welded Steel Construction (Metal Arc Welding)"
CSA W48	"Welding Electrodes"
CSA W186	"Welding of Reinforcing Bars in Reinforced Concrete Construction"

Figure 46-8. *The Canadian Standards Association (CSA) develops standards and certification requirements used throughout Canada.*

✓ **Point**

Purchase orders for materials refer to applicable materials standards and codes.

European Standards Council (CEN)

European standards are produced by the European Standards Council (CEN) and are known as Euronorms. Euronorms have the prefix letters EN. Euronorms replace the standards of the individual countries of the European Community with single documents for specific items such as various base metals and filler metals.

International Organization for Standardization (ISO)

The International Organization for Standardization (ISO) promotes the development of standards to facilitate the international exchange of goods and services. ISO publishes several standards on welding electrodes.

USING MATERIALS STANDARDS

Materials standards provide information on commercially available base metals and welding consumables. The variations between materials standards permit specification and procurement consistent with the design and service requirements of the fabrication. Certification of products ensures that materials procured for welding and fabrication meet the specifications. The relatively high temperatures and stresses experienced during welding may lead to premature failure if improper materials are used. For example, the substitution of free-machining steel for low-carbon steel in a part to be welded may lead to failure because the presence of sulfur or selenium in the free-machining steel leads to hot cracking. The specified base metal and filler metal types indicated in welding procedure qualification records must be used, with no substitutions.

Variations between Materials Standards

Variations between materials standards allow them to cover many industrial applications and meet a wide range of quality requirements. Designers select standards that meet the required quality level for the intended service. Using a higher quality than necessary adds to cost. Using a lower quality than necessary may lead to premature failure in service.

For example, ASTM A53 and A106 are specifications for two types of steel piping. ASTM A53 is specified for piping for general use and is not made for any particular steelmaking process. ASTM A106 specifies fine-grain steelmaking practices that result in seamless steels (not shaped by seam welding) less prone to exhibit leakage throughout the pipe wall. These features make ASTM A106 more appropriate for critical service applications where failure from leakage or fracture might lead to injury or significant property damage. **See Figure 46-9.** The excess cost involved in using A106 for a general purpose application would be unnecessary since A53 would be acceptable.

STANDARDS COMPARISON					
Standard	**Type**	**Title**	**Most Common**	**Description**	**Uses**
ASTM A53	Specification	"Pipe, Steel, Black and Hot Dipped, Zinc Coated, Welded and Seamless"	Type E Grade B	• Resistance welded • Slightly higher carbon contact than Grade A • Not to fine-grain steelmaking practice	General use
ASTM A106	Specification	"Seamless Carbon Steel Pipe for High-Temperature Service"	Grade B	• Seamless • Balance of strength and weldability • Made to fine-grain steelmaking practice	Critical service

Figure 46-9. *Specification ASTM A53 is a general steel piping specification; ASTM A106 is preferred for critical applications.*

Certification

A *certification* is a notarized statement provided by a supplier verifying that a product meets the specification under which it is sold. Certification types include mill test report, product analysis, certificate of compliance, and filler metal approval.

A *mill test report (MTR),* or certificate of analysis (COA), is certification issued by the primary manufacturer (mill) verifying the chemical analysis and mechanical test properties of stock obtained from a starting ingot or billet of metal. The MTR is reviewed when the order is received. An MTR allows the receiver to check that the materials meet specifications. MTRs do not cost extra when requested in the original purchase order.

Product analysis is supplementary certification that a particular product form is fabricated from a specific billet of metal. Product analysis is performed on items such as tubing or pipe fittings to ensure that substitutions have not been made during processing of the metal. Testing procedures for product analysis are usually destructive, and components that are tested in order to generate a product analysis must be discarded. Product analyses may be included in the certification as a supplemental requirement in ASTM specifications at additional cost. Product analyses are required only at the discretion of the user.

A *certificate of compliance (COC)* is a statement that a material meets the specifications to which it was purchased. A certificate of compliance has little value unless the supplier has an acceptable quality program that verifies that the acceptance steps are valid and have been performed.

Filler metal approval is the process of testing samples of as-received filler metal to certify conformance to a specification. An approved inspector witnesses welding of test plates using electrodes selected at random and mechanical property tests carried out on samples of the test weldments. Approvals are granted for filler metals based on the results of the tests. The approved inspector places the approved product on a qualified products list (QPL).

Retention of filler metal on the approved lists may be subject to annual tests. Filler metal approvals include covered electrodes; submerged arc electrode wire with flux combinations; and flux cored arc welding electrodes with gas combinations.

✓ Point

A certification is a notarized statement that a material meets specifications.

✓ Point

A mill test report is a certification that provides results of chemical and mechanical property tests to indicate the material meets specifications.

MTR Segments

MTR segments indicate the conformance of a material to the standard. These include chemical analysis, mechanical properties, method of manufacture, and special requirements. Each MTR segment is checked against the standard it references to ensure the materials are as specified. Incoming materials are examined to ensure that their markings and dimensions conform to the standard.

When required by codes or standards, it is necessary to verify that received materials conform to the relevant MTR or product analysis. MTRs and product analyses are turned over to the responsible organization after verification or maintained on file. **See Figure 46-10.** To verify conformance to the MTR or product analysis, follow the procedure:

1. Verify that heat number(s) match the heat numbers recorded on the materials.
2. Verify that the chemical compositions are within the limitations indicated by the materials specification.
3. Verify that mechanical properties are within the limitations indicated by the materials specification.
4. Verify that special tests and supplementary requirements conform to the materials specification.

Weld Filler Metals

Weld filler metals are selected in compliance with AWS A5.01, *Filler Metal Procurement Guidelines*. The purchase order must indicate the filler metal specification and embedded designation, and filler metal diameter, length, and quantity. To verify that the filler metal meets the specification, the box and accompanying paperwork are checked. **See Figure 46-11.** When the box is opened for use, secondary verification is required. Secondary verification consists of supplementary inspection techniques and may include verification of the marking or tab on each piece of filler metal and of the filler metal diameter; and if required for critical applications, may include supplementary chemical analysis.

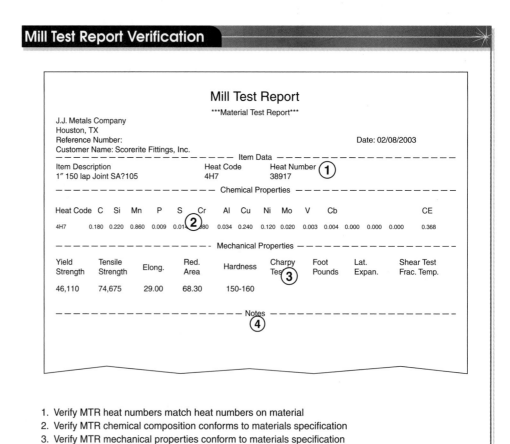

Mill Test Report Verification

Mill Test Report
Material Test Report

J.J. Metals Company
Houston, TX
Reference Number: Date: 02/08/2003
Customer Name: Scorerite Fittings, Inc.
—————————————————— Item Data ——————————————————

Item Description	Heat Code	Heat Number ①
1″ 150 lap joint SA?105	4H7	38917

—————————————— Chemical Properties ———————————————

Heat Code	C	Si	Mn	P	S	Cr ②	Al	Cu	Ni	Mo	V	Cb	CE
4H7	0.180	0.220	0.860	0.009	0.01	80	0.034	0.240	0.120	0.020	0.003	0.004 0.000 0.000 0.000	0.368

————————————— Mechanical Properties ——————————————

Yield Strength	Tensile Strength	Elong.	Red. Area	Hardness	Charpy Test ③	Foot Pounds	Lat. Expan.	Shear Test Frac. Temp.
46,110	74,675	29.00	68.30	150-160				

———————————————— Notes ————————————————
④

1. Verify MTR heat numbers match heat numbers on material
2. Verify MTR chemical composition conforms to materials specification
3. Verify MTR mechanical properties conform to materials specification
4. Verify MTR special tests and supplementary requirements conform to materials specification

Figure 46-10. *When required by codes or standards, it is necessary to verify that received materials conform to the relevant MTR or product analysis.*

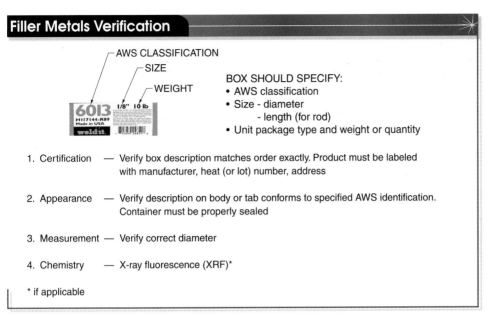

Filler Metals Verification

AWS CLASSIFICATION
SIZE
WEIGHT

6013
H117144-R89
Made in USA
weldit

1/8" 10 lb

BOX SHOULD SPECIFY:
- AWS classification
- Size - diameter
  - length (for rod)
- Unit package type and weight or quantity

1. Certification — Verify box description matches order exactly. Product must be labeled with manufacturer, heat (or lot) number, address

2. Appearance — Verify description on body or tab conforms to specified AWS identification. Container must be properly sealed

3. Measurement — Verify correct diameter

4. Chemistry — X-ray fluorescence (XRF)*

* if applicable

Figure 46-11. *Secondary verification consists of supplementary inspection techniques and may include verification of the marking or tab on each piece of filler metal, the filler metal diameter, and if required for critical applications, supplementary chemical analysis.*

FABRICATION STANDARDS AND CODES

Fabrication standards and codes are developed from many sources of experience on the reliability of weldment designs for different applications. Fabrication standard development has been driven by the need to define an adequate weld versus the perfect weld for a specific application. Ongoing field experience and research result in continuous refinement of weld quality requirements in industry codes and standards to maintain competitiveness of each segment of business.

Industry-based professional organizations write welding codes and standards. Codes are developed for regulated industries. Standards are developed for less regulated or nonregulated industries. Fabrication standards and codes cover pressure vessels and storage tanks, piping systems, construction, transportation, and heavy machinery. **See Appendix.**

Pressure Vessels and Storage Tanks

Pressure vessels and storage tanks may contain flammable, toxic, or corrosive liquids and gases. Pressure vessel or storage tank leakage or rupture may lead to significant loss of life and property damage. Pressure vessel and storage tank codes are written for the fabrication of boilers and pressure vessels, nuclear plants, storage tanks, and compressed gas containment systems. Additional in-service inspection and repair codes address repair of boilers, pressure vessels, and storage tanks that have been in service.

Boilers and Pressure Vessels. Boilers and pressure vessels, and items classified as pressure vessels, such as heat exchangers, must meet the requirements of the ASME International Boiler and Pressure Vessel Code in their design and fabrication. Many countries outside the USA and Canada recognize and accept the ASME International Boiler and Pressure Vessel Code, but an equal number of other countries accept only their own national code.

A *manufacturing data report (MDR)* is a legal document signed by the representatives of the manufacturer and the manufacturer's authorized inspection agency. An MDR certifies that all details of design, material, construction, and workmanship conform to the ASME International Boiler and Pressure Vessel Code.

> **✓ Point**
>
> *Fabrication standards and codes may be grouped into pressure vessels and storage tanks, piping systems, construction, transportation, and machinery.*

Pressure vessel fabrication requirements are typically covered under the ASME International Boiler and Pressure Vessel Code.

To etch a surface, the vessel is gently swabbed with a suitable acidic solution. After etching, excess acid is neutralized and removed by thoroughly flushing the surface with water. Sufficient water must flow over the surface to remove all traces of acid both from the nameplate and from the surface of the equipment.

By alternating grinding and etching, the nameplate stamping is made readable again. It may be necessary to experiment with the acid etching technique using a piece of aluminum or stainless steel sheet metal containing stamped identifications. The acid etching technique is a viable method of restoring damaged stamped identification tags on motors, tanks, and other items of equipment.

Repairs to boilers and pressure vessels are covered by in-service inspection and repair codes, National Board Inspection Code (NBIC), or API 510, depending on which code is recognized by the state in which the work is done. The purpose of in-service inspection and repair codes is to maintain the integrity of pressure boilers and pressure vessels after they have been placed in service by providing rules and guidelines for inspection after installation, repair, alteration, and rerating. Alteration is any repair that does not restore a mechanical component to its original design. Rerating is revision of the allowable design parameters of a mechanical component from the original design arising from formal study of its current condition. Rerating a pressure vessel results in changes to the design pressure and temperature, which must be recorded on the nameplate.

Any welding done on the pressure boundary of a pressure vessel is subject to the requirements of the applicable in-service inspection and repair code. A pressure boundary is a physical envelope that contains the working pressure of a piece of equipment. Welding in a plug or performing a weld repair to a heat exchanger tube-to-tubesheet joint is classified as a pressure vessel repair because it involves welding directly on a

Many states require that an ASME symbol stamped pressure vessel be registered with the National Board. The pressure vessel is assigned a number, known as the National Board number. The National Board number is shown on the MDR and on the vessel nameplate. The manufacturer sends two copies of the MDR to the National Board, which keeps one on file and sends the other to the state where the pressure vessel will be installed.

A pressure vessel loses its ASME International Boiler and Pressure Vessel Code identity if the MDR is missing and cannot be replaced, or if the nameplate is obliterated. Depending on the jurisdictional authority, such a vessel may need to be replaced. A nameplate must always be clearly visible. With an insulated vessel, a cutout should be made to ensure visibility. Do not paint over a nameplate or otherwise obliterate it.

Information should be restored to a nameplate should it be removed or otherwise deleted. The acid etching technique often reveals information that has been stamped on sheet metal. The acid etching technique consists of grinding, etching, and neutralizing.

Stamped information is revealed using a pencil grinder to very lightly grind the surface of the nameplate to reveal the information.

⚠ WARNING

Acids should be handled in accordance with written procedures to prevent personal injury and equipment damage.

pressure boundary of the heat exchanger. **See Figure 46-12.**

Repair organizations that make repairs or alterations will usually have an "R" or "NR" symbol stamp issued by the National Board of Boiler and Pressure Vessel Inspectors.

Figure 46-12. *Repairs made to a heat exchanger tube-to-tubesheet joint are classified as pressure vessel repairs because they involve welding directly on a pressure boundary of the heat exchanger.*

Nuclear Plants. Nuclear plant components such as nuclear reactors and materials used in nuclear plants are covered by the provisions of Section III of the ASME International Boiler and Pressure Vessel Code and the Nuclear Regulatory Commission Specification, *Quality Assurance Criteria for Nuclear Power Plants and Fuel Reprocessing Plants.* One exception is nuclear plant components developed for naval ships, which are covered by a code issued by the Department of Defense (DOD) Naval Ship Division Code, *Standard for Welding Reactor Coolant and Associated Systems and Components for Naval Nuclear Power Plants.* The DOD code is similar to Section III of the ASME International Boiler and Pressure Vessel Code. Because of the significant consequences of a failure, nuclear plant codes impose the strictest certification requirements on materials and the traceability of all materials to the point of origin.

Storage Tanks. Storage tanks consist of aboveground storage tanks and elevated storage tanks. Aboveground storage tanks usually contain various fluids, such as petroleum products and chemical solutions, and usually rest on a concrete slab or dunnage. *Dunnage* is a series of steel I-beams parallel to one another. Elevated storage tanks contain water, and may contain petroleum products, and rest on steel towers. All storage tank codes and standards refer to Section IX of the ASME International Boiler and Pressure Vessel Code for welding qualification.

Aboveground storage tanks are designed and fabricated based on the pressure of the tank. API STD 620, *Design and Construction of Large, Welded, Low Pressure Storage Tanks,* covers the design and construction of field-welded pressure tanks used for storage of petroleum intermediates and finished products under a pressure of 15 psig or less (low pressure). API STD 650, *Welded Steel Tanks for Oil Storage,* covers the material, design, fabrication, erection, and testing requirements for vertical, cylindrical welded steel storage tanks that are above ground and not subject to internal pressure.

Fabrication requirements for elevated steel tanks are described in the joint American Waterworks Association (AWWA) and American National Standards Institute (ANSI) ANSI/AWWA D100, *Standard for Welded Steel Tanks for Water Storage.* The joint standard provides a purchase specification to facilitate the manufacture and procurement of welded steel tanks for the storage of water.

Aboveground storage tank repair is covered in in-service inspection and repair code API STD 653, *Tank Inspection, Repair, Alteration, and Reconstruction.* API 653 is based on accumulated knowledge of owners, manufacturers, and repairers of steel storage tanks. API 653 provides guidance in the inspection, repair, alteration, and reconstruction of steel aboveground storage tanks used in the petroleum and chemical industries. Welding requirements are based on equivalence standard API 650.

Compressed Gas Equipment. The Compressed Gas Association (CGA) develops compressed gas equipment standards. CGA C-3, *Standards for Welding on Thin-Walled, Steel Containers,* covers welding requirements in the manufacture and repair of Department of Transportation (DOT) compressed cylinders.

Piping Systems

Piping systems, like pressure vessels, may transport flammable, toxic, or corrosive liquids. Piping systems are usually more susceptible to catastrophic failure consequences compared with pressure vessels or tanks because piping systems contain many joints and often consist of long exposed runs that may be subject to mechanical abuse. Piping system design, fabrication, and repair are covered by codes that encompass pressure piping, line piping, and water piping.

Pressure Piping. Pressure piping in thermal and nuclear power plants, refineries, and chemical plants is designed and fabricated in accordance with ASME B31, Code for Pressure Piping. Pressure piping is usually medium- to thick-wall (described by schedule) and medium- to large-size (described by diameter). The ASME Code for Pressure Piping is divided into seven Sections applicable to different end-use categories of pressure piping.

Welding procedures and qualifications vary according to the applicable Section of the ASME Code for Pressure Piping. Welding procedures and qualifications are generally in accordance with Section IX of ASME International Boiler and Pressure Vessel Code unless other codes or qualifications are referred to.

Pressure piping repair is covered by in-service inspection and repair code API 570, *Piping Inspection Code: Inspection, Repair, Alteration, and Rerating of In-Service Piping Systems.* API 570 is also applicable to ASME B31.3, *Process Piping,* and other pressure piping code sections. API 570 establishes requirements and guidelines that allow owners and users of piping systems to maintain the safety and integrity of the piping systems that have been placed into service. All repair and alteration welding must be done in accordance with ASME B31.3, or the code to which the piping system was built.

Line Piping (Cross Country Piping). Line piping consists of transmission and distribution piping that transports fuel gases, crude petroleum, and petroleum products. *Transmission piping* is medium- to high-strength steel, relatively thin-wall and large-diameter, and conveys products from locations of production to intermediate facilities. *Distribution piping* is carbon-steel, standard-size pipe of small diameter that conveys products from intermediate facilities to consumers.

Transmission piping welding requires special techniques and procedures and is governed by API 1104. API 1104 applies to arc welding and oxyfuel welding of piping used in the compression, pumping, and transmission of fuel gases, crude petroleum, and petroleum products. API 1104 presents methods for the production of acceptable welds by qualified welders using qualified welding procedures, materials, and equipment. It also contains acceptability standards and standards for repair of weld defects. API 1104 also applies to distribution piping where applicable.

Line piping repair and maintenance are covered in API Recommended Practice 1107. The primary purpose of API Recommended Practice 1107 is safety. It prohibits unsafe practices and warns against practices for which caution is necessary. API Recommended Practice 1107 includes methods for the inspection and repair of welds, and for installing appurtenances on loaded piping systems.

Water Piping. Water piping is made of low-carbon steel. AWWA C206: Field Welding of Steel Water Pipe, covers the welding of circumferential joints as well as the fabrication and installation of specials and accessories. The maximum thickness of piping covered by AWWA C206 is 1¼″.

Construction

Construction applications of welding encompass structural steel and aluminum used for buildings and highway bridges; reinforcing steel for concrete; and sheet metal. Welded joint types and configurations in construction applications are critical to the integrity of the component. Catastrophic failure may cause loss of life, injury, and costly related damage.

Structural Steel. Structural steel fabrication practices for constructing buildings and edifices are comprehensively regulated to prevent unsafe conditions during or after construction. Steel buildings welded in most cities in North America are covered by codes and specifications. Many large cities publish their own specific codes, while others follow AWS D1.1, *Structural Welding Code–Steel.* AWS D1.1 covers welding requirements for any type of welded structure made from commonly used carbon and low-alloy structural steels. AWS D1.1 does not apply to base metals less than ⅛″ thick. Additionally, it contains allowable unit stresses, structural details, workmanship standards, inspection procedures, and acceptance criteria. AWS D1.1 contains sections devoted exclusively to buildings (static loading), bridges (dynamic loading), and tubular structures.

Structural Aluminum. Structural aluminum is used for its lightness coupled with its strength and atmospheric corrosion resistance. Welding requirements for structural aluminum are contained in AWS D1.2, *Structural Welding Code–Aluminum.* AWS D1.2 contains general rules for the regulation of welding in aluminum construction plus additional, supplementary

rules applicable to statically loaded structures, dynamically loaded structures, and tubular structures.

Sheet Metal. Sheet metal is metal that is ⅛″ thick or less, corresponding to a gauge number of 11 or higher. The higher the number, the thinner the gauge. Under normal manual or semiautomatic welding conditions, sheet metal as thin as 0.035″ or roughly 20-gauge can be welded. There are two sheet metal welding codes, which apply to structural and nonstructural applications.

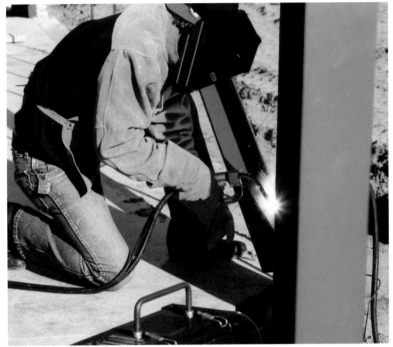

The Lincoln Electric Company

AWS D1.1, Structural Welding Code–Steel, contains sections devoted exclusively to buildings.

AWS D1.3, *Structural Welding Code–Sheet Steel,* covers requirements for welding sheet steel having a minimum specified yield point no greater than 80 ksi. AWS D1.3 covers sheet steel with or without zinc coating (galvanizing). The welding may involve connections of sheet or strip steel to thicker supporting structural members, in which case provisions of AWS D1.1, *Structural Welding Code–Steel,* also apply.

AWS D9.1, *Sheet Metal Welding Code,* covers nonstructural sheet metal requirements. AWS D9.1 provides requirements for welding carbon steel, low-alloy steel, austenitic and ferritic stainless steel, aluminum, copper, and nickel alloy sheet steels. AWS D9.1 provides requirements for nonstructural fabrication and erection of sheet metal by welding and braze welding for heating, ventilating, and air conditioning systems; architectural usage; food processing equipment; and similar applications. Where differential air pressures of more than 120″ (30 kPa) of water or structural requirements are involved, other standards are to be used.

Reinforcing Steel. Reinforcing steel is high-carbon steel rod used to reinforce concrete for structural applications and is manufactured to ASTM A615. AWS D1.4, *Structural Welding Code–Reinforcing Steel,* covers requirements for welding reinforcing steel in most reinforced concrete applications. AWS D1.4 contains regulations for welding reinforcing steel, and provides acceptable criteria for such wel . Highway bridge welding is under the jurisdiction of the state or provincial department of transportation, either by reference to, or by direct copy of AWS D1.5, *Bridge Welding Code.* AWS D1.5 is a joint standard of the American Association of State Highway and Transportation Officials (AASHTO) and the AWS. AWS D1.5 covers welding requirements for AASHTO welded highway bridges made from carbon and low-alloy steels. Failure-critical members of a bridge may require special standards of welded workmanship only by organizations having the proper personnel, experience, procedures, knowledge, and equipment. A *failure-critical member* is a tension member or component whose failure would likely result in collapse of the structure.

Many states supplement the AASHTO and AWS requirements with their own additional standards. Some states require welders to be examined yearly and be certified by the state to work on bridges. Some states maintain rosters of certified welders.

Transportation

The transportation industry represents a diverse set of end uses for welded products. Welded joints in transportation equipment are subject to tensile, compressive, torsional, bending, and shear stresses, in addition to fatigue stresses because of loading and motion. Transportation welding is not as regulated as welding in other industry segments, with the exception of certain types of transportation where there is significant opportunity for catastrophe in the event of failure. Transportation welding standards and codes cover automobiles and trucks, railroad cars and locomotives, aircraft and aerospace vehicles, ships and barges, shipping containers, and underwater welding.

Automobiles and Trucks. Automobile and truck welding is usually carried out by resistance welding and robotic arc welding. **See Figure 46-13.** For high production rates such as automobile and truck subassemblies, multiple spot welding machines are used. Welding specifications for resistance and arc welding are covered in joint standards created by the SAE and the AWS. AWS Recommended Practice D8.7, *Automotive Weld Quality–Resistance Spot Welding,* covers quality requirements for resistance spot welding of common automotive sheet steel systems, excluding high-strength low-alloy steel. AWS D8.8, *Specification for Automotive Weld Quality–Arc Welding of Steel,* defines practical tolerances for good fit-up in order to achieve satisfactory weld quality in automotive structural parts joined by robotic welding. Metal stampings and press-formed parts must be made to produce weld joint fit-up within the allowances of the specification.

Railroad Cars and Locomotives. Repair of railroad cars and locomotives is in accordance with AWS D15.1, *Railroad Welding Specification–Cars and Locomotives.* AWS D15.1 is jointly developed with the Association of American Railroads (AAR). Part I covers specific requirements for welding in the railroad industry. Part II covers specific requirements for welding on railroad freight cars other than tank cars. Welding on freight cars is performed as required in Part I except as specifically detailed in Part II. The rules for welding on tanks in tank cars are covered by the ASME International Boiler and Pressure Vessel Code. Part III of AWS D15.1 covers specific requirements for welding locomotives with emphasis on the welding of base metals less than 1/8″ thick.

Chrysler Corporation

Figure 46-13. *Automobile and truck welding is usually performed by robotic arc welding.*

Aircraft and Aerospace Vehicles. The United States Department of Defense (DOD) standard, MIL-STD-1595, *Qualifications of Aircraft, Missile, and Aerospace Fusion Welders,* establishes the procedure for welders and welding operators engaged in the fabrication of components for aircraft, missiles, and other aerospace equipment by fusion welding processes. The standard is applicable when required in the contracting documents, or when invoked in the absence of a specified welder qualification document. MIL-STD-1595 covers many welding processes, metals, and levels of proficiency for testing welders. Qualification to this standard is performed under the supervision of government inspectors.

Ships and Barges. Ship and barge welding requirements are covered by jurisdictions or insurance companies. The American Bureau of Shipping (ABS) issues *Rules for Building and Classing Steel Vessels,* one section of which covers welding requirements. These rules are required for ships registered and insured in the United States. ABS also approves specific welding consumables in *Approved Welding: Electrodes, Wire-Flux, and Wire-Gas Combinations.* Many insurance companies also publish specifications that cover welding. All United States federal government vessels are covered by codes issued by the U.S. Coast Guard or the Navships Division, Department of Defense. Their requirements are covered, respectively, in *Marine Engineering Regulations,* subchapter F, Part 57, *Welding and Brazing;* and *Fabrication, Welding, and Inspection of Ships Hulls,* Navships 0900-000-1000.

AWS D3.5, *Guide for Steel Hull Welding,* provides information on practical methods to weld steel hulls for ships, barges, mobile offshore drilling units, and other marine vessels. The guide provides information on weldability of steel plates, shapes, castings, and forgings. Hull construction is discussed in terms of preparation of materials, erection and fitting, and distortion control.

AWS D3.7, *Guide for Aluminum Hull Welding,* provides information on welding aluminum hulls and related ship structures. It applies chiefly to the welding of aluminum hulls that are over 30′ in length and made of sheet and plate 3/16″ thick or more. The distinction is made because there are different requirements for welding thin (less than 3/16″) and thick (greater than 1/4″) aluminum.

Shipping Containers. Shipping containers are used to transport gas under high pressure and for tanks carrying liquid petroleum and similar products. The fabrication of shipping containers is under strict regulation because of the serious consequences of failure. The United States Government publishes the Code of Federal Regulations (CFR), which includes standards that govern the fabrication of shipping containers. The applicable standards are 49 CFR 178.345, *General*

Design and Construction Requirements; and 49 CFR 178.337–Specification MC 331, *Cargo Tank Motor Vehicles.*

Underwater Welding. Underwater welding can be performed in wet or dry environments. Wet underwater welding (welding in the wet) is done under fully immersed conditions and produces relatively poor quality welds that are intended for temporary applications. Dry underwater welding (welding in the dry) is achieved by creating a local underwater environment free of water in which to perform welding. High-quality welds are possible with dry underwater welding. **See Figure 46-14.**

AWS D3.6, *Specification for Underwater Welding,* covers the requirements for wet and dry underwater welding. Weld quality categories (classes) are linked to weld quality requirements. Class A is for welds comparable in quality to above-water welding. Class B is for less critical applications. Class C is for applications where load bearing is not a primary consideration. Class O is for when it is necessary to meet the requirements of another designated code or specification.

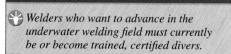

Welders who want to advance in the underwater welding field must currently be or become trained, certified divers.

Machinery

Heavy machinery is subject to rotation, vibration, sudden (impact) or slow application of large loads, and load reversals (fatigue). There are no codes that cover welding of heavy machinery. However, AWS publishes standards that cover the welding of overhead cranes and material handling equipment, machine tools, earthmoving and construction equipment, and rotating equipment. AWS standards for heavy machinery welding indicate minimum requirements for welded fabrication of the types of equipment covered.

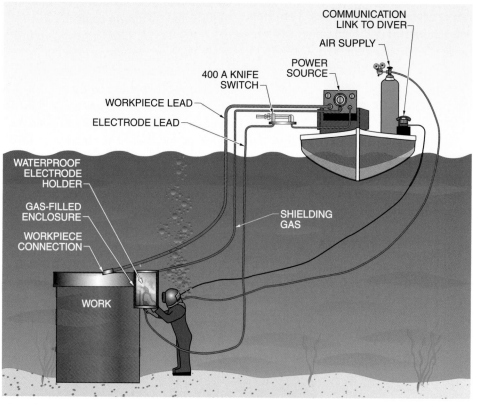

Figure 46-14. *High-quality welds are possible with dry underwater welding.*

Overhead Cranes and Material Handling Equipment. Overhead cranes and material handling equipment welding uses plate girders and other welded plate structures rather than rolled beams normally used in fabricating steel for bridges and buildings. Overhead cranes and material handling equipment are subject to vibration and moving loads. Service conditions and the associated fully reversible loading to which cranes and equipment are exposed results in a large number of load cycles in a relatively short period and local bending stresses of significant levels. AWS D14.1, *Specification for Welding Industrial Mill Cranes and Other Material Handling Equipment,* covers base metals, filler metals, joint designs, and qualification of welders and welding operators who work on overhead cranes and material handling equipment.

Machine Tools. Machine tool welding is covered in AWS D14.2, *Specification for Metal Cutting Machine Tool Weldments,* which details requirements for the manufacture and repair of machine tool components, including structures and castings. Filler metals are recommended for the applicable base metals and include carbon steels, low-alloy steels, and austenitic stainless steels. Joint designs and unit stresses are provided for fillet and groove welds.

Earthmoving and Construction Equipment. Earthmoving and construction equipment welding is covered in AWS D14.3, *Specification for Welding Earthmoving and Construction Equipment,* which applies to all structural welds used in such equipment. AWS D14.3 reflects welding practices used by manufacturers within the industry and incorporates various methods that have been proven successful by individual manufacturers. No restrictions are placed on the use of any welding process or procedure, provided the weld produced meets the qualification requirements of the specification.

Rotating Equipment. Rotating equipment welding, such as on fans, pumps, and compressors, is covered in AWS D14.6, *Specification for Welding of Rotating Elements of Equipment.* The standard covers base metals; welding processes; filler metals; welding procedure and performance qualification; fabrication requirements; inspection and quality control; and modification and repair.

General Machinery. Machine and equipment welding is covered in AWS D14.4, *Specification for Welded Joints in Machinery and Equipment.* The standard covers weld joint design, base metals, and welding procedure qualification and test methods.

 Refer to Quick Quiz® on CD-ROM

- Materials standards and codes are developed by consensus (agreement) among parties representing producers, end users, and general interest groups.
- Codes are mandatory standards that have been adopted by a jurisdictional body.
- Standards types include specifications, recommended practices, and codes.
- Two types of activity in standards creation are new standards development and existing standards revision.
- Various industry groups write materials standards and codes, but the largest set of standards is produced by ASTM International (ASTM).
- Standards pertaining to welding are published by AWS and cover welding processes, filler metals, and health.
- Purchase orders for materials refer to applicable materials standards and codes.
- A certification is a notarized statement that a material meets specifications.
- A mill test report is a certification that provides results of chemical and mechanical property tests to indicate the material meets specifications.
- Fabrication standards and codes may be grouped into pressure vessels and storage tanks, piping systems, construction, transportation, and machinery.

Questions for Study and Discussion

1. What types of groups must interact in order to create an effective industry standard?
2. What is the difference between a specification and a recommended practice?
3. What is the difference between a standard and a code?
4. What organization is the largest source of materials standards?
5. Explain each of the components for an ASTM material designated as A193-97 grade B7 (i.e., A, 193, 97, and grade B7).
6. Why is it necessary to indicate not only the ASTM standard number for a material but also the embedded grade, type, or class?
7. What are the AWS prefixes for rod, electrode, rod or wire, and electrode rod or wire?
8. How is an ASME material identified compared with an equivalent ASTM material?
9. How is an ASME filler metal identified compared with an equivalent AWS material?
10. What is the difference between a certification and a mill test report?
11. Does a certificate of compliance provide numerical information on analysis or properties of a material?
12. What type of information is contained in a manufacturing data report for a pressure vessel?

Refer to Chapter 46 in the *Welding Skills Workbook* for additional exercises.

Appendix

STANDARD WELDING TERMINOLOGY	
Common (Field) Terminology	**Standard AWS Terminology**
Welding Processes	
Gas welding	Oxyfuel gas welding (OFW)
Stick welding	Shielded metal arc welding (SMAW)
TIG welding	Gas tungsten arc welding (GTAW)
MIG welding	Gas metal arc welding (GMAW)
Short arc	Short circuiting transfer
Spray arc	Spray transfer
Welding Terms	
Arc gap, electrode gap	Arc length
Arc gas	Orifice gas
Back-up bar	Backing or backing bar
Blowhole, gas pocket, or wormhole	Porosity
Burn-through	Melt-through
Cap pass	Cover pass
Cold lap	Incomplete fusion
Contact tube	Contact tip
Cup or gas cup	Gas nozzle
Downhand	Flat position welding
Edge-flange weld	Edge weld in a flanged butt joint
Fill pass or filler pass	Intermediate weld pass
Filler bead	Intermediate weld bead
Flame cutting or gas cutting	Oxygen cutting
Ground clamp, Welding ground, or work connection	Workpiece connection
Ground lead or work lead	Workpiece lead
Included angle	Groove angle
Joint opening	Root opening
Land	Root face
Machine welding	Mechanized welding
Metallizing	Thermal spraying
Molten weld pool	Weld pool
Nondestructive evaluation or Nondestructive testing	Nondestructive examination
Parent metal	Base metal
Postweld heat treatment	Postheating
Puddle or weld puddle	Weld pool
Root gap	Root opening
Shoulder	Root face
Shrinkage stress	Residual stress
Skip weld	Intermittent weld
Silver soldering	Brazing
Soft solder	Solder
Hard solder	Brazing filler metal
Suck-back	Underfill
Vertical down	Downhill
Vertical up	Uphill
Wash pass	Cover pass

MASTER CHART OF WELDING AND JOINING PROCESSES

ARC WELDING (AW)

arc stud welding	SW
atomic hydrogen welding	AHW
bare metal arc welding	BMAW
carbon arc welding	CAW
gas carbon arc welding	CAW-G
shielded carbon arc welding	CAW-S
twin carbon arc welding	CAW-T
electrogas welding	EGW
flux cored arc welding	FCAW
gas-shielded flux cored arc welding	FCAW-G
self-shielded flux cored arc welding	FCAW-S
gas metal arc welding	GMAW
pulsed gas metal arc welding	GMAW-P
short circuit gas metal arc welding	GMAW-S
gas tungsten arc welding	GTAW
pulsed gas tungsten arc welding	GTAW-P
magnetically impelled arc welding	MIAW
plasma arc welding	PAW
shielded metal arc welding	SMAW
submerged arc welding	SAW
series submerged arc welding	SAW-S

RESISTANCE WELDING (RW)

flash welding	FW
pressure-controlled resistance welding	RW-PC
projection welding	PW
resistance seam welding	RSEW
high-frequency seam welding	RSEW-HF
induction seam welding	RSEW-I
mash seam welding	RSEW-MS
resistance spot welding	RSW
upset welding	UW
high-frequency	UW-HF
induction	UW-I

SOLDERING (S)

dip soldering	DS
furnace soldering	FS
induction soldering	IS
infrared soldering	IRS
iron soldering	INS
resistance soldering	RS
torch soldering	TS
ultrasonic soldering	USS
pressure gas soldering	WS

WELDING AND JOINING PROCESSES

SOLID STATE WELDING (SSW)

coextrusion welding	CEW
cold welding	CW
diffusion welding	DFW
hot isostatic pressure welding	HIPW
explosion welding	EXW
forge welding	FOW
friction welding	FRW
direct drive friction welding	FRW-DD
friction stir welding	FSW
inertia friction welding	FRW-I
hot pressure welding	HPW
roll welding	ROW
ultrasonic welding	USW

OXYFUEL GAS WELDING (OFW)

air acetylene welding	AAW
oxyacetylene welding	OAW
oxyhydrogen welding	OHW
pressure gas welding	PGW

BRAZING (B)

block brazing	BB
diffusion brazing	DFB
dip brazing	DB
exothermic brazing	EXB
furnace brazing	FB
induction brazing	IB
infrared brazing	IRB
resistance brazing	RB
torch brazing	TB
twin carbon arc brazing	TCAB

OTHER WELDING AND JOINING

adhesive bonding	AB
braze welding	BW
arc braze welding	ABW
carbon arc braze welding	CABW
electron beam braze welding	EBBW
exothermic braze welding	EXBW
flow brazing	FLB
flow welding	FLOW
laser beam braze welding	LBBW
electron beam welding	EBW
high vacuum	EBW-HV
medium vacuum	EBW-MV
nonvacuum	EBW-NW
electroslag welding	ESW
consumable guide electroslag welding	ESW-CG
induction welding	IW
laser beam welding	LBW
percussion welding	PEW
thermite welding	TW

American Welding Society

MASTER CHART OF ALLIED PROCESSES

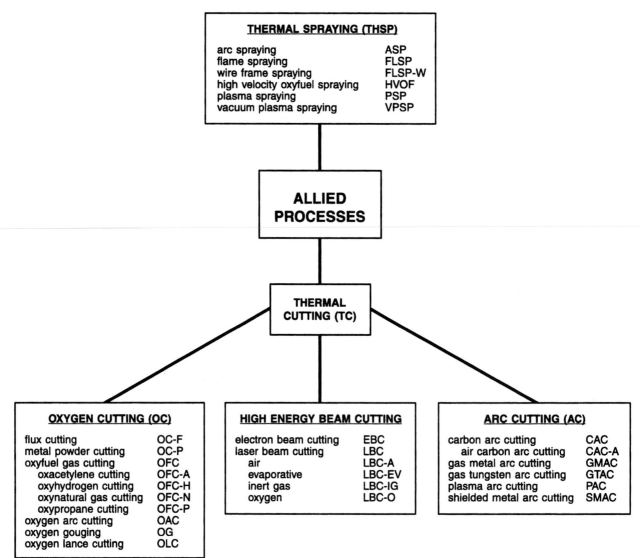

THERMAL SPRAYING (THSP)

arc spraying	ASP
flame spraying	FLSP
wire frame spraying	FLSP-W
high velocity oxyfuel spraying	HVOF
plasma spraying	PSP
vacuum plasma spraying	VPSP

ALLIED PROCESSES

THERMAL CUTTING (TC)

OXYGEN CUTTING (OC)

flux cutting	OC-F
metal powder cutting	OC-P
oxyfuel gas cutting	OFC
oxacetylene cutting	OFC-A
oxyhydrogen cutting	OFC-H
oxynatural gas cutting	OFC-N
oxypropane cutting	OFC-P
oxygen arc cutting	OAC
oxygen gouging	OG
oxygen lance cutting	OLC

HIGH ENERGY BEAM CUTTING

electron beam cutting	EBC
laser beam cutting	LBC
air	LBC-A
evaporative	LBC-EV
inert gas	LBC-IG
oxygen	LBC-O

ARC CUTTING (AC)

carbon arc cutting	CAC
air carbon arc cutting	CAC-A
gas metal arc cutting	GMAC
gas tungsten arc cutting	GTAC
plasma arc cutting	PAC
shielded metal arc cutting	SMAC

American Welding Society

COMMON WELDING SYMBOLS

OTHER SIDE / ARROW SIDE	Arrow Side/Other Side	SMAW	Welding Process or Specification
	Fillet Weld Arrow Side		Weld-All-Around
	Fillet Weld Other Side		Scarf for Brazed Joint
	Fillet Weld Both Sides		Field Weld
	Single-V-Groove Weld	1/16	Melt-Thru Dimension
	Square-Groove Weld	1/8	Surface Weld Dimension
	Double-V-Groove Weld		Flush or Flat Contour of Weld
	Single Bevel Groove Weld		Convex Contour of Weld
	Double Bevel Groove Weld		Concave Contour of Weld
	J-Groove Weld	3	Consumable Insert
	U-Groove Weld	PT	Liquid Penetrant Examination Examine-All-Around
	Flare-V-Groove Weld	RT	Radiographic Examination
	Flare-Bevel-Groove Weld		
	Plug or Slot Weld	VT	Visual Examination Examine in Field
	Spot Weld		
	Seam Weld	(4) MT (4)	Magnetic Examination on All Four Spot Welds
	Back Weld		
	Melt-Thru Weld	G	Multiple Reference Lines
	Surfacing Weld		
	Edge Weld	3/8 6-10	Intermittent Weld Pitch
3/8	Fillet Weld Leg Size	NOTE: 3/8" LEG ON PART B (3/16 x 3/8) SEE NOTE	Single Fillet Weld Unequal Leg Size
(.38)	Single-V-Groove Penetration	1/4 1/4	Double Fillet Equal Leg Size
	Combined Welding Symbols	.25 (.38)	Depth of Bevel

AMERICAN WELDING SOCIETY
Welding Symbol Chart

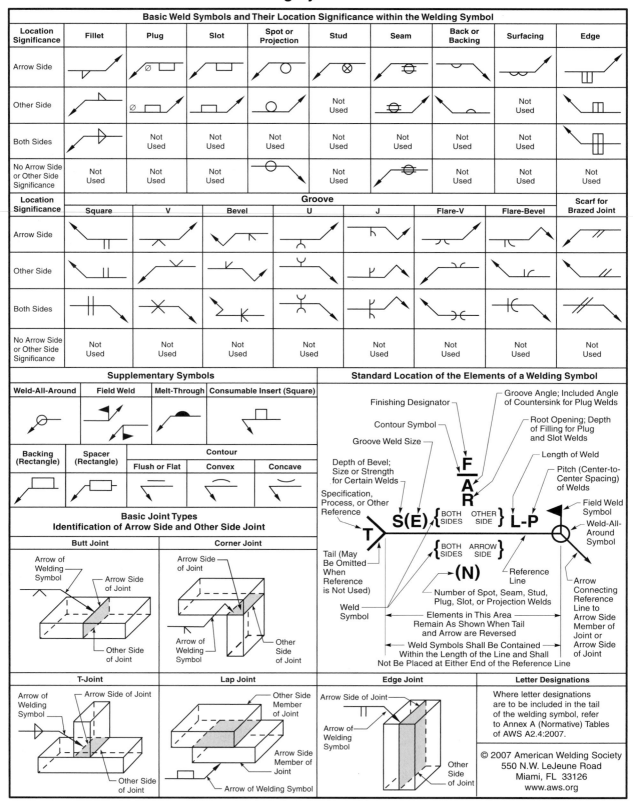

AMERICAN WELDING SOCIETY
Welding Symbol Chart

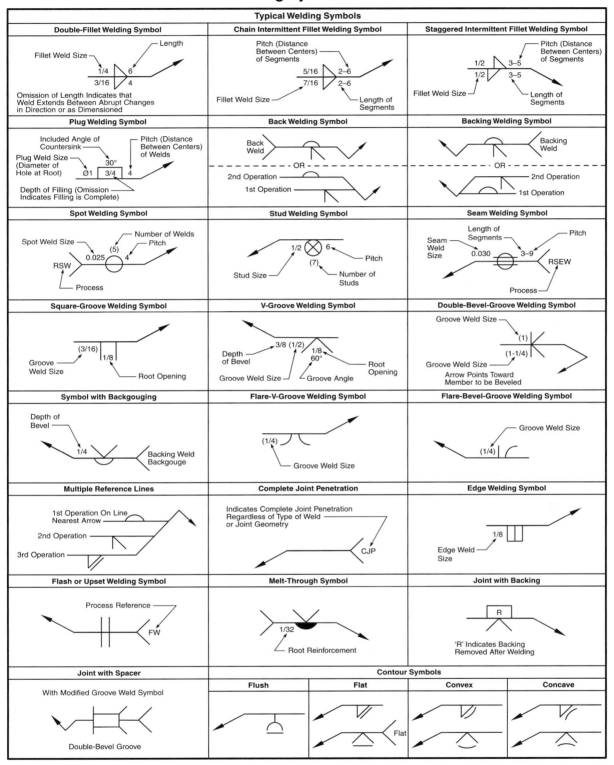

Typical Welding Symbols

Double-Fillet Welding Symbol
Length
Fillet Weld Size — 1/4 6
3/16 4
Omission of Length Indicates that Weld Extends Between Abrupt Changes in Direction or as Dimensioned

Chain Intermittent Fillet Welding Symbol
Pitch (Distance Between Centers) of Segments
5/16 2–6
7/16 2–6
Fillet Weld Size — Length of Segments

Staggered Intermittent Fillet Welding Symbol
Pitch (Distance Between Centers) of Segments
1/2 3–5
1/2 3–5
Fillet Weld Size — Length of Segments

Plug Welding Symbol
Included Angle of Countersink
Plug Weld Size (Diameter of Hole at Root) — Ø1 3/4 4 30°
Pitch (Distance Between Centers) of Welds
Depth of Filling (Omission Indicates Filling is Complete)

Back Welding Symbol
Back Weld
— OR —
2nd Operation
1st Operation

Backing Welding Symbol
Backing Weld
— OR —
2nd Operation
1st Operation

Spot Welding Symbol
Spot Weld Size — 0.025 (5) 4
Number of Welds
Pitch
RSW
Process

Stud Welding Symbol
1/2 ⊗ 6
(7)
Stud Size
Pitch
Number of Studs

Seam Welding Symbol
Length of Segments
Seam Weld Size — 0.030 3–9
Pitch
RSEW
Process

Square-Groove Welding Symbol
(3/16)
1/8
Groove Weld Size
Root Opening

V-Groove Welding Symbol
3/8 (1/2) 1/8
60°
Depth of Bevel
Groove Weld Size
Groove Angle
Root Opening

Double-Bevel-Groove Welding Symbol
Groove Weld Size
(1)
(1-1/4)
Groove Weld Size
Arrow Points Toward Member to be Beveled

Symbol with Backgouging
Depth of Bevel
1/4
Backing Weld Backgouge

Flare-V-Groove Welding Symbol
(1/4)
Groove Weld Size

Flare-Bevel-Groove Welding Symbol
Groove Weld Size
(1/4)

Multiple Reference Lines
1st Operation On Line Nearest Arrow
2nd Operation
3rd Operation

Complete Joint Penetration
Indicates Complete Joint Penetration Regardless of Type of Weld or Joint Geometry
CJP

Edge Welding Symbol
1/8
Edge Weld Size

Flash or Upset Welding Symbol
Process Reference
FW

Melt-Through Symbol
1/32
Root Reinforcement

Joint with Backing
R
'R' Indicates Backing Removed After Welding

Joint with Spacer
With Modified Groove Weld Symbol
Double-Bevel Groove

Contour Symbols

Flush	Flat	Convex	Concave
	Flat		

*It should be understood that these charts are intended only as shop aids. The only complete and official presentation of the standard welding symbols is in A2.4.

WELD TYPE AND JOINT APPLICATIONS

JOINT / TYPE	BUTT	LAP	T	EDGE	CORNER
FILLET	—	✓	✓	—	✓
SQUARE-GROOVE	✓	—	✓	✓	✓
BEVEL-GROOVE	✓	✓	✓	✓	✓
V-GROOVE	✓	—	—	✓	✓
U-GROOVE	✓	—	—	✓	✓
J-GROOVE	✓	✓	✓	✓	✓
FLARE-BEVEL-GROOVE	✓	✓	✓	✓	✓
FLARE-V-GROOVE	✓	—	—	✓	✓
PLUG	—	✓	✓	—	✓
SLOT	—	✓	✓	—	✓
EDGE	✓	✓	✓	✓	✓
SPOT	—	✓	✓	—	✓
PROJECTION	—	✓	✓	—	✓
SEAM	—	✓	✓	✓	✓
BRAZE	✓	✓	✓	—	✓

WELD JOINTS AND POSITIONS

	BUTT	LAP	T	EDGE	CORNER
FLAT					
HORIZONTAL					
VERTICAL					
OVERHEAD					

WELDING SYMBOL

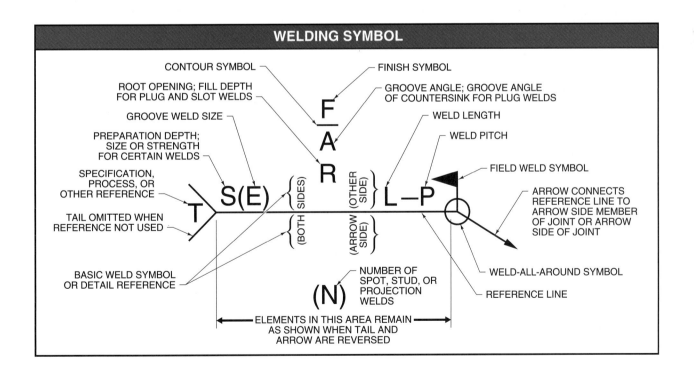

CONTOUR SYMBOL

FINISH SYMBOL

ROOT OPENING; FILL DEPTH FOR PLUG AND SLOT WELDS

GROOVE ANGLE; GROOVE ANGLE OF COUNTERSINK FOR PLUG WELDS

GROOVE WELD SIZE

WELD LENGTH

PREPARATION DEPTH; SIZE OR STRENGTH FOR CERTAIN WELDS

WELD PITCH

SPECIFICATION, PROCESS, OR OTHER REFERENCE

FIELD WELD SYMBOL

TAIL OMITTED WHEN REFERENCE NOT USED

ARROW CONNECTS REFERENCE LINE TO ARROW SIDE MEMBER OF JOINT OR ARROW SIDE OF JOINT

BASIC WELD SYMBOL OR DETAIL REFERENCE

NUMBER OF SPOT, STUD, OR PROJECTION WELDS

WELD-ALL-AROUND SYMBOL

REFERENCE LINE

ELEMENTS IN THIS AREA REMAIN AS SHOWN WHEN TAIL AND ARROW ARE REVERSED

$\frac{F}{A}$ R S(E) (SIDES) (OTHER SIDE) (ARROW SIDE) (BOTH) L–P (N)

Form A.7.1

SUGGESTED
WELDING PROCEDURE SPECIFICATION (WPS)

Identification _____

Date _____ Revision _____

Company name _____

Supporting PQR no.(s) _____ Type - Manual ( )   Semiautomatic ( )

Welding process(es) _____ Mechanized ( )   Automatic ( )

Backing: Yes ( )   No ( )

Backing material (type) _____

Material number _____ Group _____ To material number _____ Group _____

Material spec. type and grade _____ To material spec. type and grade_____

Base metal thickness range: Groove _____ Fillet _____

Deposited weld metal thickness range _____

Filler metal F no. _____ A no. _____

Spec. no. (AWS) _____ Flux tradename _____

Electrode-flux (Class) _____ Type _____

Consumable insert: Yes ( )   No ( )   Classifications _____

Shape _____

Position(s) of joint _____ Size _____

Welding progression: Up ( )   Down ( )   Ferrite number (when reqd.) _____

PREHEAT: **GAS:**

Preheat temp., min _____ Shielding gas(es) _____

Interpass temp., max _____ Percent composition _____
(continuous or special heating, where
applicable, should be recorded)   Flow rate _____
                                             Root shielding gas _____

POSTWELD HEAT TREATMENT: Trailing gas composition _____

Temperature range _____ Trailing gas flow rate _____

Time range _____

Tungsten electrode, type and size _____

Mode of metal transfer for GMAW: Short-circuiting ( )   Globular ( )   Spray ( )

Electrode wire feed speed range: _____

Stringer bead ( )   Weave bead ( )   Peening: Yes ( )   No ( )

Oscillation _____

Standoff distance

Multiple ( )   or single electrode ( )

Other _____

	Filler metal				Current			
Weld layer(s)	Process	Class	Dia.	Type & polarity	Amp range	Volt range	Travel speed range	
								e.g., Remarks, comments, hot wire addition, technique, torch angle, etc.

Approved for Production by _____
                                             Employer

Note: Those items that are not applicable should be marked N.A.

American Welding Society

WPS no. used for test _____   Welding process(es) _____

Company _____   Equipment type and model (sw) _____

JOINT DESIGN USED (2.6.1)

WELD INCREMENT SEQUENCE

Single ( )   Double weld ( )

HEAT POSTWELD TREATMENT (2.6.6)

Backing material _____

Temp. _____

Root opening _____ Root face dimension _____

Time _____

Groove angle _____ Radius (J-U) _____

Other _____

Back gouging: Yes ( ) No ( ) Method _____

GAS (2.6.7)

BASE METALS (2.6.2)

Gas type(s) _____

Material spec._____ To _____

Gas mixture percentage _____

Type or grade _____ To _____

Flow rate _____

Material no. _____ To material no. _____

Backing gas _____ Flow rate _____

Group no. _____ To group no. _____

Root shielding gas _____

Thickness _____

EBW vacuum ( ) Absolute pressure ( )

Diameter (pipe) _____

ELECTRICAL CHARACTERISTICS (2.6.8)

Surfacing: Material _____ Thickness _____

Electrode extension _____

Chemical composition _____

Other _____

Standoff distance _____

FILLER METALS (2.6.3)

Transfer mode (GMAW) _____

Weld metal analysis A no. _____

Electrode diameter tungsten _____

Filler metal F no. _____

Type tungsten electrode _____

AWS specification _____

Current: AC ( ) DCEP ( ) DCEN ( ) Pulsed ( )

AWS classification _____

Heat input _____

Flux class _____ Flux brand _____

EBW: beam focus current _____ Pulse freq. _____

Consumable insert: Spec. _____ Class. _____

Filament type _____ Shape ___ Size _____

Supplemental filler metal spec. _____ Class. _____

Other _____

Non-classified filler metals _____

TECHNIQUE (2.6.9)

Consumable guide (ESW) Yes ( ) No ( )

Oscillation frequency _____Weave width _____

Supplemental deoxidant (EBW) _____

Dwell time _____

POSITION (2.6.4)

String or weave bead _____ Weave width _____

Position of groove _____ Fillet _____

Multiple-pass or single pass (per side) _____

Vertical progression: Up ( ) Down ( )

Number of electrodes _____

Peening _____

Electrode spacing _____

PREHEAT (2.6.5)

Arc timing (SW) _____ Lift ( )

Preheat temp., actual min. _____

PAW: Conventional ( ) Key hole ( )

Interpass temp., actual max. _____

Interpass cleaning:

Pass no.	Filler metal size	Amps	Volts	Travel speed (ipm)	Filler metal wire (ipm)	Slope induction	Special notes (process, etc.)

Note: Those items that are not applicable should be marked N.A.

Form A.7.2

TENSILE TEST SPECIMENS: SUGGESTED PROCEDURE QUALIFICATION RECORD   PQR No.

Type: _____ Tensile specimen size: _____ Area: _____

Groove (   )   Reinforcing bar (   )   Stud welds (   )

Tensile test results: (Minimum required UTS _____ psi)

Specimen no.	Width, in.	Thickness, in.	Area, in.^{2}	Max load lb	UTS, psi	Type failure and location

GUIDED BEND TEST SPECIMENS - SPECIMEN SIZE: _____

Type	Result	Type	Result

MACRO-EXAMINATION RESULTS: Reinforcing bar (   )   Stud (   )

1. _____   4. _____
2. _____   5. _____
3. _____

SHEAR TEST RESULTS - FILLETS: 1. _____   3. _____
2. _____   4. _____

IMPACT TEST SPECIMENS

Type: _____ Size: _____

Test temperature: _____

Specimen location: WM = weld metal;   BM = base metal;   HAZ = heat-affected zone

Test results:

Welding position	Specimen location	Energy absorbed (ft.-lb)	Ductile fracture area (percent)	Lateral expansion (mils)

IF APPLICABLE   **RESULTS**

Hardness tests: (   )   Values _____   Acceptable (   )   Unacceptable (   )

Visual (special weldments 2.4.2) (   )   Acceptable (   )   Unacceptable (   )

Torque (   ) psi   Acceptable (   )   Unacceptable (   )

Proof test (   )   Method _____   Acceptable (   )   Unacceptable (   )

Chemical analysis (   )   Acceptable (   )   Unacceptable (   )

Nondestructive exam (   )   Process _____   Acceptable (   )   Unacceptable (   )

Other _____   Acceptable (   )   Unacceptable (   )

Mechanical testing by (Company) _____ Lab No. _____

We certify that the statements in this Record are correct and that the test welds were prepared, welded, and tested in accordance with the requirements of the American Welding Society Standard for Welding Procedure and Performance Qualification (AWS B2-83)

Qualifier: _____   Reviewed by: _____

Date: _____   Approved by: _____
                                                        Employer

American Welding Society

Form A.7.3

SUGGESTED
WELDER PERFORMANCE QUALIFICATION TEST RECORD

Name _____ Identification _____ . Welder ( )   Operator ( )

Social security number: _____ Qualified to WPS no. _____

Process(es) _____ Manual ( )   Semiautomatic ( )   Automatic ( )   Mechanized ( )

Test base metal specification _____ To _____

Material number _____ To _____

Fuel gas (OFW) _____

AWS filler metal classification _____ F no. _____

Backing: Yes   ( )   No   ( )   Double ( ) or Single side ( )
Current: AC   ( )   DC   ( )   Short circuiting transfer (GMAW) Yes (..) No (..)
Consumable insert: Yes   ( )   No   ( )
Root shielding: Yes   ( )   No   ( )

TEST WELDMENT	POSITION TESTED	WELDMENT THICKNESS (T)

GROOVE:

Pipe   1G ( )   2G ( )   5G ( )   6G ( )   6GR ( )   Diameter(s) _____ (T) _____
Plate   1G ( )   2G ( )   3G ( )   4G ( )   (T) _____
Rebar   1G ( )   2G ( )   3G ( )   4G ( )   Bar size _____ Butt   ( )
                                                                              Spliced butt ( )

FILLET:

Pipe ( )   1F ( )   2F ( )   3F ( )   5F ( )   Diameter _____ (T) _____
Plate ( ) 1F ( )   2F ( )   3F ( )   4F ( )   (T) _____

Other (describe) _____

Test results: Remarks

Visual test	N/A ( )	Pass ( )	Fail ( )
Bend test	N/A ( )	Pass ( )	Fail ( )
Macro test	N/A ( )	Pass ( )	Fail ( )
Tension test	N/A ( )	Pass ( )	Fail ( )
Radiographic test	N/A ( )	Pass ( )	Fail ( )
Penetrant test	N/A ( )	Pass ( )	Fail ( )

QUALIFIED FOR:
PROCESSES
GROOVE: **THICKNESS**

Pipe   1G ( )   2G ( )   5G ( )   6G ( )   6GR ( )   (T) Min _____ Max. _____ Dia _____
Plate 1G ( )   2G ( )   3G ( )   4G ( )   (T) Min _____ Max. _____
Rebar 1G ( )   2G ( )   3G ( )   4G ( )   Bar size   Min _____ Max. _____

FILLET:

Pipe   1F ( )   2F ( )   4F ( )   5F ( )   (T) Min _____ Max. _____
Plate 1F ( )   2F ( )   3F ( )   4F ( )   (T) Min _____ Max. _____
Rebar 1F ( )   2F ( )   3F ( )   4F ( )   Bar size   Min _____ Max. _____

Weld cladding ( )   Position(s) _____ T Min _____ Max. _____ Clad Min _____

Consumable insert ( )   Backing type ( )
Uphill ( )   Downhill ( )
Single side ( )   Double side ( )   No backing ( )
Short circuiting ( ) ( )   Spray ( )   Pulsed Spray ( )
Reinforcing bar - butt ( ) or Spliced butt ( )

The above named person is qualified for the welding process(es) used in this test within the limits of essential variables including materials and filler metal variables of the AWS Standard for Welding Procedure and Performance Qualification (AWS B2.1).

Date tested _____   Signed by _____
                                                                   Qualifier

ELECTRODE SELECTION CHART*

Variables	Electrode Class†										
	E6010	E6011	E6012	E6013	E6027	E7014	E7024	E7016	E7018	E7028	E6020
Groove butt welds, flat (< ¼″)	5	5	3	8	10	9	9	7	9	10	10
Groove butt welds, all positions (< ¼″)	10	9	5	8	(b)	6	(b)	7	6	(b)	(b)
Fillet welds, flat or horizontal	2	3	8	7	9	9	10	5	9	9	10
Fillet welds, all positions	10	9	6	7	(b)	7	(b)	8	6	(b)	(b)
Current (C)‡	DCEP	DCEP AC	DCEN AC	DC AC	DC AC	DC AC	DC AC	DCEP AC	DCEP AC	DCEP AC	DC AC
Thin material (¼″)	5	7	8	9	(b)	8	7	2	2	(b)	(b)
Heavy plate or highly restrained joint	8	8	8	8	8	8	7	10	9	9	8
High-sulfur or off-analysis steel	(b)	(b)	5	3	(b)	3	5	9	9	9	(b)
Deposition rate	4	4	5	5	10	6	10	4	6	8	6
Depth of penetration	10	9	6	5	8	6	4	7	7	7	8
Appearance, undercutting	6	6	8	9	10	9	10	7	10	10	9
Soundness	6	6	3	5	9	7	8	10	9	9	9
Ductility	6	7	4	5	10	6	5	10	10	10	10
Low-temperature impact strength	8	8	4	5	9	8	9	10	10	10	8
Low spatter loss	1	2	6	7	10	9	10	6	8	9	9
Poor fit-up	6	7	10	8	(b)	9	8	4	4	4	(b)
Welder appeal	7	6	8	9	10	10	10	6	8	9	9
Slag removal	9	8	6	8	9	8	9	4	7	8	9

* Rating is on a comparative basis of same-size electrodes with 10 as the highest value. Ratings may change with size
† AWS
‡ DCEP–direct current electrode positive; DCEN–direct current electrode negative; AC–alternating current; DC–direct current, either polarity
(b) Not recommended

FILLER METALS FOR GTAW PLAIN CARBON AND LOW ALLOY STEELS

Base Alloy	Filler Metal	AWS Specification
A36 A529 A570 A573 A53 AI 06 A501 A242 A441 A588 A572, Grade 42 A633 Grades A, B, C, D	ER70S-2 ER70S-3 ER70S-6 ER70S-7	A5.18
A572, Grade 60, 65 A633, Grade E A242 A588	ER80S-G ER80S-B2 ER80S-B2L ER80S-Nil ER80S-Ni2 ER80S-Ni3	A5.28
A514/A517	ER100S-1 ER110S-1 ER126S-1	A5.28
A533, Grade B	ER100S-1 ER110S-1	A5.28
A537	ER80S-Nil ER80S-Ni2 ER80S-Ni3	A5.28
A543, Grade B	ER110S-1 ER120S-1	A5.28
A678, Grade C	ER100S-1	A5.28
HY80 HY100	ER110S-1 ER120S-1	A5.28

FILLER METALS FOR GTAW AUSTENITIC STAINLESS STEELS

Base Alloy	Filler Metal	AWS Specification
201 202	ER209, ER219 ER308	A5.9
301, 302 304, 305	ER308	A5.9
304L	ER308L ER347	A5.9
309	ER309	A5.9
309S	ER309L	A5.9
310, 314, 3105	ER310	A5.9
316	ER316	A5.9
316L	ER316L	A5.9
316H	ER16-8-2 ER316H	A5.9
317	ER317	A5.9
317L	ER317L	A5.9
321	ER321	A5.9
330	ER330	A5.9
347, 348	ER347	A5.9
349	ER349	A5.9

FILLER METALS FOR GTAW MARTENSITIC, FERRITIC PRECIPITATION HARDENING AND DUPLEX STAINLESS STEELS

Base Alloy	Filler Metal	AWS Specification
410	ER410	A5.9
410 NiMo	ER410 NiMo	A5.9
420	ER420	A5.9
430	ER430	A5.9
630	ER630	A5.9
16-8-2	ER16-8-2	A5.9
17-4 PH 15-5 PH	AMS 5826 (17-4 PH) or ER 308	A5.9
Stainless W	AMS 5805C (A286) or ERNiMo-3	A5.14
17-7 PH	AMS 5824A (17-7 PH)	
PH15-7Mo	AMS 5812C (PH15-7Mo)	
AM350	AMS 5774-B (AM350)	
AM355	AMS 5780A (AM355)	
A286	ERNiCrFe-6 or ERNiMo-3	A5.14
2205	ER 2209	
255	ER 2553	

FILLER METALS FOR GTAW ALUMINUM ALLOYS

Base Alloy	Filler Metal		AWS Specification
	Strength	Ductility	
1100	ER4043	ER1100	A5.10
2219	ER2319	ER2319	A5.10
3003	ER4043	ER1100	A5.10
5052	ER5356	ER5654	A5.10
5083	ER5183	ER5356	A5.10
5086	ER5356	ER5356	A5.10
5454	ER5356	ER5554	A5.10
5456	ER5556	ER5356	A5.10
6061	ER5356	ER5356	A5.10
6063	ER5356	ER5356	A5.10
7005	ER5556	ER5356	A5.10
7039	ER5556	ER5356	A5.10

ACETYLENE CUTTING TIP SIZES

Metal Thickness (in.)	Tip Size	Cutting Oxygen Pressure (PSI)	Preheat Oxygen (PSI)	Acetylene Pressure (PSI)	Kerf Width
⅛	000	20–25	3–5	3–5	.04
¼	00	20–25	3–5	3–5	.05
⅜	0	25–30	3–5	3–5	.06
½	0	30–35	3–6	3–5	.006
¾	1	30–35	4–7	3–5	.07
1	2	35–40	7–8	3–6	.09
2	3	40–45	5–10	4–8	.11
3	4	40–50	5–10	5–11	.12
4	5	45–55	6–12	6–13	.15

WELDING CABLE SIZES

Amps	Length of Circuit*						
	100′	150′	200′	250′	300′	350′	400′
100	4	4	2	2	1	1/0	1/0
150	4	2	1	1/0	2/0	3/0	3/0
200	2	1	1/0	2/0	3/0	4/0	4/0
250	1	1/0	2/0	3/0	4/0		
300	1/0	2/0	3/0	4/0			
350	1/0	3/0	4/0				
400	2/0	3/0					
450	2/0	4/0					
500	3/0	4/0					
550	3/0	4/0					
600	4/0						

* includes both welding and ground leads (based on 4 V drop), 60% duty cycle

MISCELLANEOUS WELD DEFECTS

Problem	Cause	Remedy
Undercutting	Improper Electrode Manipulation	Pause at each side of the weld bead when using a weaving technique; use proper electrode angles
	Welding current set too high	Reduce welding current (use proper current for electrode size and welding position)
	Arc length too long	Reduce arc length
	Travel speed too fast	Reduce travel speed
	Arc blow	Reduce effects of arc blow; reset workpiece connections
Overlap	Travel speed too slow	Increase travel speed
	Incorrect electrode angle	Use proper electrode angle
	Too large electrode	Use smaller electrode
Spatter	Arc blow	Reduce effects of arc blow; reset workpiece connections
	Welding current set too high	Reduce welding current (use proper current for electrode size and welding position)
	Arc length too long	Reduce arc length
	Wet, dirty, or damaged electrode	Properly maintain and store electrodes

POTENTIAL EFFECTS OF OXYGEN-DEFICIENT ATMOSPHERES*

Oxygen Content†	Effects and Symptoms‡
19.5	Minimum permissible oxygen level
15–19.5	Decreased ability to work strenuously. May impair condition and induce early symptoms in persons with coronary, pulmonary, or circulatory problems
12–14	Respiration exertion and pulse increases. Impaired coordination, perception, and judgment
10–11	Respiration further increases in rate and depth, poor judgment, lips turn blue
8–9	Mental failure, fainting, unconsciousness, ashen face, blue lips, nausea, and vomiting
6–7	8 min, 100% fatal; 6 min, 50% fatal; 4 min–5 min, recovery with treatment
4–5	Coma in 40 sec, convulsions, respiration ceases, death

* values are approximate and vary with state of health and physical activities
† % by volume
‡ at atmospheric pressure

WELD DEFECT EVALUATION GUIDE . . .

Types of Defect	Pressure Vessels (per ASME, Section VIII)		Chemical Plant and Petroleum Refinery Piping (per ANSI Piping Code)	
	100% X-Ray*	Spot X-Ray*	100% X-Ray*	Random X-Ray*
Cracks	None Allowed	None Allowed	None Allowed	None Allowed
Incomplete Penetration at root pass	None Allowed	None Allowed	None Allowed	• Maximum of 1/32″ or 20% of wall thickness, whichever is smaller • Maximum length of 1½″ in 6″ weld • None allowed for longitudinal welds
Incomplete Penetration due to high-low fit-up	None Allowed	None Allowed	None Allowed	• Maximum of 1/32″ or 20% of wall thickness, whichever is smaller • Maximum length of 1½″ in 6″ weld • None allowed for longitudinal welds
Lack of Fusion at root pass	None Allowed	None Allowed	None Allowed	None Allowed
Lack of Fusion at groove face or between beads, "cold lap"	None Allowed	None Allowed	None Allowed	None Allowed
Melt-through	Not Covered	Not Covered	Not Covered	Not Covered
Internal Concavity	Shall not reduce weld thickness to less than thinner material. Contour of concavity shall be smooth	Shall not reduce weld thickness to less than thinner material. Contour of concavity shall be smooth	Shall not reduce weld thickness to less than thinner material	Shall not reduce weld thickness to less than thinner material
Undercut at root pas or cover pass	1/32″ or 10%t†, whichever is less	1/32″ or 10%t†, whichever is less	• Maximum depth of 1/32″ or 25% of wall thickness, whichever is smaller • None allowed for longitudinal butt joints	• Maximum depth of 1/32″ or 25% of wall thickness, whichever is smaller • None allowed for longitudinal butt joints
Slag Inclusions elongated, except as noted	**Material Thickness** — **Maximum Slag Length** less than or equal to ¾″ — ¼″ ¾″ to 2¼″ — 1/3t† greater than 2¼″ — ¾″	• Maximum length of 2/3T‡ where T‡ is thickness of material, with ¾″ Maximum • Maximum total length of T‡ in 6″ weld length	• Maximum length of 1/3T‡ and width lesser of 3/32″ or 1/3T‡ • Maximum total length of T‡ in 12T‡ of weld	• Maximum length of 2T‡ and width lesser of ⅛″ or ½T‡ • Maximum total length of 4T‡ in 6″ weld
Porosity	• Maximum individual size shall be smaller of ¼t† or 5/32″ (or 1/3t† or ¼″ if 1″ separation) • The length of an acceptable cluster shall not exceed the lesser of 1″ or 2t†	Any size or amount is acceptable	• Maximum 1/3T‡ or ⅛″, whichever is less, greatest dimension of individual pore • Maximum total area 3X area of maximum single allowable pore for any square inch of weld (For piping or elongated porosity use slag inclusion criteria)	• Maximum ½T‡ or ⅛″, whichever is less, greatest dimension of individual pore • Maximum total area 3X area of maximum single allowable pore for any square inch of weld
Excess Weld Reinforcement	**Material Thickness** — **Maximum Height** 3/16″ to less than 1″ — 3/32″ 1″ to less than 2″ — ⅛″ 2″ to less than 3″ — 5/32″ 3″ to less than 4″ — 7/32″	**Material Thickness** — **Maximum Slag Length** 3/16″ to less than 1″ — 3/32″ 1″ to less than 2″ — ⅛″ 2″ to less than 3″ — 5/32″ 3″ to less than 4″ — 7/32″	**Thinner Mat'l Thickness. (T‡)** — **Maximum Height** less than or equal to ¼″ — 1/16″ greater than ¼″ to ½″ — ⅛″ greater than ½″ to 1″ — 5/32″ greater than 1″ — 3/16″	**Thinner Mat'l Thickness (T‡)** — **Maximum Height** less than or equal to ¼″ — 1/16″ greater than ¼″ to ½″ — ⅛″ greater than ½″ to ½″ — 5/32″ greater than 1″ — 3/16″
Excessive Root Penetration	Same as Excess Weld Reinforcement	Same as Excess Weld Reinforcement	Same as Excess Weld Reinforcement	Same as Excess Weld Reinforcement
Misalignment	**Material‖ Thickness** — **Long# Maximum** — **Circum.** Maximum** less than equal to ½″ — ¼t† — ¼t† greater than ½″ to ¾″ — ⅛″ — ¼t† greater than ¾″ to 1½″ — ⅛″ — 3/16″ greater than 1½″ to 2″ — ⅛″ — ⅛t† greater than 2″ lesser of: — 1/16t† or ⅜″ — ⅛t† or ¾″	**Material‖ Thickness** — **Long# Maximum** — **Circum.** Maximum** less than or equal to ½″ — ¼t† — ¼t† greater than ½″ to ¾″ — ⅛″ — ¼t† greater than ¾″ to 1½″ — ⅛″ — 3/16″ greater than 1½″ to 2″ — ⅛″ — ⅛t† greater than 2″ lesser of: — 1/16t† or ⅜″ — ⅛t† or ¾″	Inside diameters of components at ends to be joined must be aligned within engineering design and welding procedure. If the external surfaces of the two components are not aligned, the weld shall be tapered between the two surfaces	Inside diameters of components at ends to be joined must be aligned within engineering design and welding procedure. If the external surfaces of the two components are not aligned, the weld shall be tapered between the two surfaces
Accumulation of Discontinuities	Not Covered	Not Covered	Not Covered	Not Covered
General	No coarse ripples, grooves, overlaps, abrupt ridges or valleys	No coarse ripples, grooves, overlaps abrupt ridges or valleys	Longitudinal butt welds same as 100% X-Ray, except as noted	

* 100% X-Ray, Random X-Ray, and Spot X-Ray are quality level designations used by the ASME pressure vessel and ANSI piping codes and are also used when other NDE methods of evaluation are used
† t = weld thickness
‡ T = thinner material thickness
§ w = weld width
‖ see UHT-20 for special heat-treated ferritic steels
\# joint category A
** joint categories B, C, and D

... WELD DEFECT EVALUATION GUIDE

Types of Defect	Pipelines (per API Std. 1104)	Storage Tanks (per API Std. 650)	Power Boilers (per ASME Section 1)
Cracks	None allowed (except shallow crater cracks in the cover pass with maximum length of 5⁄32″)	None Allowed	None Allowed
Incomplete Penetration at root pass	• Maximum of 1″ in length in 12″ of weld, or 8% of weld length if less than 12″ • Maximum individual length of 1″	None Allowed	None Allowed
Incomplete Penetration due to high-low fit-up	• Maximum individual length of 2″ • Maximum accumulated length of 3″ in 12″ of continuous weld	None Allowed	None Allowed
Lack of Fusion at root pass	• Maximum of 1″ in length in 12″ of weld, or 8% of weld length if less than 12″ • Maximum individual length of 1″	None Allowed	None Allowed
Lack of Fusionat sidewall or between beads, "cold lap"	• Maximum individual length of 2″ • Maximum accumulated length of 2″ in 12″ of continuous weld	None Allowed	None Allowed
Melt-through	Pipe Diameter / Maximum Defect / Maximum Total less than 2¾″ OD — ¼″ — 1″ greater than or equal to 2¾″ OD — ¼″ — ½″ in 2″	Not Covered	Not Covered
Internal Concavity	If density of radiographic image of internal concavity is less than base metal, any length is allowable. If more dense, then see burn-through above	Shall not reduce weld thickness to less than thinner material. Contour of concavity shall be smooth	Not Covered
Undercut at root pass or cover pass	• Maximum depth 1⁄32″ or 12½% wall thickness, whichever is smaller. • Maximum 2″ length or 1⁄6 wall thickness, whichever is less, for depth of 1⁄64″ to 1⁄32″ or 6% to 12½% of wall thickness, whichever is less	• For horizontal butt joints: maximum depth 1⁄32″ • For vertical butt joints: maximum depth 1⁄64″	1⁄32″ or 10% t†, whichever is less
Slag Inclusions elongated, except as noted	• Maximum length is 2″ and width 1⁄16″ • Maximum total length 2″ in 12″ of weld. Parallel slag lines are considered separate if width of either exceeds 1⁄32″. Isolated Slag Inclusions: • Maximum width 1⁄8″ and ½″ length in 12″ of weld. • No more than 4 isolated inclusions of 1⁄8″ maximum width.	Material Thickness / Maximum Slag Length less than or equal to ¾″ — ¼″ ¾″ to 2¼″ — 1⁄3t† greater than 2¼″ — ¾″ Maximum length of t† in 12t† length	Material Thickness / Maximum Slag Length less than or equal to ¾″ — ¼″ ¾″ to 2¼″ — 1⁄3t† greater than 2¼″ — ¾″ Maximum total length of t† in 12t† length
Porosity	Spherical: Maximum dimension 1⁄8″ or 25% of wall thickness, whichever is less Cluster: Maximum area of ½″ diameter with maximum individual pore dimension of 1⁄16″. Maximum ½″ length in 12″ weld Hollow Bead: Maximum length ½″. Maximum 2″ length in 12″ weld with individual discontinuities exceeding ¼″ in length separated by at least 2″	• For aligned rounded indications, the summation of diameters less than t in 12t† length • Maximum individual size shall be the smaller of ¼″ or 5⁄32″ ; or 1⁄3t† or ¼ if 1″ separation • The length of an acceptable cluster shall not exceed the lesser of 1″ or 2t†	• For aligned rounded indications, the summation of diameters less than t in 12t† length • Maximum individual size shall be the smaller of ¼″ or 5⁄32″ ; or 1⁄3t† or ¼″ if 1″ separation • The length of an acceptable cluster shall not exceed the lesser of 1″ or 2t†
Excess Weld Reinforcement	Maximum height of 1⁄16″	Material Thickness / Vertical Joint Maximum / Horizontal Joint Maximum less than or equal to ½″ — 3⁄32″ — 1⁄8″ greater than ½″ to 1″ — 1⁄8″ — 3⁄16″ greater than 1″ — 3⁄16″ — ¼″	Material Thickness / Pipe/Tube Maximum / Other Weld Maximum greater than 3⁄16″ to ½″ — 5⁄32″ — 3⁄32″ greater than ½″ to 1″ — 3⁄16″ — 3⁄32″ greater than 1″ to 2″ — ¼″ — 1⁄8″ greater than 2″ to 3″ — ¼″or 1⁄8w§ — 5⁄32″ greater than 3″ to 4″ — ¼″or 1⁄8w§ — 7⁄32″
Excessive Root Penetration	Not Covered	Same as Excess Weld Reinforcement	Same as Excess Weld Reinforcement
Misalignment	Maximum 1⁄16″ Any greater offset, provided it is caused by dimensional variations, shall be equally distributed around the circumference of the pipe	Vertical misalignment less than or equal to 10%t† or 1⁄16″, whichever is larger Horizontal misalignment less than or equal to 20%t† of upper plate, with 1⁄8″ maximum	Material Thickness / Long Maximum / Circum. Maximum less than or equal to ½″ — ¼t† — ¼t† greater than ½″ to ¾″ — 1⁄8″ — ¼t† greater than ¾″ to 1½″ — 1⁄8″ — 3⁄16″ greater than 1½″ to 2 — 1⁄8″ — 1⁄8t† 2″ lesser of: — 1⁄16t† or 3⁄8″ — 1⁄8t† or ¾″
Accumulation of Discontinuities	Maximum of 2″ in any 12″ or 8% of weld length excluding high-low condition	Not Covered	Not Covered
General	Rights of Rejection—"Since NDE methods give limited indications, the Company may reject welds which appear to meet these standards of acceptability, if in its opinion the depth of the defect may be detrimental to the strength of weld."		No ripples, grooves, abrupt ridges, and valleys to avoid stress risers

* 100% X-Ray, Random X-Ray, and Spot X-Ray are quality level designations used by the ASME pressure vessel and ANSI piping codes and are also used when other NDE methods of evaluation are used
† t = weld thickness
‡ T = thinner material thickness
§ w = weld width
‖ see UHT-20 for special heat-treated ferritic steels
joint category A
** joint categories B, C, and D

NONDESTRUCTIVE EXAMINATION

Method	Letter Designation
Acoustic emission	AET
Electromagnetic	ET
Leak	LT
Magnetic particle	MT
Neutron radiographic	NRT
Penetrant*	PT
Proof*	PRT
Radiographic*	RT
Ultrasonic*	UT
Visual*	VT

*methods used for testing pipe welds

NDE EXAMINATION SYMBOL

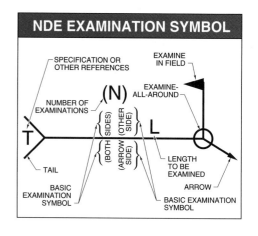

SPARK CHART

Metal	Stream Volume	Relative Length*	Color of Stream	Color of Bursts	Quantity of Bursts	Nature of Bursts
1. Wrought Iron	Large	65	Straw	White	Very few	Forked
2. Machine Steel (AISI 1020)	Large	70	White	White	Few	Forked
3. Carbon Tool Steel	Moderately large	55	White	White	Very many	Fine, repeating
4. Gray Cast Iron	Small	25	Red	Straw	Many	Fine, repeating
5. White Cast Iron	Very small	20	Red	Straw	Few	Fine, repeating
6. Annealed Mall. Iron	Moderate	30	Red	Straw	Many	Fine, repeating
7. High-Speed Steel (18-4-1)	Small	60	Red	Straw	Extremely few	Forked
8. Austenitic Manganese Steel	Moderately large	45	White	White	Many	Fine, repeating
9. Stainless Steel (Type 410)	Moderate	50	Straw	White	Moderate	Forked
10. Tungsten-Chromium Die Steel	Small	35	Red	Straw†	Many	Fine, repeating†
11. Nitrided Nitralloy	Large (curved)	55	White	White	Moderate	Forked
12. Stellite®	Very small	10	Orange	Orange	None	
13. Cemented Tungsten Carbide	Extremely small	2	Light Orange	Light Orange	None	
14. Nickel	Very small‡	10	Orange	Orange	None	
15. Copper, Brass, and Aluminum	None				None	

* actual length varies with grinding wheel, pressure, etc.
† blue-white spurts
‡ some wavy streaks

BRINELL HARDNESS NUMBERS*

Impression Diameter (mm)	Brinell Hardness Number For a Load of Kg.						Impression Diameter (mm)	Brinell Hardness Number For a Load of Kg.				
	500	1000	1500	2000	2500	3000		500	1000	1500	2000	2500
2.00	158	316	473	632	788	945	4.25	33.6	67.2	101	134	167
2.05	150	300	450	600	750	899	4.30	32.8	65.6	98.3	131	164
2.10	143	286	428	572	714	856	4.35	32.0	64.0	95.9	128	160
2.15	136	272	408	544	681	817	4.40	31.2	62.4	93.6	125	156
2.20	130	260	390	520	650	780	4.45	30.5	61.0	91.4	122	153
2.25	124	248	372	496	621	745	4.50	29.8	59.6	89.3	119	149
2.30	119	238	356	476	593	712	4.55	29.1	58.2	87.2	116	145
2.35	114	228	341	456	568	682	4.60	28.4	56.8	85.2	114	142
2.40	109	218	327	436	545	653	4.65	27.8	55.6	83.3	111	139
2.45	104	208	312	416	522	627	4.70	27.1	54.2	81.4	108	136
2.50	100	200	301	400	500	601	4.75	26.5	53.0	79.6	106	133
2.55	96.3	193	289	385	482	578	4.80	25.9	51.8	77.8	104	130
2.60	92.6	185	278	370	462	555	4.85	25.4	50.8	76.1	102	127
2.65	89.0	178	267	356	445	534	4.90	24.8	49.6	74.4	99.2	124
2.70	85.7	171	257	343	429	514	4.95	24.3	48.6	72.8	97.2	122
2.75	82.6	165	248	330	413	495	5.00	23.8	47.6	71.3	95.2	119
2.80	79.6	159	239	318	398	477	5.05	23.3	46.6	69.8	93.2	117
2.85	76.8	154	230	307	384	461	5.10	22.8	45.6	68.3	91.2	114
2.90	74.1	148	222	296	371	444	5.15	22.3	44.6	66.9	89.2	112
2.95	71.5	143	215	286	358	429	5.20	21.8	43.6	65.5	87.2	109
3.00	69.1	138	207	276	346	415	5.25	21.4	42.8	64.1	85.6	107
3.05	66.8	134	200	267	334	401	5.30	20.9	41.8	62.8	83.6	105
3.10	64.6	129	194	258	324	388	5.35	20.5	41.0	61.5	82.0	103
3.15	62.5	125	188	250	313	375	5.40	20.1	40.2	60.3	80.4	101
3.20	60.5	121	182	242	303	363	5.45	19.7	39.4	59.1	78.8	98.5
3.25	58.6	117	176	234	293	352	5.50	19.3	38.6	57.9	77.2	96.5
3.30	56.8	114	170	227	284	341	5.55	18.9	37.8	56.8	75.6	95.0
3.35	55.1	110	165	220	276	331	5.60	18.6	37.2	55.7	74.4	92.5
3.40	53.4	107	160	214	267	321	5.65	18.2	36.4	54.6	72.8	90.8
3.45	51.8	104	156	207	259	311	5.70	17.8	35.6	53.5	71.2	89.2
3.50	50.3	101	151	201	252	302	5.75	17.5	35.0	52.5	70.0	87.5
3.55	48.9	97.8	147	196	244	293	5.80	17.2	34.4	51.5	68.8	85.5
3.60	47.5	95.0	142	190	238	285	5.85	16.8	33.6	50.5	67.2	84.2
3.65	46.1	92.2	138	184	231	277	5.90	16.5	33.0	49.6	66.0	82.5
3.70	44.9	89.8	135	180	225	269	5.95	16.2	32.4	48.7	64.8	81.2
3.75	43.6	87.2	131	174	218	262	6.00	15.9	31.8	47.7	63.6	79.5
3.80	42.4	84.8	127	170	212	255	6.05	15.6	31.2	46.8	62.4	78.0
3.85	41.3	82.6	124	165	207	248	6.10	15.3	30.6	46.0	61.2	76.7
3.90	40.2	80.4	121	161	201	241	6.15	15.1	30.2	45.2	60.4	75.3
3.95	39.1	78.2	117	156	196	235	6.20	14.8	29.6	44.3	59.2	73.8

* diameter of ball=10mm

Tinius Olsen Testing Machine Company, Inc.

LOW-ALLOY, HIGH-STRENGTH ELECTRODES

Type/Suffix	Welding Application
E-7018-A1	Carbon and molybdenum steels
E-8016-B2 E-8018-B2L	Chromium—molybdenum steels
E-8016-C1 E-8018-C1 E-8018-C2 E-8018-C3	Nickel steels
E-9016-B3 E-9018-B3L	Chromium—molybdenum steels
E-10016-D2	Manganese—molybdenum steels

UNIT PREFIXES

PREFIX	UNIT	SYMBOL	NUMBER
Other larger multiples			
Mega	Million	M	$1,000,000 = 10^6$
Kilo	Thousand	k	$1,000 = 10^3$
Hecto	Hundred	h	$100 = 10^2$
Deka	Ten	d	$10 = 10^1$
			Unit $1 = 10^0$
Deci	Tenth	d	$0.1 = 10^{-1}$
Centi	Hundredth	c	$0.01 = 10^{-2}$
Milli	Thousandth	m	$0.001 = 10^{-3}$
Micro	Millionth	μ	$0.000001 = 10^{-6}$
Other smaller multiples			

ROCKWELL HARDNESS CONVERSION TABLE . . .

Rockwell			Superficial Rockwell			Vickers	Knoop	Brinell	Tensile Strength	Brinell
B 100 kgf 1/16" ball	**A** 60 kgf diamond	**E** 100 kgf 1/8" ball	**15T** 15 kgf 1/16" ball	**30T** 30 kgf 1/16" ball	**45T** 45 kgf 1/16" ball	**Hardness**	**Hardness** 500 gf and over	**Hardness** 3000 gf 10mm ball	**1000 lbs square inch**	**Hardness** 500 kgf 10mm ball
100	61.5	—	93.1	83.1	729	240	251	240	116	201
99	60.9	—	92.8	82.5	719	234	246	234	114	195
98	60.2	—	92.5	81.8	709	228	241	228	109	189
97	59.5	—	92.1	81.1	699	222	236	222	105	184
96	58.9	—	91.8	80.4	689	216	231	216	102	179
95	58.3	—	91.5	79.8	67.9	210	226	210	100	175
94	57.6	—	91.2	79.1	66.9	205	221	205	98	171
93	57.0	—	90.8	78.4	65.9	200	216	200	94	167
92	56.4	—	90.5	77.8	64.8	195	211	195	92	163
91	55.8	—	90.2	77.1	63.8	190	206	190	90	160
90	55.2	—	89.9	76.4	62.8	185	201	185	89	157
89	54.6	—	89.5	75.8	61.8	180	196	180	88	154
88	54.0	—	89.2	75.1	60.8	176	192	176	86	151
87	53.4	—	88.9	74.4	59.8	172	188	172	84	148
86	52.8	—	88.6	73.8	58.8	169	184	169	83	145
85	52.3	—	88.2	73.1	57.8	165	180	165	82	142
84	51.7	—	87.9	72.4	56.8	162	176	162	81	140
83	51.1	—	87.6	71.8	55.8	159	173	159	80	137
82	50.6	—	87.3	71.1	54.8	156	170	156	76	135
81	50.0	—	86.9	70.4	53.8	153	167	153	73	133
80	49.5	—	86.6	69.7	52.8	150	164	150	72	130
79	48.9	—	86.3	69.1	51.8	147	161	147	70	128
78	48.4	—	86.0	68.4	50.8	144	158	144	69	126
77	47.9	—	85.6	67.7	49.8	141	155	141	68	124
76	47.3	—	85.3	67.1	48.8	139	152	139	67	122
75	46.8	—	85.0	66.4	47.8	137	150	137	66	120
74	46.3	—	84.7	65.7	46.8	135	147	135	65	118
73	45.8	—	84.3	65.1	45.8	132	145	132	64	116
72	45.3	—	84.0	64.4	44.8	130	143	130	63	114
71	44.8	100	83.7	63.7	43.8	127	141	127	62	112
70	44.3	99.5	83.4	63.1	42.8	125	139	125	61	110
69	43.8	99.0	83.0	62.4	41.8	123	137	123	60	109
68	43.3	98.0	82.7	61.7	40.8	121	135	121	59	108
67	42.8	97.5	82.4	61.0	39.8	119	133	119	58	106
66	42.3	97.0	82.1	60.4	38.7	117	131	117	57	104
65	41.8	96.0	81.8	59.7	37.7	116	129	116	56	102
64	41.4	95.5	81.4	59.0	36.7	114	127	114	—	100
63	40.9	95.0	81.1	58.4	35.7	112	125	112	—	99
62	40.4	94.5	80.8	57.7	34.7	110	124	110	—	98
61	40.0	93.5	80.5	57.0	33.7	108	122	108	—	96
60	39.5	93.0	80.1	56.4	32.7	107	120	107	—	95
59	39.0	92.5	79.8	55.7	31.7	106	118	106	—	94
58	38.6	92.0	79.5	55.0	30.7	104	117	104	—	92
57	38.1	91.0	79.2	54.4	29.7	103	115	103	—	91
56	37.7	90.5	78.8	53.7	28.7	101	114	101	—	90
55	37.2	90.0	78.5	53.0	27.7	100	112	100	—	89
54	36.8	89.5	78.2	52.4	26.7	—	111	—	—	87
53	36.3	89.0	77.9	51.7	25.7	—	110	—	—	86
52	35.9	88.0	77.5	51.0	24.7	—	109	—	—	85
51	35.5	87.5	77.2	50.3	23.7	—	108	—	—	84
50	35.0	87.0	76.9	49.7	22.7	—	107	—	—	83
49	34.6	86.5	76.6	49.0	21.7	—	106	—	—	82
48	34.1	85.5	76.2	48.3	20.7	—	105	—	—	81
47	33.7	85.0	75.9	47.7	19.7	—	104	—	—	80
46	33.3	84.5	75.6	47.0	18.7	—	103	—	—	80

...ROCKWELL HARDNESS CONVERSION TABLE										
Rockwell			Superficial Rockwell			Vickers	Knoop	Brinell	Tensile Strength	Brinell
B 100 kgf 1/16" ball	A 60 kgf diamond	E 100 kgf 1/8" ball	15T 15 kgf 1/16" ball	30T 30 kgf 1/16" ball	45T 45 kgf 1/16" ball	Hardness	Hardness 500 gf and over	Hardness 3000 gf 10mm ball	1000 lbs square inch	Hardness 500 kgf 10mm ball
45	32.9	84.0	75.3	46.3	17.7	—	102	—	—	79
44	32.4	83.5	74.9	45.7	16.7	—	101	—	—	78
43	32.0	82.5	74.6	45.0	15.7	—	100	—	—	77
42	31.6	82.0	74.3	44.3	14.7	—	99	—	—	76
41	31.2	81.5	74.0	43.7	13.6	—	98	—	—	75
40	30.7	81.0	73.6	43.0	12.6	—	97	—	—	75
39	30.3	80.0	73.3	42.3	11.6	—	96	—	—	74
38	29.9	79.5	73.0	41.6	10.6	—	95	—	—	73
37	29.5	79.0	72.7	41.0	9.6	—	94	—	—	72
36	29.1	78.5	72.3	40.3	8.6	—	93	—	—	72
35	28.7	78.0	72.0	39.6	7.6	—	92	—	—	71
34	28.2	77.0	71.7	39.0	6.6	—	91	—	—	70
33	27.8	76.5	71.4	38.3	5.6	—	90	—	—	69
32	27.4	76.0	71.0	37.6	4.6	—	89	—	—	69
31	27.0	75.5	70.7	37.0	3.6	—	88	—	—	68
30	26.6	75.0	70.4	36.3	2.6	—	87	—	—	67

PIPE							
PIPE SIZE (NOMINAL)*	OUTSIDE DIAMETER (ACTUAL SIZE)*	INSIDE DIAMETER*			SCHEDULE–WALL THICKNESS*		
		STD	XS	XXS	SCHEDULE 40	SCHEDULE 60	SCHEDULE 80
1/8	0.405	0.269	0.215	—	0.068	0.095	—
1/4	0.540	0.364	0.302	—	0.088	0.119	—
3/8	0.675	0.493	0.423	—	0.091	0.126	—
1/2	0.840	0.622	0.546	0.252	0.109	0.147	0.294
3/4	1.050	0.824	0.742	0.434	0.113	0.154	0.308
1	1.315	1.049	0.957	0.599	0.133	0.179	0.358
1 1/4	1.660	1.380	1.278	0.896	0.140	0.191	0.382
1 1/2	1.900	1.610	1.500	1.100	0.145	0.200	0.400
2	2.375	2.067	1.939	1.503	0.154	0.218	0.436
2 1/2	2.875	2.469	2.323	1.771	0.203	0.276	0.552
3	3.500	3.068	2.900	2.300	0.216	0.300	0.600
3 1/2	4.000	3.548	3.364	2.728	0.226	0.318	—
4	4.500	4.026	3.826	3.152	0.237	0.337	0.674
5	5.563	5.047	4.813	4.063	0.258	0.375	0.750
6	6.625	6.065	5.761	4.897	0.280	0.432	0.864
8	8.625	7.981	7.625	6.875	0.322	0.500	0.875
10	10.750	10.020	9.750	8.750	0.365	0.500	—
12	12.750	12.000	11.750	10.750	0.406	0.500	—

*In in.

AISI-SAE DESIGNATION SYSTEM

Numbers and Digits	Type of steel and/or nominal alloy content

Carbon steels

10xx — Plain carbon (1% Mn max)

11xx — Resulfurized

12xx — Resulfurized and rephosphorized

15xx — Plain carbon (1.00% Mn to 1.65% Mn max)

Manganese steels

13xx — 1.75% Mn

Nickel steels

23xx — 3.5% Ni

25xx — 5% Ni

Nickel-chromium steels

31xx — 1.25% Ni; .65% Cr and .80% Cr

32xx — 1.75% Ni; 1.07% Cr

33xx — 3.50% Ni; 1.50% Cr and 1.57% Cr

34xx — 3.00% Ni; .77% Cr

Molybdenum steels

40xx — .20% Mo and .25% Mo

44xx — .40% Mo and .52% Mo

Chromium-molybdenum steels

41xx — .50% Cr, .80% Cr, and .95% Cr; .12% Mo, .20% Mo, .25% Mo, and .30% Mo

Nickel-chromium-molybdenum steels

43xx — 1.82% Ni; .50% Cr and .80% Cr; .25% Mo

43BVxx — 1.82% Ni; .50% Cr; .12% Mo and .25% Mo; .03% V min

47xx — 1.05% Ni; .45% Cr; .20% Mo and .35% Mo

81xx — .30% Ni; .40% Cr; .12% Mo

86xx — .55% Ni; .50% Cr; .20% Mo

87xx — .55% Ni; .50% Cr; .25% Mo

88xx — .55% Ni; .50% Cr; .35% Mo

93xx — 3.25% Ni; 1.20% Cr; .12% Mo

94xx — .45% Ni; .40% Cr; .12% Mo

97xx — .55% Ni; .20% Cr; .20% Mo

98xx — 1.00% Ni; .80% Cr; .25% Mo

Nickel-molybdenum steels

46xx — .85% Ni and 1.82% Ni; .20% Mo and .25% Mo

48xx — 3.50% Ni; .25% Mo

Chromium steels

50xx — .27% Cr, .40% Cr, .50% Cr, and .65% Cr

51xx — .80% Cr, .87% Cr, .92% Cr, .95% Cr, 1.00% Cr, and 1.05% Cr

Chromium steels

50xxx — .50% Cr

51xxx — 1.02% Cr   } C 1.00% min

52xxx — 1.45% Cr

Chromium-vanadium steels

61xx — .60% Cr, .80% Cr, and .95% Cr; .10% V and .15% V min

Tungsten-chromium steel

72xx — 1.75% W; 0.75% Cr

Silicon-manganese steels

92xx — 1.40% Si and 2.00% Si; .65% Mn, .82% Mn, and .85% Mn; 0% Cr and .65% Cr

High-strength low-alloy steels

9xx — Various SAE grades

Boron steels

xxBxx — B denotes boron steel

Leaded steels

xxLxx — L denotes leaded steel

ASTM SPECIFICATIONS FOR CHROME-MOLY STEEL PRODUCTS					
Type	Forgings	Tubes	Pipe	Castings	Plate
½Cr-½Mo	A182-F2	A213-T2	A335-P2 A369-FP2 A426-CP2	A356-GR5	A387-Gr2
1Cr-½Mo	A182-F12 A336-F12	A213-T12	A335-P12 A369-FP12 A426-CP12	—	A387-Gr2
1¼Cr-½Mo	A182-F11/F11A A336-F11/F11A	A199-T11 A200-T11 A213-T11	A335-P11 A369-FP11 A426-CP11	A217-WC6/11 A356-Gr6 A389-C23	A387-Gr11
2¼Cr-1Mo	A182-F22/F22a A336-F22/F22A	A199-T22 A200-T22 A213-T22	A335-P22 A369-FP22 A426-CP22	A217-WC9 A356-Gr10	A387-Gr22
3Cr-1Mo	A182-F21 A336-F21/F21A	A199-T21 A200-T21 A213-T21	A335-P21 A369-FP21 A426-CP21	—	A387-Gr21
5Cr-½Mo	A182-F5/F5a A336-F5/F5A	A199-T5 A200-T5 A213-T5	A335-P5 A369-FP5 A426-CP5	A217-C5	A387-Gr5
5Cr-½MoSi	—	A213-T5b	A335-P5b A426-CP5b	—	—
5Cr-½MoTi	—	A213-T5c	A335-P5c	—	—
7Cr-½Mo	A182-F7	A199-T7 A200-T7 A213-T7	A335-P7 A369-FP7 A426-CP7	—	A387-Gr7
9Cr-1Mo	A182-F9 A336-F9	A199-T9 A200-T9 A213-T9	A335-P9 A369-FP9 A426-CP9	A217-C12	A387-Gr9
9Cr-1Mo and V+Nb+N	A182-F91	A199-T91 A200-T91 A213-T91	A335-P91 A369-FP91	—	A387-Gr91

MILLIMETER EQUIVALENTS OF AN INCH

Inches		mm	Inches		mm	Inches		mm
	0.00004	0.001		0.11811	3		0.550	13.970
	0.00039	0.01	1/8	0.1250	3.175		0.55118	14
	0.00079	0.02		0.13780	3.5	9/16	0.56250	14.2875
	0.001	0.025	9/64	0.14063	3.5719		0.57087	14.5
	0.00118	0.03		0.150	3.810	37/64	0.57813	14.6844
	0.00157	0.04		0.15625	3.9688		0.59055	15
	0.00197	0.05		0.15748	4	19/32	0.59375	15.0812
	0.002	0.051	11/64	0.17188	4.3656		0.600	15.24
	0.00236	0.06		0.1750	4.445	39/64	0.60938	15.4781
	0.00276	0.07		0.17717	4.5		0.61024	15.5
	0.003	0.0762	3/16	0.18750	4.7625	5/8	0.6250	15.875
	0.00315	0.08		0.19685	5		0.62992	16
	0.00354	0.09		0.20	5.08	41/64	0.64063	16.2719
	0.00394	0.1	13/64	0.20313	5.1594		0.64961	16.5
	0.004	0.1016		0.21654	5.5		0.650	16.51
	0.005	0.1270	7/32	0.21875	5.5562	21/32	0.65625	16.6688
	0.006	0.1524		0.2250	5.715		0.66929	17
	0.007	0.1778	15/64	0.23438	5.9531	43/64	0.67188	17.0656
	0.00787	0.2		0.23622	6	11/16	0.68750	17.4625
	0.008	0.2032	1/4	0.250	6.35		0.68898	17.5
	0.009	0.2286					0.700	17.78
	0.00984	0.25		0.255591	6.5	45/64	0.70313	17.8594
	0.01	0.254	17/64	0.26563	6.7469		0.70866	18
	0.01181	0.3		0.275	6.985	23/32	0.71875	18.2562
1/64	0.01563	0.3969		0.27559	7		0.72835	18.5
	0.01575	0.4	9/32	0.28125	7.1438	47/64	0.73438	18.6531
	0.01969	0.5		0.29528	7.5		0.74803	19
	0.02	0.508	19/64	0.29688	7.5406	3/4	0.750	19.050
	0.02362	0.6		0.30	7.62			
	0.025	0.635	5/16	0.3125	7.9375	49/64	0.76563	19.4469
	0.02756	0.7		0.31496	8		0.76772	19.5
	0.0295	0.75	21/64	0.32813	8.3344	25/32	0.78125	19.8438
	0.03	0.762		0.33465	8.5		0.78740	20
1/32	0.03125	0.7938	11/32	0.34375	8.7375	51/64	0.79688	20.2406
	0.0315	0.8		0.350	8.89		0.800	20.320
	0.03543	0.9		0.35433	9		0.80709	20.5
	0.03937	1	23/64	0.35938	9.1281	13/16	0.81250	20.6375
	0.04	1.016		0.37402	9.5		0.82677	21
3/64	0.04687	1.191	3/8	0.375	9.525	53/64	0.82813	21.0344
	0.04724	1.2	25/64	0.39063	9.9219	27/32	0.84375	21.4312
	0.05	1.27		0.39370	10		0.84646	21.5
	0.05512	1.4		0.400	10.16		0.850	21.590
	0.05906	1.5	13/32	0.40625	10.3188	55/64	0.85938	21.8281
	0.06	1.524		0.41339	10.5		0.86614	22
1/16	0.06250	1.5875	27/64	0.42188	10.7156	7/8	0.875	22.225
	0.06299	1.6		0.43307	11		0.88583	22.5
	0.06693	1.7	7/16	0.43750	11.1125	57/64	0.89063	22.6219
	0.07	1.778		0.450	11.430		0.900	22.860
	0.07087	1.8		0.45276	11.5		0.90551	23
	0.075	1.905	29/64	0.45313	11.5094	29/32	0.90625	23.0188
5/64	0.07813	1.9844	15/32	0.46875	11.9062	59/64	0.92188	23.4156
	0.07874	2		0.47244	12		0.92520	23.5
	0.08	2.032	31/64	0.48438	12.3031	15/16	0.93750	23.8125
	0.08661	2.2		0.49213	12.5		0.94488	24
	0.09	2.286	1/2	0.50	12.7		0.950	24.130
	0.09055	2.3				61/64	0.95313	21.2094
3/32	0.09375	2.3812		0.51181	13		0.96457	24.5
	0.09843	2.5	33/64	0.51563	13.0969	31/32	0.96875	24.6062
	0.1	2.54	17/32	0.53125	13.4938		0.98425	25
	0.10236	2.6		0.53150	13.5	63/64	0.98438	25.0031
7/64	0.10937	2.7781	35/64	0.54688	13.8906	1	1.0000	25.4

LIST OF MICROETCHANTS*

Alloy Family	Common Name for Etchant	ASTM E407 No.[†]
Carbon and low alloy steels	Nital or Picral	74a, 76
Tool steels	Nital	74a
Cast irons	Nital	74a
Austenitic stainless steels	Oxalic	13b
Precipitation hardening stainless steels	Fry's	79
Ferritic and martensitic stainless steels	Viella's	80
Heat resistant castings	Glyceregia	87
Ni, Ni-Cu, and Ni-Fe	Acetic-nitric-water	134
Ni-Mo	Chromic-HCl	143
Ni-Cr-Mo (Alloy C-276)	Oxalic	13c
Ni-Cr-Mo (all other)	Hydrochloric-methanol	23
Ni-Fe-Cr-Mo (Alloy 20 Cb·3)	Hydrochloric-nitric	88
Ni-Fe-Cr-Mo (all other)	Hydrochloric-copper sulfate	25
Ni and Fe base superalloys	Kalling's or Glyceregia	94, 87
W, Mo	Murakami's	98c
Ta, Cb	Sulfuric-HF-peroxide	163
Ti	Kroll's	192
Zr	Nitric-HF-hydrochloric	66
Al	Keller's	3
Mg	Acetic-glycol	119
Pb	Acetic-nitric-glycerin	113
Sn, Sn-Pb	Nital	74d
Zinc	Chromic-sodium sulfate	200
Cu alloys	Ammonium hydroxide-peroxide	44
Cu-Zn	Phosphoric-water	8b

* Exercise extreme caution in handling all chemicals, especially HF. Follow safety precautions described in ASTM E407.
† See numerical list of etchants

NUMERICAL LIST OF ETCHANTS . . .

Etchant	Composition	Procedure
3	2 ml HF 3 ml HCl 5 ml HNO$_3$ 190 ml water	(a) Immerse 10 sec to 20 sec. Wash in stream of warm water. Reveals general structure. (b) Dilute with four parts water—color constituents—mix fresh.
8	10 ml H$_3$PO$_4$ 90 ml water	(b) Electrolytic at 1 V to 8 V for 5 sec to 10 sec.

13	10 g oxalic acid
	100 ml water
		Electrolytic at 6 V.
		  (b) 1 min
		  (c) 2 sec to 3 sec

23	5 ml HCl 95 ml ethanol (95%) or methanol (95%)	Electrolytic at 6 V for 10 sec to 20 sec.
44	50 ml NH$_4$OH 20 to 50 ml H$_2$O$_2$ (3%) 0 to 50 ml water	Use fresh. Peroxide content varies directly with copper content of alloy to be etched. Swab or immerse up to 1 min. Film on etched aluminum bronze removed by No. 82.
66	30 ml HF 15 ml HNO$_3$ 30 ml HCl	Swab 3 sec to 10 sec or immerse to 2 min.
74	1 to 5 ml HNO3 100 ml ethanol (95%) or methanol (95%)	Etching rate is increased, selectivity decreased with increased percentage of HNO$_3$. (a) Immerse a few seconds to a minute. (b) Swab or immerse several minutes.
75	5 g picric acid 8 g CuCl$_2$ 20 ml HCl 6 ml HNO$_3$ 200 ml ethanol (95%) or methanol (95%)	Immerse 1 sec to 2 sec at a time and immediately rinse with methanol. Repeat as often as necessary. Long immersion times result in copper deposition on surface.
76	10 g picric acid 10 ml ethanol (95%) or methanol (95%)	Composition given will saturate the solution with picric acid. Immerse a few seconds to a minute or more.
79	40 ml HCl 5 g CuCl$_2$ 30 ml water 25 ml ethanol (95%) or methanol (95%)	Swab a few seconds to a minute.
80	5 ml HCl 1 g picric acid 100 ml ethanol (95%) or methanol (95%)	Swab or immerse a few seconds to 15 min. Reaction may be accelerated by adding a few drops of 3% H$_2$O$_2$.
82	5 g FeCl$_3$ 5 drops HCl 100 ml water	Immerse 5 sec to 10 sec.
87	10 ml HNO$_3$ 20 to 50 ml HCl 30 ml glycerin	**Warning:** Nitrogen dioxide gas given off. Use hood. Mix HCl and glycerin thoroughly before adding HNO3. Do not store. Discard before solution attains a dark orange color. Swab or immerse a few seconds to a few minutes. Higher percentage of HCl minimizes pitting. A hot water rinse just prior to etching may be used to activate the reaction. Sometimes a few passes on the final polishing wheel is also necessary to remove a passive surface.

. . . NUMERICAL LIST OF ETCHANTS		
Etchant	**Composition**	**Procedure**
88	10 ml HNO$_3$ 20 ml HCl 30 ml water	**Warning:** Nitrogen dioxide gas given off. Use hood. Discard before solution attains a dark orange color. Immerse a few seconds to a minute. Much stronger reaction than No. 87.
94	2 g CuCl$_2$ 40 ml HCl 40 to 80 ml ethanol (95%) or methanol (95%)	Submerged swabbing for a few seconds to several minutes.
98	10 g K$_3$Fe(CN)$_6$ 10 g KOH or NaOH 100 ml water	**Warning:** Extremely poisonous hydrogen cyanide given off. Use hood. Poisonous by ingestion as well as contact. To discard, neutralize (or turn basic) with ammonia and flush down acid drain with water. Use fresh. (c) Swab 5 to 60 sec. Immersion will produce a stain etch. Follow with water rinse, alcohol rinse, dry.
113	15 ml acetic acid 15 ml HNO$_3$ 60 ml glycerin	Use fresh solution at 80°C (175°F).
119	1 ml HNO$_3$ 20 ml acetic acid 60 ml diethylene glycol 20 ml water	Swab 1 to 3 sec for F and T6, 10 sec for T4 and 0 temper.
134	70 ml H$_3$PO$_4$ 30 ml water	Electrolytic for 5 V to 10 V for 5 to 60 sec. Polishes at high currents.
143	0.01 to 1 g CrO$_3$ 100 ml HCl	Allow solution to age a few minutes before using. Swab or immerse a few seconds to a few minutes.
163	30 ml H$_2$SO$_4$ 30 ml HF 3 to 5 drops H$_2$O$_2$ (30%) 30 ml water	Immerse 5 to 60 sec. Use this solution for alternate etch and polishing.
192	1 to 3 ml HF 2 to 6 ml HNO$_3$ 100 ml water	Swab 3 to 10 sec or immerse 10 to 30 sec. HF attacks and HNO$_3$ brightens the surface of titanium. Make concentation changes on this basis.

200	A– 40 g CrO$_3$
	    3 g NaSO$_4$
	    200 ml water
	
	B– 40 g CrO$_3$
	    200 ml water
		Immerse in Solution A with gentle agitation for several seconds. Rinse in Solution B.

LETTER SIZES*									
SIZE	DRILL DIAMETER	SIZE	DRILL DIAMETER	SIZE	DRILL DIAMETER	SIZE	DRILL DIAMETER	SIZE	DRILL DIAMETER
A	.234	G	.261	L	.290	Q	.332	V	.377
B	.238	H	.266	M	.295	R	.339	W	.386
C	.242	I	.272	N	.302	S	.348	X	.397
D	.246	J	.277	O	.316	T	.358	Y	.404
E	.250	K	.281	P	.323	U	.368	Z	.413
F	.257								

*in in.

PIPE FITTINGS AND VALVES

	FLANGED	SCREWED	BELL & SPIGOT		FLANGED	SCREWED	BELL & SPIGOT		FLANGED	SCREWED	BELL & SPIGOT
BUSHING		⊐	∈	REDUCING FLANGE	⊐			AUTOMATIC BY-PASS VALVE			
CAP		⊣	⊃	BULL PLUG	⊐		○	AUTOMATIC REDUCING VALVE			
REDUCING CROSS				PIPE PLUG		◁	⊏	STRAIGHT CHECK VALVE			
STRAIGHT-SIZE CROSS				CONCENTRIC REDUCER	▷	▷	≫	COCK			
CROSSOVER				ECCENTRIC REDUCER							
45° ELBOW				SLEEVE	⊩---⊩	⊣---⊢	⊃---∈	DIAPHRAGM VALVE			
90° ELBOW				STRAIGHT-SIZE TEE				FLOAT VALVE			
ELBOW – TURNED DOWN	⊖⊩	⊖⊣	⊖∈	TEE – OUTLET UP	⊩⊙⊩	⊣⊙⊣	⊃⊙∈	GATE VALVE			
ELBOW – TURNED UP	⊙⊩	⊙⊣	⊙→	TEE – OUTLET DOWN	⊩⊖⊩	⊣⊖⊣	⊃⊖∈	MOTOR-OPERATED GATE VALVE			
BASE ELBOW				DOUBLE-SWEEP TEE				GLOBE VALVE			
DOUBLE-BRANCH ELBOW				REDUCING TEE				MOTOR-OPERATED GLOBE VALVE			
LONG-RADIUS ELBOW				SINGLE-SWEEP TEE				ANGLE HOSE VALVE			
REDUCING ELBOW				SIDE OUTLET TEE – OUTLET DOWN				GATE HOSE VALVE			
SIDE OUTLET ELBOW – OUTLET DOWN	⊙⊩	⊙⊣	⊙→	SIDE OUTLET TEE – OUTLET UP				GLOBE HOSE VALVE			
SIDE OUTLET ELBOW – OUTLET UP	⊙⊩	⊙⊣	⊙→	UNION	⊣⊩	⊣⊢		LOCKSHIELD VALVE			
STREET ELBOW				ANGLE CHECK VALVE				QUICK-OPENING VALVE			
CONNECTING PIPE JOINT	⊣⊢	⊣	←	ANGLE GATE VALVE – ELEVATION				SAFETY VALVE			
EXPANSION JOINT				ANGLE GATE VALVE – PLAN							
LATERAL				ANGLE GLOBE VALVE – ELEVATION				GOVERNOR-OPERATED AUTOMATIC VALVE			
ORIFICE FLANGE	⊣⊪⊩			ANGLE GLOBE VALVE – PLAN							

AMSE International

DRILLED HOLES

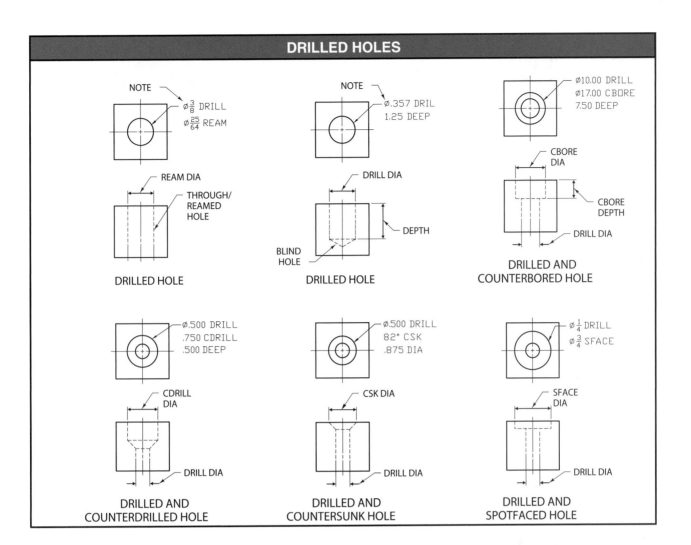

| | DRILLED HOLE | | DRILLED HOLE | | DRILLED AND COUNTERBORED HOLE |

| | DRILLED AND COUNTERDRILLED HOLE | | DRILLED AND COUNTERSUNK HOLE | | DRILLED AND SPOTFACED HOLE |

DRILL SIZES*

SIZE	DRILL DIAMETER	SIZE	DRILL DIAMETER	SIZE	DRILL DIAMETER	SIZE	DRILL DIAMETER	SIZE	DRILL DIAMETER	SIZE	DRILL DIAMETER
1	.2280	17	.1730	33	.1130	49	.0730	65	.0350	81	.0130
2	.2210	18	.1695	34	.1110	50	.0700	66	.0330	82	.0125
3	.2130	19	.1660	35	.1100	51	.0670	67	.0320	83	.0120
4	.2090	20	.1610	36	.1065	52	.0635	68	.0310	84	.0115
5	.2055	21	.1590	37	.1040	53	.0595	69	.0292	85	.0110
6	.2040	22	.1570	38	.1015	54	.0550	70	.0280	86	.0105
7	.2010	23	.1540	39	.0995	55	.0520	71	.0260	87	.0100
8	.1990	24	.1520	40	.0980	56	.0465	72	.0250	88	.0095
9	.1960	25	.1495	41	.0960	57	.0430	73	.0240	89	.0091
10	.1935	26	.1470	42	.0935	58	.0420	74	.0225	90	.0087
11	.1910	27	.1440	43	.0890	59	.0410	75	.0210	91	.0083
12	.1890	28	.1405	44	.0860	60	.0400	76	.0200	92	.0079
13	.1850	29	.1360	45	.0820	61	.0390	77	.0180	93	.0075
14	.1820	30	.1285	46	.0810	62	.0380	78	.0160	94	.0071
15	.1800	31	.1200	47	.0785	63	.0370	79	.0145	95	.0067
16	.1770	32	.1160	48	.0760	64	.0360	80	.0135	96	.0063

*in in.

STANDARD SERIES THREADS — GRADED PITCHES

NOMINAL DIAMETER	UNC		UNF		UNEF	
	TPI	TAP DRILL	TPI	TAP DRILL	TPI	TAP DRILL
0 (.0600)			80	3/64		
1 (.0730)	64	No. 53	72	No. 53		
2 (.0860)	56	No. 50	64	No. 50		
3 (.0990)	48	No. 47	56	No. 45		
4 (.1120)	40	No. 43	48	No. 42		
5 (.1250)	40	No. 38	44	No. 37		
6 (.1380)	32	No. 36	40	No. 33		
8 (.1640)	32	No. 29	36	No. 29		
10 (.1900)	24	No. 25	32	No. 21		
12 (.2160)	24	No. 16	28	No. 14	32	No.13
1/4 (.2500)	20	No. 7	28	No. 3	32	7/32
5/16 (.3125)	18	F	24	I	32	9/32
3/8 (.3750)	16	5/16	24	Q	32	11/32
7/16 (.4375)	14	U	20	25/64	28	13/32
1/2 (.5000)	13	27/64	20	29/64	28	15/32
9/16 (.5625)	12	31/64	18	33/64	24	33/64
5/8 (.6250)	11	17/32	18	37/64	24	37/64
11/16 (.6875)					24	41/64
3/4 (.7500)	10	21/32	16	11/16	20	45/64
13/16 (.8125)					20	49/64
7/8 (.8750)	9	49/64	14	13/16	20	53/64
15/16 (.9375)					20	57/64
1 (1.000)	8	7/8	12	59/64	20	61/64

PREFERRED METRIC SCREW THREADS

COARSE (GENERAL PURPOSE)				FINE			
Nominal Size and Thd Pitch	Tap Drill Diameter*	Nominal Size and Thd Pitch	Tap Drill Diameter*	Nominal Size and Thd Pitch	Tap Drill Diameter*	Nominal Size and Thd Pitch	Tap Drill Diameter*
M1.6 x 0.35	1.25	M20 x 2.5	17.5	—	—	M20 x 1.5	18.5
M2 x 0.4	1.6	M24 x 3	21.0	—	—	M24 x 2	22.0
M2.5 x 0.45	2.05	M30 x 3.5	26.5	—	—	M30 x 2	28.0
M3 x 0.5	2.5	M36 x 4	32.0	—	—	M36 x 2	33.0
M4 x 0.7	3.3	M42 x 4.5	37.5	—	—	M42 x 2	39.0
M5 x 0.8	4.2	M48 x 5	43.0	—	—	M48 x 2	45.0
M6 x 1	5.0	M56 x 5.5	50.5	—	—	M56 x 2	52.0
M8 x 1.25	6.8	M64 x 6	58.0	M8 x 1	7.0	M64 x 2	60.0
M10 x 1.5	8.5	M72 x 6	66.0	M10 x 1.25	8.75	M72 x 2	68.0
M12 x 1.75	10.30	M80 x 6	74.0	M12 x 1.25	10.5	M80 x 2	76.0
M16 x 2	14.00	M90 x 6	84.0	M16 x 1.5	14.5	M90 x 2	86.0
—	—	M100 x 6	94.0	—	—	M100 x 2	96.0

Glossary

acetylene: A colorless gas with a very distinctive, nauseating odor that is highly combustible when mixed with oxygen. Unstable at pressures above 15 psi. Used in oxyacetylene welding. See *oxyacetylene welding.*

acoustic emission testing (AE): A proof test that consists of detecting acoustic signals produced by plastic deformation or crack formation during mechanical loading or thermal stressing of metals.

acrylic: A one-part UV (heat-cure) or two-part adhesive that can be used on a variety of materials.

actual throat: The shortest distance from the face of a fillet weld to the weld root after welding. See *weld face.*

adhesive bonding: The joining of parts with an adhesive placed between the faying (mating) surfaces, which produces an adhesive bond.

adhesive wear: The removal of metal from a surface by welding together and subsequent shearing of minute areas of two surfaces that slide across each other under pressure.

air carbon arc cutting (CAC-A): A cutting process in which the cutting of metals is accomplished by melting with the heat of an arc between a carbon electrode and the base metal.

air cut time: The time that a piece of equipment spends in the nonproductive activity of moving from one weld to another.

alignment marker: A center punch mark made across the joint in various locations.

alloy: Metal that consists of more than one chemical element, with at least one of the elements being a pure metal.

alteration: Any repair that does not restore a mechanical component to its original design.

alternating current (AC): An electrical current that has alternating positive and negative values. See *current.*

ammeter: Electrical instrument that measures amperage.

amperage: The quantity of electricity measured.

ampere (A): The basic unit of measure for electrical current. See *conductor.*

amplitude: The height of the EP and EN portions of the AC cycle above or below zero.

anaerobic adhesive: A one-part adhesive or sealant that cures due to the absence of air which has been displaced between mated parts.

angle beam: A vibrating pulse wave traveling other than perpendicular to the surface.

annealing: Heat treatment process that softens a metal by heating it to a suitable temperature, holding it at that temperature, and cooling it at a suitable rate.

arc blow: The deflection of the welding arc by magnetic forces that occur due to current flow.

arc length: The distance from the tip of the electrode to the weld pool.

arc strike: A discontinuity that results from arcing of the electrode and consists of any localized remelted metal, heat-affected metal, or change in the surface profile of any base metal.

arc voltage (working voltage): The voltage present after an arc is struck and maintained.

arc welding (AW): A group of welding processes that produce coalescence of metals by heating them with an electric arc.

armature: The part of the generator that rotates with the shaft and delivers the electricity.

arrowhead: A termination of the carrier line in the shape of an arrowhead.

arrow side: The surface that is in the direct line of vision of the welder.

artifact: A nonrelevant indication that appears on a radiograph.

A-scan presentation: A method of data presentation using a horizontal base line that indicates distance or time, and a vertical deflection from the base line that indicates relative amplitude of the returning signal.

autogenous weld: A fusion weld made without filler metal.

automatic welding: A welding process that requires minimal observation by the operator and no manual adjustment of the controls.

auto-refrigeration: Cooling that occurs when gas expands, as in the sudden release of gas from a pipe or piece of equipment.

axis: Straight line around which a geometric figure is generated.

backfire: A quick recession of the flame into the welding tip, typically followed by extinction of the flame.

backgouging: The removal of weld metal and base metal from the weld root side of a welded joint to facilitate complete fusion and complete joint penetration when welding on that side is completed.

background current: The current setting for the low pulse, which maintains the arc.

backhand welding: A welding technique in which the torch points toward the weld and away from the direction of travel.

backing symbol: Supplementary symbol indicated by a rectangle on the opposite side of the groove weld symbol on the reference line. See *supplementary symbol*.

backing weld: A weld made at the back of a single-groove weld, which is deposited before any welding on the opposite side is done.

backlighting: A lighting method that uses a diffused light source to eliminate or soften shadow detail.

back (transverse) pitch: Distance from the center of one row of rivets to the center of the adjacent row of rivets. See *rivet*.

back-step welding: A welding process in which weld passes are made in the direction opposite to the progress of welding.

back-to-back positioning: A mechanical restraint method that places identical weldments back-to-back and clamps them together.

back weld: A weld made in the weld root opposite the face of the weld.

bake-out: A temperature-control process used on a casting to remove hydrogen and other contaminants that could cause cracking during welding.

balloting: A formal method of documenting and voting upon the reviewers' suggestions.

bar: Round-, square-, or rectangular-shaped structural steel. See *structural steel*.

base metal: The metal or alloy that is to be welded.

base metal material specification: The chemical composition or industry specification of the base metal.

base metal thickness range: A procedure qualification variable that indicates the range of base metal thicknesses covered in the procedure qualification record.

base metal weldability classification: An alphanumeric system that groups base metals with similar welding characteristics.

bead: Narrow layer or layers of metal deposited on the base metal as an electrode melts. See *base metal* and *electrode*.

beam: I-shaped structural steel. See *structural steel*.

bending strength: A combination of tensile and compressive forces, and is a property that measures resistance to bending or deflection in the direction that the load is applied.

bending stress: See *flexural stress*.

bend test: A destructive test used to determine the ductility of a weld by bending a welded specimen around a standardized mandrel.

bevel: Sloped edge of an object running from surface to surface.

binocular microscope: A light microscope that provides a low-magnification, three-dimensional view of the surface.

biprism: Two uniaxial double-refracting crystals.

bird nesting: A tangle of wire that forms in a wire feeder when welding wire is restricted in the liner or by a burnback condition.

blend grinding: A mechanical repair method in which a thinned, pitted, or cracked region of a part is smoothed to create a gentle transition with the unaffected surface.

blind hole: Drilled hole that does not pass through.

blind rivet: Rivet with a hollow shank that joins two parts with access from one side. See *rivet* and *shank*.

bloom: A slight haze that appears on the surface of the specimen and is evidence of the first appearance of the microstructure.

bourdon tube: A coiled fluid-containing tube that straightens out as the internal pressure on the fluid is increased.

brazed joint tension shear test: A shear test that determines the strength of filler metal in a brazed joint.

braze welding (BW): A joining process that produces a coalescence of metals with filler metals that begin to melt at temperatures above 840°F, below the melting point of the metals joined, and in which the filler metal is not distributed into the joint by capillary action.

brazing (B): A group of joining processes that produce a coalescence of metals using nonferrous filler metals that have a melting point below that of the base metal.

brazing symbol: Graphic symbol that shows braze locations and specifications on prints.

brazing temperature range: The temperature range within which the base metal is heated to enable the filler metal to wet the base metal and form a brazed joint.

break line: Line that shows internal features or avoids showing continuous features.

brightfield illumination: An illumination process in which the surface features perpendicular to the optical axis of the microscope appear the brightest.

Brinell hardness test: An indentation hardness test that uses a machine to press a 10 mm diameter, hardened steel ball into the surface of a test specimen.

brinelling: Localized plastic deformation or surface denting caused by repeated local impact or overload.

brittleness: Lack of ductility in a metal. See *ductility.*

broken-out section: Partial section view which appears to have been broken out of the object. See *section view.*

buildup lighting: A lighting method that combines (adding or deleting) light sources to achieve the desired lighting effect.

burnback: A condition that occurs when welding wire is restricted, and fuses to the end of the contact tip.

burst: A complex branching of the carrier line.

buttering: A surfacing weld technique that applies weld metal on one or more joint surfaces to provide compatible base metal for subsequent completion of the weld.

butt joint: A weld joint in which two workpieces are set approximately level to each other and are positioned edge-to-edge. See *weld joint.*

C

calibration block: A piece of material of specified composition, heat treatment, geometric form, and surface finish, by which ultrasonic equipment can be assessed and calibrated for the examination of material of the same general condition.

calibration standard: A calibration block or a reference block.

Canadian Center for Occupational Health and Safety: Federal agency that standardizes safety practices for most types of work.

capacitor discharge (CD) start: An arc starting method that uses a burst of high voltage from a bank of capacitors in the welding power supply.

capillary action: The force that distributes liquid filler metal through surface tension between the faying surfaces of the joint.

carbon equivalent: A formula based on the chemical composition of a steel, which provides a numerical value to indicate whether preheat and postheating are required.

carburizing: Case-hardening process for low-carbon steels that uses an environment with sufficient carbon potential and a temperature above the upper critical temperature. See *case hardening.*

carburizing flame: A reducing flame in which there is an excess of fuel gas.

carrier line: An incandescent (glowing) streak that traces the trajectory (path) of each particle (spark).

cartesian coordinate system: A system of locating points in space defined by perpendicular planes.

case hardening: Process of hardening low-carbon or mild steels by adding carbon, nitrogen, or a combination of carbon and nitrogen to the outer surface, forming a hard, thin outer shell.

cast: Metal heated to its liquid state and poured into a mold, where it cools and resolidifies.

casting alloy: Alloy poured into a sand or permanent metal mold. See *alloy.*

cavitation: Surface damage caused by collapsing vapor bubbles in a flowing liquid.

certificate of analysis (COA): See *mill test report.*

certificate of compliance (COC): A statement by a manufacturer, without supporting documentation, that the supplied metal meets specifications.

certification: A notarized statement provided by a supplier verifying that a product meets the specification under which it is sold.

chamfer: Sloped edge of an object running from surface to side. See *edge.*

channel: C-shaped structural steel used in conjunction with other structural shapes as support members or combined to serve as an I beam. See *structural steel*.

charpy: Impact test specimen supported horizontally between two anvils with the pendulum allowed to strike opposite the notch.

Charpy V-notch test: A toughness test that uses the energy produced by a dynamic load, and measures the energy needed to break a small machine-notched test specimen.

check valve: A valve that allows the flow of liquid or gas in one direction only.

chemical analysis: A destructive quantitative identification method that requires removal of a small sample (1 g to 2 g) of metal for chemical analysis of its constituent elements.

chemical inhomogeneity: Any disturbance in the chemical composition gradient of a metal.

chemical polishing: A polishing process that uses chemical reactions to remove the rough peaks on the specimen surface.

chemical properties: Properties of metals that are directly related to molecular composition and pertaining to the chemical reactivity of metals and the surrounding environment.

chemical spot testing: A semi-quantitative identification method that uses chemicals that react when placed on certain types of metals.

chill plate: A metal plate used to prevent overheating during welding.

chisel testing: A qualitative identification method that identifies metal by the shape of the chips it produces.

circular magnetization: A concentric magnetic field produced by a straight conductor, such as a piece of wire, carrying an electrical current.

cleaning action: The removal of the oxide coating base metal by bombarding it with gas ions.

code: A type of standard that is mandatory and is used by a jurisdictional body.

coefficient of linear expansion: The change in unit dimension, such as length, caused by a 1° rise in temperature.

coefficient of volumetric expansion: The rate of change in the volume of an object per degree of temperature.

cold crack: A crack that develops after solidification is complete.

cold mechanical repair: A mechanical repair method that consists of spanning a crack in a failed part with structural repair components anchored into sound base metal on both sides of the crack.

cold welding (CW): Welding process in which a weld is produced using pressure at room temperature to cause deformation at the joint.

cold worked: Metal that is hammered, rolled, or drawn through a die.

color-coding: An identification marking that consists of colored stripes painted on one end of metal to allow for permanent storage or temporary storage and subsequent retrieval from a metal service center or a user's storeroom.

color test: Metal identification test that identifies metals by their color.

combined weld symbols: Weld symbols used when the weld joint, weld type, and welding operation require more information than can be specified with one weld symbol. See *weld symbol*, *weld joint*, and *weld type*.

commutator: The part of an armature that connects the armature to the insulated copper bars on which the brushes ride.

compression: Stress caused by two equal forces acting on the same axial line to crush an object. See *stress* and *axis*.

compressive strength: The ability of a metal to resist being crushed.

concave: Curved inward.

concave root surface: A depression in the weld extending below the surface of the adjacent base metal caused by an underfill in the root pass of a weld.

concavity: An indentation on the side of the joint opposite the weld. This is also referred to as suck-back.

conductor: Any material through which electricity flows easily.

confined space: A space large enough and so configured that an employee can physically enter and perform assigned work, has limited or restricted means for entry and exit, and is not designed for continuous employee occupancy.

constant-current welding machine: A welding machine that maintains a relatively constant current over a wide range of welding voltages caused by changes in arc length.

constant pitch: Standard screw thread series with a set number of threads per inch regardless of diameter. See *standard series*.

constant potential: Generation of a stable voltage regardless of the current output produced by the welding machine.

consumable insert: Spacer that provides proper opening of a weld joint and becomes part of the filler metal during welding. See *weld joint*.

consumable insert symbol: Supplementary symbol indicated by a square on the opposite side of a groove weld on the reference line. See *supplementary symbol* and *groove weld*.

contact tip-to-work distance (CTWD): Distance from the end of the contact tip to the work; includes arc length.

continuous magnetization method: An MT examination technique in which the magnetic particles are applied while the magnetizing force is maintained.

contour symbol: Supplementary symbol indicated by a horizontal line or arc parallel to the weld symbol, which specifies the shape of the completed weld. See *supplementary symbol* and *weld symbol*.

conventional sine wave transformer-rectifier: A welding power source that produces a sinusoidal waveform.

conventional square wave transformer-rectifier: A welding power source that produces a relatively square 60 Hz waveform.

convex: Curved outward.

cooling rate: The rate of temperature change of a weld joint over time from the welding temperature to room temperature.

corner joint: A joint formed when two members are positioned at an approximate right angle in an L shape.

corrosion: Combining metals with elements in the environment that leads to deterioration of the metal.

corrosion allowance: An additional thickness of metal above the design thickness that allows for metal loss from corrosion or wear without reducing the design thickness.

counterbored hole: Enlarged and recessed hole with square shoulders.

counterdrilled hole: Hole with a cone-shaped opening below the outer surface.

countersink: Tool that produces a countersunk hole. See *countersunk hole*.

countersunk hole: Hole with a cone-shaped opening or recess at the outer surface. See *countersink*.

couplant: A liquid substance used between the search unit and the test surface to permit or improve the transmission of ultrasonic energy.

cover pass: The final weld pass deposited.

crack: A fracture-type discontinuity characterized by a sharp tip and a high ratio of length to width, and width to opening displacement.

crater: A depression at the termination of the weld bead. See *base metal*.

cracking: The opening and closing of cylinder valves quickly to clear any debris.

creep: Slow plastic elongation that occurs during extended service under load above a specific temperature for that metal. See *strain*.

critical temperature: Temperature above which steel must be heated so it will harden when quenched.

crosschecking: A series of parallel cracks about ½″ apart that occur in brittle deposits (with hardness greater than HRC 50) as they undergo stress relief.

cryogenic properties: Ability of a metal to resist failure when subjected to very low temperatures.

crystal structure: A specific arrangement of atoms in an orderly and repeating three-dimensional pattern.

cubic foot: $1'-0'' \times 1'-0'' \times 1'-0''$ or 1728 cu in.

cubic inch: $1'' \times 1'' \times 1''$ or its equivalent.

Curie temperature: The temperature of magnetic transformation, above which a metal is nonmagnetic, and below which it is magnetic.

curing: A process that converts the adhesive from its applied condition to the final solid state.

current: The amount of electron flow through an electrical circuit. See *conductor*.

cutting plane line: Line that shows where an object is imagined to be cut in order to view internal features.

cyaniding: Process of hardening low-carbon steel by heating it in sodium cyanide or potassium cyanide.

cyanoacrylate adhesive: A one-part adhesive that cures instantly by reacting to trace surface moisture to bond mated parts.

cyclical (variable) load: A load that varies with time and rate, but without the sudden change that occurs with an impact load.

darkfield illumination: An illumination process that illuminates the specimen at sufficient obliqueness (a narrow angle to the surface) so that the contrast is completely reversed from that obtained with brightfield illumination.

defect: One or more indications whose aggregate size, shape, orientation, location, or properties fail to meet the acceptance criteria of the applicable fabrication code or standard.

demagnetization: The elimination or reduction of residual magnetism created by MT.

density testing: A semi-quantitative identification method that measures the density of an unknown metal.

depth of fusion: Distance that fusion extends into the base metal from the surface metal during welding. See *fusion face* and *weld interface*.

depth of field: The total depth of the image that can be maintained in focus within a lens.

derating: A lowering of the current output level of an AC welding machine when being used for GTAW.

design thickness: The thickness of metal required to support the load on a part.

destructive testing: Any type of testing that damages the test part (specimen).

developer: A material that is applied to the test surface to accelerate bleedout and enhance the contrast of indications.

developing time: The elapsed time between the application of the developer and the examination of the part.

diffraction: A modification of light in which the rays appear to be deflected to produce fringes of parallel light and dark colored bands.

diffused light: A lighting source that uses a semi-opaque screen (such as ground glass) to diffuse the light source, reduce glare, and soften harsh details.

dilution: A change in the chemical composition of a weld metal from the pure filler metal due to contributions of the base metals.

direct current (DC): An electrical current that flows in one direction only. See *current*.

direct current electrode negative (DCEN): Flow of current from electrode (–) to work (+). See *electrode*.

direct current electrode positive (DCEP): Flow of current from work (–) to electrode (+). See *electrode*.

dissimilar metal welding: The joining of two metals of different composition using a compatible filler metal to ensure the weld meets required properties.

distortion: The undesirable dimensional change of a fabrication.

distribution piping: Carbon-steel, standard-size pipe of small diameter that conveys products from intermediate facilities to consumers.

double-bevel-groove weld: Groove weld having joint members beveled on both sides with the weld made from both sides. See *groove weld*.

double-flare-bevel-groove weld: Groove weld having two radiused joint members with the weld made from both sides. See *groove weld*.

double-flare-V-groove weld: Groove weld having radiused joint members with the weld made from both sides. See *groove weld*.

double-groove weld: A groove weld that is made from both sides of the joint. See *groove weld*.

double-J-groove weld: Groove weld having joint members grooved in a J shape on both sides with the weld made from both sides. See *groove weld*.

double-square-groove weld: Groove weld having square-edged joint members with the weld made from both sides. See *groove weld*.

double-U-groove weld: Groove weld having joint members grooved in a U shape on both sides with the weld made from both sides. See *groove weld*.

double-V-groove weld: Groove weld having joint members angled on both sides with the weld made from both sides. See *groove weld*.

downhill welding: Welding with a downward progression. See *vertical welding*.

drag: Lag between the top of the cut and the bottom as cutting proceeds.

drag travel angle: Travel angle where the electrode points away from the direction of travel.

drill: Round hole in a material produced by a twist drill.

dry magnetization method: An MT examination technique in which the magnetic particles are in a dry powder form.

ductility: A measure of the ability of a metal to yield plastically under load, rather than fracture.

dunnage: A series of steel I-beams parallel to one another.

duty cycle: Percentage of time during a specified period that a welding machine can be operated at its rated load without exceeding the temperature limits of the insulation on the components parts.

dwell time: The total time penetrant is in contact with the component surface, including application and drain times.

dynamic electricity: Electricity in motion in an electric current. *See current.*

E

earmuffs: A device worn over the ears to reduce the level of noise reaching the eardrum.

earplugs: A device inserted into the ear canal to reduce the level of noise reaching the eardrum.

eddy current: An electrical current caused to flow in a conductor by the time or space variation, or both, of an applied magnetic field.

edge: Intersection of two surfaces.

edge joint: Weld joint formed when the edges of two or more parallel or nearly parallel workpieces are joined. See *weld joint.*

edge preparation: The preparation of the workpiece edges by cutting, cleaning, or other methods.

effective throat: The minimum distance between the weld face and the weld root, minus convexity.

elastic deformation: Ability of a metal to return to its original size and shape after loading and unloading.

elastic limit (yield): The maximum stress to which a material is subjected without any permanent strain remaining after stress is completely removed.

electrical circuit: Path taken by electric current flowing from one terminal of the welding machine, through a conductor, and to the other terminal. See *current* and *conductor.*

electrical conductivity: The rate at which electric current flows through a metal.

electrical properties: Ability of a metal to conduct or resist electricity or the flow of electrons.

electrical resistivity (resistivity): The electrical resistance of a unit volume of a material.

electrical resistivity testing: A semi-quantitative identification method that uses differences in electrical resistivity to identify metals.

electrode: A component of the welding circuit that conducts electrical current to the weld area. See *weld bead.*

electrode angles: The angles at which the electrode is held in relation to the joint during welding.

electrode extension: On a GTAW torch, the distance from the end of the collet to the tip of the electrode.

electrode holder: A handle-like tool that holds the electrode during welding. Also called a stinger. See *electrode.*

electrogas welding (EGW): A welding process that uses an arc between a filler metal electrode and the weld pool, using approximately vertical welding and a backing bar to control the molten weld metal.

electrolytic polishing: A polishing process in which the mount is the anode (connected to the positive terminal) in an electrolytic solution and current is passed from a metal cathode (connected to the negative terminal).

electromagnetic examination (ET): An NDE method that uses electromagnetic energy having frequencies less than visible light to yield information on the quality of the part being tested.

electron beam welding (EBW): A welding process that produces coalescence with a concentrated beam, composed primarily of high-velocity electrons, impinging on the joint.

electroplating: The application of a thin, hard chrome coating to repair minor damage.

embrittlement: The complete loss of ductility and toughness of a metal, so that it fractures when a small load is applied.

epoxy: A two-part adhesive that cures when resin and hardener are combined.

equipment calibration standard: A test piece that contains typical discontinuities that demonstrate that calibration equipment is detecting the discontinuities for which the part is being inspected.

erosion (low-stress abrasion): A form of abrasive wear in which the force of an abrasive and the surface causes the removal of surface material.

erosion-corrosion: The detrimental effect of velocity or turbulence in a corrosive environment.

essential variable: A welding qualification variable which, if altered, shall be considered to affect the mechanical properties of the weld.

etching: The controlled selective attack on a metal surface for revealing the microstructural detail of a polished specimen.

examiner: A person who is qualified, or qualified and certified, to conduct certain types of NDE processes.

excess weld reinforcement: Weld metal built up in excess of the quantity required to fill a joint.

explosion welding (EXW): A welding process that produces a weld by extreme impact of the metals through controlled detonation.

F

face reinforcement: Reinforcement on the same side as the welding. See *filler metal.*

failure-critical member: A tension member or component whose failure would likely result in collapse of the structure.

failure modes and effects analysis: A failure analysis process that provides a diagnosis of the technical cause of failure using experience gained from previous failures.

false indication: An NDE indication interpreted to be caused by a discontinuity at a location where no discontinuity actually exists.

fast-fill (F1 group) electrode: An electrode with a high iron powder coating that has a soft arc and high deposit rates. See *electrode.*

fast-freeze (F3 group) electrode: An electrode that produces a crisp, deep-penetrating arc and fast-freezing weld bead. See *electrode.*

fatigue: Failure of a material operating under alternating (cyclic) stresses at a value below the tensile strength of the material.

fatigue strength: Property of a metal to resist various kinds of rapidly alternating stresses. See *stress.*

faying surface: The surface of a joint member that is in contact with, or in close proximity to, the member to which it will be joined. See *capillary action.*

ferromagnetic material: A material that can be magnetized or strongly attracted by a magnetic field.

ferrous metal: Any metal with iron as a major alloying element.

field rivet: Rivet placed in the field. See *rivet.*

field weld symbol: Supplementary symbol indicated by a triangular flag rising from the intersection of the arrow and reference line, which specifies the welding operation is to be completed in the field at the location of final installation. See *supplementary symbol.*

file testing: A qualitative identification method in which a file is used to indicate the hardness of steel compared with that of the file.

filler metal: Metal deposited in a welded, brazed, or soldered joint during the welding process.

filler metal approval: The process of testing samples of as-received filler metal to certify conformance to a specification.

filler metal quantity: The deposited weld metal thickness range for groove or fillet welds.

filler metal specification: Identification of filler metal by AWS number or other specification designation.

filler metal usability classification: Alphanumeric method of grouping filler metals with similar characteristics.

fillet weld: A weld type of approximately triangular cross section joining two surfaces at approximately right angles. See *weld type.*

fillet weld break test: A break test in which the specimen is tested with the weld root in tension.

fillet weld leg: Distance from the joint root to the weld toe. See *joint root.*

fillet weld leg size: Dimension from the root of a weld to the toes of a weld after welding. See *fillet weld leg.*

fillet weld shear test: A shear test in which a tensile load is placed on a fillet weld specimen so that the load shears the fillet weld in a longitudinal or a transverse direction.

fill-freeze (F2 group) electrode: An electrode that has a moderately forceful arc and a medium deposition rate. See *electrode.*

fill-freeze (F4 group) electrode: An electrode that produces sound welds with excellent notch toughness and high ductility.

fill lighting: A lighting method that uses a small region of a brighter light to increase detail on a dark area of a subject.

fitting: Standard connection used to join two or more pieces of pipe.

fit-up: The positioning of pipe with other pipe or fittings before welding.

fixed automation system: A system that uses machines designed for a single production function.

fixture: A device used to maintain the correct positional relationship between weldment components as required by print specifications.

flame spraying: A thermal spraying process that uses an oxyfuel gas flame as a source of heat for melting the coating material.

flanged joint: An edge joint in which one of the joint members has a flanged edge at the weld joint.

flash arrestor: A safety device that prevents an explosion or a backfire in the torch or torch head from reaching the regulator and the acetylene cylinder.

flashback: A recession of the flame into or back of the mixing chamber in a flame torch or flame spray torch.

flashlight: A lighting source that provides a pulse of very intense light.

flash welding (FW): A resistance welding process that produces a weld at the faying surfaces of a joint by the intense heat of an arc that occurs when the workpieces are contacted and by the application of pressure after heating has been substantially completed.

flaw (indication): A discontinuity that can be detected through NDE techniques.

flexible automation system: A system that uses programmable movements of the torch and sometimes the workpiece.

flexural (bending) stress: A stress caused by equal forces acting perpendicular to the horizontal axis of an object.

fluidity: A measure of the viscosity or flowability of a liquid or molten solid.

fluorescence: The ability of certain atoms to emit light when they are exposed to external radiation of shorter wavelengths.

flush patch: A replacement of a thin metal component that is applied to provide a smooth transition with the component.

flux: A material that hinders or prevents the formation of oxides and other undesirable substances in molten metal.

flux-cored arc welding (FCAW): An arc welding process that uses a tubular electrode with flux in its core.

forced cooling: Rapid cooling of a solidified weld joint between passes using water.

forehand welding: A welding technique in which the torch points away from the weld in the direction of travel.

forged: Metal formed by a mechanical or hydraulic press with or without heat.

forge welding (FOW): A welding process that produces a weld by heating the metals to welding temperature and applying forceful blows to cause deformation at the faying surfaces.

fork: A simple branching of the carrier line.

foundry mark: An identification marking embossed on the exterior of castings.

fracture test: Metal identification test that breaks the metal sample to check for ductility and grain size. See *ductility.*

frequency: The number of cycles (AC sine waves) per second measured in hertz (Hz).

fretting: Surface damage between two materials, usually metal, caused by oscillatory movement between the surfaces.

friction welding (FRW): A welding process that joins two metal parts that rotate or are in relative motion with respect to one another when they are brought into contact and pressure is applied between them.

full skip: One complete reflection of the ultrasonic beam.

fusion: Melting together of filler metal and base metal. See *filler metal* and *base metal.*

fusion face: Surface of the base metal that is melted during welding. See *fusion.*

fusion welding: The melting of metal with filler metal, or filler metal only, to make a weld.

galling: A condition that occurs when excessive friction, caused by rubbing of high spots on the surface, results in localized welding with subsequent spalling (formation of surface slivers) and further roughening of the rubbing surfaces.

gas metal arc welding (GMAW): An arc welding process that uses an arc between a continuous wire electrode and the weld pool.

gas-shielded flux cored arc welding (FCAW-G): An FCAW variation in which the shielding is obtained from both the CO_2 gas flowing from the nozzle and from the flux core of the electrode.

gas tungsten arc spot welding: An arc welding process that produces localized fusion similar to resistance spot welding but does not require accessibility to both sides of the joint.

gas tungsten arc welding (GTAW): An arc welding process in which shielding gas protects the arc between a nonconsumable tungsten electrode and the base metal. See *electrode.*

globular transfer: The transfer of molten metal in large droplets from the welding wire to the workpiece across an arc.

gouging: A cutting process that removes metal by melting or burning off a portion of the base metal to form a bevel or groove.

gouging (high-stress abrasion): A severe form of abrasive wear in which the force between an abrasive body and the wearing surface is large enough to macroscopically gouge, groove, or deeply scratch the surface.

grain: An assembly of crystals having different orientations of their crystal components. See *crystal*.

grain structure: Pattern of the grains in a metal. See *grain*.

graphitization: The formation of iron carbide that results in loss of ductility.

grinding: The mechanical removal of metal from the surface using hard, brittle grains of an abrasive material.

grip: Effective holding length of a rivet. See *rivet*.

groove face: Surface of the joint member included in the groove of the weld.

groove weld: A weld type made in the groove between the two members to be joined. See *weld type*.

grounding device (ground): Connection between welding cable and weld parts in the welding circuit.

guided bend test: A bend test in which a rectangular piece of welded metal is bent around a U-shaped die and forced into a U shape.

H

hardfacing: Application of filler metals that provide a coating to protect the base metal from wear caused by impact, abrasion, erosion, or from other wear. See *filler metal* and *base metal*.

hardness: The resistance of a material to deformation, indentation, or scratching.

hardness test: a destructive test used to determine the relative hardness of the weld area as compared with the base metal.

heat-affected zone (HAZ): A narrow band of base metal adjacent to the weld joint whose properties and/or metallurgical structure are altered by the heat of welding. See *base metal* and *mechanical property*.

heating rate: The rate of temperature change of a weld joint over time from room temperature to the welding temperature.

heat input: The amount of heat applied to the filler metal and the base metal surface at the required rate to form a weld pool, plus the additional heat required to compensate for heat that is conducted away from the weld.

heat shaping: The application of localized heating to cause movement of a distorted part and restore its dimensions.

heat sink: A piece of metal that draws some of the heat generated by the arc away from the weld zone.

hertz (Hz): The international unit of frequency equal to 1 cycle per second.

high-carbon steel: Steel with a carbon range of 0.45% to 0.75%.

high frequency (HF): High-voltage, low-current pulses over 16,000 cycles per second (16 KHz).

high frequency (HF) start: An arc starting method that uses high frequency to create a path for the arc between the electrode and base metal.

hot crack: A crack formed at temperatures near the completion of solidification.

hot melt adhesive: Thermoplastic material that is applied in a molten state and cures to a solid state when cooled.

hot wire welding: A gas tungsten arc welding process in which the filler metal is preheated as it enters the weld pool.

hot work: Any operation that involves open flames, heat, and/or sparks and has the potential to cause a fire or explosion.

hydrogen-assisted cracking: Loss of toughness in steels resulting from hydrogen atoms created at the surface of the metal by corrosion that diffuse into the HAZ and the base metal.

hydrostatic testing (hydrotesting): Proof testing of closed containers such as vessels, tanks, and piping systems by filling them with water and applying a predetermined test pressure.

image quality indicator (IQI): A device or combination of devices whose demonstrated image determines radiographic quality and sensitivity.

impact damage: Removal of material from and damage to a surface caused by repetitive collisions or impact between two surfaces.

impact load: Load that is applied suddenly or intermittently. See *load*.

impact strength: Ability of a metal to resist loads that are applied suddenly and often at high velocity.

impact testing: Special testing performed on small, notched specimens, to simulate a stress concentration effect.

inclusion: Entrapped foreign solid material in deposited weld metal, such as slag or flux, tungsten, or oxide.

incomplete fusion: A lack of union (fusion) between adjacent weld passes or base metal.

incomplete penetration: A condition in a groove weld in which weld metal does not extend through the joint thickness.

inert gas: A gas that does not readily combine with other elements.

inspector: A person who is qualified, or qualified and certified, to apply the results of NDE flaw characterization to determine whether the flaws meet the acceptance criteria of the applicable fabrication code or standard.

intergranular penetration: Penetration of molten metal along the grain boundaries of the base metal that leads to embrittlement of the base metal.

intermediate weld pass: A single progression of welding subsequent to the root pass and before the cover pass.

intermittent welds: Short sections of fillet welds applied at specified intervals on the weld part. See *fillet weld*.

interpass temperature: Weld area temperature between passes of a multiple-pass weld. See *weld pass*.

interpass temperature control: Maintaining the temperature range within the weld between weld passes until welding is complete.

interstitial element: A chemical element added in small amounts, whose atomic size is significantly less than the major elements present in the metal.

inverter: 1. An electrical device that changes DC into AC. **2.** A welding power source that produces a true square AC waveform.

J

joint design: The shape, dimensions, and configuration of the joint.

joint penetration: The distance the weld metal extends from the weld face into the joint.

joint root: The portion of a weld joint where joint members are the closest to each other.

K

kerf: The width of the cut metal.

killed steel: Steel that is completely deoxidized during steel production by adding silicon or aluminum in the furnace ladle or to the mold.

L

lamellar tearing: A subsurface terrace and step-like crack pattern in wrought steel base metal oriented parallel to the base metal working direction.

laminar discontinuity: A discontinuity that is relatively thin and flat.

lamination: A defect or discontinuity that is aligned parallel to the worked surface of the metal.

lap joint: A weld joint between two overlapping members in parallel planes. See *weld joint*.

large rivets: Rivets with a shank ½″ or greater in diameter. See *rivet* and *shank*.

laser beam welding (LBW): A welding process that produces coalescence with the heat from a laser beam impinging on the joint.

liquid impingement: Progressive material removal from a surface by the striking action of a liquid.

liquid penetrant examination (PT): An NDE technique that uses dyes suspended in high-fluidity liquids to penetrate solid materials and indicate the presence of discontinuities.

liquidus: The lowest temperature at which an alloy is completely molten.

liquidus temperature: The melting temperature of a filler metal.

load: External mechanical force applied to a component. See *stress*.

load cell: A device that uses the elastic deformation of a spring or diaphragm that is calibrated to indicate the mechanical load applied to the specimen.

lock: A precision, high-strength steel member with a multi-lobed outer contour.

longitudinal crack: A crack with its major axis oriented approximately parallel to the weld axis.

longitudinal magnetization: A magnetic field produced when the current-carrying conductor is coiled and the magnetic field is parallel to the axis of the coil.

longitudinal shrinkage: Weld metal shrinkage that occurs parallel to the weld axis.

longitudinal wave: A compression wave that represents wave motion in which the particle oscillation is in the same direction as wave propagation.

low-carbon steel: Steel with a carbon range of 0.05% to 0.30%.

M

machinable electrode: Electrode whose deposits are soft and ductile enough so that they can easily be machined after welding. See *electrode*.

machining: Precise shaping to a desired profile using special tools to remove material.

macroetchants: Deep etchants that are intended to develop gross features such as weld solidification structures.

magnetic field: The space within and around a magnetized part or conductor carrying current in which a magnetic force is exerted.

magnetic leakage field: The magnetic field that leaves or enters the surface of a part at a discontinuity or change in section configuration of a magnetic circuit.

magnetic particle: A finely divided ferromagnetic material that is capable of being individually magnetized and attracted to distortion in a magnetic field.

magnetic particle examination (MT): An NDE method that uses a strong magnetizing current and a finely divided powder to detect defects.

magnetic response testing: A qualitative identification method in which a magnet is laid on the surface of an unknown metal to test for a magnetic force.

magnetism: The ability of a metal to be attracted by a magnet, or to develop residual magnetism when placed in a magnetic or electrical field.

main lighting: A primary lighting method that uses a light source at a vertical angle of 40° to 60° to the subject.

malleability: Ability of a metal to be deformed by compressive forces without developing defects such as those encountered in rolling, pressing, or forging.

manual welding: Welding with a torch, welding gun, or electrode holder, held and manipulated by hand.

manufacturing data report (MDR): A legal document signed by the representatives of the manufacturer and the manufacturer's authorized inspection agency.

margin: Distance from the edge of a plate to the centerline of the nearest row of rivets. See *rivet*.

material safety data sheet (MSDS): A document that includes data about every hazardous component comprising 1% or more of a material's content and is used by a manufacturer, importer, or distributor to relay chemical hazard information to the employee.

materials nonconformance report: A form created by the receiver of the metal to audit manufacturer paperwork regarding supplied metals.

materials test report (MTR): A certified statement issued by the primary manufacturer indicating the chemical analysis and mechanical properties of the metal.

mechanical bonding: The joining of two components by locking, compression, or surface tension.

mechanical property: A property of metal that describes the behavior of metals under applied loads.

mechanical repair: A repair weld process that consists of methods that do not create a metallurgical bond between the restored parts or at the restored surface.

mechanical restraint: A device used to restrict movement and counteract shrinkage stresses that occur during welding.

mechanized welding: A welding process in which the welding is automatic, but the operator must make process adjustments manually.

medium-carbon steel: Steel with a carbon range of 0.30% to 0.45%.

melting point: The temperature at which a metal passes from a solid state to a liquid (molten) state.

melt-through: A discontinuity that occurs in butt welds when the arc melts through the bottom of the weld.

metallograph: A metallurgical microscope equipped to photograph microstructures and produce photomicrographs.

metallurgical bond: The joining of two components by atomic fusion.

metallurgical structure: The arrangement of atoms in repeating patterns within a metal.

metallurgy: The study of the influence of crystal and grain structure of metals on the mechanical, physical, and chemical properties of metals.

metal transfer: Manner in which molten metal transfers from the end of the electrode across the welding arc to the weld pool.

meter: A device used to measure and indicate the flow of a gas, liquid, or current through a system.

metric equivalent standard: A version of a standard in which all the units are indicated in metric (SI) values.

microanalysis: Chemical analysis of extremely small regions of the specimen surface using tools such as energy-dispersive X-ray analysis or electron probe microanalysis.

microhardness test: A microhardness test is a type of indentation hardness test that uses light loads of less than 200 g.

microstructure: The appearance of the metallurgical structure of metals when they are specially prepared to reveal their features.

mill test report (MTR): Certification issued by the primary manufacturer (mill) verifying the chemical analysis and mechanical test properties of stock obtained from a starting ingot or billet of metal. See *certificate of analysis (COA)*.

mock-up: A simulation of the repair area on which the welder performs work in the expected position of the repair.

modulus of elasticity: A measure of the stiffness of an object under tension or compression.

multiple-impulse welding: A form of resistance welding in which welds are made with repeated electrical impulses.

N

neutral flame: A flame that has neither oxidizing nor carburizing characteristics.

nil ductility transition (NDT) temperature test: A toughness test that measures the temperature at which the fracture behavior of a metal changes from ductile to brittle in the presence of a stress raiser.

noise reduction rating number (NRR): A number that indicates the noise level reduction in decibels (dB).

Nomarski illumination: An illumination process that illuminates the specimen using polarized light that is separated into two beams by a biprism.

nonconsumable electrode: An electrode that does not melt and become part of the weld.

nondestructive examination (NDE): The development and application of technical methods to examine materials or components in ways that do not impair their future usefulness and serviceability.

nonessential variable: A qualification variable that may be changed in a WPS without requalification of the WPS.

nonferrous metal: A metal that contains no iron.

non-permit confined space: Confined space that does not contain, or have the potential to contain, any hazards capable of causing death or serious physical harm.

nonrelevant indication: An NDE indication caused by a discontinuity that, after evaluation, does not need to be rejected.

notch effect: A stress-concentrating condition caused by an abrupt change in section thickness or in continuity of the structure.

nugget: The weld metal that joins the member in spot, seam, and projection welds.

null point method: An alternative method of thermoelectric potential sorting.

Occupational Safety and Health Administration (OSHA): Federal agency that standardizes safety practices for most types of work environments.

ohm: The basic unit of measurement of resistance.

open-circuit voltage: Voltage produced when a welding machine is ON but no welding is being done.

open root joint: An unwelded joint that does not use backing or consumable inserts.

optical emission spectrometer: An instrument used for optical emission spectroscopy that is placed on the surface of an unknown metal.

optical emission spectroscopy: A semi-quantitative identification method that separates and analyzes the light emitted from an unknown metal surface when it is arced by an electric current.

other side: The opposite surface of the joint.

overheating: Microstructural damage or change caused by cutting operations.

overlapping: Extending the weld metal beyond the weld toes. Also called *cold lapping*.

oxidation: The combination of a metal with oxygen in the air to form metal oxide.

oxidizing flame: A flame in which there is an excess of oxygen.

oxyacetylene welding (OAW): An oxyfuel welding process that uses acetylene as the fuel gas.

oxyfuel cutting (OFC): A group of cutting processes that use high heat temperatures generated by burning a fuel gas in oxygen to accelerate the chemical reaction between oxygen and the base metal to sever and remove the metal.

oxyfuel welding (OFW): A type of welding that uses heat from the combustion of a mixture of oxygen and a fuel gas such as acetylene, methylacetylene-propadiene (MAPP), propane, natural gas, hydrogen, or propylene.

P

paperwork: Physical certification or documentation provided by a product manufacturer or supplier.

pass: Each layer of bead deposited on the base metal.

peak current: The high current setting for the pulse cycle.

peel test: A shear test in which a specimen is gripped in a vise and then bent and peeled apart with pincers to reveal the weld.

peening: The mechanical working of weld metal using impact blows.

penetrant: A solution or suspension of dye.

percent on time: The length of time that peak current is maintained before it drops to background current. Also referred to as peak time.

permit-required confined space: A confined space that contains or has the potential to contain a hazardous atmosphere; contains a material that has the potential to engulf the entrant; has the internal configuration that the entrant could be trapped or asphyxiated; or contains any other recognized serious safety or health hazards.

personal protective equipment (PPE): Any device worn by welders to prevent burns and protect against hazardous radiation, fumes, and gases, as well as eye injuries and hearing loss.

photomacrography: The documentation of macroetched samples using photography.

physical failure analysis: A failure analysis process that provides a diagnosis of the technical cause of failure using rigorous analytical methods.

pipe jig: A device that holds sections of pipe or fittings before tack welding.

pitting (spalling): The forming of localized cavities in metal resulting from corrosion, repetitive sliding or rolling surface stresses, or poor electroplating.

plane-strain fracture toughness test: A toughness test that measures the resistance of metals to brittle fracture propagation in the presence of stress raisers such as weld defects.

plasma arc cutting (PAC): A cutting process that uses a constricted arc to remove molten metal with a high-velocity jet of ionized gas.

plasma arc welding (PAW): An arc welding process that uses a constricted arc between a nonconsumable tungsten electrode and the weld pool (transferred arc), or between the electrode and constricting nozzle (non-transferred arc).

plasma spraying: A thermal spraying process in which a plasma torch is used as a heat source for melting and propelling the surfacing material to the workpiece.

plastic strain: Strain that remains permanent after the stress is removed.

plug weld: A weld made in a circular hole in one member to fuse it to the other member.

pneumatic testing: A proof test in which air is pressurized inside a closed vessel to reveal leaks.

polarity: The positive (+) or negative (−) state of an object.

polarized illumination: An illumination process that reveals microstructural features in metals that are optically anisotropic.

polarizer: A device into which normal light passes and from which polarized light emerges.

polysulfide adhesive: A one- or two-part adhesive or sealant that cures by evaporation or catalyst.

polyurethane: A one- or two-part adhesive with excellent flexibility that cures by evaporation, catalyst, or heat.

positioner: A mechanical device that supports and moves weldments for maximum loading, welding, and unloading efficiency.

postheat: The reheating of the weld area to a high temperature, holding for a predetermined time at temperature, and cooling at a specified rate.

prebending: A mechanical restraint method that relies on locating workpieces out of position before welding so that welding shrinkage stresses pull the workpieces back into position.

postflow: The flow of shielding gas after the arc has been extinguished.

preheat: The heating of the joint area to a predetermined temperature in order to slow the cooling rate.

prequalified WPS: A welding procedure specification that complies with the stipulated conditions of a particular fabrication standard or code and is acceptable for use under that code without requiring additional qualification testing.

primary weld: A weld that is an integral part of a structure and that directly transfers the load. See *load*.

prod: A set of hand-held electrodes used to transmit the magnetizing current from the source to the material being inspected.

prod method: A wet or dry continuous method in which portable prod-type electrical contacts are pressed against the areas to be examined to magnetize them.

product analysis: A chemical report that a particular metal, such as tubing or piping, is made from a particular heat of metal.

projection weld: A resistance weld type produced by the heat obtained from the resistance to the flow of welding current. See *weld type, fusion,* and *base metal.*

projection welding (PW): A welding process that produces a weld using heat obtained from resistance of the workpiece to the welding current.

proof testing: The application of specific loads to welded structures, without failure or permanent deformation, to assess their mechanical integrity.

proportional limit: The maximum stress at which stress is directly proportional to strain. See *stress*.

pulsed GTAW (GTAW-P): A gas tungsten arc welding variation in which direct current is pulsed between a peak current (high pulse) and a background current (low pulse).

pulsed spray transfer: A spray transfer mode in which current is cycled from low to high, at which point spray transfer occurs.

pulse-echo mode: A UT inspection method in which the presence and position of a reflector are indicated by the echo amplitude and time.

pulses per second: The number of times per second that the current achieves peak current.

pure metal: Metal that consists of one chemical element.

push travel angle: Travel angle where the electrode points toward the direction of travel.

qualitative identification: Metal identification by a qualified person to confirm the identity of an unknown metal.

radiograph: A permanent, visible image on a recording medium produced by penetrating radiation passing through a material being tested.

radiographic examination: The use of X rays or nuclear radiation (gamma rays) to detect various types of internal and external discontinuities in material.

reaming: Enlarging and improving the surface quality of a hole.

recommended practice: A type of standard that provides instructions for performing one or more repetitive technical functions.

rectifier: An electrical device that changes AC current into DC current.

red hardness: The capacity to resist softening in the red heat temperature range.

reducing flame: See *carburizing flame*.

reduction: Loss or removal of oxygen during the welding process.

reference block: A test piece of the same material, shape, and significant dimensions as a particular object under examination, and which may contain natural or artificial discontinuities or defects.

reflected light: A lighting source that bounces light off a white card, wall, or ceiling.

reinforcement: Amount of weld metal that is piled up above the surface of the pieces being joined.

relevant indication: An NDE indication caused by a discontinuity that requires evaluation.

rerating: Revision of the allowable design parameters of a mechanical component from the original design arising from formal study of its current condition.

residual magnetization method: An MT examination technique in which magnetic particles are applied after the magnetizing force has been disconnected.

residual stress: Stress that occurs in a joint member or material after welding has been completed, resulting from thermal or mechanical conditions.

resistance: The opposition of the material in a conductor to the passage of electric current, causing the electrical energy to be transformed into heat.

resistance welding (RW): A group of welding processes in which fusion occurs from the heat obtained by resistance to the flow of current through the metals joined.

retentivity: The ability of a material to retain a portion of the applied magnetic field after the magnetizing force has been removed.

right angle: Angle that contains 90°.

rimmed steel: Steel with little or no deoxidizer addition.

ripple: The shape within the deposited bead caused by the movement of the welding heat source. See *bead*.

rivet: Cylindrical metal pin with a preformed head.

rivet pitch: Distance from the center of one rivet to the center of the next rivet in the same row. See *rivet*.

robot: A programmed path device used to position the torch and at times the workpiece.

Rockwell hardness test: An indentation hardness test that uses two loads, supplied sequentially, to form an indentation on a metal test specimen to determine hardness.

roll welding: A welding procedure that applies heat and pressure to interlock the faying surfaces of the weld.

root bead: A weld bead that extends into or includes part or all of the joint root.

root cause failure analysis: A failure analysis process that determines how to prevent a failure from recurring by understanding how the actions of humans or systems may have led to the technical cause of the failure.

root edge: Weld face that comes to a point and has no width. See *weld face.*

root face: The portion of the groove face within the joint root.

root opening: The distance between joint members at the root of the weld before welding.

root pass: The initial weld pass that provides complete penetration through the thickness of the joint member. See *weld pass.*

root reinforcement: Reinforcement on the side opposite the one on which welding took place. See *filler metal.*

root surface: Surface of the weld on the opposite side of the joint on which welding was done.

rough polishing: A polishing process that is performed on a series of rotating wheels covered with a low-nap cloth (cloth containing a small amount of fiber).

run-off tab: A piece of metal of the same composition and thickness as the base metal that is tacked to the metal to allow the weld to be completed.

S

scan presentation: A method of data presentation using a horizontal base line that indicates distance or time, and a vertical deflection from the base line that indicates relative amplitude of the returning signal.

scratch start: An arc starting method that uses the high open-circuit voltage of the constant-current (CC) welding power source to start the arc when the electrode comes into contact with the base metal.

screw thread series: Groupings of diameter-pitch combinations.

sealant: A product used to seal, fill voids, and waterproof parts.

seam weld: A continuous weld made between or upon overlapping members that produces a continuous seam or series of overlapping spot welds. See *weld type, fusion,* and *spot weld.*

search unit (probe): An electroacoustic device for transmitting or receiving ultrasonic energy, or both.

secondary weld: A weld used to hold joint members and subassemblies together.

section view: Interior view of an object through which a cutting plane has been passed. See *cutting plane line.*

segregation: Any concentration of alloying chemical elements in a specific region of a metal. See *base metal.*

selective plating: A form of electroplating used for touch-up repairs on worn or damaged parts.

self-shielded flux-cored arc welding (FCAW-S): An FCAW variation in which shielding gas is provided exclusively by the flux within the electrode core.

semiautomatic welding: Manual welding with equipment that controls one or more welding conditions automatically.

semikilled steel: Steel in which deoxidizers only partially kill the oxygen-carbon reaction.

semi-quantitative identification: Metal identification by applying a physical stimulus to an unknown metal to produce a signal that is interpreted against a set of standards.

sensitization: Precipitation of chromium carbides in stainless steels from exposure to high temperatures, as in welding, typically in the HAZ.

servomotor: An AC or DC motor with encoder feedback to indicate how far the motor has rotated.

shank: Cylindrical body of a rivet. See *rivet.*

sheared plate: Plate that is rolled between horizontal and vertical rollers and trimmed on all edges.

shearing: The parting of material when one blade forces the material past an opposing blade.

shear strength: Ability of a metal to withstand two equal forces acting in opposite directions.

shear stress (shear): Stress caused by two equal and parallel forces acting upon an object from opposite directions. See *stress.*

shear wave: A transverse wave that represents wave motion in which the particle oscillation is perpendicular to wave propagation direction.

sheet: Structural steel $\frac{3}{16}''$ or less used to cover large expanses of a structure. See *structural steel.*

sheet lining (wallpapering): A weld repair method that uses thin, usually $\frac{1}{16}''$, sheets of corrosion-resistant material that are welded to a corroded surface.

shielded metal arc welding (SMAW): An arc welding process that produces an arc between a consumable, coated electrode and the workpiece, creating a weld pool.

shop rivet: Rivet placed in the shop. See *rivet*.

short circuiting transfer: A metal transfer mode in which molten metal from consumable welding wire is deposited during repeated short circuits. See *electrode*.

shrinkage stress: Stress that occurs in weld filler metal as it cools, contracts, and solidifies.

silicone: A one- or two-part adhesive or sealant that cures by evaporation or catalyst.

single-bevel-groove weld: Groove weld having one joint member beveled, with the weld made from that side. See *groove weld*.

single-flare-bevel-groove weld: Groove weld having one straight and one radiused joint member, with the weld made from one side. See *groove weld*.

single-flare-V-groove weld: Groove weld having radiused joint members, with the weld made from one side. See *groove weld*.

single-groove weld: A groove weld made from one side only. See *groove weld*.

single-J-groove weld: Groove weld having joint members grooved in a J shape on one side, with the weld made from the grooved side. See *groove weld*.

single-square-groove weld: Groove weld having square-edged joint members, with the weld made from one side. See *groove weld*.

single-U-groove weld: Groove weld having joint members grooved in a U shape on one side, with the weld made from the grooved side. See *groove weld*.

single-V-groove weld: Groove weld having both joint members angled on the same side, with the weld made from the grooved side. See *groove weld*.

sinusoidal waveform: A 60 cycle per second (60 Hz) sine wave.

slag inclusions: Small particles of slag (cooled flux) trapped in the weld metal which prevent complete penetration.

sleeving: A weld repair method that applies surfacing to badly worn shafts by welding snug-fitting semicircular forms to cover the shaft surface.

slope: The shape of the volt-amp curve on a GMAW welding machine. Slope is an important function of GMAW-S because it controls the magnitude of the short-circuit current when the welding wire is in contact with the work. Slope is represented by the volt-ampere curve.

slot weld: A weld made in an elongated hole in one member to fuse it to the other member. See *weld type*.

slurry: A mixture of solid particles in a liquid.

slurry erosion: The progressive loss of material from a surface caused by slurry moving over the surface.

soldering (S): A group of joining processes that produce a coalescence of metals with nonferrous filler metals having a melting point below that of the base metals. See *filler metal* and *base metal*.

soldering copper: A tool that consists of a copper or steel heating tip fastened to a rod with a wooden handle.

solidification temperature: Temperature at which the atoms of a metal assume their characteristic crystal structure. See *crystal structure*.

solid particle impingement: Wearing away of a surface by repeated impacts from solid particles.

solidus: The highest temperature at which an alloy is completely solid.

solidus temperature: The highest temperature that a metal can reach and remain in a solid state.

solvent-base adhesive: A one-part adhesive with a rubber or plastic base that cures by solvent evaporation.

space lattice: Uniform pattern produced by lines connected through the atoms.

spacer symbol: Supplementary symbol indicated by a rectangle centered on reference line. See *supplementary symbol*.

spark testing: A semi-quantitative identification method that identifies metals by the shape, length, and color of the spark produced when the metal is held against a grinding wheel rotating at high speed.

spatter: A discontinuity that occurs when metal particles are expelled during fusion welding and do not form part of the weld.

special series: Screw thread series with combinations of diameter and pitch not in the standard screw thread series. See *screw thread series*.

specification: A type of standard that indicates the technical and commercial requirements for a product.

specific heat: The ratio of the quantity of heat required to increase the temperature of a unit mass of metal by 1°, compared with the amount of heat required to raise the same mass of water by the same temperature.

spin testing: Proof testing of rotating machinery done by spinning it at speeds above design values to develop desired stresses from centrifugal forces.

splat: A flattened particle that cools rapidly and solidifies as it strikes a metal surface.

spotface: Flat surface machined at a right angle to a drilled hole. See *right angle*.

spotlight: An intense lighting source that uses a single bulb in a reflector.

spot weld: A weld with an approximately circular cross section made between or upon overlapping members. See *weld type* and *fusion*.

spot-weld tension shear test: A shear test that determines the strength of arc welds and resistance spot welds.

spray and fuse (spraywelding): A two step thermal spray process in which a thermal spray coating is deposited and subsequently fused by heating with a torch or by placing the part in a furnace.

spray transfer: A metal transfer mode in which molten welding wire is propelled axially across the arc in small droplets.

staggered intermittent fillet welds: Intermittent fillet welds that have a staggered pitch and are applied to both sides of a weld joint.

standard: A document that, by agreement, serves as a model for the measurement of a property or the establishment of a procedure.

standard series: Screw thread series of coarse (UNC/UNRC), fine (UNF/UNRF), and extra-fine (UNEF/UNREF) graded pitches and eight series with constant pitches. See *screw thread series*.

starved joint: A joint that contains insufficient adhesive to create an optimum bond.

static electricity: Electricity at rest or electricity that is not moving.

static load: A load that remains constant. See *load*.

stator: The stationary part of a generator that produces a rotating magnetic field.

steel deoxidation: The process of removing a controlled amount of oxygen from steel during steelmaking.

stencil marking: An identification marking that consists of continuous or repeated ink markings on the metal.

stickout: 1. The amount of unmelted electrode extending beyond the end of the gas nozzle when using GMAW and FCAW as the welding process. **2.** The distance from the end of the nozzle to the tip of the electrode in GTAW.

stopoff: A material used to outline areas that are not to be brazed.

straight bead: A type of weld bead made without any appreciable weaving motion.

straight beam: A vibrating pulse wave traveling perpendicular to the surface.

strain: The accompanying change in dimensions when a load induces stress in a material. See *stress*.

strength: The ability of a metal to resist deformation from mechanical forces exerted on it.

stress: The internal resistance of a material to an externally applied load. See *strain*.

stress relieving: Process of heating a metal to a suitable temperature, holding it at that temperature to reduce residual stresses, and cooling it slowly to minimize the development of new residual stresses. See *stress*.

strongback: A mechanical restraint device that is attached to one side of a weld joint to hold workpieces in alignment during welding.

structural steel: Steel used in the erection of a structure.

structural weld repair: Restoration of a load-bearing structure by welding to meet performance requirements.

stud weld: A weld produced by joining a metal stud or similar part to a member. See *weld type*.

submerged arc welding (SAW): An arc welding process that uses an arc between a bare metal electrode and the weld pool.

subresonant vibration: Vibration frequency less than the resonant frequency of the weld.

subsurface deformation: Microstructural damage or change produced by cutting and that occurs below the surface of the specimen.

supplementary essential variable: A qualification variable, for metals where impact testing is required, that requires a new welding procedure specification.

supplementary symbol: Symbol used on welding symbols to further define the operation to be completed.

surface feature: A part of a surface where change occurs.

surfacing: The application of a layer or layers of material to a surface to obtain desired properties or dimensions.

surfacing weld: A weld applied to a surface, as opposed to a joint, to obtain desired properties or dimensions.

surfacing weld repair: The application of a layer, or layers, of specially formulated weld metal to restore worn or corroded components to extend their useful life.

sweat soldering: A process whereby two surfaces are soldered together without allowing the solder to be seen.

T

tack weld: A weld used to hold workpieces in proper alignment until the final welds are made.

tail: Part of a welding symbol included when a specific welding process, specification, or procedure must be indicated. See *welding symbol.*

tap start: An arc starting method that requires the electrode to touch the work.

teach pendant: The input method that the robot programmer uses to move the robot and create robot programs.

tee: T-shaped structural steel made of I beams cut to specifications by mill or suppliers. See *structural steel.*

tensile strength: A measure of the maximum stress that a material can resist under tensile stress. See *load.*

tensile test: A destructive test that measures the effects of a tensile force on a material.

tensile test machine: A testing machine composed of two major components that are the means of applying the load to the specimen and the means of measuring the applied load.

tension (tensile stress): Stress caused by two equal forces acting on the same axial line to pull an object apart.

tension shear test: A shear test in which a prepared specimen is pulled to failure in a tensile testing machine.

theoretical throat: Distance from the joint root, perpendicular to the hypotenuse of the largest right-angle triangle that can be inscribed within the cross section of a fillet weld. See *fillet weld, weld face,* and *weld root.*

thermal conductivity: The rate which metal transmits heat.

thermal equilibrium: A steady-state condition in which time is available for the diffusion of atoms.

thermal expansion: A measure of the change in dimension of a member caused by heating or cooling.

thermal properties: One of the physical properties of metal. Includes melting point, thermal conductivity, and thermal expansion and contraction.

thermal spraying (THSP): A group of processes in which finely divided metallic or nonmetallic materials are deposited in a molten or semimolten condition to form a coating.

thermoelectric potential sorting: A semi-quantitative identification method that uses measurement of the electric potential generated when two metals are heated.

threaded fasteners: Devices such as nuts and bolts that join or fasten parts together with threads.

through hole: Drilled hole passing completely through the material.

T-joint: A weld joint formed when two workpieces are positioned at approximately 90° to one another in the form of a T.

torch positioner: A fixed-path mechanical apparatus that moves the torch in a specified path.

torch testing: A qualitative identification method that identifies a metal by the melting rate, the appearance of the metal when heat is applied, and the action of the molten metal.

torque: Product of the applied force (P) times the distance (L) from the center of application.

torsion (torsional stress): Stress caused by two forces acting in opposite twisting directions. See *stress.*

torsional strength: The measure of a material's ability to withstand forces that cause it to twist.

toughness: The ability of a metal to absorb energy, such as impact loads, and deform rather than crack or fail catastrophically. See *ductility.*

toughness test: A dynamic test in which a specimen is broken by a single blow and the energy absorbed in breaking the piece is measured in foot-pounds (ft-lb).

transformer: An electrical device that changes AC voltage from one level to another.

transmission piping: Medium- to high-strength steel, relatively thin-wall and large-diameter, that conveys products from locations of production to intermediate facilities.

transverse crack: A crack with its major axis oriented approximately perpendicular to the weld axis.

transverse shrinkage: Weld metal shrinkage that occurs perpendicular to the weld axis.

travel angle: An angle less than 90° between the electrode axis and a line perpendicular to the workpiece and in a plane determined by the electrode axis and the weld axis.

travel speed: The rate at which the electrode moves along the weld joint. See *electrode.*

tubing: Round-, square-, or rectangular-shaped structural steel. See *structural steel.*

U

ultimate tensile strength: A measure of the maximum stress (load) that a metal can withstand.

ultrasonic examination (UT): An NDE method that introduces ultrasonic waves (vibrations) into, through, or onto the surface of a part and determines various attributes of the material from its effects on the ultrasonic waves.

ultrasonic welding (USW): A welding process that produces a weld by applying high-frequency vibratory energy to workpieces that are held together under pressure.

undercutting: Creating a groove in the base metal that is not completely filled by weld metal during the welding process.

underfill: A discontinuity in which the weld face or root surface extends below the adjacent surface of the base metal.

undesirable microstructure: The creation, through the heat of welding, of microstructures that are preferentially attacked in a corrosive environment.

unified numbering system (UNS): A common embedded designation system that unifies all families of metals and alloys.

union: Fitting consisting of three parts having threads and flanges which draw together when tightened.

universal plate: Plate that is rolled between horizontal and vertical rollers and trimmed only on the ends.

uphill welding: Welding with an upward progression. See *vertical weld.*

upset welding (UW): A resistance welding process that produces a weld on the faying surfaces by the heat obtained from resistance to the flow of current through the surface contact areas while under constant pressure.

user enquiry: A formal procedure developed by standards committees and code-creating organizations to help users interpret issues and offer suggestions.

vacuum box testing: The application of a partial vacuum to one side of a structure and examining for the presence of leaks.

variable load: Load that varies with time and rate, but without the sudden change that occurs with an impact load. See *impact load.*

variable voltage control: A control that spans a range of voltages and is used to set the open-circuit voltage on a welding machine.

vertical weld: A weld with the axis of the weld approximately vertical. See *downhill welding* and *uphill welding.*

very-high carbon steel: Steel with a carbon range of 0.75% to 1.7%.

vibratory stress relief: The application of subresonant vibration during welding to control distortion, or after cooling to provide stress relief.

Vickers hardness test: An indentation hardness test that uses an indenter with a 136° square-base diamond cone, and that may be used to test hardness in the base metal, weld metal, and HAZ.

viscosity: The resistance of a substance to flow in a fluid or semi-fluid state.

visual examination (VT): Application of the naked eye, assisted as necessary by low-power magnification and measuring devices, to monitor welding quality.

visual identification: Metal identification that consists of checking the appearance of the base metal or filler metal for key features that identify the metal type.

volt (V): Unit of measure for electricity that expresses the electrical pressure differential between two points in a conductor. See *conductor.*

voltage: The amount of electrical pressure in a circuit.

voltage drop: The voltage decrease across a component due to resistance to the flow of current. See *current* and *resistance.*

volt-ampere (VA) curve: Curve showing the relationship between output voltage and output current.

voltmeter: An instrument used to measure voltage.

water-base adhesive: A one-part adhesive that cures by water evaporation.

weave bead: A type of weld bead made with transverse oscillation.

weaving: A welding technique in which the energy source is moved transversely as it progresses along the weld joint.

weld-all-around symbol: Supplementary symbol indicated by a circle at the intersection of the arrow and reference line, which specifies that the weld extends completely around the joint. See *supplementary symbol.*

weld bead: Weld that results from a weld pass. See *weld pass.*

weld contour: Cross-sectional shape of the completed weld face. See *weld face.*

weld cracks: Linear discontinuities that occur in the base metal, weld interface, or the weld metal. See *base metal* and *weld interface.*

weld defects: Undesirable characteristics of a weld which may cause the weld to be rejected.

weld discontinuity: An interruption in the typical structure of a weld.

welder certification: A written statement that the welder has produced welds meeting a prescribed standard of welding performance.

welder performance qualification: A test that demonstrates a welder's ability to produce welds that meet required standards.

welder registration: The act of approving a copy of the welder's certification document by an appropriate authority.

weld face: The exposed surface of the weld, bounded by the weld toes on the side on which welding was done. See *weld toe.*

weld finish: Method used to achieve the surface finish. See *base metal.*

weld gauge: A device for measuring the size and shape of welds.

welding: The coalescence or joining together of metals, with or without a filler metal, using heat, pressure, or heat and pressure.

welding procedure qualification record (WPQR): Documentation of the welding variables used to produce an acceptable test weld and the results of tests conducted on the weld to qualify a WPS.

welding procedure qualification variable: A condition (parameter) that affects the integrity of a weld joint.

welding procedure specification (WPS): A document providing the required welding variables for a specific application to ensure repeatability by properly trained welders and welding operators.

welding symbol: A graphical representation of the specifications for producing a welded joint. (See Appendix)

weld interface: The boundary between the weld metal and the base metal in a fusion weld.

weld joint: The physical configuration at the juncture of the members to be welded.

weld leg: The distance from the joint root to the toe of a fillet weld.

weld metal: The portion of a fusion weld that is completely melted during welding.

weld overlay: The application of surfacing using a welding process that creates a metallurgical bond with the base metal through melting of the surfacing metal.

weld pass: A single progression of welding along the weld joint.

weld reinforcement: The amount of weld metal in excess of that required to fill the joint.

weld repair: A repair weld process that consists of methods that join failed parts or restore their surface using a welding process.

weld root: The area where filler metal intersects base metal and extends the furthest into the weld joint.

weld symbol: A graphic symbol connected to the reference line of a welding symbol specifying the weld type.

weld throat: Distance through the center of the weld from the face to the root. See *weld face* and *weld root.*

weld toe: The junction of the base metal and the weld face.

weld type: The cross-sectional shape of the weld after filler metal is added to the joint.

weld width: The distance from toe to toe across the face of the weld.

wet magnetization method: An MT examination technique in which the magnetic particles are suspended in a liquid medium.

whipping: A manual welding technique in which the arc is moved quickly forward about one electrode diameter and back about one-half an electrode diameter as it progresses along the weld joint.

wire feeder: A welding machine accessory that holds a filler metal spool and allows it to be fed to the hot wire torch as welding progresses.

work angle: An angle less than 90° in a line perpendicular to the workpiece and in a plane determined by the electrode axis and the weld axis.

working voltage: See *arc voltage.*

workmanship standard: A section of a joint similar to the one in manufacture in which portions of each successive weld pass are shown.

wraparound guided bend test: A bend test in which a specimen is bent around a stationary mandrel a specified amount to expose weld discontinuities.

x-ray fluorescence spectrography (XRF): A nondestructive quantitative identification method that uses a gamma ray beam to identify an unknown metal.

yield point: The location on the stress-strain curve where an increase in strain occurs without an increase in stress.

yield strength: The level of stress within a metal that is sufficient to cause plastic flow.

yoke: A temporary horseshoe magnet made of soft, low-retentivity iron that is magnetized by a small coil wound around the horizontal bar.

yoke method: A dry continuous method of MT for detection of surface discontinuities.

Index

full-open corner joint, *38,* 38
full skip, 460
furnace heating, *277,* 277
fusion welding, 32
FW, 367, *368*

V

vacuum box testing, 474
variable loads, *34*
variables, 235. 513
ventilation, 12–13, *13*
ventilation system, *104,* 105
vertical position, 42–43, *43*
vertical position gouging, *318*
vertical position, OAW, *87,* 87
vertical position, SMAW, 157–158
vertical welds, 157
V heating, *660,* 660, *661*
vibratory stress relief, 556
Vickers hardness test, 425–426, *436*
viscosity, 408
visual identification, 562–566
  appearance, 562
  color, 563, *564*
    characteristic color groupings, 564
    color-coding, *565*
  markings, *564,* 564–566
visual nondestructive examination, 442–445
  after welding, 444–445
  before welding, 443
  during welding, 443–444, *444*
voltage, *94,* 94–96, *95*
voltage drop, 94
volt-ampere curve, 97
volumetric expansion, 545
  coefficients, *545*
VT. *See* visual nondestructive examination

W

wallpapering, 333, *334*
warpage, 272
washing, 319
water-base adhesive, 409
water piping specifications, 697
wear, types of, 288–291
weave bead, 520
weaving, 141, *142, 152*
  during uphill welding, *158,* 158
weld
  appearance after welding, 445
  dimensional accuracy after welding, *445,* 445
weld-all-around symbol, *674,* 674
weld beads, *31,* 31
  depositing, 73
  quality during welding, 444
weld defects, 74, 441, 493, *494*

weld discontinuities. *See* discontinuities
welder certification, 531
welder performance qualifications, 531
  for brazing, 535, *537*
  for pipe, 535
  for plate, 534–535
  positions, 533
  for sheet steel, 535–536, *536*
  members, 534–535
welder registration, 531
weld face, 32, *33*
weld gauge, *445,* 445
welding
  applications, *1*
  check valves, *58,* 58
  codes, 532–533
  equipment, 54, 102
    assembly, 63–65, *64*
    automatic, 396–397
    for SMAW, 122–123, *123*
  goggles, 59
  hoses, 59
  processes, 2–5, *3*
  regulators, 56–58, *57*
  sparklighters, *59,* 59
  tips, *56,* 56
  torches, 54–55, *55*
welding arc, sustaining, 121, *122*
welding helmets. *See* helmets
welding guns, 216–218, *217*
  EBW, *375*
  GMAW, *372*
  FCAW, 261–262, *262*
welding leads, 102
welding machines
  constant-current, 97–98, *98*
  engine-driven, 100–101, *101*
  inverter arc, 100
  ratings, 101–102
  robotic, *6,* 6
  static, 98–100
  transformer, 98
  transformer-rectifier, *99,* 99
welding occupations, 5–7
  employment outlook, 6
  job classifications, 7
  training, 6–7
welding positions. *See* positions, welding
welding procedure qualification record, 526–527
welding procedures
  qualification record, 532
  specification, 523
welding processes
  for aluminum alloys, 629
  for copper alloys, 625
  for magnesium alloys, 632
  for nickel alloys, 622–623

  as procedure qualification variable, 513–514
  for titanium alloys, 634–635
welding screen, *104,* 105
welding symbols. *See* symbols, welding
welding technique
  as procedure qualification variable, 519
  and dilution, 293
welding variables, 235, 513
welding wire, *226, 227*
  diameters, *228*
weld interface, 543
weld joints. *See* joints
weld leg, 32, 33
weld metal, 542, 543
  deposition, 654–655
  excess, *655*
  shrinkage, 649
weld overlays, 288
  welding processes, 294–300
weld passes. *See* passes
weld pool, 71–72
weld reinforcement, *32,* 32
weld repair, 332. *See also* repair welding
  methods, 332–334
    sleeving, 334
    structural, *332,* 332
    surfacing, 332–333
    wallpapering, 333, *334*
  plans, 334–342, *335*
    confined space entry permit, *341*
    determining repair necessity, 334–336
    distortion control, 337
    identifying base metal, 337
    oxygen-deficient atmospheres, *340*
    procedures, 338–342
weld root, 32, *33*
welds
  cleaning, 132, *133*
weld stresses. *See* stress
weld toe, *32,* 32, *33*
weld types, 39, 514
  back weld, 42
  fillet weld, 39
  groove weld, 39, 41
  plug weld, 41
  projection weld, 42
  seam weld, 41
  slot weld, 41
  spot weld, 41
  stud weld, 41
  surfacing weld, 41
weld width, *32,* 32
wet magnetization method, 455
whipping, 158
whiskers, 242

wire brushes, *104,* 105
wire feeder, 196, 218–221, *219*
  drive rolls, *220*
  for FCAW, 265
work angles, *129,* 129
  overhead position, *164,* 164
work boots, 18
working voltage, 94, *95*
work leads, 103
workmanship standard, *444,* 444
workpiece positioner, 390, *394,*
      394–395
wormholes, 503
WPQ. *See* welder performance
      qualifications
WPQR, 526–527
WPS, 523
wraparound guided bend test, 420–422,
      421, 422
  bend locations, *421*
  calculation of strain, 422

X-ray fluorescence spectrography,
      576–577, *577*
XRF, 576–577, *577*

yield point, *415,* 415
yield strength, 555
yoke, 454
yoke method, 455

USING THE *WELDING SKILLS* INTERACTIVE CD-ROM

Before removing the Interactive CD-ROM from the protective sleeve, please note that the book cannot be returned for refund or credit if the CD-ROM sleeve seal is broken.

System Requirements

To use this Windows®-compatible CD-ROM, your computer must meet the following minimum system requirements:

- Microsoft® Windows Vista™, Windows 2000®, or Windows NT® operating system
- Intel® 1.3 GHz processor (or equivalent)
- 128MB of available RAM (256MB recommended)
- 335MB of available hard disc space
- 1024 × 768 monitor resolution
- CD-ROM drive (or equivalent optical drive)
- Sound output capability and speakers
- Microsoft® Internet Explorer® 6.0 or Firefox® 2.0 web browser
- Active Internet connection required for Internet links

Opening Files

Insert the Interactive CD-ROM into the computer CD-ROM drive. Within a few seconds, the home screen will be displayed allowing access to all features of the CD-ROM. Information about the usage of the CD-ROM can be accessed by clicking on Using This Interactive CD-ROM. The Quick Quizzes®, Illustrated Glossary, Flash Cards, Welding Resources, Media Clips, and ATPeResources.com can be accessed by clicking on the appropriate button on the home screen. Clicking on the American Tech web site button (www.go2atp.com) accesses information on related educational products. Unauthorized reproduction of the material on this CD-ROM is strictly prohibited.